Volume Two of

THE WORLD OF MATHEMATICS

A small library of the literature of mathematics from A'h-mosé the Scribe to Albert Einstein, presented with commentaries and notes by JAMES R. NEWMAN

LONDON
GEORGE ALLEN AND UNWIN LTD

First Published in Great Britain in 1960

ACKNOWLEDGEMENTS

THE EDITOR wishes to express his gratitude for permission to reprint material from the following sources:

American Association for the Advancement of Science for "What Is Calculus of Variations and What Are Its Applications?", by Karl Menger, from *Scientific Monthly*, Vol. 45.

Messrs. Appleton-Century-Crofts, Inc., for "Gustav Theodor Fechner", from *A History of Experimental Psychology*, by E. G. Boring.

Messrs. G. Bell & Sons Ltd., for "The Theory of Relativity" from *Readable Relativity* by Clement V. Durell.

Cambridge University Press for "The Constants of Nature", from *The Expanding Universe*, and "The New Law of Gravitation and the Old Law", from *Space Time and Gravitation*, by Sir Arthur Stanley Eddington; for "Causality and Wave Mechanics", from *Science and Humanism*, and "Heredity and the Quantum Theory", from *What Is Life?* by Erwin Schrodinger; for selection from *John Couch Adams and the Discovery of Neptune*, by Sir Harold Spencer Jones; for "Mathematics of Heredity" from *Mendel's Principle of Heredity*, by Gregor Mendel, edited by W. Bateson; and for "On Magnitude", from *On Growth and Form*, by D'Arcy Wentworth Thompson.

Commentary for "Sociology Learns the Language of Mathematics" by Abraham Kaplan.

Northwestern University Press for "Mathematics of Motion", from *Dialogues Concerning Two New Sciences*, by Galileo Galilei, translated by Henry Crew and Alfonso De Salvio.

Messrs. Longmans, Green & Co. Ltd., for "Mathematics of Natural Selection", from *The Causes of Evolution*, by J. B. S. Haldane; and Messrs. Chatto & Windus Ltd., for "On Being the Right Size" from *Possible Worlds* by J. B. S. Haldane.

The American Economic Association for "The Theory of Economic Behavior" by Leonid Hurwicz, from *The American Economic Review XXXV*, 1945.

Journal of the American Statistical Association for "The Meaning of Probability", by Ernest Nagel, March, 1936.

Little, Brown & Company, for "The Longitude", from *The Story of Maps*, by Lloyd Arnold Brown.

Macmillan & Co. Ltd., London, for "The Application of Probability to Conduct", from *A Treatise on Probability*, by John Maynard Keynes. By permission of the Executors of Lord Keynes.

Messrs. G. Bell & Sons Ltd., and the Executors of Sir William Bragg for "The Röntgen Rays", from *The Universe of Light*, by Sir William Bragg.

Harvard University Press for "Kinetic Theory of Gases" by Daniel Bernoulli, from *Source Book in Physics* by W. F. Magie.

Penguin Books Ltd., for "Theory of Games", by S. Vajda, from *Penguin Science News*, 28.

Messrs. Hutchinson & Co. Ltd., for "Mathematics of War and Foreign Politics" and "Statistics of Deadly Quarrels" from *Psychological Factors of Peace and War*, by Lewis Fry Richardson and P. Pear.

Oxford University Press for "Plateau's Problem", from *What Is Mathematics?* by Richard Courant and Herbert Robbins.

The Science Press, Lancaster, Pennsylvania, for "Chance", from *Foundations of Science*, by Henri Poincaré, translated by George Bruce Halsted.

Scientific American for "Crystals and the Future of Physics" by Philippe Le Corbeiller.

Messrs. Hutchinson & Co. Ltd., for "Mendeléeff", from *Crucibles: The Story of Chemistry* by Bernard Jaffe.

Society for Promoting Christian Knowledge for selection from *Soap Bubbles* by C. V. Boys.

Messrs. Taylor & Francis Ltd., and *The Philosophical Magazine* for "Atomic Numbers" by H. G. J. Moseley.

Cambridge University Press for "The Uncertainty Principle" from *The Physical Principles of the Quantum Theory*, by Dr. Werner Heisenberg.

Dover Publications Inc., for "Concerning Probability" from *Philosophical Essay on Probabilities*, by Pierre Simon de Laplace.

Printed in Great Britain
Novello & Co. Ltd.,
Soho, London

Table of Contents

VOLUME TWO

PART V: Mathematics and the Physical World

Galileo Galilei: Commentary 726

1. Mathematics of Motion *by* GALILEO GALILEI 734

The Bernoullis: Commentary 771

2. Kinetic Theory of Gases *by* DANIEL BERNOULLI 774

A Great Prize, a Long-Suffering Inventor and the First Accurate Clock: Commentary 778

3. The Longitude *by* LLOYD A. BROWN 780

John Couch Adams: Commentary 820

4. John Couch Adams and the Discovery of Neptune *by* SIR HAROLD SPENCER JONES 822

H. G. J. Moseley: Commentary 840

5. Atomic Numbers *by* H. G. J. MOSELEY 842

The Small Furniture of Earth: Commentary 851

6. The Röntgen Rays *by* SIR WILLIAM BRAGG 854
7. Crystals and the Future of Physics *by* PHILIPPE LE CORBEILLER 871

Queen Dido, Soap Bubbles, and a Blind Mathematician: Commentary 882

8. What Is Calculus of Variations and What Are Its Applications? *by* KARL MENGER 886
9. The Soap-bubble *by* C. VERNON BOYS 891
10. Plateau's Problem *by* RICHARD COURANT *and* HERBERT ROBBINS 901

The Periodic Law and Mendeléeff: Commentary 910
11. Periodic Law of the Chemical Elements *by* DMITRI MENDELÉEFF 913
12. Mendeléeff *by* BERNARD JAFFE 919

Gregor Mendel: Commentary 932
13. Mathematics of Heredity by GREGOR MENDEL 937

J. B. S. Haldane: Commentary 950
14. On Being the Right Size *by* J. B. S. HALDANE 952
15. Mathematics of Natural Selection *by* J. B. S. HALDANE 958

Erwin Schrödinger: Commentary 973
16. Heredity and the Quantum Theory *by* ERWIN SCHRÖDINGER 975

D'Arcy Wentworth Thompson: Commentary 996
17. On Magnitude *by* D'ARCY WENTWORTH THOMPSON 1001

Uncertainty: Commentary 1047
18. The Uncertainty Principle *by* WERNER HEISENBERG 1051
19. Causality and Wave Mechanics *by* ERWIN SCHRÖDINGER 1056

Sir Arthur Stanley Eddington: Commentary 1069
20. The Constants of Nature *by* SIR ARTHUR STANLEY EDDINGTON 1074
21. The New Law of Gravitation and the Old Law *by* SIR ARTHUR STANLEY EDDINGTON 1094

Commentary 1105
22. The Theory of Relativity *by* CLEMENT V. DURELL 1107

PART VI: Mathematics and Social Science

The Founder of Psychophysics: Commentary 1146
1. Gustav Theodor Fechner *by* EDWIN G. BORING 1148

Sir Francis Galton: Commentary 1167
2. Classification of Men According to Their Natural Gifts *by* SIR FRANCIS GALTON 1173

Thomas Robert Malthus: Commentary 1189
3. Mathematics of Population and Food *by* THOMAS ROBERT MALTHUS 1192

Cournot, Jevons, and the Mathematics of Money: Commentary 1200
4. Mathematics of Value and Demand *by* AUGUSTIN COURNOT 1203
5. Theory of Political Economy *by* WILLIAM STANLEY JEVONS 1217

A Distinguished Quaker and War: Commentary 1238
6. Mathematics of War and Foreign Politics *by* LEWIS FRY RICHARDSON 1240
7. Statistics of Deadly Quarrels *by* LEWIS FRY RICHARDSON 1254

The Social Application of Mathematics: Commentary 1264
8. The Theory of Economic Behavior *by* LEONID HURWICZ 1267
9. Theory of Games *by* S. VAJDA 1285
10. Sociology Learns the Language of Mathematics *by* ABRAHAM KAPLAN 1294

PART VII: The Laws of Chance

Pierre Simon de Laplace: Commentary 1316
1. Concerning Probability *by* PIERRE SIMON DE LAPLACE 1325
2. The Red and the Black *by* CHARLES SANDERS PEIRCE 1334
3. The Probability of Induction *by* CHARLES SANDERS PEIRCE 1341

Lord Keynes: Commentary 1355
4. The Application of Probability to Conduct *by* JOHN MAYNARD KEYNES 1360

An Absent-minded Genius and the Laws of Chance: Commentary 1374
5. Chance *by* HENRI POINCARÉ 1380

Ernest Nagel and the Laws of Probability: Commentary 1395
6. The Meaning of Probability *by* ERNEST NAGEL 1398

INDEX [IN VOLUME FOUR] 2471

PART V

Mathematics and the Physical World

1. Mathematics of Motion *by* GALILEO GALILEI
2. Kinetic Theory of Gases *by* DANIEL BERNOULLI
3. The Longitude *by* LLOYD A. BROWN
4. John Couch Adams and the Discovery of Neptune *by* SIR HAROLD SPENCER JONES
5. Atomic Numbers *by* H. G. J. MOSELEY
6. The Röntgen Rays *by* SIR WILLIAM BRAGG
7. Crystals and the Future of Physics *by* PHILIPPE LE CORBEILLER
8. What Is Calculus of Variations and What Are Its Applications? *by* KARL MENGER
9. The Soap-bubble *by* C. VERNON BOYS
10. Plateau's Problem *by* RICHARD COURANT *and* HERBERT ROBBINS
11. Periodic Law of the Chemical Elements *by* DMITRI MENDELÉEFF
12. Mendeléeff *by* BERNARD JAFFE
13. Mathematics of Heredity *by* GREGOR MENDEL
14. On Being the Right Size *by* J. B. S. HALDANE
15. Mathematics of Natural Selection *by* J. B. S. HALDANE
16. Heredity and the Quantum Theory *by* ERWIN SCHRÖDINGER
17. On Magnitude *by* D'ARCY WENTWORTH THOMPSON
18. The Uncertainty Principle *by* WERNER HEISENBERG
19. Causality and Wave Mechanics *by* ERWIN SCHRÖDINGER
20. The Constants of Nature *by* SIR ARTHUR STANLEY EDDINGTON
21. The New Law of Gravitation and the Old Law *by* SIR ARTHUR STANLEY EDDINGTON
22. The Theory of Relativity *by* CLEMENT V. DURELL

COMMENTARY ON

GALILEO GALILEI

MODERN science was founded by men who asked more searching questions than their predecessors. The essence of the scientific revolution of the sixteenth and seventeenth centuries is a change in mental outlook rather than a flowering of invention and Galileo, more than any other single thinker, was responsible for that change.

Galileo has been called the first of the moderns. "As we read his writings we instinctively feel at home; we know that we have reached the method of physical science which is still in use." [1] Galileo's primary interest was to discover *how* rather than *why* things work. He did not depreciate the role of theory and was himself unrivaled in framing bold hypotheses. But he recognized that theory must conform to the results of observation, that the schemes of Nature are not drawn up for our easy comprehension. "Nature nothing careth," he says, "whether her abstruse reasons and methods of operating be or be not exposed to the capacity of men." He insisted on the supremacy of the "irreducible and stubborn facts" however "unreasonable" they might seem.[2] "I know very well," says Salviati, a character representing Galileo himself in the *Dialogues Concerning the Two Principal Systems of the World,* "that one sole experiment, or concludent demonstration, produced on the contrary part, sufficeth to batter to the ground . . . a thousand . . . probable Arguments."

The origins of modern science can of course be traced much further back—at least to the thirteenth- and fourteenth-century philosophers, Robert Grosseteste, Adam Marsh, Nicole Oresme, Albertus Magnus, William of Occam.[3] Recent historical researches have broadened our understanding of the evolution of scientific thought, proved its continuity (history, like nature, evidently abhors making jumps) and helped to kill the already tottering myth that the science of the Middle Ages was little more than commentary and sterile exegesis.[4] The enlarged perspective

[1] Sir William Dampier, *A History of Science, and Its Relations with Philosophy and Religion,* Fourth Edition, Cambridge (England), 1949, p. 129.

[2] It is a great mistake, as Whitehead points out, "to conceive this historical revolt as an appeal to reason. On the contrary it was through and through an anti-intellectualist movement. It was the return to the contemplation of brute fact; and it was based on a recoil from the inflexible rationality of medieval thought." *Science and the Modern World,* Chapter I.

[3] See for example Herbert Butterfield, *The Origins of Modern Science,* London, 1949; A. C. Crombie, *Robert Grosseteste and the Origins of Experimental Science,* Oxford, 1953; A. C. Crombie, *Augustine to Galileo, the History of Science,* A.D. *400–1650,* London, 1952; A. R. Hall, *The Scientific Revolution, 1500–1800,* London, 1954.

[4] In this discussion I have drawn on material of mine published in the pages of *Scientific American*; in particular on a review of the Butterfield book cited in the preceding note (*Scientific American,* July 1950, p. 56 *et seq.*) I am much indebted to *The Origins of Modern Science* for its masterly presentation of the period treated above.

does not, however, diminish one's admiration for the stupendous achievements of Galileo. It is in his approach to the problems of motion that his imaginative powers are most wonderfully exhibited. Let us examine briefly the ideas he had to overthrow and the system he had to create in order to found a rational science of mechanics. According to Aristotle, all heavy bodies had a "natural" motion toward the center of the universe, which, for medieval thinkers was the center of the earth. All other motion was "violent" motion, because it required a constant motive force, and because it contravened the tendency of bodies to sink to their natural place. The acceleration of falling bodies was explained on the ground that they moved more "jubilantly"—somewhat like a horse—as they got nearer home. The planetary spheres, seen to be exempt from the "natural" tendency, were kept wheeling in their great arcs by the labors of a sublime Intelligence or Prime Mover. Except for falling bodies things moved only when and as long as effort was expended to keep them moving. They moved fast when the mover worked hard; their motion was impeded by friction; they stopped when the mover stopped. For the motion of terrestrial objects Aristotle had in mind the example of the horse and cart; in the celestial regions his mechanics left "the door halfway open for spirits already."

On the whole, Aristotle's theory of motion squared well enough with common experience, and his teachings prevailed for more than fifteen centuries. Eventually, however, men began to discover small but disturbing inconsistencies between experimental data and the Aristotelian dictates. There was the anomaly of the misbehaving arrow which, according to the horse and cart concept of motion, should have fallen to earth the instant it lost contact with the bowstring. Nor was the traditional explanation of the acceleration of falling bodies swallowed forever without protest. In each case the paradox was met by an ingenious modification of the accepted system (this was known, from Plato's celebrated phrase, as "saving the phenomena"); yet every such tailoring, however skillful, was a source of controversy and raised new suspicions as to the validity of all Aristotle's teachings.

In the fourteenth century Jean Buridan and others at the University of Paris developed a "theory of impetus" which proved to be a major factor in dethroning Aristotelian mechanics. This theory, later picked up by Leonardo da Vinci, held that a projectile kept moving by virtue of a something "inside the body itself" which it had acquired in the course of getting under way. Falling bodies accelerated because "impetus" was continually being added to the constant fall produced by the original weight. The importance of the theory lay in the fact that men for the first time were presented with the idea of motion as a lingering aftereffect derived from an initial impulse. This was midway to the modern view, pretty clearly expressed in Galileo, that a body "continues its motion in a straight line until something intervenes to halt or slacken or deflect it."

What was needed to complete the journey was an extraordinary transposition of ideas from the real to an imaginary world. The ghosts of Plato and Pythagoras returned triumphantly to point the way. Modern mechanics describes quite well how real bodies behave in the real world; its principles and laws are derived, however, from a nonexistent conceptual world of pure, clean, empty, boundless Euclidean space, in which perfect geometric bodies execute perfect geometric figures. Until the great thinkers, operating, in Butterfield's words, "on the margin of contemporary thought," were able to establish the mathematical hypothesis of this ideal Platonic world, and to draw their mathematical consequences, it was impossible for them to construct a rational science of mechanics applicable to the physical world of experience. This was the forward leap of imagination required—the new look at familiar things in an unfamiliar way, to see what in fact there was to be seen, rather than what some classical or medieval writer had said ought to be seen. Buridan, Nicole Oresme and Albert of Saxony with their theory of impetus; Galileo with his beautiful systematization of everyday mechanical occurrences and his ability to picture such situations as the behavior of perfectly spherical balls moving on perfectly horizontal planes; Tycho Brahe with his immense and valuable observational labors in astronomy; Copernicus with the *De Revolutionibus Orbium* and heliocentric hypothesis; Kepler with his laws of planetary motion and his passionate search for harmony and "sphericity"; Descartes with his discourse on method, his determination to have all science as closely knit as mathematics, his wedding of algebra to geometry; Huygens with his mathematical analysis of circular motion and centrifugal force; Gilbert with his *terella*, his theory of magnetism and gravitation; Viète, Stevin and Napier with their aids to simplicity of mathematical notation and operations: each took a part in the grand renovation not only of the physical sciences but of the whole manner of thinking about the furniture of the outside world. They made possible the culminating intellectual creation of the seventeenth century, the clockwork universe of Newton in which marbles and planets rolled about as a result of the orderly interplay of gravitational forces, in which motion was as "natural" as rest, and in which God, once having wound the clock, had no further duties.

Galileo was the principal figure in this drama of changing ideas.[5] He

[5] "Taking his achievements in mechanics as a whole, we must admit that in the progress from the pre-Galileon to the post-Newtonian period, Galileo's contribution extends more than halfway. And it must be remembered that he was the pioneer. Newton said no more than the truth when he declared that if he saw further than other men it was because he stood on the shoulders of giants. Galileo in these matters had no giants on whom to mount; the only giants he encountered were those who had first to be destroyed before vision of any kind became possible." Herbert Dingle, *The Scientific Adventure* ("Galileo Galilei (1564 1642)"), London, 1952, p. 106.

was the first to grasp the importance of the concept of *acceleration* in dynamics. Acceleration means change of velocity, in magnitude or direction. As against Aristotle's view that motion required a force to maintain it, Galileo held that it is not motion but rather "the creation or destruction" of motion or a change in its direction—i.e. acceleration, which requires the application of external force. He discovered the law of falling bodies. This law, as Bertrand Russell remarks, given the concept of "acceleration," is of "the utmost simplicity." [6] A falling body moves with constant acceleration except for the resistance of the air. At first Galileo supposed that the speed of a falling body was proportional to the distance fallen through. When this hypothesis proved unsatisfactory, he modified it to read that speed was proportional to the time of fall. The mathematical consequences of this assumption he was able partially to verify by experiment.

Because free-falling bodies attained a velocity beyond the capacity of the measuring instruments then available, Galileo approached the problem of verification by experiments in which the effect of gravity was "diluted." He proved that a body moving down an inclined plane of given height attains a velocity independent of the angle of slope, and that its terminal speed is the same as if it had fallen through the same vertical height. The trials on the inclined plane thus confirmed his law. The famous story about his dropping different weights from the Tower of Pisa to disprove Aristotle's contention that heavy objects fall faster than light ones is probably untrue; but it follows from Galileo's law that all bodies, heavy or light, are subject to the same acceleration. He himself had no doubt on this score—whether or not he actually corroborated the principle by experiment—but it was not until after his death, when the air pump had been invented, that a complete proof was given by causing bodies to fall in a vacuum.

Experimenting with a pendulum, Galileo obtained further evidence to sustain the principle of "persistence of motion." The swinging bob of a pendulum is analogous to a body falling down an inclined plane. A ball rolling down a plane, assuming negligible friction, will climb another plane to a height equal to that of its starting point. And so, as Galileo found, with the bob. If released at one horizontal level C (as in the diagram), it will ascend to the same height DC, whether it moves by the arc BD or, when the string is caught by nails at E or F, by the steeper arcs BG or BI.[7] It was a short step from his work on the problem of persistence of motion to Newton's first law of motion, also known as the law of inertia.

Another of Galileo's important discoveries in dynamics resulted from his study of the path of projectiles. That a cannon ball moves forward and

[6] Bertrand Russell, *A History of Western Philosophy*, London, 1945.

[7] See Crombie (*Augustine to Galileo*), *op. cit.*, pp. 299–300.

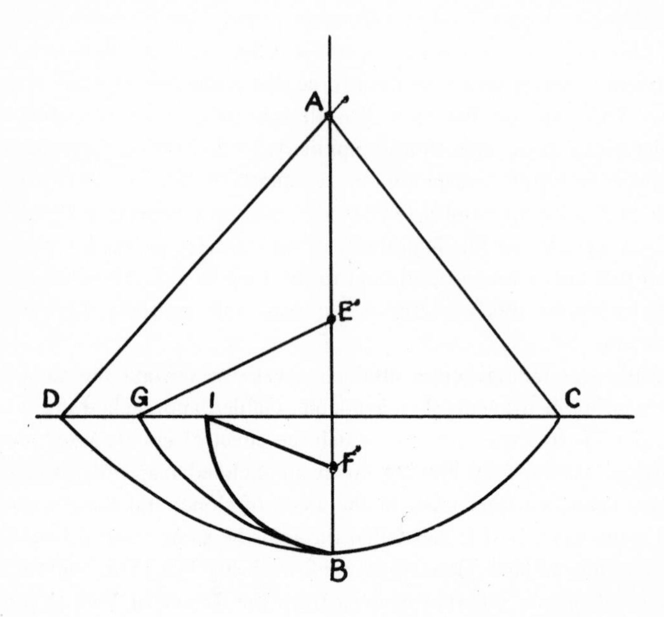

also falls was obvious, but how these motions were combined was not understood. Galileo showed that the trajectory of the projectile could be resolved into two simultaneous motions: one horizontal, with the velocity (disregarding the small air resistance) remaining unchanged, the other vertical, conforming to the law of falling bodies, i.e., 16 feet in the first second, 48 feet in the second second, 80 feet in the third, and so on. Combining the two motions makes the path a parabola. Galileo's principle of the persistence of motion, and his method of dissecting compound motions solved the apparent anomalies gleefully urged by those opposed to the Copernican system. It could now be explained why an object dropped from the mast of a ship fell to the foot of the mast and was not left behind by the ship's motion; why a stone dropped from a tower landed at its base and not to the west of it, even though the rotating earth had moved towards the east while the stone fell. The stone, as Galileo realized, shares the velocity of rotation of the earth, and retains it on the way down.[8]

It was Galileo's way to turn back and forth from hypothesis and deduc-

[8] "In fact if the tower were high enough, there would be the opposite effect to that expected by the opponents of Copernicus. The top of the tower, being further from the center of the earth than the bottom, is moving faster, and therefore the stone should fall slightly to the east of the foot of the tower. This effect, however, would be too slight to be measurable." Russell, *op. cit.*, p. 534.

tion to experiment: no one before him attained a comparable skill in blending experiments with mathematical abstractions. In all his investigations he followed the procedure epitomized in a famous passage of Francis Bacon: "to educe and form axioms from experience . . . to deduce and derive new experiments from axioms. . . . For our road does not lie on a level, but ascends and descends; first ascending to axioms, then descending to works." In the true Platonic tradition he was convinced that the mathematical models which led him to observations were the "enduring reality, the substance, underlying phenomena." [9]

"Philosophy," he wrote in his polemic treatise, *Il Saggiatore*, "is written in that vast book which stands forever open before our eyes, I mean the universe; but it cannot be read until we have learnt the language and become familiar with the characters in which it is written. It is written in mathematical language, and the letters are triangles, circles and other geometrical figures, without which means it is humanly impossible to comprehend a single word."

* * * * *

The facts of Galileo's life are well known and I shall restrict myself to a bare summary. He was born in 1564, the year Michelangelo died, and he died in 1642, the year in which Newton was born. ("I commend these facts," says Bertrand Russell, "to those (if any) who still believe in metempsychosis.") His father was a noble of Florence and Galileo was well educated, first in medicine, at the University of Pisa, then in mathematics and physics. One of his first discoveries was that of the isochronism of the pendulum; he was seventeen when the sight of a lamp set swinging in the cathedral of Pisa—which he measured by his pulse beats—inspired this conjecture. Another of his early works was the invention of a hydrostatic balance. For eighteen years (1592–1610) he held the chair of mathematics at Padua. He was an enormously popular lecturer, and "such was the charm of his demonstrations that a hall capable of containing 2000 people had eventually to be assigned for the accommoda-

[9] "Galileo's Platonism was of the same kind as that which had led to Archimedes being known in the sixteenth century as the 'Platonic philosopher,' and with Galileo mathematical abstractions got their validity as statements about Nature by being solutions of particular physical problems. By using this method of abstracting from immediate and direct experience, and by correlating observed events by means of mathematical relations which could not themselves be observed, he was led to experiments of which he could not have thought in terms of the old commonsense empiricism. A good example of this is his work on the pendulum.

"By abstracting from the inessentials of the situation, 'the opposition of the air, and line, or other accidents,' he was able to demonstrate the law of the pendulum, that the period of oscillation is independent of the arc of swing and simply proportional to the square-root of the length. This having been proved, he could then reintroduce the previously excluded factors. He showed, for instance, that the reason why a real pendulum, of which the thread was not weightless, came to rest, was not simply because of air resistance, but because each particle of the thread acted as a small pendulum. Since they were at different distances from the point of suspension, they had different frequencies and therefore inhibited each other." Crombie, *op. cit.*, (*Augustine to Galileo*), pp. 295–296.

tion of the overflowing audiences which they attracted." [10] Galileo made first-class contributions to hydrostatics, to the mechanics of fluids and to acoustics. He designed the first pendulum clock, invented the first thermometer (a glass bulb and tube, filled with air and water, with the end of the open tube dipping in a vessel of water), and constructed, from his own knowledge of refraction, one of the first telescopes and a compound microscope.[11] Although he was early drawn to the Copernican system, it was not until 1604, on the appearance of a new star, that he publicly renounced the Aristotelian axiom of the "incorruptibility of the heavens"; a short time later he entirely abandoned the Ptolemaic principles. Having greatly improved on his first telescope, Galileo made a series of discoveries which opened a new era in the history of astronomy. He observed the mountains in the moon and roughly measured their height; the visibility of "the old moon in the new moon's arms" he explained by earth-shine; [12] he discovered four of the eleven satellites of Jupiter,[13] innumerable stars and nebulae, sun spots, the phases of Venus predicted by Copernicus, the librations of the moon.

In 1615, after Galileo had removed from Padua to Florence, his advocacy of the Copernican doctrines began to bring him into conflict with the Church. At first the warnings were mild. "Write freely," he was told by a high ecclesiastic, Monsignor Dini, "but keep outside the sacristy." He made two visits to Rome to explain his position; the second on the accession of Pope Urban VIII, who received him warmly. But the publication in 1632 of his powerfully argued, beautifully written masterpiece, *Dialogue on the Great World Systems*, an evaluation, "sparkling with malice," of the comparative merits of the old and new theories of celestial motion, brought a head-on collision with the Inquisition.[14] Galileo was

[10] *Encyclopaedia Britannica*, Eleventh Edition, article on Galileo.

[11] It is the Dutchman, Jan Lippershey, to whom priority in the invention of these instruments (1600) is usually attributed. See H. C. King, *The History of the Telescope*, Cambridge, 1955.

A small excursion on Galileo's work with the microscope will perhaps be permitted. "When the Frenchman, Jean Tarde, called on Galileo in 1614, he said 'Galileo told me that the tube of a telescope for observing the stars is no more than two feet in length; but to see objects well, which are very near, and which on account of their smaller size are hardly visible to the naked eye, the tube must be two or three times longer. He tells me that with this long tube he has seen flies which look as big as a lamb, are covered all over with hair, and have very pointed nails, by means of which they keep themselves up and walk on glass although hanging feet upwards.'" *Galileo, Opera, Ed. Naz.* Vol. 19, p. 589, as quoted in Crombie, *op. cit.*, p. 352.

[12] Sir Oliver Lodge, *Pioneers of Science*, London, 1928, p. 100.

[13] It is not always possible to prove to a philosopher the existence of a thing by bringing it into plain view. The professor of philosophy at Padua "refused to look through Galileo's telescope, and his colleague at Pisa labored before the Grand Duke with logical arguments, 'as if with magical incantations to charm the new planets out of the sky.'" Dampier, *op. cit.*, p. 130.

[14] In 1953 appeared two new excellent editions of the Dialogue, last translated into English by Sir Thomas Salusbury in 1661: (1) *Galileo's Dialogue on the Great World Systems*, edited by Giorgio de Santillana, Chicago, 1953 (based on the Salusbury translation); (2) *Galileo Galilei—Dialogue Concerning the Two Chief World*

summoned to Rome and tried for heresy. Before his spirit was broken he observed: "In these and other positions certainly no man doubts but His Holiness the Pope hath always an absolute power of admitting or condemning them; but it is not in the power of any creature to make them to be true or false or otherwise than of their own nature and in fact they are." Long questioning—though it is unlikely physical torture was applied—brought him to his knees. He was forced to recant, to recite penitential psalms, and was sentenced to house imprisonment for the rest of his life. He retired to a villa at Arcetri, near Florence, where he continued, though much enfeebled and isolated, to write and meditate. In 1637 he became blind and thereafter the rigor of his confinement was relaxed so as to permit him to have visitors. Among those who came, it is said, was John Milton. He died aged seventy-eight.

I have taken a substantial excerpt from the *Dialogues Concerning Two New Sciences* (*Discorsi e Dimostriazioni Matematiche Intorno a due nuove scienze*),[15] a work completed in 1636. It presents his mature and final reflections on the science of mechanics and is a monument of literature and science. On the margin of Galileo's own copy of the *Dialogue on the Great World Systems* appears a note in his handwriting which sums up his lifelong, passionate and courageous dedication to the unending struggle of reason against authority:

"In the matter of introducing novelties. And who can doubt that it will lead to the worst disorders when minds created free by God are compelled to submit slavishly to an outside will? When we are told to deny our senses and subject them to the whim of others? When people devoid of whatsoever competence are made judges over experts and are granted authority to treat them as they please? These are the novelties which are apt to bring about the ruin of commonwealths and the subversion of the state."

Systems—Ptolemaic and Copernican, translated by Stillman Drake, Foreword by Albert Einstein, Berkeley and Los Angeles, 1953. For the most detailed and searching modern account of Galileo's clash with the Church, and trial for heresy, see Giorgio de Santillana, *The Crime of Galileo*, Chicago, 1955.

[15] The standard English translation is by Henry Crew and Alfonso de Salvio, The Macmillan Company, New York, 1914. A reprint has recently (1952) been issued by Dover Publications, Inc., New York.

[My uncle Toby] proceeded next to Galileo and Torricellius, wherein, by certain Geometrical rules, infallibly laid down, he found the precise part to be a "Parabola"—or else an "Hyperbola,"—and that the parameter, or "latus rectum," of the conic section of the said path, was to the quantity and amplitude in a direct ratio, as the whole line to the sine of double the angle of incidence, formed by the breech upon an horizontal line;—and that the semiparameter,—stop! my dear uncle Toby—stop!

—LAWRENCE STERNE

In questions of science the authority of a thousand is not worth the humble reasoning of a single individual. —GALILEO GALILEI

1 Mathematics of Motion

By GALILEO GALILEI

THIRD DAY

CHANGE OF POSITION [*De Motu Locali*]

MY purpose is to set forth a very new science dealing with a very ancient subject. There is, in nature, perhaps nothing older than motion, concerning which the books written by philosophers are neither few nor small; nevertheless I have discovered by experiment some properties of it which are worth knowing and which have not hitherto been either observed or demonstrated. Some superficial observations have been made, as, for instance, that the free motion [*naturalem motum*] of a heavy falling body is continuously accelerated; [1] but to just what extent this acceleration occurs has not yet been announced; for so far as I know, no one has yet pointed out that the distances traversed, during equal intervals of time, by a body falling from rest, stand to one another in the same ratio as the odd numbers beginning with unity.[2]

It has been observed that missiles and projectiles describe a curved path of some sort; however no one has pointed out the fact that this path is a parabola. But this and other facts, not few in number or less worth knowing, I have succeeded in proving; and what I consider more important, there have been opened up to this vast and most excellent science, of which my work is merely the beginning, ways and means by which other minds more acute than mine will explore its remote corners. . . .

NATURALLY ACCELERATED MOTION

The properties belonging to uniform motion have been discussed in the preceding section; but accelerated motion remains to be considered.

[1] "Natural motion" of the author has here been translated into "free motion"—since this is the term used to-day to distinguish the "natural" from the "violent" motions of the Renaissance. [*Trans.*]

[2] A theorem demonstrated in Corollary I, p. 746. [*Trans.*]

And first of all it seems desirable to find and explain a definition best fitting natural phenomena. For anyone may invent an arbitrary type of motion and discuss its properties; thus, for instance, some have imagined helices and conchoids as described by certain motions which are not met with in nature, and have very commendably established the properties which these curves possess in virtue of their definitions; but we have decided to consider the phenomena of bodies falling with an acceleration such as actually occurs in nature and to make this definition of accelerated motion exhibit the essential features of observed accelerated motions. And this, at last, after repeated efforts we trust we have succeeded in doing. In this belief we are confirmed mainly by the consideration that experimental results are seen to agree with and exactly correspond with those properties which have been, one after another, demonstrated by us. Finally, in the investigation of naturally accelerated motion we were led, by hand as it were, in following the habit and custom of nature herself, in all her various other processes, to employ only those means which are most common, simple and easy.

For I think no one believes that swimming or flying can be accomplished in a manner simpler or easier than that instinctively employed by fishes and birds.

When, therefore, I observe a stone initially at rest falling from an elevated position and continually acquiring new increments of speed, why should I not believe that such increases take place in a manner which is exceedingly simple and rather obvious to everybody? If now we examine the matter carefully we find no addition or increment more simple than that which repeats itself always in the same manner. This we readily understand when we consider the intimate relationship between time and motion; for just as uniformity of motion is defined by and conceived through equal times and equal spaces (thus we call a motion uniform when equal distances are traversed during equal time-intervals), so also we may, in a similar manner, through equal time-intervals, conceive additions of speed as taking place without complication; thus we may picture to our mind a motion as uniformly and continuously accelerated when, during any equal intervals of time whatever, equal increments of speed are given to it. Thus if any equal intervals of time whatever have elapsed, counting from the time at which the moving body left its position of rest and began to descend, the amount of speed acquired during the first two time-intervals will be double that acquired during the first time-interval alone; so the amount added during three of these time-intervals will be treble; and that in four, quadruple that of the first time-interval. To put the matter more clearly, if a body were to continue its motion with the same speed which it had acquired during the first time-interval and were to retain this same uniform speed, then its motion would be twice as slow as that which it would have if its velocity had been acquired during *two* time-intervals.

And thus, it seems, we shall not be far wrong if we put the increment of speed as proportional to the increment of time; hence the definition of motion which we are about to discuss may be stated as follows: A motion is said to be uniformly accelerated, when starting from rest, it acquires, during equal time-intervals, equal increments of speed.

SAGREDO. Although I can offer no rational objection to this or indeed to any other definition, devised by any author whomsoever, since all definitions are arbitrary, I may nevertheless without offense be allowed to doubt whether such a definition as the above, established in an abstract manner, corresponds to and describes that kind of accelerated motion which we meet in nature in the case of freely falling bodies. And since the Author apparently maintains that the motion described in his definition is that of freely falling bodies, I would like to clear my mind of certain difficulties in order that I may later apply myself more earnestly to the propositions and their demonstrations.

SALVIATI. It is well that you and Simplicio raise these difficulties. They are, I imagine, the same which occurred to me when I first saw this treatise, and which were removed either by discussion with the Author himself, or by turning the matter over in my own mind.

SAGR. When I think of a heavy body falling from rest, that is, starting with zero speed and gaining speed in proportion to the time from the beginning of the motion; such a motion as would, for instance, in eight beats of the pulse acquire eight degrees of speed; having at the end of the fourth beat acquired four degrees; at the end of the second, two; at the end of the first, one: and since time is divisible without limit, it follows from all these considerations that if the earlier speed of a body is less than its present speed in a constant ratio, then there is no degree of speed however small (or, one may say, no degree of slowness however great) with which we may not find this body travelling after starting from infinite slowness, i.e., from rest. So that if that speed which it had at the end of the fourth beat was such that, if kept uniform, the body would traverse two miles in an hour, and if keeping the speed which it had at the end of the second beat, it would traverse one mile an hour, we must infer that, as the instant of starting is more and more nearly approached, the body moves so slowly that, if it kept on moving at this rate, it would not traverse a mile in an hour, or in a day, or in a year or in a thousand years; indeed, it would not traverse a span in an even greater time; a phenomenon which baffles the imagination, while our senses show us that a heavy falling body suddenly acquires great speed.

SALV. This is one of the difficulties which I also at the beginning, experienced, but which I shortly afterwards removed; and the removal was effected by the very experiment which creates the difficulty for you. You say the experiment appears to show that immediately after a heavy body starts from rest it acquires a very considerable speed: and I say that the same experiment makes clear the fact that the initial motions of a falling body, no matter how heavy, are very slow and gentle. Place a heavy body upon a yielding material, and leave it there without any pressure except that owing to its own weight; it is clear that if one lifts this body a cubit or two and allows it to fall upon the same material, it will, with this impulse, exert a new and greater pressure than that caused by its mere weight; and this effect is brought about by the [weight of the] falling body together with the velocity acquired during the fall, an effect which will be greater and greater according to the height of the fall, that is according as the velocity of the falling body becomes greater. From the quality and intensity of the blow we are thus enabled to accurately estimate the speed of a falling body. But tell me, gentlemen, is it not true that if a block be allowed to fall upon a stake from a height of four cubits and drives it into the earth, say, four finger-breadths, that coming from a height of two cubits it will drive the stake a much less distance, and from the height of one cubit a still less distance; and finally if the block be lifted only one finger-breadth how much more will it accomplish than if merely laid on top of the stake without percussion? Certainly very little. If it be lifted only the thickness of a leaf, the effect will be altogether imperceptible. And since the effect of the blow depends upon the velocity of this striking body, can any one doubt the motion is very slow and the speed more than small whenever the effect [of the blow] is imperceptible? See now the power of truth; the same experiment which at first glance seemed to show one thing, when more carefully examined, assures us of the contrary.

But without depending upon the above experiment, which is doubtless very conclusive, it seems to me that it ought not to be difficult to establish such a fact by reasoning alone. Imagine a heavy stone held in the air at rest; the support is removed and the stone set free; then since it is heavier than the air it begins to fall, and not with uniform motion but slowly at the beginning and with a continuously accelerated motion. Now since velocity can be increased and diminished without limit, what reason is there to believe that such a moving body starting with infinite slowness, that is, from rest, immediately acquires a speed of ten degrees rather than one of four, or of two, or of one, or of a half, or of a hundredth; or, indeed, of any of the infinite number of small values [of speed]? Pray listen. I hardly think you will refuse to grant that the gain of speed of the stone falling from rest follows the same sequence as the diminution and loss of this same speed when, by some impelling force, the stone is thrown to its former elevation: but even if you do not grant this, I do not see how you can doubt that the ascending stone, diminishing in speed, must before coming to rest pass through every possible degree of slowness.

SIMPLICIO. But if the number of degrees of greater and greater slowness is limitless, they will never be all exhausted, therefore such an ascending heavy body will never reach rest, but will continue to move without limit always at a slower rate; but this is not the observed fact.

SALV. This would happen, Simplicio, if the moving body were to maintain its speed for any length of time at each degree of velocity; but it merely passes each point without delaying more than an instant: and since each time-interval however small may be divided into an infinite number of instants, these will always be sufficient [in number] to correspond to the infinite degrees of diminished velocity.

That such a heavy rising body does not remain for any length of time at any given degree of velocity is evident from the following: because if, some time-interval having been assigned, the body moves with the same speed in the last as in the first instant of that time-interval, it could from this second degree of elevation be in like manner raised through an equal height, just as it was transferred from the first elevation to the second, and by the same reasoning would pass from the second to the third and would finally continue in uniform motion forever. . . .

SALV. The present does not seem to be the proper time to investigate the cause of the acceleration of natural motion concerning which various opinions have been expressed by various philosophers, some explaining it by attraction to the center, others to repulsion between the very small parts of the body, while still others attribute it to a certain stress in the surrounding medium which closes in behind the falling body and drives it from one of its positions to another. Now, all these fantasies, and others too, ought to be examined; but it is not really worth while. At present it is the purpose of our Author merely to investigate and to demonstrate some of the properties of accelerated motion (whatever the cause of this acceleration may be)—meaning thereby a motion, such that the momentum of its velocity [*i momenti della sua velocità*] goes on increasing after departure from rest, in simple proportionality to the time, which is the same as saying that in equal time-intervals the body receives equal increments of velocity; and if we find the properties [of accelerated motion] which will be demonstrated later are realized in freely falling and accelerated bodies, we may conclude that the assumed definition includes such a motion of falling bodies and that their speed [*accelerazione*] goes on increasing as the time and the duration of the motion.

SAGR. So far as I see at present, the definition might have been put a little more clearly perhaps without changing the fundamental idea, namely, uniformly accelerated motion is such that its speed increases in proportion to the space traversed; so that, for example, the speed acquired by a body in falling four cubits would be double that acquired in falling two cubits and this latter speed would be double that acquired in the first cubit. Because there is no doubt but that a heavy body falling from the height of six cubits has, and strikes with, a momentum [*impeto*] double that it had at the end of three cubits, triple that which it had at the end of one.

SALV. It is very comforting to me to have had such a companion in error; and moreover let me tell you that your proposition seems so highly probable that our Author himself admitted, when I advanced this opinion to him, that he had for some time shared the same fallacy. But what most surprised me was to see two propositions so inherently probable that they commanded the assent of everyone to whom they were presented, proven in a few simple words to be not only false, but impossible.

SIMP. I am one of those who accept the proposition, and believe that a falling body acquires force [*vires*] in its descent, its velocity increasing in proportion to the space, and that the momentum [*momento*] of the falling body is doubled when it falls from a doubled height; these propositions, it appears to me, ought to be conceded without hesitation or controversy.

SALV. And yet they are as false and impossible as that motion should be completed instantaneously; and here is a very clear demonstration of it. If the velocities are in proportion to the spaces traversed, or to be traversed, then these spaces are traversed in equal intervals of time; if, therefore, the velocity with which the falling body traverses a space of eight feet were double that with which it covered the first four feet (just as the one distance is double the other) then the time-intervals required for these passages would be equal. But for one and the same body to fall eight feet and four feet in the same time is possible only in the case of instantaneous [discontinuous] motion; but observation shows us that the motion of a falling body occupies time, and less of it in covering a distance of four feet than of eight feet; therefore it is not true that its velocity increases in proportion to the space.

The falsity of the other proposition may be shown with equal clearness. For if we consider a single striking body the difference of momentum in its blows can depend only upon difference of velocity; for if the striking body falling from a double height were to deliver a blow of double momentum, it would be necessary for this body to strike with a doubled velocity; but with this doubled speed it would traverse a doubled space in the same time-interval; observation however shows that the time required for fall from the greater height is longer.

SAGR. You present these recondite matters with too much evidence and ease; this great facility makes them less appreciated than they would be had they been presented in a more abstruse manner. For, in my opinion, people esteem more lightly that knowledge which they acquire with so little labor than that acquired through long and obscure discussion.

SALV. If those who demonstrate with brevity and clearness the fallacy of many popular beliefs were treated with contempt instead of gratitude the injury would be quite bearable; but on the other hand it is very unpleasant and annoying to see men, who claim to be peers of anyone in a certain field of study, take for granted certain conclusions which later are quickly and easily shown by another to be false. I do not describe such a feeling as one of envy, which usually degenerates into hatred and anger against those who discover such fallacies; I would call it a strong desire to maintain old errors, rather than accept newly discovered truths. This desire at times induces them to unite against these truths, although at heart believing in them, merely for the purpose of lowering the esteem in which certain others are held by the unthinking crowd. Indeed, I have heard from our Academician many such fallacies held as true but easily refutable; some of these I have in mind.

SAGR. You must not withhold them from us, but, at the proper time, tell us about them even though an extra session be necessary. But now, continuing the thread of our talk, it would seem that up to the present we have established the definition of uniformly accelerated motion which is expressed as follows:

> A motion is said to be equally or uniformly accelerated when, starting from rest, its momentum (*celeritatis momenta*) receives equal increments in equal times.

SALV. This definition established, the Author makes a single assumption, namely,

> The speeds acquired by one and the same body moving down planes of different inclinations are equal when the heights of these planes are equal.

By the height of an inclined plane we mean the perpendicular let fall from the upper end of the plane upon the horizontal line drawn through the lower end of the same plane. Thus, to illustrate, let the line AB be horizontal, and let the planes CA and CD be inclined to it; then the Author calls the perpendicular CB the "height" of the planes CA and CD; he supposes that the speeds acquired by one and the same body, descending along the planes CA and CD to the terminal points A and D are equal since the heights of these planes are the same, CB; and also it must be understood that this speed is that which would be acquired by the same body falling from C to B.

SAGR. Your assumption appears to me so reasonable that it ought to be conceded without question, provided of course there are no chance or outside resistances, and that the planes are hard and smooth, and that

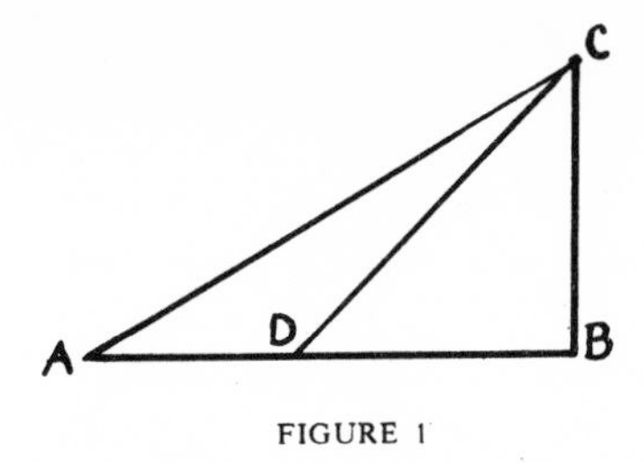

FIGURE 1

the figure of the moving body is perfectly round, so that neither plane nor moving body is rough. All resistance and opposition having been removed, my reason tells me at once that a heavy and perfectly round ball descending along the lines CA, CD, CB would reach the terminal points A, D, B, with equal momenta [*impeti eguali*].

SALV. Your words are very plausible; but I hope by experiment to increase the probability to an extent which shall be little short of a rigid demonstration.

Imagine this page to represent a vertical wall, with a nail driven into it; and from the nail let there be suspended a lead bullet of one or two ounces by means of a fine vertical thread, AB, say from four to six feet long, on this wall draw a horizontal line DC, at right angles to the vertical thread AB, which hangs about two finger-breadths in front of the wall. Now bring the thread AB with the attached ball into the position AC and set it free; first it will be observed to descend along the arc CBD, to pass the point B, and to travel along the arc BD, till it almost reaches the horizontal CD, a slight shortage being caused by the resistance of the air and the string; from this we may rightly infer that the ball in its descent through the arc CB acquired a momentum [*impeto*] on reaching B, which was just sufficient to carry it through a similar arc BD to the same height. Having repeated this experiment many times, let us now drive a nail into the wall close to the perpendicular AB, say at E or F, so that it projects out some five or six finger-breadths in order that the thread, again carrying the bullet through the arc CB, may strike upon the nail E when the bullet reaches B, and thus compel it to traverse the arc BG, described about E as center. From this we can see what can be done by the same momentum [*impeto*] which previously starting at the same point B carried the same body through the arc BD to the horizontal CD. Now, gentlemen, you will observe with pleasure that the ball swings to the point G in the horizontal, and you would see the same thing happen if the obstacle were placed at some lower point, say at F, about which the ball would describe the arc BI, the rise of the ball always terminating exactly on the line CD. But when the nail is placed so low that the remainder of the thread below it will not reach to the height CD

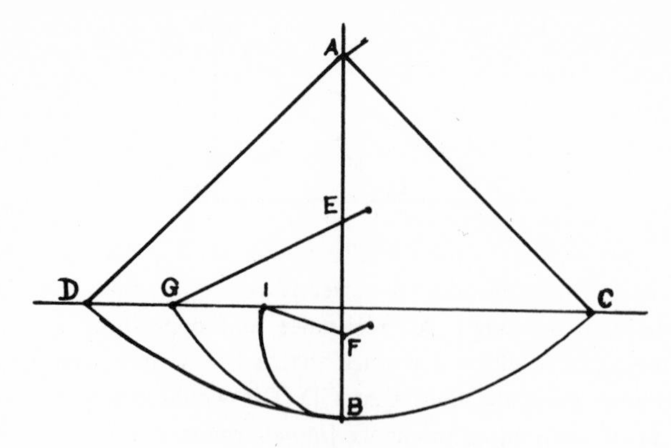

FIGURE 2

(which would happen if the nail were placed nearer B than to the intersection of AB with the horizontal CD) then the thread leaps over the nail and twists itself about it.

This experiment leaves no room for doubt as to the truth of our supposition; for since the two arcs CB and DB are equal and similarly placed, the momentum [*momento*] acquired by the fall through the arc CB is the same as that gained by fall through the arc DB; but the momentum [*momento*] acquired at B, owing to fall through CB, is able to lift the same body [*mobile*] through the arc BD; therefore, the momentum acquired in the fall BD is equal to that which lifts the same body through the same arc from B to D; so, in general, every momentum acquired by fall through an arc is equal to that which can lift the same body through the same arc. But all these momenta [*momenti*] which cause a rise through the arcs BD, BG, and BI are equal, since they are produced by the same momentum, gained by fall through CB, as experiment shows. Therefore all the momenta gained by fall through the arcs DB, GB, IB are equal.

SAGR. The argument seems to me so conclusive and the experiment so well adapted to establish the hypothesis that we may, indeed, consider it as demonstrated.

SALV. I do not wish, Sagredo, that we trouble ourselves too much about this matter, since we are going to apply this principle mainly in motions which occur on plane surfaces, and not upon curved, along which acceleration varies in a manner greatly different from that which we have assumed for planes.

So that, although the above experiment shows us that the descent of the moving body through the arc CB confers upon it momentum [*momento*] just sufficient to carry it to the same height through any of the arcs BD, BG, BI, we are not able, by similar means, to show that the event would be identical in the case of a perfectly round ball descending along planes whose inclinations are respectively the same as the chords of these arcs. It seems likely, on the other hand, that, since these planes form angles at the point B, they will present an obstacle to the ball which has descended along the chord CB, and starts to rise along the chord BD, BG, BI.

In striking these planes some of its momentum [*impeto*] will be lost and it will not be able to rise to the height of the line CD; but this obstacle, which interferes with the experiment, once removed, it is clear that the momentum [*impeto*] (which gains in strength with descent) will be able to carry the body to the same height. Let us then, for the present, take this as a postulate, the absolute truth of which will be established when we find that the inferences from it correspond to and agree perfectly with experiment. The Author having assumed this single principle passes next to the propositions which he clearly demonstrates; the first of these is as follows:

THEOREM I, PROPOSITION I

The time in which any space is traversed by a body starting from rest and uniformly accelerated is equal to the time in which that same space would be traversed by the same body moving at a uniform speed whose value is the mean of the highest speed and the speed just before acceleration began.

Let us represent by the line AB the time in which the space CD is traversed by a body which starts from rest at C and is uniformly accelerated; let the final and highest value of the speed gained during the interval AB be represented by the line EB drawn at right angles to AB; draw the line AE, then all lines drawn from equidistant points on AB and parallel to BE will represent the increasing values of the speed, beginning with the instant A. Let the point F bisect the line EB; draw FG parallel to BA, and GA parallel to FB, thus forming a parallelogram AGFB which will be equal in area to the triangle AEB, since the side GF bisects the side AE at the point I; for if the parallel lines in the triangle AEB are extended to GI, then the sum of all the parallels contained in the quadrilateral is equal to the sum of those contained in the triangle AEB; for those in the triangle IEF are equal to those contained in the triangle GIA, while those included in the trapezium AIFB are common. Since each and every instant of time in the time-interval AB has its corresponding point on the line AB, from which points parallels drawn in and limited by the triangle AEB represent the increasing values of the growing velocity, and since parallels contained within the rectangle represent the values of a speed which is not increasing, but constant, it appears, in like manner, that the momenta [*momenti*] assumed by the moving body may also be

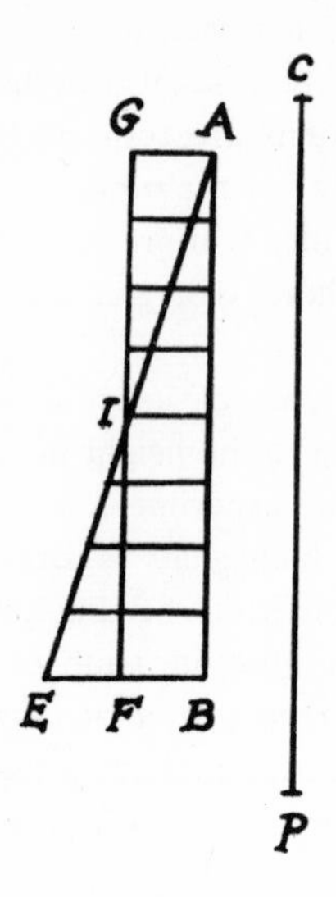

FIGURE 3

represented, in the case of the accelerated motion, by the increasing parallels of the triangle AEB, and, in the case of the uniform motion, by the parallels of the rectangle GB. For, what the momenta may lack in the first part of the accelerated motion (the deficiency of the momenta being represented by the parallels of the triangle AGI) is made up by the momenta represented by the parallels of the triangle IEF.

Hence it is clear that equal spaces will be traversed in equal times by two bodies, one of which, starting from rest, moves with a uniform acceleration, while the momentum of the other, moving with uniform speed, is one-half its maximum momentum under accelerated motion.

Q. E. D.

THEOREM II, PROPOSITION II

The spaces described by a body falling from rest with a uniformly accelerated motion are to each other as the squares of the time-intervals employed in traversing these distances.

Let the time beginning with any instant A be represented by the straight line AB in which are taken any two time-intervals AD and AE. Let HI represent the distance through which the body, starting from rest at H, falls with uniform acceleration. If HL represents the space traversed during the time-interval AD, and HM that covered during the interval AE, then the space MH stands to the space LH in a ratio which is the square of the ratio of the time AE to the time AD; or we may say simply that the distances HM and HL are related as the squares of AE and AD.

Draw the line AC making any angle whatever with the line AB; and

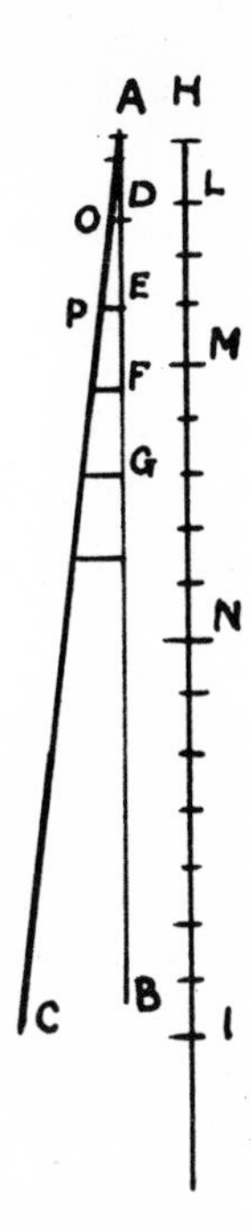

FIGURE 4

from the points D and E, draw the parallel lines DO and EP; of these two lines, DO represents the greatest velocity attained during the interval AD, while EP represents the maximum velocity acquired during the interval AE. But it has just been proved that so far as distances traversed are concerned it is precisely the same whether a body falls from rest with a uniform acceleration or whether it falls during an equal time-interval with a constant speed which is one-half the maximum speed attained during the accelerated motion. It follows therefore that the distances HM and HL are the same as would be traversed, during the time-intervals AE and AD, by uniform velocities equal to one-half those represented by DO and EP respectively. If, therefore, one can show that the distances HM and HL are in the same ratio as the squares of the time-intervals AE and AD, our proposition will be proven.

But in the fourth proposition of the first book it has been shown that the spaces traversed by two particles in uniform motion bear to one another a ratio which is equal to the product of the ratio of the velocities by the ratio of the times. But in this case the ratio of the velocities is the same as the ratio of the time-intervals (for the ratio of AE to AD is the same as that of ½ EP to ½ DO or of EP to DO). Hence the ratio of the spaces traversed is the same as the squared ratio of the time-intervals. Q. E. D.

Evidently then the ratio of the distances is the square of the ratio of the final velocities, that is, of the lines EP and DO, since these are to each other as AE to AD.

COROLLARY I

Hence it is clear that if we take any equal intervals of time whatever, counting from the beginning of the motion, such as AD, DE, EF, FG, in which the spaces HL, LM, MN, NI are traversed, these spaces will bear to one another the same ratio as the series of odd numbers, 1, 3, 5, 7; for this is the ratio of the differences of the squares of the lines [which represent time], differences which exceed one another by equal amounts, this excess being equal to the smallest line [viz. the one representing a single time-interval]: or we may say [that this is the ratio] of the differences of the squares of the natural numbers beginning with unity.

While, therefore, during equal intervals of time the velocities increase as the natural numbers, the increments in the distances traversed during these equal time-intervals are to one another as the odd numbers beginning with unity. . . .

SIMP. I am convinced that matters are as described, once having accepted the definition of uniformly accelerated motion. But as to whether this acceleration is that which one meets in nature in the case of falling bodies, I am still doubtful; and it seems to me, not only for my own sake but also for all those who think as I do, that this would be the proper moment to introduce one of those experiments—and there are many of them, I understand—which illustrate in several ways the conclusions reached.

SALV. The request which you, as a man of science, make, is a very reasonable one; for this is the custom—and properly so—in those sciences where mathematical demonstrations are applied to natural phenomena, as is seen in the case of perspective, astronomy, mechanics, music, and others where the principles, once established by well-chosen experiments, become the foundations of the entire superstructure. I hope therefore it will not appear to be a waste of time if we discuss at considerable length this first and most fundamental question upon which hinge numerous consequences of which we have in this book only a small number, placed there by the Author, who has done so much to open a pathway hitherto closed to minds of speculative turn. So far as experiments go they have not been neglected by the Author; and often, in his company, I have attempted in the following manner to assure myself that the acceleration actually experienced by falling bodies is that above described.

A piece of wooden moulding or scantling, about 12 cubits long, half a cubit wide, and three finger-breadths thick, was taken; on its edge was cut a channel a little more than one finger in breadth; having made this groove very straight, smooth, and polished, and having lined it with parchment, also as smooth and polished as possible, we rolled along it a hard, smooth, and very round bronze ball. Having placed this board in a sloping position, by lifting one end some one or two cubits above the other, we rolled the ball, as I was just saying, along the channel, noting, in a manner presently to be described, the time required to make the descent. We repeated this experiment more than once in order to measure the time with an accuracy such that the deviation between two observations never exceeded one-tenth of a pulse-beat. Having performed this operation and having assured ourselves of its reliability, we now rolled the ball only one-quarter the length of the channel; and having measured the time of its descent, we found it precisely one-half of the former. Next we tried other distances, comparing the time for the whole length with that for the half, or with that for two-thirds, or three-fourths, or indeed for any fraction; in such experiments, repeated a full hundred times, we always found that the spaces traversed were to each other as the squares of the times, and this was true for all inclinations of the plane, i.e., of the channel, along which we rolled the ball. We also observed that the times of descent, for various inclinations of the plane, bore to one another precisely that ratio which, as we shall see later, the Author had predicted and demonstrated for them.

For the measurement of time, we employed a large vessel of water placed in an elevated position; to the bottom of this vessel was soldered a pipe of small diameter giving a thin jet of water, which we collected in a small glass during the time of each descent, whether for the whole length of the channel or for a part of its length; the water thus collected was weighed, after each descent, on a very accurate balance; the differences and ratios of these weights gave us the differences and ratios of the times, and this with such accuracy that although the operation was repeated many, many times, there was no appreciable discrepancy in the results.

SIMP. I would like to have been present at these experiments; but feeling confidence in the care with which you performed them, and in the fidelity with which you relate them, I am satisfied and accept them as true and valid.

SALV. Then we can proceed without discussion.

COROLLARY II

Secondly, it follows that, starting from any initial point, if we take any two distances, traversed in any time-intervals whatsoever, these time-intervals bear to one another the same ratio as one of the distances to the mean proportional of the two distances.

For if we take two distances ST and SY measured from the initial

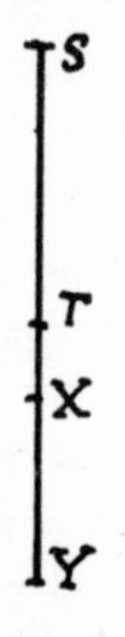

FIGURE 5

point S, the mean proportional of which is SX, the time of fall through ST is to the time of fall through SY as ST is to SX; or one may say the time of fall through SY is to the time of fall through ST as SY is to SX. Now since it has been shown that the spaces traversed are in the same ratio as the squares of the times; and since, moreover, the ratio of the space SY to the space ST is the square of the ratio SY to SX, it follows that the ratio of the times of fall through SY and ST is the ratio of the respective distances SY and SX.

SCHOLIUM

The above corollary has been proven for the case of vertical fall; but it holds also for planes inclined at any angle; for it is to be assumed that along these planes the velocity increases in the same ratio, that is, in proportion to the time, or, if you prefer, as the series of natural numbers. . . .

THEOREM III, PROPOSITION III

If one and the same body, starting from rest, falls along an inclined plane and also along a vertical, each having the same height, the times of descent will be to each other as the lengths of the inclined plane and the vertical.

Let AC be the inclined plane and AB the perpendicular, each having the same vertical height above the horizontal, namely, BA; then I say, the time of descent of one and the same body along the plane AC bears a ratio to the time of fall along the perpendicular AB, which is the same as the ratio of the length AC to the length AB. Let DG, EI and LF be any lines parallel to the horizontal CB; then it follows from what has preceded that a body starting from A will acquire the same speed at the point G as at D, since in each case the vertical fall is the same; in like manner the speeds at I and E will be the same; so also those at L and F. And in general the speeds at the two extremities of any parallel drawn from any point on AB to the corresponding point on AC will be equal.

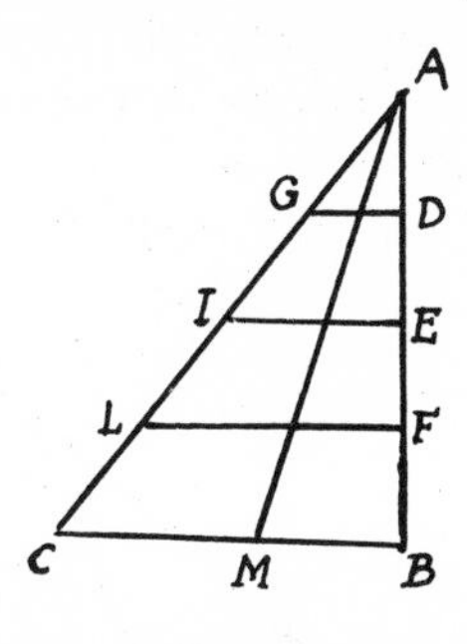

FIGURE 6

Thus the two distances AC and AB are traversed at the same speed. But it has already been proved that if two distances are traversed by a body moving with equal speeds, then the ratio of the times of descent will be the ratio of the distances themselves; therefore, the time of descent along AC is to that along AB as the length of the plane AC is to the vertical distance AB. Q. E. D.

SAGR. It seems to me that the above could have been proved clearly and briefly on the basis of a proposition already demonstrated, namely, that the distance traversed in the case of accelerated motion along AC or AB is the same as that covered by a uniform speed whose value is one-half the maximum speed, CB; the two distances AC and AB having been traversed at the same uniform speed it is evident, from Proposition I, that the times of descent will be to each other as the distances.

COROLLARY

Hence we may infer that the times of descent along planes having different inclinations, but the same vertical height stand to one another in the same ratio as the lengths of the planes. For consider any plane AM extending from A to the horizontal CB; then it may be demonstrated in the same manner that the time of descent along AM is to the time along AB as the distance AM is to AB; but since the time along AB is to that along AC as the length AB is to the length AC, it follows, *ex æquali*, that as AM is to AC so is the time along AM to the time along AC.

THEOREM IV, PROPOSITION IV

The times of descent along planes of the same length but of different inclinations are to each other in the inverse ratio of the square roots of their heights.

From a single point B draw the planes BA and BC, having the same length but different inclinations; let AE and CD be horizontal lines drawn to meet the perpendicular BD; and let BE represent the height of the plane AB, and BD the height of BC; also let BI be a mean proportional to BD and BE; then the ratio of BD to BI is equal to the square root of the ratio of BD to BE. Now, I say, the ratio of the times of descent along BA and BC is the ratio of BD to BI; so that the time of descent along BA is related to the height of the other plane BC, namely BD as the time

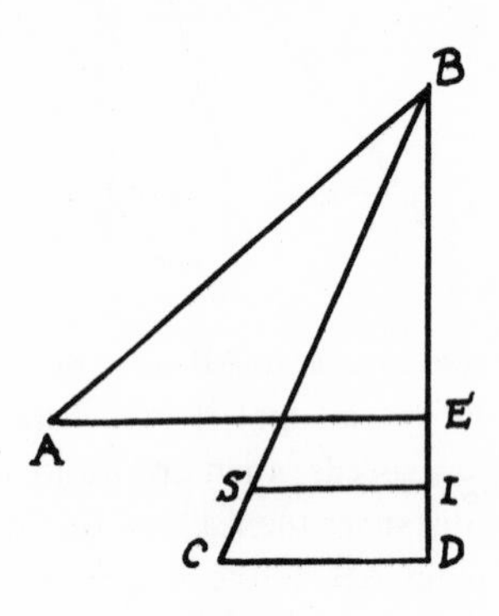

FIGURE 7

along BC is related to the height BI. Now it must be proved that the time of descent along BA is to that along BC as the length BD is to the length BI.

Draw IS parallel to DC; and since it has been shown that the time of fall along BA is to that along the vertical BE as BA is to BE; and also that the time along BE is to that along BD as BE is to BI; and likewise that the time along BD is to that along BC as BD is to BC, or as BI to BS; it follows, *ex æquali*, that the time along BA is to that along BC as BA to BS, or BC to BS. However, BC is to BS as BD is to BI; hence follows our proposition.

THEOREM V, PROPOSITION V

The times of descent along planes of different length, slope and height bear to one another a ratio which is equal to the product of the ratio of the lengths by the square root of the inverse ratio of their heights.

Draw the planes AB and AC, having different inclinations, lengths, and heights. My theorem then is that the ratio of the time of descent along AC to that along AB is equal to the product of the ratio of AC to AB by the square root of the inverse ratio of their heights.

For let AD be a perpendicular to which are drawn the horizontal lines BG and CD; also let AL be a mean proportional to the heights AG and AD; from the point L draw a horizontal line meeting AC in F; accordingly AF will be a mean proportional between AC and AE. Now since the time of descent along AC is to that along AE as the length AF is to AE;

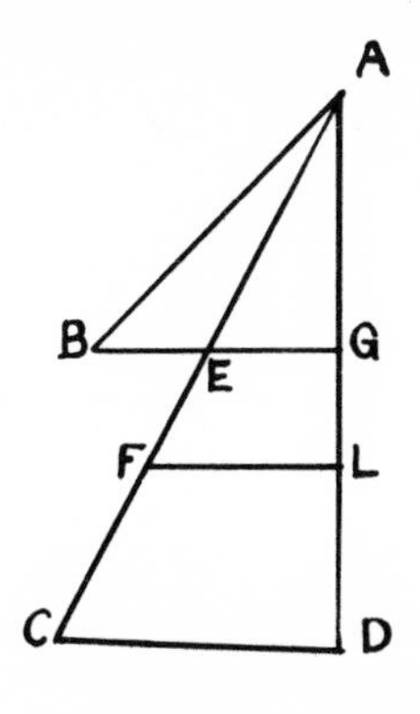

FIGURE 8

and since the time along AE is to that along AB as AE is to AB, it is clear that the time along AC is to that along AB as AF is to AB.

Thus it remains to be shown that the ratio of AF to AB is equal to the product of the ratio of AC to AB by the ratio of AG to AL, which is the inverse ratio of the square roots of the heights DA and GA. Now it is evident that, if we consider the line AC in connection with AF and AB, the ratio of AF to AC is the same as that of AL to AD, or AG to AL which is the square root of the ratio of the heights AG and AD; but the ratio of AC to AB is the ratio of the lengths themselves. Hence follows the theorem.

THEOREM VI, PROPOSITION VI

If from the highest or lowest point in a vertical circle there be drawn any inclined planes meeting the circumference the times of descent along these chords are each equal to the other.

On the horizontal line GH construct a vertical circle. From its lowest point—the point of tangency with the horizontal—draw the diameter FA and from the highest point, A, draw inclined planes to B and C, any points whatever on the circumference; then the times of descent along these are equal. Draw BD and CE perpendicular to the diameter; make AI a mean proportional between the heights of the planes, AE and AD; and since the rectangles FA.AE and FA.AD are respectively equal to the squares of AC and AB, while the rectangle FA.AE is to the rectangle FA.AD as AE is to AD, it follows that the square of AC is to the square of AB as the length AE is to the length AD. But since the length AE is to AD as the square of AI is to the square of AD, it follows that the squares on the lines AC and AB are to each other as the squares on the lines AI and AD, and hence also the length AC is to the length AB as AI

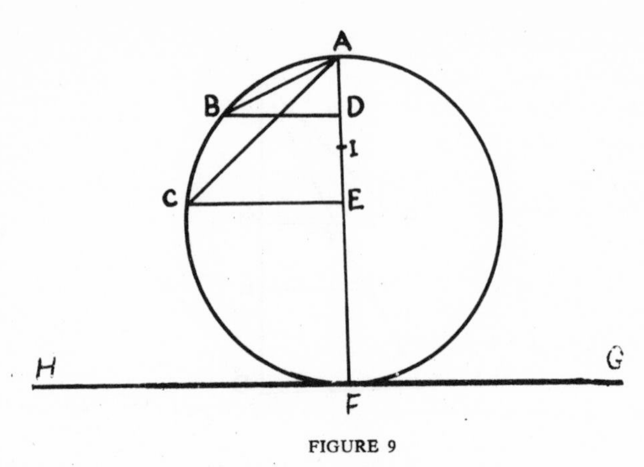

FIGURE 9

is to AD. But it has previously been demonstrated that the ratio of the time of descent along AC to that along AB is equal to the product of the two ratios AC to AB and AD to AI; but this last ratio is the same as that of AB to AC. Therefore the ratio of the time of descent along AC to that along AB is the product of the two ratios, AC to AB and AB to AC. The ratio of these times is therefore unity. Hence follows our proposition.

By use of the principles of mechanics [*ex mechanicis*] one may obtain the same result. . . .

SCHOLIUM

We may remark that any velocity once imparted to a moving body will be rigidly maintained as long as the external causes of acceleration or retardation are removed, a condition which is found only on horizontal planes; for in the case of planes which slope downwards there is already present a cause of acceleration, while on planes sloping upward there is retardation; from this it follows that motion along a horizontal plane is perpetual; for, if the velocity be uniform, it cannot be diminished or slackened, much less destroyed. Further, although any velocity which a body may have acquired through natural fall is permanently maintained so far as its own nature [*suapte natura*] is concerned, yet it must be remembered that if, after descent along a plane inclined downwards, the body is deflected to a plane inclined upward, there is already existing in this latter plane a cause of retardation; for in any such plane this same body is subject to a natural acceleration downwards. Accordingly we have here the superposition of two different states, namely, the velocity acquired during the preceding fall which if acting alone would carry the body at a uniform rate to infinity, and the velocity which results from a natural acceleration downwards common to all bodies. It seems altogether reasonable, therefore, if we wish to trace the future history of a body which has descended along some inclined plane and has been deflected along some plane inclined upwards, for us to assume that the maximum speed acquired during descent is permanently maintained during the ascent. In the ascent, however, there supervenes a natural inclination downwards, namely, a motion which, starting from rest, is accelerated at the usual rate. If perhaps this discussion is a little obscure, the following figure will help to make it clearer.

Let us suppose that the descent has been made along the downward sloping plane AB, from which the body is deflected so as to continue its motion along the upward sloping plane BC; and first let these planes be of equal length and placed so as to make equal angles with the horizontal line GH. Now it is well known that a body, starting from rest at A, and descending along AB, acquires a speed which is proportional to the time,

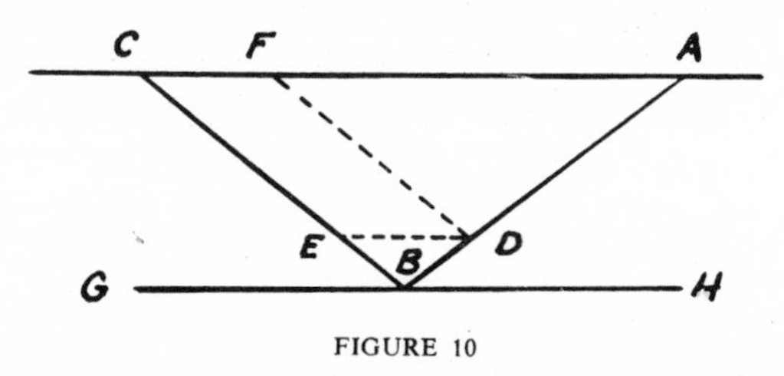

FIGURE 10

which is a maximum at B, and which is maintained by the body so long as all causes of fresh acceleration or retardation are removed; the acceleration to which I refer is that to which the body would be subject if its motion were continued along the plane AB extended, while the retardation is that which the body would encounter if its motion were deflected along the plane BC inclined upwards; but, upon the horizontal plane GH, the body would maintain a uniform velocity equal to that which it had acquired at B after fall from A; moreover this velocity is such that, during an interval of time equal to the time of descent through AB, the body will traverse a horizontal distance equal to twice AB. Now let us imagine this same body to move with the same uniform speed along the plane BC so that here also during a time-interval equal to that of descent along AB, it will traverse along BC extended a distance twice AB; but let us suppose that, at the very instant the body begins its ascent it is subjected, by its very nature, to the same influences which surrounded it during its descent from A along AB, namely, it descends from rest under the same acceleration as that which was effective in AB, and it traverses, during an equal interval of time, the same distance along this second plane as it did along AB; it is clear that, by thus superposing upon the body a uniform motion of ascent and an accelerated motion of descent, it will be carried along the plane BC as far as the point C where these two velocities become equal.

If now we assume any two points D and E, equally distant from the vertex B, we may then infer that the descent along BD takes place in the same time as the ascent along BE. Draw DF parallel to BC; we know that, after descent along AD, the body will ascend along DF; or, if, on reaching D, the body is carried along the horizontal DE, it will reach E with the same momentum [*impetus*] with which it left D; hence from E the body will ascend as far as C, proving that the velocity at E is the same as that at D.

From this we may logically infer that a body which descends along any inclined plane and continues its motion along a plane inclined upwards will, on account of the momentum acquired, ascend to an equal height above the horizontal; so that if the descent is along AB the body will be carried up the plane BC as far as the horizontal line ACD: and this is true whether the inclinations of the planes are the same or different, as in the case of the planes AB and BD. But by a previous postulate the speeds acquired by fall along variously inclined planes having the same vertical

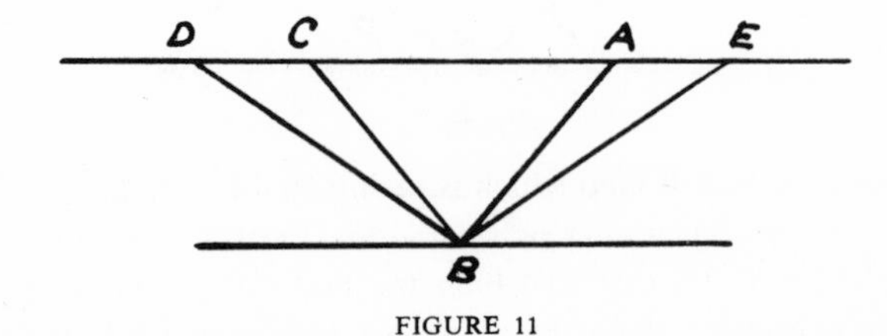

FIGURE 11

height are the same. If therefore the planes EB and BD have the same slope, the descent along EB will be able to drive the body along BD as far as D; and since this propulsion comes from the speed acquired on reaching the point B, it follows that this speed at B is the same whether the body has made its descent along AB or EB Evidently then the body will be carried up BD whether the descent has been made along AB or along EB. The time of ascent along BD is however greater than that along BC, just as the descent along EB occupies more time than that along AB; moreover it has been demonstrated that the ratio between the lengths of these times is the same as that between the lengths of the planes. . . .

FOURTH DAY

SALV. Once more, Simplicio is here on time; so let us without delay take up the question of motion. The text of our Author is as follows:

The Motion of Projectiles

In the preceding pages we have discussed the properties of uniform motion and of motion naturally accelerated along planes of all inclinations. I now propose to set forth those properties which belong to a body whose motion is compounded of two other motions, namely, one uniform and one naturally accelerated; these properties, well worth knowing, I propose to demonstrate in a rigid manner. This is the kind of motion seen in a moving projectile; its origin I conceive to be as follows:

Imagine any particle projected along a horizontal plane without friction; then we know, from what has been more fully explained in the preceding pages, that this particle will move along this same plane with a motion which is uniform and perpetual, provided the plane has no limits. But if the plane is limited and elevated, then the moving particle, which we imagine to be a heavy one, will on passing over the edge of the plane acquire, in addition to its previous uniform and perpetual motion, a downward propensity due to its own weight; so that the resulting motion which I call projection [*projectio*], is compounded of one which is uniform and horizontal and of another which is vertical and naturally accelerated. We now proceed to demonstrate some of its properties, the first of which is as follows:

THEOREM I, PROPOSITION I

A projectile which is carried by a uniform horizontal motion compounded with a naturally accelerated vertical motion describes a path which is a semi-parabola.

SAGR. Here, Salviati, it will be necessary to stop a little while for my sake and, I believe, also for the benefit of Simplicio; for it so happens that I have not gone very far in my study of Apollonius and am merely aware of the fact that he treats of the parabola and other conic sections, without an understanding of which I hardly think one will be able to follow the proof of other propositions depending upon them. Since even in this first beautiful theorem the author finds it necessary to prove that the path of a projectile is a parabola, and since, as I imagine, we shall have to deal with only this kind of curves, it will be absolutely necessary to have a thorough acquaintance, if not with all the properties which Apollonius has demonstrated for these figures, at least with those which are needed for the present treatment.

SALV. You are quite too modest, pretending ignorance of facts which not long ago you acknowledged as well known—I mean at the time when we were discussing the strength of materials and needed to use a certain theorem of Apollonius which gave you no trouble.

SAGR. I may have chanced to know it or may possibly have assumed it, so long as needed, for that discussion; but now when we have to follow all these demonstrations about such curves we ought not, as they say, to swallow it whole, and thus waste time and energy.

SIMP. Now even though Sagredo is, as I believe, well equipped for all his needs, I do not understand even the elementary terms; for although our philosophers have treated the motion of projectiles, I do not recall their having described the path of a projectile except to state in a general way that it is always a curved line, unless the projection be vertically upwards. But if the little Euclid which I have learned since our previous discussion does not enable me to understand the demonstrations which are to follow, then I shall be obliged to accept the theorems on faith without fully comprehending them.

SALV. On the contrary, I desire that you should understand them from the Author himself, who, when he allowed me to see this work of his, was good enough to prove for me two of the principal properties of the parabola because I did not happen to have at hand the books of Apollonius. These properties, which are the only ones we shall need in the present discussion, he proved in such a way that no prerequisite knowledge was required. These theorems are, indeed, given by Apollonius, but after many preceding ones, to follow which would take a long while. I wish to shorten our task by deriving the first property purely and simply from the mode of generation of the parabola and proving the second immediately from the first.

Beginning now with the first, imagine a right cone, erected upon the circular base *ibkc* with apex at *l*. The section of this cone made by a plane drawn parallel to the side *lk* is the curve which is called a *parabola*. The base of this parabola *bc* cuts at right angles the diameter *ik* of the circle *ibkc*, and the axis *ad* is parallel to the side *lk*; now having taken any

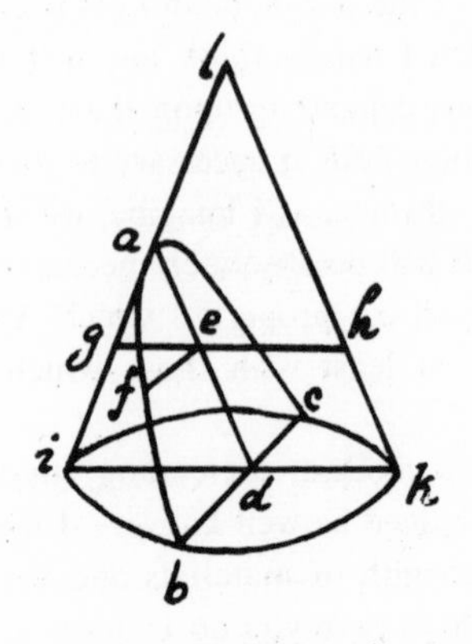

FIGURE 12

point *f* in the curve *bfa* draw the straight line *fe* parallel to *bd*; then, I say, the square of *bd* is to the square of *fe* in the same ratio as the axis *ad* is to the portion *ae*. Through the point *e* pass a plane parallel to the circle *ibkc*, producing in the cone a circular section whose diameter is the line *geh*. Since *bd* is at right angles to *ik* in the circle *ibk*, the square of *bd* is equal to the rectangle formed by *id* and *dk*; so also in the upper circle which passes through the points *gfh* the square of *fe* is equal to the rectangle formed by *ge* and *eh*; hence the square of *bd* is to the square of *fe* as the rectangle *id.dk* is to the rectangle *ge.eh*. And since the line *ed* is parallel to *hk*, the line *eh*, being parallel to *dk*, is equal to it; therefore the rectangle *id.dk* is to the rectangle *ge.eh*. And since the line *ed* is parallel to *hk*, the line *eh*, being parallel to *dk*, is equal to it; therefore the rectangle *id.dk* is to the rectangle *ge.eh* as *id* is to *ge*, that is, as *da* is to *ae*; whence also the rectangle *id.dk* is to the rectangle *ge.eh*, that is, the square of *bd* is to the square of *fe*, as the axis *da* is to the portion *ae*. Q. E. D.

The other proposition necessary for this discussion we demonstrate as follows. Let us draw a parabola whose axis *ca* is prolonged upwards to a point *d*; from any point *b* draw the line *bc* parallel to the base of the parabola; if now the point *d* is chosen so that $da = ca$, then, I say, the straight line drawn through the points *b* and *d* will be tangent to the parabola at *b*. For imagine, if possible, that this line cuts the parabola above or that its prolongation cuts it below, and through any point *g* in it draw the straight

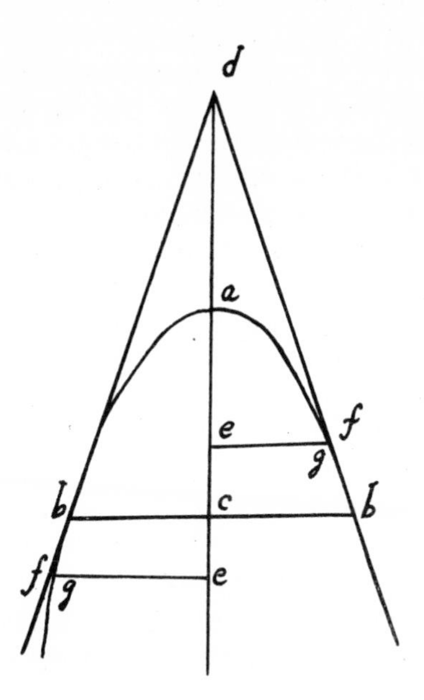

FIGURE 13

line *fge*. And since the square of *fe* is greater than the square of *ge*, the square of *fe* will bear a greater ratio to the square of *bc* than the square of *ge* to that of *bc*; and since, by the preceding proposition, the square of *fe* is to that of *bc* as the line *ea* is to *ca*, it follows that the line *ea* will bear to the line *ca* a greater ratio than the square of *ge* to that of *bc*, or, than the square of *ed* to that of *cd* (the sides of the triangles *deg* and *dcb* being proportional). But the line *ea* is to *ca*, or *da*, in the same ratio as four times the rectangle *ea.ad* is to four times the square of *ad*, or, what is the same, the square of *cd*, since this is four times the square of *ad*; hence four times the rectangle *ea.ad* bears to the square of *cd* a greater ratio than the square of *ed* to the square of *cd*; but that would make four times the rectangle *ea.ad* greater than the square of *ed*; which is false, the fact being just the opposite, because the two portions *ea* and *ad* of the line *ed* are not equal. Therefore the line *db* touches the parabola without cutting it. Q. E. D.

SIMP. Your demonstration proceeds too rapidly and, it seems to me, you keep on assuming that all of Euclid's theorems are as familiar and available to me as his first axioms, which is far from true. And now this fact which you spring upon us, that four times the rectangle *ea.ad* is less than the square of *de* because the two portions *ea* and *ad* of the line *de* are not equal brings me little composure of mind, but rather leaves me in suspense.

SALV. Indeed, all real mathematicians assume on the part of the reader perfect familiarity with at least the elements of Euclid; and here it is necessary in your case only to recall a proposition of the Second Book in which he proves that when a line is cut into equal and also into two unequal parts, the rectangle formed on the unequal parts is less than that formed on the equal (i.e., less than the square on half the line), by an amount which is the square of the difference between the equal and unequal segments. From this it is clear that the square of the whole line which is equal to four times the square of the half is greater than four times the rectangle of the unequal parts. In order to understand the following portions of this treatise it will be necessary to keep in mind the two elemental theorems from conic sections which we have just demonstrated; and these two theorems are indeed the only ones which the Author uses. We can now resume the text and see how he demonstrates his first proposition in which he shows that a body falling with a motion compounded of a uniform horizontal and a naturally accelerated [*naturale descendente*] one describes a semi-parabola.

Let us imagine an elevated horizontal line or plane *ab* along which a body moves with uniform speed from *a* to *b*. Suppose this plane to end abruptly at *b*; then at this point the body will, on account of its weight, acquire also a natural motion downwards along the perpendicular *bn*.

Draw the line *be* along the plane *ba* to represent the flow, or measure, of time; divide this line into a number of segments, *bc, cd, de*, representing equal intervals of time; from the points *b, c, d, e*, let fall lines which are parallel to the perpendicular *bn*. On the first of these lay off any distance *ci*, on the second a distance four times as long, *df*; on the third, one nine times as long, *eh*; and so on, in proportion to the squares of *cb, db, eb*,

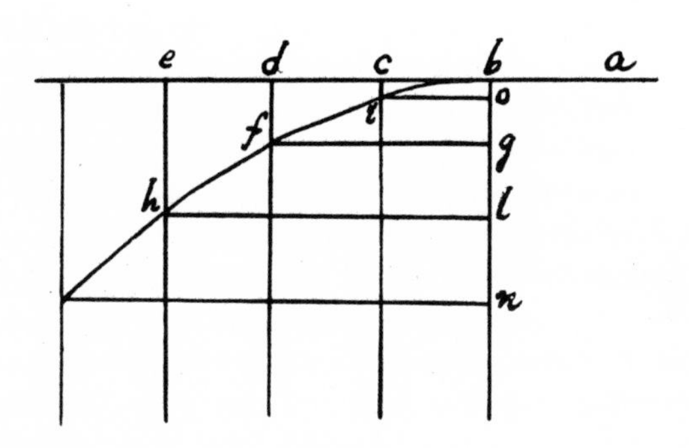

FIGURE 14

or, we may say, in the squared ratio of these same lines. Accordingly we see that while the body moves from *b* to *c* with uniform speed, it also falls perpendicularly through the distance *ci*, and at the end of the time-interval *bc* finds itself at the point *i*. In like manner at the end of the time-interval *bd*, which is the double of *bc*, the vertical fall will be four times the first distance *ci*; for it has been shown in a previous discussion that the distance traversed by a freely falling body varies as the square of the time; in like manner the space *eh* traversed during the time *be* will be nine times *ci*; thus it is evident that the distances *eh, df, ci* will be to one another as the squares of the lines *be, bd, bc*. Now from the points *i, f, h* draw the straight lines *io, fg, hl* parallel to *be*; these lines *hl, fg, io* are equal to *eb, db* and *cb*, respectively; so also are the lines *bo, bg, bl* respectively equal to *ci, df*, and *eh*. The square of *hl* is to that of *fg* as the line *lb* is to *bg*; and the square of *fg* is to that of *io* as *gb* is to *bo*; therefore the points *i, f, h*, lie on one and the same parabola. In like manner it may be shown that, if we take equal time-intervals of any size whatever, and if we imagine the particle to be carried by a similar compound motion, the positions of this particle, at the ends of these time-intervals, will lie on one and the same parabola. Q. E. D.

SALV. This conclusion follows from the converse of the first of the two propositions given above. For, having drawn a parabola through the points *b* and *h*, any other two points, *f* and *i*, not falling on the parabola must lie either within or without; consequently the line *fg* is either longer or shorter than the line which terminates on the parabola. Therefore the square of *hl* will not bear to the square of *fg* the same ratio as the line *lb* to *bg*, but a greater or smaller; the fact is, however, that the square of *hl* does bear this same ratio to the square of *fg*. Hence the point *f* does lie on the parabola, and so do all the others.

SAGR. One cannot deny that the argument is new, subtle and conclusive, resting as it does upon this hypothesis, namely, that the horizontal motion remains uniform, that the vertical motion continues to be accelerated downwards in proportion to the square of the time, and that such motions and velocities as these combine without altering, disturbing, or hindering each other,[3] so that as the motion proceeds the path of the projectile does not change into a different curve: but this, in my opinion, is impossible. For the axis of the parabola along which we imagine the natural motion of a falling body to take place stands perpendicular to a horizontal surface and ends at the center of the earth; and since the parabola deviates more and more from its axis no projectile can ever reach the center of the earth or, if it does, as seems necessary, then the path of the projectile must transform itself into some other curve very different from the parabola.

SIMP. To these difficulties, I may add others. One of these is that we suppose the horizontal plane, which slopes neither up nor down, to be represented by a straight line as if each point on this line were equally distant from the center, which is not the case; for as one starts from the middle [of the line] and goes toward either end, he departs farther and farther from the center [of the earth] and is therefore constantly going uphill. Whence it follows that the motion cannot remain uniform through any distance whatever, but must continually diminish. Besides, I do not see how it is possible to avoid the resistance of the medium which must destroy the uniformity of the horizontal motion and change the law of acceleration of falling bodies. These various difficulties render it highly improbable that a result derived from such unreliable hypotheses should hold true in practice.

SALV. All these difficulties and objections which you urge are so well founded that it is impossible to remove them; and, as for me, I am ready to admit them all, which indeed I think our Author would also do. I grant that these conclusions proved in the abstract will be different when applied in the concrete and will be fallacious to this extent, that neither will the horizontal motion be uniform nor the natural acceleration be in the ratio assumed, nor the path of the projectile a parabola, etc. But, on the other hand, I ask you not to begrudge our Author that which other eminent men have assumed even if not strictly true. The authority of Archimedes alone will satisfy everybody. In his Mechanics and in his first quadrature of the parabola he takes for granted that the beam of a balance

[3] A very near approach to Newton's Second Law of Motion. [*Trans.*]

or steelyard is a straight line, every point of which is equidistant from the common center of all heavy bodies, and that the cords by which heavy bodies are suspended are parallel to each other.

Some consider this assumption permissible because, in practice, our instruments and the distances involved are so small in comparison with the enormous distance from the center of the earth that we may consider a minute of arc on a great circle as a straight line, and may regard the perpendiculars let fall from its two extremities as parallel. For if in actual practice one had to consider such small quantities, it would be necessary first of all to criticise the architects who presume, by use of a plumbline, to erect high towers with parallel sides. I may add that, in all their discussions, Archimedes and the others considered themselves as located at an infinite distance from the center of the earth, in which case their assumptions were not false, and therefore their conclusions were absolutely correct. When we wish to apply our proven conclusions to distances which, though finite, are very large, it is necessary for us to infer, on the basis of demonstrated truth, what correction is to be made for the fact that our distance from the center of the earth is not really infinite, but merely very great in comparison with the small dimensions of our apparatus. The largest of these will be the range of our projectiles—and even here we need consider only the artillery—which, however great, will never exceed four of those miles of which as many thousand separate us from the center of the earth; and since these paths terminate upon the surface of the earth only very slight changes can take place in their parabolic figure which, it is conceded, would be greatly altered if they terminated at the center of the earth.

As to the perturbation arising from the resistance of the medium this is more considerable and does not, on account of its manifold forms, submit to fixed laws and exact description. Thus if we consider only the resistance which the air offers to the motions studied by us, we shall see that it disturbs them all and disturbs them in an infinite variety of ways corresponding to the infinite variety in the form, weight, and velocity of the projectiles. For as to velocity, the greater this is, the greater will be the resistance offered by the air; a resistance which will be greater as the moving bodies become less dense [*men gravi*]. So that although the falling body ought to be displaced [*andare accelerandosi*] in proportion to the square of the duration of its motion, yet no matter how heavy the body, if it falls from a very considerable height, the resistance of the air will be such as to prevent any increase in speed and will render the motion uniform; and in proportion as the moving body is less dense [*men grave*] this uniformity will be so much the more quickly attained and after a shorter fall. Even horizontal motion which, if no impediment were offered, would be uniform and constant is altered by the resistance of the air and finally ceases; and here again the less dense [*piu leggiero*] the body the quicker the process. Of these properties [*accidenti*] of weight, of velocity, and also of form [*figura*], infinite in number, it is not possible to give any exact description; hence, in order to handle this matter in a scientific way, it is necessary to cut loose from these difficulties; and having discovered and demonstrated the theorems, in the case of no resistance, to use them and apply them with such limitations as experience will teach. And the advantage of this method will not be small; for the material and shape of the projectile may be chosen, as dense and round as possible, so that it will encounter the least resistance in the medium. Nor will the spaces and velocities in general be so great but that we shall be easily able to correct them with precision.

In the case of those projectiles which we use, made of dense [*grave*] material and round in shape, or of lighter material and cylindrical in shape, such as arrows, thrown from a sling or crossbow, the deviation from an exact parabolic path is quite insensible. Indeed, if you will allow me a little greater liberty, I can show you, by two experiments, that the dimensions of our apparatus are so small that these external and incidental resistances, among which that of the medium is the most considerable, are scarcely observable.

I now proceed to the consideration of motions through the air, since it is with these that we are now especially concerned; the resistance of the air exhibits itself in two ways: first by offering greater impedance to less dense than to very dense bodies, and secondly by offering greater resistance to a body in rapid motion than to the same body in slow motion.

Regarding the first of these, consider the case of two balls having the same dimensions, but one weighing ten or twelve times as much as the other; one, say, of lead, the other of oak, both allowed to fall from an elevation of 150 or 200 cubits.

Experiment shows that they will reach the earth with slight difference in speed, showing us that in both cases the retardation caused by the air is small; for if both balls start at the same moment and at the same elevation, and if the leaden one be slightly retarded and the wooden one greatly retarded, then the former ought to reach the earth a considerable distance in advance of the latter, since it is ten times as heavy. But this does not happen; indeed, the gain in distance of one over the other does not amount to the hundredth part of the entire fall. And in the case of a ball of stone weighing only a third or half as much as one of lead, the difference in their times of reaching the earth will be scarcely noticeable. Now since the speed [*impeto*] acquired by a leaden ball in falling from a height of 200 cubits is so great that if the motion remained uniform the ball would, in an interval of time equal to that of the fall, traverse 400 cubits, and since this speed is so considerable in comparison with those which, by use of bows or other machines except fire arms, we are able to give to our projectiles, it follows that we may, without sensible error, regard as absolutely true those propositions which we are about to prove without considering the resistance of the medium.

Passing now to the second case, where we have to show that the resistance of the air for a rapidly moving body is not very much greater than for one moving slowly, ample proof is given by the following experiment. Attach to two threads of equal length—say four of five yards—two equal leaden balls and suspend them from the ceiling; now pull them aside from the perpendicular, the one through 80 or more degrees, the other through not more than four or five degrees; so that, when set free, the one falls, passes through the perpendicular, and describes large but slowly decreasing arcs of 160, 150, 140 degrees, etc.; the other swinging through small and also slowly diminishing arcs of 10, 8, 6 degrees, etc.

In the first place it must be remarked that one pendulum passes through its arcs of 180°, 160°, etc., in the same time that the other swings through its 10°, 8°, etc., from which it follows that the speed of the first ball is 16 and 18 times greater than that of the second. Accordingly, if the air offers more resistance to the high speed than to the low, the frequency of vibration in the large arcs of 180° or 160°, etc., ought to be less than in the small arcs of 10°, 8°, 4°, etc., and even less than in arcs of 2°, or 1°; but this prediction is not verified by experiment; because if two persons start to count the vibrations, the one the large, the other the small, they will discover that after counting tens and even hundreds they will not differ by a single vibration, not even by a fraction of one.

This observation justifies the two following propositions, namely, that vibrations of very large and very small amplitude all occupy the same time and that the resistance of the air does not affect motions of high speed more than those of low speed, contrary to the opinion hitherto generally entertained.

SAGR. On the contrary, since we cannot deny that the air hinders both of these motions, both becoming slower and finally vanishing, we have to admit that the retardation occurs in the same proportion in each case. But how? How, indeed, could the resistance offered to the one body be greater than that offered to the other except by the impartation of more momentum and speed [*impeto e velocità*] to the fast body than to the slow? And if this is so the speed with which a body moves is at once the cause and measure [*cagione e misura*] of the resistance which it meets. Therefore, all motions, fast or slow, are hindered and diminished in the same proportion; a result, it seems to me, of no small importance.

SALV. We are able, therefore, in this second case to say that the errors, neglecting those which are accidental, in the results which we are about to demonstrate are small in the case of our machines where the velocities employed are mostly very great and the distances negligible in comparison with the semi-diameter of the earth or one of its great circles.

SIMP. I would like to hear your reason for putting the projectiles of fire arms, i.e., those using powder, in a different class from the projectiles employed in bows, slings, and crossbows, on the ground of their not being equally subject to change and resistance from the air.

SALV. I am led to this view by the excessive and, so to speak, supernatural violence with which such projectiles are launched; for, indeed, it appears to me that without exaggeration one might say that the speed of a ball fired either from a musket or from a piece of ordnance is supernatural. For if such a ball be allowed to fall from some great elevation its speed will, owing to the resistance of the air, not go on increasing indefinitely; that which happens to bodies of small density in falling through short distances—I mean the reduction of their motion to uniformity—will also happen to a ball of iron or lead after it has fallen a few thousand cubits; this terminal or final speed [*terminata velocità*] is the maximum which such a heavy body can naturally acquire in falling through the air. This speed I estimate to be much smaller than that impressed upon the ball by the burning powder.

An appropriate experiment will serve to demonstrate this fact. From a height of one hundred or more cubits fire a gun [*archibuso*] loaded with a lead bullet, vertically downwards upon a stone pavement; with the same gun shoot against a similar stone from a distance of one or two cubits, and observe which of the two balls is the more flattened. Now if the ball which has come from the greater elevation is found to be the less flattened of the two, this will show that the air has hindered and diminished the speed initially imparted to the bullet by the powder, and that the air will not permit a bullet to acquire so great a speed, no matter from what height it falls; for if the speed impressed upon the ball by the fire does not exceed that acquired by it in falling freely [*naturalmente*] then its downward blow ought to be greater rather than less.

This experiment I have not performed, but I am of the opinion that a musket-ball or cannon-shot, falling from a height as great as you please, will not deliver so strong a blow as it would if fired into a wall only a few cubits distant, i.e., at such a short range that the splitting or rending of the air will not be sufficient to rob the shot of that excess of supernatural violence given it by the powder.

The enormous momentum [*impeto*] of these violent shots may cause some deformation of the trajectory, making the beginning of the parabola flatter and less curved than the end; but, so far as our Author is concerned, this is a matter of small consequence in practical operations, the main one of which is the preparation of a table of ranges for shots of high elevation, giving the distance attained by the ball as a function of the angle of elevation; and since shots of this kind are fired from mortars [*mortari*] using small charges and imparting no supernatural momentum [*impeto sopranaturale*] they follow their prescribed paths very exactly.

But now let us proceed with the discussion in which the Author invites us to the study and investigation of the motion of a body [*impeto del mobile*] when that motion is compounded of two others; and first the case in which the two are uniform, the one horizontal, the other vertical.

THEOREM II, PROPOSITION II

When the motion of a body is the resultant of two uniform motions, one horizontal, the other perpendicular, the square of the resultant momentum is equal to the sum of the squares of the two component momenta.

Let us imagine any body urged by two uniform motions and let *ab* represent the vertical displacement, while *bc* represents the displacement which, in the same interval of time, takes place in a horizontal direction.

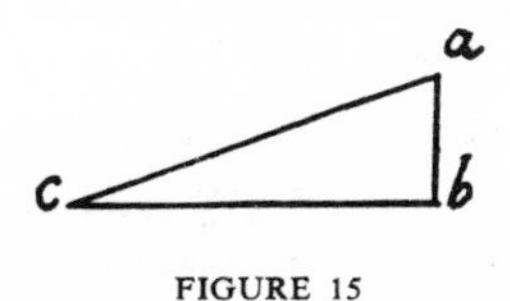

FIGURE 15

If then the distances *ab* and *bc* are traversed, during the same time-interval, with uniform motions the corresponding momenta will be to each other as the distances *ab* and *bc* are to each other; but the body which is urged by these two motions describes the diagonal *ac*; its momentum is proportional to *ac*. Also the square of *ac* is equal to the sum of the squares of *ab* and *bc*. Hence the square of the resultant momentum is equal to the sum of the squares of the two momenta *ab* and *bc*. Q. E. D.

SIMP. At this point there is just one slight difficulty which needs to be cleared up; for it seems to me that the conclusion just reached contradicts a previous proposition in which it is claimed that the speed [*impeto*] of a body coming from *a* to *b* is equal to that in coming from *a* to *c*; while now you conclude that the speed [*impeto*] at *c* is greater than that at *b*.

SALV. Both propositions, Simplicio, are true, yet there is a great difference between them. Here we are speaking of a body urged by a single motion which is the resultant of two uniform motions, while there we were speaking of two bodies each urged with naturally accelerated motions, one along the vertical *ab* the other along the inclined plane *ac*. Besides the time-intervals were there not supposed to be equal, that along the incline *ac* being greater than that along the vertical *ab*; but the motions of which we now speak, those along *ab*, *bc*, *ac*, are uniform and simultaneous.

SIMP. Pardon me; I am satisfied; pray go on.

SALV. Our Author next undertakes to explain what happens when a body is urged by a motion compounded of one which is horizontal and uniform and of another which is vertical but naturally accelerated; from these two components results the path of a projectile, which is a parabola. The problem is to determine the speed [*impeto*] of the projectile at each point. With this purpose in view our Author sets forth as follows the manner, or rather the method, of measuring such speed [*impeto*] along the path which is taken by a heavy body starting from rest and falling with a naturally accelerated motion.

THEOREM III, PROPOSITION III

Let the motion take place along the line *ab*, starting from rest at *a*, and in this line choose any point *c*. Let *ac* represent the time, or the measure of the time, required for the body to fall through the space *ac*; let *ac* also represent the velocity [*impetus seu momentum*] at *c* acquired by a fall through the distance *ac*. In the line *ab* select any other point *b*. The problem now is to determine the velocity at *b* acquired by a body in falling through the distance *ab* and to express this in terms of the velocity at *c*, the measure of which is the length *ac*. Take *as* a mean proportional between *ac* and *ab*. We shall prove that the velocity at *b* is to that at *c* as the length *as* is to the length *ac*. Draw the horizontal line *cd*, having twice the length of *ac*, and *be*, having twice the length of *ba*. It then fol-

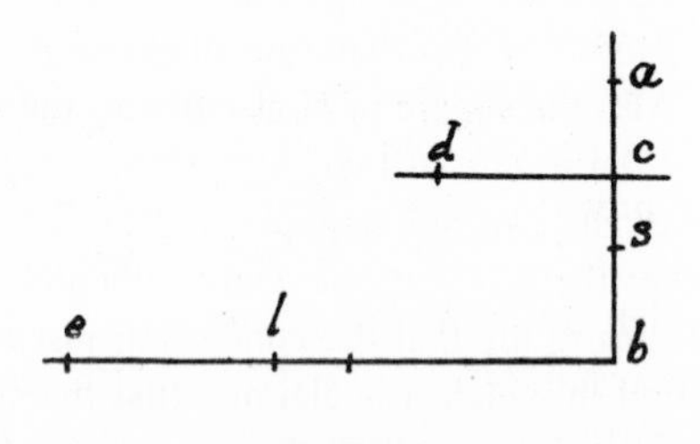

FIGURE 16

lows, from the preceding theorems, that a body falling through the distance *ac*, and turned so as to move along the horizontal *cd* with a uniform speed equal to that acquired on reaching *c* will traverse the distance *cd* in the same interval of time as that required to fall with accelerated motion from *a* to *c*. Likewise *be* will be traversed in the same time as *ba*. But the time of descent through *ab* is *as*; hence the horizontal distance *be* is also traversed in the time *as*. Take a point *l* such that the time *as* is to the time *ac* as *be* is to *bl*; since the motion along *be* is uniform, the distance *bl*, if traversed with the speed [*momentum celeritatis*] acquired at *b*, will occupy the time *ac*; but in this same time-interval, *ac*, the distance *cd* is traversed with the speed acquired in *c*. Now two speeds are to each other as the distances traversed in equal intervals of time. Hence the speed at *c* is to the speed at *b* as *cd* is to *bl*. But since *dc* is to *be* as their halves, namely, as *ca* is to *ba*, and since *be* is to *bl* as *ba* is to *sa*; it follows that *dc* is to *bl* as *ca* is to *sa*. In other words, the speed at *c* is to that at *b* as *ca* is to *sa*, that is, as the time of fall through *ab*.

The method of measuring the speed of a body along the direction of its fall is thus clear; the speed is assumed to increase directly as the time. . . .

PROBLEM I, PROPOSITION IV

SALV. Concerning motions and their velocities or momenta [*movimenti e lor velocità o impeti*] whether uniform or naturally accelerated, one cannot speak definitely until he has established a measure for such velocities and also for time. As for time we have the already widely adopted hours, first minutes and second minutes. So for velocities, just as for intervals of time, there is need of a common standard which shall be understood and accepted by everyone, and which shall be the same for all. As has already been stated, the Author considers the velocity of a freely falling body adapted to this purpose, since this velocity increases according to the same law in all parts of the world; thus for instance the speed acquired by a leaden ball of a pound weight starting from rest and falling vertically through the height of, say, a spear's length is the same in all places; it is therefore excellently adapted for representing the momentum [*impeto*] acquired in the case of natural fall.

It still remains for us to discover a method of measuring momentum in the case of uniform motion in such a way that all who discuss the subject will form the same conception of its size and velocity [*grandezza e velocità*]. This will prevent one person from imagining it larger, another smaller, than it really is; so that in the composition of a given uniform motion with one which is accelerated different men may not obtain different values for the resultant. In order to determine and represent such a momentum and particular speed [*impeto e velocità particolare*] our Author has found no better method than to use the momentum acquired by a body in naturally accelerated motion. The speed of a body which has in this manner acquired any momentum whatever will, when converted into uniform motion, retain precisely such a speed as, during a time-interval equal to that of the fall, will carry the body through a distance equal to twice that of the fall. But since this matter is one which is fundamental in our discussion it is well that we make it perfectly clear by means of some particular example.

Let us consider the speed and momentum acquired by a body falling through the height, say, of a spear [*picca*] as a standard which we may use in the measurement of other speeds and momenta as occasion demands; assume for instance that the time of such a fall is four seconds [*minuti secondi d'ora*]; now in order to measure the speed acquired from a fall through any other height, whether greater or less, one must not conclude that these speeds bear to one another the same ratio as the heights of fall; for instance, it is not true that a fall through four times a given height confers a speed four times as great as that acquired by descent through the given height; because the speed of a naturally accelerated motion does not vary in proportion to the time. As has been shown above, the ratio of the spaces is equal to the square of the ratio of the times.

If, then, as is often done for the sake of brevity, we take the same limited straight line as the measure of the speed, and of the time, and also of the space traversed during that time, it follows that the duration of fall and the speed acquired by the same body in passing over any other distance, is not represented by this second distance, but by a mean proportional between the two distances. This I can better illustrate by an example.

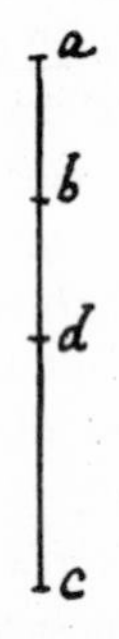

FIGURE 17

In the vertical line *ac*, lay off the portion *ab* to represent the distance traversed by a body falling freely with accelerated motion: the time of fall may be represented by any limited straight line, but for the sake of brevity, we shall represent it by the same length *ab*; this length may also be employed as a measure of the momentum and speed acquired during the motion; in short, let *ab* be a measure of the various physical quantities which enter this discussion.

Having agreed arbitrarily upon *ab* as a measure of these three different quantities, namely, space, time, and momentum, our next task is to find the time required for fall through a given vertical distance *ac*, also the momentum acquired at the terminal point *c*, both of which are to be expressed in terms of the time and momentum represented by *ab*. These two required quantities are obtained by laying off *ad*, a mean proportional between *ab* and *ac*; in other words, the time of fall from *a* to *c* is represented by *ad* on the same scale on which we agreed that the time of fall from *a* to *b* should be represented by *ab*. In like manner we may say that the momentum [*impeto o grado di velocità*] acquired at *c* is related to that acquired at *b*, in the same manner that the line *ad* is related to *ab*, since the velocity varies directly as the time, a conclusion, which although employed as a postulate in Proposition III, is here amplified by the Author.

This point being clear and well-established we pass to the consideration of the momentum [*impeto*] in the case of two compound motions, one of which is compounded of a uniform horizontal and a uniform vertical motion, while the other is compounded of a uniform horizontal and a naturally accelerated vertical motion. If both components are uniform, and one at right angles to the other, we have already seen that the square of the resultant is obtained by adding the squares of the components [p. 765] as will be clear from the following illustration.

Let us imagine a body to move along the vertical *ab* with a uniform momentum [*impeto*] of 3, and on reaching *b* to move toward *c* with a momentum [*velocità ed impeto*] of 4, so that during the same time-interval it will traverse 3 cubits along the vertical and 4 along the horizontal. But

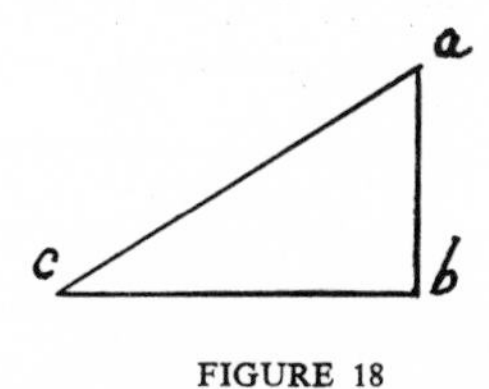

FIGURE 18

a particle which moves with the resultant velocity [*velocità*] will, in the same time, traverse the diagonal *ac*, whose length is not 7 cubits—the sum of *ab* (3) and *bc* (4)—but 5, which is *in potenza* equal to the sum of 3 and 4, that is, the squares of 3 and 4 when added make 25, which is the square of *ac*, and is equal to the sum of the squares of *ab* and *bc*. Hence *ac* is represented by the side—or we may say the root—of a square whose area is 25, namely 5.

As a fixed and certain rule for obtaining the momentum which results from two uniform momenta, one vertical, the other horizontal, we have therefore the following: take the square of each, add these together, and extract the square root of the sum, which will be the momentum resulting from the two. Thus, in the above example, the body which in virtue of its vertical motion would strike the horizontal plane with a momentum [*forza*] of 3, would owing to its horizontal motion alone strike at *c* with a momentum of 4; but if the body strikes with a momentum which is the resultant of these two, its blow will be that of a body moving with a momentum [*velocità e forza*] of 5; and such a blow will be the same at all points of the diagonal *ac*, since its components are always the same and never increase or diminish.

Let us now pass to the consideration of a uniform horizontal motion compounded with the vertical motion of a freely falling body starting from rest. It is at once clear that the diagonal which represents the motion compounded of these two is not a straight line, but, as has been demonstrated, a semi-parabola, in which the momentum [*impeto*] is always increasing because the speed [*velocità*] of the vertical component is always increasing. Wherefore, to determine the momentum [*impeto*] at any given point in the parabolic diagonal, it is necessary first to fix upon the uniform horizontal momentum [*impeto*] and then, treating the body as one falling freely, to find the vertical momentum at the given point; this latter can be determined only by taking into account the duration of fall, a consideration which does not enter into the composition of two uniform motions where the velocities and momenta are always the same; but here where one of the component motions has an initial value of zero and increases its speed [*velocità*] in direct proportion to the time, it follows that the time must determine the speed [*velocità*] at the assigned point. It only remains to obtain the momentum resulting from these two components (as in the case of uniform motions) by placing the square of the resultant equal to the sum of the squares of the two components. . . .

COMMENTARY ON
THE BERNOULLIS

IN EIGHT generations the Bach family produced at least two dozen eminent musicians and several dozen more of sufficient repute to find their way into musical dictionaries. So numerous and so eminent were they that, according to the *Britannica*, musicians were known as "Bachs" in Erfurt even when there were no longer any members of the family in the town. What the Bachs were to music, the Bernoulli clan was to science. In the course of a century eight of its members pursued mathematical studies, several attaining the foremost rank in various branches of this science as well as in related disciplines. From this group came a "swarm of descendants about half of whom were gifted above the average and nearly all of whom, down to the present day, have been superior human beings." [1]

The Bernoullis were a Protestant family driven from Antwerp in the last quarter of the sixteenth century by religious persecution. In 1583 they found asylum in Frankfurt; after a few years they moved to Basel in Switzerland. Nicolaus Bernoulli (1623–1708) was a wealthy merchant and a town councilor. This in itself is not a noteworthy achievement; he deserves rather to be remembered for his three sons Jacob, Nicolaus and John, and their descendants.[2] It is peculiar that nothing is ever said about the women the Bernoullis married; they must have made at least a genetic contribution to this illustrious spawn.

Jacob (I), for eighteen years professor of mathematics at Basel, had started out at his father's insistence as a theologian, but soon succumbed to his passion for science. He became a master of the calculus, developing

[1] E. T. Bell, *Men of Mathematics*, N. Y., 1937, p. 131. "No fewer than 120 of the descendants of the mathematical Bernoullis have been traced genealogically, and of this considerable posterity the majority achieved distinction—sometimes amounting to eminence—in the law, scholarship, science, literature, the learned professions, administration and the arts. None were failures."

[2] There is a confusion, understandable, about the Bernoulli genealogical lines, and another, less understandable, about their names. Jacob, for example, is also known as Jakob, Jacques and James; Johannes, as Johann, John and Jean. I shall use the familiar forms in the following table:

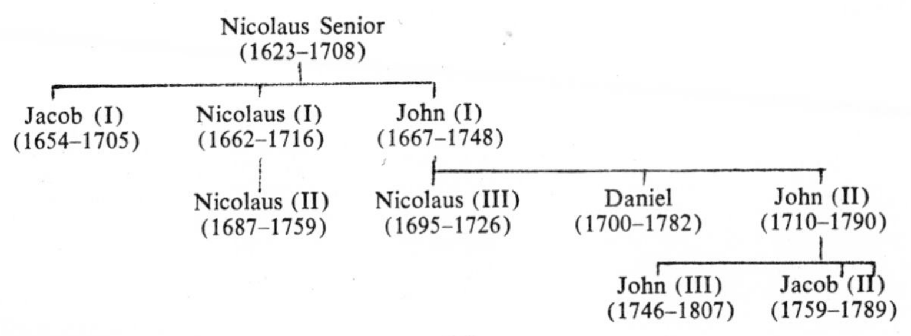

and applying it successfully to a considerable number of problems. Among his more celebrated investigations were those into the properties of the curve known as the catenary (it is formed by a heavy chain hanging freely from its two extremities), into isoperimetrical figures (those enclosing, for any given perimeter, the greatest area) and into various spiral curves. His other works include a great treatise on probability, the *Ars Conjectandi* (a selection from it appears elsewhere in these pages: see pp. 1452–1455), *A Method of Teaching Mathematics to the Blind*, based on his experience teaching the elements of science to a blind girl at Geneva, and many verses in Latin, German and French, regarded as "elegant" in their time but now forgotten. Jacob, according to Francis Galton, suffered from "a bilious and melancholic temperament"; [3] his brother John did nothing to soothe it. John also was an exceptional mathematician. He was more prolific than Jacob, made many independent and important mathematical discoveries, and enlarged scientific knowledge in chemistry, physics and astronomy. The Bernoullis, Galton says, "were mostly quarrelsome and unamiable"; John was a prime example. He was violent, abusive, jealous, and, when necessary, dishonest. He claimed a reward Jacob had offered for a solution of the isoperimetrical problem. The solution he presented was wrong; he waited until Jacob died and then published another solution which he knew to be wrong—a fact he admitted seventeen years later. His son Daniel, again a brilliant mathematician, had the temerity to win a French Academy of Sciences prize which his father had sought. John gave him a special reward by throwing him out of the house.[4] These agreeable traits were "lived out," as psychoanalysts might observe, and thus did nothing to shorten John's life. He died at the age of eighty, retaining his powers and his meanness to the end.

It is with Daniel Bernoulli that we are here mainly concerned. He was a second son, born at Groningen—where his father was then professor of mathematics—in 1700. His father did everything possible to turn him from mathematical pursuits. The program consisted of cruel mistreatment, when Daniel was a child, to destroy his self-confidence, and of later attempts to force him into business. John should have known this wouldn't work; the Bernoullis were tough as well as dedicated. When he was eleven, Daniel got instruction in geometry from his brother Nicolaus.[5] He studied

[3] Francis Galton, *Hereditary Genius*; London, reprint of 1950, p. 195.

[4] E. T. Bell, *op. cit.*, p. 134.

[5] Nicolaus Bernoulli (1695–1726) was no exception to the Bernoulli rule of extraordinariness. At the age of eight he could speak German, Dutch, French and Latin; at sixteen he became a Doctor of Philosophy at the University of Basel; he was appointed professor of mathematics at St. Petersburg at the same time as Daniel. His early death (of a "lingering fever") prevented him from developing his evident powers. As the eldest son he was better treated by his father than Daniel; at any rate he was permitted, even encouraged, to study mathematics. When he was twenty-one his father pronounced him "worthy of receiving the torch of science from his own hands."

medicine, became a physician and finally, at the age of twenty-five, accepted an appointment as professor of mathematics at St. Petersburg. In 1733 he returned to Basel to become professor of anatomy, botany, and later, "experimental and speculative philosophy," i.e., physics. He remained at this post until he was almost eighty, publishing a large number of important memoirs on physical problems and doing first-rate work in probability theory, calculus, differential equations and related fields. He won, or divided equally, no less than ten prizes put up by the French Academy of Sciences, including the one which so infuriated his father. Late in life, he particularly enjoyed recalling that, in his youth, a stranger once answered his self-introduction, "I am Daniel Bernoulli," with an "incredulous and mocking" "and I am Isaac Newton."

Bernoulli's most famous book is the *Hydrodynamica*, in which he laid the foundations, theoretical and practical, for the "equilibrium, pressure, reaction and varied velocities" of fluids. The *Hydrodynamica* is notable also for presenting the first formulation of the kinetic theory of gases, a keystone of modern physics. Bernoulli showed that, if a gas be imagined to consist of "very minute corpuscles," "practically infinite in number," "driven hither and thither with a very rapid motion," their myriad collisions with one another and impact on the walls of the containing vessel would explain the phenomenon of pressure. Moreover, if the volume of the container were slowly decreased by sliding in one end like a piston, the gas would be compressed, the number of collisions of the corpuscles would be increased per unit of time, and the pressure would rise. The same effect would follow from heating the gas; heat, as Bernoulli perceived, being nothing more than "an increasing internal motion of the particles." This "astonishing prevision of a state of physics which was not actually reached for 110 years" (notably by Joule, who calculated the statistical averages of the enormous number of molecular collisions and thus derived Boyle's law—pressure $\times$ volume $=$ constant—from the laws of impact) was fully sustained by Bernoulli's remarkable experimental and theoretical labors.[6] He provided an algebraic formulation of the relation between impacts and pressure; he even calculated the magnitude of the pressure increase resulting from decreased volume and found it corresponded to the hypothesis Boyle had confirmed by experiment, that "the pressures and expansions are in reciprocal proportions." [7]

The following selection covers these topics; it is from the tenth section of the *Hydrodynamica*, (1738).

[6] Lloyd W. Taylor, *Physics—The Pioneer Science*, Boston, 1941, p. 109.

[7] "Bernoulli in effect had made in his thinking two enormous jumps, for which the temper of his time was not ready for over three generations: first the equivalence between heat and energy, and, secondly, the idea that a well-defined relationship, such as Boyle's simple law, could be deduced from the chaotic picture of randomly moving particles." Gerald Holton, *Introduction to Concepts and Theories in Physical Science*, Cambridge (Mass.), 1952, p. 376.

Daniel Bernoulli has been called the father of mathematical physics.
—ERIC TEMPLE BELL

So many of the properties of matter, especially when in the gaseous form, can be deduced from the hypothesis that their minute parts are in rapid motion, the velocity increasing with the temperature, that the precise nature of this motion becomes a subject of rational curiosity. Daniel Bernoulli, Herapath, Joule, Krönig, Clausius, &c., have shewn that the relations between pressure, temperature and density in a perfect gas can be explained by supposing the particles to move with uniform velocity in straight lines, striking against the sides of the containing vessel and thus producing pressure. —JAMES CLERK MAXWELL (*Illustrations of the Dynamical Theory of Gases*)

2 Kinetic Theory of Gases

By DANIEL BERNOULLI

1. IN the consideration of elastic fluids we may assign to them such a constitution as will be consistent with all their known properties, that so we may approach the study of their other properties, which have not yet been sufficiently investigated. The particular properties of elastic fluids are as follows: 1. They are heavy; 2. they expand in all directions unless they are restrained; and 3. they are continually more and more compressed when the force of compression increases. Air is a body of this sort, to which especially the present investigation pertains.

2. Consider a cylindrical vessel *ACDB* (Figure 44) set vertically, and a movable piston *EF* in it, on which is placed a weight *P*: let the cavity *ECDF* contain very minute corpuscles, which are driven hither and thither with a very rapid motion; so that these corpuscles, when they strike against the piston *EF* and sustain it by their repeated impacts, form an elastic fluid which will expand of itself if the weight *P* is removed or diminished, which will be condensed if the weight is increased, and which gravitates toward the horizontal bottom *CD* just as if it were endowed with no elastic powers: for whether the corpuscles are at rest or are agitated they do not lose their weight, so that the bottom sustains not only the weight but the elasticity of the fluid. Such therefore is the fluid which we shall substitute for air. Its properties agree with those which we have already assumed for elastic fluids, and by them we shall explain other properties which have been found for air and shall point out others which have not yet been sufficiently considered.

3. We consider the corpuscles which are contained in the cylindrical cavity as practically infinite in number, and when they occupy the space *ECDF* we assume that they constitute ordinary air, to which as a standard

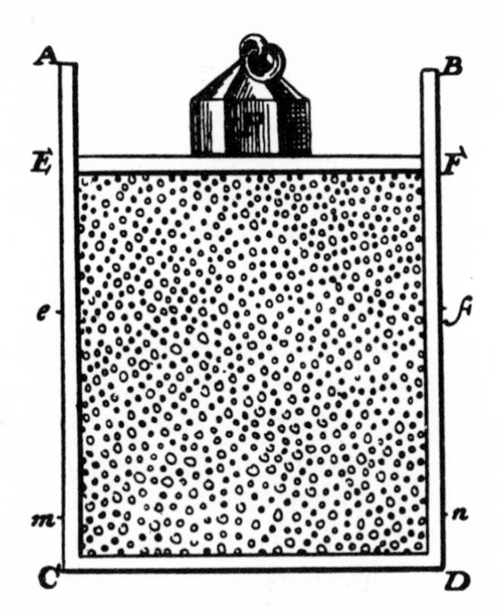

all our measurements are to be referred: and so the weight P holding the piston in the position EF does not differ from the pressure of the superincumbent atmosphere, which therefore we shall designate by P in what follows.

It should be noticed that this pressure is not exactly equal to the absolute weight of a vertical cylinder of air resting on the piston EF, as hitherto most authors have asserted without sufficient consideration; rather it is equal to the fourth proportional to the surface of the earth, to the size of the piston EF, and to the weight of all the atmosphere on the surface of the earth.

4. We shall now investigate the weight π, which is sufficient to condense the air $ECDF$ into the space $eCDf$, on the assumption that the velocity of the particles is the same in both conditions of the air, the natural condition as well as the condensed. Let $EC = 1$ and $eC = s$. When the piston EF is moved to ef, it appears that a greater effort is made by the fluid for two reasons: first, because the number of particles is now greater in the ratio of the space in which they are contained, and secondly, because each particle repeats its impacts more often. That we may properly calculate the increment which depends on the first cause we may consider the particles as if they were at rest. We shall set the number of them which are contiguous to the piston in the position $EF = n$; then the like number when the piston is in the position ef will be $= n : \left(\dfrac{eC}{EC}\right)^{2/3}$ or $= n : s^{2/3}$.

It should be noticed that the fluid is no more condensed in the lower part than in the upper part, because the weight P is infinitely greater than the weight of the fluid itself: hence it is plain that for this reason the force of the fluid is in the ratio of the numbers n and $n : s^{2/3}$ that is, as $s^{2/3}$ is to 1. Now in reference to the other increment arising from the second cause, this is found by considering the motion of the particles, and it appears that their impacts are made more often by as much as the particles are closer together: therefore the numbers of the impacts will be reciprocally as the mean distances between the surfaces of the particles, and these mean distances will be thus determined.

We assume that the particles are spheres. We represent by D the mean distance between the centers of the spheres when the piston is in the position EF, and by d the diameter of a sphere. Then the mean distance between the surfaces of the spheres will be $D - d$. But it is evident that when the piston is in the position ef, the mean distance between the centers of the spheres $= D\sqrt[3]{s}$ and therefore the mean distance between the surfaces of the spheres $= D\sqrt[3]{s} - d$. Therefore, with respect to the second cause, the force of the natural air in $ECDF$ will be to the force of the compressed air in $eCDf$ as $\dfrac{1}{D-d}$ to $\dfrac{1}{D\sqrt[3]{s}-d}$, or as $D\sqrt[3]{s} - d$ to $D - d$. When both causes are joined the predicted forces will be as $s^{2/3} \times (D\sqrt[3]{s} - d)$ to $D - d$.

For the ratio of D to d we may substitute one which is easier to understand: for if we think of the piston EF as depressed by an infinite weight, so that it descends to the position mn, in which all the particles are in contact, and if we represent the line mC by m, we shall have D is to d as 1 is to $\sqrt[3]{m}$. If we substitute this in the ratio above, we shall find that the force of the natural air in $ECDF$ is to the force of the compressed air in $eCDf$ as $s^{2/3} \times (\sqrt[3]{s} - \sqrt[3]{m})$ is to $1 - \sqrt[3]{m}$, or as $s - \sqrt[3]{mss}$ is to $1 - \sqrt[3]{m}$. Therefore $\pi = \dfrac{1 - \sqrt[3]{m}}{s - \sqrt[3]{mss}} \times P$.

5. From all the facts known we may conclude that natural air can be very much condensed and compressed into a practically infinitely small space; so that we may set $m = 0$, and hence $\pi = P/s$; so that the compressing weights are almost in the inverse ratio of the spaces which air occupies when compressed by different amounts. This law has been proved by many experiments. It certainly may be safely adopted for air that is less dense than natural air; whether it holds for considerably denser air I have not sufficiently investigated: nor have there yet been experiments instituted with the accuracy which is necessary in this case. There is special need of an experiment to find the value of m, but this experiment must be most accurately carried out and with air under very high pressure;

and the temperature of the air while it is being compressed must be carefully kept constant.

6. The elasticity of air is not only increased by condensation but by heat supplied to it, and since it is admitted that heat may be considered as an increasing internal motion of the particles, it follows that if the elasticity of air of which the volume does not change is increased, this indicates a more intense motion in the particles of air; which fits in well with our hypothesis; for it is plain that so much the greater weight P is needed to keep the air in the condition $ECDF$, as the aerial particles are agitated by the greater velocity. It is not difficult to see that the weight P should be in the duplicate ratio of this velocity because, when the velocity increases, not only the number of impacts but also the intensity of each of them increases equally, and each of them is proportional to the weight P.

Therefore, if the velocity of the particles is called v, the weight which is able to sustain the piston in the position $EF = vvP$ and in the position

$$ef = \frac{1 - \sqrt[3]{m}}{s - \sqrt[3]{mss}} \times vvP, \text{ or approximately } = \frac{vvP}{s},$$ because as we have seen

the number m is very small in comparison with unity or with the number s.

7. This theorem, as I have presented it in the preceding paragraph, in which it is shown that in air of any density but at a fixed temperature, the elasticities are proportional to the densities, and further that the increments of elasticity which are produced by equal changes of temperature are proportional to the densities, this theorem, I say, D. Amontons discovered by experiment and presented it in the Memoirs of the Royal Academy of Sciences of Paris in 1702.

COMMENTARY ON

A Great Prize, a Long-Suffering Inventor and the First Accurate Clock

THE reckoning of latitude—the distance north or south from the equator of a point on the earth's surface—was well understood by the ancients. Since the Pole Star holds approximately the same position in the heavens throughout every night, and since the earliest sailors observed that it dipped toward the horizon as they sailed south, they measured its angle with the horizon (its altitude), using an astrolabe, cross-staff or other angle-measuring device, and thus fixed their position with reference to the equator. Over the centuries the ancient methods were modified and improved, but their basic features are retained by sea and air navigators to the present day.[1]

Longitude, the measure of distance east or west from an arbitrary line, presented much greater difficulties and defied exact calculation until the eighteenth century. The combined efforts of astronomers, physicists, mathematicians and clock makers were required to solve this important problem upon which scientific cartography, sound navigation and systematic exploration and discovery depended. Curiously enough, it was not until almost half a century after Columbus had made his "long voyage" that anyone connected the fixing of longitude with the construction of reliable and portable timekeepers. The relationship between clocks and longitude is doubtless obvious to many readers, but I had better be safe and explain what is involved. Longitude is determined, in effect, by translating space into time. Take as a baseline from which east-west distances are to be measured a meridian (half a great circle included between the poles) passing through a convenient place; the modern convention fixes on Greenwich. Designate this line as the zero or prime meridian and imagine other meridians marking off the globe at intervals of 15°. Since it takes the earth twenty-four hours to complete a rotation of 360°, each meridian may be regarded as separated from its immediate neighbors to the east and west by one hour. Finding your longitude then, is "merely a matter

[1] "The quadrants, sextants and octants, developed throughout the centuries, were little more than segments of the ancient astrolabe, refined and adapted to meet the special requirements of surveyors and navigators. The modern Nautical Almanac, with its complex and multifarious tables that make it possible to find the latitude at any hour of the day or night, is nothing more than the sum total of ancient astrology, streamlined and perfected by astronomical instruments, including telescopes." Lloyd A. Brown, *The Story of Maps*, Boston, 1949, p. 180.

of comparing noons with Greenwich. You are just as long a distance from Greenwich as your noon is long a time from the Greenwich noon." [2] Nowadays Greenwich noon is ascertained by radio; but a clock set to run on Greenwich time will do the job almost as well. Starting out on a sea voyage, you take along a chronometer set to Greenwich time; after sailing west for a few days you observe, let us suppose, that when the sun is directly overhead (12 noon), your Greenwich chronometer says 3 P.M. This means that the sun has required three hours to "move" from directly above Greenwich to directly above the spot where you find yourself. To be accurate, it means that the earth has turned for three hours. Thus you have reached a point three times 15° west longitude. To give another illustration, if it is noon where you are when it is midnight in Greenwich, you are halfway around the globe, at longitude 180°.

It is easy to see, therefore, that precise clocks were needed to calculate longitude. The search for a reliable method of keeping time at sea produced its share of fantastic as well as sensible suggestions. Sir Kenelm Digby, for example, invented a "powder of sympathy" which, by a method I shall not attempt to repeat, caused a dog on shipboard to "yelp the hour on the dot." This was not the final answer to the problem. John Harrison, a Yorkshire carpenter, did better with his famous No. 4 chronometer, which took fifty years to make but lost only one second per month in trials at sea. Parliament had offered a prize of £20,000 for a dependable chronometer and Harrison—this "very ingenious and sober man," a contemporary called him—quite properly claimed the money. He thereupon became the victim of a series of unsurpassed chicaneries perpetrated by scientists and politicians in concert. He got his reward but only after a parliamentary crisis and direct intervention by the King.

The story of "The Longitude," a chronicle of science, politics, mathematics, human determination and brilliant craftsmanship, is admirably told by Lloyd A. Brown, a leading cartographer, in *The Story of Maps.* The following material is selected from his book.

[2] David Greenhood, *Down to Earth: Mapping for Everybody*, New York, 1951, p. 15.

The Art of Navigation is to be perfected by the Solution of this Problem. To find, at any Time, the Longitude of a Place at Sea. A Public Reward is promised for the Discovery. Let him obtain it who is able.

—BERNHARD VARENIUS (*Geographia Generalis, 1650*)

3 The Longitude

By LLOYD A. BROWN

SCIENTIFIC cartography was born in France in the reign of Louis XIV (1638–1715), the offspring of astronomy and mathematics. The principles and methods which had been used and talked about for over two thousand years were unchanged; the ideal of Hipparchus and Ptolemy, to locate each place on earth scientifically, according to its latitude and longitude, was still current. But something new had been introduced into the picture in the form of two pieces of apparatus—a telescope and a timekeeper. The result was a revolution in map making and a start towards an accurate picture of the earth. With the aid of these two mechanical contrivances it was possible, for the first time, to solve the problem of how to determine longitude, both on land and at sea.

The importance of longitude, the distance of a place east or west from any other given place, was fully appreciated by the more literate navigators and cartographers of history, but reactions to the question of how to find it varied from total indifference to complete despondency. Pigafetta, who sailed with Magellan, said that the great explorer spent many hours studying the problem of longitude, "but," he wrote, "the pilots content themselves with knowledge of the latitude, and are so proud [of themselves], they will not hear speak of the longitude." Many explorers of the time felt the same way, and rather than add to their mathematical burdens and observations, they were content to let well enough alone. However, "there be some," says an early writer, "that are very inquisitive to have a way to get the longitude, but that is too tedious for seamen, since it requireth the deep knowledge of astronomy, wherefor I would not have any man think that the longitude is to be found at sea by any instrument; so let no seamen trouble themselves with any such rule, but (according to their accustomed manner) let them keep a perfect account and reckoning of the way of their ship." What he meant was, let them keep their dead reckoning with a traverse board, setting down the ship's estimated daily speed and her course.[1]

Like the elixir of life and the pot of gold, longitude was a will-o'-the-wisp which most men refused to pursue and others talked about with awe.

[1] W. R. Martin, article "Navigation" in *Encyclopaedia Britannica*, 11th Edition.

"Some doo understand," wrote Richard Eden, "that the Knowledge of the Longitude myght be founde, a thynge doubtlesse greatly to be desyred, and hytherto not certaynly knowen, although Sebastian Cabot, on his death-bed told me that he had the knowledge thereof by divine revelation, yet so, that he myght not teache any man. But," adds Eden, with a certain amount of scorn, "I thinke that the good olde man, in that extreme age, somewhat doted, and had not yet even in the article of death, vtterly shaken off all wordly vayne glorie." [2]

Regardless of pessimism and indifference, the need for a method of finding the longitude was fast becoming urgent. The real trouble began in 1493, less than two months after Columbus returned to Spain from his first voyage to the west. On May 4 of that year, Pope Alexander VI issued the Bull of Demarcation to settle the dispute between Spain and Portugal, the two foremost maritime rivals in Europe. With perfect equanimity His Holiness drew a meridian line from pole to pole on a chart of the Western Ocean one hundred leagues from the Azores. To Spain he assigned all lands not already belonging to any other Christian prince which had been or would be discovered west of the line, and to Portugal all discoveries to the east of it; a masterful stroke of diplomacy, except for the fact that no one knew where the line fell. Naturally both countries suspected the worst, and in later negotiations each accused the other of pushing the line a little in the wrong direction. For all practical purposes, the term "100 leagues west of the Azores" was meaningless, as was the Line of Demarcation and all other meridians in the New World laid down from a line of reference in the Old.

Meanwhile armed convoys heavily laden with the wealth of the Indies ploughed the seas in total darkness so far as their longitude was concerned. Every cargo was worth a fortune and all the risk involved, but too many ships were lost. There were endless delays because a navigator was never sure whether he had overreached an island or was in imminent danger of arriving in the middle of the night without adequate preparations made for landing. The terrible uncertainty was wearing. In 1598 Philip III of Spain offered a perpetual pension of 6000 ducats, together with a life pension of 2000 ducats and an additional gratuity of 1000 more to the "discoverer of longitude." Moreover, there would be smaller sums available in advance for sound ideas that might lead to the discovery and for partially completed inventions that promised tangible results, and no questions asked. It was the clarion call for every crank, lunatic and undernourished inventor in the land to begin research on "the fixed point" or the "East and West navigation" as it was called. In a short time the

[2] See Richard Eden's "Epistle Dedicatory" in his translation of John Taisnier's *A very necessarie and profitable book concerning nauigation* . . . London, 1579 (?). (Quoted from *Bibliotheca Americana. A catalogue of books . . . in the library of the late John Carter Brown.* Providence, 1875, Part I, No. 310.)

Spanish government was so deluged with wild, impractical schemes and Philip was so bored with the whole thing that when an Italian named Galileo wrote the court in 1616 about another idea, the king was unimpressed. After a long, sporadic correspondence covering sixteen years, Galileo reluctantly gave up the idea of selling his scheme to the court of Spain.[3]

Portugal and Venice posted rewards, and drew the same motley array of talent and the same results as Spain. Holland offered a prize of 30,000 scudi to the inventor of a reliable method of finding the longitude at sea, and Willem Blaeu, map publisher, was one of the experts chosen by the States General to pass on all such inventions. In August, 1636, Galileo came forward again and offered his plan to Holland, this time through his Paris friend Diodati, as he did not care to have his correspondence investigated by the Inquisition. He told the Dutch authorities that some years before, with the aid of his telescope, he had discovered what might be a remarkable celestial timekeeper—Jupiter. He, Galileo, had first seen the four satellites, the "Cosmian Stars" (*Sidera Medicea*, as he called them), and had studied their movements. Around and around they went, first on one side of the planet, then on the other, now disappearing, then reappearing. In 1612, two years after he first saw them, he had drawn up tables, plotting the positions of the satellites at various hours of the night. These, he found, could be drawn up several months in advance and used to determine mean time at two different places at once. Since then, he had spent twenty-four years perfecting his tables of the satellites, and now he was ready to offer them to Holland, together with minute instructions for the use of any who wished to find the longitude at sea or on land.[4]

The States General and the four commissioners appointed to investigate the merits of Galileo's proposition were impressed, and requested further details. They awarded him a golden chain as a mark of respect and Hortensius, one of the commissioners, was elected to make the journey to Italy where he could discuss the matter with Galileo in person. But the Holy Office got wind of things and the trip was abandoned. In 1641 after a lapse of nearly three years, negotiations were renewed by the Dutch scientist Christian Huygens, but Galileo died a short time after and the idea of using the satellites of Jupiter was set aside.[5]

In the two thousand year search for a solution of the longitude problem

[3] J. J. Fahie, *Galileo. His life and work*, New York, 1903, pp. 172, 372 ff.; Rupert T. Gould, *The marine chronometer: Its history and development*, London, 1923, pp. 11, 12.

[4] J. J. Fahie, *op. cit.*, pp. 372 ff. Galileo named the satellites of Jupiter the "Cosmian Stars" after Cosmo Medici (Cosmo II, grandduke of Tuscany). See Galileo's *Opere* edited by Eugenio Alberi, 16 vols., Firenze, 1842 56. Tome III contains his "Sydereus Nuncius," pp. 59–99, describing his observations of Jupiter's satellites, and his suggestion that they be used in the determination of longitude.

[5] J. J. Fahie, *op. cit.*, pp. 373–75.

it was never a foregone conclusion that the key lay in the transportation of timekeepers. But among the optimistic who believed that a solution could somehow be found, it was agreed that it would have to come from the stars, especially for longitude at sea, where there was nothing else to observe. It might be found in the stars alone or the stars in combination with some terrestrial phenomenon. However, certain fundamental principles were apparent to all who concerned themselves with the problem. Assuming that the earth was a perfect sphere divided for convenience into 360 degrees, a mean solar day of 24 hours was equivalent to 360 degrees of arc, and 1 hour of the solar day was equivalent to 15 degrees of arc or 15 degrees of longitude. Likewise, 1 degree of longitude was the equivalent of 4 minutes of time. Finer measurements of time and longitude (minutes and seconds of time, minutes and seconds of arc) had been for centuries the stuff that dreams were made of. Surveys of the earth in an east-west direction, expressed in leagues, miles or some other unit of linear measure, would have no significance unless they could be translated into degrees and minutes of arc, fractional parts of the circumference of the earth. And how big was the earth?

The circumference of the earth and the length of a degree (1/360th part of it) had been calculated by Eratosthenes and others but the values obtained were questionable. Hipparchus had worked out the difference between a solar day and a sidereal day (the interval between two successive returns of a fixed star to the meridian), and had plotted a list of 44 stars scattered across the sky at intervals of right ascension equal to exactly one hour, so that one or more of them would be on the meridian at the beginning of every sidereal hour. He had gone a step further and adopted a meridian line through Rhodes, suggesting that longitudes of other places could be determined with reference to his prime meridian by the simultaneous observation of the moon's eclipses. This proposal assumed the existence of a reliable timekeeper which was doubtless nonexistent.

The most popular theoretical method of finding longitude, suggested by the voyages of Columbus, Cabot, Magellan, Tasman and other explorers, was to plot the variation or declination of the compass needle from the true north. This variation could be found by taking a bearing on the polestar and noting on the graduated compass card the number of points, half and quarter points (degree and minutes of arc) the needle pointed east or west of the pole. Columbus had noted this change of compass variation on his first voyage, and later navigators had confirmed the existence of a "line of no variation" passing through both poles and the fact that variation changed direction on either side of it. This being so, and assuming that the variation changed at a uniform rate with a change of longitude, it was logical to assume that here at last was a solution to the whole problem. All you had to do was compare the amount of variation at your place of observation with the tabulated variation at places whose longitude had already been determined. It was this fond hope that induced Edmund Halley and others to compile elaborate charts showing the supposed lines of equal variation throughout the world. However, it was by no means that simple, as Gellibrand and others discovered. Variation does not change uniformly with a change of longitude; likewise, changes in variation occur very slowly; so slowly, in fact, that precise east-west measurements are impractical, especially at sea. And, too, it was found that lines of equal variation do not always run north and south; some run nearly east and west. However, in spite of the flaws that cropped up, one by one, the method had strong supporters for many years, but finally died a painful, lingering death.[6]

In addition to discovering Jupiter's satellites, Galileo made a second important contribution to the solution of longitude by his studies of the pendulum and its behavior, for the application of the swinging weight to the mechanism of a clock was the first step towards the development of an accurate timekeeper.[7] The passage of time was noted by the ancients and their astronomical observations were "timed" with sundials, sandglasses and water clocks but little is known about how the latter were controlled. Bernard Walther, a pupil of Regiomontanus, seems to have been the first to time his observations with a clock driven by weights. He stated that on the 16th of January, 1484, he observed the rising of the planet Mercury, and immediately attached the weight to a clock having an hour-wheel with fifty-six teeth. By sunrise one hour and thirty-five teeth had passed, so that the elapsed time was an hour and thirty-seven minutes, according to his calculations. The next important phase in the development of a timekeeper was the attachment of a pendulum as a driving force. This clock was developed by Christian Huygens, Dutch physicist and astronomer, the son of Constantine Huygens. He built the first one in 1656 in order to increase the accuracy of his astronomical observations, and later presented it to the States General of Holland on the 16th of June, 1657. The following year he published a full description of the principles involved in the mechanism of his timekeeper and the physical laws governing the pendulum. It was a classic piece of writing, and established Huygens as one of the leading European scientists of the day.[8]

By 1666 there were many able scientists scattered throughout Europe. Their activities covered the entire fields of physics, chemistry, astronomy, mathematics, and natural history, experimental and applied. For the most

[6] R. T. Gould, *op. cit.*, pp. 4 and 4 *n*.

[7] See Galileo's *Dialogues concerning two new sciences, by Galileo Galilei*, translated by Henry Crew and Alfonso de Salvio, introduction by Antonio Favaro, New York, 1914, pp. 84, 95, 170, 254.

[8] See John L. E. Dreyer's article "Time, Measurement of" (in the *Encyclopaedia Britannica*, 11th edition, pp. 983d, 984a).

part they worked independently and their interests were widely diversified. Occasionally the various learned societies bestowed honorary memberships on worthy colleagues in foreign countries, and papers read in the various societies were exchanged with fellow scientists in foreign lands. The stage was set for the transition of cartography from an art to a science. The apparatus was at hand and the men to use it.

Pleading for the improvement of maps and surveys, Thomas Burnet made a useful distinction between the popular commercial map publications of the day and what he considered should be the goal of future map makers. "I do not doubt," he wrote, "but that it would be of very good use to have *natural* Maps of the Earth . . . as well as civil. . . . Our common Maps I call *Civil*, which note the distinction of Countries and of Cities, and represent the Artificial Earth as inhabited and cultivated: But natural Maps leave out all that, and represent the Earth as it would be if there were not an Inhabitant upon it, nor ever had been; the Skeleton of the Earth, as I may so say, with the site of all its parts. Methinks also every Prince should have such a Draught of his Country and Dominions, to see how the ground lies in the several parts of them, which highest, which lowest; what respect they have to one another, and to the Sea; how the Rivers flow, and why; how the Mountains lie, how Heaths, and how the Marches. Such a Map or Survey would be useful both in time of War and Peace, and many good observations might be made by it, not only as to Natural History and Philosophy, but also in order to the perfect improvement of a Country." [9]

These sentiments regarding "natural" maps were fully appreciated and shared by the powers in France, who proceeded to do something about it. All that was needed was an agency to acquire the services and direct the work of the available scientific talent, and someone to foot the bills. The agency was taken care of by the creation of the Académie Royale des Sciences, and the man who stood prepared to foot the bills for better maps was His Majesty Louis XIV, king of France.

Louis XIV ascended the throne when he was five years old, but had to wait sixteen years before he could take the reins of government. He had to sit back and watch the affairs of state being handled by the queen-mother and his minister, Cardinal Mazarin. He saw the royal authority weakened by domestic troubles and the last stages of the Thirty Years' War. Having suffered through one humiliation after another without being able to do anything about it, Louis resolved, when he reached the age of twenty-one, to rule as well as reign in France. He would be his own first minister. Foremost among his few trusted advisors was Jean Baptiste Colbert, minister for home affairs, who became, in a short time, the chief power behind the throne. Colbert, an ambitious and industrious man with

[9] Thomas Burnet: *The theory of the earth* . . . London, 1684, p. 144.

expensive tastes, contrived to indulge himself in literary and artistic extravagances while adding to the stature and glory of his monarch. As for the affairs of state over which he exercised control, there were two enterprises in particular which entitle Colbert to an important place in the history of France. The first was the establishment of the French Marine under a monarch who cared little for naval exploits or the importance of sea power in the growth and defense of his realm; the second was the founding, in 1666, of the Académie Royale des Sciences, now the Institut de France.[10]

The Académie Royale was Colbert's favorite project. An amateur scientist, he realized the potential value of a distinguished scientific body close to the throne, and with his unusual skill and seemingly unlimited funds he set out to make France pre-eminent in science as it was in the arts and the art of war. He scoured Europe in search of the top men in every branch of science. He addressed personal invitations to such figures as Gottfried Wilhelm von Leibnitz, German philosopher and mathematician; Niklaas Hartsoeker, Dutch naturalist and optician; Ehrenfried von Tschirnhausen, German mathematician and manufacturer of optical lenses and mirrors; Joannes Hévélius, one of Europe's foremost astronomers; Vincenzo Viviani, Italian mathematician and engineer; Isaac Newton, England's budding mathematical genius. The pensions that went with the invitations were without precedent, surpassing in generosity those established by Cardinal Richelieu for the members of the Académie Française, and those granted by Charles II for the Royal Society of London. Additional funds were available for research, and security and comfort were assured to those scientists who would agree to work in Paris, surrounded by the most brilliant court in Europe. Colbert's ambition to make France foremost in science was realized, though some of the invitations were declined with thanks. Christian Huygens joined the Académie in 1666 and received his pension of 6000 livres a year until 1681, when he returned to Holland. Olaus Römer, Dutch astronomer, also accepted. These celebrities were followed by Marin de la Chambre who became physician to Louis XIV; Samuel Duclos and Claude Bourdelin in chemistry; Jean Pecquet and Louis Gayant in anatomy; Nicholas Marchant in botany.[11]

In spite of the broad scope of its activities, the avowed purpose of founding the Académie Royale, according to His Majesty, was to correct and improve maps and sailing charts. And the solution of the major problems of chronology, geography and navigation, whose practical importance was incontestable, lay in the further study and application of astronomy.[12]

[10] See Charles J. E. Wolf, *Histoire de l'observatoire de Paris de sa fondation à 1793*, Paris, 1902; also *L'Institut de France* by Gaston Darboux, Henry Roujon and George Picot, Paris, 1907 ("Les Grandes Institutions de France").

[11] C. J. E. Wolf, *op. cit.*, pp. 5 ff.

[12] *Mémoires de l'Académie Royale des Sciences*, Vol. VIII, Paris, 1730.

To this end, astronomical observations and conferences were begun in January, 1667. The Abbé Jean Picard, Adrian Auzout, Jacques Buot and Christian Huygens were temporarily installed in a house near the Cordeliers, the garden of which was taken over for astronomical observations. There the scientists set up a great quadrant, a mammoth sextant and a highly refined version of the sundial. They also constructed a meridian line. Sometimes observations were made in the garden of the Louvre. On the whole, facilities for astronomical research were far from good, and there was considerable grumbling among the academicians.

As early as 1665, before the Académie was founded, Auzout had written Colbert an impassioned memorandum asking for an observatory, reminding him that the progress of astronomy in France would be as nothing without one. When Colbert finally made up his mind, in 1667, and the king approved the money, events moved rapidly. The site chosen for the observatory was at Faubourg St. Jacques, well out in the country, away from the lights and disconcerting noises of Paris. Colbert decided that the Observatory of Paris should surpass in beauty and utility any that had been built to date, even those in Denmark, England and China, one which would reflect the magnificence of a king who did things on a grand scale. He called in Claude Perrault, who had designed the palace at Versailles with accommodations for 6000 guests, and told him what he and his Académie wanted. The building should be spacious; it should have ample laboratory space and comfortable living quarters for the resident astronomers and their families.[13]

On the 21st of June, 1667, the day of the summer solstice, the members of the Académie assembled at Faubourg St. Jacques, and with great pomp and circumstance made observations for the purpose of "locating" the new observatory and establishing a meridian line through its center, a line which was to become the official meridian of Paris. The building was to have two octagonal towers flanking the southern façade, and eight azimuths were carefully computed so that the towers would have astronomical as well as architectural significance. Then, without waiting for their new quarters, the resident members of the Académie went back to work, attacking the many unsolved riddles of physics and natural history, as well as astronomy and mathematics. They designed and built much of the apparatus for the new observatory. They made vast improvements in the telescope as an astronomical tool; they solved mechanical and physical problems connected with the pendulum and what gravity does to it, helping Huygens get the few remaining "bugs" out of his pendulum timekeeper. They concentrated on the study of the earth, its size and shape and place in the universe; they investigated the nature and behavior of the moon and other celestial bodies; they worked towards the establishment of a

[13] C. J. E. Wolf, *op. cit.*, p. 4.

standard meridian of longitude for all nations, the meridian of Paris running through the middle of their observatory. They worked on the problem of establishing the linear value of a degree of longitude which would be a universally acceptable constant. In all these matters the Académie Royale des Sciences was fortunate in having at its disposal the vast resources of the court of France as well as the personal patronage of Louis XIV.

An accurate method of determining longitude was first on the agenda of the Académie Royale, for obviously no great improvement could be made in maps and charts until such a method was found. Like Spain and the Netherlands, France stood ready to honor and reward the man who could solve the problem. In 1667, an unnamed German inventor addressed himself to Louis XIV, stating that he had solved the problem of determining longitude at sea. The king promptly granted him a patent (brevet) on his invention, sight unseen, and paid him 60,000 livres in cash. More than this, His Majesty contracted to pay the inventor 8000 livres a year (Huygens was getting 6000!) for the rest of his life, and to pay him four sous on every ton of cargo moved in a ship using the new device, reserving for himself only the right to withdraw from the contract in consideration of 100,000 livres. All this His Majesty would grant, but on one condition: the inventor must demonstrate his invention before Colbert, Abraham Duquesne, Lieutenant-General of His Majesty's naval forces, and Messrs. Huygens, Carcavi, Roberval, Picard and Auzout of the Académie Royale des Sciences.[14]

The invention proved to be nothing more than a variation on an old theme, an ingenious combination of water wheel and odometer to be inserted in a hole drilled in the keel of a ship. The passage of water under the keel would turn the water wheel, and the distance traversed by the ship in a given period would be recorded on the odometer. The inventor also claimed that by some strange device best known to himself his machine would make any necessary compensations for tides and cross-currents of one kind and another; it was, in fact, an ideal and perfect solution to the longitude problem. The royal examiners studied the apparatus, praised its ingenuity and then submitted their report to the king in writing. They calmly pointed out, among other things, that if a ship were moving with a current, it might be most stationary with respect to the water under the keel and yet be carried along over an appreciable amount of longitude while the water wheel remained motionless. If, on the other hand, the ship were breasting a current, the odometer would register considerable progress when actually the ship might be getting nowhere. The German

[14] *Histoire de l'Académie Royale des Sciences*, Vol. I, pp. 45-46.

inventor departed from Paris richer by 60,000 livres and the members of the Académie went back to work.[15]

In 1669, after three years of intensive study, the scientists of the Académie Royale had gathered together considerable data on the celestial bodies, and had studied every method that had been suggested for the determination of longitude. The measurement of lunar distances from the stars and the sun was considered impractical because of the complicated mathematics involved. Lunar eclipses might be all right except for the infrequency of the phenomenon and the slowness of eclipses, which increase the chance of error in the observer. Moreover, lunar eclipses were utterly impractical at sea. Meridional transits of the moon were also tried with indifferent success. What the astronomers were looking for was a celestial body whose distance from the earth was so great that it would present the same appearance from any point of observation. Also wanted was a celestial body which would move in a constantly predictable fashion, exhibiting at the same time a changing picture that could be observed and timed simultaneously from different places on the earth. Such a body was Jupiter, whose four satellites, discovered by Galileo, they had observed and studied. The serious consideration of Jupiter as a possible solution of the longitude problem brought to mind a publication that had come out in 1668 written by an Italian named Cassini. While the members of the Académie continued their study of Jupiter's satellites with an eye to utilizing their frequent eclipses as a method of determining longitude, Colbert investigated the possibilities of luring Cassini to Paris.

Giovanni Domenico Cassini was born in Perinaldo, a village in the Comté of Nice, June 8, 1625, the son of an Italian gentleman. After completing his elementary schooling under a preceptor he studied theology and law under the Jesuits at Genoa and was graduated with honors. He developed a decided love of books, and while browsing in a library one day he came across a book on astrology. The work amused him, and after studying it he began to entertain his friends by predicting coming events. His phenomenal success as an astrologer plus his intellectual honesty made him very suspicious of his new-found talent, and he promptly abandoned the hocus-pocus of astrology for the less dramatic study of astron-

[15] Justin Winsor, *Narrative and Critical History of America*, Boston, 1889, Vol. II, pp. 98–99 has an interesting note on the "log." In Pigafetta's journal (January, 1521), he mentions the use of a chain dragged astern on Magellan's ships to measure their speed. The "log" as we know it was described in Bourne's *Regiment of the Sea*, 1573, and Humphrey Cole is said to have invented it. In Eden's translation of Taisnier he speaks of an artifice "not yet divulgate, which, placed in the pompe of a shyp, whyther the water hath recourse, and moved by the motion of the shypp, with wheels and weyghts, doth exactly shewe what space the shyp hath gone." See the article "Navigation" in the *Encyclopaedia Britannica*, 9th edition. For further comments on the history of the log, see L. C. Wroth, *The Way of a Ship*, Portland, Maine, 1937, pp. 72–74.

omy. He made such rapid progress and displayed such remarkable aptitude, that in 1650, when he was only twenty-five years old, he was chosen by the Senate of Bologna to fill the first chair of astronomy at the university, vacant since the death of the celebrated mathematician Bonaventura Cavalieri. The Senate never regretted their choice.[16]

One of Cassini's first duties was to serve as scientific consultant to the Church for the precise determination of Holy Days, an important application of chronology and longitude. He retraced the meridian line at the Cathedral of Saint Petronius constructed in 1575 by Ignazio Dante, and added a great mural quadrant which took him two years to build. In 1655 when it was completed, he invited all the astronomers in Italy to observe the winter solstice and examine the new tables of the sun by which the equinoxes, the solstices and numerous Holy Days could now be accurately determined.

Cassini was next appointed by the Senate of Bologna and Pope Alexander VII to determine the difference in level between Bologna and Ferrara, relative to the navigation of the Po and Reno rivers. He not only did a thorough job of surveying, but wrote a detailed report on the two rivers and their peculiarities as well. The Pope next engaged Cassini, in the capacity of a hydraulic engineer, to straighten out an old dispute between himself and the Duke of Tuscany relative to the diversion of the precious water of the Chiana River, alternate affluent of the Arno and the Tiber. Having settled the dispute to the satisfaction of the parties concerned, he was appointed surveyor of fortifications at Perugia, Pont Felix and Fort Urbino, and was made superintendent of the waters of the Po, a river vital to the conservation and prosperity of the country. In his spare time Cassini busied himself with the study of insects and to satisfy his curiosity repeated several experiments on the transfusion of blood from one animal to another, a daring procedure that was causing a flurry of excitement in the scientific world. But his major hobby was astronomy and his favorite planet was Jupiter. While he worked on the Chiana he spent many evenings at Citta della Piève observing Jupiter's satellites. His telescope was better than Galileo's and with it he was able to make some additional discoveries. He noted that the plane of the revolving satellites was such that the satellites passed across Jupiter's disc close to the equator; he noted the size of the orbit of each satellite. He was certain he could see a number of fixed spots on Jupiter's orb, and on the strength of his findings he began to time the rotation of the planet as well as the satellites, using a fairly reliable pendulum clock.[17]

[16] *Oeuvres de Fontenelle. Eloges.* Paris, 1825, Vol. I, p. 254.

[17] Joseph François Michaud: *Biographie Universelle*, Paris, 1854–65. Cassini gave Jupiter's rotation as $9^h\ 56^m$. The *correct* time is not yet known with certainty. Slightly different results are obtained by using different markings. The value $9^h\ 55^m$ is frequently used in modern texts.

After sixteen years of patient toil and constant observations, Cassini published his tables (*Ephemerides*) of the eclipses of Jupiter's satellites for the year 1668, giving on one page the appearance of the planet in a diagram with the satellites grouped around it and on the opposite page the time of the eclipse (immersion) of each satellite in hours, minutes and seconds, and the time of each emersion.[18]

Cassini, then forty-three years old, had become widely known as a scholar and skilled astronomer, and when a copy of his *Ephemerides* reached Paris Colbert decided he must get him for the Observatory and the Académie Royale. In this instance, however, it took considerable diplomacy as well as gold to get the man he wanted, for Cassini was then in the employ of Pope Clement IX, and neither Louis XIV nor Colbert cared to offend or displease His Holiness. Three distinguished scholars, Vaillant, Auzout and Count Graziani, were selected to negotiate with the Pope and the Senate of Bologna for the temporary loan of Cassini, who was to receive 9000 livres a year as long as he remained in France. The arrangements were finally completed, and Cassini arrived in Paris on the 4th of April, 1669. Two days later he was presented to the king. Although Cassini had no intention of staying indefinitely, Colbert was insistent, and in spite of the remonstrances of the Pope and the Senate of Bologna, Cassini became a naturalized citizen of France in 1673, and was thereafter known as Jean Domenique Cassini.[19]

Observations were in full swing when Cassini took his place among the savants of the Académie Royale who were expert mechanics as well as physicists and mathematicians. Huygens and Auzout had ground new lenses and mirrors, and had built vastly improved telescopes for the observatory. With the new instruments Huygens had already made some phenomenal discoveries. He had observed the rotation period of Saturn, discovered Saturn's rings and the first of the satellites. Auzout had built other instruments and applied to them an improved filar micrometer, a measuring device all but forgotten since its invention by Gascoyne (Gascoigne) about 1639. After Cassini's arrival, more apparatus was ordered, including the best telescopes available in Europe, made by Campani in Italy.[20]

One of the first important steps toward the correction of maps and charts was the remeasurement of the circumference of the earth and the establishment of a new value for a degree of arc in terms of linear measure. There was still a great deal of uncertainty as to the size of the earth,

[18] The first edition of Cassini's work was published under the title *Ephemerides Bononiensés Mediceorvm sydervm ex hypothesibvs, et tabvlis Io: Dominici Cassini* . . . Bononiae [Bologne], 1668.

[19] C. J. E. Wolf, *op. cit.*, p. 6.

[20] For a detailed inventory of the equipment built and purchased by the Académie Royale, see C. J. E. Wolf, *op. cit.*

and the astronomers were reluctant to base their new data on a fundamental value which might negate all observations made with reference to it. After poring over the writings of Hipparchus, Poseidonius, Ptolemy and later authorities such as Snell, and after studying the methods these men had used, the Académie worked out a detailed plan for measuring the earth, and in 1669 assigned Jean Picard to do the job.

The measurement of the earth at the equator, from east to west, was out of the question; no satisfactory method of doing it was known. Therefore the method used by Eratosthenes was selected, but with several important modifications and with apparatus that the ancients could only have dreamed of. Picard was to survey a line by triangulation running approximately north and south between two terminal points; he would then measure the arc between the two points (that is, the difference in latitude) by astronomical observations. After looking over the country around Paris, Picard decided he could run his line nearly northward to the environs of Picardy without encountering serious obstructions such as heavy woods and high hills.[21]

Picard selected as his first terminal point the "Pavillon" at Malvoisine near Paris, and for his second point the clock tower in Sourdon near Amiens, a distance of about thirty-two French leagues. Thirteen great triangles were surveyed between the two points, and for the purpose Picard used a stoutly reinforced iron quadrant with a thirty-eight inch radius fixed on a heavy standard. The usual pinhole alidades used for sighting were replaced by two telescopes with oculars fitted with cross hairs, an improved design of the instrument used by Tycho Brahe in Denmark. The limb of the quadrant was graduated into minutes and seconds by transversals. For measuring star altitudes involving relatively acute angles, Picard used a tall zenith sector made of copper and iron with an amplitude of about 18°. Attached to one radius of the sector was a telescope ten feet long. Also part of his equipment were two pendulum clocks, one regulated to beat seconds, the other half-seconds. For general observations and for observing the satellites of Jupiter, he carried three telescopes: a small one about five feet long and two larger ones, fourteen and eighteen feet long. Picard was well satisfied with his equipment. In describing his specially fitted quadrant, he said it did the work so accurately that during the two years it took to measure the arc of the meridian, there was never an error of more than a minute of arc in any of the angles measured on the entire circumference of the horizon, and that in many cases, on checking the instrument for accuracy, it was found to be absolutely true. And as for the

[21] For a complete account of Picard's measurement of the earth, including tables of data and a historical summary, see the *Mémoires de l'Académie Royale des Sciences*, Vol. VII, Pt. I, Paris, 1729. See also the article "Earth, Figure of the" by Alexander Ross Clarke and Frederick Robert Helmert in the *Encyclopaedia Britannica*, 11th edition, p. 801.

pendulum clocks he carried, Picard was pleased to report that they "marked the seconds with greater accuracy than most clocks mark the half hours." [22]

When the results of Picard's survey were tabulated, the distance between his two terminal points was found to be 68,430 toises 3 pieds. The difference in latitude between them was measured, not by taking the altitude of the sun at the two terminal points, but by measuring the angle between the zenith and a star in the kneecap of Cassiopeia, first at Malvoisine and then at Sourdon. The difference was 1° 11′ 57″. From these figures the value of a degree of longitude was calculated as 57,064 toises 3 pieds. But on checking from a second base line of verification which was surveyed in the same general direction as the first, this value was revised to 57,060 toises, and the diameter of the earth was announced as 6,538,594 toises. All measurements of longitude made by the Académie Royale were based on this value, equivalent to about 7801 miles, a remarkably close result.[23]

In 1676, after the astronomers had revised and enlarged his *Ephemerides* of 1668, Cassini suggested that the corrected data might now be used for the determination of longitude, and Jupiter might be given a trial as a celestial clock. The idea was approved by his colleagues and experimental observations were begun, based on a technique developed at the Observatory and on experience acquired by a recent expedition to Cayenne for the observation of the planet Mars. The scientists were unusually optimistic as the work began, and in a rare burst of enthusiasm one of them wrote, *"Si ce n'est pas -là le véritable secret des Longitudes, au-moins en approche-t-il de bien près."*

Because of his tremendous energy, skill and patience, Cassini had by this time assumed the leadership of the scientists working at the observatory, even though he did not have the title of Director. He carried on an extensive correspondence with astronomers in other countries, particularly in Italy where the best instruments were available and where he and his work were well known. Astronomers in foreign parts responded with enthusiasm when they learned of the work that was being done at the Paris Observatory. New data began to pour in faster than the resident astronomers could appraise and tabulate it. Using telescopes and the satellites of Jupiter, hundreds of cities and towns were now being located for the first time with reference to a prime meridian and to each other. All of the standard maps of Europe, it seemed, would have to be scrapped.[24]

[22] *Mémoires de l'Académie Royale des Sciences,* Vol. VII, Pt. I. Pagination varies widely in different editions of this series.

[23] *Ibid.*, pp. 306–07, gives a table of the linear measures used by Picard in making his computations. Measurements were made between the zenith and a star in the kneecap of Cassiopeia, probably δ (Al Rukbah). See the *Mémoires de l'Académie Royale des Sciences,* Vol. VII, Pt. II, Paris, 1730, p. 305.

[24] For Cassini's status in the Académie Royale, see C. J. E. Wolf, *op. cit.*

With so much new information available, Cassini conceived the idea of compiling a large-scale map of the world (planisphere) on which revised geographical information could be laid down as it came in from various parts of the world, especially the longitudes of different places, hitherto unknown or hopelessly incorrect. For this purpose the third floor of the west tower of the Observatory was selected. There was plenty of space, and the octagonal walls of the room had been oriented by compass and quadrant when the foundation of the building was laid. The planisphere, on an azimuthal projection with the North Pole at the center, was executed in ink by Sédileau and de Chazelles on the floor of the tower under the watchful eye of Cassini. The circular map was twenty-four feet in diameter, with meridians radiating from the center to the periphery, like the spokes of a wheel, at intervals of 10°. The prime meridian of longitude (through the island of Ferro) was drawn from the center at any angle "half way between the two south windows of the tower" to the point where it bisected the circumference of the map. The map was graduated into degrees from zero to 360 in a counterclockwise direction around the circle. The parallels of latitude were laid down in concentric circles at intervals of 10°, starting with zero at the equator and numbering both ways. For convenient and rapid "spotting" of places, a cord was attached to a pin fastened to the center of the map with a small rider on it, so that by swinging the cord around to the proper longitude and the rider up or down to the proper latitude, a place could be spotted very quickly.

On this great planisphere the land masses were of course badly distorted, but it did not matter. What interested the Académie was the precise location, according to latitude and longitude, of the important places on the earth's surface, places that could be utilized in the future for bases of surveying operations. For this reason it was much more important to have the names of a few places strategically located and widely distributed, according to longitude, than it was to include a great many places that were scientifically unimportant. For the same reason, most of the cities and towns that boasted an astronomical observatory, regardless of how small, were spotted on the map.

The planisphere was highly praised by all who saw it. The king came to see it with Colbert and all the court. His Majesty graciously allowed Cassini, Picard and de la Hire to demonstrate the various astronomical instruments used by the members of the Académie to study the heavens and to determine longitude by remote control, as it were. They showed him their great planisphere and explained how the locations of different places were being corrected on the basis of data sent in from the outside world. It was enough to make even Louis pause.[25]

[25] *Ibid.*, pp. 62–65. For an account of the great planisphere, see the *Histoire de l'Académie Royale des Sciences*, Vol. I, pp. 225–26; C. J. E. Wolf, *op. cit.* A facsimile

What effect the king's visit had on future events it is difficult to estimate, but in the next few years a great deal of surveying was done. Many surveying expeditions were sent out from the Observatory, and the astronomers went progressively further afield. Jean Richer led an expedition to Cayenne and Jean Mathieu de Chazelles went to Egypt. Jesuit missionaries observed at Madagascar and in Siam. Edmund Halley, who was in close touch with the work going on in France, made a series of observations at the Cape of Good Hope. Thevenot, the historian and explorer, communicated data on several lunar eclipses observed at Goa. About this time, Louis-Abel Fontenay, a Jesuit professor of mathematics at the College of Louis le Grand, was preparing to leave for China. Hearing of the work being done by Cassini and his colleagues, Fontenay volunteered to make as many observations as he could without interfering with his missionary duties. Cassini trained him and sent him on his way prepared to contribute data on the longitudes to the Orient. Thus the importance of remapping the world and the feasibility of the method devised by the Académie Royale began to dawn on the scholars of Europe, and many foreigners volunteered to contribute data. Meanwhile Colbert raised more money and Cassini sent more men into the field.

One of the longest and most difficult expeditions organized by the Académie Royale was led by Messrs. Varin and des Hayes, two of His Majesty's engineers for hydrography, to the island of Gorée and the West Indies. It was also one of the most important for the determination of longitudes in the Western Hemisphere, involving as it did the long jump across the Atlantic Ocean, a span where some of the most egregious errors in longitude had been made. Cassini's original plan, approved by the king, was to launch the expedition from Ferro, on the extreme southwest of the Canary Islands, an island frequently used by cartographers as a prime meridian of longitude. But as there was some difficulty about procuring passage for the expedition, it was decided to take a departure from Gorée, a small island off Cape Verde on the west coast of Africa, where a French colony had recently been established by the Royal Company of Africa.[26]

Before their departure, Varin and des Hayes spent considerable time at the Observatory, where they were thoroughly trained by Cassini and where they could make trial observations to perfect their technique. They received their final instructions in the latter part of 1681, and set out for

---

of one of the printed versions of the map, reduced, was issued with Christian Sandler's *Die Reformation der Kartographie um 1700*, Munich, 1905; a second one, colored, was published in 1941 by the University of Michigan from the original in the William L. Clements Library. For bibliographical notes regarding the publication of the map, see L. A. Brown, *Jean Domenique Cassini* . . . pp. 62–73.

[26] The island of Ferro, the most southwest of the Canary Islands, was a common prime meridian among cartographers as late as 1880. It was considered the dividing line between the Eastern and Western hemispheres! (See Lippincott's *A complete pronouncing gazetteer or geographical dictionary of the world*. Philadelphia, 1883.)

Rouen equipped with a two and a half foot quadrant, a pendulum clock, and a nineteen foot telescope. Among the smaller pieces of apparatus they carried were a thermometer, a barometer and a compass. From Rouen they moved to Dieppe, where they were held up more than a month by storm weather and contrary winds. With time on their hands, they made a series of observations to determine the latitude and longitude of the city. The two men finally arrived at Gorée in March, 1682, and there they were joined by M. de Glos, a young man trained and recommended by Cassini. De Glos brought along a six foot sextant, an eighteen foot telescope, a small zenith sector, an astronomical ring and another pendulum clock. Although the primary object of the expedition was to determine longitudes by observing the eclipses of the satellites of Jupiter, the three men had orders to observe the variation of the compass at every point in their travels, especially during the ocean voyage, and to make thermometrical and barometric observations whenever possible; in short, they were to gather all possible scientific data that came their way. From Gorée the expedition sailed for Guadeloupe and Martinique, and for the next year extensive observations were made. The three men returned to Paris in March, 1683.[27]

Cassini's instructions to the party were given in writing. They furnish a clear picture of the best seventeenth century research methods and at the same time explain just how terrestrial longitudes were determined by timing the eclipses of the satellites of Jupiter. The object was simple enough: to find the difference in *mean* or *local* time between a prime meridian such as Ferro or Paris and a second place such as Guadeloupe, the difference in time being equivalent to the difference in longitude. Two pendulum clocks were carried on the expedition, and before leaving they were carefully regulated at the Observatory. The pendulum of one was adjusted so that it would keep *mean* time, that is twenty-four hours a day. The second clock was set to keep *sidereal* or *star* time ($23^h\ 56^m\ 4^s$).[28] The rate of going for the two timekeepers was carefully tabulated over a long period so that the observers might know in advance what to expect when the temperature, let us say, went up or down ten degrees in a twenty-four hour period. These adjustments were made by raising or lowering the pendulum bob to speed up or slow down the clock. After the necessary adjustments were made, the position of the pendulum bobs on the rods was marked and the clocks were taken apart for shipping.

Having arrived at the place where observations were to be made, the

[27] L. A. Brown: *Jean Domenique Cassini* . . . pp. 42-44.

[28] Cassini suggested two methods of adjusting a clock to *mean* time. The first was to make a series of observations of the sun (equal altitudes) and afterwards correct them with tables of the equation of time, and second, to regulate one clock to keep sidereal time by observing two successive transits of a star and correct the second clock from it.

astronomers selected a convenient, unobstructed space and set up their instruments. They fixed their pendulum bobs in position and started their clocks, setting them at the approximate hour of the day. The next operation was to establish a meridian line, running true north and south, at the place of observation. This was done in several ways, each method being used as a check against the accuracy of the others. The first was to take a series of equal altitude observations of the sun, a process which would also give a check on the accuracy of the clock which was to keep mean time. To do this, the altitude of the sun was taken with a quadrant or sextant approximately three (or four) hours before apparent noon. At the moment the sight was taken the hour, minute and second were recorded in the log. An afternoon sight was taken when the sun had descended to precisely the same angle recorded in the morning observation. Again the time was taken at the instant of observation. The difference in time between the two observations divided by two and added to the morning time gave the hour, minute and second indicated at apparent noon. This observation was repeated two days in succession, and the difference in minutes and seconds recorded by the clock on the two days (always different because of the declination of the sun) divided by two and added to the first gave the observers what the clock did in twenty-four hours; in other words, it gave them mean time. A very simple check on the arrival of apparent noon, when the sun reaches the meridian, was to drop a plumb line from the fixed quadrant and note the shadow on the ground as each observation was taken. These observations were repeated daily so that the observers always knew their local time.

The second pendulum clock was much simpler to adjust. All they had to do was set up a telescope in the plane of the meridian, sight it on a fixed star and time two successive transits of the star. When the pendulum was finally adjusted so that $23^h\ 56^m\ 4^s$ elapsed between two successive transits, the job was done. The latitude of the place of observation was equally simple to determine. The altitude of the sun at apparent noon was taken with a quadrant and the angle, referred to the tables of declination, gave the observers their latitude. A check on the latitude was made at night, by observing the height of the polestar.

With the meridian line established, and a clock regulated to keep mean time, the next thing was to observe and time the eclipses of the satellites of Jupiter, at least two of which are eclipsed every two days. As Cassini pointed out, this was not always a simple matter, because not all eclipses are visible from the same place and because bad weather often vitiates the observations. Observations called for a very fussy technique.

The most satisfactory time observations of Jupiter, in Cassini's opinion, could be made of the immersions and emersions of the first satellite. Six phases of the eclipse should be timed: during the immersion of the satellite

(1) when the satellite is at a distance from the limb of Jupiter equal to its own diameter; (2) when the satellite just touches Jupiter; (3) when it first becomes entirely hidden by Jupiter's disc. During the emersion of the satellite (4) the instant the satellite begans to reappear; (5) when it becomes detached from Jupiter's disc; (6) when the satellite has moved away from Jupiter a distance equal to its own diameter. To observe and time these phases was a two-man job: one to observe and one to keep a record of the time in minutes and seconds. If an observer had to work alone, Cassini recommended the "eye and ear" method of timing observations, which is still good observational practice. The observer begins to count out loud "one-five-hundred, two-five-hundred, three-five-hundred" and so on, the instant the eclipse begins, and he continues to count until he can get to his clock and note the time. Then by subtracting his count from the clock reading, he has the time at which the observation was made.

The emersion of the satellite, Cassini warned, always requires very careful observation, because you see nothing while you are waiting for it. At the instant you see a faint light in the region where the satellite should reappear, you should begin counting without leaving the telescope until you are sure you are seeing the actual emersion. You may make several false starts before you actually see and can time the actual emersion. Other observations worth using, according to Cassini, were the conjunctions of two satellites going in opposite directions. A conjunction was said to occur when the centers of the two satellites were in a straight perpendicular line. In all important observations requiring great accuracy, Cassini recommended a dress rehearsal the day before and at the same hour, so that if the instruments did not behave or the star was found to be in a difficult position, all necessary adjustments would have been made in advance.[29]

In addition to the observations for the determination of longitude, all expeditions sent out from the Paris Observatory were cautioned to note any variation in the functioning of their pendulum clocks. This did not mean normal variations caused by changes in temperature. Such variations could be predicted in advance by testing the metal pendulum rods: determining the coefficient of expansion at various temperatures. What they were watching for was a change caused by a variation in gravity. There were two reasons involved, one practical and one theoretical. The pendulum was an extremely important engine, since it was the driving force of the best clocks then in use. And too, the whole subject of gravitation, whose leading exponents were Christian Huygens, Isaac Newton and

[29] Cassini's "Instructions" were printed in full in the *Mémoires de l'Académie Royale des Sciences*, Vol. VIII, Paris, 1730. For a translation into English see L. A. Brown, *Jean Domenique Cassini. . .* pp. 48–60.

The planet Jupiter showing the six positions of the first satellite used by seventeenth-century astronomers to determine the difference in longitude between two places.

Robert Hooke, was causing a stir in the scientific world. The idea of using the pendulum experimentally for studying gravitation came from Hooke, and the theories of Newton and Huygens might well be proved or disproved by a series of experiments in the field. What no one knew was that these field trials would result in the discovery that the earth is not a perfect sphere but an oblate spheroid, a sphere flattened at the poles.

What effect, if any, did a change of latitude produce in the oscillations of a pendulum if the temperature remained unchanged? Many scientists said none, and experiments seemed to prove it. Members of the Académie had transported timekeepers to Copenhagen and The Hague to try them at different latitudes, and a series of experiments had been conducted in London. The results were all negative; at every place a pendulum of a given length (39.1 inches) beat seconds or made 3600 oscillations an hour. However, there was one exception. In 1672, Jean Richer had made an expedition to Cayenne (4° 56′ 5″ N.) to observe the opposition of Mars. On the whole the expedition was a success, but Richer had had trouble with his timekeeper. Although the length of the pendulum had been carefully adjusted at the Observatory before he sailed, Richer found that in Cayenne his clock lost about two minutes and a half a day, and that in order to get it to keep mean time he had to shorten the pendulum (raise the bob) by more than a "ligne" (about 1⁄12 of an inch). All this was very trying to Cassini, who was a meticulous observer. "It is suspected," he wrote, "that this resulted from some error in the observation."

Had he not been a gentleman as well as a scholar, he would have said that Richer was just plain careless.[30]

The following year, 1673, Huygens published his masterpiece on the oscillation of the pendulum, in which he set down for the first time a sound theory on the subject of centrifugal force, principles which Newton later applied to his theoretical investigation of the earth.[31] The first opportunity to confirm the fallacy of Richer's observations on the behavior of his timekeeper came when Varin and des Hayes sailed for Martinique (14° 48′ N.) and Guadeloupe (between 15° 47′ and 16° 30′ N.). Cassini cautioned them to check their pendulums with the greatest possible care and they did. But, unfortunately, their clocks behaved badly, and they, too, had to shorten the pendulums in order to make them beat mean time. Cassini was still dubious, but not Isaac Newton. In the third book of his *Principia* he concluded that this variation of the pendulum in the vicinity of the equator must be caused either by a diminution of gravity resulting from a bulging of the earth at the equator, or from the strong, counteracting effect of centrifugal force in that region.[32]

The discoveries made by the Académie Royale des Sciences set a fast pace in the scientific world and pointed the way towards many others. The method of finding longitude by means of the eclipses of Jupiter's satellites had proved to be feasible and accurate, but it was not accepted by foreign countries without a struggle. Tables of Jupiter's satellites were finally included in the English *Nautical Almanac* and remained there in good standing for many years, along with tables of lunar distances and other star data associated with rival methods of finding longitude. It was generally conceded, however, that Jupiter could not be used for finding longitude at sea, in spite of Galileo's assertions to the contrary. Many inventors besides the great Italian had come up with ingenious and wholly impractical devices to provide a steady platform on shipboard from which astronomical observations could be made. But the fact remained that the sea was too boisterous and unpredictable for astronomers and their apparatus.

England made her official entry in the race for the longitude when Charles II ordered the construction of a Royal Observatory, for the advancement of navigation and nautical astronomy, in Greenwich Park, overlooking the Thames and the plain of Essex.[33] In England things moved slowly at first, but they moved. The king was determined to have the tables of the heavenly bodies corrected for the use of his seamen and

[30] L. A. Brown: *Jean Domenique Cassini . . .* p. 57.

[31] Christian Huygens: *Horologivm oscillatorivm; sive de motu pendulorvm ad horologia aptato demonstrationes geometricae*, Paris, 1673.

[32] Isaac Newton: *Philosophiae Naturalis Principia Mathematica*, 3 vols., London, 1687.

[33] R. T. Gould, *op. cit.*, p. 9; also Henry S. Richardson's *Greenwich: its history, antiquities, improvements and public buildings*, London, 1834.

so appointed John Flamsteed "astronomical observator" by a royal warrant dated March 4, 1675, at the handsome salary of £100 a year, out of which he paid £10 in taxes. He had to provide his own instruments, and as an additional check to any delusions of grandeur he might have, he was ordered to give instruction to two boys from Christ's Hospital. Stark necessity made him take several private pupils as well. Dogged by ill health and the irritations common to the life of a public servant, Flamsteed was nevertheless buoyed up by the society of Newton, Halley, Hooke and the scientists of the Académie Royale, with whom he corresponded. A perfectionist of the first magnitude, Flamsteed was doomed to a life of unhappiness by his unwillingness to publish his findings before he had had a chance to check them for accuracy. To Flamsteed, no demand was sufficiently urgent to justify such scientific transgression.

Flamsteed worked under constant pressure. Everybody, it seemed, wanted data of one sort of another, and wanted it in a hurry. Newton needed full information on "places of the moon" in order to perfect his lunar theory. British scientists, as a group, had set aside the French method of finding longitude and all other methods requiring the use of sustained observations at sea. They were approaching the problem from another angle, and demanded complete tables of lunar distances and a complete catalogue of star places. Flamsteed did as he was told, and for fifteen years (1689–1704) spent most of his time at the pedestrian task of compiling the first Greenwich star catalogue and tables of the moon, meanwhile reluctantly doling out to his impatient peers small doses of what he considered incomplete if not inaccurate data.[34]

The loudest clamors for information came from the Admiralty and from the waterfront. In 1689 war broke out with France. In 1690 (June 30) the English fleet was defeated by the French at the battle of Beachy Head. Lord Torrington, the English admiral, was tried by court-martial and acquitted, but nevertheless dismissed from the service. In 1691 several ships of war were lost off Plymouth because the navigators mistook the Deadman for Berry Head. In 1707 Sir Cloudesley Shovel, returning with his fleet from Gibraltar, ran into dirty weather. After twelve days of groping in a heavy overcast, all hands were in doubt as to the fleet's position. The Admiral called for the opinion of his navigators, and with one exception they agreed that the fleet was well to the west of Ushant, off the Brittany peninsula. The fleet stood on, but that night, in a heavy fog, they ran into the Scilly Islands off the southwest coast of England. Four ships and two thousand men were lost, including the Admiral. There was a story current, long after, that a seaman on the flagship had estimated from his own dead reckoning that the fleet was in a dangerous position. He had the temerity to point this out to his superiors, who

[34] See Francis Baily's *Account of the Rev. John Flamsteed*, London, 1835.

sentenced him forthwith to be hanged at the yardarm for mutiny. The longitude had to be found! [35]

There was never a shortage of inventive genius in England, and many fertile minds were directed towards the problem of finding longitude at sea. In 1687 two proposals were made by an unknown inventor which were novel, to say the least. He had discovered that a glass filled to the brim with water would run over at the instant of new and full moon, so that the longitude could be determined with precision at least twice a month. His second method was far superior to the first, he thought, and involved the use of a popular nostrum concocted by Sir Kenelm Digby called the "powder of sympathy." This miraculous healer cured open wounds of all kinds, but unlike ordinary and inferior brands of medicine, the powder of sympathy was applied, not to the wound but to the weapon that inflicted it. Digby used to describe how he made one of his patients jump sympathetically merely by putting a dressing he had taken from the patient's wound into a basin containing some of his curative powder. The inventor who suggested using Digby's powder as an aid to navigation proposed that before sailing every ship should be furnished with a wounded dog. A reliable observer on shore, equipped with a standard clock and a bandage from the dog's wound, would do the rest. Every hour, on the dot, he would immerse the dog's bandage in a solution of the powder of sympathy and the dog on shipboard would yelp the hour.[36]

Another serious proposal was made in 1714 by William Whiston, a clergyman, and Humphrey Ditton, a mathematician. These men suggested that a number of lightships be anchored in the principal shipping lanes at regular intervals across the Atlantic ocean. The lightships would fire at regular intervals a star shell timed to explode at 6440 feet. Sea captains could easily calculate their distance from the nearest lightship merely by timing the interval between the flash and the report. This system would be especially convenient in the North Atlantic, they pointed out, where the depth never exceeded 300 fathoms! For obvious reasons, the proposal of Whiston and Ditton was not carried out, but they started something. Their plan was published, and thanks to the publicity it received in various periodicals, a petition was submitted to Parliament on March 25, 1714, by "several Captains of Her Majesty's Ships, Merchants of London, and Commanders of Merchantmen," setting forth the great importance of finding the longitude and praying that a public reward be offered for some practicable method of doing it.[37] Not only the petition but the

[35] R. T. Gould, *op. cit.*, p. 2.

[36] See *Curious Enquiries*, London, 1687; R. T. Gould, *op. cit.*, p. 11. The title of Sir Kenelm Digby's famous work, which appeared in French as well as English, was *A late discourse . . . touching the cure of wounds by the powder of sympathy; with instructions how to make the said powder . . .* Second edition, augmented, London, 1658.

[37] See Whiston and Ditton's *A new method for discovering the longitude*. London,

proposal of Whiston and Ditton were referred to a committee, who in turn consulted a number of eminent scientists including Newton and Halley.

That same year Newton prepared a statement which he read to the committee. He said, "That, for determining the Longitude at Sea, there have been several Projects, true in the Theory, but difficult to execute." Newton did not favor the use of the eclipses of the satellites of Jupiter, and as for the scheme proposed by Whiston and Ditton, he pointed out that it was rather a method of "keeping an Account of the Longitude at Sea, than for finding it, if at any time it should be lost." Among the methods that are difficult to execute, he went on. "One is, by a Watch to keep time exactly: But, by reason of the Motion of a Ship, the Variation of Heat and Cold, Wet and Dry, and the Difference of Gravity in Different Latitudes, such a Watch hath not yet been made." That was the trouble: such a watch had not been made.[38]

The idea of transporting a timekeeper for the purpose of finding longitude was not new, and the futility of the scheme was just as old. To the ancients it was just a dream. When Gemma Frisius suggested it in 1530 there were mechanical clocks, but they were a fairly new invention, and crudely built, which made the idea improbable if not impossible.[39] The idea of transporting "some true Horologie or Watch, apt to be carried in journeying, which by an Astrolabe is to be rectified . . ." was again stated by Blundeville in 1622, but still there was no watch which was "true" in the sense of being accurate enough to use for determining longitude.[40] If a timekeeper was the answer, it would have to be very accurate indeed. According to Picard's value, a degree of longitude was equal to about sixty-eight miles at the equator, or four minutes, by the clock. One minute of time meant seventeen miles—towards or away from danger. And if on a six weeks' voyage a navigator wanted to get his longitude within half a degree (thirty-four miles) the rate of his timekeeper must not gain or lose more than two minutes in forty-two days, or *three seconds a day.*

Fortified by these calculations, which spelled the impossible, and the report of the committee, Parliament passed a bill (1714) "for providing a publick reward for such person or persons as shall discover the Longitude." It was the largest reward ever offered, and stated that for any practical invention the following sum would be paid: [41]

£10,000 for any device that would determine the longitude within 1 degree.

1714. The petition appeared in various periodicals: *The Guardian*, July 14; *The Englishman*, Dec. 19, 1713 (R. T. Gould, *op. cit.*, p. 13).

[38] R. T. Gould, *op. cit.*, p. 13.

[39] Gemma Frisius: *De principiis astronomiae et cosmographiae*, Antwerp, 1530.

[40] Thomas Blundeville, *M. Blundeville his exercises . . .*, Sixth Edition, London, 1622, p. 390.

[41] 12 Anne, Cap. 15; R. T. Gould, *op. cit.*, p. 13.

£15,000 for any device that would determine the longitude within 40 minutes.

£20,000 for any device that would determine the longitude within 30 minutes (2 minutes of time or 34 miles).

As though aware of the absurdity of their terms, Parliament authorized the formation of a permanent commission—the Board of Longitude—and empowered it to pay one half of any of the above rewards as soon as a majority of its members were satisfied that any proposed method was practicable and useful, and that it would give security to ships within eighty miles of danger, meaning land. The other half of any reward would be paid as soon as a ship using the device should sail from Britain to a port in the West Indies without erring in her longitude more than the amounts specified. Moreover, the Board was authorized to grant a smaller reward for a less accurate method, provided it was practicable, and to spend a sum not to exceed £2000 on experiments which might lead to a useful invention.

For fifty years this handsome reward stood untouched, a prize for the impossible, the butt of English humorists and satirists. Magazines and newspapers used it as a stock cliché. The Board of Longitude failed to see the joke. Day in and day out they were hounded by fools and charlatans, the perpetual motion lads and the geniuses who could quarter a circle and trisect an angle. To handle the flood of crackpots, they employed a secretary who handed out stereotyped replies to stereotyped proposals. The members of the Board met three times a year at the Admirality, contributing their services and their time to the Crown. They took their responsibilities seriously and frequently called in consultants to help them appraise a promising invention. They were generous with grants-in-aid to struggling inventors with sound ideas, but what they demanded was results.[42] Neither the Board nor any one else knew exactly what they were looking for, but what everyone knew was that the longitude problem had stopped the best minds in Europe, including Newton, Halley, Huygens, von Leibnitz and all the rest. It was solved, finally, by a ticking machine in a box, the invention of an uneducated Yorkshire carpenter named John Harrison. The device was the marine chronometer.

Early clocks fell into two general classes: nonportable timekeepers driven by a falling weight, and portable timekeepers such as table clocks and crude watches, driven by a coiled spring. Gemma Frisius suggested the latter for use at sea, but with reservations. Knowing the unreliable temperament of spring-driven timekeepers, he admitted that sand and water clocks would have to be carried along to check the error of a

[42] R. T. Gould, *op. cit.*, p. 16. According to the Act of 1712 the Board was comprised of: "The Lord High Admiral or the First Lord of the Admiralty; The Speaker of the House of Commons; The First Commissioner of the Navy; The First Commissioner of Trade; The Admirals of the Red, White and Blue Squadrons; The Master of the Trinity House; The President of the Royal Society; The Astronomer-Royal; The Savilian, Lucasian, and Plumian Professors of Mathematics."

spring-driven machine. In Spain, during the reign of Philip II, clocks were solicited which would run exactly twenty-four hours a day, and many different kinds had been invented. According to Alonso de Santa Cruz there were "some with wheels, chains and weights of steel: some with chains of catgut and steel: others using sand, as in sandglasses: others with water in place of sand, and designed after many different fashions: others again with vases or large glasses filled with quicksilver: and, lastly, some, the most ingenious of all, driven by the force of the wind, which moves a weight and thereby the chain of the clock, or which are moved by the flame of a wick saturated with oil: and all of them adjusted to measure twenty-four hours exactly." [43]

Robert Hooke became interested in the development of portable timekeepers for use at sea about the time Huygens perfected the pendulum clock. One of the most versatile scientists and inventors of all time, Hooke was one of those rare mechanical geniuses who was equally clever with a pen. After studying the faults of current timekeepers and the possibility of building a more accurate one, he slyly wrote a summary of his investigations, intimating that he was completely baffled and discouraged. "All I could obtain," he said, "was a Catalogue of Difficulties, *first* in the doing of it, *secondly* in the bringing of it into publick use, *thirdly*, in making advantage of it. Difficulties were proposed from the alteration of *Climates, Airs, heats* and *colds*, temperature of *Springs*, the nature of *Vibrations*, the wearing of Materials, the motion of the Ship, and divers others." Even if a reliable timekeeper were possible, he concluded, "it would be difficult to bring it to use, for Sea-men know their way already to any Port. . . ." As for the rewards: "the Praemium for the Longitude," there never was any such thing, he retorted scornfully. "No King or State would pay a farthing for it."

In spite of his pretended despondency, Hooke nevertheless lectured in 1664 on the subject of applying springs to the balance of a watch in order to render its vibrations more uniform, and demonstrated, with models, twenty different ways of doing it. At the same time he confessed that he had one or two other methods up his sleeve which he hoped to cash in on at some future date. Like many scientists of the time, Hooke expressed the principle of his balance spring in a Latin anagram; roughly: *Ut tensio, sic vis,* "as the tension is, so is the force," or, "the force exerted by a spring is directly proportional to the extent to which it is tensioned." [44]

The first timekeeper designed specifically for use at sea was made by Christian Huygens in 1660. The escapement was controlled by a pendu-

[43] *Ibid.*, p. 20; this translation is from a paraphrase by Duro in his *Disquisiciones Nauticas.*

[44] *Ibid.*, p. 25. The anagram was a device commonly used in the best scientific circles of the time to establish priority of invention or discovery without actually disclosing anything that might be seized upon by a zealous colleague.

lum instead of a spring balance, and like many of the clocks that followed, it proved useless except in a flat calm. Its rate was unpredictable; when tossed around by the sea it either ran in jerks or stopped altogether. The length of the pendulum varied with changes of temperature, and the rate of going changed in different latitudes, for some mysterious reason not yet determined. But by 1715 every physical principal and mechanical part that would have to be incorporated in an accurate timekeeper was understood by watchmakers. All that remained was to bridge the gap between a good clock and one that was nearly perfect. It was that half degree of longitude, that two minutes of time, which meant the difference between conquest and failure, the difference between £20,000 and just another timekeeper.[45]

One of the biggest hurdles between watchmakers and the prize money was the weather: temperature and humidity. A few men included barometric pressure. Without a doubt, changes in the weather did things to clocks and watches, and many suggestions were forthcoming as to how this principal source of trouble could be overcome. Stephen Plank and William Palmer, watchmakers, proposed keeping a timekeeper close to a fire, thus obviating errors due to change in temperature. Plank suggested keeping a watch in a brass box over a stove which would always be hot. He claimed to have a secret process for keeping the temperature of the fire uniform. Jeremy Thacker, inventor and watchmaker, published a book on the subject of the longitude, in which he made some caustic remarks about the efforts of his contemporaries.[46] He suggested that one of his colleagues, who wanted to test his clock at sea, should first arrange to have two consecutive Junes equally hot at every hour of every day. Another colleague, referred to as Mr. Br . . . e, was dubbed the Corrector of the Moon's Motion. In a more serious vein, Thacker made several sage observations regarding the physical laws with which watchmakers were struggling. He verified experimentally that a coiled spring loses strength when heated and gains it when cooled. He kept his own clock under a kind of bell jar connected with an exhaust pump, so that it could be run in a partial vacuum. He also devised an auxiliary spring which kept the clock going while the mainspring was being wound. Both springs were wound outside the bell by means of rods passed through stuffing boxes, so that neither the vacuum nor the clock mechanism

[45] *Ibid.*, pp. 27–30; Huygens described his pendulum clock in his *Horologium Oscillatorium*, Paris, 1673.

[46] Ibid., pp. 32, 33; Jeremy Thacker wrote a clever piece entitled: *The Longitudes Examined, beginning with a short epistle to the Longitudinarians and ending with the description of a smart, pretty Machine of my Own which I am (almost) sure will do for the Longitude and procure me The Twenty Thousand Pounds.* By Jeremy Thacker, of Beverly in Yorkshire. ". . . *quid non mortalia pectora cogis Auri sacra Fames* . . ." London. Printed for J. Roberts at the Oxford Arms in Warwick Lane, 1714. Price Sixpence.

would have to be disturbed. In spite of these and other devices, watchmakers remained in the dark and their problems remained unsolved until John Harrison went to work on the physical laws behind them. After that they did not seem so difficult.[47]

Harrison was born at Foulby in the parish of Wragby, Yorkshire, in May, 1693. He was the son of a carpenter and joiner in the service of Sir Rowland Winn of Nostell Priory. John was the oldest son in a large family. When he was six years old he contracted smallpox, and while convalescing spent hours watching the mechanism and listening to the ticking of a watch laid on his pillow. When his family moved to Barrow in Lincolnshire, John was seven years old. There he learned his father's trade and worked with him for several years. Occasionally he earned a little extra by surveying and measuring land, but he was much more interested in mechanics, and spent his evenings studying Nicholas Saunderson's published lectures on mathematics and physics. These he copied out in longhand including all the diagrams. He also studied the mechanism of clocks and watches, how to repair them and how they might be improved. In 1715, when he was twenty-two, he built his first grandfather clock or "regulator." The only remarkable feature of the machine was that all the wheels except the escape wheel were made of oak, with the teeth, carved separately, set into a groove in the rim.[48]

Many of the mechanical faults in the clocks and watches that Harrison saw around him were caused by the expansion and contraction of the metals used in their construction. Pendulums, for example, were usually made of an iron or steel rod with a lead bob fastened at the end. In winter the rod contracted and the clock went fast, and in summer the rod expanded, making the clock lose time. Harrison made his first important contribution to clockmaking by developing the "gridiron" pendulum, so named because of its appearance. Brass and steel, he knew, expand for a given increase in temperature in the ratio of about three to two (100 to

[47] The invention of a "maintaining power" is erroneously attributed to Harrison. See R. T. Gould, *op. cit.*, p. 34, who says that Thacker antedates Harrison by twenty years on the invention of an auxiliary spring to keep a machine going while it was being wound. In spite of the magnitude of John Harrison's achievement, the inventive genius of Pierre Le Roy of Paris produced the prototype of the modern chronometer. As Rupert Gould points out, Harrison's Number Four was a remarkable piece of mechanism, "a satisfactory marine timekeeper, one, too, which was of permanent usefullness, and which could be duplicated as often as necessary. But No. 4, in spite of its fine performance and beautiful mechanism, cannot be compared, for efficiency and design, with Le Roy's wonderful machine. The Frenchman, who was but little indebted to his precessors, and not at all to his contemporaries, evolved, by sheer force of genius, a timekeeper which contains all the essential mechanism of the modern chronometer." (See R. T. Gould, *op. cit.*, p. 65.)

[48] For a biographical sketch of Harrison see R. T. Gould, *op. cit.*, pp. 40 ff. Saunderson (1682–1739) was Lucasian professor of mathematics at Cambridge. Harrison's first "regulator" is now in the museum of the Clockmakers' Co. of London. "The term 'regulator' is used to denote any high-class pendulum clock designed for use solely as an accurate time-measurer, without any additions such as striking mechanism, calendar work, &c." (Gould, p. 42 *n.*)

62). He therefore built a pendulum with nine alternating steel and brass rods, so pinned together that expansion or contraction caused by variation in the temperature was eliminated, the unlike rods counteracting each other.[49]

The accuracy of a clock is no greater than the efficiency of its escapement, the piece which releases for a second, more or less, the driving power, such as a suspended weight or a coiled mainspring. One day Harrison was called out to repair a steeple clock that refused to run. After looking it over he discovered that all it needed was some oil on the pallets of the escapement. He oiled the mechanism and soon after went to work on a design for an escapement that would not need oiling. The result was an ingenious "grasshopper" escapement that was very nearly frictionless and also noiseless. However, it was extremely delicate, unnecessarily so, and was easily upset by dust or unnecessary oil. These two improved parts alone were almost enough to revolutionize the clockmaking industry. One of the first two grandfather clocks he built that were equipped with his improved pendulum and grasshopper escapement did not gain or lose more than a second a month during a period of fourteen years.

Harrison was twenty-one years old when Parliament posted the £20,000 reward for a reliable method of determining longitude at sea. He had not finished his first clock, and it is doubtful whether he seriously aspired to winning such a fortune, but certainly no young inventor ever had such a fabulous goal to shoot at, or such limited competition. Yet Harrison never hurried his work, even after it must have been apparent to him that the prize was almost within his reach. On the contrary, his real goal was the perfection of his marine timekeeper as a precision instrument and a thing of beauty. The monetary reward, therefore, was a foregone conclusion.

His first two fine grandfather clocks were completed by 1726, when he was thirty-three years old, and in 1728 he went to London, carrying with him full-scale models of his gridiron pendulum and grasshopper escapement, and working drawings of a marine clock he hoped to build if he could get some financial assistance from the Board of Longitude. He called on Edmund Halley, Astronomer Royal, who was also a member of the Board. Halley advised him not to depend on the Board of Longitude, but to talk things over with George Graham, England's leading horologist.[50] Harrison called on Graham at ten o'clock one morning, and together they talked pendulums, escapements, remontoires and springs until eight o'clock in the evening, when Harrison departed a happy man. Graham had advised him to build his clock first and then apply to the

[49] *Ibid.*, pp. 40–41. Graham had experimented with a gridiron pendulum and in 1725 had produced a pendulum with a small jar of mercury attached to the bob which was supposed to counteract the expansion of the rod caused by a rise in temperature.

[50] *Ibid.*, pp. 42, 43; Graham and Tompion are the only two horologists buried in Westminster Abbey.

Board of Longitude. He had also offered to loan Harrison the money to build it with, and would not listen to any talk about interest or security of any kind. Harrison went home to Barrow and spent the next seven years building his first marine timekeeper, his "Number One," as it was later called.

In addition to heat and cold, the archenemies of all watchmakers, he concentrated on eliminating friction, or cutting it down to a bare minimum, on every moving part, and devised many ingenious ways of doing it; some of them radical departures from accepted watchmaking practice. Instead of using a pendulum, which would be impractical at sea, Harrison designed two huge balances weighing about five pounds each, that were connected by wires running over brass arcs so that their motions were always opposed. Thus any effect on one produced by the motion of the ship would be counteracted by the other. The "grasshopper" escapement was modified and simplified and two mainsprings on separate drums were installed. The clock was finished in 1735.

There was nothing beautiful or graceful about Harrison's Number One. It weighed seventy-two pounds and looked like nothing but an awkward, unwieldly piece of machinery. However, everyone who saw it and studied its mechanism declared it a masterpiece of ingenuity, and its performance certainly belied its appearance. Harrison mounted its case in gimbals and for a while tested it unofficially on a barge in the Humber River. Then he took it to London where he enjoyed his first brief triumph. Five members of the Royal Society examined the clock, studied its mechanism and then presented Harrison with a certificate stating that the principles of this timekeeper promised a sufficient degree of accuracy to meet the requirements set forth in the Act of Queen Anne. This historic document, which opened for Harrison the door to the Board of Longitude, was signed by Halley, Smith, Bradley, Machin and Graham.

On the strength of the certificate, Harrison applied to the Board of Longitude for a trial at sea, and in 1736 he was sent to Lisbon in H.M.S. *Centurion*, Captain Proctor. In his possession was a note from Sir Charles Wager, First Lord of the Admiralty, asking Proctor to see that every courtesy be given the bearer, who was said by those who knew him best to be "a very ingenious and sober man." Harrison was given the run of the ship, and his timekeeper was placed in the Captain's cabin where he could make observations and wind his clock without interruption. Proctor was courteous but skeptical. "The difficulty of measuring Time truly," he wrote, "where so many unequal Shocks and Motions stand in Opposition to it, gives me concern for the honest Man, and makes me feel he has attempted Impossibilities." [51]

No record of the clock's going on the outward voyage is known, but

[51] *Ibid.*, pp. 45–46.

after the return trip, made in H.M.S. *Orford*, Robert Man, Harrison was given a certificate signed by the master (that is, navigator) stating: "When we made the land, the said land, according to my reckoning (and others), ought to have been the Start; but before we knew what land it was, John Harrison declared to me and the rest of the ship's company, that according to his observations with his machine, it ought to be the Lizard—the which, indeed, it was found to be, his observation showing the ship to be more west than my reckoning, above one degree and twenty-six miles." It was an impressive report in spite of its simplicity, and yet the voyage to Lisbon and return was made in practically a north and south direction; one that would hardly demonstrate the best qualities of the clock in the most dramatic fashion. It should be noted, however, that even on this well-worn trade route it was not considered a scandal that the ship's navigator should make an error of 90 miles in his landfall.

On June 30, 1737, Harrison made his first bow to the mighty Board of Longitude. According to the official minutes, "Mr. John Harrison produced a new invented machine, in the nature of clockwork, whereby he proposes to keep time at sea with more exactness than by any other instrument or method hitherto contrived . . . and proposes to make another machine of smaller dimensions within the space of two years, whereby he will endeavour to correct some defects which he hath found in that already prepared, so as to render the same more perfect . . ." The Board voted him £500 to help defray expenses, one half to be paid at once and the other half when he completed the second clock and delivered same into the hands of one of His Majesty's ship's captains.[52]

Harrison's Number Two contained several minor mechanical improvements and this time all the wheels were made of brass instead of wood. In some respects it was even more cumbersome than Number One, and it weighed one hundred and three pounds. Its case and gimbal suspension weighed another sixty-two pounds. Number Two was finished in 1739, but instead of turning it over to a sea captain appointed by the Board to receive it, Harrison tested it for nearly two years under conditions of "great heat and motion." Number Two was never sent to sea because by the time it was ready, England was at war with Spain and the Admiralty had no desire to give the Spaniards an opportunity to capture it.

In January, 1741, Harrison wrote the Board that he had begun work on a third clock which promised to be far superior to the first two. They voted him another £500. Harrison struggled with it for several months, but seems to have miscalculated the "moment of inertia" of its balances. He thought he could get it going by the first of August, 1741, and have it ready for a sea trial two years later. But after five years the Board learned "that it does not go well, at present, as he expected it would, yet he plainly

[52] *Ibid.*, p. 47.

perceived the Cause of its present Imperfection to lye in a certain part [the balances] which, being of a different form from the corresponding part in the other machines, had never been tried before." Harrison had made a few improvements in the parts of Number Three and had incorporated in it the same antifriction devices he had used on Number Two, but the clock was still bulky and its parts were far from delicate; the machine weighed sixty-six pounds and its case and gimbals another thirty-five.[53]

Harrison was again feeling the pinch, even though the Board had given him several advances to keep him going, for in 1746, when he reported on Number Three, he laid before the Board an impressive testimonial signed by twelve members of the Royal Society including the President, Martin Folkes, Bradley, Graham, Halley and Cavendish, attesting the importance and practical value of his inventions in the solution of the longitude problem. Presumably this gesture was made to insure the financial support of the Board of Longitude. However, the Board needed no prodding. Three years later, acting on its own volition, the Royal Society awarded Harrison the Copley medal, the highest honor it could bestow. His modesty, perseverance and skill made them forget, at least for a time, the total lack of academic background which was so highly revered by that august body.[54]

Convinced that Number Three would never satisfy him, Harrison proposed to start work on two more timekeepers, even before Number Three was given a trial at sea. One would be pocketsize and the other slightly larger. The Board approved the project and Harrison went ahead. Abandoning the idea of a pocketsize chronometer, Harrison decided to concentrate his efforts on a slightly larger clock, which could be adapted to the intricate mechanism he had designed without sacrificing accuracy. In 1757 he began work on Number Four, a machine which "by reason alike of its beauty, its accuracy, and its historical interest, must take pride of place as the most famous chronometer that ever has been or ever will be made." It was finished in 1759.[55]

Number Four resembled an enormous "pair-case" watch about five inches in diameter, complete with pendant, as though it were to be worn. The dial was white enamel with an ornamental design in black. The hour and minute hands were of blued steel and the second hand was polished. Instead of a gimbal suspension, which Harrison had come to distrust, he used only a soft cushion in a plain box to support the clock. An adjustable

[53] *Ibid.*, pp. 47–49, for details of the technical improvements made in No. 2 and No. 3.

[54] *Ibid.*, p. 49. Some years later he was offered the honor of Fellow of the Royal Society, but he declined it in favor of his son William.

[55] *Ibid.*, pp. 50, 53; for a full description of No. 4 see H. M. Frodsham in the *Horological Journal* for May, 1878—with drawings of the escapement, train and remontoire taken from the duplicate of No. 4 made by Larcum Kendall.

outer box was fitted with a divided arc so that the timekeeper could be kept in the same position (with the pendant always slightly above the horizontal) regardless of the lie of the ship. When it was finished, Number Four was not adjusted for more than this one position, and on its first voyage it had to be carefully tended. The watch beat five to the second and ran for thirty hours without rewinding. The pivot holes were jeweled to the third wheel with rubies and the end stones were diamonds. Engraved in the top-plate were the words "John Harrison & Son, A.D. 1759." Cunningly concealed from prying eyes beneath the plate was a mechanism such as the world had never seen; every pinion and bearing, each spring and wheel was the end product of careful planning, precise measurement and exquisite craftsmanship. Into the mechanism had gone "fifty years of self-denial, unremitting toil, and ceaseless concentration." To Harrison, whose singleness of purpose had made it possible for him to achieve the impossible, Number Four was a satisfactory climax to a lifetime of effort. He was proud of this timekeeper, and in a rare burst of eloquence he wrote, "I think I may make bold to say, that there is neither any other Mechanical or Mathematical thing in the World that is more beautiful or curious in texture than this my watch or Time-keeper for the Longitude . . . and I heartily thank Almighty God that I have lived so long, as in some measure to complete it." [56]

After checking and adjusting Number Four with his pendulum clock for nearly two years, Harrison reported to the Board of Longitude, in March 1761, that Number Four was as good as Number Three and that its performance greatly exceeded his expectations. He asked for a trial at sea. His request was granted, and in April, 1761, William Harrison, his son and right-hand man, took Number Three to Portsmouth. The father arrived a short time later with Number Four. There were numerous delays at Portsmouth, and it was October before passage was finally arranged for young Harrison aboard H.M.S. *Deptford*, Dudley Digges, bound for Jamaica. John Harrison, who was then sixty-eight years old, decided not to attempt the long sea voyage himself; and he also decided to stake everything on the performance of Number Four, instead of sending both Three and Four along. The *Deptford* finally sailed from Spithead with a convoy, November 18, 1761, after first touching at Portland and Plymouth. The sea trial was on.

Number Four had been placed in a case with four locks, and the four keys were given to William Harrison, Governor Lyttleton of Jamaica, who was taking passage on the *Deptford*, Captain Digges, and his first lieutenant. All four had to be present in order to open the case, even for winding. The Board of Longitude had further arranged to have the longitude of Jamaica determined *de novo* before the trial, by a series of obser-

[56] *Ibid.*, p. 63.

vations of the satellites of Jupiter, but because of the lateness of the season it was decided to accept the best previously established reckoning. Local time at Portsmouth and at Jamaica was to be determined by taking equal altitudes of the sun, and the difference compared with the time indicated by Harrison's timekeeper.

As usual, the first scheduled port of call on the run to Jamaica was Madeira. On this particular voyage, all hands aboard the *Deptford* were anxious to make the island on the first approach. To William Harrison it meant the first crucial test of Number Four; to Captain Digges it meant a test of his dead reckoning against a mechanical device in which he had no confidence; but the ship's company had more than a scientific interest in the proceedings. They were afraid of missing Madeira altogether, "the consequence of which, would have been Inconvenient." To the horror of all hands, it was found that the beer had spoiled, over a thousand gallons of it, and the people had already been reduced to drinking water. Nine days out from Plymouth the ship's longitude, by dead reckoning, was 13° 50′ west of Greenwich, but according to Number Four and William Harrison it was 15° 19′ W. Captain Digges naturally favored his dead reckoning calculations, but Harrison stoutly maintained that Number Four was right and that if Madeira were properly marked on the chart they would sight it the next day. Although Digges offered to bet Harrison five to one that he was wrong, he held his course, and the following morning at 6 A.M. the lookout sighted Porto Santo, the northeastern island of the Madeira group, dead ahead.

The *Deptford's* officers were greatly impressed by Harrison's uncanny predictions throughout the voyage. They were even more impressed when they arrived at Jamaica three days before H.M.S. *Beaver*, which had sailed for Jamaica ten days before them. Number Four was promptly taken ashore and checked. After allowing for its rate of going (2⅔ seconds per day losing at Portsmouth), it was found to be 5 seconds slow, an error in longitude of 1¼′ only, or 1¼ nautical miles.[57]

The official trial ended at Jamaica. Arrangements were made for William Harrison to make the return voyage in the *Merlin*, sloop, and in a burst of enthusiasm Captain Digges placed his order for the first Harrison-built chronometer which should be offered for sale. The passage back to England was a severe test for Number Four. The weather was extremely rough and the timekeeper, still carefully tended by Harrison, had to be moved to the poop, the only dry place on the ship, where it was pounded unmercifully and "received a number of violent shocks." However, when it was again checked at Portsmouth, its total error for the five months' voyage, through heat and cold, fair weather and foul (after allowing for its rate of going), was only $1^{m}\ 53\frac{1}{2}^{s}$, or an error in longitude of 28½′

[57] *Ibid.*, pp. 55–56 and 56 *n.*

(28½ nautical miles). This was safely within the limit of half a degree specified in the Act of Queen Anne. John Harrison and son had won the fabulous reward of £20,000.

The sea trial had ended, but the trials of John Harrison had just begun. Now for the first time, at the age of sixty-nine, Harrison began to feel the lack of an academic background. He was a simple man; he did not know the language of diplomacy, the gentle art of innuendo and evasion. He had mastered the longitude but he did not know how to cope with the Royal Society or the Board of Longitude. He had won the reward and all he wanted now was his money. The money was not immediately forthcoming.

Neither the Board of Longitude nor the scientists who served it as consultants were at any time guilty of dishonesty in their dealings with Harrison; they were only human. £20,000 was a tremendous fortune, and it was one thing to dole out living expenses to a watchmaker in amounts not exceeding £500 so that he might contribute something or other to the general cause. But it was another thing to hand over £20,000 in a lump sum to one man, and a man of humble birth at that. It was most extraordinary. Moreover, there were men on the Board and members of the Royal Society who had designs on the reward themselves or at least a cut of it. James Bradley and Johann Tobias Mayer had both worked long and hard on the compilation of accurate lunar tables. Mayer's widow was paid £3000 for his contribution to the cause of longitude, and in 1761 Bradley told Harrison that he and Mayer would have shared £10,000 of the prize money between them if it had not been for his blasted watch. Halley had struggled long and manfully on the solution of the longitude by compass variation, and was not in a position to ignore any part of £20,000. The Reverend Nevil Maskelyne, Astronomer Royal, and compiler of the *Nautical Almanac*, was an obstinate and uncompromising apostle of "lunar distances" or "lunars" for finding the longitude, and had closed his mind to any other method whatsoever. He loved neither Harrison nor his watch. In view of these and other unnamed aspirants, it was inevitable that the Board should decide that the amazing performance of Harrison's timekeeper was a fluke. They had never been allowed to examine the mechanism, and they pointed out that if a gross of watches were carried to Jamaica under the same conditions, one out of the lot might perform equally well—at least for one trip. They accordingly refused to give Harrison a certificate stating that he had met the requirements of the Act until his timekeeper was given a further trial, or trials. Meanwhile, they did agree to give him the sum of £2500 as an interim reward, since his machine had proved to be a rather useful contraption, though mysterious beyond words. An Act of Parliament (February, 1763) enabling him to receive £5000 as soon as he disclosed the secret of his invention, was completely nullified by the absurdly rigid conditions set up by the Board. He was finally granted a new trial at sea.[58]

The rules laid down for the new trial were elaborate and exacting. The difference in longitude between Portsmouth and Jamaica was to be determined *de novo* by observations of Jupiter's satellites. Number Four was to be rated at Greenwich before sailing, but Harrison balked, saying "that he did not chuse to part with it out of his hands till he shall have reaped some advantage from it." However, he agreed to send his own rating, sealed, to the Secretary of the Admiralty before the trial began. After endless delays the trial was arranged to take place between Portsmouth and Barbados, instead of Jamaica, and William Harrison embarked on February 14, 1764, in H.M.S. *Tartar*, Sir John Lindsay, at the Nore. The *Tartar* proceeded to Portsmouth, where Harrison checked the rate of Number Four with a regulator installed there in a temporary observatory. On March 28, 1764, the *Tartar* sailed from Portsmouth and the second trial was on.

It was the same story all over again. On April 18, twenty-one days out, Harrison took two altitudes of the sun and announced to Sir John that they were forty-three miles east of Porto Santo. Sir John accordingly steered a direct course for it, and at one o'clock the next morning the island was sighted, "which exactly agreed with the Distance mentioned above." They arrived at Barbados May 13, "Mr. Harrison all along in the Voyage declaring how far he was distant from that Island, according to the best settled longitude thereof. The Day before they made it, he declared the Distance: and Sir John sailed in Consequence of this Declaration, till Eleven at Night, which proving dark he thought proper to lay by. Mr. Harrison then declaring they were no more than eight or nine Miles from the Land, which accordingly at Day Break they saw from that Distance." [59]

When Harrison went ashore with Number Four he discovered that none other than Maskelyne and an assistant, Green, had been sent ahead to check the longitude of Barbados by observing Jupiter's satellites. Moreover, Maskelyne had been orating loudly on the superiority of his own method of finding longitude, namely, by lunar distances. When Harrison heard what had been going on he objected strenuously, pointing out to Sir John that Maskelyne was not only an interested party but an active and avid competitor, and should not have anything to do with the trials. A compromise was arranged, but, as it turned out, Maskelyne was suddenly indisposed and unable to make the observations.

[58] *Ibid.*, pp. 57 ff. On August 17, 1762, the Board refused to give Harrison a certificate stating that he had complied with the terms set forth in the Act of Queen Anne.

[59] *Ibid.*, p. 59; see also *A narrative of the proceedings relative to the discovery of the longitude at sea; by Mr. John Harrison's time-keeper* . . . [By James Short] London: printed for the author, 1765, pp. 7, 8.

After comparing the data obtained by observation with Harrison's chronometer, Number Four showed an error of 38.4 seconds over a period of seven weeks, or 9.6 miles of longitude (at the equator) between Portsmouth and Barbados. And when the clock was again checked at Portsmouth, after 156 days, elapsed time, it showed, after allowing for its rate of going, a total gain of only 54 seconds of time. If further allowance were made for changes of rate caused by variations in temperature, information posted beforehand by Harrison, the rate of Number Four would have been reduced to an error of 15 seconds of loss in 5 months, or less than 1/10 of a second a day.[60]

The evidence in favor of Harrison's chronometer was overwhelming, and could no longer be ignored or set aside. But the Board of Longitude was not through. In a Resolution of February 9, 1765, they were unanimously of the opinion that "the said timekeeper has kept its time with sufficient correctness, without losing its longitude in the voyage from Portsmouth to Barbados beyond the nearest limit required by the Act 12th of Queen Anne, but even considerably within the same." Now, they said, all Harrison had to do was demonstrate the mechanism of his clock and explain the construction of it, "by Means whereof other such Timekeepers might be framed, of sufficient Correctness to find the Longitude at Sea. . . ." In order to get the first £10,000 Harrison had to submit, on oath, complete working drawings of Number Four; explain and demonstrate the operation of each part, including the process of tempering the springs; and finally, hand over to the Board his first three timekeepers as well as Number Four.[61]

Any foreigner would have acknowledged defeat at this juncture, but not Harrison, who was an Englishman and a Yorkshireman to boot. "I cannot help thinking," he wrote the Board, after hearing their harsh terms, "but I am extremely ill used by gentlemen who I might have expected different treatment from. . . . It must be owned that my case is very hard, but I hope I am the first, and for my country's sake, shall be the last that suffers by pinning my faith on an English Act of Parliament." The case of "Longitude Harrison" began to be aired publicly, and several of his friends launched an impromptu publicity campaign against the Board and against Parliament. The Board finally softened their terms and Harrison reluctantly took his clock apart at his home for the edification of a committee of six, nominated by the Board; three of them, Thomas Mudge, William Matthews and Larcum Kendall, were watchmakers. Harrison then received a certificate from the Board (October 28, 1765) entitling him to £7500, or the balance due him on the first half of the reward. The second half did not come so easily.

[60] R. T. Gould, *op. cit.*, pp. 59, 60.

[61] *Ibid.*, pp. 60 ff.; Act of Parliament 5 George III, Cap. 20; see also James Short, *op. cit.*, p. 15.

Number Four was now in the hands of the Board of Longitude, held in trust for the benefit of the people of England. As such, it was carefully guarded against prying eyes and tampering, even by members of the Board. However, that learned body did its humble best. First they set out to publicize its mechanism as widely as possible. Unable to take the thing apart themselves, they had to depend on Harrison's own drawings, and these were redrawn and carefully engraved. What was supposed to be a full textual description was written by the Reverend Nevil Maskelyne and printed in book form with illustrations appended: *The Principles of Mr. Harrison's Time-Keeper, with Plates of the Same.* London, 1767. Actually the book was harmless enough, because no human being could have even begun to reproduce the clock from Maskelyne's description. To Harrison it was just another bitter pill to swallow. "They have since published all my Drawings," he wrote, "without giving me the last Moiety of the Reward, or even paying me and my Son for our Time at a rate as common Mechanicks; an Instance of such Cruelty and Injustice as I believe never existed in a learned and civilised Nation before." Other galling experiences followed.[62]

With great pomp and ceremony Number Four was carried to the Royal Observatory at Greenwich. There it was scheduled to undergo a prolonged and exhaustive series of trials under the direction of the Astronomer Royal, the Reverend Nevil Maskelyne. It cannot be said that Maskelyne shirked his duty, although he was handicapped by the fact that the timekeeper was always kept locked in its case, and he could not even wind it except in the presence of an officer detailed by the Governor of Greenwich to witness the performance. Number Four, after all, was a £10,000 timekeeper. The tests went on for two months. Maskelyne tried the watch in various positions for which it was not adjusted, dial up and dial down. Then for ten months it was tested in a horizontal position, dial up. The Board published a full account of the results with a preface written by Maskelyne, in which he gave it as his studied opinion "That Mr. Harrison's Watch cannot be depended upon to keep the Longitude within a Degree, in a West-India Voyage of six weeks, nor to keep the Longitude within Half a Degree for more than a Fortnight, and then it must be kept in a Place where the Thermometer is always some Degrees above freezing." (There was still £10,000 prize money outstanding.)[63]

The Board of Longitude next commissioned Larcum Kendall, watchmaker, to make a duplicate of Number Four. They also advised Harrison that he must make Number Five and Number Six and have them tried at

[62] R. T. Gould, *op. cit.*, pp. 61–62; 62 *n*; the unauthorized publication of Harrison's drawings with a preface by the Reverend Maskelyne, appeared under the title: *The principles of Mr. Harrison's time-keeper, with plates of the same,* London, 1767.

[63] R. T. Gould, *op. cit.*, p. 63; see also Nevil Maskelyne's *An account of the going of Mr. John Harrison's watch* . . . London, 1767.

sea, intimating that otherwise he would not be entitled to the other half of the reward. When Harrison asked if he might use Number Four for a short time to help him build two copies of it, he was told that Kendall needed it to work from and that it would be impossible. Harrison did the best he could, while the Board laid plans for an exhaustive series of tests for Number Five and Number Six. They spoke of sending them to Hudson's Bay and of letting them toss and pitch in the Downs for a month or two as well as sending them out to the West Indies.

After three years (1767–1770) Number Five was finished. In 1771, just as the Harrisons were finishing the last adjustments on the clock, they heard that Captain Cook was preparing for a second exploring cruise, and that the Board was planning to send Kendall's duplicate of Number Four along with him. Harrison pleaded with them to send Number Four and Number Five instead, telling them he was willing to stake his claim to the balance of the reward on their performance, or to submit "to any mode of trial, by men not already proved partial, which shall be definite in its nature." The man was now more than ever anxious to settle the business once and for all. But it was not so to be. He was told that the Board did not see fit to send Number Four out of the kingdom, nor did they see any reason for departing from the manner of trial already decided upon.

John Harrison was now seventy-eight years old. His eyes were failing and his skilled hands were not as steady as they were, but his heart was strong and there was still a lot of fight left in him. Among his powerful friends and admirers was His Majesty King George the Third, who had granted Harrison and his son an audience after the historic voyage of the *Tartar*. Harrison now sought the protection of his king, and "Farmer George," after hearing the case from start to finish, lost his patience. "By God, Harrison, I'll see you righted," he roared. And he did. Number Five was tried at His Majesty's private observatory at Kew. The king attended the daily checking of the clock's performance, and had the pleasure of watching the operation of a timekeeper whose total error over a ten week's period was 4½ seconds.[64]

Harrison submitted a memorial to the Board of Longitude, November 28, 1772, describing in detail the circumstances and results of the trial at Kew. In return, the Board passed a resolution to the effect that they were not the slightest bit interested; that they saw no reason to alter the manner of trial they had already proposed and that no regard would be paid for a trial made under any other conditions. In desperation Harrison decided to play his last card—the king. Backed by His Majesty's personal interest in the proceedings, Harrison presented a petition to the House of Commons with weight behind it. It was heralded as follows: "The Lord North, by His Majesty's Command, acquainted the House that His Maj-

[64] R. T. Gould, *op. cit.*, pp. 64–65.

esty, having been informed of the Contents of the said Petition, recommended it to the Consideration of the House." Fox was present to give the petition his full support, and the king was willing, if necessary, to appear at the Bar of the House under an inferior title and testify in Harrison's behalf. At the same time, Harrison circulated a broadside, *The Case of Mr. John Harrison*, stating his claims to the second half of the reward.[65]

The Board of Longitude began to squirm. Public indignation was mounting rapidly and the Speaker of the House informed the Board that consideration of the petition would be deferred until they had an opportunity to revise their proceedings in regard to Mr. Harrison. Seven Admiralty clerks were put to work copying out all the Board's resolutions concerning Harrison. While they worked day and night to finish the job, the Board made one last desperate effort. They summoned William Harrison to appear before them; but the hour was late. They put him through a catechism and tried to make him consent to new trials and new conditions. Harrison stood fast, refusing to consent to anything they might propose. Meanwhile a money bill was drawn up by Parliament in record time; the king gave it the nod and it was passed. The Harrisons had won their fight.

[65] *Ibid.*, p. 66; see also the *Journal of the House of Commons*, 6. 5. 1772.

COMMENTARY ON

JOHN COUCH ADAMS

THE discovery in 1846 of the planet Neptune was a dramatic and spectacular achievement of mathematical astronomy. The very existence of this new member of the solar system, and its exact location, were demonstrated with pencil and paper; there was left to observers only the routine task of pointing their telescopes at the spot the mathematicians had marked. It is easy to understand the enthusiastic appraisal of Sir John Herschel, who, on the occasion of the presentation of gold medals of the Royal Astronomical Society to Leverrier and Adams, said that their codiscovery of Neptune "surpassed by intelligible and legitimate means, the wildest pretensions of clairvoyance." [1]

As early as about 1820 it had been recognized that irregularities in the motion of Uranus, deviations of its observed orbit from its calculated positions, could be accounted for only by an outside disturbing force. The German astronomer Bessel at that time remarked to Humboldt that sooner or later the "mystery of Uranus" would "be solved by the discovery of a new planet." [2] The problem was solved by two astronomers working entirely independently of each other: Urban Jean Leverrier, director of the Paris Observatory, and John Couch Adams, a twenty-six-year-old Fellow of St. John's College in Cambridge. The conquest of this intricate and incredibly laborious problem of "inverse perturbations"—i.e., given the perturbations, to find the planet—was an intellectual feat deservedly acclaimed as "sublime"; but the personal behavior of various participants in the event falls short of sublimity. The affair was marked by episodes of confusion and fecklessness, of donnish hairpulling and Gallic backbiting, of stuffiness, jealousy and general academic blight. Sir Harold Spencer Jones gives a balanced account; as the present Astronomer Royal he is understandably charitable to Sir George Airy, the Astronomer Royal in Adams' day, who almost succeeded—more or less innocently—in doing Adams out of proper recognition for his work.

I add a few biographical details about Adams not covered in Jones' lecture. While the discovery of Neptune was his most sensational achievement, it came at the beginning of a long and distinguished service to astronomy and mathematics. Adams succeeded George Peacock, the noted algebraist, as professor of astronomy and geometry at Cambridge. This post, as well as a fellowship in mathematics at Pembroke College, he held until his death. His researches are represented by important memoirs on

[1] *Nature*, Vol. XXXIV, p. 565.
[2] Giorgio Abetti, *The History of Astronomy*, 1952, New York, p. 214.

lunar motion, on the effect of planetary perturbations on the orbits and periods of certain meteors, on various problems of pure mathematics. Adams was a stylish craftsman in his mathematical work; even the examination questions he set in prize competitions were admired for their finish. Like Euler and Gauss, he found pleasure in undertaking immense numerical calculations, which he carried out with consummate accuracy. But he was not chained to his specialty. He read widely in history, biology, geology and general literature; he took a deep interest in political questions. During the Franco-Prussian war, he was so moved "that he could scarcely work or sleep." [3] Adams was a shy, gentle and unaffected man. He refused knighthood in 1847, just as he had refused to be drawn into the bitter controversy over the question of who was first to discover Neptune. The honor was tendered in a foolish attempt to settle a foolish question. The entire business was beneath Adams. "He uttered no complaint; he laid no claim to priority: Leverrier had no warmer admirer." [4] He died after a long illness on January 21, 1892.[5]

[3] J. W. L. Glaisher, biographical memoir in *The Scientific Papers of John Couch Adams*, ed. by William Grylls Adams, Cambridge, 1896, p. XLIV.

[4] *The Dictionary of National Biography*; Vol. XXII, *Supplement*, article on Adams, p. 16.

[5] For further information on the Adams-Airy affair see J. E. Littlewood, *A Mathematician's Miscellany*, London, 1953, pp. 116–134.

There are seven windows in the head, two nostrils, two eyes, two ears, and a mouth; so in the heavens there are two favorable stars, two unpropitious, two luminaries, and Mercury alone undecided and indifferent. From which and many other similar phenomena of nature, such as the seven metals, etc., which it were tedious to enumerate, we gather that the number of planets is necessarily seven.—FRANCESCO SIZZI (argument against Galileo's discovery of the satellites of Jupiter)

4 John Couch Adams and the Discovery of Neptune

By SIR HAROLD SPENCER JONES

ON the night of 13 March 1781 William Herschel, musician by profession but assiduous observer of the heavens in his leisure time, made a discovery that was to bring him fame. He had for some time been engaged upon a systematic and detailed survey of the whole heavens, using a 7 in. telescope of his own construction; he carefully noted everything that appeared in any way remarkable. On the night in question, in his own words:

'In examining the small stars in the neighbourhood of H Geminorum I perceived one that appeared visibly larger than the rest; being struck with its uncommon appearance I compared it to H Geminorum and the small star in the quartile between Auriga and Gemini, and finding it so much larger than either of them, I suspected it to be a comet.'

Most observers would have passed the object by without noticing anything unusual about it, for the minute disk was only about 4 sec. in diameter. The discovery was made possible by the excellent quality of Herschel's telescope, and by the great care with which his observations were made.

The discovery proved to be of greater importance than Herschel suspected, for the object he had found was not a comet, but a new planet, which revolved round the Sun in a nearly circular path at a mean distance almost exactly double that of Saturn; it was unique, because no planet had ever before been discovered; the known planets, easily visible to the naked eye, did not need to be discovered.

After the discovery of Uranus, as the new planet was called, it was ascertained that it had been observed as a star and its position recorded on a score of previous occasions. The earliest of these observations was made by Flamsteed at Greenwich in 1690. Lemonnier in 1769 had observed its transit six times in the course of 9 days and, had he compared the observations with one another, he could not have failed to anticipate Herschel in the discovery. As Uranus takes 84 years to make a complete revolution round the Sun, these earlier observations were of special value for the investigation of its orbit.

The positions of the planet computed from tables constructed by Delambre soon began to show discordances with observation, which became greater as time went on. As there might have been error or incompleteness in Delambre's theory and tables, the task of revision was undertaken by Bouvard, whose tables of the planet appeared in 1821. Bouvard found that, when every correction for the perturbations in the motion of Uranus by the other planets was taken into account, it was not possible to reconcile the old observations of Flamsteed, Lemonnier, Bradley, and Mayer with the observations made subsequently to the discovery of the planet in 1781.

'The construction of the tables, then,' said Bouvard, 'involves this alternative: if we combine the ancient observations with the modern, the former will be sufficiently well represented, but the latter will not be so, with all the precision which their superior accuracy demands; on the other hand, if we reject the ancient observations altogether, and retain only the modern, the resulting tables will faithfully conform to the modern observations, but will very inadequately represent the more ancient. As it was necessary to decide between these two courses, I have adopted the latter, on the ground that it unites the greatest number of probabilities in favour of the truth, and I leave to the future the task of discovering whether the difficulty of reconciling the two systems is connected with the ancient observations, or whether it depends on some foreign and unperceived cause which may have been acting upon the planet.'

Further observations of Uranus were for a time found to be pretty well represented by Bouvard's Tables, but systematic discordances between observations and the tables gradually began to show up. As time went on, observations continued to deviate more and more from the tables. It began to be suspected that there might exist an unknown distant planet, whose gravitational attraction was disturbing the motion of Uranus. An alternative suggestion was that the inverse square law of gravitation might not be exact at distances as great as the distance of Uranus from the Sun.

The problem of computing the perturbations in the motion of one planet by another moving planet, when the undisturbed orbits and the masses of the planets are known is fairly straightforward, though of some mathematical complexity. The inverse problem, of analysing the perturbations in the motion of one planet in order to deduce the position, path and mass of the planet which is producing these perturbations, is of much greater complexity and difficulty. A little consideration will, I think, show that this must be so. If a planet were exposed solely to the attractive influence of the Sun, its orbit would be an ellipse. The attractions of the other planets perturb its motion and cause it to deviate now on the one side and now on the other side of this ellipse. To determine the elements of the elliptic orbit from the positions of the planet as assigned by observation, it is necessary first to compute the perturbations produced by the other planets and to subtract them from the observed positions.

The position of the planet in this orbit at any time, arising from its undisturbed motion, can be calculated; if the perturbations of the other planets are then computed and added, the true position of the planet is obtained. The whole procedure is, in practice, reduced to a set of tables. But if Uranus is perturbed by a distant *unknown* planet, the observed positions when corrected by the subtraction of the perturbations caused by the *known* planets are not the positions in the true elliptic orbit; the perturbations of the unknown planet have not been allowed for. Hence when the corrected positions are analysed in order to determine the elements of the elliptic orbit, the derived elements will be falsified. The positions of Uranus computed from tables such as Bouvard's would be in error for two reasons; in the first place, because they are based upon incorrect elements of the elliptic orbit; in the second place, because the perturbations produced by the unknown planet have not been applied. The two causes of error have a common origin and are inextricably entangled in each other, so that neither can be investigated independently of the other. Thus though many astronomers thought it probable that Uranus was perturbed by an undiscovered planet, they could not prove it. No occasion had arisen for the solution of the extremely complicated problem of what is termed inverse perturbations, starting with the perturbed positions and deducing from them the position and motion of the perturbing body.

The first solution of this intricate problem was made by a young Cambridge mathematician, John Couch Adams. As a boy at school Adams had shown conspicuous mathematical ability, an interest in astronomy, and skill and accuracy in numerical computation. At the age of 16 he had computed the circumstances of an annular eclipse of the Sun, as visible from Lidcot, near Launceston, where his brother lived. He entered St John's College in October, 1839, at the age of 20, and in 1843 graduated Senior Wrangler, being reputed to have obtained more than double the marks awarded to the Second Wrangler. In the same year he became first Smith's Prizeman and was elected Fellow of his College.

Whilst still an undergraduate his attention had been drawn to the irregularities in the motion of Uranus. After his death there was found among his papers this memorandum, written at the beginning of his second long vacation:

Memoranda.

1841. July 3. Formed a design, in the beginning of this week, of investigating, as soon as possible after taking my degree, the irregularities in the motion of Uranus, wh. are yet unaccounted for; in order to find whether they may be attributed to the action of an undiscovered planet beyond it; and if possible thence to determine the elements of its orbit, &c. approximately, wh. wd. probably lead to its discovery.

'1841, July 3. Formed a design in the beginning of this week, of investigating, as soon as possible after taking my degree, the irregularities in the motion of Uranus, which are yet unaccounted for; in order to find whether they may be attributed to the action of an undiscovered planet beyond it; and if possible thence to determine the elements of its orbit, etc. approximately, which would probably lead to its discovery.'

As soon as Adams had taken his degree he attempted a first rough solution of the problem, with the simplifying assumptions that the unknown planet moved in a circular orbit, in the plane of the orbit of Uranus, and that its distance from the Sun was twice the mean distance of Uranus, this being the distance to be expected according to the empirical law of Bode. This preliminary solution gave a sufficient improvement in the agreement between the corrected theory of Uranus and observation to encourage him to pursue the investigation further. In order to make the observational data more complete application was made in February 1844 by Challis, the Plumian Professor of Astronomy, to Airy, the Astronomer Royal, for the errors of longitude of Uranus for the years 1818–26. Challis explained that he required them for a young friend, Mr Adams of St John's College, who was working at the theory of Uranus. By return of post, Airy sent the Greenwich data not merely for the years 1818–26 but for the years 1754–1830.

Adams now undertook a new solution of the problem, still with the assumption that the mean distance of the unknown planet was twice that of Uranus but without assuming the orbit to be circular. During term-time he had little opportunity to pursue his investigations and most of the work was undertaken in the vacations. By September 1845, he had completed the solution of the problem, and gave to Challis a paper with the elements of the orbit of the planet, as well as its mass and its position for 1 October 1845. The position indicated by Adams was actually within 2° of the position of Neptune at that time. A careful search in the vicinity of this position should have led to the discovery of Neptune. The comparison between observation and theory was satisfactory and Adams, confident in the validity of the law of gravitation and in his own mathematics, referred to the 'new planet'.

Challis gave Adams a letter of introduction to Airy, in which he said that 'from his character as a mathematician, and his practice in calculation, I should consider the deductions from his premises to be made in a trustworthy manner'. But the Astronomer Royal was in France when Adams called at Greenwich. Airy, immediately on his return, wrote to Challis saying: 'would you mention to Mr Adams that I am very much interested with the subject of his investigations, and that I should be delighted to hear of them by letter from him?'

Towards the end of October Adams called at Greenwich, on his way from Devonshire to Cambridge, on the chance of seeing the Astronomer Royal. At about that time Airy was occupied almost every day with meetings of the Railway Gauge Commission and he was in London when Adams called. Adams left his card and said that he would call again. The card was taken to Mrs Airy, but she was not told of the intention of Adams to call later. When Adams made his second call, he was informed that the Astronomer Royal was at dinner; there was no message for him and he went away feeling mortified. Airy, unfortunately, did not know of this second visit at the time. Adams left a paper summarizing the results which he had obtained and giving a list of the residual errors of the mean longitude of Uranus, after taking account of the disturbing action of the new planet. These errors were satisfactorily small, except for the first observation by Flamsteed in 1690. A few days later Airy wrote to Adams acknowledging the paper and enquiring whether the perturbations would explain the errors of the radius vector of Uranus as well as the errors of longitude; in the reduction of the Greenwich observations, Airy had shown that not only the longitude of Uranus but also its distance from the Sun (called the radius vector) showed discordances from the tabular values. Airy said at a later date that he waited with much anxiety for the answer to this query, which he looked upon as an *experimentum crucis*, and that if Adams had replied in the affirmative, he would at once have exerted all his influence to procure the publication of Adams's theory. It should be emphasized that neither Challis nor Airy knew anything about the details of Adams's investigation. Adams had attacked this difficult problem entirely unaided and without guidance. Confident in his own mathematical ability he sought no help and he needed no help.

Adams never replied to the Astronomer Royal's query; but for this failure to reply, he would almost certainly have had the sole glory of the discovery of Neptune. Airy and Adams were looking at the same problem from different points of view; Adams was so convinced that the discordances between the theory of Uranus and observation were due to the perturbing action of an unknown planet that no alternative hypothesis was considered by him; Airy, on the other hand, did not exclude the possi-

---

1845 October

According to my calculations, the obs? irregularities in the motion of Uranus may be accounted for, by supposing the existence of an ext? planet, the mass & orbit of which are as follows

Mean Dist. (assumed nearly in accordance with Bode's law)

38.4

Mean sid? motn in 365.25 days

1°.30'.9

Mean Long. 1st Octr 1845

323°.34'

Long. Perihn

315.55

Eccenty

0.1610

Mass (that of Sun being unity)

0.00016.56

For the modern obsns I have used the method of Normal places, taking the mean of the Tabular errors as given by obsns near 3 consecutive oppns to correspond with the mean of the times & the Greenw. obsns have been used down to 1830, since wh. the Cambridge & Greenwich obsns and those given in the Astron. Nachr. have been made use of. The follg are the rem. errors of mean Longitude

	Obsn − Theory		Obsn − Theory		Obsn − Theory
1780	+0″.27	1801	−0″.04	1822	+0″.30
1783	−0.23	1804	+1.76	1825	+1.92
1786	−0.96	1807	−0.21	1828	+2.25
1789	+1.82	1810	+0.56	1831	−1.06
1792	−0.91	1813	−0.94	1834	−1.44
1795	+0.09	1816	−0.31	1837	−1.62
1798	−0.99	1819	−2.00	1840	+1.73

The error for 1780 is concluded from that for 1781 given by obsn compared with those of 4 or 5 following years & also with Lemonnier's obsns in 1769 & 1771.

---

bility that the law of gravitation might not apply exactly at great distances. The purpose of his query, to which he attached great importance, was to decide between the two possibilities. As he later wrote to Challis (21 December 1846):

'There were two things to be explained, which might have existed each independently of the other, and of which one could be ascertained independently of the other: viz. error of longitude and error of radius vector. And there is no *a priori* reason for thinking that a hypothesis which will

For the ancient obsns. the follg. are the remg. errors

	Obsn.–Theory		Obsn.–Theory
1690	+44.4	1756	−4.0
1712	+6.7	1763	−5.1
1715	−6.8	1769	+0.6
1750	−1.6	1771	+11.8
1753	+5.7		

The errors are small except for Flamsteed's obsn. of 1690. This being an isolated obsn. very distant from the rest, I thought it best not to use it in forming the eqns. of condn. It is not improbable however that this error might be destroyed by a small change in the assumed mean motion of the new planet.

J. C. Adams

---

explain the error of longitude will also explain the error of radius vector. If, after Adams had satisfactorily explained the error of longitude he had (with the numerical values of the elements of the two planets so found) converted his formula for perturbation of radius vector into numbers, and if these numbers had been discordant with the *observed* numbers of discordances of radius vector, then *the theory would have been false*, NOT from any error of Adams's BUT from a failure in the law of gravitation. On this question therefore turned the continuance or fall of the law of gravitation.'

What were the reasons for Adams's failure to reply? There were several; he gave them himself at a later date (18 November 1846) in a letter to Airy. He wrote as follows:

'I need scarcely say how deeply I regret the neglect of which I was guilty in delaying to reply to the question respecting the Radius Vector of Uranus, in your note of November 5th, 1845. In palliation, though not in excuse of this neglect, I may say that I was not aware of the importance which you attached to my answer on this point and I had not the smallest notion that you felt any difficulty on it. . . . For several years past, the observed place of Uranus has been falling more and more rapidly behind its tabular place. In other words, the real angular motion of Uranus is considerably *slower* than that given by the Tables. This appeared to me to show clearly that the Tabular Radius Vector would be considerably increased by any Theory which represented the motion in Longitude. . . . Accordingly, I found that if I simply corrected the elliptic elements, so as to satisfy the modern observations as nearly as possible without taking into account any additional perturbations, the corresponding increase in the Radius Vector would not be very different from that given by my actual Theory. Hence it was that I waited to defer writing to you till I could find time to draw up an account of the method employed to obtain the results which I had communicated to you. More than once I commenced writing with this object, but unfortunately did not persevere. I was also much pained at not having been able to see you when I called at the Royal Observatory the second time, as I felt that the whole matter might be better explained by half an hour's conversation than by several letters, in writing which I have always experienced a strange difficulty. I entertained from the first the strongest conviction that the observed anomalies were due to the action of an exterior planet; no other hypothesis appeared to me to possess the slightest claims to attention. Of the accuracy of my calculations I was quite sure, from the care with which they were made and the number of times I had examined them. The only point which appeared to admit of any doubt was the assumption as to the mean distance and this I soon proceeded to correct. The work however went on very slowly throughout, as I had scarcely any time to give to these investigations, except during the vacations.

'I could not expect, however, that practical astronomers, who were already fully occupied with important labours, would feel as much confidence in the results of my investigation, as I myself did; and I therefore had our instruments put in order, with the express purpose, if no one else took up the subject, of undertaking the search for the planet myself, with the small means afforded by our observatory at St John's.'

Airy was a man with a precise and orderly mind, extremely methodical and prompt in answering letters. Another person might have followed the matter up, but not Airy. In a letter of later date to Challis, he said that 'Adams's silence . . . was so far unfortunate that it interposed an effectual barrier to all further communication. It was clearly impossible for me to write to him again.'

Meanwhile, another astronomer had turned his attention to the problem of accounting for the anomalies in the motion of Uranus. In the summer of 1845 Arago, Director of the Paris Observatory, drew the attention of his friend and protégé, Le Verrier, to the importance of investigating the theory of Uranus. Le Verrier was a young man, 8 years older than Adams, with an established reputation in the astronomical world, gained by a brilliant series of investigations in celestial mechanics. In contrast, Adams was unknown outside the circle of his Cambridge friends and he had not yet published anything.

Le Verrier decided to devote himself to the problem of Uranus and laid aside some researches on comets, on which he had been engaged. His investigations received full publicity, for the results were published, as the work proceeded, in a series of papers in the *Comptes Rendus* of the French Academy. In the first of these, communicated in November 1845 (a month after Adams had left his solution of the problem with the Astronomer Royal), Le Verrier recomputed the perturbations of Uranus by Jupiter and Saturn, derived new orbital elements for Uranus, and showed that these perturbations were not capable of explaining the observed irregularities of Uranus. In the next paper, presented in June 1846, Le Verrier discussed possible explanations of the irregularities and concluded that none was admissible, except that of a disturbing planet exterior to Uranus. Assuming, as Adams had done, that its distance was twice the distance of Uranus and that its orbit was in the plane of the ecliptic, he assigned its true longitude for the beginning of 1847; he did not obtain the elements of its orbit nor determine its mass.

The position assigned by Le Verrier differed by only 1° from the position which Adams had given seven months previously. Airy now felt no doubt about the accuracy of both calculations; he still required to be satisfied about the error of the radius vector, however, and he accordingly addressed to Le Verrier the query that he had addressed to Adams, but this time in a more explicit form. He asked whether the errors of the tabular radius vector were the consequence of the disturbance produced by an exterior planet, and explained why, by analogy with the moon's variation, this did not seem to him necessarily to be so. Le Verrier replied a few days later giving an explanation which Airy found completely satisfactory. The errors of the tabular radius vector, said Le Verrier, were not produced actually by the disturbing planet; Bouvard's orbit required correction, because it had been based on positions which were not true elliptic positions, including, as they did, the perturbations by the outer planet; the correction of the orbit, which was needed on this account, removed the discordance between the observed and tabular radius vector.

Airy was a man of quick and incisive action. He was now fully convinced that the true explanation of the irregularities in the motion of

Uranus had been provided and he felt confident that the new planet would soon be found. He had already, a few days before receiving the reply from Le Verrier, informed the Board of Visitors of the Royal Observatory, at their meeting in June, of the extreme probability of discovering a new planet in a very short time. It was in consequence of this strongly expressed opinion of Airy that Sir John Herschel (a member of the Board) in his address on 10 September to the British Association, at its meeting at Southampton, said: 'We see it [a probable new planet] as Columbus saw America from the shores of Spain. Its movements have been felt, trembling along the far-reaching line of our analysis, with a certainty hardly inferior to that of ocular demonstration.'

Airy considered that the most suitable telescope with which to make the search for the new planet was the Northumberland telescope of the Cambridge Observatory, which was larger than any telescope at Greenwich and more likely to detect a planet whose light might be feeble. Airy offered to lend Challis one of his assistants, if Challis was too busy to undertake the search himself. He pointed out that the most favourable time for the search (when the undiscovered planet would be at opposition) was near at hand. A few days later, Airy sent Challis detailed directions for carrying out the search and in a covering letter said that, in his opinion, the importance of the inquiry exceeded that of any current work, which was of such a nature as not to be totally lost by delay.

Challis decided to prosecute the search himself and began observing on 29 July 1846, three weeks before opposition. The method adopted was to make three sweeps over the area to be searched, mapping the positions of all the stars observed, and completing each sweep before beginning the next. If the planet was observed it would be revealed, when the different sweeps were compared, by its motion relative to the stars.

What followed was not very creditable to Challis. He started by observing in the region indicated by Adams: the first four nights on which observations were made were 29 July, 30 July, 4 August and 12 August. But no comparison was made, as the search proceeded, between the observations on different nights. He did indeed make a partial comparison between the nights of 30 July and 12 August, merely to assure himself that the method of observation was adequate. He stopped short at No. 39 of the stars observed on 12 August; as he found that all these had been observed on 30 July, he felt satisfied about the method of observation. If he had continued the comparison for another ten stars he would have found that a star of the 8th magnitude observed on 12 August was missing in the series of 30 July. This was the planet: it had wandered into the zone between the two dates. Its discovery was thus easily within his grasp. But 12 August was not the first time on which Challis had observed the planet; he had already observed it on 4 August and if he had compared the observations of 4 August with the observations of either 30 July or of 12 August, the planet would have been detected.

When we recall Airy's strong emphasis on carrying on the search in preference to any current work, Challis's subsequent excuses to justify his failure were pitiable. He had delayed the comparisons, he said, partly from being occupied with comet reductions (which could well have waited), and partly from a fixed impression that a long search was required to ensure success. He confessed that, in the whole of the undertaking, he had too little confidence in the indications of theory. Oh! man of little faith! If only he had shared Airy's conviction of the great importance of the search.

But we have anticipated somewhat. While Challis was laboriously continuing his search, Adams wrote on 2 September an important letter to Airy who, unknown to Adams, was then in Germany. He referred to the assumption in his first calculations that the mean distance of the supposed disturbing planet was twice that of Uranus. The investigation, he said, could scarcely be considered satisfactory while based on anything arbitrary. He had therefore repeated his calculations, assuming a somewhat smaller mean distance. The result was very satisfactory in that the agreement between theory and observations was somewhat improved and, at the same time, the eccentricity of the orbit, which in the first solution had an improbably large value, was reduced. He gave the residuals for the two solutions, and remarked that the comparison with recent Greenwich observations suggested that a still better agreement could be obtained by a further reduction in the mean distance. He asked for the results of the Greenwich observations for 1844 and 1845. He then gave the corrections to the tabular radius vector of Uranus and remarked that they were in close agreement with those required by the Greenwich observations.

Two days earlier, on 31 August, Le Verrier had communicated a third paper to the French Academy which was published in a number of the *Comptes Rendus* that reached England near the end of September. Challis received it on 29 September. Le Verrier gave the orbital elements of the hypothetical planet, its mass, and its position. From the mass and distance of the planet he inferred, on the reasonable assumption that its mean density was equal to the mean density of Uranus, that it should show a disk with an angular diameter of about 3·3 sec. Le Verrier went on to remark as follows:

'It should be possible to see the new planet in good telescopes and also to distinguish it by the size of its disk. This is a very important point. For if the planet could not be distinguished by its appearance from the stars it would be necessary, in order to discover it, to examine all the small stars in the region of the sky to be explored, and to detect a proper motion of one of them. This work would be long and wearisome. But if, on the contrary, the planet has a sensible disk which prevents it from being confused with a star, if a simple study of its physical appearance can replace the rigorous determination of the positions of all the stars, the search will proceed much more rapidly.'

After reading this memoir on 29 September, Challis searched the same night in the region indicated by Le Verrier (which was almost identical with that indicated by Adams, in the first instance, a year earlier), looking out particularly for a visible disk. Of 300 stars observed he noted one and one only as seeming to have a disk. This was, in actual fact, the planet. Its motion might have been detected in the course of a few hours, but Challis waited for confirmation until the next night, when no observation was possible because the Moon was in the way. On 1 October he learnt that the planet had been discovered at Berlin on 23 September. His last chance of making an independent discovery had gone.

For on 23 September Galle, Astronomer at the Berlin Observatory, had received a letter from Le Verrier suggesting that he should search for the unknown planet, which would probably be easily distinguished by a disk. D'Arrest, a keen young volunteer at the Observatory, asked to share in the search, and suggested to Galle that it might be worth looking among the star charts of the Berlin Academy, which were then in course of publication, to verify whether the chart for Hour 21 was amongst those that were finished. It was found that this chart had been printed at the beginning of 1846, but had not yet been distributed; it was therefore available only to the astronomers at the Berlin Observatory. Galle took his place at the telescope, describing the configurations of the stars he saw, while d'Arrest followed them on the map, until Galle said: 'And then there is a star of the 8th magnitude in such a such a position', whereupon d'Arrest exclaimed: 'That star is not on the map.' An observation the following night showed that the object had changed its position and proved that it was the planet. Had this chart been available to Challis, as it would have been but for the delay in distribution, he would undoubtedly have found the planet at the beginning of August, some weeks before Le Verrier's third memoir was presented to the French Academy.

On 1 October, Le Verrier wrote to Airy informing him of the discovery of the planet. He mentioned that the Bureau des Longitudes had adopted the name Neptune, the figure a trident, and that the name Janus (which had also been suggested) would have the inconvenience of making it appear that the planet was the last in the solar system, which there was no reason to believe.

The discovery of the planet, following the brilliant researches of Le Verrier, which were known to the scientific world through their publication by the French Academy, was received with admiration and delight, and was acclaimed as one of the greatest triumphs of the human intellect. The prior investigations of Adams, his prediction of the position of the planet, the long patient search by Challis were known to only a few people in England. Adams had published nothing; he had communicated his results to Challis and to Airy, but neither of them knew anything of the details of his investigations; his name was unknown in astronomical circles outside his own country. Adams had actually drawn up a paper to be read at the meeting of the British Association at Southampton early in September, but he did not arrive in sufficient time to present it, as Section A closed its meetings one day earlier than he had expected.

The first reference in print to the fact that Adams had independently reached conclusions similar to those of Le Verrier was made in a letter from Sir John Herschel, published in the *Athenaeum* of 3 October. It came as a complete surprise to the French astronomers and ungenerous aspersions were cast upon the work of Adams. It was assumed that his solution was a crude essay which would not stand the test of rigorous examination and that, as he had not published any account of his researches, he could not establish a claim to priority or even to a share in the discovery. Some justification seemed to be afforded by an unfortunate letter from Challis to Arago, of 5 October, stating that he had searched for the planet, in conformity with the suggestions of Le Verrier, and had observed an object on 29 September which appeared to have a disk and which later was proved to have been the planet. No reference was made in this letter to the investigations of Adams or to his own earlier searches during which the planet had twice been observed. Airy, moreover, wrote to Le Verrier on 14 October, mentioning that collateral researches, which had led to the same result as his own, had been made in England. He went on to say: 'If in this I give praise to others I beg that you will not consider it as at all interfering with my acknowledgment of your claims. You are to be recognized, without doubt, as the real predictor of the planet's place. I may add that the English investigations, as I believe, were not so extensive as yours. They were known to me earlier than yours.' It is difficult to understand why Airy wrote in these terms; he had expressed the highest admiration for the manner in which the problem had been solved by Le Verrier, but he was not in a position to express any opinion about the work of Adams, which he had not yet seen.

At the meeting of the French Academy on 12 October, Arago made a long and impassioned defence of his protégé, Le Verrier, and a violent attack on Adams, referring scornfully to what he described as his *clandestine* work. 'Le Verrier is to-day asked to share the glory, so loyally, so rightly earned, with a young man who has communicated nothing to the public and whose calculations, more or less incomplete, are, with two exceptions, totally unknown in the observatories of Europe! No! no! the friends of science will not allow such a crying injustice to be perpetrated. He concluded by saying that Adams had no claim to be mentioned, in the history of Le Verrier's planet, by a detailed citation nor even by the slightest allusion. National feeling ran very high in France. The paper *Le National* asserted that the three foremost British astronomers (Herschel, Airy and Challis) had organized a miserable plot to steal the discovery from M. Le Verrier and that the researches of Adams were merely a myth invented for this purpose.

In England opinion was divided; some English astronomers contended that because Adams's results had not been publicly announced he could claim no share in the discovery. But for the most part it was considered that the credit for the successful prediction of the position of the unknown planet should be shared equally between Adams and Le Verrier. Adams himself took no part in the heated discussions which went on for some time with regard to the credit for the discovery of the new planet; he never uttered a single word of criticism or blame in connexion with the matter.

The controversy was lifted to a higher plane by a letter from Sir John Herschel to *The Guardian* in which he said:

'The history of this grand discovery is that of *thought* in one of its highest manifestations, of science in one of its most refined applications. So viewed, it offers a deeper interest than any personal question. In proportion to the importance of the step, it is surely interesting to know that more than one mathematician has been found capable of taking it. The fact, thus stated, becomes, so to speak, a measure of the maturity of our science; nor can I conceive anything better calculated to impress the general mind with a respect for the mass of accumulated facts, laws, and methods, as they exist at present, and the reality and efficiency of the forms into which they have been moulded, than such a circumstance. We need some reminder of this kind in England, where a want of faith in the higher theories is still to a certain degree our besetting weakness.'

At the meeting of the Royal Astronomical Society on 13 November 1846, three important papers were read. The first, by the Astronomer Royal, was an 'Account of some Circumstances historically connected with the Discovery of the Planet Exterior to Uranus'. All the correspondence with Adams, Challis and Le Verrier was given, as well as the two memoranda from Adams, the whole being linked together by Airy's own comments. The account made it perfectly clear that Adams and Le Verrier had independently solved the same problem, that the positions which they had assigned to the new planet were in close agreement, and that Adams had been the first to solve the problem. The second was Challis's 'Account of Observations undertaken in search of the Planet discovered at Berlin on September 23, 1846', which showed that in the course of the search for the planet, he had twice observed it before its discovery at Berlin, and that he had observed it a third time before the news of this discovery reached England. The third paper was by Adams and was entitled 'An Explanation of the observed Irregularities in the Motion of Uranus, on the Hypothesis of Disturbances caused by a more Distant Planet; with a determination of the Mass, Orbit, and Position of the Disturbing Body'.

Adams's memoir was a masterpiece; it showed a thorough grasp of the problem; a mathematical maturity which was remarkable in one so young; and a facility in dealing with complex numerical computations. Lieut. Stratford, Superintendent of the Nautical Almanac, reprinted it as an Appendix to the *Nautical Almanac* for 1851, then in course of publication, and sent sufficient copies to Schumacher, editor of the *Astronomische Nachrichten*, for distribution with that periodical. Hansen, the foremost exponent of the lunar theory, wrote to Airy to say that, in his opinion, Adams's investigation showed more mathematical genius than Le Verrier's. Airy, a competent judge, gave his own opinion in a letter to Biot, who had sent to Airy a paper he had written about the new planet. He sent it, he said, with some diffidence because he had expressed a more favourable opinion of the work of Adams than Airy had given. In reply, Airy wrote: 'On the whole I think his [Adams's] mathematical investigation superior to M. Le Verrier's. However, both are so admirable that it is difficult to say.' He went on to state that 'I believe I have done more than any other person to place Adams in his proper position'.

With the independent investigations of both men published, there was no difficulty in agreeing that each was entitled to an equal share of the honour. The verdict of history agrees with that of Sir John Herschel who, in addressing the Royal Astronomical Society in 1848, said:

'As genius and destiny have joined the names of Le Verrier and Adams, I shall by no means put them asunder; nor will they ever be pronounced apart so long as language shall celebrate the triumphs of science in her sublimest walks. On the great discovery of Neptune, which may be said to have surpassed, by intelligible and legitimate means, the wildest pretensions of clairvoyance, it would now be quite superfluous for me to dilate. That glorious event and the steps which led to it, and the various lights in which it has been placed, are already familiar to everyone having the least tincture of science. I will only add that as there is not, nor henceforth ever can be, the slightest rivalry on the subject of these two illustrious men—as they have met as brothers, and as such will, I trust, ever regard each other—we have made, we could make, no distinction between them on this occasion. May they both long adorn and augment our science, and add to their own fame, already so high and pure, by fresh achievements.'

Although on 1 October, Le Verrier had informed Airy that the Bureau des Longitudes had assigned the name Neptune to the new planet, Arago announced to the French Academy on 5 October that Le Verrier had delegated to him the right of naming the planet and that he had decided, in the exercise of this right, to call it Le Verrier. 'I pledge myself', he said, 'never to call the new planet by any other name than Le Verrier's Planet.' As though to justify this name, Le Verrier's collected memoirs on the perturbations of Uranus, which were published in the *Connaissance des Temps* for 1849 were given the title 'Recherches sur les mouvements de la planète Herschel (dite Uranus)' with a footnote to say that Le Verrier considered it as a strict duty, in future publications, to ignore the name of Uranus entirely and to call the planet only by the name of Herschel!

The name Le Verrier for the planet was not welcomed outside France. It was not in accordance with the custom of naming planets after mythological deities, and it ignored entirely the claims of Adams. Moreover, it might set a precedent. As Smyth said to Airy: 'Mythology is neutral ground. Herschel is a good name enough. Le Verrier somehow or other suggests a Fabriquant and is therefore not so good. But just think how awkward it would be if the next planet should be discovered by a German, by a Bugge, a Funk, or your hirsute friend Boguslawski!'

The widespread feeling against the name of Le Verrier was shared in Germany by Encke, Gauss and Schumacher and in Russia by Struve. Airy therefore wrote to Le Verrier:

'From my conversation with lovers of astronomy in England and from my correspondence with astronomers in Germany, I find that the name assigned by M. Arago is not well received. They think, in the first place, that the character of the name is at variance with that of the names of all the other planets. They think in the next place that M. Arago, as your delegate, could do only what you could do, and that you would not have given the name which M. Arago has given. They are all desirous of receiving a mythological name selected by you. In these feelings I do myself share. It was believed at first that you approved of the name Neptune, and in that supposition we have used the name Neptune when it was necessary to give a name. Now if it was understood that you still approve of the name Neptune (or Oceanus as some English mythologists suggested—or any other of the same class), I am sure that all England and Germany would adopt it at once. I am not sure that they will adopt the name which M. Arago has given.'

Airy might have added, but did not, that there was a general feeling, not merely in England, but also in Germany and Russia, that the name Le Verrier by implication denied any credit to the work of Adams and that, for this reason also, it was inappropriate.

Le Verrier, in reply, said that since one spoke of Comet Halley, Comet Encke, etc., he saw nothing inappropriate in Planet Le Verrier; that the

Bureau des Longitudes had given the name Neptune without his consent and had now withdrawn it; [1] and that, since he had delegated the selection of the name to Arago, it was a matter that no longer concerned him. At a later date, when relations between Arago and Le Verrier had become strained, the true story was told by Arago. It appears that Arago had at first agreed to the name Neptune, but Le Verrier had implored him, in order to serve him as a friend and as a countryman, to adopt the name Le Verrier. Arago had in the end agreed, but on condition that Uranus should always be called Planet Herschel, a name which Arago himself had frequently used. The greatest men are liable to human weaknesses and failings; Le Verrier was described by his friends as a *mauvais coucheur,* an uncomfortable bedfellow. By the general consensus of astronomers the name Neptune was adopted for the new planet; the name Le Verrier did not long survive.

In the history of the discovery of Neptune so many chances were missed which might have changed completely the course of events, that it is perhaps not surprising to find that the planet might have been discovered 50 years earlier. When sufficient observations of Neptune had been obtained to enable a fairly accurate orbit to be computed, a search was made to find out whether the planet had been observed as a star before its discovery. It was discovered that a star recorded in the *Histoire Céleste* of Lalande as having been observed on 10 May 1795 was missing in the sky; its position was marked as uncertain but was in close agreement with the position to be expected for Neptune. The original manuscripts of Lalande at the Paris Observatory were consulted; it was found that Lalande had, in fact, observed Neptune not only on 10 May but also on 8 May. The two positions, being found discordant and thought to refer to one and the same star, Lalande rejected the observation of 8 May and printed in the *Histoire Céleste* only the observation of 10 May marking it as doubtful, although it was not so marked in the manuscript. The change in position in the two days agreed closely with the motion of Neptune in the interval. If Lalande had taken the trouble to make a further observation to check the other two, he could scarcely have failed to discover the planet. Airy's comment, when sending the information about the two observations to Adams, was 'Let no one after this blame Challis'.

[1] This statement by Le Verrier was not correct. The minutes of the Bureau des Longitudes show that the Bureau had not considered assigning a name by 1 October, when Le Verrier had written not only to Airy but also to various other astronomers in Germany and Russia informing them that the Bureau des Longitudes 'had adopted the name Neptune, the figure a trident'. The Bureau neither assigned the name Neptune nor subsequently withdrew it. The minutes of the Bureau des Longitudes show that Le Verrier's statements were repudiated by the Bureau at a subsequent meeting. It seems that the name Neptune was Le Verrier's own choice in the first instance but that he soon decided that he would like the planet to be named Le Verrier. There is no explanation of his reasons for stating that the name Neptune had been assigned by the Bureau des Longitudes; it was, in fact, outside the competence of the Bureau to assign a name to a newly discovered planet.

COMMENTARY ON
H. G. J. MOSELEY

ON August 10, 1915, the twenty-eight-year-old British physicist, Henry Gwyn-Jeffreys Moseley, died in the trenches of Gallipoli. This extraordinarily gifted young scientist had left the laboratory of Sir Ernest Rutherford in Manchester to join the army in 1914. He was the victim not only of a Turkish bullet but of an incomparably stupid recruitment program which permitted a scientist of such proven talent to become a soldier. For Moseley's brilliance was known. He had worked with Rutherford for some years, and at the age of twenty-seven had made discoveries in physics as important as any achieved in this century. Thus was squandered a life of the greatest promise to the history of science. It was from such spectacular instances of waste that both Great Britain and the U. S. learned to keep their best scientists out of the firing lines of the Second World War.

Moseley's researches on atomic structure are reported in two communications to the *Philosophical Magazine*, from which the excerpts below have been selected. Although comparatively easy to explain, they are written in the usual terse and difficult idiom of the physicist. Therefore, I had better summarize their content and say a few words about how Moseley's work fitted into the scene of contemporary physics.

After Roentgen discovered X-rays in 1895 a number of scientists devoted themselves to comparing these radiations with light waves. Attempts were made to reflect and refract X-rays, to determine whether they produced interference phenomena as does light. The attempts failed. If a parallel beam of light is allowed to fall on a grating, a surface on which many thousands of fine lines have been drawn to the inch, the transmitted or reflected light is broken up into its component colors, or wave lengths, and appears as a spectrum. When this procedure was applied to X-rays it also failed. These experiments indicated either that X-rays were not waves or that the waves were so short that the methods of study used for light were unsuitable. In 1912 Max von Laue suggested that, while a man-made grating was too crude for the job, a grating provided by nature in the form of crystals might catch waves less than a thousandth as long as the shortest light waves. Von Laue worked out the mathematics of his theory and Friedrich and Knipping confirmed it in a series of beautiful experiments. The array of regular lines of atoms in the crystal served as a supergrating ("of very minute dimensions") and yielded characteristic diffraction spectra for X-rays. These spectra could be photographed and their lines analyzed. Thus, the wave nature of X-rays was established and, incidentally, a way was opened into the great new field of the structure of crystals.[1]

Now we come to Moseley's experiments. X-rays are produced by permitting a stream of electrons emitted by the negative pole (cathode) in a vacuum tube to impinge on a solid target. Roentgen had used platinum for the target but it soon became clear that if other elements were substituted they would also respond to cathode bombardment by emitting X-rays. Moseley successively used forty-two different elements as targets. He passed each set of X-rays through a crystal, photographed the resulting diffraction patterns and analyzed the waves emanating from each sample. He noticed that his figures were falling into a remarkable order. As he moved element by element up the Periodic Table—in which the elements are arranged by atomic weight and bracketed in groups according to their chemical properties—he found a regular increase in the square roots of the frequency of vibration of the characteristic spectrum lines. If this square root is multiplied by an appropriate constant so as to convert the regular increase to unity, one gets what is known as the series of atomic numbers ranging from 1 for hydrogen to 98 for californium. (Moseley went only as far as 79, for gold, when he was called away on more urgent business. The remaining places in the table were gradually filled in by other physicists.)

What are these numbers which match so smoothly the order of the Periodic Table? (A few irregularities crop up but they have been explained and do not shake the theory.) The little arithmetic trick of converting the increase to unity should not mislead anyone to believe that the numbers merely represent a superimposed order. For the fact is that the atomic number of each element, representing the square root of frequency of an atom "suitably excited" to emit X radiation, assigns to the element its true place in the Table because it is the number of the unit positive charges in the nucleus of the atom. It is this charge which determines the element's chemical behavior. All subsequent advances in nuclear physics depend upon and are an outgrowth of the fundamental insight gained by Moseley. The beauty and simplicity of his theory, its revelation of the existence of an almost uncanny step-by-step order in the arrangement of the basic structures of matter, would have pleased the ancient Greek philosophers who held that number ruled the universe. "In our own day," says Dampier, "Aston with his integral atomic weights, Moseley with his atomic numbers, Planck with his quantum theory, and Einstein with his claim that physical facts such as gravitation are exhibitions of local space-time properties, are reviving ideas that, in older, cruder forms, appear in Pythagorean philosophy."

[1] For a further discussion of X-rays, von Laue's experiment, and the structure of crystals, see the selections by Bragg and Le Corbeiller, and the introduction preceding them, pp. 851–881.

The most important discoveries of the laws, methods and progress of Nature have nearly always sprung from the examination of the smallest objects which she contains. —J. B. LAMARCK

5 Atomic Numbers

By H. G. J. MOSELEY

THE HIGH FREQUENCY SPECTRA OF THE ELEMENTS

IN the absence of any available method of spectrum analysis, the characteristic types of X radiation, which an atom emits if suitably excited, have hitherto been described in terms of their absorption in aluminium. The interference phenomena exhibited by X-rays when scattered by a crystal have now, however, made possible the accurate determination of the frequencies of the various types of radiation. This was shown by W. H. and W. L. Bragg, who by this method analysed the line spectrum emitted by the platinum target of an X-ray tube. C. G. Darwin and the author extended this analysis and also examined the continuous spectrum, which in this case constitutes the greater part of the radiation. Recently Prof. Bragg has also determined the wave-lengths of the strongest lines in the spectra of nickel, tungsten, and rhodium. The electrical methods which have hitherto been employed are, however, only successful where a constant source of radiation is available. The present paper contains a description of a method of photographing these spectra, which makes the analysis of the X-rays as simple as any other branch of spectroscopy. The author intends first to make a general survey of the principal types of high-frequency radiation, and then to examine the spectra of a few elements in greater detail and with greater accuracy. The results already obtained show that such data have an important bearing on the question of the internal structure of the atom, and strongly support the views of Rutherford and of Bohr.

Kaye has shown that an element excited by a stream of sufficiently fast cathode rays emits its characteristic X radiation. He used as targets a number of substances mounted on a truck inside an exhausted tube. A magnetic device enabled each target to be brought in turn into the line of fire. This apparatus was modified to suit the present work. The cathode stream was concentrated on to a small area of the target, and a platinum plate furnished with a fine vertical slit placed immediately in front of the part bombarded. The tube was exhausted by a Gaede mercury pump, charcoal in liquid air being also sometimes used to remove water vapour. The X-rays, after passing through the slit marked S in Figure 1, emerged

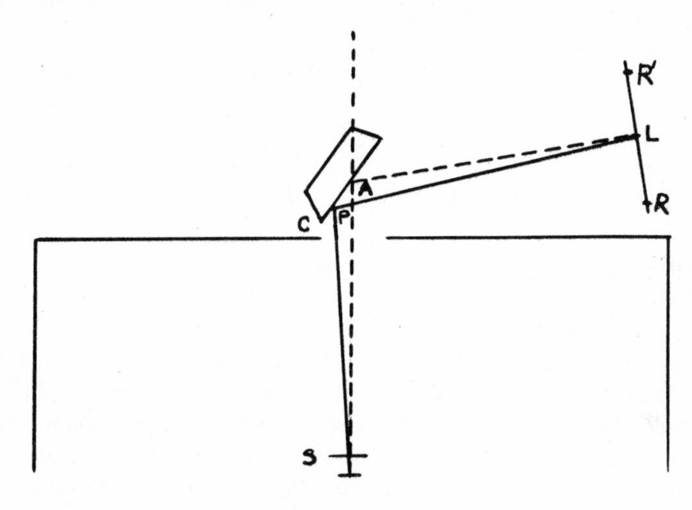

FIGURE 1

through an aluminum window of ·02 mm. thick. The rest of the radiation was shut off by a lead box which surrounded the tube. The rays fell on the cleavage face, C, of a crystal of potassium ferrocyanide which was mounted on the prism-table of a spectrometer. The surface of the crystal was vertical and contained the geometrical axis of the spectrometer.

Now it is known that X-rays consist in general of two types, the heterogeneous radiation and characteristic radiations of definite frequency. The former of these is reflected from such a surface at all angles of incidence, but at the large angles used in the present work the reflexion is of very little intensity. The radiations of definite frequency, on the other hand, are reflected only when they strike the surface at definite angles, the glancing angle of incidence θ, the wave-length λ, and the "grating constant" d of the crystal being connected by the relation

$$n\lambda = 2d \sin \theta, \qquad \ldots\ldots\ldots\ldots\ldots (1)$$

where n, an integer, may be called the "order" in which the reflexion occurs. The particular crystal used, which was a fine specimen with face 6 cm. square, was known to give strong reflexions in the first three orders, the third order being the most prominent.

If then a radiation of definite wave-length happens to strike any part P of the crystal at a suitable angle, a small part of it is reflected. Assuming for the moment that the source of the radiation is a point, the locus of P is obviously the arc of a circle, and the reflected rays will travel along the generating lines of a cone with apex at the image of the source. The effect on a photographic plate L will take the form of the arc of an hyperbola, curving away from the direction of the direct beam. With a fine slit at S, the arc becomes a fine line which is slightly curved in the direction indicated.

The photographic plate was mounted on the spectrometer arm, and both the plate and the slit were 17 cm. from the axis. The importance of this arrangement lies in a geometrical property, for when these two distances are equal the point L at which a beam reflected at a definite angle strikes the plate is independent of the position of P on the crystal surface. The angle at which the crystal is set is then immaterial so long as a ray can strike some part of the surface at the required angle. The angle θ can be obtained from the relation $2\theta = 180° - SPL = 180° - SAL$.

The following method was used for measuring the angle SAL. Before taking a photograph a reference line R was made at both ends of the plate by replacing the crystal by a lead screen furnished with a fine slit which coincided with the axis of the spectrometer. A few seconds' exposure to the X-rays then gave a line R on the plate, and so defined on it the line joining S and A. A second line RQ was made in the same way after turning the spectrometer arm through a definite angle. The arm was then turned to the position required to catch the reflected beam and the angles LAP for any lines which were subsequently found on the plate deduced from the known value of RAP and the position of the lines on the plate. The angle LAR was measured with an error of not more than $0°{\cdot}1$, by superposing on the negative a plate on which reference lines had been marked in the same way at intervals of 1°. In finding from this the glancing angle of reflexion two small corrections were necessary in practice, since neither the face of the crystal nor the lead slit coincided accurately with the axis of the spectrometer. Wave-lengths varying over a range of about 30 per cent. could be reflected for a given position of the crystal.

In almost all cases the time of exposure was five minutes. Ilford X-ray plates were used and were developed with rodinal. The plates were mounted in a plate-holder, the front of which was covered with black paper. In order to determine the wave-length from the reflexion angle θ it is necessary to know both the order n in which the reflexion occurs and the grating constant d. n was determined by photographing every spectrum both in the second order and the third. This also gave a useful check on the accuracy of the measurements; d cannot be calculated directly for the complicated crystal potassium ferrocyanide. The grating constant of this particular crystal had, however, previously been accurately compared with d', the constant of a specimen of rock-salt. It was found that

$$d = 3d' \frac{{\cdot}1988}{{\cdot}1985}.$$

Now W. L. Bragg has shown that the atoms in a rock-salt crystal are in simple cubical array. Hence the number of atoms per c.c.

$$2\frac{N\sigma}{M} = \frac{1}{(d')^3}:$$

N, the number of molecules in a gram-mol., $= 6{\cdot}05 \times 10^{23}$ assuming the charge (e) on an electron to be $4{\cdot}89 \times 10^{-10}$; σ, the density of this crystal of rock-salt, was 2·167, and M the molecular weight $= 58{\cdot}46$.

This gives $d' = 2{\cdot}814 \times 10^{-8}$ and $d = 8{\cdot}454 \times 10^{-8}$ cm. It is seen that the determination of wave-length depends on $e^{1/3}$ so that the effect of uncertainty in the value of this quantity will not be serious. Lack of homogeneity in the crystal is a more likely source of error, as minute inclusions of water would make the density greater than that found experimentally.

Twelve elements have so far been examined. . . .

The Plate shows the spectra in the third order placed approximately in register. Those parts of the photographs which represent the same angle of reflexion are in the same vertical line. . . . It is to be seen that the spectrum of each element consists of two lines. Of these the stronger has been called α in the table, and the weaker β. The lines found on any of the plates besides α and β were almost certainly all due to impurities. Thus in both the second and third order the cobalt spectrum shows Niα very strongly and Feα faintly. In the third order the nickel spectrum shows Mnα_2 faintly. The brass spectra naturally show α and β both of Cu and of Zn, but Znβ_2 has not yet been found. In the second order the ferro-vanadium and ferro-titanium spectra show very intense third-order Fe lines, and the former also shows Cuα_3 faintly. The Co contained Ni and 0·8 per cent. Fe, the Ni 2·2 per cent. Mn, and the V only a trace of Cu. No other lines have been found; but a search over a wide range of wave-lengths has been made only for one or two elements, and perhaps prolonged exposures, which have not yet been attempted, will show more complex spectra. The prevalence of lines due to impurities suggests that this may prove a powerful method of chemical analysis. Its advantage over ordinary spectroscopic methods lies in the simplicity of the spectra and the impossibility of one substance masking the radiation from another. It may even lead to the discovery of missing elements, as it will be possible to predict the position of their characteristic lines. . . .

Phil. Mag. (1914), p. 703.

The first part of this paper dealt with a method of photographing X-ray spectra, and included the spectra of a dozen elements. More than thirty other elements have now been investigated, and simple laws have been found which govern the results, and make it possible to predict with

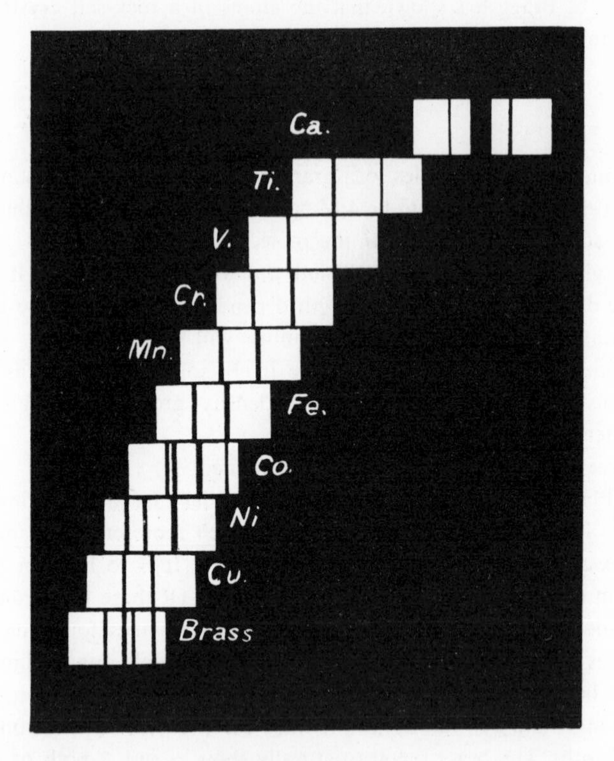

FIGURE 2—X-ray spectra.

confidence the position of the principal lines in the spectrum of any element from aluminium to gold. The present contribution is a general preliminary survey, which claims neither to be complete nor very accurate. . . .

The results obtained for radiations belonging to Barkla's K series are given in Table I, and for convenience the figures already given in Part I are included. The wave-length λ has been calculated from the glancing angle of reflexion θ by means of the relation $n\lambda = 2d \sin \theta$, where d has been taken to be $8{\cdot}454 \times 10^{-8}$ cm. As before, the strongest line is called α and the next line β. The square root of the frequency of each line is plotted in Figure 3, and the wave-lengths can be read off with the help of the scale at the top of the diagram.

The spectrum of Al was photographed in the first order only. The very light elements give several other fainter lines, which have not yet been fully investigated, while the results for Mg and Na are quite complicated, and apparently depart from the simple relations which connect the spectra of the other elements. In the spectra from yttrium onwards only the α line has so far been measured, and further results in these directions will be

TABLE I

	α line $\lambda \times 10^8$ cm.	Q_K	N Atomic Number	β line $\lambda \times 10^8$ cm.
Aluminium	8·364	12·05	13	7·912
Silicon	7·142	13·04	14	6·729
Chlorine	4·750	16·00	17	—
Potassium	3·759	17·98	19	3·463
Calcium	3·368	19·00	20	3·094
Titanium	2·758	20·99	22	2·524
Vanadium	2·519	21·96	23	2·297
Chromium	2·301	22·98	24	2·093
Manganese	2·111	23·99	25	1·818
Iron	1·946	24·99	26	1·765
Cobalt	1·798	26·00	27	1·629
Nickel	1·662	27·04	28	1·506
Copper	1·549	28·01	29	1·402
Zinc	1·445	29·01	30	1·306
Yttrium	0·838	38·1	39	—
Zirconium	0·794	39·1	40	—
Niobium	0·750	40·2	41	—
Molybdenum	0·721	41·2	42	—
Ruthenium	0·638	43·6	44	—
Palladium	0·584	45·6	46	—
Silver	0·560	46·6	47	—

TABLE II

	α line $\lambda \times 10^8$ cm.	Q_L	N Atomic Number	β line $\lambda \times 10^8$ cm.	φ line $\lambda \times 10^8$ cm.	γ line $\lambda \times 10^8$ cm.
Zirconium	6·091	32·8	40	—	—	—
Niobium	5·749	33·8	41	5·507	—	—
Molybdenum	5·423	34·8	42	5·187	—	—
Ruthenium	4·861	36·7	44	4·660	—	—
Rhodium	4·622	37·7	45	—	—	—
Palladium	4·385	38·7	46	4·168	—	3·928
Silver	4·170	39·6	47	—	—	—
Tin	3·619	42·6	50	—	—	—
Antimony	3·458	43·6	51	3·245	—	—
Lanthanum	2·676	49·5	57	2·471	2·424	2·313
Cerium	2·567	50·6	58	2·366	2·315	2·209
Praseodymium	(2·471)	51·5	59	2·265	—	—
Neodymium	2·382	52·5	60	2·175	—	—
Samarium	2·208	54·5	62	2·008	1·972	1·893
Europium	2·130	55·5	63	1·925	1·888	1·814
Gadolinium	2·057	56·5	64	1·853	1·818	—
Holmium	1·914	58·6	66	1·711	—	—
Erbium	1·790	60·6	68	1·591	1·563	—
Tantalum	1·525	65·6	73	1·330	—	1·287
Tungsten	1·486	66·5	74	—	—	—
Osmium	1·397	68·5	76	1·201	—	1·172
Iridium	1·354	69·6	77	1·155	—	1·138
Platinum	1·316	70·6	78	1·121	—	1·104
Gold	1·287	71·4	79	1·092	—	1·078

given in a later paper. The spectra both of K and Cl were obtained by means of a target of KCl, but it is very improbable that the observed lines have been attributed to the wrong elements. The α line for elements from Y onwards appeared to consist of a very close doublet, an effect previously observed by Bragg in the case of rhodium.

The results obtained for the spectra of the L series are given in Table II and plotted in Figure 3. These spectra contain five lines, α, β, γ, δ, ϵ, reckoned in order of decreasing wave-length and decreasing intensity. There is also always a faint companion α' on the long wave-length side of α, a rather faint line ϕ between β and γ for the rare earth elements at least, and a number of very faint lines of wave-length greater than α. Of these, α, β, ϕ, and γ have been systematically measured with the object of finding out how the spectrum alters from one element to another. The fact that often values are not given for all these lines merely indicates the incompleteness of the work. The spectra, so far as they have been examined, are so entirely similar that without doubt α, β, and γ at least always exist. Often γ was not included in the limited range of wave-lengths which can be photographed on one plate. Sometimes lines have not been measured, either on account of faintness or of the confusing proximity of lines due to impurities. . . .

CONCLUSIONS

In Figure 3 the spectra of the elements are arranged on horizontal lines spaced at equal distances. The order chosen for the elements is the order of the atomic weights, except in the cases of A, Co, and Te, where this clashes with the order of the chemical properties. Vacant lines have been left for an element between Mo and Ru, an element between Nd and Sa, and an element between W and Os, none of which are yet known, while Tm, which Welsbach has separated into two constituents, is given two lines. This is equivalent to assigning to successive elements a series of successive characteristic integers. On this principle the integer N for Al, the thirteenth element, has been taken to be 13, and the values of N then assumed by the other elements are given on the left-hand side of Figure 3. This proceeding is justified by the fact that it introduces perfect regularity into the X-ray spectra. Examination of Figure 3 shows that the values of $\nu^{1/2}$ for all the lines examined both in the K and the L series now fall on regular curves which approximate to straight lines. The same thing is shown more clearly by comparing the values of N in Table I with those of

$$Q_K = \sqrt{\frac{\nu}{\frac{3}{4}\nu_0}},$$

ν being the frequency of the line and ν_0 the fundamental Rydberg frequency. It is here plain that $Q_K = N - 1$ very approximately, except for

High-Frequency Spectra of the Elements.

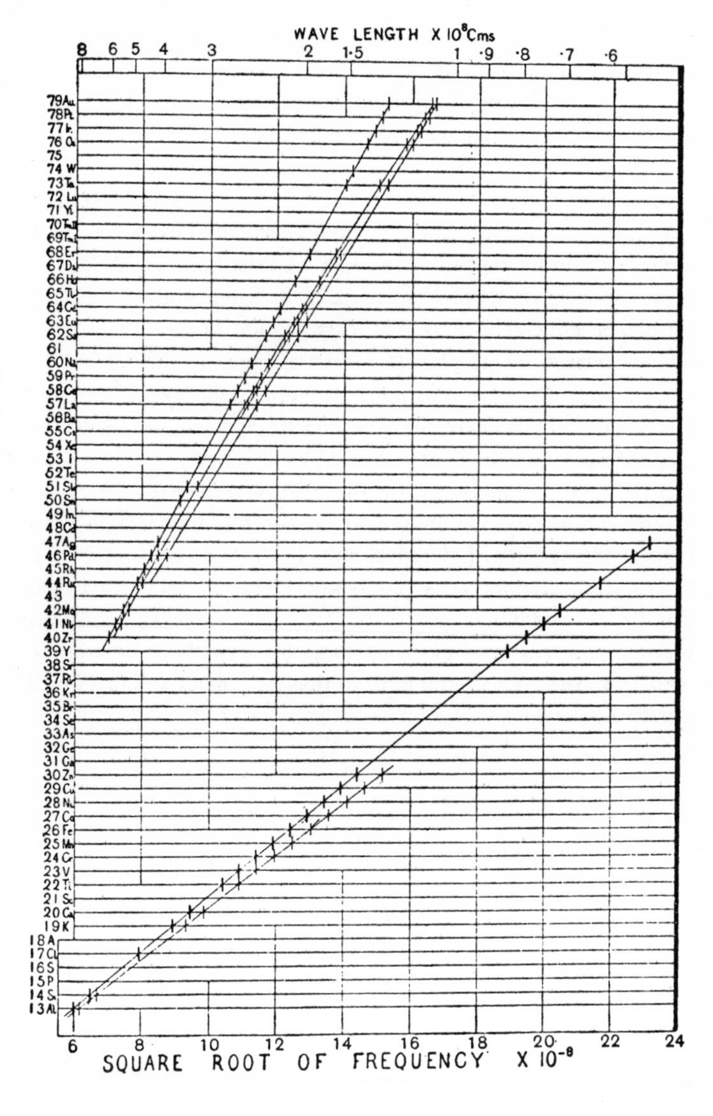

FIGURE 3

the radiations of very short wave-length which gradually diverge from this relation. Again, in Table II a comparison of N with

$$Q_L = \sqrt{\frac{\nu}{\frac{5}{36}\nu_0}},$$

where ν is the frequency of the La line, shows that $Q_L = N - 7{\cdot}4$ approximately, although a systematic deviation clearly shows that the relation is not accurately linear in this case.

Now if either the elements were not characterized by these integers, or any mistake had been made in the order chosen or in the number of places left for unknown elements, these regularities would at once disappear. We can therefore conclude from the evidence of the X-ray spectra alone, without using any theory of atomic structure, that these integers are really characteristic of the elements. Further, as it is improbable that two different stable elements should have the same integer, three, and only three, more elements are likely to exist between Al and Au. As the X-ray spectra of these elements can be confidently predicted, they should not be difficult to find. The examination of keltium would be of exceptional interest, as no place has been assigned to this element.

Now Rutherford has proved that the most important constituent of an atom is its central positively charged nucleus, and van den Broek has put forward the view that the charge carried by this nucleus is in all cases an integral multiple of the charge on the hydrogen nucleus. There is every reason to suppose that the integer which controls the X-ray spectrum is the same as the number of electrical units in the nucleus, and these experiments therefore give the strongest possible support to the hypothesis of van den Broek. Soddy has pointed out that the chemical properties of the radio-elements are strong evidence that this hypothesis is true for the elements from thallium to uranium, so that its general validity would now seem to be established.

COMMENTARY ON

The Small Furniture of Earth

THE next two selections deal with the wave theory of X-rays and the atomic theory of crystals. The bringing together and verification of these theories in a single famous experiment was one of the major events of the twentieth-century renaissance in physics.

Crystals were the subject of much attention in the eighteenth and nineteenth centuries, their optical properties and geometric relations being carefully studied by mineralogists, crystallographers and mathematicians. As early as 1824 the hypothesis was advanced that crystals consist of layers of atoms distributed in regular patterns; [1] in the 1890s mathematicians had worked out fully the number of possible ways the atoms inside a crystal could be distributed. However, experimental proof of these fundamental ideas was still lacking in 1910. Attempts to confirm the theory of crystals by diffraction experiments with light were unsuccessful because the waves were too long for the job; it was as if one were to try to measure the inside of a thimble with a foot-rule.

Another important hypothesis, also in a doubtful status at that time, related to X-rays. The majority of physicists were convinced that X-rays were waves of a length much shorter than light rays, but again early attempts to prove the conjecture were unsuccessful.

The crucial experiment testing both hypotheses was proposed by the German physicist Von Laue, who was then assistant lecturer in Professor Sommerfeld's department at the University of Munich.[2] He reasoned that if X-rays were short waves and crystals three-dimensional lattices of atoms, a pencil of X-rays passed through a crystal would produce a characteristic pattern on a photographic plate. The X-rays, in other words, could be made to report the arrangement of the tiny furniture encountered inside the crystal. The experiment, performed in 1912, was an extraordinary triumph.[3] The X-rays portrayed the interior of the crystals, and the crystals repaid the favor by disclosing the form of the X-rays. Von Laue

[1] Max von Laue, *History of Physics*, New York, 1950, p. 119: "The first scientist to combine the newly created concept of the chemical atom with this idea [of the building-block structure of crystals] and to assume that space lattices are made up of chemical atoms was the physicist Ludwig August Seeber. . . . He published his ideas in 1824, i.e., thirty-two years prior to the entry of atomistics into modern physics in the form of the kinetic theory of gases."

[2] See also the introduction to the paper by H. G. J. Moseley, pp. 840–841.

[3] The experiment, though conceived by Laue, was actually performed by two young Munich research students, Friedrich and Knipping, who had just taken their doctorates under Röntgen. See Kathleen Lonsdale, *Crystals and X-Rays*, London, 1948, pp. 1–22, for an authoritative historical introduction to the subject; also the standard work for the advanced student: Sir Lawrence Bragg, *The Crystalline State, a General Survey*, London, 1949.

described this reciprocal disclosure as "one of those surprising events to which physics owes its power of conviction."

The first of the selections below, on Von Laue's experiment and its relation to the geometry of X-rays and crystals, is by the great physicist Sir William Bragg. The history of modern crystallography is in large part the history of his researches. Bragg was born in Cumberland in 1862 and after a brilliant school career accepted a professorship in physics at the University of Adelaide, Australia. His first research paper did not appear, remarkably enough, until 1904, when he was forty-two. It is unusual for a physicist so long to defer his original investigations, but the paper itself—it was concerned with alpha particles—was immediately recognized as a first-class achievement and marked the beginning of a prolific output of creative studies. In 1907 Bragg was elected a Fellow of the Royal Society, and a year later returned to England to take the Cavendish chair at Leeds. It was there that he became interested in X-rays, which he then regarded, contrary to prevailing opinion, as particles rather than waves. The Laue experiment of 1912 convinced him he was mistaken, and in the same year Bragg and his son took up the research on X-rays and crystals for which in 1915 they jointly received the Nobel prize in physics. Their work "laid the foundation of one of the most beautiful structures of modern science"; it was used for fundamental advances in both inorganic and organic chemistry, metallurgy is deeply indebted to it, as are other branches of pure and applied science.[4]

Bragg had a crowded and happy career during which he worked on many other subjects besides X-rays and raised a crop of distinguished pupils who enriched many parts of physics. In 1915 he held a chair at University College, London; during the First World War he directed acoustic research on submarine detection; in 1923 he became director of the Royal Institution and of the Davy-Faraday Research Laboratory. Not the least of his gifts was in popular exposition. He enjoyed nothing more than to give lectures and experimental demonstrations to youngsters and general audiences; his connection with the Royal Institution (he served as president from 1935 to 1940) fortunately facilitated his exercise of this art.

The article which follows is a chapter from Bragg's book, *The Universe of Light*, based on Christmas Lectures delivered in 1931.

The second selection is a simple and attractive account by Philippe Le Corbeiller of the mathematics of crystals. A suitable complement to Bragg's discussion, it emphasizes the experimental proof of the theory of space groups. In the development of crystallography the mathematics of group theory and of symmetry has, as I remarked earlier, played a remarkable part. Just as Adams and Leverrier, for example, decreed the motion

[4] E. N. daC. Andrade, "Sir William Bragg" (obituary), *Nature*, March 28, 1942.

and position of the planet Neptune before it was discovered, so mathematicians by an exhaustive logical analysis of certain properties of space and of the possible transformations (motions) within space, decreed the permissible variations of internal structure of crystals before observers were able to discover their actual structure. Mathematics, in other words, not only enunciated the applicable physical laws, but provided an invaluable syllabus of research to guide future experimenters. The history of the physical sciences contains many similar instances of mathematical prevision. Models, concepts, theories are initially expressed in mathematical form; later they are tested by observation and either confirmed, or disproved and discarded. Not uncommonly, of course, the model is overhauled to conform more closely to experimental data and the tests are then repeated. The theory of space groups and crystal classes is among the most successful and striking examples of mathematical model making.

Mr. Le Corbeiller is professor of general education and of applied physics at Harvard University. He was born and educated in France, has served as a member of the engineering staff of the *Département des Communications*, as an official of the French Government Broadcasting System and in other administrative and academic posts. His publications include papers on algebra, number theory, oscillating generators, electroacoustics and other scientific topics. The article on crystals, cast in the congenial framework of a mock-Socratic dialogue, first appeared in *Scientific American*, January 1953.

Why think? Why not try the experiment? —JOHN HUNTER (*Letter to Edward Jenner*)

The true worth of an experimenter consists in his pursuing not only what he seeks in his experiment, but also what he did not seek.
—CLAUDE BERNARD

6 The Röntgen Rays

By SIR WILLIAM BRAGG

. . . X-RAYS are generally produced as a consequence of the electric spark or discharge in a space where the pressure of the air or other gas is extremely low. The electric spark has for centuries been a subject of interested observation, but no great step forward was made until it was arranged that the discharge should take place in a glass tube or bulb from which the air had been pumped out more or less completely. The spark became longer, wider, and more highly coloured as the pressure

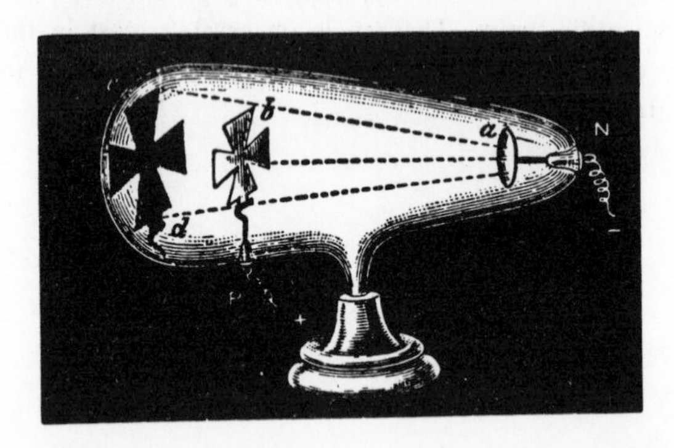

FIGURE 1—The cathode or negative terminal is on the right at *a*. The rays proceed in straight lines across the tube and excite fluorescence on the opposite wall. A metal cross *b* casts a sharp shadow.

diminished. When Crookes so improved the air pump that pressures of the order of the millionth part of atmospheric pressure became attainable a phenomenon appeared which had not been previously observed. The negative terminal became the source of a radiation which shot in a straight line across the bulb and had mechanical effects. It generated heat whenever it struck the opposite wall or some body placed to intercept it: it excited vivid fluorescence in glass and many minerals: it could turn a light mill wheel if it struck the vanes. And, a most important property, the stream could be deflected by bringing a magnet near it. This was an extremely important observation for it suggested that the stream consisted of electrified particles in flight. Such a stream would be equivalent to an electric current and would therefore be susceptible to the force of a magnet. Illustrations of Crookes's experiments are given in Figures 1, 2, 3, and 4. They are taken from the original blocks used in the published

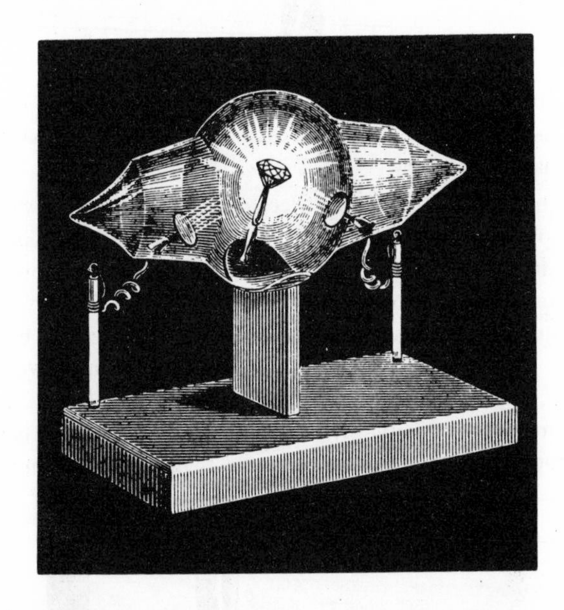

FIGURE 2—The rays excite a vivid fluorescence in the diamond mounted in the centre of the tube.

account of a discourse which he gave at the Royal Institution in April 1879. Crookes believed that the stream consisted of molecules of some kind. He argued that his air pump had attained such perfection that the comparatively few molecules left in the tube could move over distances comparable with the length of the tube without coming into collision with other molecules. Such a condition, he said, was as different from that of a gas as the latter from that of a liquid. At the end of a paper contributed to the Royal Society in the same year (1879) he wrote in a dim but interesting foreshadowing of the future which was partly to be verified:

'The phenomena in these exhausted tubes reveal to physical science a new world—a world where matter exists in a fourth state, where the corpuscular theory of light holds good, and where light does not always move in a straight line; but where we can never enter, and in which we must be content to observe and experiment from the outside.'

J. J. Thomson, Wiechert and others showed that the stream consisted of particles carrying negative charges of electricity, and that these carriers were far smaller than even the hydrogen atom. The name 'electron' was

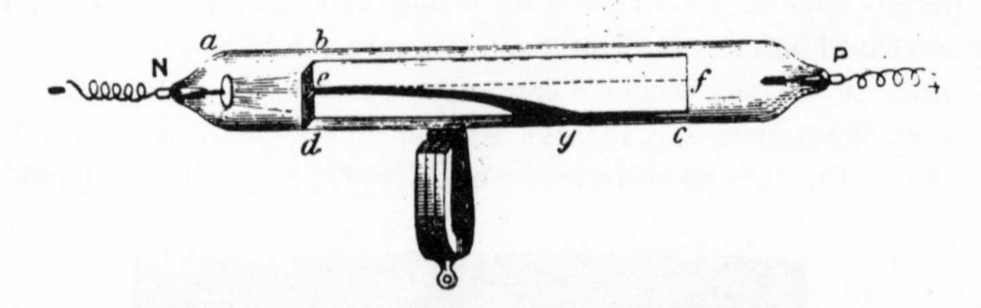

FIGURE 3—The cathode rays are limited to a narrow pencil by means of a slot placed in front of the cathode at *a*. The deflections of the stream by a horseshoe magnet are then observed easily.

given to them. It appeared that electrons could be torn away from any kind of atom, if sufficient electric force was supplied by the induction coil or other electric contrivance for producing the requisite power; and that the electrons from all sources were exactly alike. Evidently the electron was a fundamental constituent of matter. The stream of electrons was called the cathode ray because it issued from the negative or cathode terminal.

Röntgen was investigating the cathode ray phenomena when he found that photographic plates near by became fogged although no light could

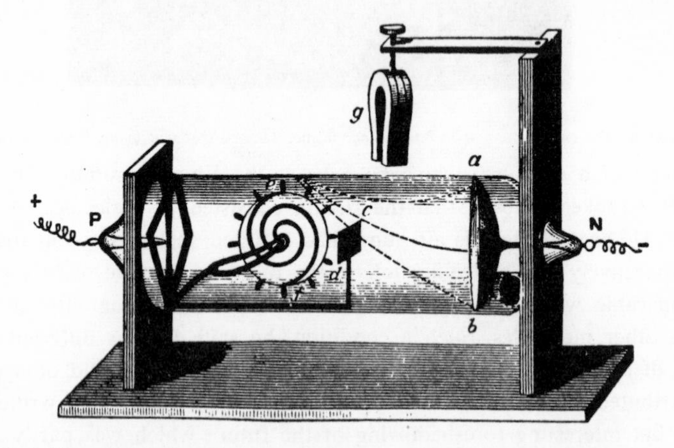

FIGURE 4—The cathode *a* is made in the shape of a saucer: this is found to have the effect of concentrating the rays to a point. Normally a screen *c* intercepts the rays, but the magnet *g* deflects them so that they get over the top of the screen and strike the vanes of the little wheel *ef*, which then spins rapidly. If the disposition of the magnet is reversed, the rays go under *c* and the wheel spins the opposite way.

have reached them. He traced the cause to a radiation issuing from his glass bulb, and in particular from the place where the cathode rays struck the wall: and proceeded to examine the general characteristics of the new rays upon which he had stumbled.

In many respects they resembled light. They moved in straight lines and cast sharp shadows, they traversed space without any obvious transference of matter, they acted on a photographic plate, they excited certain materials to fluorescence, and they could, like ultra-violet light, discharge electricity from conductors. In other ways the rays seemed to differ from light. The mirrors, prisms and lenses which deflected light had no such action on X-rays: gratings as ordinarily constructed did not diffract them: neither double refraction, nor polarisation was produced by the action of crystals. Moreover they had an extraordinary power of penetrating matter. Nothing seemed to hold them up entirely, though everything exerted some power of absorption: heavier atoms were more effective than lighter. Hence arose the quickly observed power of revealing the inner constitution of bodies opaque to light: bones cast shadows much deeper than those due to the surrounding flesh.

If the velocity of X-rays could have been shown without question to have been the same as that of light it would have established their identity: but the experiment though attempted was too difficult. Barkla showed that a pencil of X-rays could have 'sides' or be polarised if the circumstances of their origin were properly arranged, but the polarisation differed in some of its aspects from that which light could be made to exhibit. Laue's experiment brought the controversy to an end, by proving that a diffraction of X-rays could be produced which was in every way parallel to the diffraction of light: if the diffraction phenomena could be depended upon to prove the wave theory of light, exactly the same evidence existed in favour of a wave theory of X-rays.

LAUE'S EXPERIMENT

Let us now consider the details of Laue's famous experiment which has had such striking consequences. The plan of it was very simple. A fine pencil of X-rays was to be sent through a crystal and a photographic

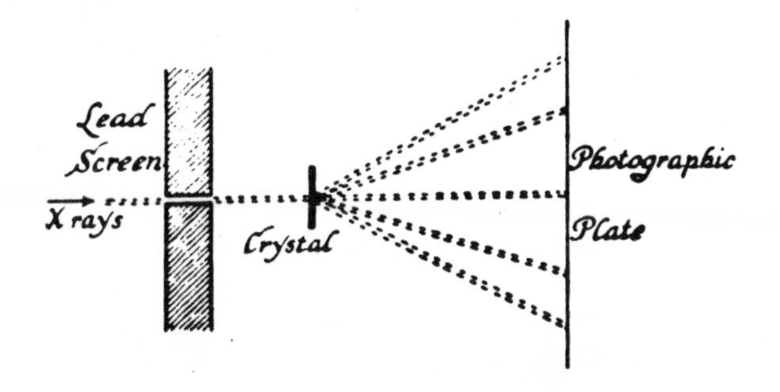

FIGURE 5—The X-rays pass through a fine opening in a lead screen and impinge upon the crystal as shown. Laue's diffraction pattern is formed upon the photographic plate.

plate was to be placed to receive the pencil after it had emerged on the other side, Figure 5. Laue surmised that beside the main image on the plate which would be caused by the incidence of the pencil there might be other subsidiary images. He based his forecast on a consideration of effects of this kind, which occur in the case of light. When a train of ether waves falls upon a plate on which parallel lines are ruled, or is transmitted through such a plate, or passes through an atmosphere in which fine particles of uniform size are suspended, there are regular deflections of the energy in various directions constituting 'diffracted' pencils. . . . In all cases of this kind it is a necessity that there should be no great disparity between the length of the wave on the one hand, and that of the regular spacing or particle diameter on the other. Laue thought that previous failures to find diffraction in the case of X-rays might be due to a want of observance of this condition. If the X-ray wave-lengths were thousands of times shorter than those of light, as he had reason for believing, it was useless to look for diffraction effects with ordinary gratings in the ordinary way. One ought to employ gratings in which the lines were drawn thousands of times closer together than in the usual practice. This is not practicable: no one can draw millions of parallel lines to the inch.

It was possible, however, that Nature had already provided the tool which could not be constructed in the workshop. The crystal might be the appropriate grating for the X-rays, because its atoms were supposed to be in regular array, and the distances that separated them were, so far as could be calculated, of the same order as the wave-length of the X-ray. Whether these anticipations were well or ill founded, they became of little consequence when the experiment was made in 1912 by Laue's colleagues, Friedrich and Knipping, and was completely successful. A complicated but symmetrical pattern of spots appeared upon the photographic plate, which, though unlike any diffraction pattern due to light, was clearly of the same nature. It was soon found that every crystal produced its own pattern and that the experiment opened up not only a new method for the investigation of the nature of X-rays, but also a new means of analysis of the structure of crystals. Examples of these patterns are given in Plate I A, B; they may be compared with Plate II C (pp. 866, 867).

In order that these points may be clear it is necessary to examine, at not too great length, the details of the experiment and its implications. We have already examined certain phenomena of crystalline structure in the case of Iceland spar; but it will be convenient to reconsider the subject and to discuss it more generally.

The most striking and characteristic features of a crystal are its regularity of form, the polished evenness of its faces and the sharpness of its edges. If we compare crystals of the same composition we find that the angles between faces are always exactly the same from crystal to crystal: while the relative values of the areas of the faces may vary considerably. In technical terms, the faces of different specimens may be unequally developed. It is natural to infer that there is an underlying regularity of structure, involving the repetition in space of a unit which is too small to be visible. As a simple analogy we might take a piece of material woven, as is customary, with warp and weft at right angles to each other. However it might be torn, it would form pieces with right angles at all the corners: but the pieces would not necessarily be square. There would be two principal directions at right angles to each other: and all tears would take place at right angles to one or other of them. If the two directions were exactly alike, in all the characteristics that could be examined, if for example, they tore with equal ease, and if the frayed edges were the same on all sides, the warp and weft must be identical: they must be composed of the same threads and have the same spacings. We could rightly say that the material is founded on a 'square' pattern. This would still be the case even if both warp and weft were not simple but complex: if for example, each of them contained coloured threads at various intervals. So long as the scheme of repetition was the same in both we should still say the pattern was square: as for example a tartan might be.

The analogy of a woven material is insufficient to represent all the complications of a crystal structure; because warp and weft cannot be inclined to each other at any angle except ninety degrees: but it illustrates the important point that in any structure built of repetitions in space, the angles of the whole must be always the same; while no such restriction applies to the areas of the faces. The unit that is repeated in every direction determines the angular form. If, for example, the unit of a composition in a plane had the form of the small unit (Figure 6), the whole might have various shapes such as are illustrated in the same figure. The edges need not always include the same angles as the edges of A but they would necessarily be inclined to one another at angles which from specimen to specimen would be invariable.

So also in the solid crystal various faces might be developed which would, in regard to the angles made with each other, display the same constancy of mutual inclination. As this is what we should expect if the crystal is composed of units repeated regularly in all directions, and as the facts agree with expectations, we assume that our preliminary conceptions of crystal structure are correct.

What will happen when a train of ether waves meets such a crystalline arrangement?

A crystal can be thought of as a series of layers spaced at regular intervals, just as in two dimensions a regular assemblage of points can be thought of as a set of rows equally spaced. Also, just as in the simpler case

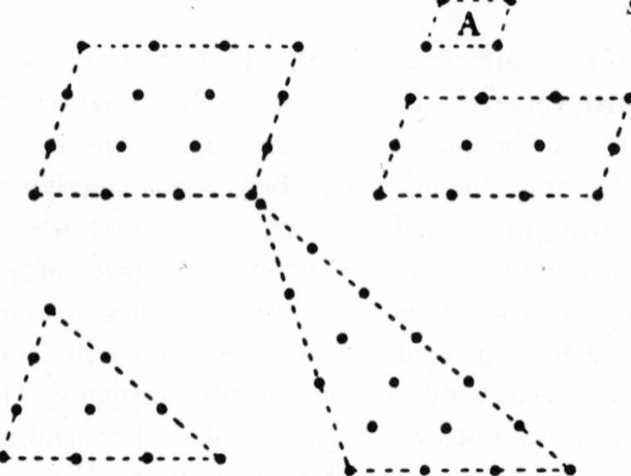

FIGURE 6—The unit of a planar design is included in the outline A. The multiplication of the unit may assume various forms, a few of which are shown. The mutual inclinations of the edges are limited to certain definite angles.

the rows might be made up in various ways as in Figure 7, so also any one crystal can be divided up in an infinite number of ways into parallel sheets.

It is convenient to consider the problem of the diffraction of X-rays in stages: first taking the scattering by a single unit, then by a sheet of units, then by the whole crystal which is made up of a succession of sheets.

The unit in a crystal is made up of a certain number of atoms arranged in a certain way: the composition and arrangement vary from crystal to crystal. When the train of waves meets the unit each atom in it scatters and can be regarded as the centre of a series of ripples spreading outwards in spherical form. At a little distance these melt into one another, and in the end there is but one spherical wave having its centre somewhere within the unit. There is however this peculiarity about the wave, that it is not equally strong in all directions. To take a simple case, imagine the unit to consist of two atoms A and B, separated by a distance

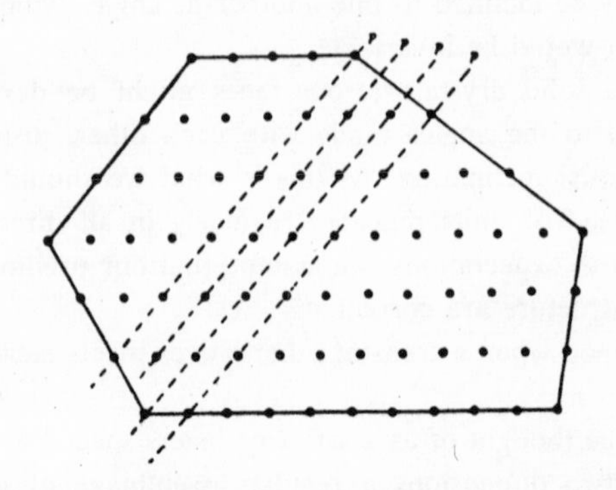

FIGURE 7—The points in the figure can be arranged in rows in various ways.

equal to half the length of a wave, as in Figure 8. The oncoming waves arrive simultaneously at the two. The waves scattered by A and B start off together. In the direction ABC the two systems are always in opposition: a crest of one set fits into a hollow in the other. In this direction they destroy each other's effects. The same happens in the reverse direction BAD. But in every other direction there is no such complete interference: in a direction such as that marked by the arrow P they support each other to some extent, and more so as the direction P is separated from C or D. In this case the combination of the scattered waves will have a spherical form, but it will not be equally intense all round. There will

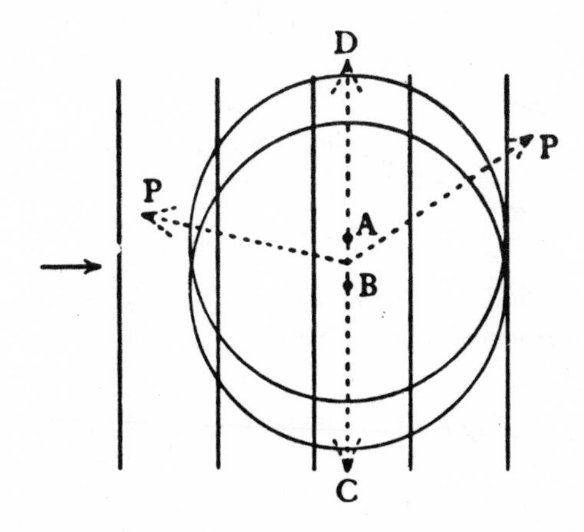

FIGURE 8—Waves represented by the vertical lines are moving upon the two atoms A and B which scatter a small fraction of their energy. The two are half a wave-length distant from one another. The spherical waves that spread away annul one another in the direction A B C or B A D, because the crest of one fits into the hollow of the other. But energy is scattered in all other directions such as are indicated by the arrows marked P.

be points C and D where the wave vanishes as the figure shows.

Other arrangements of the atoms lead to other distributions of intensity on the spherical surface: and the more complicated the arrangement the more complicated the distribution.

This complexity has however no effect on the development of our argument and is described only to make the picture more real. The one important point is that whatever the composition and arrangement of the unit, all the units behave alike. As far as we are concerned for the moment we may disregard the inequality of the distribution of the energy on the surface of the scattered wave, and remember only that the waves are in the end spherical, and that they may be regarded as originating from regularly arranged points which represent the positions of the units.

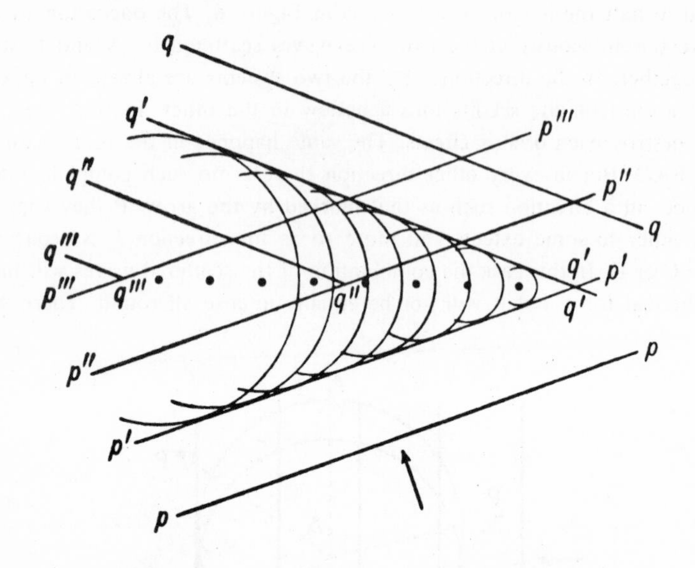

FIGURE 9—Waves, 'pp,' 'p'p'' . . . sweep over a row of points where some scattering takes place. The most of the energy goes on, but a fraction is reflected in the form of waves 'qq,' 'q'q'.' . . .

We now go on to consider the combined effect of the units in a sheet. Suppose that the dots in Figure 9 represent some of the units of pattern in a sheet which is at right angles to the paper. A set of waves is shown in section by the straight lines *pp*, *p'p'*, etc. As each wave sweeps over the points, spherical waves spread away from the point in turn, and a reflected wave is formed by their combination. This is in fact another instance of the application of the Huygens principle. We have a case of simple reflection, differing only from reflection by a mirror in the fact that only a portion of the original energy is carried away by the reflected wave. We know by experiment that in the case of a single sheet this portion is extremely small: the X-rays often sweep over millions of sheets before they are finally spent.

An analogous effect is frequently to be observed in the case of sound. A regular reflection can take place at a set of iron railings, the bulk of the energy going through. We hear such a reflection when we pass the railings in a car.

It is to be observed that the even spacing of the units, and of the dots which stand for them in Figure 9 is not necessary so far as the effect of a sheet is concerned. Nor need the railings be regularly arranged, in order to produce an echo: a reflection can even be observed from a hedge. Regularity is not important until we consider the possibility of reflection by many different sets of planes within the crystal.

In the diagram Figure 10 we represent a section of these sheets by the lines S_1 S_2 S_3, and so on; we draw them as full lines and not as rows of dots because it is of no importance where the units or the representative dots lie in each sheet. For convenience also we show by straight lines the directions in which the waves are travelling instead of the waves themselves. Thus $a\mathrm{P}a_1$ represents a case of reflection in a single sheet which we have just been considering. Besides the set represented by $a\mathrm{P}q_1$ there is another case of reflection represented by $b\mathrm{Q}b_1$, another by $c\mathrm{R}c_1$, and so on. The dimensions of the figure are grossly exaggerated in certain directions so as to show the argument more clearly. The distances between the layers are in actuality minute compared to the width of the pencil. Each ray, like $b\mathrm{Q}b_1$ represents a train of waves moving on so broad a

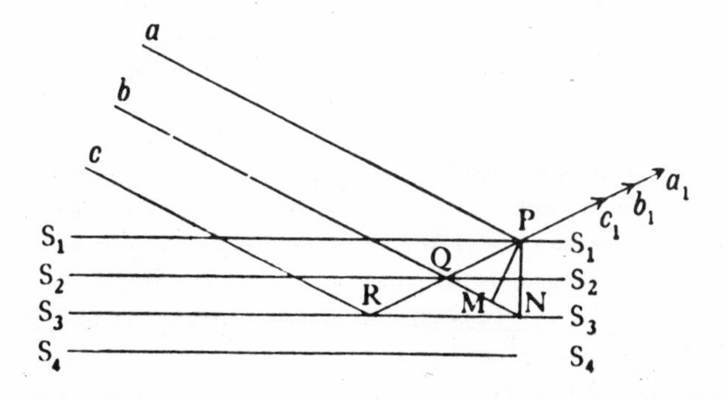

FIGURE 10—The diagram illustrates the reflection of a train of waves by a series of regular spaced sheets, each capable of reflecting a small fraction of the energy of the train.

front that the various reflected trains overlap one another sideways.

The reflected set represented by $b\mathrm{Q}b_1$ has had further to go than $a\mathrm{P}a_1$, before it re-appears and joins the latter. If we draw perpendicular distances PM and PN the extra distance is MN.

And again the set $c\mathrm{R}c_1$ lags behind $b\mathrm{Q}b_1$ just as much again as this lags behind $a\mathrm{P}a_1$, because the sheets are spaced at even distances. Behind this other reflections follow at regular intervals. The wave reflected by the crystal is the sum of all these. We may represent the summation by Figure 11 in which curves representing the reflected sets of waves are put one below another: each lagging behind the one above it by the distance MN. These waves are to be added up; for instance, along the vertical line shown in Figure 11 we are to add together Oa, Ob, Oc, etc., giving positive signs to those that are above the horizontal line and negative to those that are below. Usually the sum of those quantities will be zero, because they are just as likely to be above as below the line, and in their millions every possible size up to a maximum is to be found. The meaning of this is that there is no reflected pencil: its constituents have

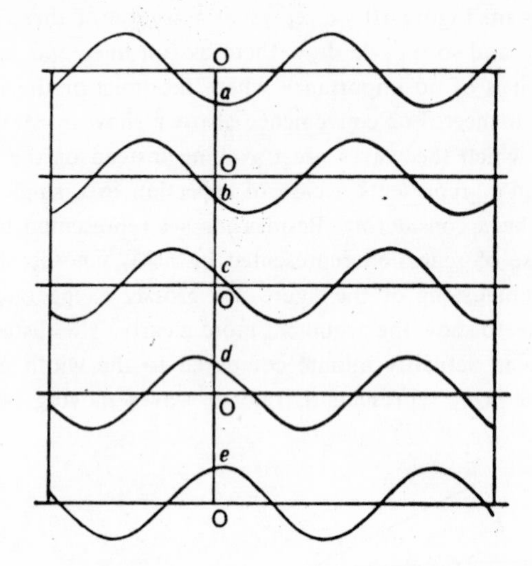

FIGURE 11—A diagram to show how the reflected waves of Figure 10 add up to nothing unless they are all exactly in phase with one another. The quantity $Oa + Ob + Oc + \ldots$ is zero because there are as many positive as negative terms in the millions that must be added together. The exception is when the reflections are all in phase, the crests of one set being exactly above or below (in this kind of figure) crests of all the other sets.

destroyed one another. There is one exception to this rule. If the lag is exactly one wave-length or two, or three, or any whole number of wave-lengths, so that the curves shown in Figure 11 lie exactly below one another, then the sum of them all is just a multiple of one of them, and as the multiplier is large the reflection is large also. The reflected energy cannot of course be greater than the incident, but calculation shows that over a very small range on either side of the reflecting angle the reflection is complete.

The amount of the lag depends on two things, the angle at which the rays strike the crystal and the spacing of the sheets. If they are nearly perpendicular to the sheets, the lag is twice the distance between two sheets that are neighbours: and this is its maximum value. The more oblique the incidence the smaller the lag; at glancing incidence it becomes very small. Thus provided the wave-length is not too great there must always be some particular angle of incidence at which the lag is exactly one wave-length or even more wave-lengths: and at these angles the reflection leaps out strongly.

If our primary rays are of one wave-length we must turn the crystal round until the angle is right. If the angle of incidence has a fixed value we may get a reflection by throwing a mixed beam upon the crystal, out of which rays of the right wave-length will be selected for reflection while the rest pass on. The late Lord Rayleigh once showed at a lecture in the Royal Institution an analogous experiment in acoustics. As its dimensions are on a scale so much larger than those of X-rays and crystals, it helps to an appreciation of the latter case. Sound waves are produced by a whistle of very high pitch, known as a bird-call. The waves are only an inch or so in length, much shorter than the waves of ordinary speech, but hundreds of millions of times longer than the ether waves of X-rays. The note is so high that many ears, especially those of older people cannot perceive it. A set of muslin screens, about a foot square are arranged in parallel sequence, upon a system of lazy-tongs which allows the common spacing to be varied. The screens may for our purpose be taken to correspond to the sheets of Figure 10: each can reflect a small fraction of an incident sound wave, but allows the bulk of it to go on.

If now the whistle and the set of screens are arranged as in the photograph, Plate I C, the sound may be reflected by the screen. The reflected sound, if it exists, is very easily detected by means of the 'sensitive flame.' This is a luminous gas jet issuing under great pressure from a long narrow tube with a fine nozzle. The pressure is adjusted until the flame is on the point of flaring, under which circumstances a high pitched sound causes it to duck and flare in a most striking way. The rapid alternations of pressure in the sound wave are the direct cause of the effect. The sensitive flame is so placed that it can detect the reflected sound if there is any, but it is screened from any direct action of the bird-call.

It is then found that if, by means of the lazy-tongs, the common spacing of the screens is altered gradually and continuously the flame passes through successive phases of flaring and silence. The explanation is the same as that of the X-ray effect just described. If there is flaring it means that the reflections from the successive screens all conspire, and this happens if the spacing is so adjusted that the lag of one reflection behind another is a whole number of wave-lengths.

Rayleigh used the analogy for the purpose of explaining the brilliant colours of crystals of chlorate of potash. These crystals have a peculiar formation, being composed of alternating layers of the crystalline material differing only in the orientation of their crystalline axes. The thickness of the layer is thousands of times smaller than the spacing of the muslin screen, but again thousands of times larger than the spacings of the layers in which the crystalline units are disposed. It is of the order of the wave-length of visible light. The same argument holds in all three cases.

We have one more point to consider before we can appreciate Laue's experiment. We have to remember that there is not only one way, there are an infinite number of ways in which a crystal can be divided into parallel sheets. Suppose that Figure 12 represents for example the disposi-

PLATE I

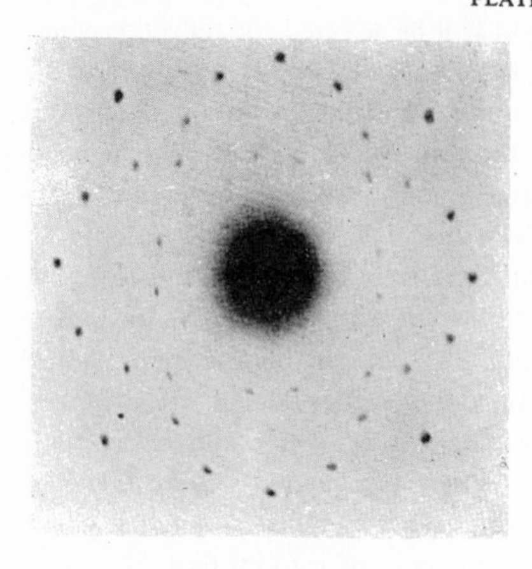

A. X-ray diffraction spectrum of rocksalt.

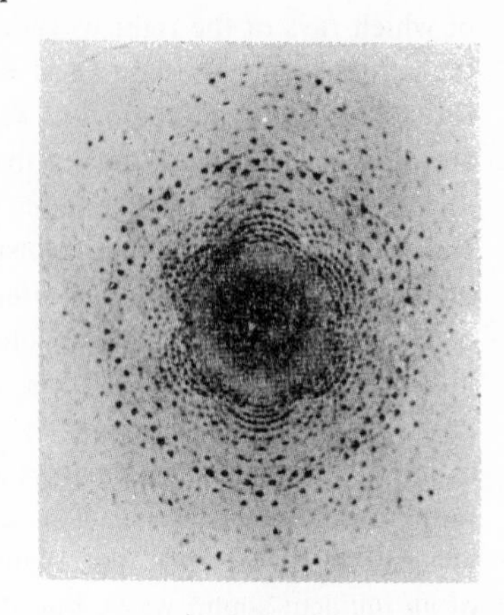

B. An X-ray diffraction spectrum of kaliophilite.

Bannister

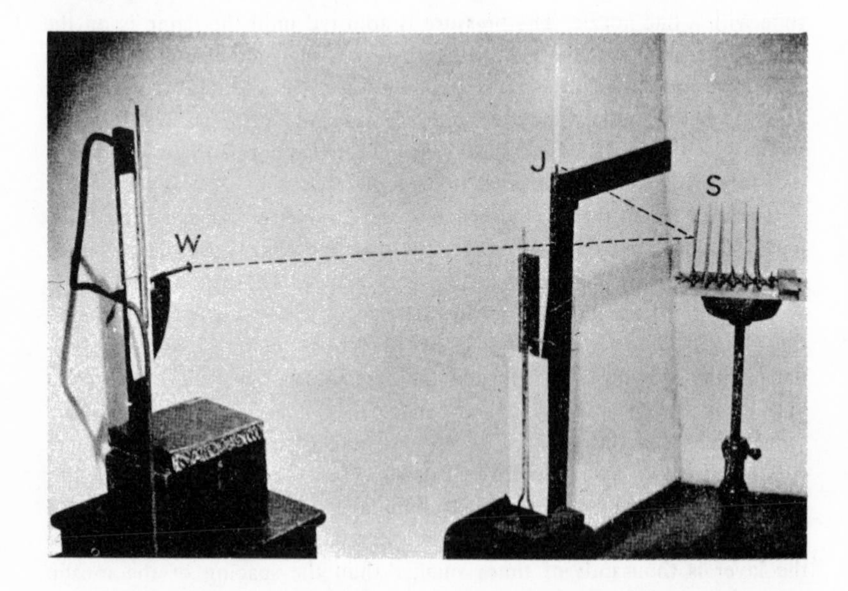

C. A photograph of the apparatus used by Lord Rayleigh and described on pp. 864–865. The whistle is at W, the set of screens at S, and the luminous gas jet at J. The dotted lines show roughly the course of the sound waves that affect the jet. The photograph shows the appearance of the jet when there is no sound, or when the screens (seen edgeways in the picture) are not placed so that their reflections reinforce one another. Owing to a peculiarity in the form of the nozzle at J, the jet does not respond to sound proceeding directly from W to J. When the screens are properly spaced the jet broadens and ducks to a fraction of its normal height.

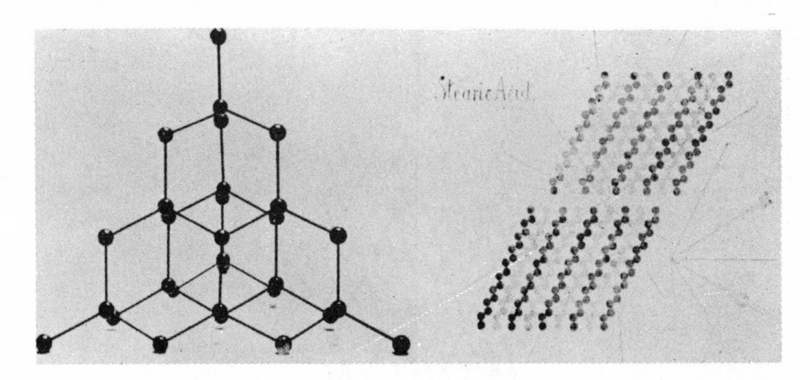

A. A model showing the structure of diamond. Each ball represents a carbon atom in respect to position only, but not in respect to form or size. The distance between the centres of two neighbouring atoms is 1.54 Angstrom Units. (One unit is the hundred-millionth of a centimetre.)

B. The arrangement of the molecules in a crystal of stearic acid. Each zig-zag represents the carbon atoms of a single molecule. The distance between the centres of two neighbouring carbon atoms is 1.54 Angstrom Units, as in diamond. The two models A and B are built on different scales. The terminal groups and the hydrogens are not shown. (From a diagram due to A. Muller, Davy Faraday Laboratory.)

C. A set of optical diffraction spectra, due to Fraunhöfer—from Guillemin's "Forces of Nature." The light is diffracted in each case by a screen containing a regular arrangement of fine openings, e.g. a part of a feather. The original is coloured, but the differences in colour are not shown in this reproduction. These optical spectra may be compared with the X-ray spectra of I A, B.

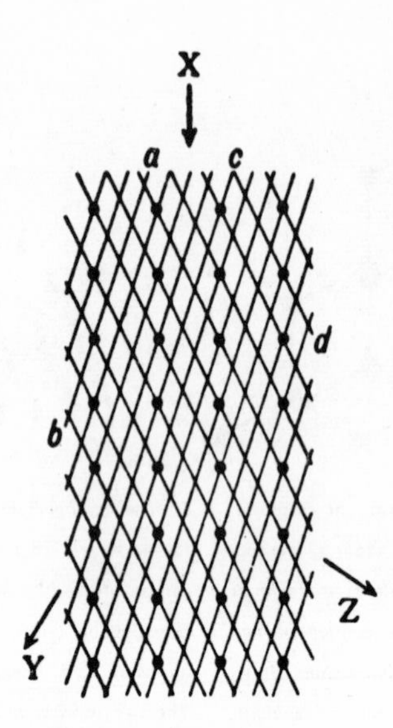

FIGURE 12—The dots represent the units of a cubic crystal. The scale would be about 100,000,000 to one. The direction marked X is that of the incident X-rays. There are reflections in the direction Y due to planes parallel to 'ab,' and in the direction Z due to planes parallel to 'cd'; also in other directions not indicated.

tion of the units in a cubic crystal: a section by the plane of the paper is all that can conveniently be shown in such a diagram. We may consider the division into sheets to be made parallel to *ab* and as far as this set of sheets is concerned, there will be a partial reflection of X-rays. Out of the original pencil which must contain a variety of wave-lengths a sharply defined selection will be reflected in the direction Y consisting of those for which the lag is equal to the wave-length: the lag depends as already said on the spacing and on the angle of incidence.

The crystal can also be divided into sheets which cut the plane of the diagram in lines parallel to *cd*. In this case reflection takes place at a different angle, and the rays reflected are of a different wave-length. The reflected ray moves away in the direction Z and makes another spot on the photographic plate.

Thus many reflections occur simultaneously and make their impressions on the photographic plate. If the crystal is of a cubic design and the incident rays are parallel to one edge the pattern that is formed will be symmetrical about two straight lines at right angles as in Plate I A. If, on the other hand, the crystal is of a hexagonal design we have a six-sided figure as in Plate I B. All these agreements between observed results and calculated expectations are perfect, and it is obvious that the fundamental

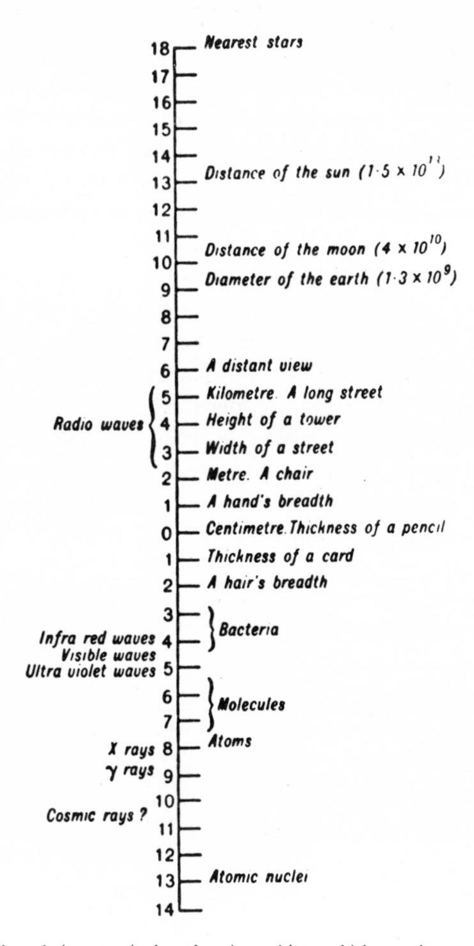

This table shows the relative magnitudes of various objects which we observe and measure. It is like a set of shelves on which we place specimens of objects and magnitudes from the very great to the very small. On a middle shelf marked zero, we have the centimetre, and the thickness of a pencil to represent objects of that order of magnitude. On the shelf above we place an object of about ten centimetres in size; the width of a hand will serve. The shelf above takes objects of about a hundred centimetres, for example smaller objects of furniture. The width of a street will represent the thousand centimetres, the height of a tower might be ten thousand centimetres or a hundred metres, and so on. Below the zero shelf comes first a shelf holding something of the order of a millimetre in thickness, as a card; then the hair's breadth on the next shelf and so on. Bacteria are at various heights on the third and fourth shelves down; molecules on the sixth and seventh, atoms nearly down to the eighth. On the other side of the vertical line the various wavelengths are shown in the same way. Distances are sometimes given in figures. The sun's distance is fifteen million million centimetres or in symbols 1.5×10^{13}. This goes therefore on the thirteenth shelf up.

hypothesis is correct. The X-rays have as much right to be spoken of as ether waves as light itself.

The two illustrations of Laue photographs (Plate I) differ much from each other: their difference illustrates the remarkable diversity that is found in such photographs. Every crystal writes its own signature. It is in some cases easy to deduce the structure of the crystal from the picture which is characteristic of it. In others the task is accomplished with difficulty: in a still greater number of cases the solution is more than present technique and skill can accomplish. Since every solid substance contains parts that are crystalline, and since in many of them the whole is an aggregation of crystals, it will readily be understood that a knowledge of crystal structure often affords an explanation of the properties of the substance. One or two examples are given in Plate II A, B, but it would be out of place to consider this subject in detail, and indeed impossible, for it has grown so rapidly that a considerable treatise is now required to give any reasonable account of it. We must be content with the proof that the X-rays can be considered as ether waves.

All things began in order, so shall they end, and so shall they begin again; according to the ordainer of order and mystical mathematics of the city of heaven.
—SIR THOMAS BROWNE

7 Crystals and the Future of Physics

By PHILIPPE LE CORBEILLER

MY PHYSICIST friend Empeiros walked into my office in a very bad mood. He had just read an article, and in a very respectable monthly, which claimed that the field of physics was limited, and moreover would soon be completely mapped out.

"Such nonsense makes me mad," said Empeiros. "Physics is being renewed all the time by someone discovering something completely unexpected. Look at X-rays, look at radium, look at cosmic rays. How could a physicist ever sit back and say that from now on nothing new will ever be found?"

"That does not seem so impossible to me," I replied. "I agree that physics seems today to be without boundaries, but that may be because its field is not yet organized. Look at the history of the great explorations. After Columbus had discovered America, one might have expected the discovery of any number of continents. Yet by the end of the 18th century the whole surface of the globe was sufficiently well mapped out to preclude the discovery of any new continent. I think exploration in physics might well repeat this pattern."

"That is begging the question," said Empeiros. "At the time of Columbus all competent people knew that the earth was round. They must therefore have known that the amount of ground to be discovered was limited. It is absurd to compare the earth to physics, which is limitless."

"The comparison is *not* absurd," I replied. "People did not always know that the earth was a globe, and as long as the earth was thought of as flat its surface could be either finite or limitless; there was no telling. Intellectually speaking the argument was clinched not when Magellan's ship circumnavigated the earth, but when your ancestor Eratosthenes gave the first estimate of the circumference of the spherical earth. The check came 2,000 years later, but the argument was not concerned with the time interval."

"You sound like a Platonist," said Empeiros, obviously soothed by my allusion to his Greek descent.

"Of course I am," I answered. "My Ph.D. work was in the theory of numbers and arithmetical groups."

"Very fine," said Empeiros, "but that isn't physics."

"It isn't physics?" I exclaimed indignantly. "Look at crystals!"

"What have crystals to do with the future of physics?" inquired Empeiros.

"Everything," I said with glee. "Crystallography is the image of the

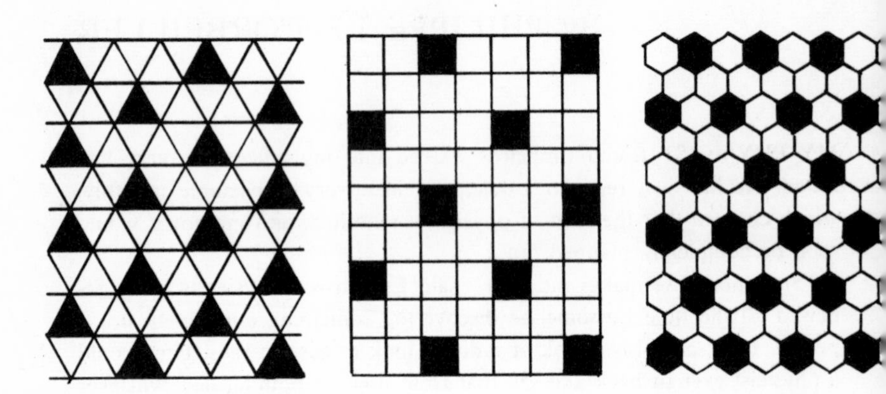

FIGURE 1—Regular tiles, i.e., tiles of which each side has the same length, will pave a floor in only three shapes: the triangle, the square and the hexagon. Tiles of other shapes will not fit to cover the entire area of the floor.

physics of tomorrow. Now I know just how to convince you. Sit down and be convinced."

Bullied, Empeiros sat down.

"There are theorems in mathematics," I began, "which tell us that there are just so many ways of combining certain things, and no more. Here is an example. Square tiles are common enough, and so are six-sided tiles. Long before Euclid, geometers asked themselves: Can we pave a floor with regular tiles of any number of sides? The answer is no. Only three kinds of regular tiles can be used: the triangle, the square and the hexagon. Regular tiles of other shapes will not fit together to cover the whole area.

"This first result may not be very exciting, but the next one is—at least Plato found it so. A regular cube is a solid limited by six equal square faces. Can we build solids with any number of equal, regular faces? Again the answer is no. Only five regular solids are possible, and no more. Some day I want to show you how to build these solids out of a sheet of strong paper; it is a fascinating game."

"I like games," said Empeiros, "but I wish you would come to physics. In mathematics we know the rules of the game, since we make them up ourselves. Here, for instance, the rule is to use nothing but regular polygons to limit our solids. I see nothing surprising in the fact that there should be exactly five solids obeying that arbitrary rule. But in physics we don't make the rules; therefore we shall never be able to say that a chapter of physics is closed."

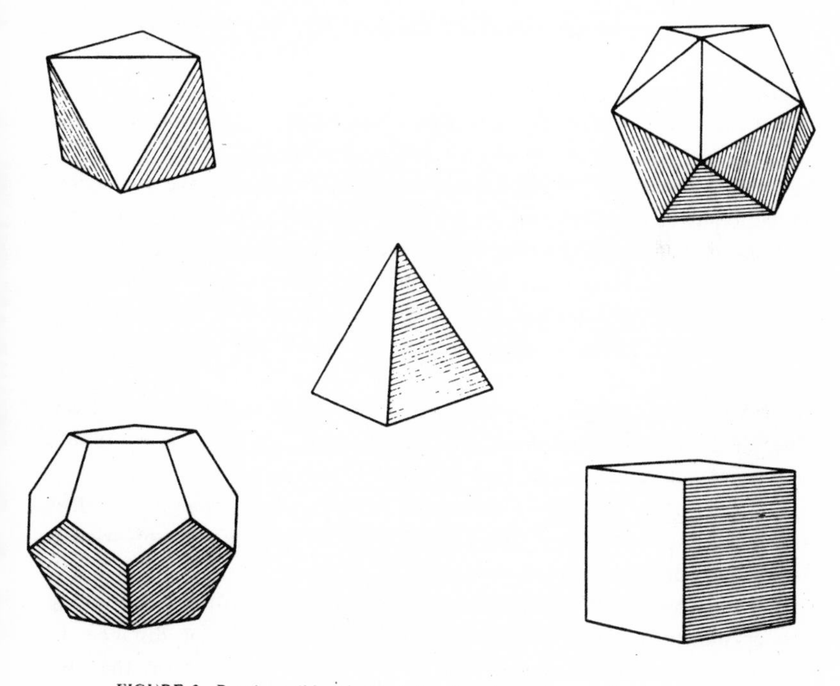

FIGURE 2—Regular solids, of which each face is regular, are five in number.

"Your argument," I answered, "is perfectly plausible. But it does not fit the case. There are 'regular solids' in physics also; we call them crystals. So here we have geometry again, and arithmetic, but this time it does not all take place in our heads. Here nature invents the rules of the mathematical game, and of course it finds itself bound by the consequences of the rules."

"Sounds like nonsense to me," said Empeiros.

"The ancients," I continued, disregarding the comment, "noticed several beautifully shaped crystals, such as quartz, growing, as it were, out of shapeless rocks. The first scientific crystallographer was a Danish bishop called Steno, who in 1669 published a dissertation *Concerning Solids Naturally Contained within Solids*. Other naturalists carried on his research, and by 1782 a Frenchman, the Abbé René Just Haüy, had found the basic rule of the game of crystals: he had found how to obtain the shape of any crystal from that of some standard simple form.

"Ever since Steno, mineralogists had been measuring angles between the faces of hundreds and hundreds of crystals and the accumulation of all these data was a holy mess. Haüy noticed that a very simple rule held for those angles relating to a specific mineral. Since it is difficult to describe geometrical things in space, allow me to use a flat plane to tell you what Haüy's Law says and what it implies. Assume that in a flat crystal a corner seemed cut off by the face BC, so that the contour of the crystal was XBCY [*see diagram on p. 876.*] On edge AX Haüy marked off equal distances AB, BD, DE, EF, and on edge AY also equal distances AC, CP, PQ, QR. Then, joining any of the points marked off on edge AX, say E, to any of the points marked off on edge AY, say P, he found the angles formed by the cut EP and the two edges equal to angles actually observed on other specimens of the same mineral.

"This, Empeiros, was a momentous event in the history of science. For the good Abbé Haüy thus became the first experimental atomist. His successors were John Dalton, Gregor Johann Mendel, J. J. Thomson, Max Planck, Albert Einstein—each discovering a new type of 'atom.' From such geniuses . . ."

"Now wait a moment," interrupted Empeiros good-humoredly. "I am interested in your story, and I want information, not enthusiasm. Would you mind telling me quietly how atomism enters here?"

"That is plain enough," I answered. "Don't you see that nature does not make cuts at random? She limits herself to a specific set of orientations; those in between are forbidden, so to speak. This is what we call space quantization. Why 'quantization'? Because we had to lay out along the crystal edges equal chunks of length, equal quanta of distance, to reproduce the observed angles. Now of course one is free to say that this is just a geometrical construction without physical significance. Haüy himself did not think so. He pictured a crystal as made up of an enormous number of very small, *identical* bricks. Later there came objections to the idea of solid matter consisting of stacked-up little bricks with no free space between. However, the essential point in Haüy's system was not the little bricks themselves, but the regular distances defined by them—the quantization of space. So that if we assume that a crystal is made of minute atoms, all identical, distributed in space in some regular lattice, we shall have retained the basic reason why a crystal face, cutting off a whole number of atoms, must necessarily cut the edges of the simple standard form in whole-number ratios and satisfy Haüy's law in an automatic way."

"I follow all that," said Empeiros, "and am ready to accept this model (without believing in it literally) since it fits the angle measurements. But I note that nature is not so limited as you say, since around each crystal vertex it has an infinity of directions to choose from. Hence we have here

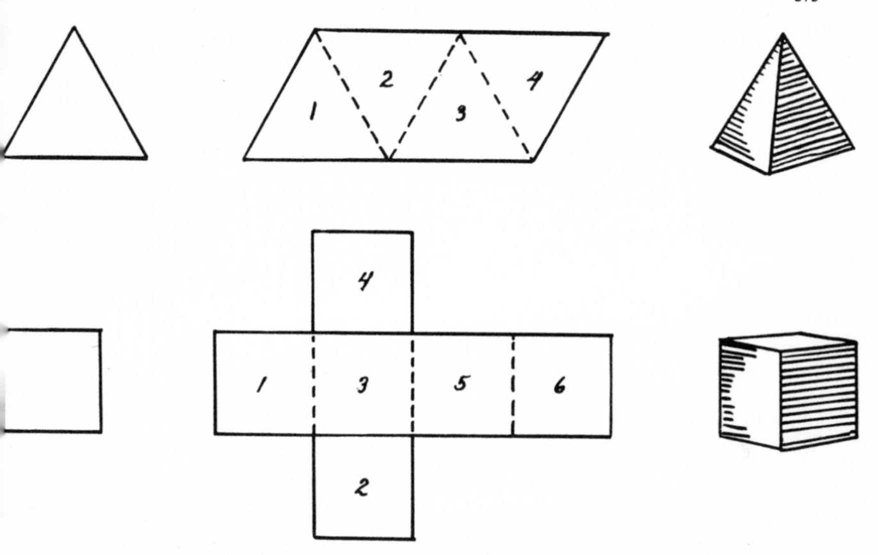

FIGURE 3—Regular solids may be made from paper to demonstrate the limitation of their class. If four regular triangles are pasted together as shown at the top of this illustration, they will form a regular tetrahedron. The same can be done to make a regular hexahedron (cube), dodecahedron, octahedron and icosahedron.

no such drastic limitations as in the mathematical problem of the regular solids."

"That's what you think," I replied, "but you are already in a mathematical trap. From the simple fact that the atoms are arranged into a regular lattice it follows that certain crystal symmetries are allowed—and no others. For instance, given a square made up of little dots, it is impossible to turn it into a regular octagon—a polygon of eight equal sides. To make an octagon of it we would have to divide each side of the square into three parts, with lengths in the ratios $1:\sqrt{2}:1$. And this is impossible to achieve with evenly spaced dots, because $\sqrt{2}$ is not the ratio of two integers—as Pythagoras found to his amazement some 2,500 years ago.

"And now the final step: If we turn the original square around by one right angle, or one fourth of one full turn, the appearance of the square will not be changed; crystallographers say that a square has a rotation axis of order four. A regular octagon could be turned around one eighth of a turn and still look the same; we say it has a rotation axis of order eight. But we have just found that a regular square array of points can never be made into a regular octagon. So we conclude that crystals may have a rotation axis of order four but not of order eight."

"I must admit," said Empeiros, "that this is amazing. Tell me one thing: How is it that I was taught a good deal of mathematics, and that I never heard of anything like that?"

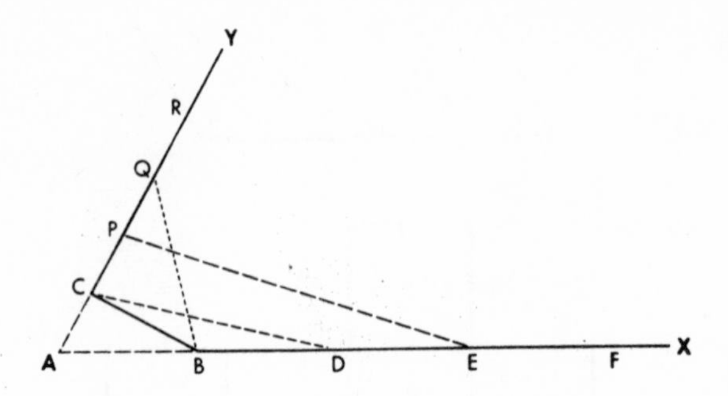

FIGURE 4—Crystallographic law, here depicted in two dimensions instead of three, states that the facets of any one crystal type can have only such angles as are formed by cuts from equidistant points on the axes AX and AY.

"That," I answered, "is because you were taught the geometry of Euclid, and the crystals follow the geometry of Pythagoras. Euclid's geometry is one of continuous lines; Pythagoras', of arrays of points. After Euclid invented his geometry, the earlier geometry of Pythagoras was dropped. Later scientists—Archimedes, Descartes, Newton, Maxwell—went deeper and deeper into the mathematics and the physics of the continuous, and brought out such things as the calculus, universal gravitation and electromagnetism.

"In the last 60 years, however, a new revolution has taken place, and everywhere we look we find that what seems to be continuous is really composed of atoms. Atoms of all kinds: chemical atoms, molecules, ions, electrons, quanta, photons, neutrons, chromosomes, genes. And everywhere the presence of atoms is revealed by the empirical discovery of laws expressed in terms of whole numbers. So you see that Haüy's Law—the Law of Rational Indices—was a forerunner of all the other atomic laws."

"But are not modern mathematicians interested in such things?" asked Empeiros.

"They are," I answered, "but they give them other names. They call them Number Theory and the Theory of Discontinuous Groups. Actually they have found much more than we can use as yet in physics, but we have in crystals illustrations of some of their simpler theorems. Mathematicians have proved, for example, that the symmetry elements of crystals can be grouped in 32 different ways, and no others. Crystallographers have classified every known crystal into one of these 32 crystal classes. It is true that examples of two of these 32 classes have not yet been found in nature. But mathematics is the science of the logically possible. It tells us that these 32 classes are possible, and while we do not happen to know

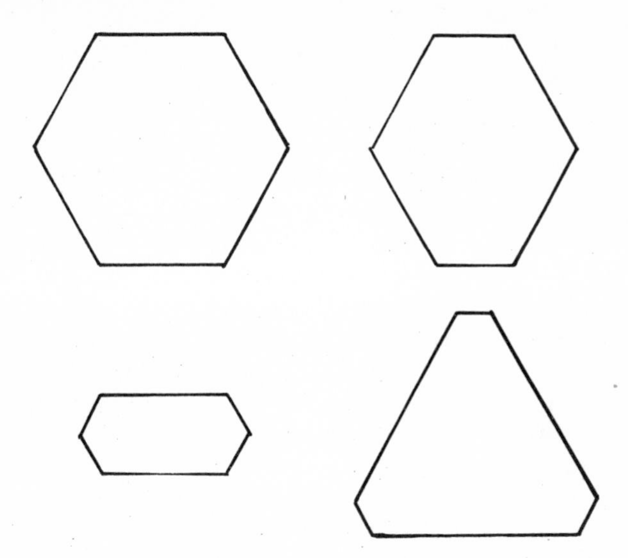

FIGURE 5—Size and shape of crystals are unimportant in establishing their identity; the angles between facets are what count. These four hexagons, all of whose angles equal 120 degrees, are thus equivalent in this sense.

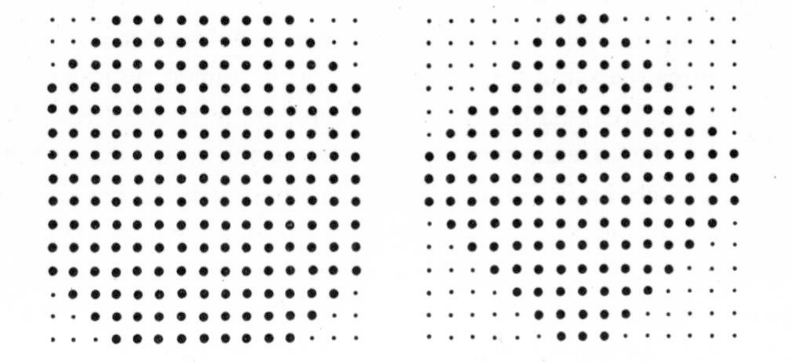

FIGURE 6—Square made up of dots cannot be changed into a regular octagon. Here two octagons have been made, but neither is regular. This illustrates a limitation of crystal form (see Figure 7).

any example of two of them, we may find them next year, or in 10 years. Whether or not we shall ever find them is quite immaterial."

"Is the science of crystals, then, entirely closed?" asked Empeiros.

"By no means," I answered. "Crystallographers have been digging deeper than this first set of symmetries which is reflected in the outward shapes of the crystals. Until now we have imagined the atoms inside the crystal to be simple little dots. If we assume that the atoms have a certain shape, the question arises in how many ways this unique shape can be twisted around in space as we jump from one corner of the lattice to the next. Again one finds that only certain possibilities are open. This very difficult problem was solved in the 1890s by three scientists working independently—Fedorov, a Russian; Schoenflies, a German, and Barlow, an Englishman. All three got the same result: there are just 230 different ways of distributing identical objects of arbitrary shape regularly in space. We say that there exist 230 different space-groups.

"These 230 space-groups fit into the 32 crystal classes in a relatively simple way. Each commands a certain external symmetry, and therefore belongs to a specific class. The various space-groups (*i.e.*, internal arrangements) within a given class cannot be distinguished from one another by investigating the crystal faces. But in 1912 the German physicist Max von Laue found a means of exploring this internal structure—a beam of X-rays.

"From that moment the 230 space-groups, which previously had been considered a mere mathematical game, assumed great practical importance. X-ray analysis enables us to say to which space-group a crystal belongs. It even gives us clues from which we may find out how the atoms are arranged in a crystal. For a simple substance such as common salt this is relatively easy; you simply try out a few likely arrangements until you hit upon the one which in all respects fits the pattern of dots made on the photographic plate by the X-ray beam passed through the crystal. But with complicated molecules the problem becomes very difficult. It requires the same tedious kind of work, guided by much ingenuity and flair, as the breaking of an intricate diplomatic code. X-ray analysis is now part of the arsenal with which the problem of the structure of proteins, of such basic importance in biochemistry, is being attacked."

"All this is fascinating," said Empeiros, "but I don't see that we have advanced a bit in the question we were going to discuss. You believe that the field of physics is finite. Are not crystals the only regular arrays in nature? They appear to me, more so now than ever, as a beautiful exception."

"They are, and they are not," I answered. "The tiles, the regular solids, the spatial arrays we have been talking about seemed to be problems in geometry. However, this geometry, the geometry of Pythagoras, merely illustrates theorems from Number Theory, the name we give to higher arithmetic. When we say there are three kinds of regular tiles, five regular solids, 32 crystal classes, 230 space-groups, we are only giving geometrical illustrations of four arithmetical problems, which have respectively 3, 5, 32 and 230 solutions. You see, then, that the essential circumstance about our problems was not that they were about geometrical arrays: it was that they were dealing with integers.

"By now we have found so many other types of particles besides those

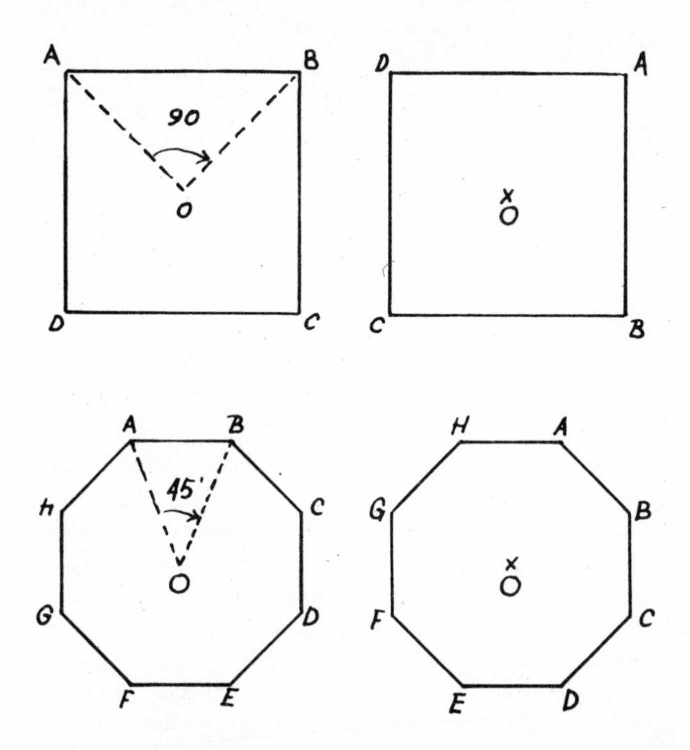

FIGURE 7—Rotation axis of a square has the order four: it can be turned 90 degrees, or four times in 360 degrees, and still look the same. An octagon has a rotation axis of the order eight. Crystals cannot have such a rotation axis because an octagon cannot be made out of a regular array of points (see Figure 6).

that are built up into crystals that the field is wide open for other applications of Number Theory and of an allied chapter of mathematics, Group Theory. Indeed quite a few have already come to light. The quantum theory of the atom is full of such relations, and the enumeration of the successive elements in the periodic table, is very similar to the enumeration of the crystal classes. Just as it would be impossible to discover tomorrow a crystal with some 33rd type of symmetry, so are we assured that no new chemical element can ever insert itself anywhere within the succession which today stretches from hydrogen to californium: it is a system of 98 elements filling all the steps from 1 to 98 electrons.

"The implications of this have crept almost unnoticed upon the chemists and physicists. Placing such limitations upon nature runs quite contrary to the traditional tenets of empiricism. That it has not received much notice shows that we are not yet adjusted to thinking in terms of

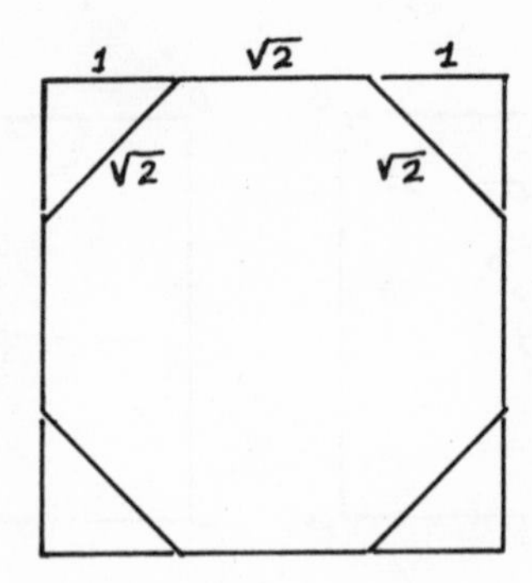

FIGURE 8—Octagon can be made out of square only by dividing each side of the latter in the ratio $1 : \sqrt{2} : 1$.

atoms and quanta, and are unaccustomed to the absolute character of the counting of combinations of integers."

"You have marshaled," said Empeiros, "a collection of results which I admit is quite impressive. You may be right in thinking that this line of thought makes a dent in the solid doctrine of empiricism. But again I remind you of your statement that the field of physics is finite. You have not yet proved that."

"And I cannot prove it," I answered. "It is a hunch and a prediction. All I can do is to show you why I think it is the trend along which physical science has been developing these last 350 years. In any new science purely descriptive knowledge is the first stage of advance. Next comes the establishment of quantitative laws, such as Boyle's law for the 'spring' of air, Newton's law of universal gravitation, Maxwell's equation of electromagnetism. In this stage we learn that things in nature are related by such and such numerical laws. But we don't understand why. Why should crystal faces always obey Haüy's Law, the Law of Rational Indices? That understanding came in 1912 when the periodic space-structure of crystals was experimentally established. We could then conclude, not from observation but led by mathematical necessity, that there are exactly 32 possible crystal classes and 230 possible types of space arrangements of atoms in a crystal.

"In the third stage of scientific knowledge, which we might call deductive or axiomatic, the natural laws obtained by observation are shown to be necessary logical consequences of a few hypotheses or assumptions. The surprising thing about the examples of deductive knowledge which we know today is the extreme simplicity of the assumptions, and how rich and far-removed are their consequences. For instance, the single assumption that any crystal is a periodic space lattice carries with it the whole theory of crystal classes and of space groups.

"At the present time some of our leading scientists, following early attempts by Einstein, are trying to merge together the experimental laws of gravitation and of electromagnetism. It is unthinkable that these two fields should be unrelated. We know from measurement that the electric repulsion of two electrons is a certain number of times greater than their Newtonian attraction. This is the empirical stage. We want to know why this ratio is that number and no other. The recent discovery of several new types of elementary particles, which just now are in meaningless disorder, makes the problem all the more pressing.

"We cannot rest satisfied with numerical statements without logical justification. Crystallography is a perfect example of how a numerical statement becomes justified. It may not be very long before we obtain deductive knowledge, on the basis of some few fundamental assumptions, of the main features of the physical universe."

"You may surely project anything you wish into the future," said Empeiros. "But you must admit that the realization of your daydream would be without precedent in the development of science."

"I admit no such thing," I retorted. "Geometry was once quite as empirical as physics is today, and the first 'theorems' were found experimentally. It took the Greeks 300 years, from Thales to Euclid, to establish a set of assumptions upon which geometry could be built logically. Now consider that practically all we know in physics has been found in 350 years, and that in the last 50 years we have traveled all the way from the phenomena of common experience to fissions in atomic nuclei. Wouldn't you think that in the next 50 years—taking us to the year 2000, a nice round figure—we should have a sporting chance of establishing in our turn a set of assumptions on which physics can be built?"

Empeiros shook his head skeptically.

COMMENTARY ON

Queen Dido, Soap Bubbles, and a Blind Mathematician

A VARIETY of natural phenomena exhibit what is called the minimum principle. The principle is displayed where the amount of energy expended in performing a given action is the least required for its execution, where the path of a particle or wave in moving from one point to another is the shortest possible, where a motion is completed in the shortest possible time, and so on. A famous example of this economy of physical behavior was discovered by Heron of Alexandria. He found that the equality of the angles of incidence and reflection formed by a light ray meeting a plane mirror assures the shortest possible path of the ray in moving from its source to a reflected point by way of the mirror.[1] Sixteen hundred years later Fermat showed that the minimum principle also defined the law of the refraction of light.[2] Other important instances arise in mechanics (e.g., a flexible chain suspended freely from its two ends "assumes a form in which its potential energy is a minimum"), electrodynamics, relativity and quantum theory.

The minimum property and its inverse twin, the maximum property, find expression in certain simple statements of geometry (suggested by practical experience); such as that a straight line is the shortest distance between two points in the plane, or, that of all closed curves of equal length the circle encloses the largest area. Many of these "self-evident" truths were also known to the ancients. The Phoenician princess Dido obtained from a native North African chief a grant of as much land as she could enclose with an ox-hide. A clever girl, she cut the ox-hide into long thin strips, tied the ends together, and staked out a large and valuable territory on which she built Carthage.[3] Horatio—he who made his reputation defending the bridge—was rewarded by a gift of as much land as he could plough round in a day, another illustration of an isoperimetrical problem. The second law of thermodynamics provides a more modern

[1] For a proof of this theorem, regarded as the "germ of the theory of geometrical optics," see Courant and Robbins, *What Is Mathematics?* New York, 1941, pp. 329–332. Also, for a general discussion of the development of the minimal principle of physics, see Wolfgang Yourgrau and Stanley Mandelstam, *Variational Principles in Dynamics and Quantum Theory*, New York, 1955.

[2] Fermat's general principle states that where light travels from any point M to any point M′ in another medium, i.e., in all cases of refraction through prisms, lenses, etc., "the ray pursues that path which requires least time." See Thomas Preston, *The Theory of Light*, Fifth Edition, London, 1928, pp. 102–103.

[3] Lord Kelvin, *Popular Lectures and Essays*, London, 1894, Vol. 2, pp. 571–572.

(and a more discouraging) example of the maximum principle: the entropy (disorder) of the universe tends toward a maximum.

The search for maximum and minimum properties played an important part in the development of modern science. Fermat's discoveries in optics, James Bernoulli's work on the path of quickest descent, were among the labors that led to the conviction that physical laws "are most adequately expressed in terms of a minimum principle that provides a natural access to a more or less complete solution of particular problems." [4] Nor can one disregard the extent to which animistic tendencies and mysticism motivated the search for a unifying principle, even among the outstanding contributors to a rational system of mechanics—Euler, Lagrange, Hamilton, Gauss, to name a few.[5]

Pierre de Maupertuis (1698–1759), the French mathematician and astronomer who enunciated the Principle of Least Action, was convinced that this comprehensive law of conservation demonstrated "God's intention to regulate physical phenomena by a general principle of highest perfection." [6] Other scientists and philosophers shared this view and even went so far as to suggest that a single rule of economy, a rule both grand and simple, would embrace all the phenomena of nature. Mathematicians, fortunately, were able to bring some clarity and order to the problem of least and most, thus preventing physics from abandoning the real world and wandering into a swamp of ideals and noble conjectures.

The calculus of variations is the branch of mathematics which deals with the type of problem we have been considering. Euler was the first to give the subject systematic treatment, though the name itself followed the notation introduced a little later by Lagrange. The method of this calculus was "to find the change caused in an expression containing any number of variables when one lets all or any of the variables change." It deals not only with the maximum and minimum properties discernible in physical events, such as the behavior of light rays, the equilibrium of a mechanical system, the resistance encountered by a bullet moving through air or by a boat through water; it treats also of economic, engineering and operational research problems in which it is sought to maximize (e.g., production, profit) or minimize (e.g., cost, time) certain critical variables.

In the first of the selections which follow, Karl Menger, a noted Austrian mathematician, now professor of mathematics at Illinois Institute of Technology, presents a general survey of the main topics and methods of

[4] Courant and Robbins, *op. cit.*, p. 330.

[5] Philip E. B. Jourdain, *The Principle of Least Action*, Chicago, 1913, p. 1.

[6] Courant and Robbins, *op. cit.*, p. 383; Jourdain, *op. cit.*, p. 10 *et seq.* Maupertuis, it should be noted, was discriminating in his choice of evidence as to the existence of God. When it was suggested that proof of divine wisdom lay in the fact that there were folds in the skin of a rhinoceros—without which that unhappy creature would have been unable to move—he asked: "What would be said of a man who should deny a Providence because the shell of a tortoise has neither folds nor joints?"

the calculus of variations. Dr. Menger is an able expositor and in his brief article gives a very satisfactory popular sketch of a difficult subject. The second selection is taken from a series of lectures on soap bubbles, delivered to "juvenile and popular audiences" at the beginning of the century by the British physicist C. Vernon Boys.[7] Boys was a talented scientist, though not known to a wide public. He was highly skilled as an experimentalist and as a designer of delicate measuring apparatus. His most celebrated piece of work was weighing the earth, which is to say measuring the gravitation constant. In 1895, following a method used by the great eighteenth-century chemist and physicist Henry Cavendish, he measured the gravitational force exerted by two lead balls of 4½ inches diameter on two tiny gold balls (a quarter of an inch in diameter) suspended from a metal beam by fine quartz threads. Comparing the pull of the lead balls on the gold balls (as measured by the deflection of the beam) with the pull of the earth on the gold balls (i.e., their weight), he was able to establish a ratio of about 1 : 1,000,000,000 and thus to fix with remarkable accuracy the mass of the earth.[8] These experiments had to be conducted in a cellar, in the middle of the night to eliminate the vibration caused by passing traffic at ordinary times. Besides making the equipment for this experiment, he designed a camera, which bears his name, to photograph the speed of lightning flashes, and he measured by photography the speed of flying bullets. Except for a brief period as an assistant professor of physics at Imperial College, Boys held no academic post. His income was derived from being a "Gas Referee"—someone who prescribes the method of testing the quality of gas—and an expert witness in patent cases. He died in 1944 at the age of eighty-nine.[9]

The excerpt from Boys' delightful little book discusses some of the brilliant work on bubbles and liquid films by the blind Belgian physicist J. Plateau (1801–1883).[10] Plateau's results bring to notice a particularly fruitful aspect of the co-operation between mathematics and experimental research. That mathematics is a handmaiden of science is a commonplace; but it is less well understood that experiments stimulate mathe-

[7] C. V. Boys, *Soap-Bubbles: Their Colours and the Forces Which Mould Them*, New and Enlarged Edition, London, 1931.

[8] "He found that two point masses of 1 gramme each, 1 centimetre apart, would attract one another with a force of 6.6576×10^{-8} dynes, which makes the density of the Earth 5.5270 times that of water." Sir William Dampier, *A History of Science*, Fourth Edition, Cambridge, 1949, p. 178.

[9] Lord Rayleigh gives an interesting account of Boys' life in *Obituary Notices of Fellows of the Royal Society*, Vol. 4, No. 13, November 1944, pp. 771–788.

[10] J. Plateau, "Sur les Figures d'Équilibre d'une Masse Liquide Sans Pésanteur," Mémoires de l'Académie Royale de Belgique, Nouvelle Série, XXIII, 1849; also, by the same author, a comprehensive treatise, *Statique Expérimentale et Théoretique des Liquides*, Paris, 1873. More recent accounts of research on soap films are given in Courant and Robbins, *op. cit.*, pp. 385–397; R. Courant, "Soap Film Experiments with Minimal Surfaces." *American Mathematical Monthly*, XLVII (1940), pp. 167–174; and in D'Arcy Wentworth Thompson's classic, *On Growth and Form*, Second Edition, Cambridge, 1952.

matical imagination, aid in the formulation of concepts and shape the direction and emphasis of mathematical studies. One of the most remarkable features of the relationship is the successful use of physical models and experiments to solve problems arising in mathematics. In some cases a physical experiment is the only means of determining whether a solution to a specific problem exists; once the existence of a solution has been demonstrated, it may then be possible to complete the mathematical analysis, even to move beyond the conclusions furnished by the model—a sort of boot-strap procedure. It is interesting to point out that what counts in this action and reaction is as much the "physical way of thinking," the turning over in imagination of physical events, as the actual doing of the experiment. Thus, in the nineteenth century, "many of the fundamental theorems of function theory were discovered by Riemann [merely] by thinking of simple experiments concerning the flow of electricity in thin sheets"—without even approaching the laboratory.[11] The celebrated minimum problem associated with Plateau's name [12]—in its simplest version: to find the surface of smallest area with a given boundary—is connected with the solution of a system of partial differential equations. As described in the selection from Courant and Robbins, Plateau found a physical solution for "very general contours" by dipping wire frames of various shapes into liquids of low surface tension.[13] A film immediately spans the frame with a surface of least area—a fact you can discover for yourself by repeating Plateau's experiments with simple home-made equipment. The general mathematical solution of Plateau's problem was somewhat harder to derive. The human brain being less agile than a soap film, the problem was not solved until 1931.

[11] Courant and Robbins, *op. cit.*, pp. 385–386. The discussion above is largely based on the excellent treatment in the Courant book.

[12] The problem was actually first proposed by the famous French mathematician J. L. Lagrange (1736–1813) and can, if you like, be traced back to Dido.

[13] For biographical data on Courant and Robbins see p. 571.

For up and down and round, says he,
Goes all appointed things,
And losses on the roundabouts
Means profits on the swings.

—PATRICK R. CHALMERS

8 What Is Calculus of Variations and What Are Its Applications?

By KARL MENGER

THE calculus of variations belongs to those parts of mathematics whose details it is difficult to explain to a non-mathematician. It is possible, however, to explain its main problems and to sketch its principal methods for everybody.

The first human being to solve a problem of calculus of variations seems to have been Queen Dido of Carthage. When she was promised as much land as might lie within the boundaries of a bull's hide, she cut the hide into many thin strips, put them together into one long strip, the ends of which she united, and then she tried to secure as extensive a territory as possible within this boundary. History does not describe the form of the territory she chose, but if she was a good mathematician she covered the territory in the form of a circle, for to-day we know: Of all surfaces bounded by curves of a given length, the circle is the one of largest area. The branch of mathematics which establishes a rigorous proof of this statement is the calculus of variations.

Newton was the first mathematician to publish a result in this field. If a body moves in the air, it meets with a certain resistance, which depends on the shape of the body. The problem Newton studied was, what shape of body would guarantee the least possible resistance? Applications of this problem are obvious. The rifle bullet is designed in such shape as to meet with a minimum resistance in the air. Newton published a correct answer to a special case of this problem, namely, that the surface of the solid considered is obtained by revolving a curve around an axis. But he did not give the proof or the calculations that had led him to the answer. So Newton's solution had no great effect on the development of mathematics.

A new branch of mathematics started with another problem formulated and studied by the brothers Bernoulli in the seventeenth century. If a small body moves under the influence of gravity along a given curve from one point to another, then the time required naturally depends on the form of the curve. Whether the body moves along a straight line (on an inclined plane) or along a circle makes a difference. Bernoulli's question was: which path takes the shortest time? One might think that the motion along the straight line is the quickest, but already Galileo had noticed that the time required along some curves is less than along a straight line. The brothers Bernoulli determined the form of the curve which takes the shortest possible time. It is a curve which was already well known in geometry for other interesting properties and had been called cycloid.

What is common to all these problems is this: A number is associated with each curve of a certain family of curves. In the first example (that of Queen Dido) the family consists of all closed curves with a given length, and the associated number is the area of the inclosed surface; in the second example (that of Newton) the number is the resistance which a body somehow associated with the curve meets in the air; in the third example (that of the brothers Bernoulli) the family of curves consists of all curves joining two given points, and the number associated with each curve is the time it takes a body to fall along this curve. The problem consists in finding the curve for which the associated number attains a maximum or a minimum—this is the largest or the smallest possible value; in Dido's example, the maximum area; in Newton's example, the minimum resistance; in Bernoulli's example, the shortest time.

Some problems concerning maxima and minima are studied in differential calculus, taught in college. They may be formulated in the following way: Given a single curve, where is its lowest and where is the highest point? or given a single surface, where are its peaks and where are its pits? With each point of the curve or the surface, there is associated a certain number, namely, the height of the point above a horizontal axis or a horizontal plane. We are looking for those points at which this height is greatest or least. In differential calculus we deal thus with maxima and minima of so-called functions of points, *i.e.*, of numbers associated with points; in calculus of variations, however, with maxima and minima of so-called functions of curves, that is, of numbers associated with curves or of numbers associated with still more complicated geometric entities, like surfaces.

A famous question concerning surfaces is the following problem, the so-called problem of Plateau: if a closed curve in our three-dimensional space is given, we can span into it many different surfaces, all of them bounded by the given curve, *e.g.*, if the given curve is a circle we can span into it a plane circular area or a hemisphere or other surfaces bounded by the circle. Each of these surfaces has an area. Which of all these surfaces has the smallest area? If the given curve lies in a plane, like a circle, then, obviously, the plane surface inscribed has the minimum area. If the given curve, however, does not lie in the plane, like a knotted curve in the three-dimensional space, then the problem of finding the surface of minimal area bounded by the curve is very complicated. The question, which was solved some years ago by T. Radó (Ohio State University) and in a still more general way by J. Douglas (Massachusetts Institute of Technology), has applications to physics, for if the curve is made of a thin wire and we try to span into it a thin soap film, then this film will assume just the form of the surface of minimum area.

We frequently find that nature acts in such a way as to minimize certain magnitudes. The soap film will take the shape of a surface of smallest area. Light always follows the shortest path, that is, the straight line, and, even when reflected or broken, follows a path which takes a minimum of time. In mechanical systems we find that the movements actually take place in a form which requires less effort in a certain sense than any other possible movement would use. There was a period, about 150 years ago, when physicists believed that the whole of physics might be deduced from certain minimizing principles, subject to calculus of variations, and these principles were interpreted as tendencies—so to say, economical tendencies of nature. Nature seems to follow the tendency of economizing certain magnitudes, of obtaining maximum effects with given means, or to spend minimal means for given effects.

In this century Einstein's general theory of relativity has as one of its basic hypotheses such a minimal principle: that in our space-time world, however complicated its geometry be, light rays and bodies upon which no force acts move along shortest lines.

If we speak of tendencies in nature or of economic principles of nature, then we do so in analogy to our human tendencies and economic principles. A producer most often will adopt a way of production which will require a minimum of cost, compared with other ways of equal results; or which, compared with other methods of equal cost, will promise a maximum return. It is obvious that for this reason the mathematical theory of economics is to a large extent application of calculus of variations. Such applications have been considered by G. C. Evans (University of California) and in particular by Charles F. Roos (New York City). A simple but interesting example, due to the economist H. Hotelling (Columbia University), is to find the most economic way of production in a mine. We may start with a great output and decrease the output later or we may increase the output in time or we may produce with a constant rate of output. Each way of production can be represented by a curve. If we have conjectures concerning the development of the price of the produced metal, then we may associate a number with each of these curves—the possible profit. The problem is to find the way of production which will probably yield the greatest profit.

In the mathematical theory of the maximum and minimum problems in calculus of variations, different methods are employed. The old classical method consists in finding criteria as to whether or not for a given curve the corresponding number assumes a maximum or minimum. In order to find such criteria a considered curve is a little varied, and it is from this method that the name "calculus of variations" for the whole branch of mathematics is derived. The first result of this method, which to-day is represented by G. A. Bliss (University of Chicago) and his school, was the equation of Euler-Lagrange, which states: A curve which minimizes or maximizes the corresponding number must in each of its points have a certain curvature which can be determined for each problem.

Another method consists in finding out quite in general whether or not a given problem is soluble at all. For example, we consider the two following extremely simple problems: two given points may be joined by all possible curves; which of them has the shortest length, and which of them has the greatest length? The first problem is soluble: The straight line segment joining the two points is the shortest line joining them. The second problem is not soluble: There is no longest curve joining two given points, for no matter how long a curve joining them may be, there is always one which is still longer. The length is a number associated with each curve which for no curve assumes a finite maximum.

This second method of calculus of variations was initiated by the German mathematician Hilbert at the beginning of the century. The Italian mathematician Tonelli found out twenty years ago that the deeper reason for the solubility of the minimum problem concerning the length, that is, for the existence of a shortest line between every two points, is the following property of the length: A curve between two fixed points being given, there are always other curves as near as you please to it, and yet much longer than the given curve (*e.g.*, some zigzag lines near the given curve). But there is no curve very near to the given curve and joining the same two points, which is much shorter than the given curve. This property of the length is called the semi-continuity of the length. Contributions to this Hilbert-Tonelli method are due to E. J. McShane (University of Virginia), L. M. Graves (University of Chicago) and to the author.

Another way of calculus of variations was started in this country. G. D. Birkhoff (Harvard University) was the first to consider so-called minimax problems dealing with "stationary" curves which are minimizing with respect to certain neighboring curves and at the same time maximizing with respect to other curves. While the minimum and maximum problems of calculus of variations correspond to the problem in the ordinary calculus of finding peaks and pits of a surface, the minimax problems correspond to the problem of finding the saddle points of the surface (the passes of a mountain). The simplest example of such a stationary curve is obtained in the following way: if we consider two points of the equator of the earth, then their shortest connection on the surface of the earth is the minor arc of the equator between them. There is, as we have seen, no longest curve joining the two points. But there is one curve on the surface of the earth which, though it is neither the shortest nor the longest one, plays a special rôle in some respects, namely, the major arc of the equator between the two points.

One of the greatest advances of calculus of variations in recent times has been the development of a complete and systematic theory of stationary curves due to Marston Morse (Institute of Advanced Study). The most simple example of this theory, which calculates the number of minimizing and maximizing curves as well as of stationary curves, is the following "geographical" theorem quoted by Morse: If we add the number of peaks and the number of pits on the surface of the earth, and subtract the number of passes, then the result will be the number 2, whatever the shape of the mountains may be (highlands excluded).

There are many technical details of calculus of variations which are hardly available to a non-mathematician. They are the type of theory which frequently leads to the belief that mathematical theories are remote from the urgent problems of the world and useless. Real mathematicians do not worry too much about these reproaches which are engendered by a lack of knowledge of the history of science. Mathematicians study their problems on account of their intrinsic interest, and develop their theories on account of their beauty. History shows that some of these mathematical theories which were developed without any chance of immediate use later on found very important applications. Certainly this is true in the case of calculus of variations: If the cars, the locomotives, the planes, etc., produced to-day are different in form from what they used to be fifteen years ago, then a good deal of this change is due to calculus of variations. For we use streamline form in order to decrease to the minimum possible the resistance of the air in driving. It is through physics that we learn the actual laws of this resistance. But if we wish to discover the form which guarantees the least resistance, then we need calculus of variations.

Lex perpetua naturae est ut agat minimo labore, mediis et modis simplicissimis, facillimis, certis et tutis: evitando, quam maxime fieri potest, incommoditates et prolixitates. —GIOVANNI BORELLI (*1608–1679*)

9 The Soap-bubble

By C. VERNON BOYS

IT can only be our familiarity with soap-bubbles from our earliest recollections, causing us to accept their existence as a matter of course, that prevents most of us from being seriously puzzled as to why they can be blown at all. And yet it is far more difficult to realize that such things ought to be possible than it is to understand anything that I have put before you as to their actions or their form. In the first place, when people realize that the surface of a liquid is tense, that it acts like a stretched skin, they may naturally think that a soap-bubble can be blown because in the case of soap-solution the "skin" is very strong. Now the fact is just the opposite. Pure water, with which a bubble cannot be blown in air and which will not even froth, has a "skin" or surface tension three times as strong as soap-solution, as tested in the usual way, *e.g.*, by the rise in a capillary tube. Even with a minute amount of soap present the surface tension falls off from about 3¼ grains to the linear inch to 1¼ grains, as calculated from experiments with bubbles by Plateau. The liquid rises but little more than one-third of the height in a capillary tube. The soap-film has two surfaces each with a strength of 1¼ grains to the inch and so pulls with a strength of 2½ grains to the inch. Many liquids will froth that will not blow bubbles. Lord Rayleigh has shown that a pure liquid will not froth though a mixture of two pure liquids, *e.g.*, water and alcohol, will. Whatever the property is which enables a liquid to froth must be well developed for it to allow bubbles to be blown. I have repeatedly spoken of the tension of a soap-film as if it were constant, and so it is very nearly, and yet, as Prof. Willard Gibbs pointed out, it cannot be exactly so. For, consider any large bubble or, for convenience, a plane vertical film stretched in a wire ring. If the tension of two grains and a half to the inch were really identically the same in all parts, the middle parts of the film being pulled upwards and downwards to an equal extent by the rest of the film above and below it would in effect not be pulled by them at all and like other unsupported things would fall, starting like a stone with the acceleration due to gravity. Now the middle part of such a film does nothing of the kind. It appears to be at rest, and if there is any downward movement it is too slow to be noticed. The upper part therefore of the film must be more tightly stretched than the lower part, the difference being the weight of the intervening film. If the ring is turned over to invert the film then the conditions are reversed, and yet the middle part does not fall. The bubble therefore has the remarkable property within small limits of adjusting its tension to the load. Willard Gibbs put forward the view that this was due to the surface material not being identical with the liquid within the thickness of the film. That the surface was contaminated by material which lowered its surface tension and which by stretching of the film became diluted, making the film stronger, or by contraction became concentrated, making the film weaker. His own words are so apt and so much better than mine that I shall quote from his *Thermodynamics*, p. 313: "For, in a thick film (as contrasted with a black film), the increase of tension with the extension, which is necessary for its stability with respect to extension, is connected with an excess of the soap (or of some of its components) at the surface as compared with the interior of the film."

This is analogous to the effect of oil on water. Lord Rayleigh has by a beautiful experiment supported the contamination theory, for he measured the surface tension of the surface of a soap-solution within the first hundredth of a second of its existence. He then found it to be the same

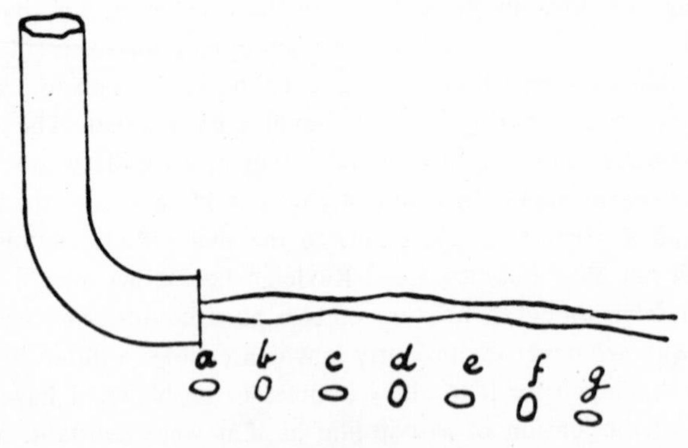

FIGURE 1

as that of water, for the surface contamination had no time to form. He allowed the liquid to issue from a small elliptical hole in a thin plate covering the end of a tube which came from a reservoir of the solution. When liquid issues from such a hole as at *a*, Figure 1, the cross section of the stream being elliptical, as shown below, tends in virtue of the surface tension to become circular, but when it gets circular the movement already set up in the section cannot be suddenly arrested and so the liquid continues its movement until it is elliptical in the other direction as at *b*, and this process is continued at a definite rate depending upon the surface tension and density of the liquid and size of the jet. At the same time the liquid is issuing at a definite rate depending on the depth of the orifice below the free surface, and so when the conditions are well chosen the liquid travels from *a* to *c* while the ellipse goes through its complete evolution, and this is repeated several times. Now if the surface tension were less the evolution would take longer and the distance between the nodes *a c e g* would be greater. With the same liquid head the distance between the nodes is the same with pure water and with soap-solution, showing that their surface tensions at first are the same, but with alcohol, which has its own surface tension from the beginning, the distance between the nodes is greater, as the surface tension is lower in a higher proportion than that in which the density is lower. Professor Donnan has quite recently shown by direct experiment of surpassing delicacy, that there is a surface concentration of the kind and amount required by Gibbs's theory.

The following experiment also indicates the existence of a surface concentration. If a soap-bubble is blown on a horizontal ring so that the diameter of the ring is very little less than that of the bubble, and the wetted stopper from a bottle of ammonia is brought close to the upper side of the bubble, it will immediately shrink away from the stopper and slip through the ring as though annoyed by the smell of the ammonia. Or, if below, it will retire to the upper side of the ring if the stopper is held below it. What really happens is that the ammonia combines with some of the constituents of the soap which are concentrated on the surface, and so raises the tension of the film on one side of the ring; it therefore contracts and blows out the film on the other side which has not yet been influenced by the ammonia. That part of the film influenced by the ammonia also becomes thicker and the rest thinner, as shown by the colours, which are then far more brilliant and variegated.

Going back now to the soap-film we see then that whatever its shape the upper parts are somewhat more tightly stretched than the lower parts, and in the case of a vertical film the difference is equal to that which will support the intervening film. There is, however, a limit beyond which this process will not go; there is a limit to the size of a soap-bubble. I do not know what this is. I have blown spherical bubbles up to 2½ feet in diameter, and others no doubt have blown bubbles larger still. I have also taken a piece of thin string ten feet long and tied it into a loop after wetting it with soap solution and letting it untwist. Holding a finger of each hand in the loop and immersing it in soap-solution I have drawn it out and pulled it tight, forming a film in this way five feet long. On holding the loop vertical the film remained unbroken, showing that five feet is less than the limit with even a moderately thick bubble. With a thin bubble the limit should be greater still. Judging by the colour and using certain other information one concludes that the average thickness of the film must have been about thirty millionths of an inch and its weight about $\frac{8}{1000}$ of a grain per square inch. Taking a film an inch wide, the five feet or sixty inches would weigh close upon half a grain, that is about one-fifth of the total load that the film can carry, showing at least 20% capacity in the soap-film for adjusting its strength to necessity, which is far more than could have been expected.

I have also found that with the application of increased forces the bubble rapidly thins to a straw colour or white, so that the 20% increase of load is not exceeded, but a film of this colour might be thirty-three feet high, or a black film ten times as much.

The feeble tension of $2\frac{1}{2}$ grains to the inch in a soap-film is quite enough in the case of the five-foot loop to make it require some exertion to keep it pulled so that the two threads are not much nearer in the middle than at the ends. In fact, this experiment provides the means by which the feeble tension of $2\frac{1}{2}$ grains to the inch may be measured by means of a seven-pound weight. If for instance the threads are seventy inches long, equidistant at their ends, and are $\frac{1}{16}$ inch nearer in the middle than at their ends when placed horizontally and stretched with seven-pound weights, then the tension of the film works out at $2\frac{1}{2}$ grains to the inch exactly. This is obtained as follows: The diameter of the circle of which the curved thread is a part is equal to half the length of the thread multiplied by itself and divided by the deviation of its middle point. The diameter of the circle then is $35 \times 35 \times 32$ inches. The tension of the thread, made equal to 7 lbs. or 7×7000 grains, is equal to the tension of the film in grains per inch multiplied by half the diameter of the circle of curvature of the thread. In other words the tension of the film in grains per inch is equal to the tension of the thread divided by half the diameter of the circle of curvature of the thread. The film tension then is equal to $\dfrac{7 \times 7000}{35 \times 35 \times 16}$ grains per inch. The fraction cancels at once to $2\frac{1}{2}$ grains per inch. As the upper parts of a vertical film are stretched more tightly than the lower parts, the pair of threads will be drawn together in the middle, rather more than $\frac{1}{16}$ inch, as the pair of threads are gradually tilted from the horizontal towards a vertical position, and then as the film drains into the threads and become lighter the threads will separate slightly.

Large bubbles are short-lived not only because, if it were a matter of chance, a bubble a foot in diameter with a surface thirty-six times as great as that of one two inches in diameter would be thirty-six times as likely to break if all the film were equally tender, but because the upper parts being in a state of greater tension have less margin of safety than the lower parts. These large bubbles however by no means necessarily break at the top. When a large, free-floating bubble is seen in bright sunlight on a dark background it is almost possible to follow the process of breaking. What is really seen, however, is a shower of spray moving in the opposite direction to that in which the hole is first made, while the air which cannot be seen blows out in the opposite direction to that of the spray. By the time it is done, and it does not take long, the momentum given to the moving drops in one direction and to the moving air in the opposite direction are equal to one another, but as the air weighs far more than the water the spray is thrown the more rapidly.

The breaking of a bubble is itself an interesting study. Duprée long ago showed that the whole of the work done in extending a bubble is to be found in the velocity given to the spray as it breaks, and thence he deduced a speed of breaking of 105 feet a second or seventy-two miles an hour for a thin bubble, and Lord Rayleigh for a thicker bubble found a speed of forty-eight feet a second or thirty-three miles an hour. The speed of breaking of a soap-bubble is curious in that it does not get up speed and keep going faster all the time, as most mechanical things do, it starts full speed at once, and its speed only changes in accordance with its thickness, the thick parts breaking more slowly. It may be worth while to show how the speed is arrived at theoretically. Take in imagination a film between two parallel and wetted wires an inch apart and extend it by drawing a wetted edge of card, india-rubber or celluloid along the wires. Then the work done in extending it for a foot, say, not counting of course the friction of the moving edge, is with a tension of 2½ grains to the inch equal to 2½ × 12 inch-grains, or if it is extended a yard to 2½ × 36 inch-grains. Let us pull it out to such a length that the weight of the film drawn out is itself equal to 2½ grains, that is to the weight its own tension will just carry. By way of example let us consider a film not very thick or very thin, but of the well-defined apple-green colour. This, it can be calculated, is just under twenty millionths of an inch thick, and it weighs $\frac{5}{1000}$ or $\frac{1}{200}$ of a grain to the square inch. The length of this film one inch wide that will weigh 2½ grains is therefore 500 inches or forty-two feet nearly. The work done in stretching a film of this area is 2½ × 500 inch-grains. The work contained in the flying spray must also be 2½ × 500 inch-grains. Now the work contained in a thing moving at any speed is the same as that which would be needed to lift it to such a height that it would if falling without obstruction acquire that speed. The work done in lifting the 2½ grains through 500 inches is exactly the same as that done in pulling out the film 500 inches horizontally against its own tension, as the force and the distance are the same in both cases. The velocity of the spray, and therefore of the edge from which it is scattered is the same as that of a stone falling through a distance equal to the length of the film, the weight of which is equal to the tension at its end. Completing the figures a stone in this latitude falling forty-two feet acquires a velocity of fifty-two feet a second, which therefore is the speed of the breaking edge of the film. I conclude therefore, as the speed found in this example is intermediate between those found by Duprée and Lord Rayleigh, that I have chosen a film intermediate in thickness between those chosen by these philosophers. I would only add that a bubble has to be reduced to one quarter of its thickness to make it break twice as fast, then the corresponding length will be four times as great, and it requires a fall from four times the height to acquire twice the speed. A black film is about one thirty-sixth of the thickness of the apple-green film, it should therefore break six times as fast or 312 feet a second, or 212 miles an hour. The extra black film of half the thickness should break at the enormous speed of 300 miles an hour. These speeds would hardly be realized in practice as the viscosity of the liquid would reduce them.

Lord Rayleigh photographed the breaking soap-film by placing a ring on which it was stretched in an inclined position, and then dropping through it a shot wet with alcohol, and about a thousandth of a second later photographing it. For this purpose he arranged two electromagnets, one to drop the wet shot and the other to drop another shot at the same instant. The second shot was allowed a slightly longer fall, so as to take a thousandth of a second, or whatever interval he wanted, longer than the first; it then by passing between two knobs in the circuit of a charged Leyden jar let this off, and the electric spark provided the light and the sufficiently short exposure to give a good and sharp photograph. The sharp retreating edge is seen with minute droplets either just detached or leaving the film. Following Lord Rayleigh I photographed a breaking film by piercing it with a minute electric spark between two needles, one on either side of the film, and by means of a piano-wire spring like a mousetrap determining the existence of this spark, and then a ten-thousandth of a second or more later letting off the spark by the light of which it was taken. My electrical arrangements were akin to those by which I photographed bullets in their flight in one thirteen-millionth of a second, but my optical arrangements were similar to Lord Rayleigh's. One photograph of a vertical film which had become a great deal thinner in its upper part, is interesting as whereas all the lower part has a circular outline, the upper part breaks into a bay showing the speed of breaking upwards in the thin film to be much greater. In all the photographs the needle spark appears by its own light, so the point at which the break first occurs is seen as well as the circular retreating edge.

* * * * *

COMPOSITE BUBBLES

A single bubble floating in the air is spherical, and as we have seen this form is assumed because of all shapes that exist this one has the smallest surface in relation to its content, that is, there is so much air within, and the elastic soap-film, trying to become as small as possible, moulds the air to this shape. If the bubble were of any other shape the film could become of less surface still by becoming spherical. When however the bubble is not single, say two have been blown in real contact with one another, again the bubbles must together take such a form that the total surface of the two spherical segments and of the part common to both, which I shall call the interface, is the smallest possible surface which will contain the two quantities of air and keep them separate. As the soap-bubble provides such a simple and pleasing way of demonstrating the solution of this problem, which is really a mathematical problem, it will be worth while to devote a little time to its consideration. Let us suppose that the two bubbles which are joined by an interface are not equal, and that

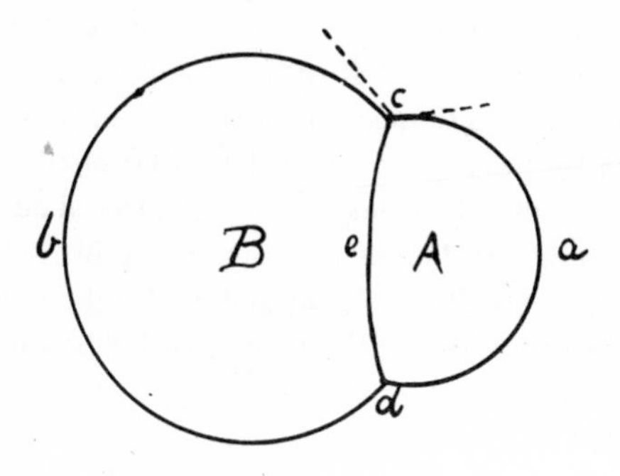

FIGURE 2

Figure 2 represents a section through the centres of both, *A* being the smaller and *B* the bigger bubble. In the first place we have seen that the pressure within a bubble is proportional to its curvature or to 1 divided by the radius of the bubble. The pressure in *A*, by which I mean the excess over atmospheric pressure, will therefore be greater than the pressure in *B* in the proportion in which the radius of *B* is greater than the radius of *A*, and the air can only be prevented from blowing through by the curvature of the interface. In fact this curvature balances the difference of pressure. Another way of saying the same thing is this: the curved and stretched film *dac* pushes the air in *A* to the left, and it takes the two less curved but equally stretched films *dbc* and *dec*, pushing to the right to balance the action of the more curved film *dac*. Or, most shortly of all, the curvature of *dac* is equal to the sum of the curvatures *dbc* and *dec*. Now consider the point *c* or *d* in the figure either of which represent a section of the circle where the two bubbles meet; at any point in this circle the three films meet and are all three pulling with the same force. They can only balance when the angles where they meet are equal or are each angles of 120°. Owing to the curvature of the lines, these angles do

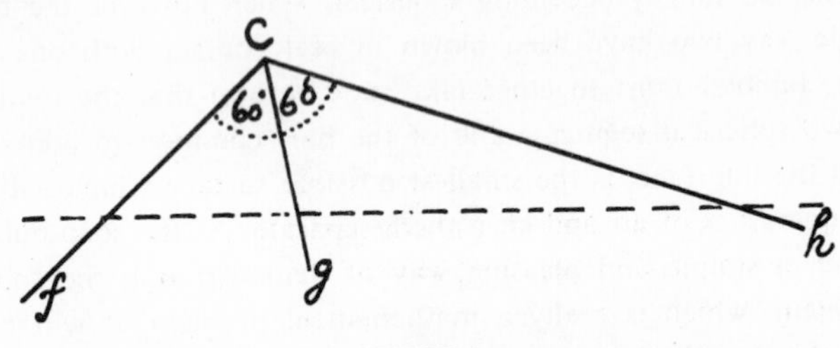

FIGURE 3

not look equal, but I have dotted in at *c* tangents to the three curves at the point *c*, and they clearly make equal angles with each other.

This equality of the angles is not an independent proposition to the last with regard to the curvatures; if either condition is fulfilled the other necessarily follows, as also does the one I opened with that the total surface must be the least possible. Plateau, the blind Belgian professor, discussed this, as he did everything that was known about the soap-bubble, in his book *Statique des Liquides*, published in Brussels, a book which is a worthy monument of the brilliant author. He there describes a simple geometrical construction by which any pair of bubbles and their interface may be drawn correctly.

From any point *C* draw three lines, *cf, cg, ch,* making angles of 60°, as shown in Figure 3. Then, on drawing any straight line across so as to cut the three lines such as the one shown dotted in the figure, the three points where it cuts the three lines will be the centres of three circles representing possible bubbles. The point where it cuts the middle line is the centre of the smaller bubble, and of the other two points, that which is nearer to *C* is the centre of the second bubble, and that which is further is the centre of the interface. Now, with the point of the compasses placed successively at each of these points, draw portions of circles passing through *C*, as shown in Figure 4, in which the construction lines of Figure 3 are shown dotted and the circular arcs are shown in full lines.

Having drawn a number of these on sheets of paper, with the curved lines very black so as to show well, lay a sheet of glass upon one, and having wetted the upper side with soap solution, blow a half bubble upon it, and then a second so as to join the first, and then with a very small pipe, or even a straw, with one end closed with sealing-wax opened afterwards

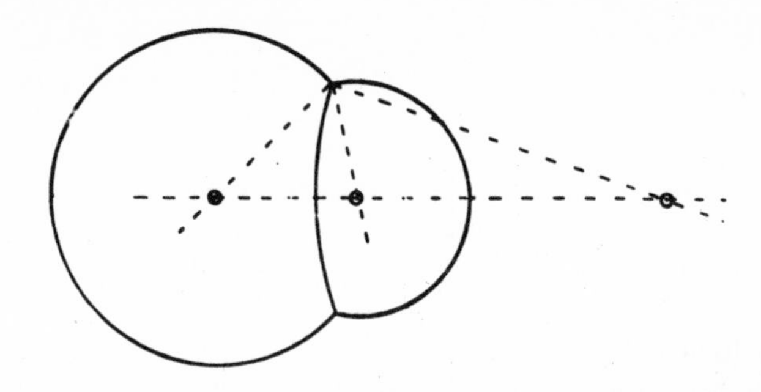

FIGURE 4

with a hot pin so as to allow air to pass very slowly, gently blow in or draw out air until the two bubbles are the same size as those in the drawing, moving the glass about so as to keep them over the figure. You will then find how the soap-bubble solves the problem automatically, and how the edges of the half bubbles exactly fit throughout their whole extent the drawings that you have made.

If the dotted line in Figure 3 cuts *cf* and *ch* at equal distances from *c*, then it will cut *cg* at half that distance from *c*, and we have the case of a bubble in contact with one of twice its diameter. In this case the interface has the same curvature as that of the larger bubble, but is reversed in direction, and each has half the curvature of the smaller bubble.

If the dotted line in Figure 3 cuts *cf* and *cg* at equal distances from *c*, it will be parallel to *ch* and will never cut it. The two bubbles will then be equal, and the interface will then have no curvature, or, in other words, it will be perfectly flat, and the line *cd*, Figure 2, which is its section, will be a straight line.

There are other cases where the same reciprocal laws apply as well as this one of the radii of curvature of joined soap bubbles. It may be written in a short form as follows: $\frac{1}{A} = \frac{1}{B} + \frac{1}{e}$, using the letters of Figure 2 to represent the lengths of the radii of the corresponding circles. For instance, a lens or mirror in optics has what is called a principal focus, *i.e.*, a distance, say A, from it at which the rays of the sun come to a focus and make it into a burning glass. If, instead of the sun, a candle flame is placed a little way off, at a distance, say B, greater than A, then the lens or mirror will produce an image of the flame at a distance, say e, such that $\frac{1}{A} = \frac{1}{B} + \frac{1}{e}$. Or, again, if the electrical resistance of a length of wire, say

A inches long, is so much, then the electrical resistance of two pieces of the same wire, B and e inches long, joined so that the current is divided between them, will be the same as that of A if $\frac{1}{A} = \frac{1}{B} + \frac{1}{e}$.

The soap-bubble then may be used to give a numerical solution of an optical and of an electrical problem.

Plateau gives one other geometrical illustration, the proof of which, however, is rather long and difficult, but which is so elegant that I cannot refrain from at least stating it. When three bubbles are in contact with one another, as shown in Figure 5, there are of course three interfaces meeting one another, as well as the three bubbles all at angles of 120°.

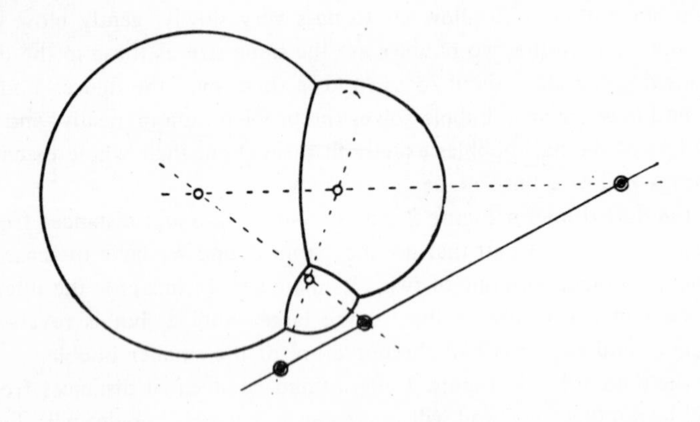

FIGURE 5

The centres of curvature also of the three bubbles and of the three interfaces, also necessarily lie in a plane, but what is not evident and yet is true is that the centres of curvature, marked by small double circles of the three interfaces, lie in a straight line. If any of you are adepts in geometry, whether Euclidean or analytical, this will be a nice problem for you to solve, as also that the surface of the three bubbles and of the three interfaces is the least possible that will confine and separate the three quantities of air. The proof that the three films drawn according to the construction of Figure 3 have the curvatures stated is much more easy, and I should recommend you to start on this first. If you want a clue, draw a line from the point where the dotted line cuts cg parallel to cf, and then consider what is before you.

I do not suppose there is anyone in this room who has not occasionally blown a common soap bubble, and while admiring the perfection of its form, and the marvelous brilliancy of its colours, wondered how it is that such a magnificent object can be so easily produced. I hope that none of you are yet tired of playing with bubbles, because, as I hope we shall see, there is more in a common bubble than those who have only played with them generally imagine. —SIR CHARLES VERNON BOYS (*Soap Bubbles*)

10 Plateau's Problem

By RICHARD COURANT
and HERBERT ROBBINS

EXPERIMENTAL SOLUTIONS OF MINIMUM PROBLEMS SOAP FILM EXPERIMENTS

INTRODUCTION

IT is usually very difficult, and sometimes impossible, to solve variational problems explicitly in terms of formulas or geometrical constructions involving known simple elements. Instead, one is often satisfied with merely proving the existence of a solution under certain conditions and afterwards investigating properties of the solution. In many cases, when such an existence proof turns out to be more or less difficult, it is stimulating to realize the mathematical conditions of the problem by corresponding physical devices, or rather, to consider the mathematical problem as an interpretation of a physical phenomenon. The existence of the physical phenomenon then represents the solution of the mathematical problem. Of course, this is only a plausibility consideration and not a mathematical proof, since the question still remains whether the mathematical interpretation of the physical event is adequate in a strict sense, or whether it gives only an inadequate image of physical reality. Sometimes such experiments, even if performed only in the imagination, are convincing even to mathematicians. In the nineteenth century many of the fundamental theorems of function theory were discovered by Riemann by thinking of simple experiments concerning the flow of electricity in thin metallic sheets.

In this section we wish to discuss, on the basis of experimental demonstrations, one of the deeper problems of the calculus of variations. This problem has been called Plateau's problem, because Plateau (1801–1883), a Belgian physicist, made interesting experiments on this subject. The problem itself is much older and goes back to the initial phases of the calculus of variations. In its simplest form it is the following: to find the surface of smallest area bounded by a given closed contour in space. We shall also discuss experiments connected with some related questions, and it will turn out that much light can thus be thrown on some of our previous results as well as on certain mathematical problems of a new type.

SOAP FILM EXPERIMENTS

Mathematically, Plateau's problem is connected with the solution of a "partial differential equation," or a system of such equations. Euler showed that all (non-plane) minimal surfaces must be saddle-shaped and that the mean curvature [1] at every point must be zero. The solution was shown to exist for many special cases during the last century, but the existence of the solution for the general case was proved only recently, by J. Douglas and by T. Radó.

Plateau's experiments immediately yield physical solutions for very general contours. If one dips any closed contour made of wire into a liquid of low surface tension and then withdraws it, a film in the form of a minimal surface of least area will span the contour. (We assume that we may neglect gravity and other forces which interfere with the tendency of the film to assume a position of stable equilibrium by attain-

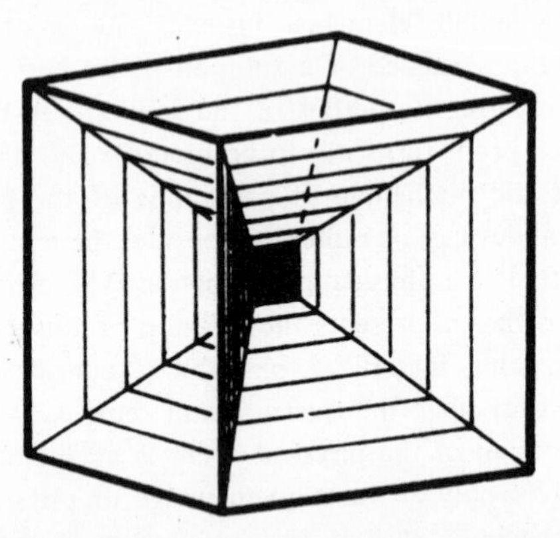

FIGURE 1—Cubic frame spanning a soap film system of 13 nearly plane surfaces.

ing the smallest possible area and thus the least possible value of the potential energy due to surface tension.) A good recipe for such a liquid is the following: Dissolve 10 grams of pure dry sodium oleate in 500 grams of distilled water, and mix 15 cubic units of the solution with 11 cubic units of glycerin. Films obtained with this solution and with frames

[1] The mean curvature of a surface at a point *P* is defined in the following way: Consider the perpendicular to the surface at *P*, and all planes containing it. These planes will intersect the surface in curves which in general have different curvatures at *P*. Now consider the curves of minimum and maximum curvature respectively. (In general, the planes containing these curves will be perpendicular to each other.) One-half the sum of these two curvatures is the mean curvature of the surface at *P*.

of brass wire are relatively stable. The frames should not exceed five or six inches in diameter.

With this method it is very easy to "solve" Plateau's problem simply by shaping the wire into the desired form. Beautiful models are obtained in polygonal wire frames formed by a sequence of edges of a regular polyhedron. In particular, it is interesting to dip the whole frame of a cube into such a solution. The result is first a system of different surfaces meeting each other at angles of 120° along lines of intersection. (If the cube is withdrawn carefully, there will be thirteen nearly plane surfaces.) Then we may pierce and destroy enough of these different surfaces so that only one surface bounded by a closed polygon remains. Several beautiful surfaces may be formed in this way. The same experiment can also be performed with a tetrahedron.

NEW EXPERIMENTS ON PLATEAU'S PROBLEM

The scope of soap film experiments with minimal surfaces is wider than these original demonstrations by Plateau. In recent years the problem of minimal surfaces has been studied when not only one but any number of contours is prescribed, and when, in addition, the topological structure of the surface is more complicated. For example, the surface might be one-sided or of genus different from zero. These more general problems produce an amazing variety of geometrical phenomena that can be exhib-

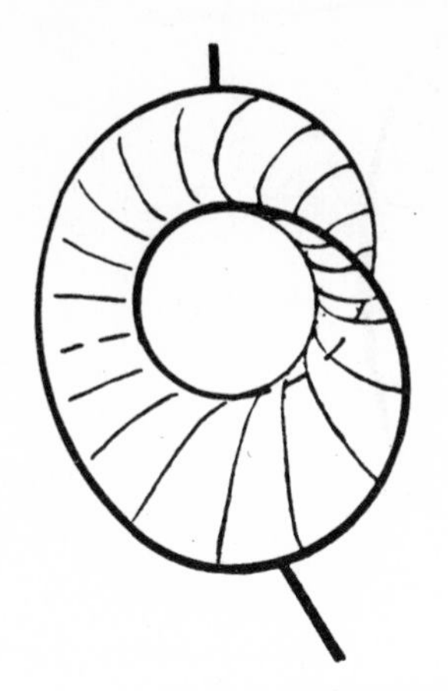

FIGURE 2—One-sided surface (Moebius strip).

FIGURE 3—Two-sided surface.

ited by soap film experiments. In this connection it is very useful to make the wire frames flexible, and to study the effect of deformations of the prescribed boundaries on the solution.

We shall describe several examples:

1) If the contour is a circle we obtain a plane circular disk. If we continuously deform the boundary circle we might expect that the minimal surface would always retain the topological character of a disk. This is not the case. If the boundary is deformed into the shape indicated by Figure 2, we obtain a minimal surface that is no longer simply connected, like the disk, but is a one-sided Moebius strip. Conversely, we might start with this frame and with a soap film in the shape of a Moebius strip. We may deform the wire frame by pulling handles soldered to it (Figure 2). In this process we shall reach a moment when suddenly the topological character of the film changes, so that the surface is again of the type of a simply connected disk (Figure 3). Reversing the deformation we again obtain a Moebius strip. In this alternating deformation process the mutation of the simply connected surface into the Moebius strip takes place at a later stage. This shows that there must be a range of shapes of the contour for which both the Moebius strip and the simply connected surface are stable, i.e., furnish relative minima. But when the Moebius strip has a much smaller area than the other surface, the latter is too unstable to be formed.

2) We may span a minimal surface of revolution between two circles. After the withdrawal of the wire frames from the solution we find, not one simple surface, but a structure of three surfaces meeting at angles of 120°,

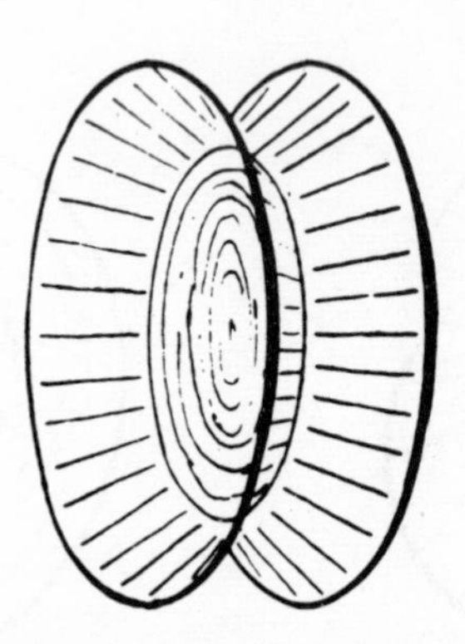

FIGURE 4—System of three surfaces.

one of which is a simple circular disk parallel to the prescribed boundary circles (Figure 4). By destroying this intermediate surface the classical catenoid is produced (the catenoid is the surface obtained by revolving a catenary about a line perpendicular to its axis of symmetry). If the two boundary circles are pulled apart, there is a moment when the doubly connected minimal surface (the catenoid) becomes unstable. At this moment the catenoid jumps discontinuously into two separated disks. This process is, of course, not reversible.

3) Another significant example is provided by the frame of Figures 5–7 in which can be spanned three different minimal surfaces. Each is bounded by the same simple closed curve; one (Figure 5) has the genus 1, while the other two are simply connected, and in a way symmetrical to each other. The latter have the same area if the contour is completely symmetrical. But if this is not the case then only one gives the absolute minimum of the area while the other will give a relative minimum, provided that the minimum is sought among simply connected surfaces. The possibility of the solution of genus 1 depends on the fact that by admitting surfaces of genus 1 one may obtain a smaller area than by requiring that the surface be simply connected. By deforming the frame we must, if the deformation is radical enough, come to a point where this is no longer true. At that moment the surface of genus 1 becomes more and more unstable and suddenly jumps discontinuously into the simply connected stable solution represented by Figure 6 or 7. If we start with one of these simply connected solutions, such as Figure 7, we may deform it in such a way that the other simply connected solution of Figure 6 becomes much more stable. The consequence is that at a certain moment a discontinuous transition from one to the other will take place. By slowly reversing the

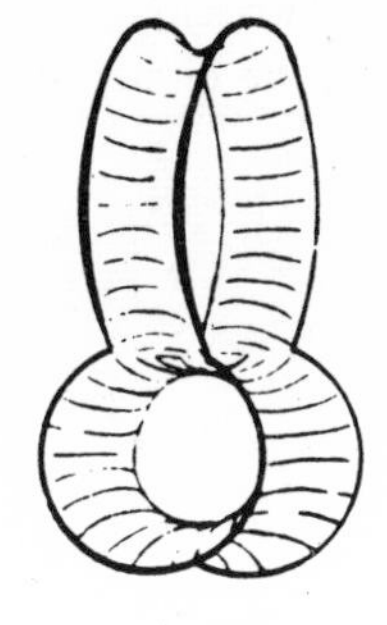

FIGURE 5

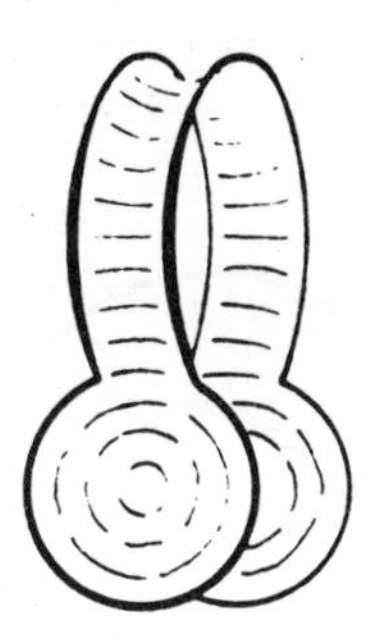

FIGURE 6

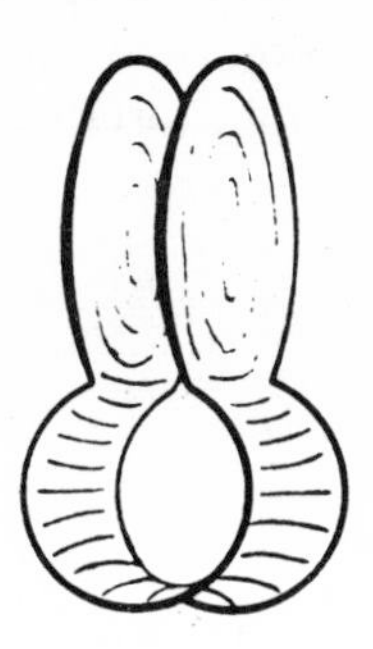

FIGURE 7

Frame spanning three different surfaces of genus 0 and 1.

deformation, we return to the initial position of the frame, but now with the other solution in it. We can repeat the process in the opposite direction, and in this way swing back and forth by discontinuous transitions between the two types. By careful handling, one may also transform discontinuously either one of the simply connected solutions into that of genus 1. For this purpose we have to bring the disk-like parts very close to each other, so that the surface of genus 1 becomes markedly more stable. Sometimes in this process intermediate pieces of film appear first and have to be destroyed before the surface of genus 1 is obtained.

This example shows not only the possibility of different solutions of the same topological type, but also of another and different type in one and

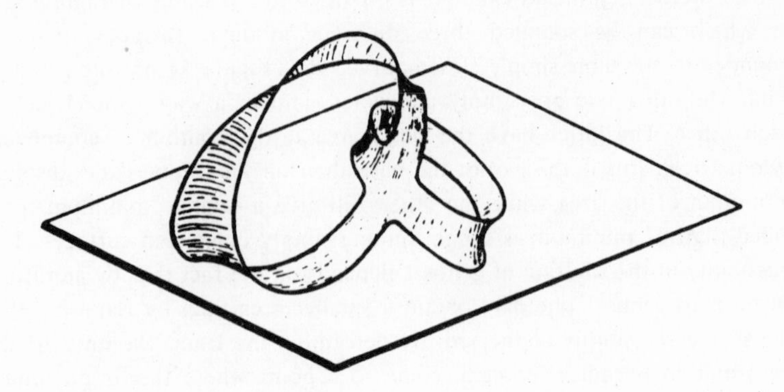

FIGURE 8—One-sided minimal surface of higher topological structure in a single contour.

the same frame; moreover, it again illustrates the possibility of discontinuous transitions from one solution to another while the conditions of the problem are changed continuously. It is easy to construct more complicated models of the same sort and to study their behavior experimentally.

An interesting phenomenon is the appearance of minimal surfaces bounded by two or more interlocked closed curves. For two circles we obtain the surface shown in Figure 9. If, in this example, the circles are

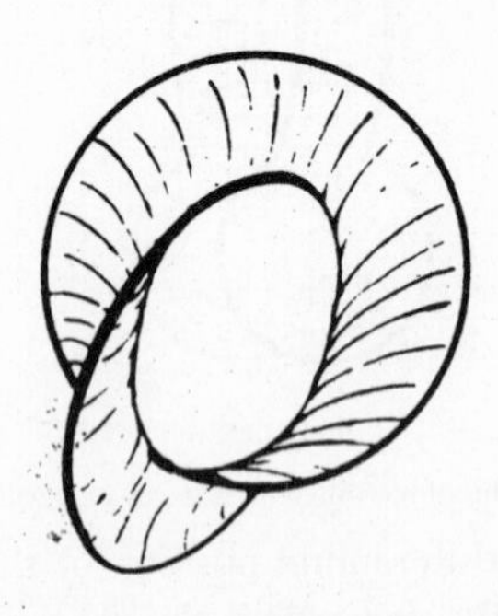

FIGURE 9—Interlocked curves.

perpendicular to each other and the line of intersection of their planes is a diameter of both circles, there will be two symmetrically opposite forms of this surface with equal area. If the circles are now moved slightly with respect to each other, the form will be altered continuously, although for each position only one form is an absolute minimum, and the other one a relative minimum. If the circles are moved so that the relative minimum is formed, it will jump over into the absolute minimum at some point. Here both of the possible minimal surfaces have the same topological character, as do the surfaces of Figures 6 and 7, one of which can be made to jump into the other by a slight deformation of the frame.

EXPERIMENTAL SOLUTIONS OF OTHER MATHEMATICAL PROBLEMS

Owing to the action of surface tension, a film of liquid is in stable equilibrium only if its area is a minimum. This is an inexhaustible source of mathematically significant experiments. If parts of the boundary of a film are left free to move on given surfaces such as planes, then on these boundaries the film will be perpendicular to the prescribed surface.

We can use this fact for striking demonstrations of Steiner's problem and its generalizations. Two parallel glass or transparent plastic plates are joined by three or more perpendiculars bars. If we immerse this object in a soap solution and withdraw it, the film forms a system of vertical planes between the plates and joining the fixed bars. The projection appearing on the glass plates is the solution of the problem of finding the shortest straight line connection between a set of fixed points.

If the plates are not parallel, the bars not perpendicular to them, or the

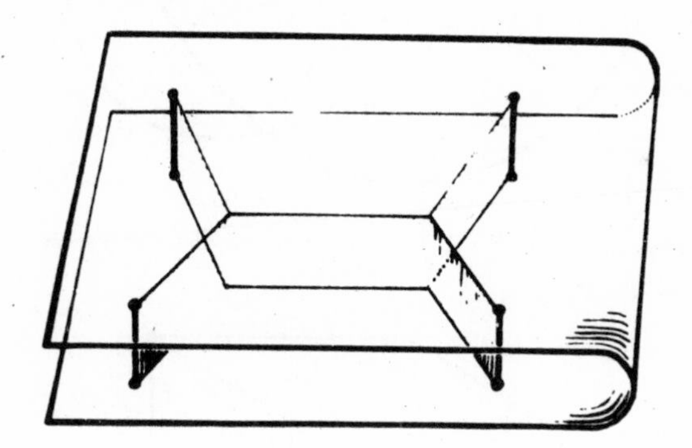

FIGURE 10—Demonstration of the shortest connection between 4 points.

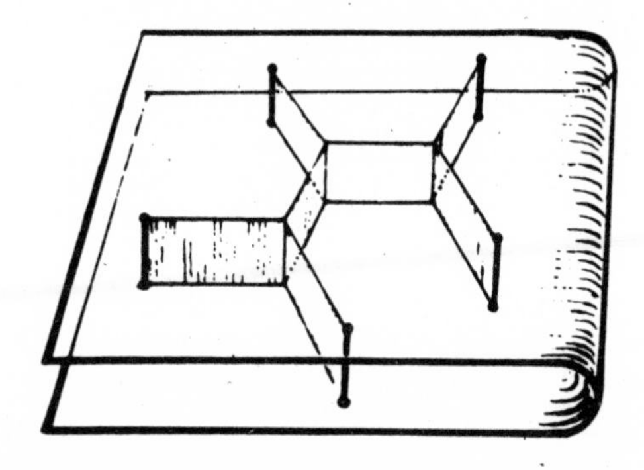

FIGURE 11—Shortest connection between 5 points.

plates curved, then the curves formed by the film on the plates will not be straight, but will illustrate new variational problems.

The appearance of lines where three sheets of a minimal surface meet at angles of 120° may be regarded as the generalization to more dimensions of the phenomena connected with Steiner's problem. This becomes clear e.g. if we join two points A, B in space by three curves, and study the corresponding stable system of soap films. As the simplest case we take for one curve the straight segment AB, and for the others two

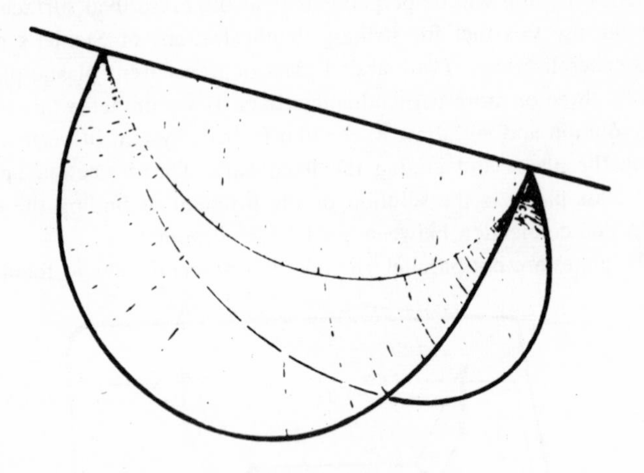

FIGURE 12—Three surfaces meeting at 120° spanned between three wires joining two points.

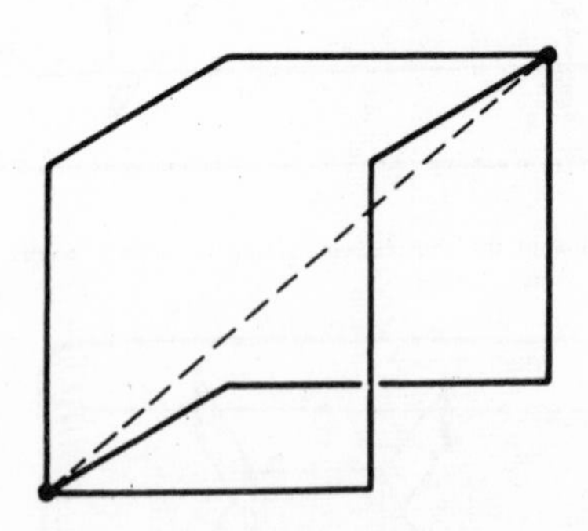

FIGURE 13—Three broken lines joining two points.

congruent circular arcs. The result is shown in Figure 12. If the planes of the arcs form an angle of less than 120°, we obtain three surfaces meeting at angles of 120°; if we turn the two arcs, increasing the included angle, the solution changes continuously into two plane circular segments.

Now let us join A and B by three more complicated curves. As an example we may take three broken lines each consisting of three edges of the same cube that join two diagonally opposite vertices: we obtain three congruent surfaces meeting in the diagonal of the cube. (We obtain this system of surfaces from that depicted in Figure 1 by destroying the films adjacent to three properly selected edges.) If we make the three broken lines joining A and B movable, we can see the line of threefold intersection become curved. The angles of 120° will be preserved (Figure 13).

All the phenomena where three minimal surfaces meet in certain lines are fundamentally of a similar nature. They are generalizations of the plane problem of joining n points by the shortest system of lines.

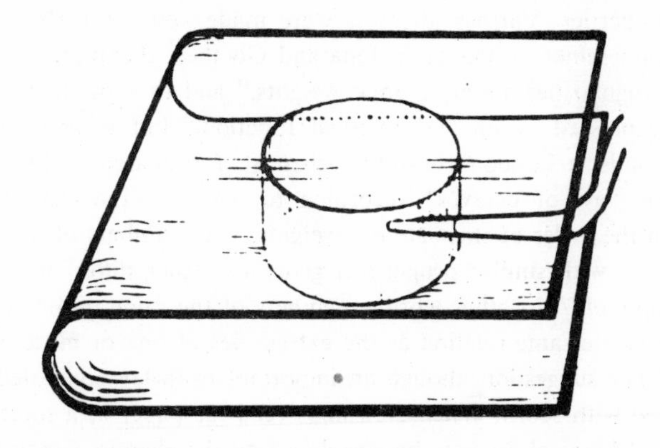

FIGURE 14—Demonstration that the circle has least perimeter for a given area.

Finally, a word about soap bubbles. The spherical soap bubble shows that among all closed surfaces including a given volume (defined by the amount of air inside), the sphere has the least area. If we consider soap bubbles of given volume which tend to contract to a minimum area but which are restricted by certain conditions, then the resulting surfaces will be not spheres, but surfaces of constant mean curvature, of which spheres and circular cylinders are special examples.

For example, we blow a soap bubble between two parallel glass plates which have previously been wetted by the soap solution. When the bubble touches one plate, it suddenly assumes the shape of a hemisphere; as soon as it also touches the other plate, it jumps into the shape of a circular cylinder, thus demonstrating the isoperimetric property of the circle in a most striking way. The fact that the soap film adjusts itself vertically to the bounding surface is the key to this experiment.

COMMENTARY ON

The Periodic Law and Mendeléeff

THE periodic law of chemistry is a perfect example of simultaneous scientific discovery. By the middle of the nineteenth century most of the atomic weights of the then-known elements had been measured with considerable accuracy, and a number of chemists, working independently, were beginning to converge on a basic relationship linking the elements by their properties. Various attempts were made—one as early as 1829 by J. W. Dobereiner, professor in Jena and Goethe's chemistry teacher—to "derive regularities among atomic weights," and to show that weights could be expressed by an "arithmetical function," but without success. John Newlands, a London consulting chemist, formulated in 1865 what he called a "law of octaves" according to which if the elements are arranged in the order of their atomic weights, "the numbers of analogous elements [i.e., with similar properties] generally either differ by 7 or by some multiple of 7; in other words, members of the same group stand to each other in the same relation as the extremities of one or more octaves in music." The suggestion, though an important truth lay concealed in it, was received with scorn. When Newlands read his paper at a meeting of the London Chemical Society, he was asked by the chemist Carey Foster "whether he had ever tried classifying the elements in the order of the initial letters of their names." [1]

Four years later the great generalization that the physical and chemical properties of the elements vary periodically with the atomic weight was perceived independently by the German chemist Julius Lothar Meyer and by the Russian chemist Dmitri Ivanovitch Mendeléeff. Meyer drew up his classification of the elements in 1868 but did not publish it until December 1869, eight months after Mendeléeff's law had appeared in Russian. The two men reached strikingly similar conclusions and chemists accord them equal recognition. But in the popular mind the periodic law is associated entirely with the name of Mendeléeff. Since Meyer is by far the less known of the two founders of this invaluable law, it is appropriate that I say one or two words about his work. Mendeléeff's achievement is admirably described in the following selections, the first an excerpt from his Faraday Lecture, delivered before the Fellows of the Chemical Society in 1889, the second a biographical essay from Bernard Jaffe's readable and authoritative book, *Crucibles: The Story of Chemistry*.

[1] Newland's contribution was ultimately recognized: in 1887 he received the Davy Medal of the Royal Society

Meyer, as others had done before him, prepared a table of the elements by groups and in the order of their atomic weights. This table in many respects resembled the present periodic table and was at some points superior to Mendeléeff's arrangement. It has been suggested that Meyer's work was what caused the Russian to revise his table and publish a second version in 1870. Meyer also made a graph in which the atomic volumes of the elements (that is, the volume of one gram multiplied by the atomic weight) were plotted against their atomic weights. The curve shown in Figure 1 was obtained by joining together the different points. It shows clearly the periodic variation of fhe atomic volume; moreover elements of similar chemical properties are seen to occur at corresponding points on the different parts of the curve.[2]

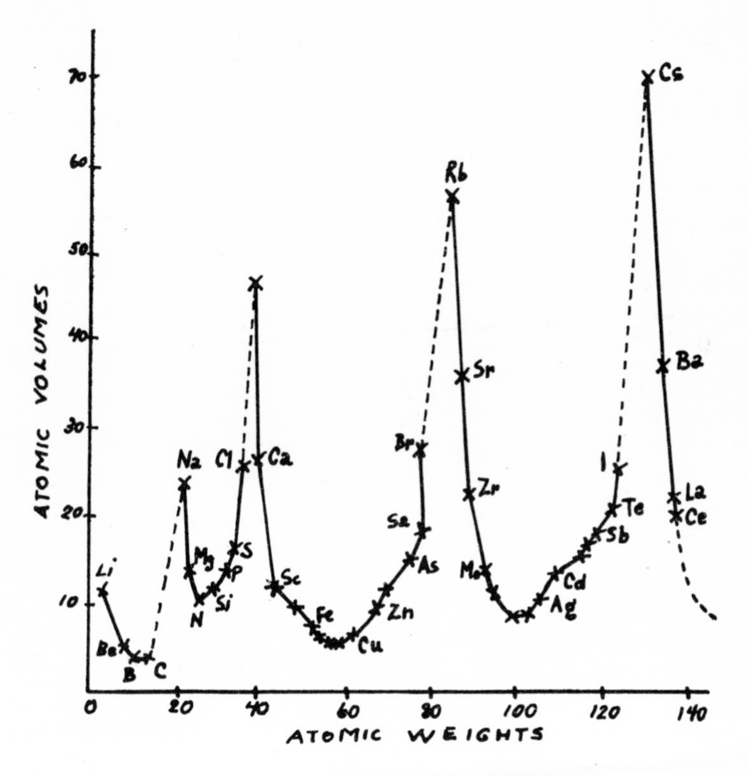

FIGURE 1

"On the abscissa, the curve is proportional to the atomic weights; the ordinates relate to the atomic volume of the elements in the solid state (except for chlorine which is in the liquid); thus the quotients from the atomic weight and the density are plotted proportionally. The atomic

[2] Alexander Findley, *A Hundred Years of Chemistry*, London, Second Edition, 1948, pp. 49–50. The figure is taken from this source.

weight of hydrogen and the density of water are taken as units. . . . Where the knowledge of the atomic volume of one or more elements is lacking, the curve is a dotted line. . . . It can be seen from the course of the curve that the atomic volume of the elements, like their chemical properties, is a periodic function of their atomic weight."

The essential fact is that all *the pictures, which science now draws of nature, and which alone seem capable of according with observational facts, are* mathematical pictures. —SIR JAMES JEANS

11 Periodic Law of the Chemical Elements

By DMITRI MENDELÉEFF

. . . BEFORE one of the oldest and most powerful of [scientific societies] I am about to take the liberty of passing in review the twenty years' life of a generalisation which is known under the name of the Periodic Law. It was in March 1869 that I ventured to lay before the then youthful Russian Chemical Society the ideas upon the same subject which I had expressed in my just written 'Principles of Chemistry.'

Without entering into details, I will give the conclusions I then arrived at in the very words I used:—

'1. The elements, if arranged according to their atomic weights, exhibit an evident *periodicity* of properties.

'2. Elements which are similar as regards their chemical properties have atomic weights which are either of nearly the same value (*e.g.* platinum, iridium, osmium) or which increase regularly (*e.g.* potassium, rubidium, cæsium).

'3. The arrangement of the elements, or of groups of elements, in the order of their atomic weights, corresponds to their so-called *valencies* as well as, to some extent, to their distinctive chemical properties—as is apparent, among other series, in that of lithium, beryllium, barium, carbon, nitrogen, oxygen, and iron.

'4. The elements which are the most widely diffused have *small* atomic weights.

'5. The *magnitude* of the atomic weight determines the character of the element, just as the magnitude of the molecule determines the character of a compound.

'6. We must expect the discovery of many yet *unknown* elements—for example, elements analogous to aluminium and silicon, whose atomic weight would be between 65 and 75.

'7. The atomic weight of an element may sometimes be amended by a knowledge of those of the contiguous elements. Thus, the atomic weight of tellurium must lie between 123 and 126, and cannot be 128.

'8. Certain characteristic properties of the elements can be foretold from their atomic weights.

'The aim of this communication will be fully attained if I succeed in drawing the attention of investigators to those relations which exist between the atomic weights of dissimilar elements, which, so far as I know, have hitherto been almost completely neglected. I believe that the solution of some of the most important problems of our science lies in researches of this kind.'

To-day, twenty years after the above conclusions were formulated, they may still be considered as expressing the essence of the now well-known periodic law.

Reverting to the epoch terminating with the sixties, it is proper to indicate three series of data without the knowledge of which the periodic law could not have been discovered, and which rendered its appearance natural and intelligible.

In the first place, it was at that time that the numerical value of atomic weights became definitely known. Ten years earlier such knowledge did not exist, as may be gathered from the fact that in 1860 chemists from all parts of the world met at Karlsruhe in order to come to some agreement, if not with respect to views relating to atoms, at any rate as regards their definite representation. Many of those present probably remember how vain were the hopes of coming to an understanding, and how much ground was gained at that Congress by the followers of the unitary theory so brilliantly represented by Cannizzaro. I vividly remember the impression produced by his speeches, which admitted of no compromise, and seemed to advocate truth itself, based on the conceptions of Avogadro, Gerhardt, and Regnault, which at that time were far from being generally recognised. And though no understanding could be arrived at, yet the objects of the meeting were attained, for the ideas of Cannizzaro proved, after a few years, to be the only ones which could stand criticism, and which represented an atom as—'the smallest portion of an element which enters into a molecule of its compound.' Only such real atomic weights—not conventional ones—could afford a basis for generalisation. It is sufficient, by way of example, to indicate the following cases in which the relation is seen at once and is perfectly clear:—

$$\begin{array}{lll} K = 39 & Rb = 85 & Cs = 133 \\ Ca = 40 & Sr = 87 & Ba = 137 \end{array}$$

whereas with the equivalents then in use—

$$\begin{array}{lll} K = 39 & Rb = 85 & Cs = 133 \\ Ca = 20 & Sr = 43{\cdot}5 & Ba = 68{\cdot}5 \end{array}$$

the consecutiveness of change in atomic weight, which with the true values is so evident, completely disappears.

Secondly, it had become evident during the period 1860–70, and even during the preceding decade, that the relations between the atomic weights of analogous elements were governed by some general and simple laws. Cooke, Cremers, Gladstone, Gmelin, Lenssen, Pettenkofer, and especially Dumas, had already established many facts bearing on that view. Thus Dumas compared the following groups of analogous elements with organic radicles:—

	Diff.		Diff.		Diff.		Diff.
		Mg = 12		P = 31		O = 8	
			8		44		8
Li = 7		Ca = 20		As = 75		S = 16	
	16		3 × 8		44		3 × 8
Na = 23		Sr = 44		Sb = 119		Se = 40	
	16		3 × 8		2 × 44		3 × 8
K = 39		Ba = 68		Bi = 207		Te = 64	

and pointed out some really striking relationships, such as the following:—

$$\begin{aligned} F &= 19. \\ Cl &= 35 \cdot 5 = 19 + 16 \cdot 5. \\ Br &= 80 = 19 + 2 \times 16 \cdot 5 + 28. \\ I &= 127 = 2 \times 19 + 2 \times 16 \cdot 5 + 2 \times 28. \end{aligned}$$

A. Strecker, in his work 'Theorien und Experimente zur Bestimmung der Atomgewichte der Elemente' (Braunschweig, 1859), after summarising the data relating to the subject, and pointing out the remarkable series of equivalents—

Cr = 26·2   Mn = 27·6   Fe = 28   Ni = 29   Co = 30
Cu = 31·7   Zn = 32·5

remarks that: It is hardly probable that all the above-mentioned relations between the atomic weights (or equivalents) of chemically analogous elements are merely accidental. We must, however, leave to the future the discovery of the *law* of the relations which appears in these figures.'

In such attempts at arrangement and in such views are to be recognised the real forerunners of the periodic law; the ground was prepared for it between 1860 and 1870, and that it was not expressed in a determinate form before the end of the decade may, I suppose, be ascribed to the fact that only analogous elements had been compared. The idea of seeking for a relation between the atomic weights of all the elements was foreign to the ideas then current, so that neither the *vis tellurique* of De Chancourtois, nor the *law of octaves* of Newlands, could secure anybody's attention. And yet both De Chancourtois and Newlands like Dumas and Strecker, more than Lenssen and Pettenkofer, had made an approach to the periodic law and had discovered its germs. The solution of the problem advanced but slowly, because the facts, but not the law, stood foremost in all attempts; and the law could not awaken a general interest so long as elements, having no apparent connection with each other, were included in the same octave, as for example:—

1st octave of Newlands.	H	F	Cl	Co & Ni	Br	Pd	I	Pt & Ir
7th Ditto . . .	O	S	Fe	Se	Rh & Ru	Te	Au	Os or Th

Analogies of the above order seemed quite accidental, and the more so as the octave contained occasionally ten elements instead of eight, and when two such elements as Ba and V, Co and Ni, or Rh and Ru, occupied one place in the octave.[1] Nevertheless, the fruit was ripening, and I now see clearly that Strecker, De Chancourtois, and Newlands stood foremost in the way towards the discovery of the periodic law, and that they merely wanted the boldness necessary to place the whole question at such a height that its reflection on the facts could be clearly seen.

A third circumstance which revealed the periodicity of chemical elements was the accumulation, by the end of the sixties, of new information respecting the rare elements, disclosing their many-sided relations to the other elements and to each other. The researches of Marignac on niobium, and those of Roscoe on vanadium, were of special moment. The striking analogies between vanadium and phosphorus on the one hand, and between vanadium and chromium on the other, which became so apparent in the investigations connected with that element, naturally induced the comparison of V = 51 with Cr = 52, Nb = 94 with Mo = 96, and Ta = 192 with W = 194; while, on the other hand, P = 31 could be compared with S = 32, As = 75 with Se = 79, and Sb = 120 with Te = 125. From such approximations there remained but one step to the discovery of the law of periodicity.

The law of periodicity was thus a direct outcome of the stock of generalisations and established facts which had accumulated by the end of the decade 1860–1870: it is an embodiment of those data in a more or less systematic expression. Where, then, lies the secret of the special importance which has since been attached to the periodic law, and has raised it to the position of a generalisation which has already given to chemistry unexpected aid, and which promises to be far more fruitful in the future and to impress upon several branches of chemical research a peculiar and original stamp? The remaining part of my communication will be an attempt to answer this question.

In the first place we have the circumstance that, as soon as the law made its appearance, it demanded a revision of many facts which were considered by chemists as fully established by existing experience. I shall return, later on, briefly to this subject, but I wish now to remind you that the periodic law, by insisting on the necessity for a revision of supposed

[1] To judge from J. A. R. Newlands's work, *On the Discovery of the Periodic Law*, London, 1884, p. 149; 'On the Law of Octaves' (from the *Chemical News*, 12, 83, August 18, 1865).

facts, exposed itself at once to destruction in its very origin. Its first requirements, however, have been almost entirely satisfied during the last 20 years; the supposed facts have yielded to the law, thus proving that the law itself was a legitimate induction from the verified facts. But our inductions from data have often to do with such details of a science so rich in facts, that only generalisations which cover a wide range of important phenomena can attract general attention. What were the regions touched on by the periodic law? This is what we shall now consider.

The most important point to notice is, that periodic functions, used for the purposes of expressing changes which are dependent on variations of time and space, have been long known. They are familiar to the mind when we have to deal with motion in closed cycles, or with any kind of deviation from a stable position, such as occurs in pendulum-oscillations. A like periodic function became evident in the case of the elements, depending on the mass of the atom. . . . The periodic law has shown that our chemical individuals display a harmonic periodicity of properties dependent on their masses. Now natural science has long been accustomed to deal with periodicities observed in nature, to seize them with the vice of mathematical analysis, to submit them to the rasp of experiment. And these instruments of scientific thought would surely, long since, have mastered the problem connected with the chemical elements, were it not for a new feature which was brought to light by the periodic law, and which gave a peculiar and original character to the periodic function.

If we mark on an axis of abscissæ a series of lengths proportional to angles, and trace ordinates which are proportional to sines or other trigonometrical functions, we get periodic curves of a harmonic character. So it might seem, at first sight, that with the increase of atomic weights the function of the properties of the elements should also vary in the same harmonious way. But in this case there is no such continuous change as in the curves just referred to, because the periods do not contain the infinite number of points constituting a curve, but a *finite* number only of such points. An example will better illustrate this view. The atomic weights—

$$Ag = 108 \qquad Cd = 112 \qquad In = 113 \qquad Sn = 118 \qquad Sb = 120$$
$$Te = 125 \qquad I = 127$$

steadily increase, and their increase is accompanied by a modification of many properties which constitutes the essence of the periodic law. Thus, for example, the densities of the above elements decrease steadily, being respectively—

$$10{\cdot}5 \qquad 8{\cdot}6 \qquad 7{\cdot}4 \qquad 7{\cdot}2 \qquad 6{\cdot}7 \qquad 6{\cdot}4 \qquad 4{\cdot}9$$

while their oxides contain an increasing quantity of oxygen—

$$Ag_2O \qquad Cd_2O_2 \qquad In_2O_3 \qquad Sn_2O_4 \qquad Sb_2O_5 \qquad Te_2O_6 \qquad I_2O_7$$

But to connect by a curve the summits of the ordinates expressing any of these properties would involve the rejection of Dalton's law of multiple proportions. Not only are there no intermediate elements between silver, which gives AgCl, and cadmium, which gives $CdCl_2$, but, according to the very essence of the periodic law, there can be none; in fact a uniform curve would be inapplicable in such a case, as it would lead us to expect elements possessed of special properties at any point of the curve. The periods of the elements have thus a character very different from those which are so simply represented by geometers. They correspond to points, to numbers, to sudden changes of the masses, and not to a continuous evolution. In these sudden changes destitute of intermediate steps or positions, in the absence of elements intermediate between, say, silver and cadmium, or aluminium and silicon, we must recognise a problem to which no direct application of the analysis of the infinitely small can be made. Therefore, neither the trigonometrical functions proposed by Ridberg and Flavitzky, nor the pendulum-oscillations suggested by Crookes, nor the cubical curves of the Rev. Mr. Haughton, which have been proposed for expressing the periodic law, from the nature of the case, can represent the periods of the chemical elements.

. . . While connecting by new bonds the theory of the chemical elements with Dalton's theory of multiple proportions, or atomic structure of bodies, the periodic law opened for natural philosophy a new and wide field for speculation. Kant said that there are in the world 'two things which never cease to call for the admiration and reverence of man: the moral law within ourselves, and the stellar sky above us.' But when we turn our thoughts towards the nature of the elements and the periodic law, we must add a third subject, namely, 'the nature of the elementary individuals which we discover everywhere around us.' Without them the stellar sky itself is inconceivable; and in the atoms we see at once their peculiar individualities, the infinite multiplicity of the individuals, and the submission of their seeming freedom to the general harmony of Nature.

Having thus indicated a new mystery of Nature, which does not yet yield to rational conception, the periodic law, together with the revelations of spectrum analysis, have contributed to again revive an old but remarkably long-lived hope—that of discovering, if not by experiment, at least by a mental effort, the *primary matter*—which had its genesis in the minds of the Grecian philosophers, and has been transmitted, together with many other ideas of the classic period, to the heirs of their civilisation.

The great mathematicians have acted on the principle "Divinez avant de démontrer," and it is certainly true that almost all important discoveries are made in this fashion. —EDWARD KASNER (*1878–1955*)

12 Mendeléeff

By BERNARD JAFFE

SIBERIA BREEDS A PROPHET

OUT of Russia came the patriarchal voice of a prophet of chemistry. "There is an element as yet undiscovered. I have named it eka-aluminum. By properties similar to those of the metal aluminum you shall identify it. Seek it, and it will be found." Startling as was this prophecy, the sage of Russia was not through. He predicted another element resembling the element boron. He was even bold enough to state its atomic weight. And before that voice was stilled, it foretold the discovery of a third element whose physical and chemical properties were thoroughly described. No man, not even the Russian himself, had beheld these unknown substances.

This was the year 1869. The age of miracles was long past. Yet here was a distinguished scientist, holding a chair of chemistry at a famous university, covering himself with the mantle of the prophets of old. Had he gathered this information from inside the crystal glass of some sorcerer? Perhaps, like the seer of ancient times, he had gone to the top of a mountain to bring down the tablets of these new elements. But this oracle disdained the robes of a priest. Rather did he announce his predictions from the stillness of his chemical laboratory, where midst the smoke, not of a burning bush, but of the fire of his furnace, he had seen visions of a great generalization in chemistry.

Chemistry had already been the object of prophecy. When Lavoisier heated some tin in a sealed flask and found it to change in appearance and weight, he saw clearly a new truth, and foretold other changes. Lockyer a year before had looked through a new instrument—the spectroscope devised by Bunsen and Kirchhof. Through this spectroscope he had gazed at the bright colored lines of a new element ninety-three million miles away. Since it was present in the photosphere of the sun he called it *helium* and predicted its existence on our earth. Twenty-one years later, William Hillebrand of the United States Geological Survey, came across this gas in the rare mineral cleveite.

But the predictions of the Russian were more astounding. He had made no direct experiments. He had come to his conclusions seemingly out of thin air. There had gradually been born in the fertile mind of this man the germ of a great truth. It was a fantastic seed but it germinated with surprising rapidity. When the flower was mature, he ventured to startle the world with its beauty.

In 1884 Sir William Ramsay had come to London to attend a dinner given in honor of William Perkin, the discoverer of the dye mauve. "I was very early at the dinner," Ramsay recalled, "and was putting off time looking at the names of people to be present, when a peculiar foreigner, every hair of whose head acted in independence of every other, came up bowing. I said, 'We are to have a good attendance, I think?' He said, 'I do not spik English.' I said, 'Vielleicht sprechen sie Deutsch?' He replied, 'Ja, ein wenig. Ich bin Mendeléeff.' Well, we had twenty minutes or so before anyone else turned up and we talked our mutual subject fairly out. He is a nice sort of fellow but his German is not perfect. He said he was raised in East Siberia and knew no Russian until he was seventeen years old. I suppose he is a Kalmuck or one of those outlandish creatures."

This "outlandish creature" was Mendeléeff, the Russian prophet to whom the world listened. Men went in search of the missing elements he described. In the bowels of the earth, in the flue dust of factories, in the waters of the oceans, and in every conceivable corner they hunted. Summers and winters rolled by while Mendeléeff kept preaching the truth of his visions. Then, in 1875, the first of the new elements he foretold was discovered. In a zinc ore mined in the Pyrenees, Lecoq de Boisbaudran came upon the hidden eka-aluminum. This Frenchman analyzed and reanalyzed the mineral and studied the new element in every possible way to make sure there was no error. Mendeléeff must indeed be a prophet! For here was a metal exactly similar to his eka-aluminum. It yielded its secret of two new lines to the spectroscope, it was easily fusible, it could form alums, its chloride was volatile. Every one of these characteristics had been accurately foretold by the Russian. Lecoq named it *gallium* after the ancient name of his native country.

But there were many who disbelieved. "This is one of those strange guesses which by the law of averages must come true," they argued. Silly to believe that new elements could be predicted with such accuracy! One might as well predict the birth of a new star in the heavens. Had not Lavoisier, the father of chemistry, declared that "all that can be said upon the nature and number of the elements is confined to discussions entirely of a metaphysical nature? The subject only furnishes us with indefinite problems."

But then came the news that Winkler, in Germany, had stumbled over another new element, which matched the eka-silicon of Mendeléeff. The German had followed the clue of the Russian. He was looking for a dirty gray element with an atomic weight of about 72, a density of 5.5, an element which was slightly acted upon by acids. From the silver ore, argyrodite, he isolated a grayish white substance with atomic weight of 72.3 and a density of 5.5. He heated it in air and found its oxide to be exactly as heavy as had been predicted. He synthesized its ethide and found it to boil at exactly the temperature that Mendeléeff had prefigured. There was not a scintilla of doubt about the fulfilment of Mendeléeff's second prophecy. The spectroscope added unequivocal testimony. Winkler announced the new element under the name of *germanium* in honor of his fatherland. The sceptics were dumbfounded. Perhaps after all the Russian was no charlatan!

Two years later the world was completely convinced. Out of Scandinavia came the report that Nilson had isolated eka-boron. Picking up the scent of the missing element in the ore of euxenite, Nilson had tracked it down until the naked element, exhibiting every property foreshadowed for it, lay before him in his evaporating dish. The data were conclusive. The whole world of science came knocking at the door of the Russian in St. Petersburg.

Dmitri Ivanovitch Mendeléeff came of a family of heroic pioneers. More than a century before his birth, Peter the Great had started to westernize Russia. Upon a marsh of pestilence he reared a mighty city which was to be Russia's window to the West. For three-quarters of a century Russia's intellectual march eastward continued, until in 1787 in Tobolsk, Siberia, the grandfather of Dmitri opened up the first printing press, and with the spirit of a pioneer published the first newspaper in Siberia, the *Irtysch*. In this desolate spot, settled two centuries before by the Cossacks, Dmitri was born on February 7, 1834. He was the last of a family of seventeen children.

Misfortune overtook his family. His father, director of the local high school, became blind, and soon after died of consumption. His mother, Maria Korniloff, a Tartar beauty, unable to support her large family on a pension of five hundred dollars a year, reopened a glass factory which her family was the first to establish in Siberia. Tobolsk at this time was an administrative center to which Russian political exiles were taken. From one of these prisoners of the revolt of 1825, a "Decembrist" who married his sister, Dmitri learned the rudiments of natural science. When fire destroyed the glass factory, little Dmitri, pet of his aged mother—she was already fifty-seven—was taken to Moscow in the hope that he might be admitted to the University. Official red tape prevented this. Determined that her son should receive a good scientific education, his mother undertook to move to St. Petersburg, where he finally gained admittance to the Science Department of the Pedagogical Institute, a school for the training of high school teachers. Here he specialized in mathematics, physics and chemistry. The classics were distasteful to this blue-eyed boy. Years later, when he took a hand in the solution of Russia's educational problems, he wrote, "We could live at the present time without a Plato, but a double number of Newtons is required to discover the secrets of nature, and to bring life into harmony with its laws."

Mendeléeff worked diligently at his studies and graduated at the head of his class. Never very robust during these early years, his health gradually weakened, and the news of his mother's death completely unnerved him. He had come to her as she lay on her death bed. She spoke to him of his future: "Refrain from illusions, insist on work and not on words. Patiently search divine and scientific truth." Mendeléeff never forgot those words. Even as he dreamed, he always felt the solid earth beneath his feet.

His physician gave him six months to live. To regain his health, he was ordered to seek a warmer climate. He went to the south of Russia and obtained a position as science master at Simferopol in the Crimea. When the Crimean War broke out he left for Odessa, and at the age of twenty-two he was back in St. Petersburg as a privat-docent. An appointment as privat-docent meant nothing more than permission to teach, and brought no stipend save a part of the fees paid by the students who attended the lectures. Within a few years he asked and was granted permission from the Minister of Public Instruction to study in France and Germany. There was no opportunity in Russia for advanced work in science. At Paris he worked in the laboratory of Henri Regnault and, for another year, at Heidelberg in a small private laboratory built out of his meager means. Here he met Bunsen and Kirchhof from whom he learned the use of the spectroscope, and together with Kopp attended the Congress of Karlsruhe, listening to the great battle over the molecules of Avogadro. Cannizzaro's atomic weights were to do valiant service for him in the years to come. Mendeléeff's attendance at this historic meeting ended his *Wanderjahre*.

The next few years were very busy ones. He married, completed in sixty days a five-hundred page textbook on organic chemistry which earned him the Domidoff Prize, and gained his doctorate in chemistry for a thesis on *The Union of Alcohol with Water*. The versatility of this gifted teacher, chemical philosopher and accurate experimenter was soon recognized by the University of St. Petersburg, which appointed him full professor before he was thirty-two.

Then came the epoch-making year of 1869. Mendeléeff had spent twenty years reading, studying and experimenting with the chemical elements. All these years he had been busy collecting a mass of data from every conceivable source. He had arranged and rearranged this data in the hope of unfolding a secret. It was a painstaking task. Thousands of scientists had worked on the elements in hundreds of laboratories scattered over the civilized world. Sometimes he had to spend days searching for missing data to complete his tables. The number of the elements had increased since the ancient artisans fashioned instruments from their gold, silver, copper, iron, mercury, lead, tin, sulphur and carbon. The alchemists had added six new elements in their futile search for the seed of gold and the elixir of life. Basil Valentine, a German physician, in the year when Columbus was discovering America had rather fancifully described antimony. In 1530 Georgius Agricola, another German, talked about bismuth in his *De Re Metallica*, a book on mining which was translated into English for the first time by Herbert Hoover and his wife in 1912. Paracelsus was the first to mention the metal zinc to the Western World. Brandt discovered glowing phosphorus in urine, and arsenic and cobalt were soon added to the list of the elements.

Before the end of the eighteenth century, fourteen more elements were discovered. In far away Choco, Colombia, a Spanish naval officer, Don Antonio de Ulloa, had picked up a heavy nugget while on an astronomical mission, and had almost discarded it as worthless before the valuable properties of the metal platinum were recognized. This was in 1735. Then came lustrous nickel, inflammable hydrogen, inactive nitrogen, life-giving oxygen, death-dealing chlorine, manganese, used for burglar-proof safes, tungsten, for incandescent lamps, chromium, for stainless steel, molybdenum and titanium, so useful in steel alloys, tellurium, zirconium, and uranium, heaviest of all the elements. The nineteenth century had hardly opened when Hatchett, an Englishman, discovered columbium (niobium) in a black mineral that had found its way from the Connecticut Valley to the British Museum. And thus the search went on, until in 1869 sixty-three different elements had been isolated and described in the chemical journals of England, France, Germany and Sweden.

Mendeléeff gathered together all the data on these sixty-three chemical elements. He did not miss a single one. He even included fluorine whose presence was known, but which had not yet been isolated because of its tremendous activity. Here was a list of all the chemical elements, every one of them consisting of different Daltonian atoms. Their atomic weights, ranging from 1 (hydrogen) to 238 (uranium), were all dissimilar. Some, like oxygen, hydrogen, chlorine and nitrogen, were gases. Others, like mercury and bromine, were liquids under normal conditions. The rest were solids. There were some very hard metals like platinum and iridium, and soft metals like sodium and potassium. Lithium was a metal so light that it could float on water. Osmium, on the other hand, was twenty-two and a half times as heavy as water. Here was mercury, a metal which was not a solid at all, but a liquid. Copper was red, gold yellow, iodine steel gray, phosphorus white, and bromine red. Some metals, like nickel and chromium, could take a very high polish; others like lead and aluminum, were duller. Gold, on exposure to the air, never tarnished, iron rusted very easily, iodine sublimed and changed into a vapor. Some elements united with one atom of oxygen, others with two, three or four atoms. A few, like potassium and fluorine, were so active that it was dangerous to handle them with the unprotected fingers. Others could remain unchanged for ages. What a maze of varying, dissimilar, physical characteristics and chemical properties!

Could some order be found in this body of diverse atoms? Was there any connection between these elements? Could some system of evolution or development be traced among them, such as Darwin, ten years before, had found among the multiform varieties of organic life? Mendeléeff wondered. The problem haunted his dreams. Constantly his mind reverted to this puzzling question.

Mendeléeff was a dreamer and a philosopher. He was going to find the key to this heterogeneous collection of data. Perhaps nature had a simple secret to unfold. And while he believed it to be "the glory of God to conceal a thing," he was firmly convinced that it was "the honor of kings to search it out." And what a boon it would prove to his students!

He arranged all the elements in the order of increasing atomic weights, starting with the lightest, hydrogen, and completing his table with uranium, the heaviest. He saw no particular value in arranging the elements in this way; it had been done previously. Unknown to Mendeléeff, an Englishman, John Newlands, had three years previously read, before the English Chemical Society at Burlington House, a paper on the arrangement of the elements. Newlands had noticed that each succeeding eighth element in his list showed properties similar to the first element. This seemed strange. He compared the table of the elements to the keyboard of a piano with its eighty-eight notes divided into periods or octaves of eight. "The members of the same group of elements," he said, "stand to each other in the same relation as the extremities of one or more *octaves* in music." The members of the learned society of London laughed at his Law of Octaves. Professor Foster ironically inquired if he had ever examined the elements according to their initial letters. No wonder—think of comparing the chemical elements to the keyboard of a piano! One might as well compare the sizzling of sodium as it skims over water to the music of the heavenly spheres. "Too fantastic," they agreed, and J. A. R. Newlands almost went down to oblivion.

Mendeléeff was clear-visioned enough not to fall into such a pit. He took sixty-three cards and placed on them the names and properties of the elements. These cards he pinned on the walls of his laboratory. Then he carefully reexamined the data. He sorted out the similar elements and pinned their cards together again on the walls. A striking relationship was thus made clear.

Mendeléeff now arranged the elements into seven groups, starting with lithium (at. wt. 7), and followed by beryllium (at. wt. 9), boron (11), carbon (12), nitrogen (14), oxygen (16) and fluorine (19). The next element in the order of increasing atomic weight was sodium (23). This element resembled lithium very closely in both physical and chemical properties. He therefore placed it below lithium in his table. After placing five more elements he came to chlorine, which had properties very similar to fluorine, under which it miraculously fell in his list. In this way he continued to arrange the remainder of the elements. When his list was completed he noticed a most remarkable order. How beautifully the elements fitted into their places! The very active metals lithium, sodium, potassium, rubidium and caesium fell into one group (No. 1). The extremely active non-metals, fluorine, chlorine, bromine and iodine, all appeared in the seventh group.

Mendeléeff had discovered that the properties of the elements "were periodic functions of their atomic weights," that is, their properties repeated themselves periodically after each seven elements. What a simple law he had discovered! But here was another astonishing fact. All the elements in Group I united with oxygen two atoms to one. All the atoms of the second group united with oxygen atom for atom. The elements in Group III joined with oxygen two atoms to three. Similar uniformities prevailed in the remaining groups of elements. What in the realm of nature could be more simple? To know the properties of one element of a certain group was to know, in a general way, the properties of all the elements in that group. What a saving of time and effort for his chemistry students!

Could his table be nothing but a strange coincidence? Mendeléeff wondered. He studied the properties of even the rarest of the elements. He re-searched the chemical literature lest he had, in the ardor of his work, misplaced an element to fit in with his beautiful edifice. Yes, here was a mistake! He had misplaced iodine, whose atomic weight was recorded as 127, and tellurium, 128, to agree with his scheme of things. Mendeléeff looked at his Periodic Table of the Elements and saw that it was good. With the courage of a prophet he made bold to say that the atomic weight of tellurium was wrong; that it must be between 123 and 126 and not 128, as its discoverer had determined. Here was downright heresy, but Dmitri was not afraid to buck the established order of things. For the present, he placed the element tellurium in its proper position, but with its false atomic weight. Years later his action was upheld, for further chemical discoveries proved his position of tellurium to be correct. This was one of the most magnificent prognostications in chemical history.

Perhaps Mendeléeff's table was now free from flaws. Again he examined it, and once more he detected an apparent contradiction. Here was gold with the accepted atomic weight of 196.2 placed in a space which rightfully belonged to platinum, whose established atomic weight was 196.7. The fault-finders got busy. They pointed out this discrepancy with scorn. Mendeléeff made brave enough to claim that the figures of the analysts, and not his table, were inaccurate. He told them to wait. He would be vindicated. And again the balance of the chemist came to the aid of the philosopher, for the then-accepted weights were wrong and Mendeléeff was again right. Gold had an atomic weight greater than platinum. This table of the queer Russian was almost uncanny in its accuracy!

Mendeléeff was still to strike his greatest bolt. Here were places in his table which were vacant. Were they always to remain empty or had the efforts of man failed as yet to uncover some missing elements which belonged in these spaces? A less intrepid person would have shrunk from the conclusion that this Russian drew. Not this Tartar, who would not cut his hair even to please his Majesty, Czar Alexander III. He was convinced of the truth of his great generalization, and did not fear the blind, chemical sceptics.

Here in Group III was a gap between calcium and titanium. Since it occurred under boron, the missing element must resemble boron. This was his eka-boron which he predicted. There was another gap in the same group under aluminum. This element must resemble aluminum, so he called it eka-aluminum. And finally he found another vacant space between arsenic and eka-aluminum, which appeared in the fourth group. Since its position was below the element silicon, he called it eka-silicon. Thus he predicted three undiscovered elements and left it to his chemical contemporaries to verify his prophecies. Not such remarkable guesses after all—at least not to the genius Mendeléeff!

In 1869 Mendeléeff, before the Russian Chemical Society, presented his paper *On the Relation of the Properties to the Atomic Weights of the Elements.* In a vivid style he told them of his epoch-making conclusions. The whole scientific world was overwhelmed. His great discovery, however, had not sprung forth overnight full grown. The germ of this important law had begun to develop years before. Mendeléeff admitted that "the law was the direct outcome of the stock of generalizations of established facts which had accumulated by the end of the decade 1860–1870." De Chancourtois in France, Strecher in Germany, Newlands in England, and Cooke in America had noticed similarities among the properties of certain elements. But no better example could be cited of how two men, working independently in different countries, can arrive at the same generalization, than the case of Lothar Meyer, who conceived the Periodic Law at almost the same time as Mendeléeff. In 1870 there appeared in *Liebig's Annalen* a table of the elements by Lothar Meyer which was almost identical with that of the Russian. The time was ripe for this great law. Some wanted the boldness or the genius necessary "to place the whole question at such a height that its reflection on the facts could be clearly seen." This was the statement of Mendeléeff himself. Enough elements had been discovered and studied to make possible the arrangement of a table such as Mendeléeff had prepared. Had Dmitri been born a generation before, he could never, in 1840, have enunciated the Periodic Law.

"The Periodic Law has given to chemistry that prophetic power long regarded as the peculiar dignity of the sister science, astronomy." So wrote the American scientist Bolton. Mendeléeff had made places for more than sixty-three elements in his Table. Three more he had predicted. What of the other missing building blocks of the universe? Twenty-five years after the publication of Mendeléeff's Table, two Englishmen, following a clue of Cavendish, came upon a new group of elements of which even the Russian had never dreamed. These elements constituted a queer company—the Zero Group as it was later named. Its members, seven in number, are the most unsociable of all the elements. Even with that ideal mixer, potassium, they will not unite. Fluorine, most violent of all the non-metals cannot shake these hermit elements out of their inertness. Moissan tried sparking them with fluorine but failed to make them combine. Besides, they are all gases, invisible and odorless. Small wonder they had remained so long hidden.

True, the first of these noble gases, as they were called, had been observed in the sun's chromosphere during a solar eclipse in August, 1868, but as nothing was known about it except its orange yellow spectral line, Mendeléeff did not even include it in his table. Later Hillebrand described a gas expelled from cleveite. He knew enough about it to state that it differed from nitrogen but failed to detect its real nature. Then Ramsay, obtaining a sample of the same mineral, bottled the gas expelled from it in a vacuum tube, sparked it and detected the spectral line of helium. The following year Kayser announced the presence of this gas in very minute amounts, one part in 185,000, in the earth's atmosphere.

The story of the discovery and isolation of these gases from the air is one of the most amazing examples of precise and painstaking researches in the whole history of science. Ramsay had been casually introduced to chemistry while convalescing from an injury received in a football game. He had picked up a textbook in chemistry and turned to the description of the manufacture of gunpowder. This was his first lesson in chemistry. Rayleigh, his co-worker, had been urged to enter either the ministry or politics, and when he claimed that he owed a duty to science, was told his action was a lapse from the straight and narrow path. Such were the initiations of these two Englishmen into the science which brought them undying fame. They worked with gases so small in volume that it is difficult to understand how they could have studied them. Rayleigh, in 1894, wrote to Lady Frances Balfour: "The new gas has been leading me a life. I had only about a quarter of a thimbleful. I now have a more decent quantity but it has cost about a thousand times its weight in gold. It has not yet been christened. One pundit suggested 'aeron,' but when I have tried the effect privately, the answer has usually been, 'When may we expect Moses?' " It was finally christened argon, and if not Moses, there came other close relatives: neon, krypton, xenon and finally radon. These gases were isolated by Ramsay and Travers from one hundred and twenty tons of air which had been liquefied. Sir William Ramsay used a micro-balance which could detect a difference in weight of one fourteen-trillionth of an ounce. He worked with a millionth of a gram of invisible, gaseous radon—the size of a tenth of a pin's head.

Besides these six Zero Group elements, some of which are doing effective work in argon and neon incandescent lamps, in helium-filled dirigibles, in electric signs, and in replacing the nitrogen in compressed air to prevent the "bends" among caisson workers, seventeen other elements were unearthed. So that, a year after Mendeléeff died in 1907, eighty-six elements were listed in the Periodic Table, a fourfold increase since the days of Lavoisier.

Mendeléeff, besides being a natural philosopher in the widest sense of the term, was also a social reformer. He was aware of the brutality and tyranny of Czarist Russia. He had learned his first lessons from the persecuted exiles in frozen Tobolsk. As he travelled about Russia, he went third class, and engaged in intimate conversation with the peasants and small tradespeople in the trains. They hated the remorseless oppression and espionage of the government. Mendeléeff was not blind to the abuses of Russian officialdom, nor did he fear to point them out. He was often vehement in his denunciations. This was a dangerous procedure in those days. But the government needed Mendeléeff, and his radical utterances were always mildly tinged with due respect for law and order. Mendeléeff was shrewd enough not to make a frontal attack on the government. He would bide his time and wait for an opportune moment when his complaints could not easily be ignored. On more than one occasion when this scientific genius showed signs of political eruption, he was hastily sent away on some government mission. Far from the centers of unrest he was much safer and of greater value to the officials.

In 1876, Mendeléeff was commissioned by the government of Alexander II to visit the oil fields of Pennsylvania in distant America. These were the early days of the petroleum industry. In 1859, Colonel Edwin L. Drake and his partner "Uncle Billy" Smith had gone to Titusville, Pennsylvania, to drive a well sixty-nine feet deep—the first to produce oil on a commercial scale. Mendeléeff had already been of invaluable service to

Russia by making a very careful study of her extensive oil fields of Baku. Here, in the Caucasus, from a gap in the rock, burned the "everlasting flame" which Marco Polo had described centuries back. Baku was the most prolific single oil district in the world and, from earliest times, people had burned its oil which they had dipped from its springs. Mendeléeff developed an ingenious theory to explain the origin of these oil deposits. He refused to accept the prevalent idea that oil was the result of the decomposition of organic material in the earth, and postulated that energy-bearing petroleum was formed by the interaction of water and metallic carbides found in the interior of the earth.

On his return from America, Mendeléeff was again sent to study the naphtha springs in the south of Russia. He did not confine his work to the gathering of statistics and the enunciation of theories. He developed in his own laboratory a new method for the commercial distillation of these products and saved Russia vast sums of money. He studied the coal region on the banks and basin of the Donetz River and opened it to the world. He was an active propagandist for Russia's industrial development and expansion, and was called upon to help frame a protective tariff for his country.

This was a period of intense social and political unrest in Russia. Alexander II had attempted to settle the land question of his twenty-three million serfs. He tried further to ameliorate conditions by reforming the judicial system, relaxing the censorship of the press, and developing educational facilities. The young students in the universities presented a petition for a change in certain educational practices. Suddenly an insurrection against the Russian government broke out in Poland. The reactionary forces again gained control. Russia was in no mood for radical changes; the requests of the students were peremptorily turned down. Mendeléeff stepped in and presented another of their petitions to the officials of the government. He was bluntly told to go back to his laboratory and stop meddling in the affairs of the state. Proud and sensitive, Mendeléeff was insulted and resigned from the University. Prince Kropotkin, a Russian anarchist of royal blood, was one of his famous students. "I am not afraid," Mendeléeff had declared, "of the admission of foreign, even of socialistic ideas into Russia, because I have faith in the Russian people who have already got rid of the Tartar domination and the feudal system." He did not change his views even after the Czar, in 1881, was horribly mangled by a bomb thrown into his carriage.

Mendeléeff had made many enemies by his espousal of liberal movements. In 1880, the St. Petersburg Academy of Sciences refused, in spite of very strong recommendations, to elect him member of its chemical section. His liberal tendencies were an abomination. But other and greater honors came to this sage. The University of Moscow promptly made him one of its honorary members. The Royal Society of England presented him with the Davy Medal which he shared with Lothar Meyer for the Periodic Classification of the Elements.

Years later, as he was being honored by the English Chemical Society with the coveted Faraday Medal, Mendeléeff was handed a small silk purse worked in the Russian national colors and containing the honorarium, according to the custom of the Society. Dramatically he tumbled the sovereigns out on the table, declaring that nothing would induce him to accept money from a Society which had paid him the high compliment of inviting him to do honor to the memory of Faraday in a place made sacred by his labors. He was showered with decorations by the chemical societies of Germany and America, by the Universities of Princeton, Cambridge, Oxford, and Göttingen. Sergius Witte, Minister of Finance under Czar Alexander III, appointed him Director of the Bureau of Weights and Measures.

Mendeléeff broke away from the conventional attitude of Russians towards women, and treated them as equals in their struggle for work and education. While he held them to be mentally inferior to men, he did not hesitate to employ women in his office, and admitted them to his lectures at the university. He was twice married. With his first wife, who bore him two children, he led an unhappy life. She could not understand the occasional fits of temper of this queer intellect. The couple soon separated and were eventually divorced. Then he fell madly in love with a young Cossack beauty of artistic temperament, and, at forty-seven, remarried. Anna Ivanovna Popova understood his sensitive nature, and they lived very happily. She would make allowances for his flights of fancy and occasional selfishness. Extremely temperamental and touchy, he wanted everybody to think well of him. At heart he was kind and lovable. Two sons and two daughters were born to them and Mendeléeff oftimes expressed the feeling that "of all things I love nothing more in life than to have my children around me." Dressed in the loose garments which his idol, Leo Tolstoy, wore, and which Anna had sewed for him, Dmitri would sit at home for hours smoking. He made an impressive figure. His deep-set blue eyes shone out of a fine expressive face half covered by a long patriarchal beard. He always fascinated his many guests with his deep guttural utterances. He loved books, especially books of adventure. Fenimore Cooper and Byron thrilled him. The theatre did not attract him, but he loved good music and painting. Accompanied by his wife, who herself had made pen pictures of some of the great figures of science, he often visited the picture galleries. His own study was adorned by her sketches of Lavoisier, Newton, Galileo, Faraday, and Dumas.

When the Russo-Japanese War broke out in February, 1904, Mendeléeff turned out to be a strict nationalist. Old as he was, he added his strength in the hope of victory. Made advisor to the Navy, he invented pyrocollodion, a new type of smokeless powder. The destruction of the Russian fleet in the Straits of Tsushima and Russia's defeat hastened his end. His lungs had always bothered him; as a youth his doctor had given him only a few months to live. But his powerfully-set frame carried him through more than seventy years of life. Then one day in February, 1907, the old scientist caught cold, pneumonia set in, and as he sat listening to the reading of Verne's *Journey to the North Pole*, he expired. Two days later Menschutkin, Russia's eminent analytical chemist, died, and within one year Russia lost also her greatest organic chemist, Friedrich Konrad Beilstein. Staggering blows to Russian chemistry.

To the end, Mendeléeff clung to scientific speculations. He published an attempt towards a chemical conception of the ether. He tried to solve the mystery of this intangible something which pervades the whole universe. To him ether was material, belonged to the Zero Group of Elements, and consisted of particles a million times smaller than the atoms of hydrogen.

Two years after he was laid beside the grave of his mother and son, Pattison Muir declared that "the future will decide whether the Periodic Law is the long looked for goal, or only a stage in the journey: a resting place while material is gathered for the next advance." Had Mendeléeff lived a few more years, he would have witnessed the complete and final development of his Periodic Table by a young Englishman at Manchester.

The Russian peasant of his day never heard of the Periodic Law, but he remembered Dmitri Mendeléeff for another reason. One day, to photograph a solar eclipse, he shot into the air in a balloon, "flew on a bubble and pierced the sky." And to every boy and girl of the Soviet Union today Mendeléeff is a national hero.

COMMENTARY ON

GREGOR MENDEL

IF research had to be expensive to be successful, as is nowadays assumed, it is doubtful that Gregor Mendel could have improved even the method of digging for garden worms. The experiments which led to the discovery of the laws of heredity were conducted on a garden strip 120 feet long and a little more than 20 feet wide. It lay alongside a little clock-tower adjoining the library of the Augustinian monastery of St. Thomas in the city of Brünn, Austria.[1] On this small plot Mendel planted hundreds of pea plants of various kinds: with white blossoms and with violet, tall and dwarf, bearing smooth or wrinkled peas. His equipment was a pair of fine forceps to handle the keel and anthers of the plants, a camel-hair pencil to dust the pollen of one plant upon the stigma of another, a supply of small paper or calico bags (to "prevent any industrious bee or enterprising pea-weevil from transferring pollen from some other flower to the stigma thus treated, and in this way invalidating the result of the hybridization experiment") and several stout notebooks (twopenny size) in which to record his results.[2] For seven years (1856–1863) Mendel performed his thousands of crossing experiments. In 1865 he read his paper on *Experiments in Plant-Hybridization* before a meeting of the Brünn Society for the Study of Natural Science. A fair-sized audience listened, at least at first, with more than perfunctory politeness to his account of the unchanging ratios in the appearance of certain characters among the hybrids. After a time the mathematics became a little difficult and attention faltered. "The minutes of the meeting inform us that there were neither questions nor discussion." [3] In 1866 the lecture was published in the "Proceedings" of this Moravian provincial society; copies were sent to 120 other societies, universities and academies in Austria and abroad. For thirty years the paper lay dormant. Then in 1900, within a period of four months, it was "discovered" three times over: by the Dutch botanist Hugo de Vries (1848–1935), by the German C. Correns, and by the Austrian Erich Von Tschermak.

"Everything of importance," Whitehead once observed, "has been said before by somebody who did not discover it." Mendel had forerunners, as did everyone except the Creator when He made the world. Others had performed crossing experiments, others had noticed the phenomena of dominance and separation, others had foreshadowed the machinery of

[1] Now Brno, Czechoslovakia.
[2] Hugo Iltis, *Life of Mendel*, New York, 1932, pp. 107–108.
[3] Iltis, *op. cit.*, p. 179.

inheritance. Mendel alone perceived the pattern running through his results, saw how the facts were linked and conceived a comprehensive theory to explain the bewildering processes of heredity.

The main outcome of his research was the discovery that certain parental characters are transmitted unchanged, without dilution or blending (to use Julian Huxley's phrase), because they are carried by some kind of distinctive unit or particle. We call these units genes, Mendel called them factors. From the outset of his arduous investigations, he grasped clearly the need for determining "the number of different forms under which the offsprings of hybrids appear," "their statistical relations," "the numerical ratios." This emphasis on precise, quantitative definition was what set his method and his results apart from the work of his predecessors and contemporaries. The paper from which I have taken the excerpt below embodies two major conclusions. In each individual, genes are found in pairs. What is now known as Mendel's First Law holds that the genes in these gene pairs, whether dominant or recessive, are independent entities "emerging separate and unchanged in the gametes [sperm or egg cells] after their co-operation in the zygote [the cell produced by the union of the gametes]." This profoundly important law of segregation may be said in a sense to contradict "our almost instinctive belief about heredity." The expression "mixed blood" reflects the common notion that hybrids are the result of a blending or fusion which resembles the blending say, of different colors. "One expects races to mix as liquids mix." But this is wrong; a mixture of genes does not produce a tincture. The genes persist as distinct units from generation to generation "without contaminating or diluting each other"; when the sexual cells mature the pairs are dissolved and the genes separate from each other. This separation provides the opportunity for wholly new combinations of characters, because the members of each pair of genes sort themselves out independently of the other pairs (with one of every two mated genes passing from each parent to each offspring). These and related facts are subsumed in Mendel's Second Law, the Law of Independent Assortment. It says in brief that the separate gene pairs "are shuffled and dealt independently of each other." [4] Upon these comparatively simple laws, substantially con-

[4] "It follows from this Second Law of Mendel's that in a cross, so long as the strains are pure and the same genes are involved, it makes no difference how they are combined in the original strains. Instead of crossing a tall yellow and a dwarf green strain, we might have crossed a dwarf yellow and a tall green; but the result, both in the first and in all subsequent generations, would have been precisely the same. Once more it is the individual genes which count, not the way they happen to be grouped in the parents' bodies. So long as all the gene-ingredients get into the first zygote in the right proportions, it makes no difference from what abode they came, any more than it makes a difference to the taste of the stew whether the pepper-pot happens to have stood next to the rice or the potatoes on the shelf." H. G. Wells, Julian S. Huxley and G. P. Wells, *The Science of Life*, New York, 1934, p. 483. This book gives an excellent account of the fundamentals of genetics. Also recommended are Iltis' biography of Mendel, cited above; Amram Scheinfeld,

firmed by an overwhelming body of experimental evidence, rests the whole vast edifice of modern genetics.

Johann Gregor Mendel was born of peasant stock, July 22, 1822, in the Moravian village of Heinzendorf. He learned fruit growing as a youth in his father's garden and retained a fondness for it all his life. There was little money available for schooling, but Mendel's parents and sisters were willing to make sacrifices for him and by the time he reached high school he was able to fend for himself. The village schools he attended taught, besides the usual essentials, natural history and natural science. This radical broadening of the curriculum was undertaken at the desire of the lady of the manor, a Countess Waldburg; the local school inspector denounced it to his superiors as a "scandal." Had he foreseen its effect on Mendel, he would have been even more outraged.

After making good progress in the study of philosophy at Olmütz, Mendel decided that his best chance for pursuing a quiet and studious life without financial cares lay in becoming a priest. At the age of twenty-one he joined the Augustinian monastery of Brünn. The prelate was a man of discernment, with respect for science and scholarship. He liked Mendel and gave him the opportunity not only of pursuing his botanical studies in the cloister gardens but of going to Vienna to attend courses in physics, mathematics and zoölogy at the university. Mendel enjoyed the security of belonging to a religious order but had not much taste for religious exercises. The time in Vienna was happily spent, as were the many years he served as a substitute teacher at the Znaim high school and later at the Brünn Modern School. It was gratifying to work outside the monastery with children, especially for one so unfitted by temperament to discharge the ordinary duties of a parish priest. Prelate Napp observed that Mendel was seized with an "unconquerable timidity" when he had to visit a sick bed or call upon anyone ill or in pain.

One of the strange disappointments of Mendel's life was his inability to qualify for a regular high-school teacher's license. He took the examination twice and was ploughed both times. The surprising fact is that he passed the part dealing with the physical sciences but was unable to convince the examiners that he was fit to teach natural history. It must be

---

The New You and Heredity, New York, 1950; Richard B. Goldschmidt, *Understanding Heredity*, New York, 1952.

Julian Huxley's *Heredity, East and West* (New York, 1949) tells the story of the Lysenko controversy and also provides an admirable summary of neo-Mendelian theories. Huxley compares the chromosomes, in which the genes are thought to be arranged in a definite linear order, to packs of playing cards. There are two full packs in the nucleus of every cell of every higher organism. Before a gamete is formed, "a complicated process of pairing and separation takes place so that each gamete has only one pack of chromosomes" containing one chromosome of each kind. Fertilization brings the two packs together, one from the egg, one from the sperm. This makes a striking and easily understandable image.

admitted that the answers he gave to some of the questions in biology were pathetic.[5] But one must also recognize that he failed partly because he was self-taught, partly because he was stubborn and the cast of his thought original. After two attempts to obtain a license, Mendel gave up trying. He continued to fill the position of "supply" (i.e., substitute) teacher but, beginning in 1856, devoted himself increasingly to his crossing experiments.

Mendel was a superb technician. He exhibited consummate skill in handling plants—he called them his "children"—and enormous patience. Every daylit hour he labored in his little garden or tramped the countryside to collect new specimens. In later years, his botanizing expeditions were curtailed by corpulence: "long walks and especially hill climbing [he wrote a friend] are very difficult for me in a world where universal gravitation prevails." Throughout his researches Mendel had a clear sense of his goals, of the questions he expected his experiments would help to answer. Although the outside world paid scant attention to his studies, he did not therefore disregard the scientific work carried on by others. He kept abreast of the literature in his own and related fields, followed carefully the advances in evolutionary theory and bought all Darwin's books directly they were published. In his reading he "paid no heed to the *Index Librorum Prohibitorum*"; in his scientific reasoning he was untrammeled by church dogma. Independence of mind was as characteristic of his work as was his genius for perceiving the simple in the bewilderingly complex.

Mendel was a cheerful, kindly and modest man with a nice sense of humor. It was good that he had these sturdy qualities to fall back upon when his great achievements failed entirely to win recognition. Even after the disappointment over his paper on sweet peas he continued researches on other plants—notably, the hawkweed—on the habits of bees, on sunspots and in meteorology. A report he wrote on a tornado which struck the town of Brünn is an example of his superb descriptive powers and scientific imagination. It has become a small classic of meteorology.

In 1868 Mendel was elected prelate of his monastery, and thereafter the duties of office encroached more and more upon his scientific work. He became a man of affairs, a government counselor, a bank chairman, an administrator and even a stubborn political controversialist. He engaged in a long struggle with the government over certain tax assessments against Catholic monasteries. His behavior in this affair had paranoid overtones, Mendel being convinced that the Austrian Parliament had passed legislation for the express purpose of harassing him and impoverishing his establishment. One suspects that the real cause of his bitterness and intransi-

[5] See Iltis, *op. cit.*, p. 69 *et seq.*

gence was the indifferent reception accorded his scientific theories; he was bent on being a difficult and disagreeable old man and he succeeded. One sympathizes with him fully. Mendel died on January 6, 1884. "My time will come," he remarked, a short time before his death. Today he is recognized, as his biographer Iltis has said, "as one of the chief among those who have brought light into the world."

This body in which we journey across the isthmus between the two oceans is not a private carriage but an omnibus. —OLIVER WENDELL HOLMES

I am, in point of fact, a particularly haughty and exclusive person, of pre-Adamite ancestral descent. You will understand this when I tell you that I can trace my ancestry back to a protoplasmal primordial atomic globule.
—W. S. GILBERT (*The Mikado*)

Much of what we know about man is derived from the study of sweet peas and a species of vinegar fly. —AUTHOR UNKNOWN

There was a young man in Rome, that was very like Augustus Caesar; Augustus took knowledge of it and sent for the man, and asked him, "Was your mother never at Rome?" He answered, "No, sir; but my father was."
—FRANCIS BACON

13 Mathematics of Heredity

By GREGOR MENDEL

EXPERIMENTS IN PLANT-HYBRIDIZATION

INTRODUCTORY REMARKS

EXPERIENCE of artificial fertilization, such as is effected with ornamental plants in order to obtain new variations in colour, has led to the experiments which will here be discussed. The striking regularity with which the same hybrid forms always reappeared whenever fertilization took place between the same species induced further experiments to be undertaken, the object of which was to follow up the developments of the hybrids in their progeny. . . .

That, so far, no generally applicable law governing the formation and development of hybrids has been successfully formulated can hardly be wondered at by anyone who is acquainted with the extent of the task, and can appreciate the difficulties with which experiments of this class have to contend. A final decision can only be arrived at when we shall have before us the results of detailed experiments made on plants belonging to the most diverse orders.

Those who survey the work done in this department will arrive at the conviction that, among all the numerous experiments made, not one has been carried out to such an extent and in such a way as to make it possible to determine the number of different forms under which the offspring of hybrids appear, or to arrange these forms with certainty according to their separate generations, or definitely to ascertain their statistical relations. It requires indeed some courage to undertake a labour of such far-reaching extent; this appears, however, to be the only right way by which we can finally reach the solution of a question the importance of which cannot be over-estimated in connection with the history of the evolution of organic forms.

The paper now presented records the results of such a detailed experiment. This experiment was practically confined to a small plant group, and is now, after eight years' pursuit, concluded in all essentials. Whether the plan upon which the separate experiments were conducted and carried out was the best suited to attain the desired end is left to the friendly decision of the reader.

SELECTION OF THE EXPERIMENTAL PLANTS

. . . The selection of the plant group which shall serve for experiments of this kind must be made with all possible care if it be desired to avoid from the outset every risk of questionable results.

The experimental plants must necessarily—

1. Possess constant differentiating characters.

2. The hybrids of such plants must, during the flowering period, be protected from the influence of all foreign pollen, or be easily capable of such protection.

The hybrids and their offspring should suffer no marked disturbance in their fertility in the successive generations.

Accidental impregnation by foreign pollen, if it occurred during the experiments and were not recognized, would lead to entirely erroneous conclusions. Reduced fertility or entire sterility of certain forms, such as occurs in the offspring of many hybrids, would render the experiments very difficult or entirely frustrate them. In order to discover the relations in which the hybrid forms stand towards each other and also towards their progenitors, it appears to be necessary that all members of the series developed in each successive generation should be, *without exception*, subjected to observation.

At the very outset special attention was devoted to the *Leguminosæ* on account of their peculiar floral structure. Experiments which were made with several members of this family led to the result that the genus *Pisum* was found to possess the necessary qualifications.

Some thoroughly distinct forms of this genus possess characters which are constant, and easily and certainly recognizable, and when their hybrids are mutually crossed they yield perfectly fertile progeny. Furthermore, a disturbance through foreign pollen cannot easily occur, since the fertilizing organs are closely packed inside the keel and the anther bursts within the bud, so that the stigma becomes covered with pollen even before the flower opens. This circumstance is of especial importance. As additional advantages worth mentioning, there may be cited the easy culture of these plants in the open ground and in pots, and also their relatively short period of growth. Artificial fertilization is certainly a somewhat elaborate process, but nearly always succeeds. For this purpose the bud is opened before it is perfectly developed, the keel is removed, and each stamen carefully extracted by means of forceps, after which the stigma can at once be dusted over with the foreign pollen. . . .

DIVISION AND ARRANGEMENT OF THE EXPERIMENTS

If two plants which differ constantly in one or several characters be crossed, numerous experiments have demonstrated that the common characters are transmitted unchanged to the hybrids and their progeny; but each pair of differentiating characters, on the other hand, unite in the hybrid to form a new character, which in the progeny of the hybrid is usually variable. The object of the experiment was to observe these variations in the case of each pair of differentiating characters, and to deduce the law according to which they appear in the successive generations. . . .

The characters which were selected for experiment relate:

1. To the *difference in the form of the ripe seeds.* These are either round or roundish; the depressions, if any, occur on the surface, being always only shallow; or they are irregularly angular and deeply wrinkled (*P. quadratum*).

2. To the *difference in the colour of the seed albumen* (endosperm). The albumen of the ripe seeds is either pale yellow, bright yellow and orange coloured, or it possesses a more or less intense green tint. This difference of colour is easily seen in the seeds as their coats are transparent.

3. To the *difference in the colour of the seed-coat.* This is either white, with which character white flowers are constantly correlated, or it is grey, grey-brown, leather-brown, with or without violet spotting, in which case the colour of the standards is violet, that of the wings purple, and the stem in the axils of the leaves is of a reddish tint. The grey seed-coats become dark brown in boiling water.

4. To the *difference in the form of the ripe pods.* These are either simply inflated, not contracted in places; or they are deeply constricted between the seeds and more or less wrinkled (*P. saccharatum*).

5. To the *difference in the colour of the unripe pods.* They are either light to dark green, or vividly yellow, in which colouring the stalks, leaf-veins, and calyx participate.

6. To the *difference in the position of the flowers.* They are either axial, that is, distributed along the main stem; or they are terminal, that is, bunched at the top of the stem and arranged almost in a false umbel; in this case the upper part of the stem is more or less widened in section (*P. umbellatum*).

7. To the *difference in the length of the stem.* The length of the stem is very various in some forms; it is, however, a constant character for

each, in so far that healthy plants, grown in the same soil, are only subject to unimportant variations in this character.

In experiments with this character, in order to be able to discriminate with certainty, the long axis of 6 to 7 ft. was always crossed with the short one of ¾ ft. to 1½ ft.

Each two of the differentiating characters enumerated above were united by cross-fertilization. There were made for the

1st trial	60	fertilizations	on	15	plants.
2nd "	58	"	"	10	"
3rd "	35	"	"	10	"
4th "	40	"	"	10	"
5th "	23	"	"	5	"
6th "	34	"	"	10	"
7th "	37	"	"	10	"

From a larger number of plants of the same variety only the most vigorous were chosen for fertilization. Weakly plants always afford uncertain results, because even in the first generation of hybrids, and still more so in the subsequent ones, many of the offspring either entirely fail to flower or only form a few and inferior seeds.

Furthermore, in all the experiments reciprocal crossings were effected in such a way that each of the two varieties which in one set of fertilizations served as seed-bearer in the other set was used as the pollen plant. . . .

THE FORMS OF THE HYBRIDS

Experiments which in previous years were made with ornamental plants have already afforded evidence that the hybrids, as a rule, are not exactly intermediate between the parental species. . . . This is . . . the case with the Pea-hybrids. In the case of each of the seven crosses the hybrid-character resembles that of one of the parental forms so closely that the other either escapes observation completely or cannot be detected with certainty. This circumstance is of great importance in the determination and classification of the forms under which the offspring of the hybrids appear. Henceforth in this paper those characters which are transmitted entire, or almost unchanged in the hybridization, and therefore in themselves constitute the characters of the hybrid, are termed the *dominant*, and those which become latent in the process *recessive*. The expression "recessive" has been chosen because the characters thereby designated withdraw or entirely disappear in the hybrids, but nevertheless reappear unchanged in their progeny, as will be demonstrated later on.

It was furthermore shown by the whole of the experiments that it is perfectly immaterial whether the dominant character belong to the seed-bearer or to the pollen-parent; the form of the hybrid remains identical in both cases. . . .

Of the differentiating characters which were used in the experiments the following are dominant:

1. The round or roundish form of the seed with or without shallow depressions.

2. The yellow colour of the seed albumen.

3. The grey, grey-brown, or leather-brown colour of the seed-coat, in association with violet-red blossoms and reddish spots in the leaf axils.

4. The simply inflated form of the pod.

5. The green colouring of the unripe pod in association with the same colour in the stems, the leaf-veins and the calyx.

6. The distribution of the flowers along the stem.

7. The greater length of stem. . . .

THE FIRST GENERATION [BRED] FROM THE HYBRIDS

In this generation there reappear, together with the dominant characters, also the recessive ones with their peculiarities fully developed, and this occurs in the definitely expressed average proportion of three to one, so that among four plants of this generation three display the dominant character and one the recessive. This relates without exception to all the characters which were investigated in the experiments. . . . *Transitional forms were not observed in any experiment.* . . .

. . . The relative numbers which were obtained for each pair of differentiating characters are as follows:

Expt. 1. Form of seed.—From 253 hybrids 7,324 seeds were obtained in the second trial year. Among them were 5,474 round or roundish ones and 1,850 angular wrinkled ones. Therefrom the ratio 2·96 to 1 is deduced.

Expt. 2. Colour of albumen.—258 plants yielded 8,023 seeds, 6,022 yellow, and 2,001 green; their ratio, therefore, is as 3·01 to 1. . . .

Expt. 3. Colour of the seed-coats.—Among 929 plants 705 bore violet-red flowers and grey-brown seed-coats, giving the proportion 3·15 to 1.

Expt. 4. Form of pods.—Of 1,181 plants 882 had them simply inflated, and in 299 they were constricted. Resulting ratio, 2·95 to 1.

Expt. 5. Colour of the unripe pods.—The number of trial plants was 580, of which 428 had green pods and 152 yellow ones. Consequently these stand in the ratio 2·82 to 1.

Expt. 6. Position of flowers.—Among 858 cases 651 had inflorescences axial and 207 terminal. Ratio, 3·14 to 1.

Expt. 7. Length of stem.—Out of 1,064 plants, in 787 cases the stem was long, and in 277 short. Hence a mutual ratio of 2·84 to 1. In this experiment the dwarfed plants were carefully lifted and transferred to a special bed. This precaution was necessary, as otherwise they would have perished through being overgrown by their tall relatives. Even in their quite young state they can be easily picked out by their compact growth and thick dark-green foliage.

If now the results of the whole of the experiments be brought together, there is found, as between the number of forms with the dominant and recessive characters, an average ratio of 2·98 to 1, or 3 to 1.

The dominant character can here have a *double signification*—viz. that of a parental character, or a hybrid character. In which of the two significations it appears in each separate case can only be determined by the following generation. As a parental character it must pass over unchanged to the whole of the offspring; as a hybrid-character, on the other hand, it must maintain the same behaviour as in the first generation.

THE SECOND GENERATION [BRED] FROM THE HYBRIDS

Those forms which in the first generation exhibit the recessive character do not further vary in the second generation as regards this character; they remain constant in their offspring.

It is otherwise with those which possess the dominant character in the first generation (bred from the hybrids). Of these *two*-thirds yield offspring which display the dominant and recessive characters in the proportion of 3 to 1, and thereby show exactly the same ratio as the hybrid forms, while only *one*-third remains with the dominant character constant.

The separate experiments yielded the following results:

Expt. 1. Among 565 plants which were raised from round seeds of the first generation, 193 yielded round seeds only, and remained therefore constant in this character; 372, however, gave both round and wrinkled seeds, in the proportion of 3 to 1. The number of the hybrids, therefore, as compared with the constants is 1·93 to 1.

Expt. 2. Of 519 plants which were raised from seeds whose albumen was of yellow colour in the first generation, 166 yielded exclusively yellow, whilst 353 yielded yellow and green seeds in the proportion of 3 to 1. There resulted, therefore, a division into hybrid and constant forms in the proportion of 2·13 to 1.

For each separate trial in the following experiments 100 plants were selected which displayed the dominant character in the first generation, and in order to ascertain the significance of this, ten seeds of each were cultivated.

Expt. 3. The offspring of 36 plants yielded exclusively grey-brown seed-coats, while of the offspring of 64 plants some had grey-brown and some had white.

Expt. 4. The offspring of 29 plants had only simply inflated pods; of the offspring of 71, on the other hand, some had inflated and some had constricted.

Expt. 5. The offspring of 49 plants had only green pods; of the offspring of 60 plants some had green, some yellow ones.

Expt. 6. The offspring of 33 plants had only axial flowers; of the offspring of 67, on the other hand, some had axial and some terminal flowers.

Expt. 7. The offspring of 28 plants inherited the long axis, and those of 72 plants some the long and some the short axis.

In each of these experiments a certain number of the plants came constant with the dominant character. For the determination of the proportion in which the separation of the forms with the constantly persistent character results, the two first experiments are of especial importance, since in these a larger number of plants can be compared. The ratios 1·93 to 1 and 2·13 to 1 gave together almost exactly the average ratio of 2 to 1. The sixth experiment gave a quite concordant result; in the others the ratio varies more or less, as was only to be expected in view of the smaller number of 100 trial plants. Experiment 5, which shows the greatest departure, was repeated, and then, in lieu of the ratio of 60 and 40, that of 65 and 35 resulted. *The average ratio of* 2 *to* 1 *appears, therefore, as fixed with certainty.* It is therefore demonstrated that, of those forms which possess the dominant character in the first generation, two-thirds have the hybrid-character, while one-third remains constant with the dominant character.

The ratio of 3 to 1, in accordance with which the distribution of the dominant and recessive characters results in the first generation, resolves itself therefore in all experiments into the ratio of 2 : 1 : 1 if the dominant character be differentiated according to its significance as a hybrid-character or as a parental one. Since the members of the first generation spring directly from the seed of the hybrids, *it is now clear that the hybrids form seeds having one or the other of the two differentiating characters, and of these one-half develop again the hybrid form, while the other half yield plants which remain constant and receive the dominant or the recessive characters in equal numbers.*

THE SUBSEQUENT GENERATIONS [BRED] FROM THE HYBRIDS

The proportions in which the descendants of the hybrids develop and split up in the first and second generations presumably hold good for all subsequent progeny. Experiments 1 and 2 have already been carried through six generations, 3 and 7 through five, and 4, 5, and 6 through four, these experiments being continued from the third generations with a small number of plants, and no departure from the rule has been perceptible. The offspring of the hybrids separated in each generation in the ratio of 2 : 1 : 1 into hybrids and constant forms.

If A be taken as denoting one of the two constant characters, for instance the dominant, a, the recessive, and Aa the hybrid form in which both are conjoined, the expression

$$A + 2Aa + a$$

shows the terms in the series for the progeny of the hybrids of two differentiating characters. . . .

THE OFFSPRING OF HYBRIDS IN WHICH SEVERAL DIFFERENTIATING CHARACTERS ARE ASSOCIATED

In the experiments above described plants were used which differed only in one essential character. The next task consisted in ascertaining whether the law of development discovered in these applied to each pair of differentiating characters when several diverse characters are united in the hybrid by crossing. As regards the form of the hybrids in these cases, the experiments showed throughout that this invariably more nearly approaches to that one of the two parental plants which possesses the greater number of dominant characters. . . . Should one of the two parental types possess only dominant characters, then the hybrid is scarcely or not at all distinguishable from it.

Two experiments were made with a considerable number of plants. In the first experiment the parental plants differed in the form of the seed and in the colour of the albumen; in the second in the form of the seed, in the colour of the albumen, and in the colour of the seed-coats. Experiments with seed-characters give the result in the simplest and most certain way.

In addition, further experiments were made with a smaller number of experimental plants in which the remaining characters by twos and threes were united as hybrids; all yielded approximately the same results. There is therefore no doubt that for the whole of the characters involved in the experiments the principle applies that *the offspring of the hybrids in which several essentially different characters are combined exhibit the terms of a series of combinations, in which the developmental series for each pair of differentiating characters are united.* It is demonstrated at the same time that *the relation of each pair of different characters in hybrid union is independent of the other differences in the two original parental stocks.* . . .

All constant combinations which in Peas are possible by the combination of the said seven differentiating characters were actually obtained by repeated crossing. . . . Thereby is . . . given the practical proof *that the constant characters which appear in the several varieties of a group of plants may be obtained in all the associations which are possible according to the [mathematical] laws of combination, by means of repeated artificial fertilization. . . .*

If we endeavour to collate in a brief form the results arrived at, we find that those differentiating characters, which admit of easy and certain recognition in the experimental plants, all behave exactly alike in their hybrid associations. The offspring of the hybrids of each pair of differentiating characters are, one-half, hybrid again, while the other half are constant in equal proportions having the characters of the seed and pollen parents respectively. If several differentiating characters are combined by cross-fertilization in a hybrid, the resulting offspring form the terms of a combination series in which the combination series for each pair of differentiating characters are united.

The uniformity of behaviour shown by the whole of the characters submitted to experiment permits, and fully justifies, the acceptance of the principle that a similar relation exists in the other characters which appear less sharply defined in plants, and therefore could not be included in the separate experiments. . . .

THE REPRODUCTIVE CELLS OF THE HYBRIDS

The results of the previously described experiments led to further experiments, the results of which appear fitted to afford some conclusions as regards the composition of the egg and pollen cells of hybrids. An important clue is afforded in *Pisum* by the circumstance that among the progeny of the hybrids constant forms appear, and that this occurs, too, in respect of all combinations of the associated characters. So far as experience goes, we find it in every case confirmed that constant progeny can only be formed when the egg cells and the fertilizing pollen are of like character, so that both are provided with the material for creating quite similar individuals, as is the case with the normal fertilization of pure species. We must therefore regard it as certain that exactly similar factors must be at work also in the production of the constant forms in the hybrid plants. Since the various constant forms are produced in *one* plant, or even in one flower of a plant, the conclusion appears logical that in the ovaries of the hybrids there are formed as many sorts of egg cells, and in the anthers as many sorts of pollen cells, as there are possible constant combination forms, and that these egg and pollen cells agree in their internal composition with those of the separate forms.

In point of fact it is possible to demonstrate theoretically that this hypothesis would fully suffice to account for the development of the hybrids in the separate generations, if we might at the same time assume that the various kinds of egg and pollen cells were formed in the hybrids on the average in equal numbers.

In order to bring these assumptions to an experimental proof, the following experiments were designed. Two forms which were constantly different in the form of the seed and the colour of the albumen were united by fertilization.

If the differentiating characters are again indicated as *A*, *B*, *a*, *b*, we have:

AB, seed parent;	*ab*, pollen parent;
A, form round;	*a*, form wrinkled;
B, albumen yellow.	*b*, albumen green.

The artificially fertilized seeds were sown together with several seeds of both original stocks, and the most vigorous examples were chosen for the reciprocal crossing. There were fertilized:

1. The hybrids with the pollen of *AB*.
2. The hybrids " " " " *ab*.
3. *AB* " " " " the hybrids.
4. *ab* " " " " the hybrids.

For each of these four experiments the whole of the flowers on three plants were fertilized. If the above theory be correct, there must be developed on the hybrids egg and pollen cells of the forms *AB*, *Ab*, *aB*, *ab*, and there would be combined:

1. The egg cells *AB*, *Ab*, *aB*, *ab* with the pollen cells *AB*.
2. The egg cells *AB*, *Ab*, *aB*, *ab* with the pollen cells *ab*.
3. The egg cells *AB* with the pollen cells *AB*, *Ab*, *aB*, *ab*.
4. The egg cells *ab* with the pollen cells *AB*, *Ab*, *aB*, *ab*.

From each of these experiments there could then result only the following forms:

1. *AB*, *ABb*, *AaB*, *AaBb*.
2. *AaBb*, *Aab*, *aBb*, *ab*.
3. *AB*, *ABb*, *AaB*, *AaBb*.
4. *AaBb*, *Aab*, *aBb*, *ab*.

If, furthermore, the several forms of the egg and pollen cells of the hybrids were produced on an average in equal numbers, then in each experiment the said four combinations should stand in the same ratio to each other. A perfect agreement in the numerical relations was, however, not to be expected, since in each fertilization, even in normal cases, some egg cells remain undeveloped or subsequently die, and many even of the well-formed seeds fail to germinate when sown. . . .

The first and second experiments had primarily the object of proving the composition of the hybrid egg cells, while the third and fourth experiments were to decide that of the pollen cells. As is shown by the above demonstration the first and third experiments and the second and fourth should produce precisely the same combinations, and even in the second year the result should be partially visible in the form and colour of the artificially fertilized seed. In the first and third experiments the dominant characters of form and colour, A and B, appear in each union, . . . partly constant and partly in hybrid union with the recessive characters a and b, for which reason they must impress their peculiarity upon the whole of the seeds. All seeds should therefore appear round and yellow, if the theory be justified. In the second and fourth experiments, on the other hand, one union is hybrid in form and in colour, and consequently the seeds are round and yellow; another is hybrid in form, but constant in the recessive character of colour, whence the seeds are round and green; the third is constant in the recessive character of form but hybrid in colour, consequently the seeds are wrinkled and yellow; the fourth is constant in both recessive characters, so that the seeds are wrinkled and green. In both these experiments there were consequently four sorts of seed to be expected—viz. round and yellow, round and green, wrinkled and yellow, wrinkled and green.

The crop fulfilled these expectations perfectly. There were obtained in the

1st Experiment, 98 exclusively round yellow seeds;
3rd " 94 " " " "

In the 2nd Experiment, 31 round and yellow, 26 round and green, 27 wrinkled and yellow, 26 wrinkled and green seeds.

In the 4th Experiment, 24 round and yellow, 25 round and green, 22 wrinkled and yellow, 27 wrinkled and green seeds. . . .

In a further experiment the characters of flower-colour and length of stem were experimented upon. . . . For the characters of form of pod, colour of pod, and position of flowers experiments were also made on a small scale, and results obtained in perfect agreement. All combinations which were possible through the union of the differentiating characters duly appeared, and in nearly equal numbers.

Experimentally, therefore, the theory is confirmed that *the pea hybrids form pollen and egg cells which, in their constitution, represent in equal numbers all constant forms which result from the combination of the characters united in fertilization.*

The difference of the forms among the progeny of the hybrids, as well as the respective ratios of the numbers in which they are observed, find a sufficient explanation in the principle above deduced. The simplest case is afforded by the developmental series of each pair of differentiating characters. This series is represented by the expression $A + 2Aa + a$, in which A and a signify the forms with constant differentiating characters, and Aa

the hybrid form of both. It includes in three different classes four individuals. In the formation of these, pollen and egg cells of the form A and a take part on the average equally in the fertilization, hence each form occurs twice, since four individuals are formed. There participate consequently in the fertilization

The pollen cells $A + A + a + a$.
The egg cells $A + A + a + a$.

It remains, therefore, purely a matter of chance which of the two sorts of pollen will become united with each separate egg cell. According, however, to the law of probability, it will always happen, on the average of many cases, that each pollen form A and a will unite equally often with each egg cell form A and a, consequently one of the two pollen cells A in the fertilization will meet with the egg cell A, and the other with an egg cell a, and so likewise one pollen cell will unite with an egg cell A, and the other with egg cell a.

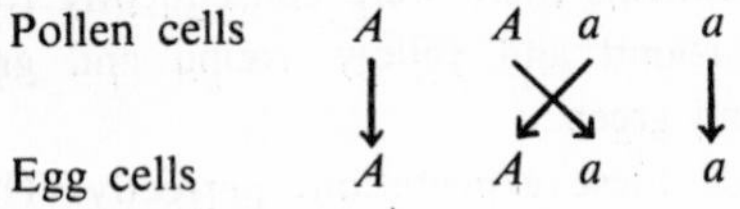

Pollen cells	A	A	a	a
	↓	↘	↙	↓
Egg cells	A	A	a	a

The results of the fertilization may be made clear by putting the signs for the conjoined egg and pollen cells in the forms of fractions, those for the pollen cells above and those for the egg cells below the line. We then have

$$\frac{A}{A} + \frac{A}{a} + \frac{a}{A} + \frac{a}{a}.$$

In the first and fourth term the egg and pollen cells are of like kind, consequently the product of their union must be constant, viz. A and a; in the second and third, on the other hand, there again results a union of the two differentiating characters of the stocks, consequently the forms resulting from these fertilizations are identical with those of the hybrid from which they sprang. *There occurs accordingly a repeated hybridization.* This explains the striking fact that the hybrids are able to produce, besides the two parental forms, offspring which are like themselves; $\frac{A}{a}$ and $\frac{a}{A}$ both give the same union Aa, since, as already remarked above, it makes no difference in the result of fertilization to which of the two characters the pollen or egg cells belong. We may write then

$$\frac{A}{A} + \frac{A}{a} + \frac{a}{A} + \frac{a}{a} = A + 2Aa + a.$$

This represents the average result of the self-fertilization of the hybrids when two differentiating characters are united in them. In individual flowers and in individual plants, however, the ratios in which the forms of the series are produced may suffer not inconsiderable fluctuations. Apart from the fact that the numbers in which both sorts of egg cells occur in the seed vessels can only be regarded as equal on the average, it remains purely a matter of chance which of the two sorts of pollen may fertilize each separate egg cell. For this reason the separate values must necessarily be subject to fluctuations, and there are even extreme cases possible, as were described earlier in connection with the experiments on the form of the seed and the colour of the albumen. The true ratios of the numbers can only be ascertained by an average deduced from the sum of as many single values as possible; the greater the number, the more are merely chance effects eliminated.

The law of combination of different characters which governs the development of the hybrids finds therefore its foundation and explanation in the principle enunciated, that the hybrids produce egg cells and pollen cells which in equal numbers represent all constant forms which result from the combinations of all the characters brought together in fertilization.

COMMENTARY ON

J. B. S. HALDANE

J. B. S. HALDANE is prodigious. In mind and body his capacity is beyond anything easily imagined. For the sake of convenient identification he may be called a geneticist; in this branch of knowledge he is among the leaders. (His regular post is professor of biometry at University College, London.) He has made contributions, many of high value, to biology, physiology, preventive medicine, botany, hematology, statistical theory, prevention of air-raid casualties and the effects of various gases and other chemical and physical agents on the human body—frequently his own. He has subjected himself to high pressures, intense cold, poisoning, disease inoculations, fevers, temporary paralysis and other unpleasantnesses—indeed he has done almost everything but put his head on a railroad track—all on behalf of science. He has so comprehensive a grasp of mathematics and physics that he has made a number of brilliant suggestions in connection with the work of E. A. Milne, the late British mathematical physicist who invented the theory of kinematic relativity, a supplement to Einstein's great concept. He is the author of several distinguished scientific books, a Fellow of the Royal Society, a rationalist, an agnostic and a Marxist. The last fact is worth mentioning because it helps to explain certain of Haldane's prolific opinions—some of them strained as well as strange—on a variety of subjects, scientific and nonscientific. There is, however, no evidence that this conviction has affected his views in the material that follows. At any rate he may have more reason than most of us to embrace the philosophy of Marx: the story is told that Haldane cured himself of stomach ulcers by reading Engels, an interesting, and hitherto perhaps unsuspected, benefit of this author's writings.

On Being the Right Size, the first of the selections below, is a sparkling little essay written more than twenty-five years ago. It deals with a problem that has attracted attention since ancient times, that cuts across various fields from biology and aeronautics to social polity and engineering. It has a moral and it has to do with mathematics; the moral is inoffensive and the mathematics is profound but simple.

The second selection consists of excerpts from Haldane's notable book, *The Causes of Evolution*. Haldane was among the first to apply mathematics to the study of natural selection and to other aspects of evolutionary theory. "Darwin," he says, "thought in words. His successors today have to think in numbers." Mathematics is used in fixing the evolutionary time scale; radioactive minerals are dated by measuring their content of a particular isotope of lead, produced gradually and at a fairly regular rate by the radioactive transformation of uranium (or thorium). Another procedure in which mathematics is needed involves meticulous measurement and comparison of fossils to determine the rate of evolutionary change. The results of such investigations are astonishing. Haldane gives the example of equine teeth which "have been getting longer for some fifty million years." [1] The process, however, is very slow: "if you measure corresponding teeth from a population of fossil horses and from their descendants two million years later, although the average values have changed, there is often still some overlap. That is to say the shortest teeth two million years later are no longer than the longest two million years past."

A complex and interesting use of mathematics is in the analysis of natural selection. One problem is "to see if you can change the character of population by exposing it to natural selection under controlled conditions." [2] Experiments along these lines with populations of flies have been conducted successfully by scientists in the U. S., England, France and Russia. The mathematical theory of these changes has drawn the attention of Haldane, S. Wright and the noted British statistical theorist R. A. Fisher, among others. The excerpts chosen have to do with the effect of selection of a given intensity and with related questions. I have cut out the more advanced mathematics so that the general reader can gain some notion of what this fascinating and important subject is about. Haldane's treatment is unfailingly clear but I will not promise that the text is so simple as to obviate the need for concentration.

[1] J. B. S. Haldane, *Everything Has a History*, London, 1951, p. 220.
[2] *Ibid.*

From what has already been demonstrated, you can plainly see the impossibility of increasing the size of structures to vast dimensions either in art or in nature; likewise the impossibility of building ships, palaces, or temples of enormous size in such a way that their oars, yards, beams, iron bolts, and, in short, all their other parts will hold together; nor can nature produce trees of extraordinary size because the branches would break down under their own weight, so also it would be impossible to build up the bony structures of men, horses, or other animals so as to hold together and perform their normal functions if these animals were to be increased enormously in height; for this increase in height can be accomplished only by employing a material which is harder and stronger than usual, or by enlarging the size of the bones, thus changing their shape until the form and appearance of the animals suggest a monstrosity. This is perhaps what our wise Poet had in mind, when he says, in describing a huge giant:

"Impossible it is to reckon his height
So beyond measure is his size." —GALILEO GALILEI

14 On Being the Right Size

By J. B. S. HALDANE

THE most obvious differences between different animals are differences of size, but for some reason the zoologists have paid singularly little attention to them. In a large textbook of zoology before me I find no indication that the eagle is larger than the sparrow, or the hippopotamus bigger than the hare, though some grudging admissions are made in the case of the mouse and the whale. But yet it is easy to show that a hare could not be as large as a hippopotamus, or a whale as small as a herring. For every type of animal there is a most convenient size, and a large change in size inevitably carries with it a change of form.

Let us take the most obvious of possible cases, and consider a giant man sixty feet high—about the height of Giant Pope and Giant Pagan in the illustrated *Pilgrim's Progress* of my childhood. These monsters were not only ten times as high as Christian, but ten times as wide and ten times as thick, so that their total weight was a thousand times his, or about eighty to ninety tons. Unfortunately the cross sections of their bones were only a hundred times those of Christian, so that every square inch of giant bone had to support ten times the weight borne by a square inch of human bone. As the human thigh-bone breaks under about ten times the human weight, Pope and Pagan would have broken their thighs every time they took a step. This was doubtless why they were sitting down in the picture I remember. But it lessens one's respect for Christian and Jack the Giant Killer.

To turn to zoology, suppose that a gazelle, a graceful little creature with long thin legs, is to become large, it will break its bones unless it does one of two things. It may make its legs short and thick, like the rhinoceros, so that every pound of weight has still about the same area of bone to support it. Or it can compress its body and stretch out its legs obliquely to gain stability, like the giraffe. I mention these two beasts because they happen to belong to the same order as the gazelle, and both are quite successful mechanically, being remarkably fast runners.

Gravity, a mere nuisance to Christian, was a terror to Pope, Pagan, and Despair. To the mouse and any smaller animal it presents practically no dangers. You can drop a mouse down a thousand-yard mine shaft; and, on arriving at the bottom, it gets a slight shock and walks away. A rat would probably be killed, though it can fall safely from the eleventh story of a building; a man is killed, a horse splashes. For the resistance presented to movement by the air is proportional to the surface of the moving object. Divide an animal's length, breadth, and height each by ten; its weight is reduced to a thousandth, but its surface only to a hundredth. So the resistance to falling in the case of the small animal is relatively ten times greater than the driving force.

An insect, therefore, is not afraid of gravity; it can fall without danger, and can cling to the ceiling with remarkably little trouble. It can go in for elegant and fantastic forms of support like that of the daddy-long-legs. But there is a force which is as formidable to an insect as gravitation to a mammal. This is surface tension. A man coming out of a bath carries with him a film of water of about one-fiftieth of an inch in thickness. This weighs roughly a pound. A wet mouse has to carry about its own weight of water. A wet fly has to lift many times its own weight and, as every one knows, a fly once wetted by water or any other liquid is in a very serious position indeed. An insect going for a drink is in as great danger as a man leaning out over a precipice in search of food. If it once falls into the grip of the surface tension of the water—that is to say, gets wet—it is likely to remain so until it drowns. A few insects, such as water-beetles, contrive to be unwettable, the majority keep well away from their drink by means of a long proboscis.

Of course tall land animals have other difficulties. They have to pump their blood to greater heights than a man and, therefore, require a larger blood pressure and tougher blood-vessels. A great many men die from burst arteries, especially in the brain, and this danger is presumably still greater for an elephant or a giraffe. But animals of all kinds find difficulties in size for the following reason. A typical small animal, say a microscopic worm or rotifer, has a smooth skin through which all the oxygen it requires can soak in, a straight gut with sufficient surface to absorb its food, and a simple kidney. Increase its dimensions tenfold in every direction, and its weight is increased a thousand times, so that if it is to use its muscles as efficiently as its miniature counterpart, it will need a thousand times as much food and oxygen per day and will excrete a thousand times as much of waste products.

Now if its shape is unaltered its surface will be increased only a hundredfold, and ten times as much oxygen must enter per minute through each square millimetre of skin, ten times as much food through each square millimetre of intestine. When a limit is reached to their absorptive powers their surface has to be increased by some special device. For example, a part of the skin may be drawn out into tufts to make gills or pushed in to make lungs, thus increasing the oxygen-absorbing surface in proportion to the animal's bulk. A man, for example, has a hundred square yards of lung. Similarly, the gut, instead of being smooth and straight, becomes coiled and develops a velvety surface, and other organs increase in complication. The higher animals are not larger than the lower because they are more complicated. They are more complicated because they are larger. Just the same is true of plants. The simplest plants, such as the green algae growing in stagnant water or on the bark of trees, are mere round cells. The higher plants increase their surface by putting out leaves and roots. Comparative anatomy is largely the story of the struggle to increase surface in proportion to volume.

Some of the methods of increasing the surface are useful up to a point, but not capable of a very wide adaptation. For example, while vertebrates carry the oxygen from the gills or lungs all over the body in the blood, insects take air directly to every part of their body by tiny blind tubes called tracheae which open to the surface at many different points. Now, although by their breathing movements they can renew the air in the outer part of the tracheal system, the oxygen has to penetrate the finer branches by means of diffusion. Gases can diffuse easily through very small distances, not many times larger than the average length travelled by a gas molecule between collisions with other molecules. But when such vast journeys—from the point of view of a molecule—as a quarter of an inch have to be made, the process becomes slow. So the portions of an insect's body more than a quarter of an inch from the air would always be short of oxygen. In consequence hardly any insects are much more than half an inch thick. Land crabs are built on the same general plan as insects, but are much clumsier. Yet like ourselves they carry oxygen around in their blood, and are therefore able to grow far larger than any insects. If the insects had hit on a plan for driving air through their tissues instead of letting it soak in, they might well have become as large as lobsters, though other considerations would have prevented them from becoming as large as man.

Exactly the same difficulties attach to flying. It is an elementary principle of aeronautics that the minimum speed needed to keep an aeroplane of a given shape in the air varies as the square root of its length. If its linear dimensions are increased four times, it must fly twice as fast. Now the power needed for the minimum speed increases more rapidly than the weight of the machine. So the larger aeroplane, which weighs sixty-four times as much as the smaller, needs one hundred and twenty-eight times its horsepower to keep up. Applying the same principles to the birds, we find that the limit to their size is soon reached. An angel whose muscles developed no more power weight for weight than those of an eagle or a pigeon would require a breast projecting for about four feet to house the muscles engaged in working its wings, while to economize in weight, its legs would have to be reduced to mere stilts. Actually a large bird such as an eagle or kite does not keep in the air mainly by moving its wings. It is generally to be seen soaring, that is to say balanced on a rising column of air. And even soaring becomes more and more difficult with increasing size. Were this not the case eagles might be as large as tigers and as formidable to man as hostile aeroplanes.

But it is time that we passed to some of the advantages of size. One of the most obvious is that it enables one to keep warm. All warm-blooded animals at rest lose the same amount of heat from a unit area of skin, for which purpose they need a food-supply proportional to their surface and not to their weight. Five thousand mice weigh as much as a man. Their combined surface and food or oxygen consumption are about seventeen times a man's. In fact a mouse eats about one quarter its own weight of food every day, which is mainly used in keeping it warm. For the same reason small animals cannot live in cold countries. In the arctic regions there are no reptiles or amphibians, and no small mammals. The smallest mammal in Spitzbergen is the fox. The small birds fly away in the winter, while the insects die, though their eggs can survive six months or more of frost. The most successful mammals are bears, seals, and walruses.

Similarly, the eye is a rather inefficient organ until it reaches a large size. The back of the human eye on which an image of the outside world is thrown, and which corresponds to the film of a camera, is composed of a mosaic of 'rods and cones' whose diameter is little more than a length of an average light wave. Each eye has about half a million, and for two objects to be distinguishable their images must fall on separate rods or cones. It is obvious that with fewer but larger rods and cones we should see less distinctly. If they were twice as broad two points would have to be twice as far apart before we could distinguish them at a given distance. But if their size were diminished and their number increased we should see no better. For it is impossible to form a definite image smaller than a wave-length of light. Hence a mouse's eye is not a small-scale model of a human eye. Its rods and cones are not much smaller than ours, and therefore there are far fewer of them. A mouse could not distinguish one human face from another six feet away. In order that they should be of any use at all the eyes of small animals have to be much larger in proportion to their bodies than our own. Large animals on the other hand only require relatively small eyes, and those of the whale and elephant are little larger than our own.

For rather more recondite reasons the same general principle holds true of the brain. If we compare the brain-weights of a set of very similar animals such as the cat, cheetah, leopard, and tiger, we find that as we quadruple the body-weight the brain-weight is only doubled. The larger animal with proportionately larger bones can economize on brain, eyes, and certain other organs.

Such are a very few of the considerations which show that for every type of animal there is an optimum size. Yet although Galileo demonstrated the contrary more than three hundred years ago, people still believe that if a flea were as large as a man it could jump a thousand feet into the air. As a matter of fact the height to which an animal can jump is more nearly independent of its size than proportional to it. A flea can jump about two feet, a man about five. To jump a given height, if we neglect the resistance of the air, requires an expenditure of energy proportional to the jumper's weight. But if the jumping muscles form a constant fraction of the animal's body, the energy developed per ounce of muscle is independent of the size, provided it can be developed quickly enough in the small animal. As a matter of fact an insect's muscles, although they can contract more quickly than our own, appear to be less efficient; as otherwise a flea or grasshopper could rise six feet into the air.

And just as there is a best size for every animal, so the same is true for every human institution. In the Greek type of democracy all the citizens could listen to a series of orators and vote directly on questions of legislation. Hence their philosophers held that a small city was the largest possible democratic state. The English invention of representative government made a democratic nation possible, and the possibility was first realized in the United States, and later elsewhere. With the development of broadcasting it has once more become possible for every citizen to listen to the political views of representative orators, and the future may perhaps see the return of the national state to the Greek form of democracy. Even the referendum has been made possible only by the institution of daily newspapers.

To the biologist the problem of socialism appears largely as a problem of size. The extreme socialists desire to run every nation as a single business concern. I do not suppose that Henry Ford would find much difficulty in running Andorra or Luxembourg on a socialistic basis. He has already more men on his pay-roll than their population. It is conceivable that a syndicate of Fords, if we could find them, would make Belgium Ltd. or Denmark Inc. pay their way. But while nationalization of certain industries is an obvious possibility in the largest of states, I find it no easier to picture a completely socialized British Empire or United States than an elephant turning somersaults or a hippopotamus jumping a hedge.

For every living creature that succeeds in getting a footing in life there are thousands or millions that perish. There is an enormous random scattering for every seed that comes to life. This does not remind us of intelligent human design. "If a man in order to shoot a hare, were to discharge thousands of guns on a great moor in all possible directions; if in order to get into a locked room, he were to buy ten thousand casual keys, and try them all; if, in order to have a house, he were to build a town, and leave all the other houses to wind and weather—assuredly no one would call such proceedings purposeful and still less would anyone conjecture behind these proceedings a higher wisdom, unrevealed reasons, and superior prudence." [*Yet so is Nature . . .*]

—J. W. N. SULLIVAN

(*The words quoted by Sullivan are from Lange, "History of Materialism."*)

15 Mathematics of Natural Selection

By J. B. S. HALDANE

WE have seen that there is no question that natural selection does occur. We must next consider what would be the effect of selection of a given intensity. The mathematical theory of natural selection where inheritance is Mendelian has been mainly developed by R. A. Fisher, S. Wright, and myself. Some of the more important results are summarised in the Appendix, but I shall deal with a few of them here. The first question which arises is how we are to measure that intensity. I shall confine myself to organisms, such as annual plants and insects, where generations do not overlap. The more general case, exemplified by man, can only be treated by means of integral equations. Suppose we have two competing types A and B, say dark and light moths or virulent and nonvirulent bacteria. Then if in one generation the ratio of A to B changes from r to $r(1+k)$ we shall call k the coefficient of selection. Of course k will not be steady. In one year an early spring will give an advantage to early maturing seeds. In the next year a late frost will reverse the process. Nor will it be constant from one locality to another, as is clear in the case of the moths just cited. We must take average values over considerable periods and areas. The value of k will increase with the proportion of individuals killed off by selection, but after selection has become intense enough to kill off about 80 per cent. of the population it increases rather slowly, roughly as the logarithm of the number killed off per survivor—sometimes even as the square root of the logarithm. In what follows I shall suppose k to be small.

The effect of selection of a given intensity depends entirely on the type of inheritance of the character selected and the system of mating. I will confine myself for the moment to characters inherited in an alternative manner, in a population either mating at random or self-fertilised. If two races do not cross, or if the inheritance is cytoplasmic, and if u_n is the ratio of A to B after n generations, then $u_n = e^{kn}u_o$, or $kn = \log_e \frac{u_n}{u_o}$. If the character is due to a single dominant gene, and u_n is the ratio of dominant to recessive genes, then

$$k_n = u_n - u_o + \log_e \frac{u_n}{u_o}.$$

This means that selection is rapid when populations contain a reasonable proportion of recessives, but excessively slow, in either direction, when recessives are very rare (see Figure 1). Thus if $k = 1/1000$, *i.e.*, 1001 of one type survive to breed for every 1000 of the other, it would take 11,739 generations to increase the number of dominants from one in a million to one in two, but 321,444 generations to increase the number of recessives in the same way. It is not surprising that the only new types which have been known to spread through a wild population under constant observation are dominants. For example, the black form of the peppered moth, *Amphidasys betularia*, which replaced the original form in the industrial districts of England and Germany during the nineteenth century, is a dominant. When the character is due to several rare genes the effect of selection is also very slow even if the genes are dominant. But however small may be the selective advantage the new character will spread, provided it is present in enough individuals of a population to prevent its disappearance by mere random extinction. Fisher has shown that it is only when k is less than the reciprocal of the number of the whole population that natural selection ceases to be effective. An average advantage of one in a million will be quite effective in most species.

A curious situation arises when two genes one at a time produce a disadvantageous type, but taken together are useful. Such a case was found by Gonzalez (1923), with three of the well-known genes in *Drosophila*. It will be seen from Table I that two of the genes, Purple and Arc, lower the expectation of life in both sexes. The third, Speck, increases the expectation of life in males, without altering it significantly in females. Purple and Arc together give considerably longer life in both sexes, but especially in the male, than either alone. The combination of all three genes restores the normal duration of life in both sexes, the increase being insignificant.

The figures for progeny in the last column are based on few families, but the fertility of Purple is significantly greater than that of the wild type.

TABLE I

Mean Life in Days, and Average Progeny per Fertile Mating of several types of Drosophila melanogaster

Type	Life of ♂	Life of ♀	Progeny
Wild	38·08 ± 0·36	40·62 ± 0·42	247
Purple (eyes)	27·42 ± 0·27	21·83 ± 0·23	325
Arc (wings)	25·20 ± 0·33	28·24 ± 0·37	127
Speck (in axilla)	46·63 ± 0·63	38·91 ± 0·65	103
Purple arc	36·00 ± 0·53	31·98 ± 0·43	230
Purple speck	23·72 ± 0·22	22·96 ± 0·19	247
Arc speck	38·41 ± 0·58	34·69 ± 0·66	106
Purple arc speck	38·38 ± 0·62	40·67 ± 0·45	118

If the percentage of fertile matings is not greatly lowered by this gene, it would tend to spread in a mixed population under Gonzalez' cultural conditions, though doubtless in a state of nature this is not so.

Of course the life-lengths of Table I do not represent selective advantages, but they only refer to five of the hundreds of genes known in *Drosophila.* No doubt none of the other common mutant genes *by itself* is advantageous in nature, or it would have spread through the species and established itself. But it is quite possible that a combination of two, three, or more would be so. The number of possible combinations of all the known genes is very large indeed. The combined mass of a population consisting of one fly of each possible type would vastly exceed that of all the known heavenly bodies, or that of the universe on the theory of general relativity. It is not an extravagant theory that at least one member of this population would be better adapted for life than the present wild type.

If we consider a case where the double dominant AB and the double recessive *aabb* are both more viable than the types A*bb* or *aa*B, then a population consisting mainly of either of the favoured types is in equilibrium, and mutation on a moderate scale is not capable of upsetting this equilibrium. But the change from one stable equilibrium to the other may take place as the result of the isolation of a small unrepresentative group of the population, a temporary change in the environment which alters the relative viability of different types, or in several other ways, one of which will be considered later.

This case seems to me very important, because it is probably the basis of progressive evolution of many organs and functions in higher animals, and of the break-up of one species into several. For an evolutionary progress to take place in a highly specialised organ such as the human eye or hand a number of changes must take place simultaneously. Thus if the eye is unusually long from back to front we get shortsightedness,

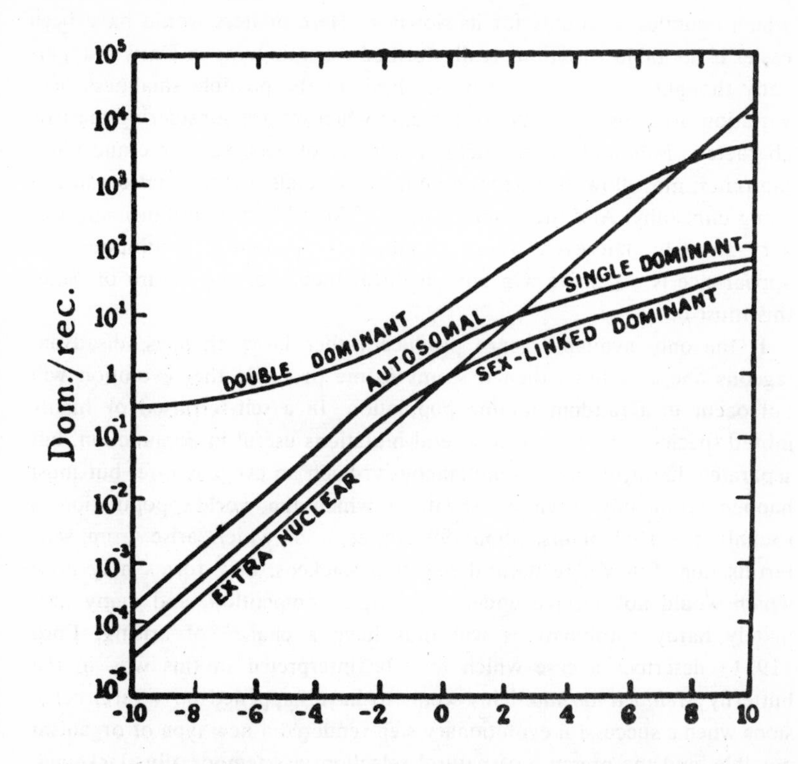

FIGURE 1—Theoretical results of selection on the composition of a population when dominants are favoured. Abscissa, number of generations multiplied by coefficient of selection. Ordinate, ratio of dominants to recessives. If the races do not interbreed the effect is the same as for an extra-nuclear factor. In the case of the double dominant, the genes are supposed to be present in equal numbers. If recessives are favoured, the sign of the abscissa is changed. For example, if $k = 0.01$, it will be seen that about 400 generations are needed for the ratio of dominants to recessives to change from 1 to 10, if autosomal single dominants are favoured.

which would not, however, occur if there were a simultaneous decrease in the curvature of the cornea or lens, which would correct the focus. As, however, abnormal eye-length is fairly common, being often inherited as a dominant, while lessened corneal curvature is rare, the usual result of the condition is shortsightedness. Actually a serious improvement in the eye would involve a simultaneous change in many of its specifications. Occasionally a single gene might produce simultaneous and harmonious changes in many at once, but this is not generally the case with new mutants, although some such genes, being almost harmless, are not eliminated, and account for much of the variation in natural populations. Evolution must have involved the simultaneous change in many genes, which doubtless accounts for its slowness. Here matters would have been easier if heritable variations really formed a continuum, as Darwin apparently thought, *i.e.*, if there were no limit to the possible smallness of a variation. But this is clearly not the case when we are considering meristic characters. Mammals have a definite number of neck vertebrae and chromosomes, most flowers a definite number of petals, exceptional organisms being unhealthy. And the atomic nature of Mendelian inheritance suggests very strongly that even where variation is apparently continuous this appearance is deceptive. On any chemical theory of the nature of genes this must be so.

If the only available genes produce rather large changes, disadvantageous one at a time, then it seems to me probable that evolution will not occur in a random mating population. In a self-fertilised or highly inbred species it may do so if several mutations useful in conjunction, but separately harmful, occur simultaneously. Such an event is rare, but must happen reasonably often in wheat, of which the world's population is roughly 5×10^{14} plants, about 99 per cent. of which arise from self-fertilisation. But where natural selection slackens, new forms may arise which would not survive under more rigid competition, and many ultimately hardy combinations will thus have a chance of arising. Ford (1931) describes a case which may be interpreted in this way in the butterfly *Melitaea aurinia*. This seems to have happened on several occasions when a successful evolutionary step rendered a new type of organism possible, and the pressure of natural selection was temporarily slackened. Thus the distinction between the principal mammalian orders seems to have arisen during an orgy of variation in the early Eocene which followed the doom of the great reptiles, and the establishment of the mammals as the dominant terrestrial group. Since that date mammalian evolution has been a slower affair, largely a progressive improvement of the types originally laid down in the Eocene.

Another possible mode of making rapid evolutionary jumps is by hybridisation. As we saw, this may lead to the immediate formation of a new species by allopolyploidy. An example of this process in Nature is given by Huskins (1930). The rice-grass *Spartina Townsendii* first appeared on the muddy foreshore of Southampton Water about 1870. It breeds true, but Stapf (1927) regards it as a hybrid of the English *S. stricta* and the (probably) American *S. alterniflora*. Huskins finds that these two latter species have chromosome numbers of 56 and 70 respectively, while *S. Townsendii* has 126 chromosomes. The basic chromosome number in the Gramineae is 7, so it would seem that a cross between an octaploid and a decaploid species gave rise to an enneaploid with 63 chromosomes, vigorous, but somewhat sterile and not breeding true. The doubling of its chromosome number gave an octocaidecaploid which combined hybrid vigour with fertility and stability. This interpretation must of course remain doubtful until the crossing has been repeated under controlled conditions, but the conjunction of morphological and cytological evidence renders it very likely.

Meanwhile the new species is proving its fitness in a true Darwinian manner by exterminating its parents, and also according to the ideas of Kropotkin, by aiding the Dutch in their struggle with the sea. Its recent origin is to be explained by the fact that its parents only hybridised as the result of human activity, *S. alterniflora* having presumably been brought on a ship from America.

Apart from this, hybridisation (where the hybrids are fertile) usually causes an epidemic of variation in the second generation which may include new and valuable types which could not have arisen within a species by slower evolution. The reason for this is that genes often exhibit quite novel behaviour in a new environment. Thus Kosswig (1929) crossed the fishes *Platypoecilus* and *Xiphophorus*, and found that some, though not all, of the genes causing abnormal colours in the latter produced very exaggerated effects when introduced into the former. Thus a gene from *Xiphophorus* for black pigment produced hybrid fish which, though quite healthy, were covered with warts of black pigment. Lotsy (1916), in particular, has emphasised the importance of hybridisation in evolution, and shown that it occurs in nature. He was able, for example, by crossing two species of *Antirrhinum* (snapdragon), namely *A. majus* and *A. glutinosum*, to obtain in the second generation plants whose flower would be ascribed by a taxonomist to the related genus *Rhinanthus*. At one time Lotsy did not believe in mutation, except by loss, and attributed all variation to hybridisation. This is certainly an exaggeration. Not only has mutation now been fully confirmed, but no such hypothesis as Lotsy's will explain the slow and steady evolution to which the geological record bears witness. Nevertheless it is difficult to doubt that hybridisation has rendered possible the coming together of certain combinations of genes which could not have arisen otherwise.

Still another possible way out of the impasse is as follows. Instead of the two or more genes changing abruptly, they may change in a number of small steps, *i.e.*, multiple allelomorphs may appear causing very slight changes in the original type of gene. Supposing blackness conferred a small advantage of about one in a thousand on the wild mouse by acting as a protective colour or otherwise, it would not be favoured by selection because it confers a definite physiological handicap. The death-rate among black mice in their first three weeks of life was shown by Detlefsen and Roberts (1918), to be decidedly larger than that of the wild type (two other colour genes had no such effect). Actually the handicap was about 4½ per cent. But it might pay the mouse to become slightly darker, changing its G gene a fraction of the way towards the gene producing black, to wait until modifying genes had accumulated which restored the physiological balance, then to proceed another step, and so on.

It will be remembered that Detlefsen (1914) crossed the ordinary guinea-pig, *Cavia porcellus*, with the smaller and darker coloured *Cavia rufescens*. He found that the dark colour was mainly due to a modification of the gene G, which behaved as a multiple allelomorph with the gene for black. Thus if *Cavia porcellus* represents the original type, which is highly probable, its gene G has changed part of the way towards producing blackness in the evolution of *C. rufescens*. Three other crosses between subspecies and geographical races in rodents give similar results. It looks as if the evolution of colour in rodents generally proceeded by rather small steps. My own quite speculative theory of orthogenetic evolution such as that described in Chapter I is that we are dealing here not only with the accumulation of numbers of genes having a similar action, but with the very slow modification of single genes, each changing in turn into a series of multiple allelomorphs. The phrase "modification of the gene" is of course a rather misleading simplification. What I mean is that mutation was constantly modifying the gene, and that at any given time natural selection acted so as to favour one particular grade of modification at the expense of the others.

One more application of mathematics, and I have done. Under what conditions can mutation overcome selection? This is quite a simple problem. Let p be the probability that a gene will mutate in a generation. We saw that p is probably usually less than a millionth, and so far always less than a thousandth. Let k be the coefficient of selection measuring the selective disadvantage of the new type, k being considerably larger than p. Then equilibrium is reached when the proportion of unfavourable to favourable phenotypes is $\frac{p}{k}$ if the mutant is recessive, $\frac{2p}{k}$ if it is dominant. The above calculations refer to a random mating population. The ratio is always $\frac{p}{k}$ in a self-fertilised population. Hence, unless k is so small as to be of the same order as p, the new type will not spread to any significant extent. Even under the extreme conditions of Muller's X-ray experiments, when mutation was a hundred and fifty times more frequent than in the normal, a disadvantage of one in two thousand would have kept any of the new recessive types quite rare. Thus until it has been shown that anywhere in nature conditions produce a mutation rate considerably higher than this, we cannot regard mutation as a cause likely by itself to cause large changes in a species. But I am not suggesting for a moment that selection alone can have any effect at all. The material on which selection acts must be supplied by mutation.

Neither of these processes alone can furnish a basis for prolonged evolution. Selection alone may produce considerable changes in a highly mixed population. A selector of sufficient knowledge and power might perhaps obtain from the genes at present available in the human species a race combining an average intellect equal to that of Shakespeare with the stature of Carnera. But he could not produce a race of angels. For the moral character or for the wings he would have to await or produce suitable mutations.

APPENDIX:

THE MEASUREMENT OF THE INTENSITY OF SELECTION

The simplest case occurs when the population consists of only two types which do not interbreed, and when generations do not overlap, as in annual plants. Suppose that for every n offspring of type A in the subsequent generation, type B gives, on an average $(1 - k)n$, we call k the coefficient of selection in favour of A. Unless k is very small it is better to use the difference of Fisher's "Malthusian parameters" for the two types. If type B gives $e^{-\kappa}n$ offspring, κ is the difference in question. When both are small, k and κ are clearly almost equal; but in general κ may have any real value, while $k = 1 - e^{-\kappa}$, or $\kappa = -\log_e (1 - k)$, so that k cannot exceed 1, but may have any negative value. Note that k in general depends both in birth-rates and death-rates. Thus in England unskilled workers have a higher death-rate, both adult and infantile, than skilled, but this is more than balanced by their greater fertility. In calculating k we may make our count at any period in the life-cycle, provided this period is the same in both generations. It is often convenient to call $1 - k$ the relative fitness of type B as compared to A.

We can generalise k in a number of ways. In general the types A and B will interbreed. If we are dealing with a difference due to a plasmon this makes no difference to our calculation. Otherwise the most satisfactory method of defining the intensity of selection would be as follows. Individuals should be counted at the moment of fertilisation. Then if for every n children of an A ♀, a B ♀ has $n(1 - k_1)$, and $n(1 - k_2)$ is the similar figure for a B ♂, we have the two coefficients k_1 and k_2. The same calculation holds for an hermaphrodite species. If selection operates mainly through death-rates, k_1 and k_2 are likely to be nearly equal. If it operates mainly through fertility this is not so. For example, male sterility is quite common in hermaphrodite plants, and in the male-sterile group

k_2 will be 1, while k_1 is small. Where (as in heterostylic plants) the success of a mating depends on the precise combination of parents, special methods must be used.

Where generations overlap, as in man, a specification of the intensity of selection involves the use of definite integrals. I showed that if the chance of a female zygote (whether alive or dead) producing offspring between the ages x and $x + \delta x$ is $K(x)\delta x$ for type A, $[K(x) - k(x)]\delta x$ for type B, then selection proceeds as if generations did not overlap, the interval between generations being $\dfrac{\int_0^\infty xK(x)\,dx}{\int_0^\infty K(x)\,dx}$, and the coefficient selection $k = \int_0^\infty k(x)\,dx$. This is true provided k is small. Similar conditions hold when selection operates on both sexes, and Norton (1928) has discussed the rather intricate mathematical problems arising in this case. In what follows we shall confine ourselves to the case where generations do not overlap, as this involves very little loss of generality, and greatly simplifies the mathematics. When generations overlap, the finite difference equations which will be developed later become integral equations.

The values of k vary greatly. For a lethal recessive $k = 1$, $\kappa = +\infty$; for many of the semi-lethal genes in *Drosophila* k exceeds 0·9, and probably for most of those studied 0·1. In many cases, *e.g.*, those characterising the races of *Taraxacum*, it is fairly large, and positive or negative according to the environment. For *Primula sinensis* we only possess data regarding mortality, as opposed to fertility. Here k varies from less than 0·01 upwards. Of 24 mutant genes 20 are neutral or nearly so, two give values of k about 0·05, one about 0·10, and one about 0·6. For some of the colour genes in mice it appears to be less than 0·05, while for the genes determining banding in *Cepea* it is 10^{-5} or less.

SELECTION OF A METRICAL CHARACTER DETERMINED BY MANY GENES

Consider an apparently continuously varying character such as human stature. The distribution of such characters is usually normal, *i.e.*, according to Gauss' error curve. When a population is in equilibrium it has been shown in several cases that mortality is higher or fertility less in those individuals which diverge most from the mean.

Fisher's (1918) analysis of Pearson's data on the correlation between relatives shows that human stature is inherited as if (apart from rather small environmental influences) it were determined by a large number of nearly completely dominant genes, each acting nearly independently on the character concerned. If there were no dominance the average stature of children would be given by that of their parents. We should get very little further information on it (as we actually do) from a knowledge of the stature of remoter ancestors.

Fisher's (1930) analysis of the effect of selection on such a population involves his theory of the evolution of dominance, which I do not myself hold. His analysis is very greatly simplified if we restrict ourselves, as I shall do here, to the case where all the genes concerned are fully dominant.

Consider a dominant gene A which is present with a genic ratio u_n, *i.e.*, the three genotypes are in the proportions u_n^2 AA : $2u_n$ A*a* : 1 *aa*.

Let α be the difference between the mean stature of the dominants and recessives. Then the average deviation of the dominants from the general mean of the population must be $\dfrac{\alpha}{(u_n+1)^2}$, that of the recessives $-\dfrac{(u_n^2+2u_n)\alpha}{(u_n+1)^2}$.

The dominants will form a normally distributed group with a mean stature exceeding the general mean by $\dfrac{\alpha}{(u_n+1)^2}$, where α may, of course, be negative. The standard deviations of the two groups will be equal, but their average divergences from the mean will differ. The group whose mean stature is nearest to that of the general population will be fittest. The two will diverge equally if dominants and recessives are present in equal numbers, *i.e.*, $u_n^2 + 2u_n = 1$, or $u_n = \sqrt{2} - 1$. In this case the population is in equilibrium. If u_n exceeds this value there will be more dominants than recessives, and the recessives will, on the whole, be more abnormal and therefore less fit than the dominants. So the proportion of dominants, and hence u_n, will increase. Similarly if u_n is less than $\sqrt{2} - 1$ it will decrease still further as the result of selection. The argument can obviously be extended to populations where complete or partial inbreeding is the rule. Fisher shows that it is also true when dominance is incomplete, in the particular case where the relative unfitness, or coefficient of selection, varies as the square of the mean deviation from the general average.

Hence *a normally distributed population cannot be in stable equilibrium as a result of selection for the characters normally distributed.* This rather sensational fact vitiates a large number of the arguments which are commonly used both for and against eugenics and Darwinism.

If the relative viability or fertility of a population whose mean stature diverges from the mean of the population by $\pm x$ be $1 - cx^2$, which follows from any of a number of simple hypotheses, then

$$k = \frac{c\alpha^2}{(u_n+1)^4}[(u_n^2+2u_n)^2-1]$$

$$= \frac{c\alpha^2(u_n^2+2u_n-1)}{(u_n+1)^2}$$

and $$\Delta u_n = \frac{c\alpha^2 u_n(u_n^2+2u_n-1)}{(u_n+1)^3}, \text{ approximately.}$$

We cannot, however, follow the course of events in such a population, because the genic ratios for a number of different genes will be varying at once, and hence the mean will vary in an unpredictable manner. In general, however, u_n will increase or decrease until its tendency to do so is checked by mutation in the opposite direction. If p be the probability of A mutating to a in one generation, q that of the reverse process, then

$$\Delta u_n = \frac{u_n(u_n^2+2u_n-1)c\alpha^2}{(u_n+1)^3} - pu_n(u_n+1) + q(u_n+1).$$

For equilibrium this must vanish. So if $x = \frac{1}{1+u}$ (the proportion of recessive genes), then

$$x^2(1-x)(1-2x^2) + \frac{p(x-1)+qx}{c\alpha^2} = 0.$$

If p and q are small compared with $c\alpha^2$ this has three roots between 0 and 1, one approximating to $\frac{1}{\sqrt{2}}$ defining an unstable equilibrium, the others near $\sqrt{\frac{p}{c\alpha^2}}$ and $1-\frac{q}{c\alpha^2}$ defining stable equilibrium. Since p and q are small, either dominants or recessives are fairly rare. Hence most of the variance is due to rare and disadvantageous genes whose supply is only kept up by mutation. But only in so far as it includes such genes does a population possess the genetic elasticity which permits it to respond to a change in environment by evolving. It must be remembered that if any gene, apart from its effect on stature, is advantageous in the heterozygous condition, it will tend to an equilibrium with u in the neighbourhood of 1. Probably some at least of the heritable stature differences are due to genes of this class. Not only the well-known vigour of hybrids, but the marked amount of heterozygosis found in selected clones, *e.g.*, of fruit trees and potatoes, makes their existence probable.

Fisher next considers what will happen if a population in equilibrium of this type is acted on by selection in favour of, say, a larger size. The numerous rare genes for small size will become still rarer, the rare genes for large size becoming commoner. Such changes, if small, will be reversed when the selection ceases. But some of the rare genes for large size will increase in numbers so much as to pass their former point of unstable equilibrium. They will therefore become very common instead of very rare.

Now if the conditions of selection change back to normal these genes will not return to their original frequency, and the mean stature will have been irreversibly increased. Again, supposing that selection increases the optimum stature of a species by a certain quantity, then when the mean stature reaches the new optimum some genes will be past their point of unstable equilibrium, but still increasing in numbers. The stature will thus, so to speak, overshoot the mark aimed at by selection. We have here, for the first time, an explanation, on strictly Darwinian lines, of useless orthogenesis.

In certain rare cases I have shown that this might occur even with regard to a character determined by a single fully dominant gene. But this is only so when selection favours dominants, and three inequalities involving k, p, and q are fulfilled. It may also happen as regards a gene (if such exist) where the heterozygote is less fit than either homozygote. But though these may be subsidiary causes of evolution beyond the optimum, they can have far less importance than the Fisher effect.

SOCIALLY VALUABLE BUT INDIVIDUALLY DISADVANTAGEOUS CHARACTERS

A study of these traits involves the consideration of small groups. For a character of this type can only spread through the population if the genes determining it are borne by a group of related individuals whose chances of leaving offspring are increased by the presence of these genes in an individual member of the group whose own private viability they lower.

Two simple cases will make this clear. Broodiness is inherited in poultry. In the wild state a broody hen is likely to live a shorter life than a non-broody one, as she is more likely to be caught by a predatory enemy while sitting. But the non-broody hen will not rear a family, so genes determining this character will be eliminated in nature. With regard to maternal instincts of this type selection will presumably strike a balance. While a mother that abandoned her eggs or young in the face of the slightest danger would be ill-represented in posterity, one who, like the average bird, does so under a sufficiently intense stimulus will live to rear another family, which a too devoted parent would not.

In the case of social insects there is no limit to the devotion and self-sacrifice which may be of biological advantage in a neuter. In a beehive the workers and young queens are samples of the same set of genotypes, so any form of behaviour in the former (however suicidal it may be) which is of advantage to the hive will promote the survival of the latter, and thus tend to spread through the species. The only bar to such a spread is the possibility that the genes in question may induce unduly altruistic behaviour in the queens. Genes causing such behaviour would tend to be eliminated.

When we pass to small social groups where every individual is a potential parent, matters are complicated. Consider a tribe or herd of N individuals mating at random, and in the ratios u_n^2 AA : $2u_n$Aa : 1 *aa*. Let the possession of the recessive character for altruistic behaviour caused by *aa* decrease the probable progeny of its possessors to $(1 - k)$ times that of the dominants. Let the presence of a fraction x of individuals in the tribe increase the probable progeny of *all* its members to $(1 + \mathrm{K}x)$ times that of a tribe possessing no recessives, which we may take to be in equilibrium. We will further suppose that a tribe composed entirely of recessives would tend to increase, hence $\mathrm{K} > k$.

Now in the next generation the number of the tribe will be increased to $\mathrm{N}\left[1 + \frac{\mathrm{K}}{(1+u_n)^2}\right]$. The number x_n of recessive genes will have changed from

$$\frac{\mathrm{N}}{1+u_n} \text{ to } \mathrm{N}\left[1 + \frac{\mathrm{K}}{(1+u_n)^2}\right]\frac{(u_n + 1 - k)}{(1+u_n)^2}$$

approximately. Hence, neglecting $k\mathrm{K}$,

$$\Delta x_n = \frac{\mathrm{N}\left(\frac{\mathrm{K}}{1+u_n} - k\right)}{(1+u_n)^2}$$

Hence the number of recessive genes will increase so long as $u_n + 1 > \frac{\mathrm{K}}{k}$, *i.e.*, recessive genes will only increase as long as they are fairly common. But meanwhile u is increasing, so this process will tend to come to an end. If altruism is dominant, we find that the numbers of the gene for it tend to increase if $(1+u_n)^2 > \frac{\mathrm{K}}{\mathrm{K}-k}$. In other words the biological advantages of altruistic conduct only outweigh the disadvantages if a substantial proportion of the tribe behave altruistically. If only a small fraction behaves in this manner, it has a very small effect on the viability of the tribe, not sufficient to counterbalance the bad effect on the individuals concerned. If $\frac{K}{k}$ be large, the proportion of altruists need not be great. If $\frac{K}{k} > N$ in the case of a dominant gene, $\frac{K}{k} > \sqrt{N}$ in the case of a recessive, a single altruistic individual will have a net biological advantage. Hence for small values of N selection is at once effective. But in large tribes the initial stages of the evolution of altruism depend not on selection, but on random survival, *i.e.*, what in physics is called fluctuation. This is quite possible when N is small, very unlikely when it is large. If any genes are common in mankind which promote conduct biologically disadvantageous to the individual in all types of society, but yet advantageous to society, they must have spread when man was divided into small endogamous groups. As many eugenists have pointed out, selection in large societies operates in the reverse direction.

But the conditions given above, though necessary for the spread of congenital altruism, are far from sufficient. Consider a tribe in which the proportion of altruists is sufficient to cause the number of the gene for it to increase. Even so the other allelomorph will increase still more rapidly. So the proportion of altruists will diminish. The tribe, however, will enlarge, and may be expected ultimately to split, like that of Abram (Genesis xiii. 11). In general this will not mend matters, but sometimes one fraction will get most of the genes for altruism, and its rate of increase be further speeded up. Finally a tribe homozygous for this gene may be produced. These events are enormously more probable if N is small, and endogamy fairly strict. Even when homozygosis is reached, however, the reverse mutation may occur, and is likely to spread. I find it difficult to suppose that many genes for absolute altruism are common in man.

At the risk of repetition I wish to add that the above analysis refers only to conduct which actually diminishes the individual's chance of leaving posterity (such a chance, though small, does exist even for worker bees). A great deal of human conduct which we call altruistic is egoistic from the point of view of natural selection. It is often correlated with well-developed parental behaviour-patterns. Moreover, altruism is commonly rewarded by poverty, and in most modern societies the poor breed quicker than the rich.

CONCLUSION

I hope that I have shown that a mathematical analysis of the effects of selection is necessary and valuable. Many statements which are constantly made, *e.g.*, "Natural selection cannot account for the origin of a highly complex character," will not bear analysis. The conclusions drawn by common sense on this topic are often very doubtful. Common sense tells us that two bodies attracting one another by gravitation tend inevitably to fall together, which would sometimes be true if the force between them varied as r^{-n}, n exceeding 2. It is not true with the inverse square law. So with selection. Unaided common sense may indicate an equilibrium, but rarely, if ever, tells us whether it is stable. If much of the investigation here summarised has only proved the obvious, the obvious is worth proving when this can be done. And if the relative importance of selection and mutation is obvious, it has certainly not always been recognised as such.

The permeation of biology by mathematics is only beginning, but unless the history of science is an inadequate guide, it will continue, and the investigations here summarised represent the beginning of a new branch of applied mathematics.

COMMENTARY ON
ERWIN SCHRÖDINGER

IN 1945 Erwin Schrödinger published a little book called *What Is Life?* The spacious title suggests an inspirational essay, but except for an epilogue on determinism and free will (discussed with great delicacy and restraint) the book is a brilliant scientific essay on the physical aspects of the living cell. The question Schrödinger considers is "How can the events in space and time which take place within the spatial boundary of a living organism be accounted for by physics and chemistry?" His "preliminary" answer is that these sciences may be able to explain life, but that they cannot do it with the resources now at their disposal.

Schrödinger draws upon experimental data as well as upon the mathematical models of atomic phenomena to make clear "a fundamental idea, which hovers between biology and physics." Adopting a conjecture already current, that a gene, or perhaps even an entire chromosome fiber, is a giant molecule, he takes the further step of supposing that its structure is that of an aperiodic solid or crystal. This would explain why its stability or "permanence" is such that a very considerable commotion is required to alter its existing pattern. Like other crystals, the chromosome can reproduce itself. But it also has another important attribute which is unique: its own complex structure (the energy state and configuration of its constituent atoms) forms a "code-script" that determines the entire future development of the organism. Though the number of atoms which make up the molecule is not very large, the number of their possible arrangements is enormous; thus "it is no longer inconceivable that the miniature code should precisely correspond with a highly complicated and specified plan of development and should somehow contain the means to put it into operation." Schrödinger now uses quantum mechanical evidence to explain what is involved in a rearrangement of structure, which must produce an alteration of the code-script and a change in the development of the organism. It will be recognized that this shift in the atomic pattern is what physicists call a quantum "jump" and biologists call a mutation. The event does not occur easily; circumstances conspire against it; nature tends to be conservative. Mutations are therefore comparatively rare. Schrödinger is concerned mainly with the physics and mathematics of these shifts from one configuration to another: when, how and why the "jumps" or mutations take place.[1]

[1] I must call the reader's attention to two papers published by Schrödinger in the August and November 1952 issues of the British *Journal of the Philosophy of Science*. These treat the question "Are there quantum 'jumps'?" Schrödinger now believes that these discontinuities in atomic behavior do *not* occur. He suggests that quantum

The inventor of this fascinating scheme is one of the leaders of modern science. Erwin Schrödinger was born at Vienna in 1887. He has occupied chairs of theoretical physics in Stuttgart, Breslau, Zurich and Berlin. At Berlin he succeeded the late Max Planck, the founder of quantum theory. He left Germany after Hitler came to power in 1933, taught for a time at Graz in Austria and then went to Dublin to become a member (for a time director, and now senior professor) of the Institute for Advanced Studies. Schrödinger's researches and publications extend over a wide range of topics including vibrations and specific heat of crystals, quantum mechanics of spectra, the "mathematical structure of the physiological color-space" and so on. His fame is based principally on his invention of wave mechanics, for which he was awarded the Nobel Prize.[2] "The series of papers, in which he developed an ingenious idea of De Broglie's to a complete theory of atomic structures, and demonstrated, moreover, the relation of his wave equation to other forms of quantum mechanics (Heisenberg-Born-Dirac) belong to the classics of theoretical physics by virtue of their depth, wealth, completeness and brilliant style." [3]

Schrödinger has the rare combination of qualities required of a great popularizer of science and the philosophy of science. He is master of his subject, original, superbly imaginative and not solemn. He makes things clear rather than easy; he does not hesitate to share perplexities with his readers. He likes to write short books, which are models of exposition, but always too short. Everything he writes—from statistical thermodynamics to free will—carries his special mark. But the energy, the play of wit, the freshness, the independence, the fondness for startling analogies and more startling conclusions recall the handicraft of William Kingdon Clifford. There is no higher praise.

"jumps" are better explained by so-called resonance properties of the atom. It is not appropriate to go into this matter further, except to say that the conjectures offered by Schrödinger in *What Is Life?* are not—at least so it seems to me—vitiated by the abandonment of the quantum "jump" concept. They can be fitted without difficulty. I think, into the modified scheme of resonance phenomena. I would add only that it is characteristic of Schrödinger to reject accepted opinions on fundamentals, and that when he puts forward one of his more daring hypotheses, he succeeds—as one suspects is his intention—in setting the scientific world on its ear.

[2] Shared with P. A. M. Dirac in 1933. For a popular exposition of wave mechanics by Schrödinger himself, see pp. 1056–1068.

[3] *Nature*, May 21, 1949, Vol. 163, p. 794.

But all that moveth doth Mutation love. —EDMUND SPENSER

In all such cases there is one common circumstance—the system has a quantity of potential energy, which is capable of being transformed into motion, but which cannot begin to be so transformed till the system has reached a certain configuration, to attain which requires an expenditure of work, which in certain cases may be infinitesimally small, and in general bears no definite proportion to the energy developed in consequence thereof. For example, the rock loosed by frost and balanced on a singular point of the mountain side, the little spark which kindles the great forest, the little word which sets the world a-fighting, the little scruple which prevents a man from doing his will, the little spore which blights all the potatoes, the little gemmule which makes us philosophers or idiots. Every existence above a certain rank has its singular points: the higher the rank the more of them. At these points, influences whose physical magnitude is too small to be taken account of by a finite being, may produce results of the greatest importance. . . . —JAMES CLERK-MAXWELL

16 Heredity and the Quantum Theory

By ERWIN SCHRÖDINGER

THE QUANTUM-MECHANICAL EVIDENCE

Und deines Geistes höchster Feuerflug
Hat schon am Gleichnis, hat am Bild genug.[1] —GOETHE

PERMANENCE UNEXPLAINABLE BY CLASSICAL PHYSICS

THUS, aided by the marvellously subtle instrument of X-rays (which, as the physicist remembers, revealed thirty years ago the detailed atomic lattice structures of crystals), the united efforts of biologists and physicists have of late succeeded in reducing the upper limit for the size of the microscopic structure, being responsible for a definite large-scale feature of the individual—the 'size of a gene'—and reducing it far below the estimates referred to earlier.[2] We are now seriously faced with the question: How can we, from the point of view of statistical physics, reconcile the facts that the gene structure seems to involve only a comparatively small number of atoms (of the order of 1000 and possibly much less), and that nevertheless it displays a most regular and lawful activity—with a durability or permanence that borders upon the miraculous.

[1] And thy spirit's fiery flight of imagination acquiesces in an image, in a parable.

[2] Earlier in the volume from which this excerpt has been taken, Schrödinger gives the estimates of maximum size of the gene (evidence obtained by breeding experiments and microscopic inspection): the maximum volume is equal to a cube of side 300 A. ("A" is the abbreviation of Ångström which is the 10^{10}th part of a meter, or in decimal notation 0.0000000001 meter.)

Let me throw the truly amazing situation into relief once again. Several members of the Habsburg dynasty have a peculiar disfigurement of the lower lip ('Habsburger Lippe'). Its inheritance has been studied carefully and published, complete with historical portraits, by the Imperial Academy of Vienna, under the auspices of the family. The feature proves to be a genuinely Mendelian 'allele' to the normal form of the lip. Fixing our attention on the portraits of a member of the family in the sixteenth century and of his descendant, living in the nineteenth, we may safely assume that the material gene structure, responsible for the abnormal feature, has been carried on from generation to generation through the centuries, faithfully reproduced at every one of the not very numerous cell divisions that lie between. Moreover, the number of atoms involved in the responsible gene structure is likely to be of the same order of magnitude as in the cases tested by X-rays. The gene has been kept at a temperature around 98° F. during all that time. How are we to understand that it has remained unperturbed by the disordering tendency of the heat motion for centuries?

A physicist at the end of the last century would have been at a loss to answer this question, if he was prepared to draw only on those laws of Nature which he could explain and which he really understood. Perhaps, indeed, after a short reflection on the statistical situation he would have answered (correctly, as we shall see): These material structures can only be molecules. Of the existence, and sometimes very high stability, of these associations of atoms, chemistry had already acquired a widespread knowledge at the time. But the knowledge was purely empirical. The nature of a molecule was not understood—the strong mutual bond of the atoms which keeps a molecule in shape was a complete conundrum to everybody. Actually, the answer proves to be correct. But it is of limited value as long as the enigmatic biological stability is traced back only to an equally enigmatic chemical stability. The evidence that two features, similar in appearance, are based on the same principle, is always precarious as long as the principle itself is unknown.

EXPLICABLE BY QUANTUM THEORY

In this case it is supplied by quantum theory. In the light of present knowledge, the mechanism of heredity is closely related to, nay, founded on, the very basis of quantum theory. This theory was discovered by Max Planck in 1900. Modern genetics can be dated from the rediscovery of Mendel's paper by de Vries, Correns and Tschermak (1900) and from de Vries's paper on mutations (1901–3). Thus the births of the two great theories nearly coincide, and it is small wonder that both of them had to reach a certain maturity before the connection could emerge. On the side of quantum theory it took more than a quarter of a century till in 1926–7 the quantum theory of the chemical bond was outlined in its general principles by W. Heitler and F. London. The Heitler-London theory involves the most subtle and intricate conceptions of the latest development of quantum theory (called 'quantum mechanics' or 'wave mechanics'). A presentation without the use of calculus is well-nigh impossible or would at least require another little volume like this. But fortunately, now that all work has been done and has served to clarify our thinking, it seems to be possible to point out in a more direct manner the connection between 'quantum jumps' and mutations, to pick out at the moment the most conspicuous item. That is what we attempt here.

QUANTUM THEORY—DISCRETE STATES—QUANTUM JUMPS

The great revelation of quantum theory was that features of discreteness were discovered in the Book of Nature, in a context in which anything other than continuity seemed to be absurd according to the views held until then.

The first case of this kind concerned energy. A body on the large scale changes its energy continuously. A pendulum, for instance, that is set swinging is gradually slowed down by the resistance of the air. Strangely enough, it proves necessary to admit that a system of the order of the atomic scale behaves differently. On grounds upon which we cannot enter here, we have to assume that a small system can by its very nature possess only certain discrete amounts of energy, called its peculiar energy levels. The transition from one state to another is a rather mysterious event, which is usually called a 'quantum jump.'

But energy is not the only characteristic of a system. Take again our pendulum, but think of one that can perform different kinds of movement, a heavy ball suspended by a string from the ceiling. It can be made to swing in a north-south or east-west or any other direction or in a circle or in an ellipse. By gently blowing the ball with a bellows, it can be made to pass continuously from one state of motion to any other.

For small-scale systems most of these or similar characteristics—we cannot enter into details—change discontinuously. They are 'quantized,' just as the energy is.

The result is that a number of atomic nuclei, including their bodyguards of electrons, when they find themselves close to each other, forming 'a system,' are unable by their very nature to adopt any arbitrary configuration we might think of. Their very nature leaves them only a very numerous but discrete series of 'states' to choose from.[3] We usually call them levels or energy levels, because the energy is a very relevant part of

[3] I am adopting the version which is usually given in popular treatment and which suffices for our present purpose. But I have the bad conscience of one who perpetuates a convenient error. The true story is much more complicated, inasmuch as it includes the occasional indeterminateness with regard to the state the system is in.

the characteristic. But it must be understood that the complete description includes much more than just the energy. It is virtually correct to think of a state as meaning a definite configuration of all the corpuscles.

The transition from one of these configurations to another is a quantum jump. If the second one has the greater energy ('is a higher level'), the system must be supplied from outside with at least the difference of the two energies to make the transition possible. To a lower level it can change spontaneously, spending the surplus of energy in radiation.

MOLECULES

Among the discrete set of states of a given selection of atoms there need not necessarily but there may be a lowest level, implying a close approach of the nuclei to each other. Atoms in such a state form a molecule. The point to stress here is, that the molecule will of necessity have a certain stability; the configuration cannot change, unless at least the energy difference, necessary to 'lift' it to the next higher level, is supplied from outside. Hence this level difference, which is a well-defined quantity, determines quantitatively the degree of stability of the molecule. It will be observed how intimately this fact is linked with the very basis of quantum theory, viz. with the discreteness of the level scheme.

I must beg the reader to take it for granted that this order of ideas has been thoroughly checked by chemical facts; and that it has proved successful in explaining the basic fact of chemical valency and many details about the structure of molecules, their binding-energies, their stabilities at different temperatures, and so on. I am speaking of the Heitler-London theory, which, as I said, cannot be examined in detail here.

THEIR STABILITY DEPENDENT ON TEMPERATURE

We must content ourselves with examining the point which is of paramount interest for our biological question, namely, the stability of a molecule at different temperatures. Take our system of atoms at first to be actually in its state of lowest energy. The physicist would call it a molecule at the absolute zero of temperature. To lift it to the next higher state or level a definite supply of energy is required. The simplest way of trying to supply it is to 'heat up' your molecule. You bring it into an environment of higher temperature ('heat bath'), thus allowing other systems (atoms, molecules) to impinge upon it. Considering the entire irregularity of heat motion, there is no sharp temperature limit at which the 'lift' will be brought about with certainty and immediately. Rather, at any temperature (different from absolute zero) there is a certain smaller or greater chance for the lift to occur, the chance increasing of course with the temperature of the heat bath. The best way to express this chance is to indicate the average time you will have to wait until the lift takes places, the 'time of expectation.'

From an investigation, due to M. Polanyi and E. Wigner,[4] the 'time of expectation' largely depends on the ratio of two energies, one being just the energy difference itself that is required to effect the lift (let us write W for it), the other one characterizing the intensity of the heat motion at the temperature in question (let us write T for the absolute temperature and kT for the characteristic energy).[5] It stands to reason that the chance for effecting the lift is smaller, and hence that the time of expectation is longer, the higher the lift itself compared with the average heat energy, that is to say, the greater the ratio $W:kT$. What is amazing is how enormously the time of expectation depends on comparatively small changes of the ratio $W:kT$. To give an example (following Delbrück): for W thirty times kT the time of expectation might be as short as $\frac{1}{10}$ sec., but would rise to 16 months when W is 50 times kT, and to 30,000 years when W is 60 times kT!

MATHEMATICAL INTERLUDE

It might be as well to point out in mathematical language—for those readers to whom it appeals—the reason for this enormous sensitivity to changes in the level step or temperature, and to add a few physical remarks of a similar kind. The reason is that the time of expectation, call it t, depends on the ratio W/kT by an exponential function, thus

$$t = \tau e^{W/kT}.$$

τ is a certain small constant of the order of 10^{-13} or 10^{-14} sec. Now, this particular exponential function is not an accidental feature. It recurs again and again in the statistical theory of heat, forming, as it were, its backbone. It is a measure of the improbability of an energy amount as large as W gathering accidentally in some particular part of the system, and it is this improbability which increases so enormously when a considerable multiple of the 'average energy' kT is required.

Actually a $W = 30kT$ (see the example quoted above) is already extremely rare. That it does not yet lead to an enormously long time of expectation (only $\frac{1}{10}$ sec. in our example) is, of course, due to the smallness of the factor τ. This factor has a physical meaning. It is of the order of the period of the vibrations which take place in the system all the time. You could, very broadly, describe this factor as meaning that the chance of accumulating the required amount W, though very small, recurs again

[4] *Zeitschrift für Physik*, Chemie (A), Haber-Band, p. 439, 1928.

[5] k is a numerically known constant, called Boltzmann's constant; $\frac{3}{2}kT$ is the average kinetic energy of a gas atom at temperature T.

and again 'at every vibration,' that is to say, about 10^{13} or 10^{14} times during every second.

FIRST AMENDMENT

In offering these considerations as a theory of the stability of the molecule it has been tacitly assumed that the quantum jump which we called the 'lift' leads, if not to a complete disintegration, at least to an essentially different configuration of the same atoms—an isomeric molecule, as the chemist would say, that is, a molecule composed of the same atoms in a different arrangement (in the application to biology it is going to represent a different 'allele' in the same 'locus' and the quantum jump will represent a mutation).

To allow of this interpretation two points must be amended in our story, which I purposely simplified to make it at all intelligible. From the way I told it, it might be imagined that only in its very lowest state does our group of atoms form what we call a molecule and that already the next higher state is 'something else.' That is not so. Actually the lowest level is followed by a crowded series of levels which do not involve any appreciable change in the configuration as a whole, but only correspond to those small vibrations among the atoms which we have mentioned on page 979. They, too, are 'quantized,' but with comparatively small steps from one level to the next. Hence the impacts of the particles of the 'heat bath' may suffice to set them up already at fairly low temperature. If the molecule is an extended structure, you may conceive these vibrations as high-frequency sound waves, crossing the molecule without doing it any harm.

So the first amendment is not very serious: we have to disregard the 'vibrational fine-structure' of the level scheme. The term 'next higher level' has to be understood as meaning the next level that corresponds to a relevant change of configuration.

SECOND AMENDMENT

The second amendment is far more difficult to explain, because it is concerned with certain vital, but rather complicated, features of the scheme of relevantly different levels. The free passage between two of them may be obstructed, quite apart from the required energy supply; in fact, it may be obstructed even from the higher to the lower state.

Let us start from the empirical facts. It is known to the chemist that the same group of atoms can unite in more than one way to form a molecule. Such molecules are called isomeric ('consisting of the same parts'; 'ἴσος = same, μέρος = part). Isomerism is not an exception, it is the rule. The larger the molecule, the more isomeric alternatives are offered. Figure 1 shows one of the simplest cases, the two kinds of propyl-alcohol, both

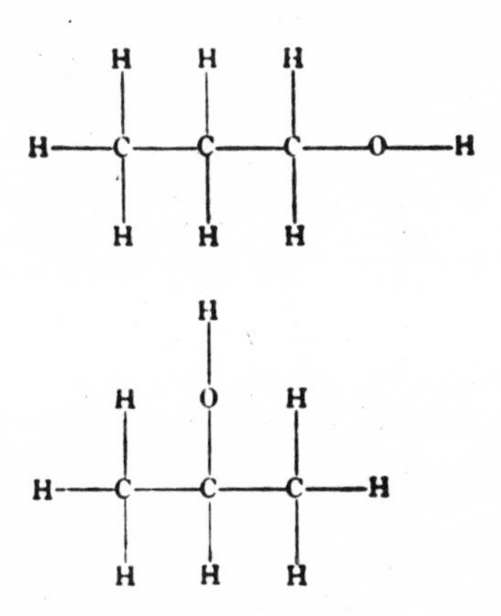

FIGURE 1—The two isomeres of propyl-alcohol.

consisting of 3 carbons (C), 8 hydrogens (H), 1 oxygen (O).[6] The latter can be interposed between any hydrogen and its carbon, but only the two cases shown in our figure are different substances. And they really are. All their physical and chemical constants are distinctly different. Also their energies are different, they represent 'different levels.'

The remarkable fact is that both molecules are perfectly stable, both behave as though they were 'lowest states.' There are no spontaneous transitions from either state towards the other.

The reason is that the two configurations are not neighbouring configurations. The transition from one to the other can only take place over intermediate configurations which have a greater energy than either of them. To put it crudely, the oxygen has to be extracted from one position and has to be inserted into the other. There does not seem to be a way of doing that without passing through configurations of considerably higher energy. The state of affairs is sometimes figuratively pictured as in Figure 2, in which 1 and 2 represent the two isomeres, 3 the 'threshold' between them, and the two arrows indicate the 'lifts,' that is to say, the energy supplies required to produce the transition from state 1 to state 2 or from state 2 to state 1, respectively.

Now we can give our 'second amendment,' which is that transitions of this 'isomeric' kind are the only ones in which we shall be interested in our biological application. It was these we had in mind when explaining 'stability' on pp. 978–979. The 'quantum jump' which we mean is the transition from one relatively stable molecular configuration to another. The energy supply required for the transition (the quantity denoted

[6] Models, in which C, H and O were represented by black, white and red wooden balls respectively, were exhibited at the lecture. I have not reproduced them here, because their likeness to the actual molecules is not appreciably greater than that of Figure 1.

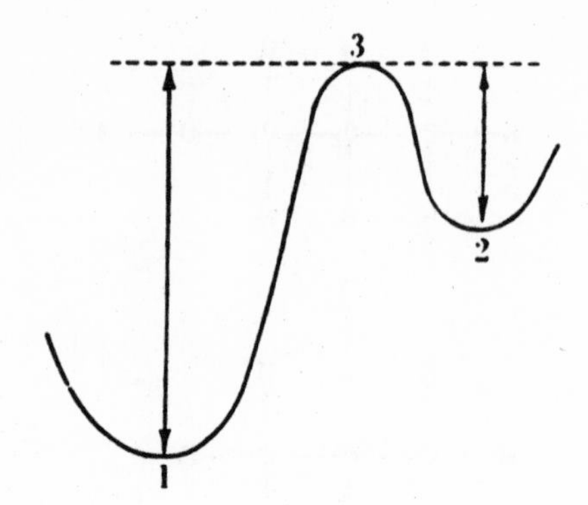

FIGURE 2—Energy threshold (3) between the isomeric levels (1) and (2). The arrows indicate the minimum energies required for transition.

by W) is not the actual level difference, but the step from the initial level up to the threshold (see the arrows in Figure 2).

Transitions with no threshold interposed between the initial and the final state are entirely uninteresting, and that not only in our biological application. They have actually nothing to contribute to the chemical stability of the molecule. Why? They have no lasting effect, they remain unnoticed. For, when they occur, they are almost immediately followed by a relapse into the initial state, since nothing prevents their return.

DELBRÜCK'S MODEL DISCUSSED AND TESTED

Sane sicut lux seipsam et tenebras manifestat, sic veritas norma sui et falsi est.[7]
—SPINOZA, *Ethics*, P. II, Prop. 43.

THE GENERAL PICTURE OF THE HEREDITARY SUBSTANCE

From these facts emerges a very simple answer to our question, namely: Are these structures, composed of comparatively few atoms, capable of withstanding for long periods the disturbing influence of heat motion to which the hereditary substance is continually exposed? We shall assume the structure of a gene to be that of a huge molecule, capable only of discontinuous change, which consists in a rearrangement of the atoms and leads to an isomeric [8] molecule. The rearrangement may affect only a small region of the gene, and a vast number of different rearrangements may be possible. The energy thresholds, separating the actual configuration from any possible isomeric ones, have to be high enough (compared

[7] Truly, as light manifests itself and darkness, thus truth is the standard of itself and of error.

[8] For convenience I shall continue to call it an isomeric transition, though it would be absurd to exclude the possibility of any exchange with the environment.

with the average heat energy of an atom) to make the change-over a rare event. These rare events we shall identify with spontaneous mutations.

The later parts of this chapter will be devoted to putting this general picture of a gene and of mutation (due mainly to the German physicist M. Delbrück) to the test, by comparing it in detail with genetical facts. Before doing so, we may fittingly make some comment on the foundation and general nature of the theory.

THE UNIQUENESS OF THE PICTURE

Was it absolutely essential for the biological question to dig up the deepest roots and found the picture on quantum mechanics? The conjecture that a gene is a molecule is to-day, I dare say, a commonplace. Few biologists, whether familiar with quantum theory or not, would disagree with it. In the opening section we ventured to put it into the mouth of a pre-quantum physicist, as the only reasonable explanation of the observed permanence. The subsequent considerations about isomerism, threshold energy, the paramount role of the ratio $W:kT$ in determining the probability of an isomeric transition—all that could very well be introduced on a purely empirical basis, at any rate without drawing explicitly on quantum theory. Why did I so strongly insist on the quantum-mechanical point of view, though I could not really make it clear in this little book and may well have bored many a reader?

Quantum mechanics is the first theoretical aspect which accounts from first principles for all kinds of aggregates of atoms actually encountered in Nature. The Heitler-London bondage is a unique, singular feature of the theory, not invented for the purpose of explaining the chemical bond. It comes in quite by itself, in a highly interesting and puzzling manner, being forced upon us by entirely different considerations. It proves to correspond exactly with the observed chemical facts, and, as I said, it is a unique feature, well enough understood to tell with reasonable certainty that 'such a thing could not happen again' in the further development of quantum theory.

Consequently, we may safely assert that there is no alternative to the molecular explanation of the hereditary substance. The physical aspect leaves no other possibility to account for its permanence. If the Delbrück picture should fail, we would have to give up further attempts. That is the first point I wish to make.

SOME TRADITIONAL MISCONCEPTIONS

But it may be asked: Are there really no other endurable structures composed of atoms except molecules? Does not a gold coin, for example, buried in a tomb for a couple of thousand years, preserve the traits of the portrait stamped on it? It is true that the coin consists of an enormous number of atoms, but surely we are in this case not inclined to attribute the mere preservation of shape to the statistics of large numbers. The same remark applies to a neatly developed batch of crystals we find embedded in a rock, where it must have been for geological periods without changing.

That leads us to the second point I want to elucidate. The cases of a molecule, a solid, a crystal are not really different. In the light of present knowledge they are virtually the same. Unfortunately, school teaching keeps up certain traditional views, which have been out of date for many years and which obscure the understanding of the actual state of affairs.

Indeed, what we have learnt at school about molecules does not give the idea that they are more closely akin to the solid state than to the liquid or gaseous state. On the contrary, we have been taught to distinguish carefully between a physical change, such as melting or evaporation in which the molecules are preserved (so that, for example, alcohol, whether solid, liquid or a gas, always consists of the same molecules, C_2H_6O), and a chemical change, as, for example, the burning of alcohol,

$$C_2H_6O + 3O_2 = 2CO_2 + 3H_2O,$$

where an alcohol molecule and three oxygen molecules undergo a re-arrangement to form two molecules of carbon dioxide and three molecules of water.

About crystals, we have been taught that they form threefold periodic lattices, in which the structure of the single molecule is sometimes recognizable, as in the case of alcohol and most organic compounds, while in other crystals, e.g. rock-salt (NaCl), NaCl molecules cannot be unequivocally delimited, because every Na atom is symmetrically surrounded by six Cl atoms, and vice versa, so that it is largely arbitrary what pairs, if any, are regarded as molecular partners.

Finally, we have been told that a solid can be crystalline or not, and in the latter case we call it amorphous.

DIFFERENT 'STATES' OF MATTER

Now I would not go so far as to say that all these statements and distinctions are quite wrong. For practical purposes they are sometimes useful. But in the true aspect of the structure of matter the limits must be drawn in an entirely different way. The fundamental distinction is between the two lines of the following scheme of 'equations':

$$\begin{aligned} &\text{molecule} = \text{solid} = \text{crystal.} \\ &\text{gas} = \text{liquid} = \text{amorphous.} \end{aligned}$$

We must explain these statements briefly. The so-called amorphous solids are either not really amorphous or not really solid. In 'amorphous' charcoal fibre the rudimentary structure of the graphite crystal has been disclosed by X-rays. So charcoal is a solid, but also crystalline. Where we find no crystalline structure we have to regard the thing as a liquid with very high 'viscosity' (internal friction). Such a substance discloses by the absence of a well-defined melting temperature and of a latent heat of melting that it is not a true solid. When heated it softens gradually and eventually liquefies without discontinuity. (I remember that at the end of the first Great War we were given in Vienna an asphalt-like substance as a substitute for coffee. It was so hard that one had to use a chisel or a hatchet to break the little brick into pieces, when it would show a smooth, shell-like cleavage. Yet, given time, it would behave as a liquid, closely packing the lower part of a vessel in which you were unwise enough to leave it for a couple of days.)

The continuity of the gaseous and liquid state is a well-known story. You can liquefy any gas without discontinuity by taking your way 'around' the so-called critical point. But we shall not enter on this here.

THE DISTINCTION THAT REALLY MATTERS

We have thus justified everything in the above scheme, except the main point, namely, that we wish a molecule to be regarded as a solid = crystal.

The reason for this is that the atoms forming a molecule, whether there be few or many of them, are united by forces of exactly the same nature as the numerous atoms which build up a true solid, a crystal. The molecule presents the same solidity of structure as a crystal. Remember that it is precisely this solidity on which we draw to account for the permanence of the gene!

The distinction that is really important in the structure of matter is whether atoms are bound together by those 'solidifying' Heitler-London forces or whether they are not. In a solid and in a molecule they all are. In a gas of single atoms (as e.g. mercury vapour) they are not. In a gas composed of molecules, only the atoms within every molecule are linked in this way.

THE APERIODIC SOLID

A small molecule might be called 'the germ of a solid.' Starting from such a small solid germ, there seem to be two different ways of building up larger and larger associations. One is the comparatively dull way of repeating the same structure in three directions again and again. That is the way followed in a growing crystal. Once the periodicity is established, there is no definite limit to the size of the aggregate. The other way is that of building up a more and more extended aggregate without the dull device of repetition. That is the case of the more and more complicated organic molecule in which every atom, and every group of atoms, plays an individual role, not entirely equivalent to that of many others (as is the case in a periodic structure). We might quite properly call that an aperiodic crystal or solid and express our hypothesis by saying: We believe a gene—or perhaps the whole chromosome fibre[9]—to be an aperiodic solid.

THE VARIETY OF CONTENTS COMPRESSED IN THE MINIATURE CODE

It has often been asked how this tiny speck of material, the nucleus of the fertilized egg, could contain an elaborate code-script involving all the future development of the organism? A well-ordered association of atoms, endowed with sufficient resistivity to keep its order permanently, appears to be the only conceivable material structure, that offers a variety of possible ('isomeric') arrangements, sufficiently large to embody a complicated system of 'determinations' within a small spatial boundary. Indeed, the number of atoms in such a structure need not be very large to produce an almost unlimited number of possible arrangements. For illustration, think of the Morse code. The two different signs of dot and dash in well-ordered groups of not more than four allow of thirty different specifications. Now, if you allowed yourself the use of a third sign, in addition to dot and dash, and used groups of not more than ten, you could form 29,524 different 'letters'; with five signs and groups up to 25, the number is 372,529,029,846,191,405.

It may be objected that the simile is deficient, because our Morse signs may have different composition (e.g. ·– – and ··–) and thus they are a bad analogue for isomerism. To remedy this defect, let us pick, from the third example, only the combinations of exactly 25 symbols and only those containing exactly 5 out of each of the supposed 5 types (5 dots, 5 dashes, etc.). A rough count gives you the number of combinations as 62,330,000,000,000, where the zeros on the right stand for figures which I have not taken the trouble to compute.

Of course, in the actual case, by no means 'every' arrangement of the group of atoms will represent a possible molecule; moreover, it is not a question of a code to be adopted arbitrarily, for the code-script must itself be the operative factor bringing about the development. But, on the other hand, the number chosen in the example (25) is still very small, and we have envisaged only the simple arrangements in one line. What we wish to illustrate is simply that with the molecular picture of the gene it is no longer inconceivable that the miniature code should precisely correspond with a highly complicated and specified plan of development and should somehow contain the means to put it into operation.

[9] That it is highly flexible is no objection; so is a thin copper wire.

COMPARISON WITH FACTS: DEGREE OF STABILITY; DISCONTINUITY OF MUTATIONS

Now let us at last proceed to compare the theoretical picture with the biological facts. The first question obviously is, whether it can really account for the high degree of permanence we observe. Are threshold values of the required amount—high multiples of the average heat energy kT—reasonable, are they within the range known from ordinary chemistry? That question is trivial; it can be answered in the affirmative without inspecting tables. The molecules of any substance which the chemist is able to isolate at a given temperature must at that temperature have a lifetime of at least minutes. (That is putting it mildly; as a rule they have much more.) Thus the threshold values the chemist encounters are of necessity precisely of the order of magnitude required to account for practically any degree of permanence the biologist may encounter; for we recall from the discussion of stability that thresholds varying within a range of about 1 : 2 will account for lifetimes ranging from a fraction of a second to tens of thousands of years.

But let me mention figures, for future reference. The ratios W/kT mentioned by way of example on page 979, viz.

$$\frac{W}{kT} = 30, \ 50, \ 60,$$

producing lifetimes of

$$\tfrac{1}{10} \text{ sec., } 16 \text{ months, } 30{,}000 \text{ years,}$$

respectively, correspond at room temperature with threshold values of

$$0{\cdot}9, \ 1{\cdot}5, \ 1{\cdot}8 \text{ electron-volts.}$$

We must explain the unit 'electron-volt,' which is rather convenient for the physicist, because it can be visualized. For example, the third number (1·8) means that an electron, accelerated by a voltage of about 2 volts, would have acquired just sufficient energy to effect the transition by impact. (For comparison, the battery of an ordinary pocket flash-light has 3 volts.)

These considerations make it conceivable that an isomeric change of configuration in some part of our molecule, produced by a chance fluctuation of the vibrational energy can actually be a sufficiently rare event to be interpreted as a spontaneous mutation. Thus we account, by the very principles of quantum mechanics, for the most amazing fact about mutations, the fact by which they first attracted de Vries's attention, namely, that they are 'jumping' variations, no intermediate forms occurring.

STABILITY OF NATURALLY SELECTED GENES

Having discovered the increase of the natural mutation rate by any kind of ionizing rays, one might think of attributing the natural rate to the radio-activity of the soil and air and to cosmic radiation. But a quantitative comparison with the X-ray results shows that the 'natural radiation' is much too weak and could account only for a small fraction of the natural rate.

Granted that we have to account for the rare natural mutations by chance fluctuations of the heat motion, we must not be very much astonished that Nature has succeeded in making such a subtle choice of threshold values as is necessary to make mutation rare. For we have, earlier in these lectures, arrived at the conclusion that frequent mutations are detrimental to evolution. Individuals which, by mutation, acquire a gene configuration of insufficient stability, will have little chance of seeing their 'ultra-radical,' rapidly mutating descendancy survive long. The species will be freed of them and will thus collect stable genes by natural selection.

THE SOMETIMES LOWER STABILITY OF MUTANTS

But, of course, as regards the mutants which occur in our breeding experiments and which we select, *qua* mutants, for studying their offspring, there is no reason to expect that they should all show that very high stability. For they have not yet been 'tried out'—or, if they have, they have been 'rejected' in the wild breeds—possibly for too high mutability. At any rate, we are not at all astonished to learn that actually some of these mutants do show a much higher mutability than the normal 'wild' genes.

TEMPERATURE INFLUENCES UNSTABLE GENES LESS THAN STABLE ONES

This enables us to test our mutability formula, which was

$$t = \tau e^{W/kT}.$$

(It will be remembered that t is the time of expectation for a mutation with threshold energy W.) We ask: How does t change with the temperature? We easily find from the preceding formula in good approximation the ratio of the value of t at temperature $T + 10$, to that at temperature T

$$\frac{t_{T+10}}{t_T} = e^{-10W/kT^2}.$$

The exponent being now negative, the ratio is, naturally, smaller than 1. The time of expectation is diminished by raising the temperature, the mutability is increased. Now that can be tested and has been tested with the fly *Drosophila* in the range of temperature which the insects will stand. The result was, at first sight, surprising. The *low* mutability of wild genes was distinctly increased, but the comparatively *high* mutability occurring with some of the already mutated genes was not, or at any rate was much less, increased. That is just what we expect on comparing our two formulae. A large value of W/kT, which according to the first formula is required to make t large (stable gene), will, according to the second one, make for a small value of the ratio computed there, that is to say for a considerable increase of mutability with temperature. (The actual values of the ratio seem to lie between about ½ and ⅕. The reciprocal, 2·5, is what in an ordinary chemical reaction we call the van 't Hoff factor.)

HOW X-RAYS PRODUCE MUTATION

Turning now to the X-ray-induced mutation rate, we have already inferred from the breeding experiments, first (from the proportionality of mutation rate, and dosage), that some single event produces the mutation; secondly (from quantitative results and from the fact that the mutation rate is determined by the integrated ionization density and independent of the wave-length), this single event must be an ionization, or similar process, which has to take place inside a certain volume of only about 10 atomic-distances-cubed, in order to produce a specified mutation. According to our picture, the energy for overcoming the threshold must obviously be furnished by that explosion-like process, ionization or excitation. I call it explosion-like, because the energy spent in one ionization (spent, incidentally, not by the X-ray itself, but by a secondary electron it produces) is well known and has the comparatively enormous amount of 30 electron-volts. It is bound to be turned into enormously increased heat motion around the point where it is discharged and to spread from there in the form of a 'heat wave,' a wave of intense oscillations of the atoms. That this heat wave should still be able to furnish the required threshold energy of 1 or 2 electron-volts at an average 'range of action' of about ten atomic distances, is not inconceivable, though it may well be that an unprejudiced physicist might have anticipated a slightly lower range of action. That in many cases the effect of the explosion will not be an orderly isomeric transition but a lesion of the chromosome, a lesion that becomes lethal when, by ingenious crossings, the uninjured partner (the corresponding chromosome of the second set) is removed and replaced by a partner whose corresponding gene is known to be itself morbid—all that is absolutely to be expected and it is exactly what is observed.

THEIR EFFICIENCY DOES NOT DEPEND ON SPONTANEOUS MUTABILITY

Quite a few other features are, if not predictable from the picture, easily understood from it. For example, an unstable mutant does not on the average show a much higher X-ray mutation rate than a stable one. Now, with an explosion furnishing an energy of 30 electron-volts you would certainly not expect that it makes a lot of difference whether the required threshold energy is a little larger or a little smaller, say 1 or 1.3 volts.

REVERSIBLE MUTATIONS

In some cases a transition was studied in both directions, say from a certain 'wild' gene to a specified mutant and back from that mutant to the wild gene. In such cases the natural mutation rate is sometimes nearly the same, sometimes very different. At first sight one is puzzled, because the threshold to be overcome seems to be the same in both cases. But, of course, it need not be, because it has to be measured from the energy level of the starting configuration, and that may be different for the wild and the mutated gene. (See Figure 2 on page 982, where '1' might refer to the wild allele, '2' to the mutant, whose lower stability would be indicated by the shorter arrow.)

On the whole, I think, Delbrück's 'model' stands the tests fairly well and we are justified in using it in further considerations.

ORDER, DISORDER AND ENTROPY

Nec corpus mentem ad cogitandum nec mens corpus ad motum, neque ad quietem nec ad aliquid (si quid est) aliud determinare potest.[10]

—SPINOZA, *Ethics*, P. III, Prop. 2

A REMARKABLE GENERAL CONCLUSION FROM THE MODEL

Let me refer to the last phrase on page 986, in which I tried to explain that the molecular picture of the gene made it at least conceivable 'that the miniature code should be in one-to-one correspondence with a highly complicated and specified plan of development and should somehow contain the means of putting it into operation.' Very well then, but how does it do this? How are we going to turn 'conceivability' into true understanding?

Delbrück's molecular model, in its complete generality, seems to contain no hint as to how the hereditary substance works. Indeed, I do not expect that any detailed information on this question is likely to come from physics in the near future. The advance is proceeding and will, I am

[10] Neither can the body determine the mind to think, nor the mind the body to move or to rest nor to anything else, if such there be.

sure, continue to do so, from biochemistry under the guidance of physiology and genetics.

No detailed information about the functioning of the genetical mechanism can emerge from a description of its structure so general as has been given above. That is obvious. But, strangely enough, there is just one general conclusion to be obtained from it, and that, I confess, was my only motive for writing this book.

From Delbrück's general picture of the hereditary substance it emerges that living matter, while not eluding the 'laws of physics' as established up to date, is likely to involve 'other laws of physics' hitherto unknown, which, however, once they have been revealed, will form just as integral a part of this science as the former.

ORDER BASED ON ORDER

This is a rather subtle line of thought, open to misconception in more than one respect. All the remaining pages are concerned with making it clear. A preliminary insight, rough but not altogether erroneous, may be found in the following considerations:

It has been explained [11] that the laws of physics, as we know them, are statistical laws.[12] They have a lot to do with the natural tendency of things to go over into disorder.

But, to reconcile the high durability of the hereditary substance with its minute size, we had to evade the tendency to disorder by 'inventing the molecule,' in fact, an unusually large molecule which has to be a masterpiece of highly differentiated order, safeguarded by the conjuring rod of quantum theory. The laws of chance are not invalidated by this 'invention,' but their outcome is modified. The physicist is familiar with the fact that the classical laws of physics are modified by quantum theory, especially at low temperature. There are many instances of this. Life seems to be one of them, a particularly striking one. Life seems to be orderly and lawful behaviour of matter, not based exclusively on its tendency to go over from order to disorder, but based partly on existing order that is kept up.

To the physicist—but only to him—I could hope to make my view clearer by saying: The living organism seems to be a macroscopic system which in part of its behaviour approaches to that purely mechanical (as contrasted with thermodynamical) conduct to which all systems tend, as the temperature approaches the absolute zero and the molecular disorder is removed.

The non-physicist finds it hard to believe that really the ordinary laws

[11] In earlier discussions of the book. ED.

[12] To state this in complete generality about 'the laws of physics' is perhaps challengeable.

of physics, which he regards as the prototype of inviolable precision, should be based on the statistical tendency of matter to go over into disorder. The general principle involved is the famous Second Law of Thermodynamics (entropy principle) and its equally famous statistical foundation. In the following sections I will try to sketch the bearing of the entropy principle on the large-scale behaviour of a living organism—forgetting at the moment all that is known about chromosomes, inheritance, and so on.

LIVING MATTER EVADES THE DECAY TO EQUILIBRIUM

What is the characteristic feature of life? When is a piece of matter said to be alive? When it goes on 'doing something,' moving, exchanging material with its environent, and so forth, and that for a much longer period than we would expect an inanimate piece of matter to 'keep going' under similar circumstances. When a system that is not alive is isolated or placed in a uniform environment, all motion usually comes to a standstill very soon as a result of various kinds of friction; differences of electric or chemical potential are equalized, substances which tend to form a chemical compound do so, temperature becomes uniform by heat conduction. After that the whole system fades away into a dead, inert lump of matter. A permanent state is reached, in which no observable events occur. The physicist calls this the state of thermodynamical equilibrium, or of 'maximum entropy.'

Practically, a state of this kind is usually reached very rapidly. Theoretically, it is very often not yet an absolute equilibrium, not yet the true maximum of entropy. But then the final approach to equilibrium is very slow. It could take anything between hours, years, centuries. . . . To give an example—one in which the approach is still fairly rapid: if a glass filled with pure water and a second one filled with sugared water are placed together in a hermetically closed case at constant temperature, it appears at first that nothing happens, and the impression of complete equilibrium is created. But after a day or so it is noticed that the pure water, owing to its higher vapour pressure, slowly evaporates and condenses on the solution. The latter overflows. Only after the pure water has totally evaporated has the sugar reached its aim of being equally distributed among all the liquid water available.

These ultimate slow approaches to equilibrium could never be mistaken for life, and we may disregard them here. I have referred to them in order to clear myself of a charge of inaccuracy.

IT FEEDS ON 'NEGATIVE ENTROPY'

It is by avoiding the rapid decay into the inert state of 'equilibrium,' that an organism appears so enigmatic; so much so, that from the earliest times of human thought some special non-physical or supernatural force (*vis viva*, entelechy) was claimed to be operative in the organism, and in some quarters is still claimed.

How does the living organism avoid decay? The obvious answer is: By eating, drinking, breathing and (in the case of plants) assimilating. The technical term is *metabolism*. The Greek word (μεταβάλλειν) means change or exchange. Exchange of what? Originally the underlying idea is, no doubt, exchange of material. (E.g. the German for metabolism is Stoffwechsel.) That the exchange of material should be the essential thing is absurd. Any atom of nitrogen, oxygen, sulphur, etc., is as good as any other of its kind; what could be gained by exchanging them? For a while in the past our curiosity was silenced by being told that we feed upon energy. In some very advanced country (I don't remember whether it was Germany or the U.S.A. or both) you could find menu cards in restaurants indicating, in addition to the price, the energy content of every dish. Needless to say, taken literally, this is just as absurd. For an adult organism the energy content is as stationary as the material content. Since, surely, any calorie is worth as much as any other calorie, one cannot see how a mere exchange could help.

What then is that precious something contained in our food which keeps us from death? That is easily answered. Every process, event, happening—call it what you will; in a word, everything that is going on in Nature means an increase of the entropy of the part of the world where it is going on. Thus a living organism continually increases its entropy—or, as you may say, produces positive entropy—and thus tends to approach the dangerous state of maximum entropy, which is death. It can only keep aloof from it, i.e. alive, by continually drawing from its environment negative entropy—which is something very positive as we shall immediately see. What an organism feeds upon is negative entropy. Or to put it less paradoxically, the essential thing in metabolism is that the organism succeeds in freeing itself from all the entropy it cannot help producing while alive.

WHAT IS ENTROPY?

What is entropy? Let me first emphasize that it is not a hazy concept or idea, but a measurable physical quantity just like the length of a rod, the temperature at any point of a body, the heat of fusion of a given crystal or the specific heat of any given substance. At the absolute zero point of temperature (roughly $-273°$ C.) the entropy of any substance is zero. When you bring the substance into any other state by slow, reversible little steps (even if thereby the substance changes its physical or chemical nature or splits up into two or more parts of different physical or chemical nature) the entropy increases by an amount which is computed by dividing every little portion of heat you had to supply in that procedure by the absolute temperature at which it was supplied—and by summing up all these small contributions. To give an example, when you melt a solid, its entropy increases by the amount of the heat of fusion divided by the temperature at the melting-point. You see from this, that the unit in which entropy is measured is cal./° C. (just as the calorie is the unit of heat or the centimetre the unit of length).

THE STATISTICAL MEANING OF ENTROPY

I have mentioned this technical definition simply in order to remove entropy from the atmosphere of hazy mystery that frequently veils it. Much more important for us here is the bearing on the statistical concept of order and disorder, a connection that was revealed by the investigations of Boltzmann and Gibbs in statistical physics. This too is an exact quantitative connection, and is expressed by

$$\text{entropy} = k \log D,$$

where k is the so-called Boltzmann constant ($= 3.2983 \,.\, 10^{-24}$ cal./° C.), and D a quantitative measure of the atomistic disorder of the body in question. To give an exact explanation of this quantity D in brief non-technical terms is well-nigh impossible. The disorder it indicates is partly that of heat motion, partly that which consists in different kinds of atoms or molecules being mixed at random, instead of being neatly separated, e.g. the sugar and water molecules in the example quoted above. Boltzmann's equation is well illustrated by that example. The gradual 'spreading out' of the sugar over all the water available increases the disorder D, and hence (since the logarithm of D increases with D) the entropy. It is also pretty clear that any supply of heat increases the turmoil of heat motion, that is to say increases D and thus increases the entropy; it is particularly clear that this should be so when you melt a crystal, since you thereby destroy the neat and permanent arrangement of the atoms or molecules and turn the crystal lattice into a continually changing random distribution.

An isolated system or a system in a uniform environment (which for the present consideration we do best to include as a part of the system we contemplate) increases its entropy and more or less rapidly approaches the inert state of maximum entropy. We now recognize this fundamental law of physics to be just the natural tendency of things to approach the chaotic state (the same tendency that the books of a library or the piles of papers and manuscripts on a writing desk display) unless we obviate it. (The analogue of irregular heat motion, in this case, is our handling those objects now and again without troubling to put them back in their proper places.)

ORGANIZATION MAINTAINED BY EXTRACTING 'ORDER' FROM THE ENVIRONMENT

How would we express in terms of the statistical theory the marvellous faculty of a living organism, by which it delays the decay into thermodynamical equilibrium (death)? We said before: 'It feeds upon negative entropy,' attracting, as it were, a stream of negative entropy upon itself, to compensate the entropy increase it produces by living and thus to maintain itself on a stationary and fairly low entropy level.

If D is a measure of disorder, its reciprocal, $1/D$, can be regarded as a direct measure of order. Since the logarithm of $1/D$ is just minus the logarithm of D, we can write Boltzmann's equation thus:

$$-\text{(entropy)} = k \log (1/D).$$

Hence the awkward expression 'negative entropy' can be replaced by a better one: entropy, taken with the negative sign, is itself a measure of order. Thus the device by which an organism maintains itself stationary at a fairly high level of orderliness (= fairly low level of entropy) really consists in continually sucking orderliness from its environment. This conclusion is less paradoxical than it appears at first sight. Rather could it be blamed for triviality. Indeed, in the case of higher animals, we know the kind of orderliness they feed upon well enough, viz. the extremely well-ordered state of matter in more or less complicated organic compounds, which serve them as foodstuffs. After utilizing it they return it in a very much degraded form—not entirely degraded, however, for plants can still make use of it. (These, of course, have their most powerful supply of 'negative entropy' in the sunlight.)

COMMENTARY ON

D'ARCY WENTWORTH THOMPSON

D'ARCY WENTWORTH THOMPSON was a huge and remarkable man who wrote a huge and remarkable book. His outlook was Pythagorean; he saw numbers in all things. But unlike Pythagoras he was neither a religious prophet nor a mystic. Thompson was a naturalist, a classical scholar and a mathematician. His friend, the late British zoölogist Clifford Dobell, said that "these three main elements in his composition were not mixed or added but as it were chemically combined to form his personality." *On Growth and Form* is a literary and scientific masterpiece, whose breadth, grace and aliveness, whose almost endless store of wonderful things uniquely reflect this personality.

Thompson was born in Edinburgh, Scotland, in 1860, the grandson of a shipmaster and the son of an able classical scholar who was born at sea off the coast of Van Dieman's Land.[1] His mother died in giving birth to him and the "adamantine" bond between father and son profoundly influenced the son's character. His father was himself a remarkable man. The elder Thompson combined a "passionate humanism"[2] with pronounced—and advanced—views on education and other controversial matters. He taught for a time at Edinburgh Academy, where his pupils included Andrew Lang and Robert Louis Stevenson, but his outspokenness and his threats to reform the curriculum did not sit well with the directors and he transferred to a post as professor of Greek at Queens College, Galway. He wrote a number of books of essays and two volumes of quite distressing nursery rhymes. In 1866 he gave the Lowell Lectures at Boston; seventy years later his son had the special joy of delivering his own series of lectures under the same auspices.

D'Arcy Thompson got his early education from his father and learned to read, write and speak Latin and Greek with "incomparable ease and fluency." After attending Edinburgh Academy, he entered the University of Edinburgh at the age of seventeen as a medical student. Two years later he won a scholarship at Trinity College, Cambridge, where he took up the study of zoölogy and natural science. Though he was an excellent

[1] For its facts this sketch of Thompson's life draws heavily on the obituary by Dobell, *Obituary Notices*, Royal Society of London 18 (1949), 599–617. (Unkeyed quotes are from this source.) Other useful sources include the obituary by H. W. Turnbull in *The Edinburgh Mathematical Notes*, 38 (1952), pp. 17–18; G. Evelyn Hutchinson, "In Memoriam: D'Arcy Wentworth Thompson," *American Scientist*; John Tyler Bonner, "D'Arcy Thompson," *Scientific American*, August 1952, pp. 60–66.

[2] G. Evelyn Hutchinson, *loc. cit.* (*see* note 1).

student he failed to get a fellowship. This was a heavy disappointment but it may have turned out for the best because it forced him to turn to a smaller university where he was more on his own and could pursue his special interests. In 1884 he was appointed professor of biology at University College, Dundee; this institution was later incorporated into the University of St. Andrews, at which time he assumed the chair in natural history. There he taught for sixty-four years.

Thompson often said and wrote, "I love variety." His publications, comprising some 300 titles, exhibit the versatility of his gifts and inclinations. Most of his literary labor was either technical or in the field of pure scholarship; such work was its own reward and he was content if only a handful of persons took notice of it. While still at Cambridge, he translated a monumental German treatise on the fertilization of flowers (Hermann Müller, *Die Befruchtung der Blumen durch Insekten*) for which Charles Darwin wrote a preface. He compiled a bibliography of world literature on protozoa, sponges, coelenterates and worms; prepared a *Glossary of Greek Birds*, and a *Glossary of Greek Fishes* (published the year before he died); spent thirty years on a translation of Aristotle's *Historia Animalium* (1910); issued papers on morphology and systematics, the bones of a fossil sea, the nervous system of cyclostome fishes, the internal ear of the sunfish, tapeworms, the bones of the parrot's skull, the cuttlefish, the arrangement of the feathers on a hummingbird (this is only a sample list); wrote several delightful essays on classical-scientific themes (e.g., "Excess and Defect," "How to Catch Cuttlefish"), and a large number of obituaries and reviews. A dozen of his classical, scientific, mathematical and literary articles and addresses appear in *Science and the Classics* (Cambridge, 1940), a book to be warmly recommended.

Despite this voluminous pen-work and research, Thompson found time for other duties. For forty-one years (1898–1939) he served as scientific adviser to the Fishery Board for Scotland, editing reports and writing papers on practical fishing problems. He was an admirable teacher and lecturer, equally at home when talking "to school children and learned societies." His pupils were counted "by the thousands"; as he grew old he was much venerated and his reputation became immense. In 1916 he was elected a fellow, and in 1931 a vice-president, of the Royal Society of London; for five years (1934–1939) he was president of the Royal Society of Edinburgh. Cambridge and Oxford conferred honorary degrees upon him and in 1937 he was knighted. He died in 1948, having only a short time before made a trip to India where he "spoke to a huge audience on the skeletal structure of birds." [3]

Thompson was magnificent in appearance and majestic in manner. His great spade beard, grown while he was still a youth, reddish and

[3] John Tyler Bonner, *loc. cit.* (*see* note 1).

later pure white, made him look like "a kindly Jupiter." He had a massive frame, blue eyes, and a large head "which he customarily covered with a black fedora that was extra high in the crown and broad of brim." [4] His talk "whether in private or at a public meeting, was easy, measured, homely, resonant and dignified." [5] His self-possession was superb: throughout the Indian lecture "he held an angry hen tucked under one arm so that he could conveniently point out on a living specimen the salient features of his description. The remonstrances of his model disturbed not at all the easy flow of his speech." [6] He married in 1901 and had three children. He had many friends, liked to be with people and teach, yet he was "at heart a lonely man, valuing most highly peace, quiet and freedom." Dobell writes that he both loved and hated the idea of death. "The world does *not* grow tedious to us," he said when he was eighty-one; "and yet we are prepared to acknowledge that the long and happy holidays have been just enough." Toward the end, like George Bernard Shaw and others who have attained a great age, he longed for release. "My day is over," he sighed; "I have done what I could, I have drawn my pay, I have had my full share of modest happiness. . . . The old world, that I knew how to live in, has passed away."

* * * * *

The first edition of *On Growth and Form* appeared in 1915. It was a modest issue of 500 copies, but after it had slowly sold out, copies were much sought and brought pretty prices in the secondhand market. The book was written in wartime; its revision, begun in 1922 but not completed until 1941, employed Thompson during another war. "It gave me solace and occupation, when service was debarred me by my years," he said in the preface to the new edition.

He wrote his book, he explained, "as an easy introduction to the study of organic Form, by methods which are the commonplaces of physical science, which are by no means novel in their application to natural history, but which nevertheless naturalists are little accustomed to employ." Its grand object—though he claimed to have achieved nothing more than a "first approximation" ("This book of mine has little need of preface, for indeed it is 'all preface' from beginning to end")—was to reduce biological phenomena to physics and if possible to mathematics. His thesis is summarized in a memorable passage: "Cell and tissue, shell and bone, leaf and flower, are so many portions of matter, and it is in obedience to the laws of physics that their particles have been moved, molded and conformed. They are no exception to the rule that θεο'ς α'εῖ γεωμετρεῖ [God always geometrizes]. Their problems of form are in the first instance math-

[4] *Ibid.*
[5] H. W. Turnbull, *loc. cit* (*see* note 1).
[6] Bonner, *loc. cit.*

ematical problems, their problems of growth are essentially physical problems, and the morphologist is, *ipso facto*, a student of physical science."

The supreme excellence of *On Growth and Form* lies, as Dobell points out, in its fresh approach to the subject and almost perfect treatment of its problems. Thompson applied himself to the "elucidation, analysis and synthesis of facts already known in classics and fishery statistics, in publications ancient, medieval and modern on zoölogy, botany, geology, chemistry, physics, astronomy, engineering, mathematics, and even mythology." He discusses magnitudes, rates of growth, the forms of cells and tissues. He presents a beautiful analysis of radiolarian skeletons which are formed, not by crystallization forces, but rather by "that symmetry of forces which results from the play and interplay of surface-tensions in the whole system." He explains the tightness of the teeth, the mathematics of the splash, the shape of bee cells, the effect of temperature on the growth of ants, the reason for the shape of blood cells, the rules governing leaf arrangements, the formation of deer's antlers, the branching of blood vessels, the building of snow crystals. By geometrical transformations based on the work of Dürer he is able to relate the shapes of apparently distinct forms of fish. The equiangular spiral is exhibited in many forms of dress: clothed as a nautilus, a molluscan, an ammonite, a univalve, an argonaut, a shrimp, an oyster, a sunflower. We are shown the mechanics of bird flight, the stress diagram of the backbone of a horse and of a dinosaur, the remarkable resemblances between various engineering artifacts, notably bridges, and reptilian skeletons, such as those of our forebears, diplodocus and stegosaurus. The description of the frame of one of these extraordinary fellows is "strictly and beautifully comparable" to the main girder of a double-armed cantilever bridge, like that crossing the Forth.[7]

It is a wise and fascinating book, at once poetic and scientific. It is written in a limpid style, strongly individual and elegant, but unaffected. As in Buffon's phrase, "le style est l'homme même." "To spin words and make pretty sentences," said Thompson, "is my one talent, and I must make the best of it." Bacon admired in Julius Caesar his "marvelous understanding of the weight and edge of words"; the same marvelous understanding pervades *On Growth and Form*.

What follows here is the second chapter of Thompson's book. The sub-

[7] "The fore-limbs, though comparatively small, are obviously fashioned for support, but the weight which they have to carry is far less than that which the hind-limbs bear. The head is small and the neck short, while on the other hand the hind quarters and the tail are big and massive. The backbone bends into a great double-armed cantilever, culminating over the pelvis and the hind-limbs, and here furnished with its highest and strongest spines to separate the tension-member from the compression member of the girder. The fore-legs form a secondary supporting pier to this great continuous cantilever, the greater part of whose weight is poised upon the hind-limbs alone." *On Growth and Form*, Second Edition, p. 1006.

ject is magnitudes (the size and speed of animals and similar matters), but the treatment of it is so broad, so discursive—though never irrelevant—that many of the topics dealt with in later chapters are brought into view. It is a long selection, longer perhaps than any other in the book, but the reader will probably find it too short. There is never enough of excellent things.

. . . indeed what reason may not go to Schoole to the wisdome of Bees, Aunts, and Spiders? what wise hand teacheth them to doe what reason cannot teach us? ruder heads stand amazed at those prodigious pieces of nature, Whales, Elephants, Dromidaries and Camels; these I confesse, are the Colossus and Majestick pieces of her hand; but in these narrow Engines there is more curious Mathematicks, and the civilitie of these little Citizens more neatly sets forth the wisedome of their Maker.

—SIR THOMAS BROWNE

17 On Magnitude

By D'ARCY WENTWORTH THOMPSON

TO terms of magnitude, and of direction, must we refer all our conceptions of Form. For the form of an object is defined when we know its magnitude, actual or relative, in various directions; and Growth involves the same concepts of magnitude and direction, related to the further concept, or "dimension," of Time. Before we proceed to the consideration of specific form, it will be well to consider certain general phenomena of spatial magnitude, or of the extension of a body in the several dimensions of space.

We are taught by elementary mathematics—and by Archimedes himself—that in similar figures the surface increases as the square, and the volume as the cube, of the linear dimensions. If we take the simple case of a sphere, with radius r, the area of its surface is equal to $4\pi r^2$, and its volume to $\frac{4}{3}\pi r^3$; from which it follows that the ratio of its volume to surface, or V/S, is $\frac{1}{3}r$. That is to say, V/S *varies* as r; or, in other words, the larger the sphere by so much the greater will be its volume (or its mass, if it be uniformly dense throughout) in comparison with its superficial area. And, taking L to represent any linear dimension, we may write the general equations in the form

$$S \propto L^2, \qquad V \propto L^3,$$

or

$$S = kL^2, \text{ and } V = k'L^3,$$

where k, k', are "factors of proportion,"

and

$$\frac{V}{S} \propto \mathrm{L}, \text{ or } \frac{V}{S} = \frac{k}{k'} L = KL.$$

So, in Lilliput, "His Majesty's Ministers, finding that Gulliver's stature exceeded theirs in the proportion of twelve to one, concluded from the similarity of their bodies that his must contain at least 1728 [or 12^3] of theirs, and must needs be rationed accordingly." [1]

From these elementary principles a great many consequences follow, all more or less interesting, and some of them of great importance. In the first place, though growth in length (let us say) and growth in volume (which is usually tantamount to mass or weight) are parts of one and the same process or phenomenon, the one attracts our *attention* by its increase very much more than the other. For instance a fish, in doubling its length, multiplies its weight no less than eight times; and it all but doubles its weight in growing from four inches long to five.

In the second place, we see that an understanding of the correlation between length and weight in any particular species of animal, in other words a determination of k in the formula $W = k.L^3$, enables us at any time to translate the one magnitude into the other, and (so to speak) to weigh the animal with a measuring-rod; this, however, being always subject to the condition that the animal shall in no way have altered its form, nor its specific gravity. That its specific gravity or density should materially or rapidly alter is not very likely; but as long as growth lasts changes of form, even though inappreciable to the eye, are apt and likely to occur. Now weighing is a far easier and far more accurate operation than measuring; and the measurements which would reveal slight and otherwise imperceptible changes in the form of a fish—slight relative differences between length, breadth and depth, for instance—would need to be very delicate indeed. But if we can make fairly accurate determinations of the length, which is much the easiest linear dimension to measure, and correlate it with the weight, then the value of k, whether it varies or remains constant, will tell us at once whether there has or has not been a tendency to alteration in the general form, or, in other words, a difference in the rates of growth in different directions. To this subject we shall return, when we come to consider more particularly the phenomenon of *rate of growth.*

We are accustomed to think of magnitude as a purely relative matter. We call a thing *big* or *little* with reference to what it is wont to be, as when we speak of a small elephant or a large rat; and we are apt accordingly to suppose that size makes no other or more essential difference,

[1] Likewise Gulliver had a whole Lilliputian hogshead for his half-pint of wine: in the due proportion of 1728 half-pints, or 108 gallons, equal to one pipe or double-hogshead. But Gilbert White of Selborne could not see what was plain to the Lilliputians; for finding that a certain little long-legged bird, the stilt, weighed 4¼ oz. and had legs 8 in. long, he thought that a flamingo, weighing 4 lbs., should have legs 10 ft. long, to be in the same proportion as the stilt's. But it is obvious to us that, as the weights of the two birds are as 1 : 15, so the legs (or other linear dimensions) should be as the cube-roots of these numbers, or nearly as 1 : 2½. And on this scale the flamingo's legs should be, as they actually are, about 20 in. long.

and that Lilliput and Brobdingnag [2] are all alike, according as we look at them through one end of the glass or the other. Gulliver himself declared, in Brobdingnag, that "undoubtedly philosophers are in the right when they tell us that nothing is great and little otherwise than by comparison": and Oliver Heaviside used to say, in like manner, that there is no absolute scale of size in the Universe, for it is boundless towards the great and also boundless towards the small. It is of the very essence of the Newtonian philosophy that we should be able to extend our concepts and deductions from the one extreme of magnitude to the other; and Sir John Herschel said that "the student must lay his account to finding the distinction of great and little altogether annihilated in nature."

All this is true of *number*, and of *relative magnitude*. The Universe has its endless gamut of great and small, of near and far, of many and few. Nevertheless, in physical science the scale of absolute magnitude becomes a very real and important thing; and a new and deeper interest arises out of the changing ratio of dimensions when we come to consider the inevitable changes of physical relations with which it is bound up. The effect of *scale* depends not on a thing in itself, but in relation to its whole environment or milieu; it is in conformity with the thing's "place in Nature," its field of action and reaction in the Universe. Everywhere Nature works true to scale, and everything has its proper size accordingly. Men and trees, birds and fishes, stars and star-systems, have their appropriate dimensions, and their more or less narrow range of absolute magnitudes. The scale of human observation and experience lies within the narrow bounds of inches, feet or miles, all measured in terms drawn from our own selves or our own doings. Scales which include light-years, parsecs, Angström units, or atomic and sub-atomic magnitudes, belong to other orders of things and other principles of cognition.

A common effect of scale is due to the fact that, of the physical forces, some act either directly at the surface of a body, or otherwise in proportion to its surface or area; while others, and above all gravity, act on all particles, internal and external alike, and exert a force which is proportional to the mass, and so usually to the volume of the body.

A simple case is that of two similar weights hung by two similar wires. The forces exerted by the weights are proportional to their masses, and these to their volumes, and so to the cubes of the several linear dimensions, including the diameters of the wires. But the areas of cross-section of the wires are as the squares of the said linear dimensions; therefore the stresses in the wires *per unit area* are not identical, but increase in the

[2] Swift paid close attention to the arithmetic of magnitude, but none to its physical aspect. See De Morgan, on Lilliput, in *N. and Q.* (2), VI, pp. 123–125, 1858. On relative magnitude see also Berkeley, in his *Essay towards a New Theory of Vision*, 1709.

ratio of the linear dimensions, and the larger the structure the more severe the strain becomes:

$$\frac{\text{Force}}{\text{Area}} \propto \frac{l^3}{l^2} \propto l,$$

and the less the wires are capable of supporting it.

In short, it often happens that of the forces in action in a system some vary as one power and some as another, of the masses, distances or other magnitudes involved; the "dimensions" remain the same in our equations of equilibrium, but the relative values alter with the scale. This is known as the "Principle of Similitude," or of dynamical similarity, and it and its consequences are of great importance. In a handful of matter cohesion, capillarity, chemical affinity, electric charge are all potent; across the solar system gravitation [3] rules supreme; in the mysterious region of the nebulae, it may haply be that gravitation grows negligible again.

To come back to homelier things, the strength of an iron girder obviously varies with the cross-section of its members, and each cross-section varies as the square of a linear dimension; but the weight of the whole structure varies as the cube of its linear dimensions. It follows at once that, if we build two bridges geometrically similar, the larger is the weaker of the two,[4] and is so in the ratio of their linear dimensions. It was elementary engineering experience such as this that led Herbert Spencer to apply the principle of similitude to biology.[5]

But here, before we go further, let us take careful note that increased weakness is no necessary concomitant of increasing size. There are exceptions to the rule, in those exceptional cases where we have to deal only with forces which vary merely with the *area* on which they impinge. If in a big and a little ship two similar masts carry two similar sails, the two sails will be similarly strained, and equally stressed at homologous places, and alike suitable for resisting the force of the same wind. Two similar umbrellas, however differing in size, will serve alike in the same weather; and the expanse (though not the leverage) of a bird's wing may be enlarged with little alteration.

[3] In the early days of the theory of gravitation, it was deemed especially remarkable that the action of gravity "is proportional to the quantity of solid matter in bodies, and not to their surfaces as is usual in mechanical causes; this power, therefore, seems to surpass mere mechanism" (Colin Maclaurin, on *Sir Isaac Newton's Philosophical Discoveries*, IV, 9).

[4] The subject is treated from the engineer's point of view by Prof. James Thomson, Comparison of similar structures as to elasticity, strength and stability, *Coll. Papers*, 1912, pp. 361–372, and *Trans. Inst. Engineers, Scotland*, 1876; also by Prof. A. Barr, *ibid.* 1899. See also Rayleigh, *Nature*, April 22, 1915; Sir G. Greenhill, On mechanical similitude, *Math. Gaz.* March 1916, *Coll. Works*, VI, p. 300. For a mathematical account, see (e.g.) P. W. Bridgman, *Dimensional Analysis* (2nd ed.), 1931, or F. W. Lanchester, *The Theory of Dimensions*, 1936.

[5] Herbert Spencer, The form of the earth, etc., *Phil. Mag.* XXX, pp. 194–6, 1847; also *Principles of Biology*, pt. II, p. 123 *seq.*, 1864.

The principle of similitude had been admirably applied in a few clear instances by Lesage,[6] a celebrated eighteenth-century physician, in an unfinished and unpublished work. Lesage argued, for example, that the larger ratio of surface to mass in a small animal would lead to excessive transpiration, were the skin as "porous" as our own; and that we may thus account for the hardened or thickened skins of insects and many other small terrestrial animals. Again, since the weight of a fruit increases as the cube of its linear dimensions, while the strength of the stalk increases as the square, it follows that the stalk must needs grow out of apparent due proportion to the fruit: or, alternatively, that tall trees should not bear large fruit on slender branches, and that melons and pumpkins must lie upon the ground. And yet again, that in quadrupeds a large head must be supported on a neck which is either excessively thick and strong like a bull's, or very short like an elephant's.[7]

But it was Galileo who, wellnigh three hundred years ago, had first laid down this general principle of similitude; and he did so with the utmost possible clearness, and with a great wealth of illustration drawn from structures living and dead.[8] He said that if we tried building ships, palaces or temples of enormous size, yards, beams and bolts would cease to hold together; nor can Nature grow a tree nor construct an animal beyond a certain size, while retaining the proportions and employing the materials which suffice in the case of a smaller structure.[9] The thing will fall to pieces of its own weight unless we either change its relative proportions, which will at length cause it to become clumsy, monstrous and inefficient, or else we must find new material, harder and stronger than was used before. Both processes are familiar to us in Nature and in art, and practical applications, undreamed of by Galileo, meet us at every turn in this modern age of cement and steel.[10]

Again, as Galileo was also careful to explain, besides the questions of pure stress and strain, of the strength of muscles to lift an increasing weight or of bones to resist its crushing stress, we have the important question of *bending moments*. This enters, more or less, into our whole

[6] See Pierre Prévost, *Notices de la vie et des écrits de Lesage*, 1805. George Louis Lesage, born at Geneva in 1724, devoted sixty-three years of a life of eighty to a mechanical theory of gravitation; see W. Thomson (Lord Kelvin), On the ultramundane corpuscles of Lesage, *Proc. R.S.E.* VII, pp. 577–589, 1872; *Phil. Mag.* XLV, pp. 321–345, 1873; and Clerk Maxwell, art. "Atom," *Encycl. Brit.* (9), p. 46.

[7] Cf. W. Walton, On the debility of large animals and trees, *Quart. Journ. of Math.* IX, pp. 179–184, 1868; also L. J. Henderson, On volume in Biology, *Proc. Amer. Acad. Sci.* II, pp. 654–658, 1916; etc.

[8] *Discorsi e Dimostrazioni matematiche, intorno à due nuove scienze attenenti alla Mecanica ed ai Muovimenti Locali:* appresso gli Elzevirii, 1638; *Opere*, ed. Favaro, VIII, p. 169 *seq.* Transl. by Henry Crew and A. de Salvio, 1914, p. 130.

[9] So Werner remarked that Michael Angelo and Bramanti could not have built of gypsum at Paris on the scale they built of travertin at Rome.

[10] The Chrysler and Empire State Buildings, the latter 1048 ft. high to the foot of its 200 ft. "mooring mast," are the last word, at present, in this brobdingnagian architecture.

range of problems; it affects the whole form of the skeleton, and sets a limit to the height of a tall tree.[11]

We learn in elementary mechanics the simple case of two similar beams, supported at both ends and carrying no other weight than their own. Within the limits of their elasticity they tend to be deflected, or to sag downwards, in proportion to the squares of their linear dimensions; if a match-stick be two inches long and a similar beam six feet (or 36 times as long), the latter will sag under its own weight thirteen hundred times as much as the other. To counteract this tendency, as the size of an animal increases, the limbs tend to become thicker and shorter and the whole skeleton bulkier and heavier; bones make up some 8 per cent of the body of mouse or wren, 13 or 14 per cent of goose or dog, and 17 or 18 per cent of the body of a man. Elephant and hippopotamus have grown clumsy as well as big, and the elk is of necessity less graceful than the gazelle. It is of high interest, on the other hand, to observe how little the skeletal proportions differ in a little porpoise and a great whale, even in the limbs and limb-bones; for the whole influence of gravity has become negligible, or nearly so, in both of these.

In the problem of the tall tree we have to determine the point at which the tree will begin to bend under its own weight if it be ever so little displaced from the perpendicular.[12] In such an investigation we have to make certain assumptions—for instance that the trunk tapers uniformly, and that the sectional area of the branches varies according to some definite law, or (as Ruskin assumed) tends to be constant in any horizontal plane; and the mathematical treatment is apt to be somewhat difficult. But Greenhill shewed, on such assumptions as the above, that a certain British Columbian pine-tree, of which the Kew flag-staff, which is 221 ft. high and 21 inches in diameter at the base, was made, could not possibly, by theory, have grown to more than about 300 ft. It is very curious that Galileo had suggested precisely the same height (*ducento braccie alta*) as the utmost limit of the altitude of a tree. In general, as Greenhill shewed, the diameter of a tall homogeneous body must increase as the power $3/2$ of its height, which accounts for the slender proportions of young trees compared with the squat or stunted appearance of old and

[11] It was Euler and Lagrange who first shewed (about 1776–1778) that a column of a certain height would merely be compressed, but one of a greater height would be bent by its own weight. See Euler, De altitudine columnarum etc., *Acta Acad. Sci. Imp. Petropol.* 1778, pp. 163–193; G. Greenhill, Determination of the greatest height to which a tree of given proportions can grow, *Cambr. Phil. Soc. Proc.* IV, p. 65, 1881, and Chree, *ibid.* VII, 1892.

[12] In like manner the wheat-straw bends over under the weight of the loaded ear, and the cat's tail bends over when held erect—not because they "possess flexibility," but because they outstrip the dimensions within which stable equilibrium is possible in a vertical position. The kitten's tail, on the other hand, stands up spiky and straight.

large ones.[13] In short, as Goethe says in *Dichtung und Wahrheit*, "Es ist dafür gesorgt dass die Bäume nicht in den Himmel wachsen."

But the tapering pine-tree is but a special case of a wider problem. The oak does not grow so tall as the pine-tree, but it carries a heavier load, and its boll, broad-based upon its spreading roots, shews a different contour. Smeaton took it for the pattern of his lighthouse, and Eiffel built his great tree of steel, a thousand feet high, to a similar but a stricter plan. Here the profile of tower or tree follows, or tends to follow, a logarithmic curve, giving equal strength throughout, according to a principle which we shall have occasion to discuss later on, when we come to treat of form and mechanical efficiency in the skeletons of animals. In the tree, moreover, anchoring roots form powerful wind-struts, and are most developed opposite to the direction of the prevailing winds; for the lifetime of a tree is affected by the frequency of storms, and its strength is related to the wind-pressure which it must needs withstand.[14]

Among animals we see, without the help of mathematics or of physics, how small birds and beasts are quick and agile, how slower and sedater movements come with larger size, and how exaggerated bulk brings with it a certain clumsiness, a certain inefficiency, an element of risk and hazard, a preponderance of disadvantage. The case was well put by Owen, in a passage which has an interest of its own as a premonition, somewhat like De Candolle's, of the "struggle for existence." Owen wrote as follows: [15] "In proportion to the bulk of a species is the difficulty of the contest which, as a living organised whole, the individual of each species has to maintain against the surrounding agencies that are ever tending to dissolve the vital bond, and subjugate the living matter to the ordinary chemical and physical forces. Any changes, therefore, in such external conditions as a species may have been originally adapted to exist in, will militate against that existence in a degree proportionate, perhaps in a geometrical ratio, to the bulk of the species. If a dry season be greatly prolonged, the large mammal will suffer from the drought sooner than the small one; if any alteration of climate affect the quantity of vegetable food, the bulky Herbivore will be the first to feel the effects of stinted nourishment."

But the principle of Galileo carries us further and along more certain lines. The strength of a muscle, like that of a rope or girder, varies with its cross-section; and the resistance of a bone to a crushing stress varies, again like our girder, with its cross-section. But in a terrestrial animal the weight

[13] The stem of the giant bamboo may attain a height of 60 metres while not more than about 40 cm. in diameter near its base, which dimensions fall not far short of the theoretical limits; A. J. Ewart, *Phil. Trans.* CXCVIII, p. 71, 1906.

[14] Cf. (*int. al.*) T. Petch, On buttress tree-roots, *Ann. R. Bot. Garden, Peradenyia*, XI, pp. 277–285, 1930. Also an interesting paper by James Macdonald, on The form of coniferous trees, *Forestry*, VI, 1 and 2, 1931/2.

[15] *Trans. Zool. Soc.* IV, p. 27, 1850.

which tends to crush its limbs, or which its muscles have to move, varies as the cube of its linear dimensions; and so, to the possible magnitude of an animal, living under the direct action of gravity, there is a definite limit set. The elephant, in the dimensions of its limb-bones, is already shewing signs of a tendency to disproportionate thickness as compared with the smaller mammals; its movements are in many ways hampered and its agility diminished: it is already tending towards the maximal limit of size which the physical forces permit.[16] The spindleshanks of gnat or daddy-long-legs have their own factor of safety, conditional on the creature's exiguous bulk and weight; for after their own fashion even these small creatures tend towards an inevitable limitation of their natural size. But, as Galileo also saw, if the animal be wholly immersed in water like the whale, or if it be partly so, as was probably the case with the giant reptiles of the mesozoic age, then the weight is counterpoised to the extent of an equivalent volume of water, and is completely counterpoised if the density of the animal's body, with the included air, be identical (as a whale's very nearly is) with that of the water around.[17] Under these circumstances there is no longer the same physical barrier to the indefinite growth of the animal. Indeed, in the case of the aquatic animal, there is, as Herbert Spencer pointed out, a distinct advantage, in that the larger it grows the greater is its speed. For its available energy depends on the mass of its muscles, while its motion through the water is opposed, not by gravity, but by "skin-friction," which increases only as the square of the linear dimensions: [18] whence, other things being equal, the bigger the ship or the bigger the fish the faster it tends to go, but only in the ratio of the square root of the increasing length. For the velocity (V) which the fish attains depends on the work (W) it can do and the resistance (R) it must overcome. Now we have seen that the dimensions of W are l^3, and of R are l^2; and by elementary mechanics

$$W \propto RV^2, \text{ or } V^2 \propto \frac{W}{R}.$$

Therefore $$V^2 \propto \frac{l^3}{l^2} = l, \text{ and } V \propto \sqrt{l}.$$

[16] Cf. A. Rauber, Galileo über Knochenformen, *Morphol. Jahrb.* VII, p. 327, 1882.

[17] Cf. W. S. Wall, *A New Sperm Whale* etc., Sydney, 1851, p. 64: "As for the immense size of Cetacea, it evidently proceeds from their buoyancy in the medium in which they live, and their being enabled thus to counteract the force of gravity."

[18] We are neglecting "drag" or "head-resistance," which, increasing as the cube of the speed, is a formidable obstacle to an unstreamlined body. But the perfect streamlining of whale or fish or bird lets the surrounding air or water behave like a perfect fluid, gives rise to no "surface of discontinuity," and the creature passes through it without recoil or turbulence. Froude reckoned skin-friction, or surface-resistance, as equal to that of a *plane* as long as the vessel's water-line, and of area equal to that of the wetted surface of the vessel.

This is what is known as *Froude's Law*, of the correspondence of speeds —a simple and most elegant instance of "dimensional theory." [19]

But there is often another side to these questions, which makes them too complicated to answer in a word. For instance, the work (per stroke) of which two similar engines are capable should vary as the cubes of their linear dimensions, for it varies on the one hand with the *area* of the piston, and on the other with the *length* of the stroke; so is it likewise in the animal, where the corresponding ratio depends on the cross-section of the muscle, and on the distance through which it contracts. But in two similar engines, the available horse-power varies as the square of the linear dimensions, and not as the cube; and this for the reason that the actual *energy* developed depends on the heating-surface of the boiler.[20] So likewise must there be a similar tendency among animals for the rate of supply of kinetic energy to vary with the surface of the lung, that is to say (other things being equal) with the *square* of the linear dimensions of the animal; which means that, *caeteris paribus*, the small animal is stronger (having more power per unit weight) than a large one. We may of course (departing from the condition of similarity) increase the heating-surface of the boiler, by means of an internal system of tubes, without increasing its outward dimensions, and in this very way Nature increases the respiratory surface of a lung by a complex system of branching tubes and minute air-cells; but nevertheless in two similar and closely related animals, as also in two steam-engines of the same make, the law is bound to hold that the rate of working tends to vary with the square of the linear dimensions, according to Froude's *law of steamship comparison*. In the case of a very large ship, built for speed, the difficulty is got over by increasing the size and number of the boilers, till the ratio between boiler-room and engine-room is far beyond what is required in an ordinary small vessel; [21] but though we find lung-space increased among

[19] Though, as Lanchester says, the great designer "was not hampered by a knowledge of the theory of dimensions."

[20] The analogy is not a very strict or complete one. We are not taking account, for instance, of the thickness of the boiler-plates.

[21] Let L be the length, S the (wetted) surface, T the tonnage, D the displacement (or volume) of a ship; and let it cross the Atlantic at a speed V. Then, in comparing two ships, similarly constructed but of different magnitudes, we know that $L = V^2$, $S = L^2 = V^4$, $D = T = L^3 = V^6$; also R (resistance) $= S.V^2 = V^6$; H (horse-power) $= R.V = V^7$; and the coal (C) necessary for the voyage $= H/V = V^6$. That is to say, in ordinary engineering language, to increase the speed across the Atlantic by 1 per cent the ship's length must be increased 2 per cent, her tonnage or displacement 6 per cent, her coal-consumption also 6 per cent, her horse-power, and therefore her boiler-capacity, 7 per cent. Her bunkers, accordingly, keep pace with the enlargement of the ship, but her boilers tend to increase out of proportion to the space available. Suppose a steamer 400 ft. long, of 2000 tons, 2000 H.P., and a speed of 14 knots. The corresponding vessel of 800 ft. long should develop a speed of 20 knots ($1 : 2 :: 14^2 : 20^2$), her tonnage would be 16,000, her H.P. 25,000 or thereby. Such a vessel would probably be driven by four propellers instead of one, each carrying 8000 H.P. See (*int. al.*) W. J. Millar, On the most economical speed to drive a steamer, *Proc. Edin. Math. Soc.* VII, pp. 27–29, 1889; Sir James R. Napier, On the most

animals where greater rate of working is required, as in general among birds, I do not know that it can be shewn to increase, as in the "over-boilered" ship, with the size of the animal, and in a ratio which outstrips that of the other bodily dimensions. If it be the case then, that the working mechanism of the muscles should be able to exert a force proportionate to the cube of the linear bodily dimensions, while the respiratory mechanism can only supply a store of energy at a rate proportional to the square of the said dimensions, the singular result ought to follow that, in swimming for instance, the larger fish ought to be able to put on a spurt of speed far in excess of the smaller one; but the distance travelled by the year's end should be very much alike for both of them. And it should also follow that the curve of fatigue is a steeper one, and the staying power less, in the smaller than in the larger individual. This is the case in long-distance racing, where neither draws far ahead until the big winner puts on his big spurt at the end; on which is based an aphorism of the turf, that a "good big 'un is better than a good little 'un." For an analogous reason wise men know that in the 'Varsity boat-race it is prudent and judicious to bet on the heavier crew.

Consider again the dynamical problem of the movements of the body and the limbs. The work done (W) in moving a limb, whose weight is p, over a distance s, is measured by ps; p varies as the cube of the linear dimensions, and s, in ordinary locomotion, varies as the linear dimensions, that is to say as the length of limb:

$$W \propto ps \propto l^3 \times l = l^4.$$

But the work done is limited by the power available, and this varies as the mass of the muscles, or as l^3; and under this limitation neither p nor s increase as they would otherwise tend to do. The limbs grow shorter, relatively, as the animal grows bigger; and spiders, daddy-long-legs and such-like long-limbed creatures attain no great size.

Let us consider more closely the actual energies of the body. A hundred years ago, in Strasburg, a physiologist and a mathematician were studying the temperature of warm-blooded animals.[22] The heat lost must, they said, be proportional to the surface of the animal: and the gain must be equal to the loss, since the temperature of the body keeps constant. It would seem, therefore, that the heat lost by radiation and that gained by oxidation vary both alike, as the surface-area, or the square of the linear dimensions, of the animal. But this result is paradoxical; for whereas the heat lost may well vary as the surface-area, that produced by oxidation ought rather to vary as the bulk of the animal: one should vary as the

profitable speed for a fully laden cargo steamer for a given voyage, *Proc. Phil. Soc., Glasgow*, VI, pp. 33 38, 1865.

[22] MM. Rameaux et Sarrus, *Bull. Acad. R. de Médecine*, III, pp. 1094 1100, 1838–39.

square and the other as the cube of the linear dimensions. Therefore the ratio of loss to gain, like that of surface to volume, ought to increase as the size of the creature diminishes. Another physiologist, Carl Bergmann,[23] took the case a step further. It was he, by the way, who first said that the real distinction was not between warm-blooded and cold-blooded animals, but between those of constant and those of variable temperature: and who coined the terms *homœothermic* and *poecilothermic* which we use today. He was driven to the conclusion that the smaller animal does produce more heat (per unit of mass) than the large one, in order to keep pace with surface-loss; and that this extra heat-production means more energy spent, more food consumed, more work done.[24] Simplified as it thus was, the problem still perplexed the physiologists for years after. The tissues of one mammal are much like those of another. We can hardly imagine the muscles of a small mammal to produce more heat (*caeteris paribus*) than those of a large; and we begin to wonder whether it be not nervous excitation, rather than quality of muscular tissue, which determines the rate of oxidation and the output of heat. It is evident in certain cases, and may be a general rule, that the smaller animals have the bigger brains; "plus l'animal est petit," says M. Charles Richet, "plus il a des échanges chimiques actifs, et plus son cerveau est volumineux."[25] That the smaller animal needs more food is certain and obvious. The amount of food and oxygen consumed by a small flying insect is enormous; and bees and flies and hawkmoths and humming-birds live on nectar, the richest and most concentrated of foods.[26] Man consumes a fiftieth part of his own weight of food daily, but a mouse will eat half its own weight in a day; its rate of living is faster, it breeds faster, and old age comes to it much sooner than to man. A warm-blooded animal much smaller than a mouse becomes an impossibility; it could neither obtain nor yet digest the food required to maintain its constant temperature, and hence no mammals and no birds are as small as the smallest frogs or fishes. The

[23] Carl Bergmann, Verhältnisse der Wärmeökonomie der Tiere zu ihrer Grösse, *Göttinger Studien*, I, pp. 594–708, 1847—a very original paper.

[24] The metabolic activity of sundry mammals, per 24 hours, has been estimated as follows:

	Weight (kilo.)	Calories per kilo.
Guinea-pig	0·7	223
Rabbit	2	58
Man	70	33
Horse	600	22
Elephant	4000	13
Whale	150000	*circa* 1·7

[25] Ch. Richet, Recherches de calorimétrie, *Arch. de Physiologie* (3), VI, pp. 237 291, 450–497, 1885. Cf. also an interesting historical account by M. Elie le Breton, Sur la notion de "masse protoplasmique active": i. Problèmes posés par la signification de la loi des surfaces, *ibid.* 1906, p. 606.

[26] Cf. R. A. Davies and G. Fraenkel, The oxygen-consumption of flies during flight, *Jl. Exp. Biol.* XVII, pp. 402–407, 1940.

disadvantage of small size is all the greater when loss of heat is accelerated by conduction as in the Arctic, or by convection as in the sea. The far north is a home of large birds but not of small; bears but not mice live through an Arctic winter; the least of the dolphins live in warm waters, and there are no small mammals in the sea. This principle is sometimes spoken of as *Bergmann's Law*.

The whole subject of the conservation of heat and the maintenance of an all but constant temperature in warm-blooded animals interests the physicist and the physiologist alike. It drew Kelvin's attention many years ago,[27] and led him to shew, in a curious paper, how larger bodies are kept warm by clothing while smaller are only cooled the more. If a current be passed through a thin wire, of which part is covered and part is bare, the thin bare part may glow with heat, while convection-currents streaming round the covered part cool it off and leave it in darkness. The hairy coat of very small animals is apt to look thin and meagre, but it may serve them better than a shaggier covering.

Leaving aside the question of the supply of energy, and keeping to that of the mechanical efficiency of the machine, we may find endless biological illustrations of the principle of similitude. All through the physiology of locomotion we meet with it in various ways: as, for instance, when we see a cockchafer carry a plate many times its own weight upon its back, or a flea jump many inches high. "A dog," says Galileo, "could probably carry two or three such dogs upon his back; but I believe that a horse could not carry even one of his own size."

Such problems were admirably treated by Galileo and Borelli, but many writers remained ignorant of their work. Linnaeus remarked that if an elephant were as strong in proportion as a stag-beetle, it would be able to pull up rocks and level mountains; and Kirby and Spence have a well-known passage directed to shew that such powers as have been conferred upon the insect have been withheld from the higher animals, for the reason that had these latter been endued therewith they would have "caused the early desolation of the world." [28]

Such problems as that presented by the flea's jumping powers,[29] though essentially physiological in their nature, have their interest for us here:

[27] W. Thomson, On the efficiency of clothing for maintaining temperature, *Nature*, XXIX, p. 567, 1884.

[28] *Introduction to Entomology*, II, p. 190, 1826. Kirby and Spence, like many less learned authors, are fond of popular illustrations of the "wonders of Nature," to the neglect of dynamical principles. They suggest that if a white ant were as big as a man, its tunnels would be "magnificent cylinders of more than three hundred feet in diameter"; and that if a certain noisy Brazilian insect were as big as a man, its voice would be heard all the world over, "so that Stentor becomes a mute when compared with these insects!" It is an easy consequence of anthropomorphism, and hence a common characteristic of fairy-tales, to neglect the dynamical and dwell on the geometrical aspect of similarity.

[29] The flea is a very clever jumper; he jumps backwards, is stream-lined accordingly, and alights on his two long hind-legs. Cf. G. I. Watson, in *Nature*, 21 May 1938.

because a steady, progressive diminution of activity with increasing size would tend to set limits to the possible growth in magnitude of an animal just as surely as those factors which tend to break and crush the living fabric under its own weight. In the case of a leap, we have to do rather with a sudden impulse than with a continued strain, and this impulse should be measured in terms of the velocity imparted. The velocity is proportional to the impulse (x), and inversely proportional to the mass (M) moved: $V = x/M$. But, according to what we still speak of as "Borelli's law," the impulse (i.e., the work of the impulse) is proportional to the volume of the muscle by which it is produced,[30] that is to say (in similarly constructed animals) to the mass of the whole body; for the impulse is proportional on the one hand to the cross-section of the muscle, and on the other to the distance through which it contracts. It follows from this that the velocity is constant, whatever be the size of the animal.

Putting it still more simply, the work done in leaping is proportional to the mass and to the height to which it is raised, $W \propto mH$. But the muscular power available for this work is proportional to the mass of muscle, or (in similarly constructed animals) to the mass of the animal, $W \propto m$. It follows that H is, or tends to be, a constant. In other words, all animals, provided always that they are similarly fashioned, with their various levers in like proportion, ought to jump not to the same relative but to the same *actual* height.[31] The grasshopper seems to be as well planned for jumping as the flea, and the actual heights to which they jump are much of a muchness; but the flea's jump is about 200 times its own height, the grasshopper's at most 20–30 times; and neither flea nor grasshopper is a better but rather a worse jumper than a horse or a man.[32]

As a matter of fact, Borelli is careful to point out that in the act of leaping the impulse is not actually instantaneous, like the blow of a hammer, but takes some little time, during which the levers are being extended by which the animal is being propelled forwards; and this interval of time will be longer in the case of the longer levers of the larger animal. To some extent, then, this principle acts as a corrective to the more general one, and tends to leave a certain balance of advantage in regard to leaping power on the side of the larger animal.[33] But on the

[30] That is to say, the available energy of muscle, in ft.-lbs. per lb. of muscle, is the same for all animals: a postulate which requires considerable qualification when we come to compare very different kinds of muscle, such as the insect's and the mammal's.

[31] Borelli, Prop. CLXXVII. Animalia minora et minus ponderosa majores saltus efficiunt respectu sui corporis, si caetera fuerint paria.

[32] The high jump is nowadays a highly skilled performance. For the jumper contrives that his centre of gravity goes *under* the bar, while his body, bit by bit, goes *over* it.

[33] See also (*int. al.*), John Bernoulli, *De Motu Musculorum*, Basil., 1694; Chabry. Mécanisme du saut, *J. de l'Anat. et de la Physiol.* XIX, 1883; Sur la longueur des membres des animaux sauteurs, *ibid.* XXI, p. 356, 1885; Le Hello. De l'action des organes locomoteurs, etc., *ibid.* XXIX, pp. 65–93, 1893; etc.

other hand, the question of strength of materials comes in once more, and the factors of stress and strain and bending moment make it more and more difficult for nature to endow the larger animal with the length of lever with which she has provided the grasshopper or the flea. To Kirby and Spence it seemed that "This wonderful strength of insects is doubtless the result of something peculiar in the structure and arrangement of their muscles, and principally their extraordinary power of contraction." This hypothesis, which is so easily seen on physical grounds to be unnecessary, has been amply disproved in a series of excellent papers by Felix Plateau.[34]

From the *impulse* of the preceding case we may pass to the *momentum* created (or destroyed) under similar circumstances by a given force acting for a given time: $mv = Ft$.

We know that $m \propto l^3$, and $t = l/v$,
so that $l^3 v = Fl/v$, or $v^2 = F/l^2$.

But whatsoever force be available, the animal may only exert so much of it as is in proportion to the strength of his own limbs, that is to say to the cross-section of bone, sinew and muscle; and all of these cross-sections are proportional to l^2, the square of the linear dimensions. The maximal force, F_{max}, which the animal *dare* exert is proportional, then, to l^2;

therefore $F_{max}/l^2 =$ constant.

And the maximal speed which the animal can safely reach, namely $V_{max} = F_{max}/l$, is also constant, or independent (*caeteris paribus*) of the dimensions of the animal.

A spurt or effort may be well within the capacity of the animal but far beyond the margin of safety, as trainer and athlete well know. This margin is a narrow one, whether for athlete or racehorse; both run a constant risk of overstrain, under which they may "pull" a muscle, lacerate a tendon, or even "break down" a bone.[35]

It is fortunate for their safety that animals do not jump to heights proportional to their own. For conceive an animal (of mass m) to jump to a certain altitude, such that it reaches the ground with a velocity v; then if c be the crushing strain at any point of the sectional area (A) of the limbs, the limiting condition is that

$$mv = cA.$$

If the animal vary in magnitude without change in the height to which it jumps (or in the velocity with which it descends), then

$$c \propto \frac{m}{A} \propto \frac{l^3}{l^2}, \text{ or } l.$$

[34] Recherches sur la force absolue des muscles des Invertébrés, *Bull. Acad. R. de Belgique* (3), VI, VII, 1883 84: see also *ibid.* (2), XX, 1865; XXII, 1866; *Ann. Mag. N.H.* XVII, p. 139, 1866; XIX, p. 95, 1867. Cf. M. Radau, Sur la force musculaire des insectes, *Revue des deux Mondes*, LXIV, p. 770, 1866. The subject had been well treated by Straus-Dürckheim, in his *Considérations générales sur l'anatomie comparée des animaux articulés*, 1828.

[35] Cf. The dynamics of sprint-running, by A. V. Hill and others, *Proc. R.S.* (B), CII, pp. 29 42, 1927; or *Muscular Movement in Man*, by A. V. Hill, New York, 1927, ch. VI, p. 41.

The crushing strain varies directly with the linear dimensions of the animal; and this, a dynamical case, is identical with the usual statical limitation of magnitude.

But if the animal, with increasing size or stature, jump to a correspondingly increasing height, the case becomes much more serious. For the final velocity of descent varies as the square root of the altitude reached, and therefore as the square root of the linear dimensions of the animal. And since, as before,

$$c \propto mv \propto \frac{l^3}{l^2} V,$$

$$c \propto \frac{l^2}{l^2} . \sqrt{l}, \text{ or } c \propto l^{1/2}.$$

If a creature's jump were in proportion to its height, the crushing strains would so increase that its dimensions would be limited thereby in a much higher degree than was indicated by statical considerations. An animal may grow to a size where it is unstable dynamically, though still on the safe side statically—a size where it moves with difficulty though it rests secure. It is by reason of dynamical rather than of statical relations that an elephant is of graver deportment than a mouse.

An apparently simple problem, much less simple than it looks, lies in the act of walking, where there will evidently be great economy of work if the leg swing with the help of gravity, that is to say, at a *pendulum-rate*. The conical shape and jointing of the limb, the time spent with the foot upon the ground, these and other mechanical differences complicate the case, and make the rate hard to define or calculate. Nevertheless, we may convince ourselves by counting our steps, that the leg does actually tend to swing, as a pendulum does, at a certain definite rate.[36] So on the same principle, but to the slower beat of a longer pendulum, the scythe swings smoothly in the mower's hands.

To walk quicker, we "step out"; we cause the leg-pendulum to describe a greater arc, but it does not swing or vibrate faster until we shorten the pendulum and begin to run. Now let two similar individuals, *A* and *B*, walk in a similar fashion, that is to say with a similar *angle* of swing (Figure 1). The *arc* through which the leg swings, or the *amplitude* of each step, will then vary as the length of leg (say as a/b), and so as the height or other linear dimension (l) of the man.[37] But the time of swing varies inversely as the square root of the pendulum-length, or $\sqrt{a}/\sqrt{b}$. Therefore the velocity, which is measured by amplitude/time, or $a/b \times \sqrt{b}/\sqrt{a}$,

[36] The assertion that the limb tends to swing in pendulum-time was first made by the brothers Weber (*Mechanik der menschl. Gehwerkzeuge*, Göttingen, 1836). Some later writers have criticised the statement (e.g., Fischer, Die Kinematik des Beinschwingens etc., *Abh. math. phys. Kl. k. Sächs. Ges.* XXV–XXVIII, 1899–1903), but for all that, with proper and large qualifications, it remains substantially true.

[37] So the stride of a Brobdingnagian was 10 yards long, or just twelve times the 2 ft. 6 in., which make the average stride or half-pace of a man.

will also vary as the square root of the linear dimensions; which is Froude's law over again.

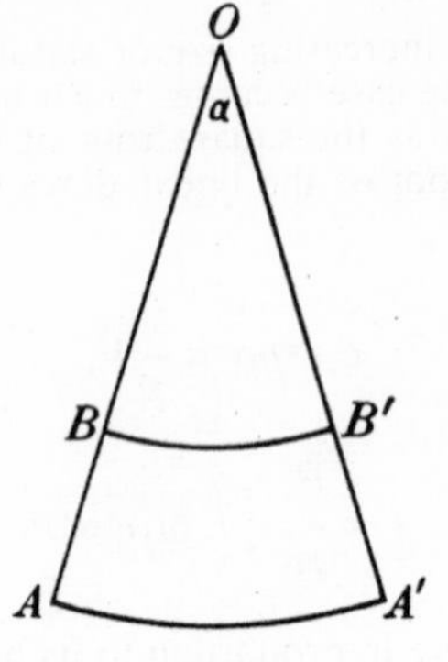

FIGURE 1

The smaller man, or smaller animal, goes slower than the larger, but only in the ratio of the square roots of their linear dimensions; whereas, if the limbs moved alike, irrespective of the size of the animal—if the limbs of the mouse swung no faster than those of the horse—then the mouse would be as slow in its gait or slower than the tortoise. M. Delisle [38] saw a fly walk three inches in half-a-second; this was good steady walking. When we walk five miles an hour we go about 88 inches in a second, or $88/6 = 14 \cdot 7$ times the pace of M. Delisle's fly. We should walk at just about the fly's pace if our stature were $1/(14 \cdot 7)^2$, or $1/216$ of our present height—say $72/216$ inches, or one-third of an inch high. Let us note in passing that the number of legs does not matter, any more than the number of wheels to a coach; the centipede runs none the faster for all his hundred legs.

But the leg comprises a complicated system of levers, by whose various exercise we obtain very different results. For instance, by being careful to rise upon our instep we increase the length or amplitude of our stride, and improve our speed very materially; and it is curious to see how Nature lengthens this metatarsal joint, or instep-lever, in horse [39] and hare and greyhound, in ostrich and in kangaroo, and in every speedy animal. Furthermore, in running we bend and so shorten the leg, in order to accommodate it to a quicker rate of pendulum-swing.[40] In short the jointed structure of the leg permits us to use it as the shortest possible lever while

[38] Quoted in Mr. John Bishop's interesting article in Todd's *Cyclopaedia*, III, p. 443.

[39] The "cannon-bones" are not only relatively longer but may even be actually longer in a little racehorse than a great carthorse.

[40] There is probably another factor involved here: for in bending and thus shortening the leg, we bring its centre of gravity nearer to the pivot, that is to say to the joint, and so the muscle tends to move it the more quickly. After all, we know that the pendulum theory is not the whole story, but only an important first approximation to a complex phenomenon.

it is swinging, and as the longest possible lever when it is exerting its propulsive force.

The bird's case is of peculiar interest. In running, walking or swimming, we consider the speed which an animal *can attain*, and the increase of speed which increasing size permits of. But in flight there is a certain necessary speed—a speed (relative to the air) which the bird *must attain* in order to maintain itself aloft, and which *must* increase as its size increases. It is highly probable, as Lanchester remarks, that Lilienthal met his untimely death (in August 1896) not so much from any intrinsic fault in the design or construction of his machine, but simply because his engine fell somewhat short of the power required to give the speed necessary for its stability.

Twenty-five years ago, when this book was written, the bird, or the aeroplane, was thought of as a machine whose sloping wings, held at a given angle and driven horizontally forward, deflect the air downwards and derive support from the upward reaction. In other words, the bird was supposed to communicate to a mass of air a downward momentum equivalent (in unit time) to its own weight, and to do so by direct and continuous impact. The downward momentum is then proportional to the mass of air thrust downwards, and to the rate at which it is so thrust or driven: the mass being proportional to the wing-area and to the speed of the bird, and the rate being again proportional to the flying speed; so that the momentum varies as the square of the bird's linear dimensions and also as the square of its speed. But in order to balance its weight, this momentum must also be proportional to the cube of the bird's linear dimensions; therefore the bird's necessary speed, such as enables it to maintain level flight, must be proportional to the square root of its linear dimensions, and the whole work done must be proportional to the power 3½ of the said linear dimensions.

The case stands, so far, as follows: m, the mass of air deflected downwards; M, the momentum so communicated; W, the work done—all in unit time; w, the weight, and V, the velocity of the bird; l, a linear dimension, the form of the bird being supposed constant.

$$M = w = l^3,\ \text{but}\ M = mV,\ \text{and}\ m = l^2V.$$

Therefore $$M = l^2V^2 = l^3,$$

and therefore $$V = \sqrt{l}$$

and $$W = MV = l^{3\frac{1}{2}}.$$

The gist of the matter is, or seems to be, that the work which *can be done* varies with the available weight of muscle, that is to say, with the mass of the bird; but the work which *has to be done* varies with mass and distance; so the larger the bird grows, the greater the disadvantage under which all its work is done.[41] The disproportion does not seem very great at first sight, but it is quite enough to tell. It is as much as to say that, every time we double the linear dimensions of the bird, the difficulty of flight, or the work which must needs be done in order to fly, is increased in the ratio of 2^3 to $2^{3½}$, or $1 : \sqrt{2}$, or say $1 : 1\cdot4$. If we take the ostrich to exceed the sparrow in linear dimensions as 25 : 1, which seems well within the mark, the ratio would be that between $25^{3½}$ and 25^3, or between 5^7 and 5^6; in other words, flight would be five times more difficult for the larger than for the smaller bird.

But this whole explanation is doubly inadequate. For one thing, it takes no account of *gliding flight*, in which energy is drawn from the wind, and neither muscular power nor engine power are employed; and we see that the larger birds, vulture, albatross or solan-goose, depend on gliding more and more. Secondly, the old simple account of the impact of the wing upon the air, and the manner in which a downward momentum is communicated and support obtained, is now known to be both inadequate and erroneous. For the science of flight, or aerodynamics, has grown out of the older science of hydrodynamics; both deal with the special properties of a fluid, whether water or air; and in our case, to be content to think of the air as a body of mass m, to which a velocity v is imparted, is to neglect all its fluid properties. How the fish or the dolphin swims, and how the bird flies, are up to a certain point analogous problems; and *stream-lining* plays an essential part in both. But the bird is much heavier than the air, and the fish has much the same density as the water, so that the problem of keeping afloat or aloft is negligible in the one, and all-important in the other. Furthermore, the one fluid is highly compressible, and the other (to all intents and purposes) incompressible; and it is this very difference which the bird, or the aeroplane, takes special advantage of, and which helps, or even enables, it to fly.

It remains as true as ever that a bird, in order to counteract gravity, must cause air to move downward and obtains an upward reaction thereby. But the air displaced downward beneath the wing accounts for a small and varying part, perhaps a third perhaps a good deal less, of the whole force derived; and the rest is generated above the wing, in a less simple way. For, as the air streams pass the slightly sloping wing, as smoothly as the stream-lined form and polished surface permit, it swirls round the front or "leading" edge,[42] and then streams swiftly over the

[41] This is the result arrived at by Helmholtz, Ueber ein Theorem geometrischähnliche Bewegungen flüssiger Körper betreffend, nebst Anwendung auf das Problem Luftballons zu lenken, *Monatsber. Akad. Berlin*, 1873, pp. 501–514. It was criticized and challenged (somewhat rashly) by K. Müllenhof, Die Grösse der Flugflächen etc., *Pflüger's Archiv*, xxxv, p. 407; xxxvi, p. 548, 1885.

[42] The arched form, or "dipping front edge" of the wing, and its use in causing a vacuum above, were first recognised by Mr. H. F. Phillips, who put the idea into a patent in 1884. The facts were discovered independently, and soon afterwards, both by Lilienthal and Lanchester.

upper surface of the wing; while it passes comparatively slowly, checked by the opposing slope of the wing, across the lower side. And this is as much as to say that it tends to be compressed below and rarefied above; in other words, that a partial vacuum is formed above the wing and follows it wherever it goes, so long as the stream-lining of the wing and its angle of incidence are suitable, and so long as the bird travels fast enough through the air.

The bird's weight is exerting a downward force upon the air, in one way just as in the other; and we can imagine a barometer delicate enough to shew and measure it as the bird flies overhead. But to calculate that force we should have to consider a multitude of component elements; we should have to deal with the stream-lined tubes of flow above and below, and the eddies round the fore-edge of the wing and elsewhere; and the calculation which was too simple before now becomes insuperably difficult. But the principle of necessary speed remains as true as ever. The bigger the bird becomes, the more swiftly must the air stream over the wing to give rise to the rarefaction or negative pressure which is more and more required; and the harder must it be to fly, so long as work has to be done by the muscles of the bird. The general principle is the same as before, though the quantitative relation does not work out as easily as it did. As a matter of fact, there is probably little difference in the end; and in aeronautics, the "total resultant force" which the bird employs for its support is said, *empirically*, to vary as the square of the air-speed: which is then a result analogous to Froude's law, and is just what we arrived at before in the simpler and less accurate setting of the case.

But a comparison between the larger and the smaller bird, like all other comparisons, applies only so long as the other factors in the case remain the same; and these vary so much in the complicated action of flight that it is hard indeed to compare one bird with another. For not only is the bird continually changing the incidence of its wing, but it alters the lie of every single important feather; and all the ways and means of flight vary so enormously, in big wings and small, and Nature exhibits so many refinements and "improvements" in the mechanism required, that a comparison based on size alone becomes imaginary, and is little worth the making.

The above considerations are of great practical importance in aeronautics, for they shew how a provision of increasing speed *must* accompany every enlargement of our aeroplanes. Speaking generally, the necessary or minimal speed of an aeroplane varies as the square root of its linear dimensions; if (*caeteris paribus*) we make it four times as long, it must, in order to remain aloft, fly twice as fast as before.[43] If a given machine

[43] G. H. Bryan, *Stability in Aviation*, 1911; F. W. Lanchester, *Aerodynamics*, 1909; cf. (*int. al.*) George Greenhill, *The Dynamics of Mechanical Flight*, 1912; F. W. Headley, *The Flight of Birds*, and recent works.

weighing, say, 500 lb. be stable at 40 miles an hour, then a geometrically similar one which weighs, say, a couple of tons has its speed determined as follows:

$$W : w :: L^3 : l^3 :: 8 : 1.$$

Therefore $$L : l :: 2 : 1.$$

But $$V^2 : v^2 :: L : l.$$

Therefore $$V : v :: \sqrt{2} : 1 = 1.414 : 1..$$

That is to say, the larger machine must be capable of a speed of 40 × 1.414, or about 56½, miles per hour.

An arrow is a somewhat rudimentary flying-machine; but it is capable, to a certain extent and at a high velocity, of acquiring "stability," and hence of actual flight after the fashion of an aeroplane; the duration and consequent range of its trajectory are vastly superior to those of a bullet of the same initial velocity. Coming back to our birds, and again comparing the ostrich with the sparrow, we find we know little or nothing about the actual speed of the latter; but the minimal speed of the swift is estimated at 100 ft. per second, or even more—say 70 miles an hour. We shall be on the safe side, and perhaps not far wrong, to take 20 miles an hour as the sparrow's minimal speed; and it would then follow that the ostrich, of 25 times the sparrow's linear dimensions, would have to fly (if it flew at all) with a minimum velocity of 5 × 20, or 100 miles an hour.[44]

The same principle of *necessary speed*, or the inevitable relation between the dimensions of a flying object and the minimum velocity at

[44] Birds have an ordinary and a forced speed. Meinertzhagen puts the ordinary flight of the swift at 68 m.p.h., which tallies with the old estimate of Athanasius Kircher (*Physiologia*, ed. 1680, p. 65) of 100 ft. per second for the swallow. Abel Chapman (*Retrospect*, 1928, ch. XIV) puts the gliding or swooping flight of the swift at over 150 m.p.h., and that of the griffon vulture at 180 m.p.h.; but these skilled flier, was seen to fly as slowly as 15 m.p.h. A hornet or a large dragonfly may reach 14 or 18 m.p.h.; but for most insects 2–4 metres per sec., say 4–9 m.p.h., is a common speed (cf. A. Magnan, *Vol des Insectes*, 1834, p. 72). The larger diptera are very cock, and 30 m.p.h. for starling, chaffinch, quail and crow. A migrating flock of lapwing travelled at 41 m.p.h., ten or twelve miles more than the usual speed of the single bird. Lanchester, on theoretical considerations, estimates the speed of the herring gull at 26 m.p.h., and of the albatross at about 34 miles. A tern, a very skilful flier, was seen to fly as slowly as 15 m.p.h. A hornet or a large dragonfly may reach 14 or 18 m.p.h.; but for most insects 2–4 metres per sec., say 4–9 m.p.h., is a common speed (cf. A. Magnan, *Vol. des Insectes*, 1834, p. 72). The larger diptera are very swift, but their speed is much exaggerated. A deerfly (*Cephenomyia*) has been said to fly at 400 yards per second, or say 800 m.p.h., an impossible velocity (Irving Langmuir, *Science*, March 11, 1938). It would mean a pressure on the fly's head of half an atmosphere, probably enough to crush the fly; to maintain it would take half a horsepower; and this would need a food-consumption of 1½ times the fly's weight *per second*! 25 m.p.h. is a more reasonable estimate. The naturalist should not forget, though it does not touch our present argument, that the aeroplane is built to the pattern of a beetle rather than of a bird; for the elytra are not wings but planes. Cf. *int. al.*, P. Amans. Géométrie . . . des ailes rigides, *C. R. Assoc. Franç. pour l'avancem. des Sc.* 1901.

which its flight is stable, accounts for a considerable number of observed phenomena. It tells us why the larger birds have a marked difficulty in rising from the ground, that is to say, in acquiring to begin with the horizontal velocity necessary for their support; and why accordingly, as Mouillard [45] and others have observed, the heavier birds, even those weighing no more than a pound or two, can be effectually caged in small enclosures open to the sky. It explains why, as Mr. Abel Chapman says, "all ponderous birds, wild swans and geese, great bustard and capercailzie, even blackcock, fly faster than they appear to do," while "light-built types with a big wing-area,[46] such as herons and harriers, possess no turn of speed at all." For the fact is that the heavy birds must fly quickly, or not at all. It tells us why very small birds, especially those as small as humming-birds, and *a fortiori* the still smaller insects, are capable of "stationary flight," a very slight and scarcely perceptible velocity relatively to the air being sufficient for their support and stability. And again, since it is in all these cases velocity relatively to the air which we are speaking of, we comprehend the reason why one may always tell which way the wind blows by watching the direction in which a bird starts to fly.

The wing of a bird or insect, like the tail of a fish or the blade of an oar, gives rise at each impulsion to a swirl or vortex, which tends (so to speak) to cling to it and travel along with it; and the resistance which wing or oar encounter comes much more from these vortices than from the viscosity of the fluid.[47] We learn as a corollary to this, that vortices form only at the edge of oar or wing—it is only the length and not the breadth of these which matters. A long narrow oar outpaces a broad one, and the efficiency of the long, narrow wing of albatross, swift or hawk-moth is so far accounted for. From the length of the wing we can calculate approximately its rate of swing, and more conjecturally the dimensions of each vortex, and finally the resistance or lifting power of the

(From V. Bjerknes)

	Weight gm.	Length of wing m.	Beats per sec.	Speed of wing-tip m./s.	Radius of vortex *	Force of wing-beat gm.	Specific force, F/W
Stork	3500	0·91	2	5·7	1·5	1480	2 : 5
Gull	1000	0·60	3	5·7	1·0	640	2 : 3
Pigeon	350	0·30	6	5·7	0·5	160	1 : 2
Sparrow	30	0·11	13	4·5	0·18	13	2 : 5
Bee	0·07	0·01	200	6·3	0·02	0·2	3½ : 1
Fly	0·01	0·007	190	4·2	0·01	0·04	4 : 1

* Conjectural.

[45] Mouillard, *L'empire de l'air; essai d'ornithologie appliquée à l'aviation*, 1881; transl. in *Annual Report of the Smithsonian Institution*, 1892.

[46] On wing-area in relation to weight of bird see Lendenfeld in *Naturw. Wochenschr.* Nov. 1904, transl. in *Smithsonian Inst. Rep.* 1904; also E. H. Hankin, *Animal Flight*, 1913; etc.

[47] Cf. V. Bjerknes, *Hydrodynamique physique*, II, p. 293, 1934.

stroke; and the result shews once again the advantages of the small-scale mechanism, and the disadvantage under which the larger machine or larger creature lies.

A bird may exert a force at each stroke of its wing equal to one-half, let us say for safety one-quarter, of its own weight, more or less; but a bee or a fly does twice or thrice the equivalent of its own weight, at a low estimate. If stork, gull or pigeon can thus carry only one-fifth, one-third, one-quarter of their weight by the beating of their wings, it follows that all the rest must be borne by *sailing-flight* between the wing-beats. But an insect's wings lift it easily and with something to spare; hence sailing-flight, and with it the whole principle of necessary speed, does not concern the lesser insects, nor the smallest birds, at all; for a humming-bird can "stand still" in the air, like a hover-fly, and dart backwards as well as forwards, if it please.

There is a little group of Fairy-flies (Mymaridae), far below the size of any small familiar insects; their eggs are laid and larvae reared within the tiny eggs of larger insects; their bodies may be no more than ½ mm. long, and their outspread wings 2 mm. from tip to tip (Figure 2). It is a peculiarity of some of these that their little wings are made of a few hairs or bristles, instead of the continuous membrane of a wing. How these act on the minute quantity of air involved we can only conjecture. It would seem that that small quantity reacts as a viscous fluid to the beat of the wing; but there are doubtless other unobserved anomalies in the mechanism and the mode of flight of these pigmy creatures.[48]

The ostrich has apparently reached a magnitude, and the moa certainly did so, at which flight by muscular action, according to the normal anatomy of a bird, becomes physiologically impossible. The same reasoning applies to the case of man. It would be very difficult, and probably absolutely impossible, for a bird to flap its way through the air were it of the bigness of a man; but Borelli, in discussing the matter, laid even greater stress on the fact that a man's pectoral muscles are so much less in proportion than those of a bird, that however we might fit ourselves out with wings, we could never expect to flap them by any power of our own weak muscles. Borelli had learned this lesson thoroughly, and in one of his chapters he deals with the proposition: *Est impossibile ut homines propriis viribus artificiose volare possint.*[49] But gliding flight, where wind-force and gravitational energy take the place of muscular power, is another story, and its limitations are of another kind. Nature has many modes and mechanisms of flight, in birds of one kind and another, in bats

[48] It is obvious that in a still smaller order of magnitude the Brownian movement would suffice to make flight impossible.

[49] Giovanni Alfonso Borelli, *De Motu Animalium*, I, Prop. CCIV, p. 243, edit. 1685. The part on *The Flight of Birds* is issued by the Royal Aeronautical Society as No. 6 of its *Aeronautical Classics*.

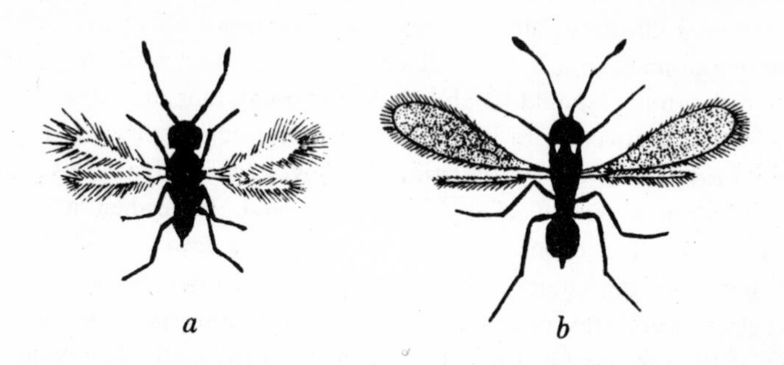

FIGURE 2—Fairy-flies ("Mymaridae"): after F. Enock. × 20.

and beetles, butterflies, dragonflies and what not; and gliding seems to be the common way of birds, and the flapping flight (*remigio alarum*) of sparrow and of crow to be the exception rather than the rule. But it were truer to say that gliding and soaring, by which energy is captured from the wind, are modes of flight little needed by the small birds, but more and more essential to the large. Borelli had proved so convincingly that we could never hope to fly *propriis viribus*, that all through the eighteenth century men tried no more to fly at all. It was in trying *to glide* that the pioneers of aviation, Cayley, Wenham and Mouillard, Langley, Lilienthal and the Wrights—all careful students of birds—renewed the attempt; [50] and only after the Wrights had learned to glide did they seek to add power to their glider. Flight, as the Wrights declared, is a matter of practice and of skill, and skill in gliding has now reached a point which more than justifies all Leonardo da Vinci's attempts to fly. Birds shew infinite skill and instinctive knowledge in the use they make of the horizontal acceleration of the wind, and the advantage they take of ascending currents in the air. Over the hot sands of the Sahara, where every here and there hot air is going up and cooler coming down, birds keep as best they can to the one, or glide quickly through the other; so we may watch a big dragonfly planing slowly down a few feet above the heated soil, and only every five minutes or so regaining height with a vigorous stroke of his wings. The albatross uses the upward current on the lee-side of a great ocean-wave; so, on a lesser scale, does the flying-fish; and the seagull flies in curves, taking every advantage of the varying wind-velocities at different levels over the sea. An Indian vulture flaps his way up for a few laborious yards,

[50] Sir George Cayley (1774–1857), father of British aeronautics, was the first to perceive the capabilities of rigid planes, and to experiment on gliding flight. He anticipated all the essential principles of the modern aeroplane, and his first paper "On Aerial Navigation" appeared in *Nicholson's Journal* for November 1809. F. H. Wenham (1824–1908) studied the flight of birds and estimated the necessary proportion of surface to weight and speed; he held that "the whole secret of success in flight depends upon a proper concave form of the supporting surface." See his paper "On Aerial Locomotion" in the *Report of the Aeronautical Society*, 1866.

then catching an upward current soars in easy spirals to 2000 feet; here he may stay, effortless, all day long, and come down at sunset. Nor is the modern sail-plane much less efficient than a soaring bird; for a skilful pilot in the tropics should be able to roam all day long at will.[51]

A bird's sensitiveness to air-pressure is indicated in other ways besides. Heavy birds, like duck and partridge, fly low and apparently take advantage of air-pressure reflected from the ground. Water-hen and dipper follow the windings of the stream as they fly up or down; a bee-line would give them a shorter course, but not so smooth a journey. Some small birds —wagtails, woodpeckers and a few others—fly, so to speak, by leaps and bounds; they fly briskly for a few moments, then close their wings and shoot along.[52] The flying-fishes do much the same, save that they keep their wings outspread. The best of them "taxi" along with only their tails in the water, the tail vibrating with great rapidity, and the speed attained lasts the fish on its long glide through the air.[53]

Flying may have begun, as in Man's case it did, with short spells of gliding flight, helped by gravity, and far short of sustained or continuous locomotion. The short wings and long tail of Archaeopteryx would be efficient as a slow-speed glider; and we may still see a Touraco glide down from his perch looking not much unlike Archeaopteryx in the proportions of his wings and tail. The small bodies, scanty muscles and narrow but vastly elongated wings of a Pterodactyl go far beyond the limits of mechanical efficiency for ordinary flapping flight; but for gliding they approach perfection.[54] Sooner or later Nature does everything which is physically possible; and to *glide* with skill and safety through the air is a possibility which she did not overlook.

Apart from all differences in the action of the limbs—apart from differences in mechanical construction or in the manner in which the mechanism is used—we have now arrived at a curiously simple and uniform result. For in all the three forms of locomotion which we have attempted to study, alike in swimming and in walking, and even in the more complex problem of flight, the general result, obtained under very different conditions and arrived at by different modes of reasoning, shews in every case that speed tends to vary as the square root of the linear dimensions of the animal.

[51] Sir Gilbert Walker, in *Nature*, Oct. 2, 1937.

[52] Why large birds cannot do the same is discussed by Lanchester, *op. cit.*, Appendix IV.

[53] Cf. Carl L. Hubbs, On the flight of . . . the Cypselurinae, and remarks on the evolution of the flight of fishes, *Papers of the Michigan Acad. of Sci.*, XVII, pp. 575–611, 1933. See also E. H. Hankin, *P.Z.S.* 1920 pp. 467–474; and C. M. Breeder, On the structural specialisation of flying fishes from the standpoint of aerodynamics, *Copeia*, 1930, pp. 114–121.

[54] The old conjecture that their flight was helped or rendered possible by a denser atmosphere than ours is thus no longer called for.

While the rate of progress tends to increase slowly with increasing size (according to Froude's law), and the rhythm or pendulum-rate of the limbs to increase rapidly with decreasing size (according to Galileo's law), some such increase of velocity with decreasing magnitude is true of all the rhythmic actions of the body, though for reasons not always easy to explain. The elephant's heart beats slower than ours,[55] the dog's quicker; the rabbit's goes pit-a-pat; the mouse's and the sparrow's are too quick to count. But the very "rate of living" (measured by the O consumed and CO_2 produced) slows down as size increases; and a rat lives so much faster than a man that the years of its life are three, instead of threescore and ten.

From all the foregoing discussion we learn that, as Crookes once upon a time remarked,[56] the forms as well as the actions of our bodies are entirely conditioned (save for certain exceptions in the case of aquatic animals) by the strength of gravity upon this globe; or, as Sir Charles Bell had put it some sixty years before, the very animals which move upon the surface of the earth are proportioned to its magnitude. Were the force of gravity to be doubled our bipedal form would be a failure, and the majority of terrestrial animals would resemble short-legged saurians, or else serpents. Birds and insects would suffer likewise, though with some compensation in the increased density of the air. On the other hand, if gravity were halved, we should get a lighter, slenderer, more active type, needing less energy, less heat, less heart, less lungs, less blood. Gravity not only controls the actions but also influences the forms of all save the least of organisms. The tree under its burden of leaves or fruit has changed its every curve and outline since its boughs were bare, and a mantle of snow will alter its configuration again. Sagging wrinkles, hanging breasts and many another sign of age are part of gravitation's slow relentless handiwork.

There are other physical factors besides gravity which help to limit the size to which an animal may grow and to define the conditions under which it may live. The small insects skating on a pool have their movements controlled and their freedom limited by the surface-tension between water and air, and the measure of that tension determines the magnitude which they may attain. A man coming wet from his bath carries a few ounces of water, and is perhaps 1 per cent. heavier than before; but a wet fly weighs twice as much as a dry one, and becomes a helpless thing. A small insect finds itself imprisoned in a drop of water, and a fly with two feet in one drop finds it hard to extricate them.

The mechanical construction of insect or crustacean is highly efficient up to a certain size, but even crab and lobster never exceed certain mod-

[55] Say 28 to 30 beats to the minute.
[56] *Proc. Psychical Soc.* XII, pp. 338–355, 1897.

erate dimensions, perfect within these narrow bounds as their construction seems to be. Their body lies within a hollow shell, the stresses within which increase much faster than the mere scale of size; every hollow structure, every dome or cylinder, grows weaker as it grows larger, and a tin canister is easy to make but a great boiler is a complicated affair. The boiler has to be strengthened by "stiffening rings" or ridges, and so has the lobster's shell; but there is a limit even to this method of counteracting the weakening effect of size. An ordinary girder-bridge may be made efficient up to a span of 200 feet or so; but it is physically incapable of spanning the Firth of Forth. The great Japanese spider-crab, *Macrocheira*, has a span of some 12 feet across; but Nature meets the difficulty and solves the problem by keeping the body small, and building up the long and slender legs out of short lengths of narrow tubes. A hollow shell is admirable for small animals, but Nature does not and cannot make use of it for the large.

In the case of insects, other causes help to keep them of small dimensions. In their peculiar respiratory system blood does not carry oxygen to the tissues, but innumerable fine tubules or tracheae lead air into the interstices of the body. If we imagine them growing even to the size of crab or lobster, a vast complication of tracheal tubules would be necessary, within which friction would increase and diffusion be retarded, and which would soon be an inefficient and inappropriate mechanism.

The vibration of vocal chords and auditory drums has this in common with the pendulum-like motion of a limb that its rate also tends to vary inversely as the square root of the linear dimensions. We know by common experience of fiddle, drum or organ, that pitch rises, or the frequency of vibration increases, as the dimensions of pipe or membrane or string diminish; and in like manner we expect to hear a bass note from the great beasts and a piping treble from the small. The rate of vibration (N) of a stretched string depends on its tension and its density; these being equal, it varies inversely as its own length and as its diameter. For similar strings, $N \propto 1/l^2$, and for a circular membrane, of radius r and thickness e, $N \propto 1/(r^2 \sqrt{e}\ )$.

But the delicate drums or tympana of various animals seem to vary much less in thickness than in diameter, and we may be content to write, once more, $N \propto 1/r^2$.

Suppose one animal to be fifty times less than another, vocal chords and all: the one's voice will be pitched 2500 times as many beats, or some ten or eleven octaves, above the other's; and the same comparison, or the same contrast, will apply to the tympanic membranes by which the vibrations are received. But our own perception of musical notes only reaches to 4000 vibrations per second, or thereby; a squeaking mouse or bat is heard by few, and to vibrations of 10,000 per second we are all of us stone-deaf. Structure apart, mere size is enough to give the lesser birds and beasts a music quite different to our own: the humming-bird, for aught we know, may be singing all day long. A minute insect may utter and receive vibrations of prodigious rapidity; even its little wings may beat hundreds of times a second.[57] Far more things happen to it in a second than to us; a thousandth part of a second is no longer negligible, and time itself seems to run a different course to ours.

The eye and its retinal elements have ranges of magnitude and limitations of magnitude of their own. A big dog's eye is hardly bigger than a little dog's; a squirrel's is much larger, proportionately, than an elephant's; and a robin's is but little less than a pigeon's or a crow's. For the rods and cones do not vary with the size of the animal, but have their dimensions optically limited by the interference-patterns of the waves of light, which set bounds to the production of clear retinal images. True, the larger animal may want a larger field of view; but this makes little difference, for but a small area of the retina is ever needed or used. The eye, in short, can never be very small and need never be very big; it has its own conditions and limitations apart from the size of the animal. But the insect's eye tells another story. If a fly had an eye like ours, the pupil would be so small that diffraction would render a clear image impossible. The only alternative is to unite a number of small and optically isolated simple eyes into a compound eye, and in the insect Nature adopts this alternative possibility.[58]

Our range of vision is limited to a bare octave of "luminous" waves, which is a considerable part of the whole range of light-heat rays emitted by the sun; the sun's rays extend into the ultra-violet for another half-octave or more, but the rays to which our eyes are sensitive are just those which pass with the least absorption through a watery medium. Some ancient vertebrate may have learned to see in an ocean which let a certain part of the sun's whole radiation through, which part is *our part* still; or perhaps the watery media of the eye itself account sufficiently for the selective filtration. In either case, the dimensions of the retinal elements are so closely related to the wave-lengths of light (or to their interference patterns) that we have good reason to look upon the retina as perfect of its kind, within the limits which the properties of light itself impose; and this perfection is further illustrated by the fact that a few light-quanta, perhaps a single one, suffice to produce a sensation.[59] The hard eyes of insects are sensitive over a wider range. The bee has two visual optima, one coincident with our own, the other and principal one high up in the

[57] The wing-beats are said to be as follows: dragonfly 28 per sec., bee 190, house-fly 330; cf. Erhard, *Verh. d. d. zool. Gesellsch.* 1913, p. 206.

[58] Cf. C. J. van der Horst, The optics of the insect eye, *Acta Zoolog.* 1933, p. 108.

[59] Cf. Niels Bohr, in *Nature*, April 1, 1933, p. 457. Also J. Joly, *Proc. R.S.* (B), XCII, p. 222, 1921.

ultra-violet.[60] And with the latter the bee is able to see that ultra-violet which is so well reflected by many flowers that flower-photographs have been taken through a filter which passes these but transmits no other rays.[61]

When we talk of light, and of magnitudes whose order is that of a wave-length of light, the subtle phenomenon of colour is near at hand. The hues of living things are due to sundry causes; where they come from chemical pigmentation they are outside our theme, but oftentimes there is no pigment at all, save perhaps as a screen or background, and the tints are those proper to a scale of wave-lengths or range of magnitude. In birds these "optical colours" are of two chief kinds. One kind include certain vivid blues, the blue of a blue jay, an Indian roller or a macaw; to the other belong the iridescent hues of mother-of-pearl, of the humming-bird, the peacock and the dove: for the dove's grey breast shews many colours yet contains but one—*colores inesse plures nec esse plus uno*, as Cicero said. The jay's blue feather shews a layer of enamel-like cells beneath a thin horny cuticle, and the cell-walls are spongy with innumerable tiny air-filled pores. These are about 0.3μ in diameter, in some birds even a little less, and so are not far from the limits of microscopic vision. A deeper layer carries dark-brown pigment, but there is no blue pigment at all; if the feather be dipped in a fluid of refractive index equal to its own, the blue utterly disappears, to reappear when the feather dries. This blue is like the colour of the sky; it is "Tyndall's blue," such as is displayed by turbid media, cloudy with dust-motes or tiny bubbles of a size comparable to the wave-lengths of the blue end of the spectrum. The longer waves of red or yellow pass through, the shorter violet rays are reflected or scattered; the intensity of the blue depends on the size and concentration of the particles, while the dark pigment-screen enhances the effect.

Rainbow hues are more subtle and more complicated; but in the peacock and the humming-bird we know for certain [62] that the colours are those of Newton's rings, and are produced by thin plates or films covering the barbules of the feather. The colours are such as are shewn by films about $\frac{1}{2}\mu$ thick, more or less; they change towards the blue end of the spectrum as the light falls more and more obliquely; or towards the red end if you soak the feather and cause the thin plates to swell. The barbules

[60] L. M. Bertholf, Reactions of the honey-bee to light, *Journ. of Agric. Res.* XLIII, p. 379; XLIV, p. 763, 1931.

[61] A. Kuhn, Ueber den Farbensinn der Bienen, *Ztschr. d. vergl. Physiol.* V, pp. 762–800, 1927; cf. F. K. Richtmeyer, Reflection of ultra-violet by flowers, *Journ. Optical Soc. Amer.* VII, pp. 151–168, 1923; etc.

[62] Rayleigh, *Phil. Mag.* (6), XXXVII, p. 98, 1919. For a review of the whole subject, and a discussion of its many difficulties, see H. Onslow, On a periodic structure in many insect scales, etc., *Phil. Trans.* (B), CCXI, pp. 1–74, 1921; also C. W. Mason, *Journ. Physic. Chemistry,* XXVII, XXX, XXXI, 1923–25–27; F. Suffert, *Zeitschr. f. Morph. u. Oekol. d. Tiere,* I, pp. 171–306, 1924 (scales of butterflies); also B. Reusch and Th. Elsasser in *Journ. f. Ornithologie,* LXXIII, 1925; etc.

of the peacock's feather are broad and flat, smooth and shiny, and their cuticular layer splits into three very thin transparent films, hardly more than 1μ thick, all three together. The gorgeous tints of the humming-birds have had their places in Newton's scale defined, and the changes which they exhibit at varying incidence have been predicted and explained. The thickness of each film lies on the very limit of microscopic vision, and the least change or irregularity in this minute dimension would throw the whole display of colour out of gear. No phenomenon of organic magnitude is more striking than this constancy of size; none more remarkable than that these fine lamellae should have their tenuity so sharply defined, so uniform in feather after feather, so identical in all the individuals of a species, so constant from one generation to another.

A simpler phenomenon, and one which is visible throughout the whole field of morphology, is the tendency (referable doubtless in each case to some definite physical cause) for mere bodily *surface* to keep pace with *volume*, through some alteration of its form. The development of villi on the lining of the intestine (which increase its surface much as we enlarge the effective surface of a bath-towel), the various valvular folds of the intestinal lining, including the remarkable "spiral value" of the shark's gut, the lobulation of the kidney in large animals,[63] the vast increase of respiratory surface in the air-sacs and alveoli of the lung, the development of gills in the larger crustacea and worms though the general surface of the body suffices for respiration in the smaller species—all these and many more are cases in which a more or less constant ratio tends to be maintained between mass and surface, which ratio would have been more and more departed from with increasing size, had it not been for such alteration of surface-form.[64] A leafy wood, a grassy sward, a piece of sponge, a reef of coral, are all instances of a like phenomenon. In fact, a deal of evolution is involved in keeping due balance between surface and mass as growth goes on.

In the case of very small animals, and of individual cells, the principle becomes especially important, in consequence of the molecular forces whose resultant action is limited to the superficial layer. In the cases just mentioned, action is *facilitated* by increase of surface: diffusion, for instance, of nutrient liquids or respiratory gases is rendered more rapid by the greater area of surface; but there are other cases in which the ratio of surface to mass may change the whole condition of the system. Iron rusts when exposed to moist air, but it rusts ever so much faster, and is soon

[63] Cf. R. Anthony, *C.R.* CLXIX, p. 1174, 1919, etc. Cf. also A. Pütter, Studien über physiologische Ähnlichkeit, *Pflüger's Archiv*, CLXVIII, pp. 209–246, 1917.

[64] For various calculations of the increase of surface due to histological and anatomical subdivision, see E. Babak, Ueber die Oberflächenentwickelung bei Organismen, *Biol. Centralbl.* XXX, pp. 225–239, 257–267, 1910.

eaten away, if the iron be first reduced to a heap of small filings; this is a mere difference of degree. But the spherical surface of the rain-drop and the spherical surface of the ocean (though both happen to be alike in mathematical form) are two totally different phenomena, the one due to surface-energy, and the other to that form of mass-energy which we ascribe to gravity. The contrast is still more clearly seen in the case of waves: for the little ripple, whose form and manner of propagation are governed by surface-tension, is found to travel with a velocity which is inversely as the square root of its length; while the ordinary big waves, controlled by gravitation, have a velocity directly proportional to the square root of their wave-length. In like manner we shall find that the form of all very small organisms is independent of gravity, and largely if not mainly due to the force of surface-tension: either as the direct result of the continued action of surface-tension on the semi-fluid body, or else as the result of its action at a prior stage of development, in bringing about a form which subsequent chemical changes have rendered rigid and lasting. In either case, we shall find a great tendency in small organisms to assume either the spherical form or other simple forms related to ordinary inanimate surface-tension phenomena, which forms do not recur in the external morphology of large animals.

Now this is a very important matter, and is a notable illustration of that principle of similitude which we have already discussed in regard to several of its manifestations. We are coming to a conclusion which will affect the whole course of our argument throughout this book, namely that there is an essential difference in kind between the phenomena of form in the larger and the smaller organisms. I have called this book a study of *Growth and Form*, because in the most familiar illustrations of organic form, as in our own bodies for example, these two factors are inseparably associated, and because we are here justified in thinking of form as the direct resultant and consequence of growth: of growth, whose varying rate in one direction or another has produced, by its gradual and unequal increments, the successive stages of development and the final configuration of the whole material structure. But it is by no means true that form and growth are in this direct and simple fashion correlative or complementary in the case of minute portions of living matter. For in the smaller organisms, and in the individual cells of the larger, we have reached an order of magnitude in which the intermolecular forces strive under favourable conditions with, and at length altogether outweigh, the force of gravity, and also those other forces leading to movements of convection which are the prevailing factors in the larger material aggregate.

However, we shall require to deal more fully with this matter in our discussion of the rate of growth, and we may leave it meanwhile, in order to deal with other matters more or less directly concerned with the magnitude of the cell.

The living cell is a very complex field of energy, and of energy of many kinds, of which surface-energy is not the least. Now the whole surface-energy of the cell is by no means restricted to its *outer* surface; for the cell is a very heterogeneous structure, and all its protoplasmic alveoli and other visible (as well as invisible) heterogeneities make up a great system of internal surfaces, at every part of which one "phase" comes in contact with another "phase," and surface-energy is manifested accordingly. But still, the external surface is a definite portion of the system, with a definite "phase" of its own, and however little we may know of the distribution of the total energy of the system, it is at least plain that the conditions which favour equilibrium will be greatly altered by the changed ratio of external surface to mass which a mere change of magnitude produces in the cell. In short, the phenomenon of division of the growing cell, however it be brought about, will be precisely what is wanted to keep fairly constant the ratio between surface and mass, and to retain or restore the balance between surface-energy and the other forces of the system.[65] But when a germ-cell divides or "segments" into two, it does not increase in mass; at least if there be some slight alleged tendency for the egg to increase in mass or volume during segmentation it is very slight indeed, generally imperceptible, and wholly denied by some.[66] The growth or development of the egg from a one-celled stage to stages of two or many cells is thus a somewhat peculiar kind of growth; it is growth limited to change of form and increase of surface, unaccompanied by growth in volume or in mass. In the case of a soap-bubble, by the way, if it divide into two bubbles the volume is actually diminished, while the surface-area is greatly increased;[67] the diminution being due to a cause which we shall have to study later, namely to the increased pressure due to the greater curvature of the smaller bubbles.

An immediate and remarkable result of the principles just described is a tendency on the part of all cells, according to their kind, to vary but little about a certain mean size, and to have in fact certain absolute limitations of magnitude. The diameter of a large parenchymatous cell is perhaps tenfold that of a little one; but the tallest phanerogams are ten thousand times the height of the least. In short, Nature has her materials of predeterminate dimensions, and keeps to the same bricks whether she

[65] Certain cells of the cucumber were found to divide when they had grown to a volume half as large again as that of the "resting cells." Thus the volumes of resting, dividing and daughter cells were as 1 : 1·5 : 0·75; and their surfaces, being as the power 2/3 of these figures, were, roughly, as 1 : 1·3 : 0·8. The ratio of S/V was then as 1 : 0·9 : 1·1, or much nearer equality. Cf. F. T. Lewis, *Anat. Record*, XLVII, pp. 59–99, 1930.

[66] Though the entire egg is not increasing in mass, that is not to say that its living protoplasm is not increasing all the while at the expense of the reserve material.

[67] Cf. P. G. Tait, *Proc. R.S.E.* V, 1866 and VI, 1868.

build a great house or a small. Even ordinary drops tend towards a certain fixed size, which size is a function of the surface-tension, and may be used (as Quincke used it) as a measure thereof. In a shower of rain the principle is curiously illustrated, as Wilding Köller and V. Bjerknes tell us. The drops are of graded sizes, *each twice as big as another*, beginning with the minute and uniform droplets of an impalpable mist. They rotate as they fall, and if two rotate in contrary directions they draw together and presently coalesce; but this only happens when two drops are falling side by side, and since the rate of fall depends on the size it always is a pair of coequal drops which so meet, approach and join together. A supreme instance of constancy or quasi-constancy of size, remote from but yet analogous to the size-limitation of a rain-drop or a cell, is the fact that the stars of heaven (however else one differeth from another), and even the nebulae themselves, are all wellnigh co-equal in *mass*. Gravity draws matter together, condensing it into a world or into a star; but ethereal pressure is an opponent force leading to disruption, negligible on the small scale but potent on the large. High up in the scale of magnitude, from about 10^{33} to 10^{35} grams of matter, these two great cosmic forces balance one another; and all the magnitudes of all the stars lie within or hard by these narrow limits.

In the living cell, Sachs pointed out (in 1895) that there is a tendency for each nucleus to gather around itself a certain definite amount of protoplasm.[68] Driesch,[69] a little later, found it possible, by artificial subdivision of the egg, to rear dwarf sea-urchin larvae, one-half, one-quarter, or even one-eighth of their usual size; which dwarf larvae were composed of only a half, a quarter or an eighth of the normal number of cells. These observations have been often repeated and amply confirmed: and Loeb found the sea-urchin eggs capable of reduction to a certain size, but no further.

In the development of *Crepidula* (an American "slipper-limpet," now much at home on our oyster-beds), Conklin[70] has succeeded in rearing dwarf and giant individuals, of which the latter may be five-and-twenty times as big as the former. But the individual cells, of skin, gut, liver, muscle and other tissues, are just the same size in one as in the other, in dwarf and in giant.[71] In like manner the leaf-cells are found to be of

[68] *Physiologische Notizen* (9), p. 425, 1895. Cf. Amelung, *Flora*, 1893; Strasbürger, Ueber die Wirkungssphäre der Kerne und die Zellgrösse, *Histol. Beitr.* (5), pp. 95–129, 1893; R. Hertwig, Ueber Korrelation von Zell- und Kerngrösse (Kernplasmarelation), *Biol. Centralbl.* XVIII, pp. 49–62, 108–119, 1903; G. Levi and T. Terni, Le variazioni dell' indice plasmatico-nucleare durante l' intercisis, *Arch. Ital. di Anat.* X, p. 545, 1911; also E. le Breton and G. Schaeffer, *Variations biochimiques du rapport nucléo-plasmatique*, Strasburg, 1923.

[69] *Arch. f. Entw. Mech.* IV, 1898, pp. 75, 247.

[70] E. G. Conklin, Cell-size and nuclear size, *Journ. Exp. Zool.* XII, pp. 1–98, 1912; Body-size and cell-size, *Journ. of Morphol.* XXIII, pp. 159–188, 1912. Cf. M. Popoff, Ueber die Zellgrösse, *Arch. f. Zellforschung*, III, 1909.

[71] Thus the fibres of the crystalline lens are of the same size in large and small dogs, Rabl, *Z. f. w. Z.* LXVII, 1899. Cf. (*int. al.*) Pearson, On the size of the blood-corpuscles

the same size in an ordinary water-lily, in the great *Victoria regia*, and in the still huger leaf, nearly 3 metres long, of *Euryale ferox* in Japan.[72] Driesch has laid particular stress upon this principle of a "fixed cell-size," which has, however, its own limitations and exceptions. Among these exceptions, or apparent exceptions, are the giant frond-like cell of a Caulerpa or the great undivided plasmodium of a Myxomycete. The flattening of the one and the branching of the other serve (or help) to increase the ratio of surface to content, the nuclei tend to multiply, and streaming currents keep the interior and exterior of the mass in touch with one another.

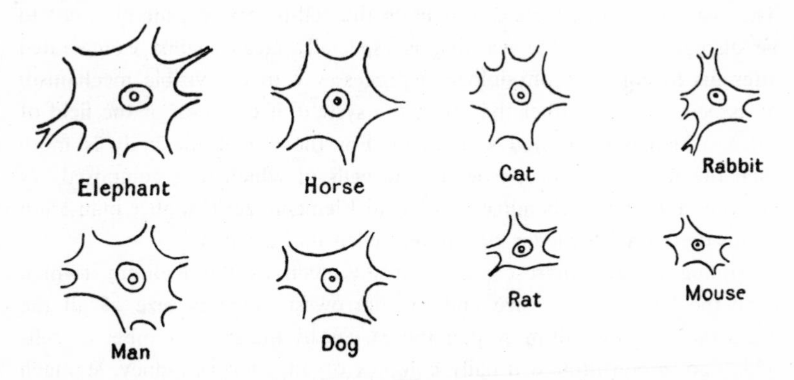

FIGURE 3—Motor ganglion-cells, from the cervical spinal cord. From Minot, after Irving Hardesty.

We get a good and even a familiar illustration of the principle of size-limitation in comparing the brain-cells or ganglion-cells, whether of the lower or of the higher animals.[73] In Figure 3 we shew certain identical nerve-cells from various mammals, from mouse to elephant, all drawn to the same scale of magnification; and we see that they are all of much the same *order* of magnitude. The nerve-cell of the elephant is about twice that of the mouse in linear dimensions, and therefore about eight times greater in volume or in mass. But making due allowance for difference of shape, the linear dimensions of the elephant are to those of the mouse

in Rana, *Biometrika*, VI, p. 403, 1909. Dr. Thomas Young caught sight of the phenomenon early in last century: "The solid particles of the blood do not by any means vary in magnitude in the same ratio with the bulk of the animal," *Natural Philosophy*, ed. 1845, p. 466; and Leeuwenhoek and Stephen Hales were aware of it nearly two hundred years before. Leeuwenhoek indeed had a very good idea of the size of a human blood-corpuscle, and was in the habit of using its diameter—about 1/3000 of an inch—as a standard of comparison. But though the blood-corpuscles shew no relation of magnitude to the size of the animal, they are related without doubt to its activity; for the corpuscles in the sluggish Amphibia are much the largest known to us, while the smallest are found among the deer and other agile and speedy animals (cf. Gulliver, *P.Z.S.* 1875, p. 474, etc.). This correlation is explained by the surface condensation or adsorption of oxygen in the blood-corpuscles, a process greatly facilitated and intensified by the increase of surface due to their minuteness.

[72] Okada and Yomosuke, in *Sci. Rep. Tohoku Univ.* III, pp. 271–278, 1928.

[73] Cf. P. Enriques, La forma come funzione della grandezza: Ricerche sui gangli nervosi degli invertebrati, *Arch. f. Entw. Mech.* XXV, p. 655, 1907–8.

as not less than one to fifty; and the bulk of the larger animal is something like 125,000 times that of the less. It follows, if the size of the nerve-cells are as eight to one, that, in corresponding parts of the nervous system, there are more than 15,000 times as many individual cells in one animal as in the other. In short we may (with Enriques) lay it down as a general law that among animals, large or small, the ganglion-cells vary in size within narrow limits; and that, amidst all the great variety of structure observed in the nervous system of different classes of animals, it is always found that the smaller species have simpler ganglia than the larger, that is to say ganglia containing a smaller number of cellular elements.[74] The bearing of such facts as this upon the cell-theory in general is not to be disregarded; and the warning is especially clear against exaggerated attempts to correlate physiological processes with the visible mechanism of associated cells, rather than with the system of energies, or the field of force, which is associated with them. For the life of the body is more than the *sum* of the properties of the cells of which it is composed: as Goethe said, "Das Lebendige ist zwar in Elemente zerlegt, aber man kann es aus diesen nicht wieder zusammenstellen und beleben."

Among certain microscopic organisms such as the Rotifera (which have the least average size and the narrowest range of size of all the Metazoa), we are still more palpably struck by the small number of cells which go to constitute a usually complex organ, such as kidney, stomach or ovary; we can sometimes number them in a few units, in place of the many thousands which make up such an organ in larger, if not always higher, animals. We have already spoken of the Fairy-flies, a few score of which would hardly weigh down one of the larger rotifers, and a hundred thousand would weigh less than one honey-bee. Their form is complex and their little bodies exquisitely beautiful; but I feel sure that their cells are few, and their organs of great histological simplicity. These considerations help, I think, to shew that, however important and advantageous the subdivision of the tissues into cells may be from the constructional, or from the dynamic, point of view, the phenomenon has less fundamental importance than was once, and is often still, assigned to it.

Just as Sachs shewed there was a limit to the amount of cytoplasm which could gather round a nucleus, so Boveri has demonstrated that the

[74] While the difference in cell-volume is vastly less than that between the volumes, and very much less also than that between the surfaces, of the respective animals, yet there *is* a certain difference; and this it has been attempted to correlate with the need for each cell in the many-celled ganglion of the larger animal to possess a more complex "exchange-system" of branches, for intercommunication with its more numerous neighbours. Another explanation is based on the fact that, while such cells as continue to divide throughout life tend to uniformity of size in all mammals, those which do not do so, and in particular the ganglion cells, continue to grow, and their size becomes, therefore, a function of the duration of life. Cf. G. Levi, Studii sulla grandezza delle cellule, *Arch. Ital. di Anat. e di Embriolog.* v, p. 291, 1906; cf. also A. Berezowski, Studien über die Zellgrösse, *Arch. f. Zellforsch.* v, pp. 375–384, 1910.

nucleus itself has its own limitations of size, and that, in cell-division after fertilisation, each new nucleus has the same size as its parent nucleus;[75] we may nowadays transfer the statement to the chromosomes. It may be that a bacterium lacks a nucleus for the simple reason that it is too small to hold one, and that the same is true of such small plants as the Cyanophyceae, or blue-green algae. Even a chromatophore with its "pyrenoids" seems to be impossible below a certain size.[76]

Always then, there are reasons, partly physiological but in large part purely physical, which define or regulate the magnitude of the organism or the cell. And as we have already found definite limitations to the increase in magnitude of an organism, let us now enquire whether there be not also a lower limit below which the very existence of an organism becomes impossible.

A bacillus of ordinary size is, say, 1μ in length. The length (or height) of a man is about a million and three-quarter times as great, i.e., 1.75 metres, or $1.75 \times 10^6\mu$; and the mass of the man is in the neighbourhood of 5×10^{18} (five million, million, million) times greater than that of the bacillus. If we ask whether there may not exist organisms as much less than the bacillus as the bacillus is less than the man, it is easy to reply that this is quite impossible, for we are rapidly approaching a point where the question of molecular dimensions, and of the ultimate divisibility of matter, obtrudes itself as a crucial factor in the case. Clerk Maxwell dealt with this matter seventy years ago, in his celebrated article *Atom*.[77] Kolli (or Colley), a Russian chemist, declared in 1893 that the head of a spermatozoon could hold no more than a few protein molecules; and Errera, ten years later, discussed the same topic with great ingenuity.[78] But it needs no elaborate calculation to convince us that the smaller bacteria or micrococci nearly approach the smallest magnitudes which we can conceive to have an organised structure. A few small bacteria are the smallest of visible organisms, and a minute species associated with influenza, *B. pneumosinter*, is said to be the least of them all. Its size is of the order of $0 \cdot 1\mu$ or rather less; and here we are in close touch with

[75] Boveri, *Zellenstudien*, V: Ueber die Abhängigkeit der Kerngrösse und Zellenzahl von der Chromosomenzahl der Ausgangszellen. Jena, 1905. Cf. also (*int. al.*) H. Voss, Kerngrössenverhältnisse in der Leber etc., *Ztschr. f. Zellforschung*, VII, pp. 187–200, 1928.

[76] The size of the nucleus may be affected, even determined, by the number of chromosomes it contains. There are giant races of *Oenothera, Primula* and *Solanum* whose cell-nuclei contain twice the normal number of chromosomes, and a dwarf race of a little freshwater crustacean, *Cyclops*, has half the usual number. The cytoplasm in turn varies with the amount of nuclear matter, the whole cell is unusually large or unusually small; and in these exceptional cases we see a direct relation between the size of the organism and the size of the cell. Cf. (*int. al.*) R. P. Gregory, *Proc. Camb. Phil. Soc.* XV, pp. 239–246, 1909; F. Keeble, *Journ. of Genetics*, II, pp. 163–188, 1912.

[77] *Encyclopaedia Britannica*, 9th edition, 1875.

[78] Leo Errera, Sur la limite de la petitesse des organismes, *Bull. Soc. Roy. des Sc. méd. et nat. de Bruxelles*, 1903; *Recueil d'œuvres* (*Physiologie générale*), p. 325.

the utmost limits of microscopic vision, for the wave-lengths of visible light run only from about 400 to 700mμ. The largest of the bacteria, *B. megatherium*, larger than the well-known *B. anthracis* of splenic fever, has much the same proportion to the least as an elephant to a guinea-pig.[79]

Size of body is no mere accident. Man, respiring as he does, cannot be as small as an insect, nor *vice versa*; only now and then, as in the Goliath beetle, do the sizes of mouse and beetle meet and overlap. The descending scale of mammals stops short at a weight of about 5 grams, that of beetles at a length of about half a millimetre, and every group of animals has its upper and its lower limitations of size. So, not far from the lower limit of our vision, does the long series of bacteria come to an end. There remain still smaller particles which the ultra-microscope in part reveals; and here or hereabouts are said to come the so-called viruses or "filter-passers," brought within our ken by the maladies, such as hydrophobia, or foot-and-mouth disease, or the mosaic diseases of tobacco and potato, to which they give rise. These minute particles, of the order of one-tenth the diameter of our smallest bacteria, have no diffusible contents, no included water—whereby they differ from every living thing. They appear to be inert colloidal (or even crystalloid) aggregates of a nucleo-protein, of perhaps ten times the diameter of an ordinary protein-molecule, and not much larger than the giant molecules of haemoglobin or haemocyanin.[80]

Bejerinck called such a virus a *contagium vivum*; "infective nucleo-protein" is a newer name. We have stepped down, by a single step, from living to non-living things, from bacterial dimensions to the molecular magnitudes of protein chemistry. And we begin to suspect that the virus-diseases are not due to an "organism, capable of physiological reproduction and multiplication, but to a mere specific chemical substance, capable of catalysing pre-existing materials and thereby producing more and more molecules like itself. The spread of the virus in a plant would then be a mere autocatalysis, not involving the transport of matter, but only a progressive change of state in substances already there." [81]

But, after all, a simple tabulation is all we need to shew how nearly the least of organisms approach to molecular magnitudes. The same table will suffice to shew how each main group of animals has its mean and

[79] Cf. A. E. Boycott, The transition from live to dead, *Proc. R. Soc. of Medicine*, XXII (*Pathology*), pp. 55–69, 1928.

[80] Cf. Svedberg, *Journ. Am. Chem. Soc.* XLVIII, p. 30, 1926. According to the Foot-and-Mouth Disease Research Committee (*5th Report*, 1937), the foot-and-mouth virus has a diameter, determined by graded filters, of 8–12mμ; while Kenneth Smith and W. D. MacClement (*Proc. R.S.* (B), CXXV, p. 296, 1938) calculate for certain others a diameter of no more than 4mμ or less than a molecule of haemocyanin.

[81] H. H. Dixon, Croonian lecture on the transport of substances in plants, *Proc. R.S.* (B), CXXV, pp. 22, 23, 1938.

characteristic size, and a range on either side, sometimes greater and sometimes less.

Our table of magnitudes is no mere catalogue of isolated facts, but goes deep into the relation between the creature and its world. A certain range, and a narrow one, contains mouse and elephant, and all whose business it is to walk and run; this is our own world, with whose dimensions our lives, our limbs, our senses are in tune. The great whales grow out of this range by throwing the burden of their bulk upon the waters; the dinosaurs wallowed in the swamp, and the hippopotamus, the sea-elephant and Steller's great sea-cow pass or passed their lives in the rivers or the sea.

LINEAR DIMENSIONS OF ORGANISMS, AND OTHER OBJECTS

	m.		
(10,000 km.)	10^7	A quadrant of the earth's circumference	
(1000 km.)	10^6	Orkney to Land's End	
	10^5		
	10^4		
		Mount Everest	
(km.)	10^3		
	10^2	Giant trees: *Sequoia*	
		Large whale	
	10^1	Basking shark	
		Elephant; ostrich; man	
(metre)	10^0		
		Dog; rat; eagle	
	10^{-1}		
		Small birds and mammals; large insects	
(cm.)	10^{-2}		
		Small insects; minute fish	
(mm.)	10^{-3}		
		Minute insects	Cells
	10^{-4}		
		Protozoa; pollen-grains	
	10^{-5}		
		Large bacteria; human blood-corpuscles	
(micron, μ)	10^{-6}		
		Minute bacteria	
	10^{-7}		
		Limit of microscopic vision	
		Viruses, or filter-passers	Colloid particles
	10^{-8}	Giant albuminoids, casein, etc.	
		Starch-molecule	
($m\mu$)	10^{-9}		
		Water-molecule	
(Ångström unit)	10^{-10}		

The things which fly are smaller than the things which walk and run; the flying birds are never as large as the larger mammals, the lesser birds and mammals are much of a muchness, but insects come down a step in the scale and more. The lessening influence of gravity facilitates flight, but makes it less easy to walk and run; first claws, then hooks and suckers and glandular hairs help to secure a foothold, until to creep upon wall or ceiling becomes as easy as to walk upon the ground. Fishes, by evading gravity, increase their range of magnitude both above and below that of terrestrial animals. Smaller than all these, passing out of our range of vision and going down to the least dimensions of living things, are protozoa, rotifers, spores, pollen-grains [82] and bacteria. All save the largest of these float rather than swim; they are buoyed up by air or water, and fall (as Stokes's law explains) with exceeding slowness.

There is a certain narrow range of magnitudes where (as we have partly said) gravity and surface tension become comparable forces, nicely balanced with one another. Here a population of small plants and animals not only dwell in the surface waters but are bound to the surface film itself—the whirligig beetles and pond-skaters, the larvae of gnat and mosquito, the duckweeds (*Lemna*), the tiny *Wolffia*, and *Azolla*; even in mid-ocean, one small insect (*Halobates*) retains this singular habitat. It would be a long story to tell the various ways in which surface-tension is thus taken full advantage of. Gravitation not only limits the magnitude but controls the form of things. With the help of gravity the quadruped has its back and its belly, and its limbs upon the ground; its freedom of motion in a plane perpendicular to gravitational force; its sense of fore-and-aft, its head and tail, its bilateral symmetry. Gravitation influences both our bodies and our minds. We owe to it our sense of the vertical, our knowledge of up-and-down; our conception of the horizontal plane on which we stand, and our discovery of two axes therein, related to the vertical as to one another; it was gravity which taught us to think of three-dimensional space. Our architecture is controlled by gravity, but gravity has less influence over the architecture of the bee; a bee might be excused, might even be commended, if it referred space to four dimensions instead of three! [83] The plant has its root and its stem; but about this vertical or gravitational axis its radiate symmetry remains, undisturbed by directional polarity, save for the sun. Among animals, radiate symmetry is confined to creatures of no great size; and some form or degree of spherical symmetry becomes the rule in the small world of the protozoon—unless gravity resume its sway through the added burden of a shell.

[82] Pollen-grains, like protozoa, have a considerable range of magnitude. The largest, such as those of the pumpkin, are about 200μ in diameter; these have to be carried by insects, for they are above the level of Stokes's law, and no longer float upon the air. The smallest pollen-grains, such as those of the forget-me-not, are about $4\frac{1}{2}\mu$ in diameter (Wodehouse).

[83] Corresponding, that is to say, to the four axes which, meeting in a point, make co-equal angles (the so-called tetrahedral angles) one with another, as do the basal angles of the honeycomb.

The creatures which swim, walk or run, fly, creep or float are, so to speak, inhabitants and natural proprietors of as many distinct and all but separate worlds. Humming-bird and hawkmoth may, once in a way, be co-tenants of the same world; but for the most part the mammal, the bird, the fish, the insect and the small life of the sea, not only have their zoological distinctions, but each has a physical universe of its own. The world of bacteria is yet another world again, and so is the world of colloids; but through these small Lilliputs we pass outside the range of living things.

What we call mechanical principles apply to the magnitudes among which we are at home; but lesser worlds are governed by other and appropriate physical laws, of capillarity, adsorption and electric charge. There are other worlds at the far other end of the scale, in the uttermost depths of space, whose vast magnitudes lie within a narrow range. When the globular star-clusters are plotted on a curve, apparent diameter against estimated distance, the curve is a fair approximation to a rectangular hyperbola; which means that, to the same rough approximation, the actual diameter is identical in them all.[84]

It is a remarkable thing, worth pausing to reflect on, that we can pass so easily and in a dozen lines from molecular magnitudes [85] to the dimensions of a Sequoia or a whale. Addition and subtraction, the old arithmetic of the Egyptians, are not powerful enough for such an operation; but the story of the grains of wheat upon the chessboard shewed the way, and Archimedes and Napier elaborated the arithmetic of multiplication. So passing up and down by easy steps, as Archimedes did when he numbered the sands of the sea, we compare the magnitudes of the great beasts and the small, of the atoms of which they are made, and of the world in which they dwell.[86]

While considerations based on the chemical composition of the organism have taught us that there must be a definite lower limit to its magnitude, other considerations of a purely physical kind lead us to the same

[84] See Harlow Shapley and A. B. Sayer, The angular diameters of globular clusters, *Proc. Nat. Acad. of Sci.* XXI, pp. 593–597, 1935. The same is approximately true of the spiral nebulae also.

[85] We may call (after Siedentopf and Zsigmondi) the smallest visible particles *microns*, such for instance as small bacteria, or the fine particles of gum-mastich in suspension, measuring $0 \cdot 5$ to $1 \cdot 0\mu$; *sub-microns* are those revealed by the ultramicroscope, such as particles of colloid gold (2–15mμ), or starch-molecules (5mμ); amicrons, under 1mμ, are not perceptible by either method. A water-molecule measures, probably, about $0 \cdot 1$mμ.

[86] Observe that, following a common custom, we have only used a logarithmic scale for the round numbers representing powers of ten, leaving the interspaces between these to be filled up, if at all, by ordinary numbers. There is nothing to prevent us from using fractional indices, if we please, throughout, and calling a blood-corpuscle, for instance, $10^{-3 \cdot 2}$ cm. in diameter, a man $10^{2 \cdot 25}$ cm. high, or Sibbald's Rorqual $10^{1 \cdot 48}$ metres long. This method, implicit in that of Napier of Merchiston, was first set forth by Wallis, in his *Arithmetica infinitorum*.

conclusion. For our discussion of the principle of similitude has already taught us that long before we reach these all but infinitesimal magnitudes the dwindling organism will have experienced great changes in all its physical relations, and must at length arrive at conditions surely incompatible with life, or what we understand as life, in its ordinary development and manifestation.

We are told, for instance, that the powerful force of surface-tension, or capillarity, begins to act within a range of about 1/500,000 of an inch, or say $0 \cdot 05\mu$. A soap film, or a film of oil on water, may be attenuated to far less magnitudes than this; the black spots on a soap bubble are known, by various concordant methods of measurement, to be only about 6×10^{-7} cm., or about $6\text{m}\mu$ thick, and Lord Rayleigh and M. Devaux have obtained films of oil of $2\text{m}\mu$, or even $1\text{m}\mu$ in thickness. But while it is possible for a fluid film to exist of these molecular dimensions, it is certain that long before we reach these magnitudes there arise conditions of which we have little knowledge, and which it is not easy to imagine. A bacillus lives in a world, or on the borders of a world, far other than our own, and preconceptions drawn from our experience are not valid there. Even among inorganic, non-living bodies, there comes a certain grade of minuteness at which the ordinary properties become modified. For instance, while under ordinary circumstances crystallisation starts in a solution about a minute solid fragment or crystal of the salt, Ostwald has shewn that we may have particles so minute that they fail to serve as a nucleus for crystallisation—which is as much as to say that they are too small to have the form and properties of a "crystal." And again, in his thin oil-films, Lord Rayleigh noted the striking change of physical properties which ensues when the film becomes attenuated to one, or something less than one, close-packed layer of molecules, and when, in short, it no longer has the properties of matter *in mass*.

These attenuated films are now known to be "monomolecular," the long-chain molecules of the fatty acids standing close-packed, like the cells of a honeycomb, and the film being just as thick as the molecules are long. A recent determination makes the several molecules of oleic, palmitic and stearic acids measure $10 \cdot 4$, $14 \cdot 1$ and $15 \cdot 1$ cm. in length, and in breadth $7 \cdot 4$, $6 \cdot 0$ and $5 \cdot 5$ cm., all by 10^{-8}: in good agreement with Lord Rayleigh and Devaux's lowest estimates (F. J. Hill, *Phil. Mag.* 1929, pp. 940–946). But it has since been shewn that in aliphatic substances the long-chain molecules are not erect, but inclined to the plane of the film; that the zig-zag constitution of the molecules permits them to interlock, so giving the film increased stability; and that the interlock may be by means of a first or second zig-zag, the measured area of the film corresponding precisely to these two dimorphic arrangements. (Cf. C. G. Lyons and E. K. Rideal, *Proc. R.S.* (A), CXXVIII, pp. 468–473, 1930.) The film may be lifted on to a polished surface of metal, or even on a sheet of paper, and one monomolecular layer so added to another; even the complex protein molecule can be unfolded to form a film one amino-acid molecule thick. The whole subject of monomolecular layers, the nature of the film, whether condensed, expanded or gaseous, its astonishing sensitiveness to the least impurities, and the manner of spreading of the one liquid over the other, has become of great interest and importance through the work of Irving Langmuir, Devaux, N. K. Adam and others, and throws new light on the whole subject of molecular magnitudes.[87]

The surface-tension of a drop (as Laplace conceived it) is the cumulative effect, the statistical average, of countless molecular attractions, but we are now entering on dimensions where the molecules are few.[88] The free surface-energy of a body begins to vary with the *radius*, when that radius is of an order comparable to inter-molecular distances; and the whole expression for such energy tends to vanish away when the radius of the drop or particle is less than $0 \cdot 01\mu$, or $10\text{m}\mu$. The qualities and properties of our particle suffer an abrupt change here; what then can we attribute, in the way of properties, to a corpuscle or organism as small or smaller than, say, $0 \cdot 05$ or $0 \cdot 03\mu$? It must, in all probability, be a homogeneous structureless body, composed of a very small number of albumenoid or other molecules. Its vital properties and functions must be extremely limited; its specific outward characters, even if we could see it, must be *nil*; its osmotic pressure and exchanges must be anomalous, and under molecular bombardment they may be rudely disturbed; its properties can be little more than those of an ion-laden corpuscle, enabling it to perform this or that specific chemical reaction, to effect this or that disturbing influence, or produce this or that pathogenic effect. Had it sensation, its experiences would be strange indeed; for if it could feel, it would regard a fall in temperature as a movement of the molecules around, and if it could see it would be surrounded with light of many shifting colours, like a room filled with rainbows.

The dimensions of a cilium are of such an order that its substance is mostly, if not all, under the peculiar conditions of a surface-layer, and surface-energy is bound to play a leading part in ciliary action. A cilium or flagellum is (as it seems to me) a portion of matter in a state *sui generis*, with properties of its own, just as the film and the jet have theirs. And just as Savart and Plateau have told us about jets and films, so will the physicist some day explain the properties of the cilium and flagellum. It is certain that we shall never understand these remarkable structures so long as we magnify them to another scale, and forget that new and peculiar physical properties are associated with the scale to which they belong.[89]

[87] Cf. (*int. al.*) Adam, *Physics and Chemistry of Surfaces*, 1930; Irving Langmuir, *Proc. R.S.* (A), CLXX, 1939.

[88] See a very interesting paper by Fred Vles, Introduction à la physique bactérienne, *Revue Scient.* 11 juin 1921. Cf. also N. Rashevsky, Zur Theorie d. spontanen Teilung von mikroskopischen Tropfen, *Ztschr. f. Physik*, XLVI, p. 578, 1928.

[89] The cilia on the gills of bivalve molluscs are of exceptional size, measuring from say 20 to 120μ long. They are thin triangular plates, rather than filaments; they are from 4 to 10μ broad at the base, but less than 1μ thick. Cf. D. Atkins, *Q.J.M.S.*, 1938, and other papers.

As Clerk Maxwell put it, "molecular science sets us face to face with physiological theories. It forbids the physiologist to imagine that structural details of infinitely small dimensions (such as Leibniz assumed, one within another, *ad infinitum*) can furnish an explanation of the infinite variety which exists in the properties and functions of the most minute organisms." And for this reason Maxwell reprobates, with not undue severity, those advocates of pangenesis and similar theories of heredity, who "would place a whole world of wonders within a body so small and so devoid of visible structure as a germ." But indeed it scarcely needed Maxwell's criticism to shew forth the immense physical difficulties of Darwin's theory of *pangenesis*: which, after all, is as old as Democritus, and is no other than that Promethean *particula undique desecta* of which we have read, and at which we have smiled, in our Horace.

There are many other ways in which, when we make a long excursion into space, we find our ordinary rules of physical behaviour upset. A very familiar case, analysed by Stokes, is that the viscosity of the surrounding medium has a relatively powerful effect upon bodies below a certain size. A droplet of water, a thousandth of an inch (25μ) in diameter, cannot fall in still air quicker than about an inch and a half per second; as its size decreases, its resistance varies as the radius, not (as with larger bodies) as the surface; and its "critical" or terminal velocity varies as the square of the radius, or as the surface of the drop. A minute drop in a misty cloud may be one-tenth that size, and will fall a hundred times slower, say an inch a minute; and one again a tenth of this diameter (say 0.25μ, or about twice as big as a small micrococcus) will scarcely fall an inch in two hours.[90] Not only do dust-particles, spores [91] and bacteria fall, by reason of this principle, very slowly through the air, but all minute bodies meet with great proportionate resistance to their movements through a fluid. In salt water they have the added influence of a larger coefficient of friction than in fresh; [92] and even such comparatively large organisms as the diatoms and the foraminifera, laden though they are with a heavy shell of flint or lime, seem to be poised in the waters of the ocean, and fall with exceeding slowness.

When we talk of one thing touching another, there may yet be a distance between, not only measurable but even large compared with the magnitudes we have been considering. Two polished plates of glass or

[90] The resistance depends on the radius of the particle, the viscosity, and the rate of fall (V); the effective weight by which this resistance is to be overcome depends on gravity, on the density of the particle compared with that of the medium, and on the mass, which varies as r^3. Resistance $= krV$, and effective weight $= k'r^3$; when these two equal one another we have the critical or terminal velocity, and $V \propto r^2$.

[91] A. H. R. Buller found the spores of a fungus (*Collybia*), measuring $5 \times 3\mu$, to fall at the rate of half a millimetre per second, or rather more than an inch a minute; *Studies on Fungi*, 1909.

[92] Cf. W. Krause, *Biol. Centralbl.* I, p. 578, 1881; Flügel, *Meteorol. Ztschr.* 1881, p. 321.

steel resting on one another are still about 4μ apart—the average size of the smallest dust; and when all dust-particles are sedulously excluded, the one plate sinks slowly down to within 0.3μ of the other, an apparent separation to be accounted for by minute irregularities of the polished surfaces.[93]

The Brownian movement has also to be reckoned with—that remarkable phenomenon studied more than a century ago by Robert Brown,[94] Humboldt's *facile princeps botanicorum*, and discoverer of the nucleus of the cell.[95] It is the chief of those fundamental phenomena which the biologists have contributed, or helped to contribute, to the science of physics.

The quivering motion, accompanied by rotation and even by translation, manifested by the fine granular particle issuing from a crushed pollen-grain, and which Brown proved to have no vital significance but to be manifested by all minute particles whatsoever, was for many years unexplained. Thirty years and more after Brown wrote, it was said to be "due, either directly to some calorical changes continually taking place in the fluid, or to some obscure chemical action between the solid particles and the fluid which is indirectly promoted by heat." [96] Soon after these words were written it was ascribed by Christian Wiener [97] to molecular movements within the fluid, and was hailed as visible proof of the atomistic (or molecular) constitution of the same. We now know that it is indeed due to the impact or bombardment of molecules upon a body so small that these impacts do not average out, for the moment, to appoximate equality on all sides.[98] The movement becomes manifest with particles of somewhere about 20μ, and is better displayed by those of about 10μ, and especially well by certain colloid suspensions or emulsions

[93] Cf. Hardy and Nottage, *Proc. R.S.* (A), CXXVIII, p. 209, 1928; Baston and Bowden, *ibid.* CXXXIV, p. 404, 1931.

[94] *A Brief Description of Microscopical Observations . . . on the Particles contained in the Pollen of Plants; and on the General Existence of Active Molecules in Organic and Inorganic Bodies*, London, 1828. See also *Edinb. New Philosoph. Journ.* V, p. 358, 1828; *Edinb. Journ. of Science*, I, p. 314, 1829; *Ann. Sc. Nat.* XIV, pp. 341–362, 1828; etc. The Brownian movement was hailed by some as supporting Leibniz's theory of Monads, a theory once so deeply rooted and so widely believed that even under Schwann's cell-theory Johannes Müller and Henle spoke of the cells as "organische Monaden"; cf. Emil du Bois Reymond, Leibnizische Gedanken in der neueren Naturwissenschaft *Monatsber. d. k. Akad. Wiss.*, Berlin, 1870.

[95] The "nucleus" was first seen in the epidermis of Orchids; but "this areola, or nucleus of the cell as perhaps it might be termed, is not confined to the epidermis," etc. See his paper on Fecundation in Orchideae and Asclepiadae, *Trans. Linn. Soc.* XVI, 1829–33, also *Proc. Linn. Soc.* March 30, 1832.

[96] Carpenter, *The Microscope*, edit. 1862, p. 185.

[97] In *Poggendorff's Annalen*, CXVIII, pp. 79–94, 1863. For an account of this remarkable man, see *Naturwissenschaften*, XV, 1927; cf. also Sigmund Exner, Ueber Brown's Molecularbewegung, *Sitzungsber. kk. Akad. Wien*, LVI, p. 116, 1867.

[98] Perrin, Les preuves de la réalité moléculaire, *Ann. de Physique*, XVII, p. 549, 1905; XIX, p. 571, 1906. The actual molecular collisions are unimaginably frequent; we see only the residual fluctuations.

whose particles are just below 1μ in diameter.[99] The bombardment causes our particles to behave just like molecules of unusual size, and this behaviour is manifested in several ways.[100] Firstly, we have the quivering movement of the particles; secondly, their movement backwards and forwards, in short, straight disjointed paths; thirdly, the particles rotate, and do so the more rapidly the smaller they are: and by theory, confirmed by observation, it is found that particles of 1μ in diameter rotate on an average through $100°$ a second, while particles of 13μ turn through only $14°$ a minute. Lastly, the very curious result appears, that in a layer of fluid the particles are not evenly distributed, nor do they ever fall under the influence of gravity to the bottom. For here gravity and the Brownian movement are rival powers, striving for equilibrium; just as gravity is opposed in the atmosphere by the proper motion of the gaseous molecules. And just as equilibrium is attained in the atmosphere when the molecules are so distributed that the density (and therefore the number of molecules per unit volume) falls off in geometrical progression as we ascend to higher and higher layers, so is it with our particles within the narrow limits of the little portion of fluid under our microscope.

It is only in regard to particles of the simplest form that these phenomena have been theoretically investigated,[101] and we may take it as certain that more complex particles, such as the twisted body of a Spirillum, would shew other and still more complicated manifestations. It is at least clear that, just as the early microscopists in the days before Robert Brown never doubted but that these phenomena were purely vital, so we also may still be apt to confuse, in certain cases, the one phenomenon with the other. We cannot, indeed, without the most careful scrutiny, decide whether the movements of our minutest organisms are intrinsically "vital" (in the sense of being beyond a physical mechanism, or working model) or not. For example, Schaudinn has suggested that the undulating movements of *Spirochaete pallida* must be due to the presence of a minute, unseen, "undulating membrane"; and Doflein says of the same species that "sie verharrt oft mit eigenthümlich zitternden Bewegungen zu einem Orte." Both movements, the trembling or quivering

[99] Wiener was struck by the fact that the phenomenon becomes conspicuous just when the size of the particles becomes comparable to that of a wave-length of light.

[100] For a full, but still elementary, account, see J. Perrin, *Les Atomes*; cf. also Th. Svedberg, *Die Existenz der Moleküle*, 1912; R. A. Millikan, *The Electron*, 1917, etc. The modern literature of the Brownian movement (by Einstein, Perrin, de Broglie, Smoluchowski and Millikan) is very large, chiefly owing to the value which the phenomenon is shewn to have in determining the size of the atom or the charge on an electron, and of giving, as Ostwald said, experimental proof of the atomic theory.

[101] Cf. R. Gans, Wie fallen Stäbe und Scheiben in einer reibenden Flüssigkeit? *Münchener Bericht*, 1911, p. 191; K. Przibram, Ueber die Brown'sche Bewegung nicht kugelförmiger Teilchen, *Wiener Bericht*, 1912, p. 2339; 1913, pp. 1895–1912.

movement described by Doflein, and the undulating or rotating movement described by Schaudinn, are just such as may be easily and naturally interpreted as part and parcel of the Brownian phenomenon.

While the Brownian movement may thus simulate in a deceptive way the active movements of an organism, the reverse statement also to a certain extent holds good. One sometimes lies awake of a summer's morning watching the flies as they dance under the ceiling. It is a very remarkable dance. The dancers do not whirl or gyrate, either in company or alone; but they advance and retire; they seem to jostle and rebound; between the rebounds they dart hither or thither in short straight snatches of hurried flight, and turn again sharply in a new rebound at the end of each little rush.[102] Their motions are erratic, independent of one another, and devoid of common purpose.[103] This is nothing else than a vastly magnified picture, or simulacrum, of the Brownian movement; the parallel between the two cases lies in their complete irregularity, but this in itself implies a close resemblance. One might see the same thing in a crowded market-place, always provided that the bustling crowd had no *business* whatsoever. In like manner Lucretius, and Epicurus before him, watched the dust-motes quivering in the beam, and saw in them a mimic representation, *rei simulacrum et imago*, of the eternal motions of the atoms. Again the same phenomenon may be witnessed under the microscope, in a drop of water swarming with Paramoecia or such-like Infusoria; and here the analogy has been put to a numerical test. Following with a pencil the track of each little swimmer, and dotting its place every few seconds (to the beat of a metronome), Karl Przibram found that the mean successive distances from a common base-line obeyed with great exactitude the "Einstein formula," that is to say the particular form of the "law of chance" which is applicable to the case of the Brownian movement.[104] The phenomenon is (of course) merely analogous, and by no means identical with the Brownian movement; for the range of motion of the little active organisms, whether they be gnats or infusoria, is vastly greater than that of the minute particles which are passive under bombardment; nevertheless Przibram is inclined to think that even his comparatively large infusoria are small enough for the molecular bombardment to be a stim-

[102] As Clerk Maxwell put it to the British Association at Bradford in 1873, "We cannot do better than observe a swarm of bees, where every individual bee is flying furiously, first in one direction and then in another, while the swarm as a whole is either at rest or sails slowly through the air."

[103] Nevertheless there may be a certain amount of bias or direction in these seemingly random divagations: cf. J. Brownlee, *Proc. R.S.E.* xxxi, p. 262, 1910–11; F. H. Edgeworth, *Metron*, i, p. 75, 1920; Lotka, *Elem. of Physical Biology*, 1925, p. 344.

[104] That is to say, the mean square of the displacements of a particle, in any direction, is proportional to the interval of time. Cf. K. Przibram, Ueber die ungeordnete Bewegung niederer Tiere, *Pflüger's Archiv*, cliii, pp. 401–405, 1913; *Arch. f. Entw. Mech.* xliii, pp. 20–27, 1917.

ulus, even though not the actual cause, of their irregular and interrupted movements.[105]

George Johnstone Stoney, the remarkable man to whom we owe the name and concept of the *electron*, went further than this; for he supposed that molecular bombardment might be the source of the life-energy of the bacteria. He conceived the swifter moving molecules to dive deep into the minute body of the organism, and this in turn to be able to make use of these importations of energy.[106]

We draw near the end of this discussion. We found, to begin with, that "scale" had a marked effect on physical phenomena, and that increase or diminution of magnitude might mean a complete change of statical or dynamical equilibrium. In the end we begin to see that there are discontinuities in the scale, defining phases in which different forces predominate and different conditions prevail. Life has a range of magnitude narrow indeed compared to that with which physical science deals; but it is wide enough to include three such discrepant conditions as those in which a man, an insect and a bacillus have their being and play their several roles. Man is ruled by gravitation, and rests on mother earth. A water-beetle finds the surface of a pool a matter of life and death, a perilous entanglement or an indispensable support. In a third world, where the bacillus lives, gravitation is forgotten, and the viscosity of the liquid, the resistance defined by Stokes's law, the molecular shocks of the Brownian movement, doubtless also the electric charges of the ionised medium, make up the physical environment and have their potent and immediate influence on the organism. The predominant factors are no longer those of our scale; we have come to the edge of a world of which we have no experience, and where all our preconceptions must be recast.

[105] All that is actually proven is that "pure chance" has governed the movements of the little organism. Przibram has made the analogous observation that infusoria, when not too crowded together, spread or diffuse through an aperture from one vessel to another at a rate very closely comparable to the ordinary laws of molecular diffusion.

[106] *Phil. Mag.* April 1890.

COMMENTARY ON

Uncertainty

WHAT are the limits of physical knowledge? The question would have been answered more optimistically half a century ago than today. The fact is we are confronted with a paradox: despite the immense progress in physics during the past fifty years, it is now generally believed that we cannot find out certain things about a physical system which any self-respecting scientist of the eighteenth or nineteenth century would have confidently asserted could be found out. The most secure part of our present knowledge is knowing what it is that we cannot know.

Let me illustrate the difference between the older and the more modern views. "We ought then," Laplace wrote, "to regard the present state of the universe as the effect of its anterior state and as the cause of the one which is to follow. Given for one instant an intelligence which could comprehend all the forces by which nature is animated and the respective situation of the beings who compose it—an intelligence sufficiently vast to submit these data to analysis—it would embrace in the same formula the movements of the greatest bodies of the universe and those of the lightest atom; for it, nothing would be uncertain and the future, as the past, would be present to its eyes." [1] Laplace did not suggest that the human brain was capable of such an analysis, but he implied no theoretical barrier blocked the way. Contrast his outlook with that of the distinguished contemporary mathematician and physicist, Sir Edmund Whittaker. Whittaker gathers together a number of statements, each of which asserts "the impossibility of achieving something, even though there may be an infinite number of ways of trying to achieve it." [2] These "Postulates of Impotence" include the assertion, in relativity (as a general law of nature), of "the impossibility of recognizing absolute velocity"; the postulate that it is impossible "at any instant to assert that a particular electron is identical with some particular electron which had been observed at an earlier instant"; the principle, essential to the late E. A. Milne's cosmological theory, that "it is impossible to tell where one is in the universe"; and the celebrated uncertainty principle of Heisenberg—a basic postulate of quantum mechanics—that it is impossible "to measure precisely the momentum of a particle at the same time as a precise measurement of its position is made." Whittaker's postulates, it should be observed, are neither statements of experimental fact nor of logical necessity. Each one

[1] Pierre Simon Marquis de Laplace, *A Philosophical Essay on Probabilities*, translated from the Sixth French Edition by F. W. Truscott and F. L. Emory, New York, 1917, p. 4.

[2] Sir Edmund T. Whittaker, *From Euclid to Eddington*, Cambridge, 1949, pp. 58–60.

simply expresses a conviction that any attempt to do a certain thing will fail. At the very least this is a more modest stand than Laplace's. He was not a modest man.

Heisenberg's principle concerns the accuracy with which certain physical quantities can be measured, the *theoretical*, not merely the practical limits of precision. It had long been supposed that the precision of measurements was limited only by the instruments and methods used, and that as these were improved a corresponding improvement in accuracy would result—without limit. Heisenberg showed that this faith in the perfectability of measurement was unjustified as regards the observation of very small particles. The very act of observing the position and the velocity of an electron interferes with it sufficiently to produce errors of measurement. It turns out that the more sharply one specifies the position of a particle, the less sharply can its velocity or momentum be determined; and vice versa. At best, according to Heisenberg, the product of these errors cannot be less than $h/2\pi$, where h is Planck's quantum constant, the minimum packet of energy encountered in nature.*

Sir William Bragg has written a very picturesque description of the uncertainty fiasco. "Given the most delicate apparatus in the world, the collection of information about the gas in some headquarters means that each molecule has sent a message, necessarily an ether wave because that is the only way in which a message can travel from one point to another. Forcing it to send such a wave by throwing light on it is again of necessity a brutal cataclysmic process. The molecule has a nasty shock, and it is not the molecule it was before. Further, we cannot say exactly how the shock has affected it, except by making it send another message at a future time. This gives it another shock and though we have succeeded in discovering what it has been doing in the past we are as badly off as before as regards predicting its future. It is important to realize that the shock is received because the molecule having had light directed upon it, sends a message betraying its whereabouts; the shock has nothing whatever to do with the apparatus receiving the message, which we may make as delicate as we like." [3]

Bragg makes it clear that we are faced here with an essential principle of physics. It is doubtful that it is also an essential principle of philosophy, as Eddington and others have concluded. Eddington called Heisenberg's result the principle of indeterminacy.[4] By "indeterminacy" he meant not only the fact that we as observers cannot tell exactly what is happening in the world of tiny dimensions—that "every time we determine something in the present we spoil something else" (as Bragg said)—but also the fact

* "Packet of energy" is a figure of speech; the dimensions of h are energy times time momentum times length.

[3] Sir William Bragg, "The Physical Sciences," *Science*, March 16, 1934.

[4] Sir William Dampier, *A History of Science*, Fourth Edition, Cambridge, 1949, p. 397.

that the particles of that world are not ruled by the law of cause and effect. If one cannot describe a causal chain—well, then there simply is none; a curious sort of reasoning. Its appeal lay in providing an escape from the inflexible determinism of the Laplacean universe. Clergymen, as Bertrand Russell and Susan Stebbing have remarked, were particularly grateful for this solace. The announcement that electrons enjoyed free will —whatever that might mean—was cheerful news; it almost offset the series of blows that formal religion suffered in the generations from Galileo to Darwin.

Russell was right that too much fuss has been made about the uncertainty principle.[5] This is not to pooh-pooh its importance to physics. But it is essential not to confuse matters, and to keep in mind that this postulate of impotence pertains to the observer and not to the observed. To say that a psychological examination disturbs the person being examined, no matter how carefully the test is conducted, is not the same as saying that his mental behavior is not determined. Similarly Heisenberg's principle "does nothing whatever to show that the course of nature is not determined. It shows merely that the old space-time apparatus is not quite adequate to the needs of modern physics, which, in any case, is known on other grounds." [6]

The two following selections deal with the problems we have been considering. The first is a short excerpt from Heisenberg's *The Physical Principles of the Quantum Theory*, a book based on lectures given at the University of Chicago in 1929.[7] The discussion is for advanced students but the excerpt I have selected will give the average reader a pretty good idea of indeterminacy, and of another radical concept, that of complementarity, proposed by the great Danish physicist, Niels Bohr. The second selection is taken from a little book by Erwin Schrödinger.[8] (I have referred to Schrödinger at greater length elsewhere in these pages—see pp. 973–974. The excerpt deals in lively and lucid fashion with a variety of the profound changes which experimental and mathematical physics have brought about

[5] Bertrand Russell, *Human Knowledge, Its Scope and Limits*, London, 1948. For an exceptionally searching discussion of the philosophical questions involved. I recommend also W. H. Watson, *On Understanding Physics*, Cambridge, 1938. As may be seen from Watson's book and later studies, Heisenberg's own formulation of the uncertainty relation is not the only one current.

[6] This passage continues: "Space and time were invented by the Greeks, and served their purpose admirably until the present century. Einstein replaced them by a kind of centaur which he called 'space-time,' and this did well enough for a couple of decades, but modern quantum mechanics has made it evident that a more fundamental reconstruction is necessary. The Principle of Indeterminacy is merely an illustration of this necessity, not of the failure of physical laws to determine the course of nature." Bertrand Russell, *The Scientific Outlook*, London, 1931, pp. 108–9. See also L. Susan Stebbing, *Philosophy and the Physicists*, Penguin Books, Middlesex, England, 1944.

[7] Werner Heisenberg, *The Physical Principles of the Quantum Theory*, Chicago, 1930.

[8] Erwin Schrödinger, *Science and Humanism; Physics in Our Time*, Cambridge, 1951.

in our picture of the physical world. Schrödinger discusses the nature of our physical models, the concepts of continuity and causality—or rather what is now left of the older, comfortable versions of these ideas—and the mathematical discipline known as wave-mechanics, of which he was a founder. It is a brilliant performance; when you have gone through it your picture of the world will, I promise, be both clearer and more topsy-turvy than ever before.

A brief note about Heisenberg: He was born in Munich in 1902, and there received his early education. At the University of Munich it was his good fortune to study with Arnold Sommerfeld, the noted German physicist who made significant contributions to the quantum and relativity theories. Heisenberg later worked at Göttingen under Max Born, and there first met Niels Bohr. For a period they worked closely together; in the years when Heisenberg held a chair in physics at Leipzig he spent a good part of his time with the Danish physicist in his laboratory at Copenhagen. Heisenberg's new theory of quantum mechanics, "based only on what can be observed, that is, on the radiation absorbed and emitted by the atom," was framed in 1925.[9] This paper and later ones which proved the validity of the classic equations of mechanics formulated by the Irish scientist William Rowan Hamilton (1805–1865) were major factors in the modern revolution in physics. His paper on indeterminacy was published in 1927;[10] he was awarded the Nobel prize in physics in 1932. Heisenberg openly fought the Nazi excesses in the 1930s and was attacked for his views; he was offered many positions in the U. S., but refused to emigrate. During the war he worked on the German atom bomb project.[11]

[9] For an admirable nonmathematical summary of this and related achievements see A. S. Eddington, *The Nature of the Physical World*, Cambridge, 1928, p. 206. Be careful, however, about accepting all of Eddington's irresistibly persuasive philosophical conjectures.

[10] *Zeitschrift für Physik*, Vol. 43, p. 172 *et seq.*

[11] For an interesting account of Heisenberg's relations with the Nazis, and also his fiasco in bomb making, see Samuel A. Goudsmit, *Alsos*, New York, 1947.

A particle can have a position or it can have a velocity, but in the strict sense it cannot have both. . . . Nature puts up with our probings into its mysteries only on conditions. The more we clarify the secret of position, the more deeply hidden becomes the secret of velocity. It reminds one of the man and woman in the weather house: if one comes out, the other goes in. . . . The product of the two unknowns is always an integral multiple of an elementary quantum of action. We can distribute the uncertainty as we wish, but we can never get away from it. —WERNER HEISENBERG

Oh, let us never, never doubt
What nobody is sure about. —HILAIRE BELLOC

18 The Uncertainty Principle

By WERNER HEISENBERG

ILLUSTRATIONS OF THE UNCERTAINTY RELATIONS

THE uncertainty principle refers to the degree of indeterminateness in the possible present knowledge of the simultaneous values of various quantities with which the quantum theory deals; it does not restrict, for example, the exactness of a position measurement alone or a velocity measurement alone. Thus suppose that the velocity of a free electron is precisely known, while the position is completely unknown. Then the principle states that every subsequent observation of the position will alter the momentum by an unknown and undeterminable amount such that after carrying out the experiment our knowledge of the electronic motion is restricted by the uncertainty relation. This may be expressed in concise and general terms by saying that every experiment destroys some of the knowledge of the system which was obtained by previous experiments. This formulation makes it clear that the uncertainty relation does not refer to the past; if the velocity of the electron is at first known and the position then exactly measured, the position for times previous to the measurement may be calculated. Then for these past times $\Delta p \Delta q$ is smaller than the usual limiting value, but this knowledge of the past is of a purely speculative character, since it can never (because of the unknown change in momentum caused by the position measurement) be used as an initial condition in any calculation of the future progress of the electron and thus cannot be subjected to experimental verification. It is a matter of personal belief whether such a calculation concerning the past history of the electron can be ascribed any physical reality or not.

(*a*) *Determination of the position of a free particle.*—As a first example of the destruction of the knowledge of a particle's momentum by an apparatus determining its position, we consider the use of a micro-

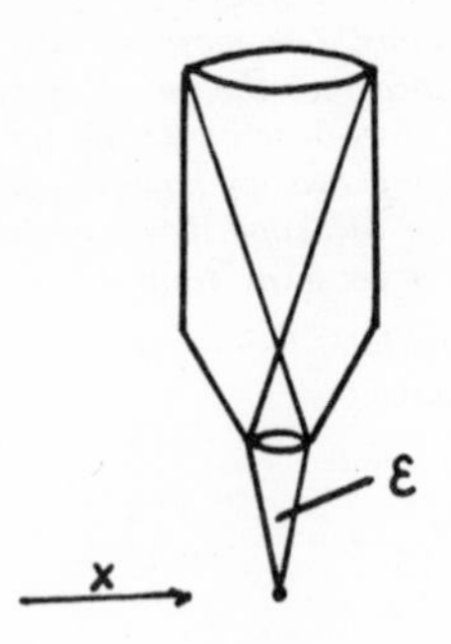

FIGURE 1

scope.[1] Let the particle be moving at such a distance from the microscope that the cone of rays scattered from it through the objective has an angular opening ϵ. If λ is the wave-length of the light illuminating it, then the uncertainty in the measurement of the x-co-ordinate (see Figure 1) according to the laws of optics governing the resolving power of any instrument is:

$$\Delta x = \frac{\lambda}{\sin \epsilon}. \tag{1}$$

But, for any measurement to be possible at least one photon must be scattered from the electron and pass through the microscope to the eye of the observer. From this photon the electron receives a Compton recoil of order of magnitude h/λ. The recoil cannot be exactly known, since the direction of the scattered photon is undetermined within the bundle of rays entering the microscope. Thus there is an uncertainty of the recoil in the x-direction of amount

$$\Delta p_x \sim \frac{h}{\lambda} \sin \epsilon, \tag{2}$$

and it follows that for the motion after the experiment

$$\Delta p_x \Delta x \sim h. \tag{3}$$

Objections may be raised to this consideration; the indeterminateness of the recoil is due to the uncertain path of the light quantum within the bundle of rays, and we might seek to determine the path by making the microscope movable and measuring the recoil it receives from the light quantum. But this does not circumvent the uncertainty relation, for it immediately raises the question of the position of the microscope, and its position and momentum will also be found to be subject to equation (3). The position of the microscope need not be considered if the electron and a fixed scale be simultaneously observed through the moving microscope,

[1] Niels Bohr, *Nature*, 121, 580, 1928.

and this seems to afford an escape from the uncertainty principle. But an observation then requires the simultaneous passage of at least two light quanta through the microscope to the observer—one from the electron and one from the scale—and a measurement of the recoil of the microscope is no longer sufficient to determine the direction of the light scattered by the electron. And so on *ad infinitum*.

BOHR'S CONCEPT OF COMPLEMENTARITY [2]

With the advent of Einstein's relativity theory it was necessary for the first time to recognize that the physical world differed from the ideal world conceived in terms of everyday experience. It became apparent that ordinary concepts could only be applied to processes in which the velocity of light could be regarded as practically infinite. The experimental material resulting from modern refinements in experimental technique necessitated the revision of old ideas and the acquirement of new ones, but as the mind is always slow to adjust itself to an extended range of experience and concepts, the relativity theory seemed at first repellently abstract. None the less, the simplicity of its solution for a vexatious problem has gained it universal acceptance. As is clear from what has been said, the resolution of the paradoxes of atomic physics can be accomplished only by further renunciation of old and cherished ideas. Most important of these is the idea that natural phenomena obey exact laws—the principle of causality. In fact, our ordinary description of nature, and the idea of exact laws, rests on the assumption that it is possible to observe the phenomena without appreciably influencing them. To co-ordinate a definite cause to a definite effect has sense only when both can be observed without introducing a foreign element disturbing their interrelation. The law of causality, because of its very nature, can only be defined for isolated systems, and in atomic physics even approximately isolated systems cannot be observed. This might have been foreseen, for in atomic physics we are dealing with entities that are (so far as we know) ultimate and indivisible. There exist no infinitesimals by the aid of which an observation might be made without appreciable perturbation.

Second among the requirements traditionally imposed on a physical theory is that it must explain all phenomena as relations between objects existing in space and time. This requirement has suffered gradual relaxation in the course of the development of physics. Thus Faraday and Maxwell explained electromagnetic phenomena as the stresses and strains of an ether, but with the advent of the relativity theory, this ether was dematerialized; the electromagnetic field could still be represented as a set of vectors in space-time, however. Thermodynamics is an even better example of a theory whose variables cannot be given a simple geometric

[2] *Nature*, 121, 580, 1928.

interpretation. Now, as a geometric or kinematic description of a process implies observation, it follows that such a description of atomic processes necessarily precludes the exact validity of the law of causality—and conversely. Bohr [3] has pointed out that it is therefore impossible to demand that both requirements be fulfilled by the quantum theory. They represent complementary and mutually exclusive aspects of atomic phenomena. This situation is clearly reflected in the theory which has been developed. There exists a body of exact mathematical laws, but these cannot be interpreted as expressing simple relationships between objects existing in space and time. The observable predictions of this theory can be approximately described in such terms, but not uniquely—the wave and the corpuscular pictures both possess the same approximate validity. This indeterminateness of the picture of the process is a direct result of the interdeterminateness of the concept "observation"—it is not possible to decide, other than arbitrarily, what objects are to be considered as part of the observed system and what as part of the observer's apparatus. In the formulas of the theory this arbitrariness often makes it possible to use quite different analytical methods for the treatment of a single physical experiment. Some examples of this will be given later. Even when this arbitrariness is taken into account the concept "observation" belongs, strictly speaking, to the class of ideas borrowed from the experiences of everyday life.[4] It can only be carried over to atomic phenomena when due regard is paid to the limitations placed on all space-time descriptions by the uncertainty principle.

The general relationships discussed here may be summarized in the following diagrammatic form:

CLASSICAL THEORY

CAUSAL RELATIONSHIPS OF PHENOMENA DESCRIBED IN TERMS OF SPACE AND TIME

QUANTUM THEORY

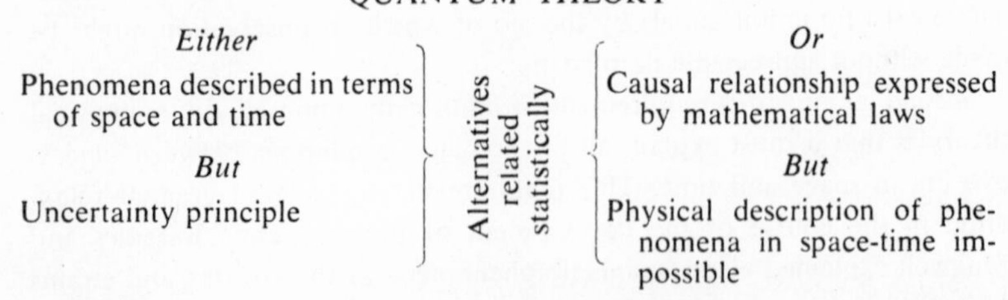

It is only after attempting to fit this fundamental complementarity of space-time description and causality into one's conceptual scheme that one

[3] *Ibid.*

[4] It need scarcely be remarked that the term "observation" as here used does not refer to the observation of lines on photographic plates, etc., but rather to the observation of "the electrons in a single atom," etc.

is in a position to judge the degree of consistency of the methods of quantum theory (particularly of the transformation theory). To mold our thoughts and language to agree with the observed facts of atomic physics is a very difficult task, as it was in the case of the relativity theory. In the case of the latter, it proved advantageous to return to the older philosophical discussions of the problems of space and time. In the same way it is now profitable to review the fundamental discussions, so important for epistemology, of the difficulty of separating the subjective and objective aspects of the world. Many of the abstractions that are characteristic of modern theoretical physics are to be found discussed in the philosophy of past centuries. At that time these abstractions could be disregarded as mere mental exercises by those scientists whose only concern was with reality, but today we are compelled by the refinements of experimental art to consider them seriously.

It is impossible to trap modern physics into predicting anything with perfect determinism because it deals with probabilities from the outset.
—SIR ARTHUR STANLEY EDDINGTON

19 Causality and Wave Mechanics

By ERWIN SCHRÖDINGER

. . . AS our mental eye penetrates into smaller and smaller distances and shorter and shorter times, we find nature behaving so entirely differently from what we observe in visible and palpable bodies of our surrounding that *no* model shaped after our large-scale experiences can ever be 'true.' A completely satisfactory model *of this type* is not only practically inaccessible, but not even thinkable. Or, to be precise, we can, of course, think it, but however we think it, it is wrong; not perhaps quite as meaningless as a 'triangular circle,' but much more so than a 'winged lion.'

CONTINUOUS DESCRIPTION AND CAUSALITY

I shall try to be a little clearer about this. From our experiences on a large scale, from our notion of geometry and of mechanics—particularly

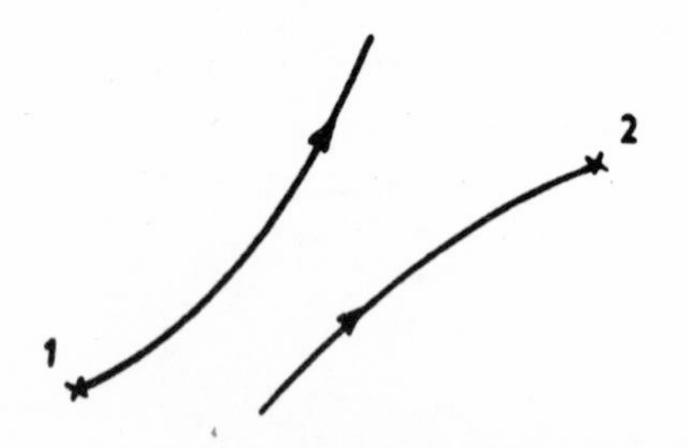

FIGURE 1

the mechanics of the celestial bodies—physicists had distilled the one clear-cut demand that a truly clear and complete description of any physical happening has to fulfil: it ought to inform you precisely of what happens at any point in space at any moment of time—of course, within the spatial domain and the period of time covered by the physical events you wish to describe. We may call this demand the *postulate of continuity of the description.* It is this postulate of continuity that appears to be unfulfillable! There are, as it were, gaps in our picture.

This is intimately connected with what I called earlier the lack of individuality of a particle, or even of an atom. If I observe a particle here and now, and observe a similar one a moment later at a place very near the former place, not only cannot I be sure whether it is 'the same,' but this statement has no absolute meaning. This *seems* to be absurd. For we are so used to thinking that at every moment between the two observations the first particle must have been *somewhere*, it must have followed a *path*, whether we know it or not. And similarly the second particle must have come from somewhere, it must have *been* somewhere at the moment of our first observation. So in principle it must be decided, or decidable, whether these two paths are the same or not—and thus whether it *is* the same particle. In other words we assume—following a habit of thought that applies to palpable objects—that we could have kept our particle under *continuous observation*, thereby ascertaining its identity.

This habit of thought we must dismiss. *We must not admit the possibility of continuous observation.* Observations are to be regarded as discrete, disconnected events. Between them there are gaps which we cannot fill in. These are cases where we should upset everything if we admitted the possibility of continuous observation. That is why I said it is better to regard a particle not as a permanent entity but as an instantaneous event. Sometimes these events form chains that give the illusion of permanent beings—but only in particular circumstances and only for an extremely short period of time in every single case.

Let us go back to the more general statement I made before, namely that the classical physicist's naïve ideal cannot be fulfilled, his demand that in principle information about every point in space at every moment of time should at least be *thinkable*. That this ideal breaks down has a very momentous consequence. For in the times when this ideal of continuity of

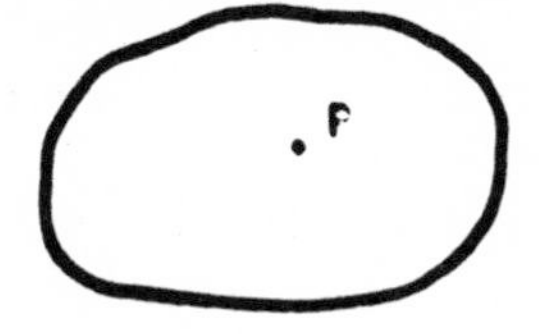

FIGURE 2

description was not doubted, the physicists had used it to formulate the *principle of causality* for the purposes of their science in a very clear and precise fashion—the only one in which they could use it, the ordinary enouncements being much too ambiguous and imprecise. It includes in this form, the principle of 'close action' (or the absence of *actio in distans*) and runs as follows: The exact physical situation at *any* point P at a given moment t is unambiguously determined by the exact physical situation within a certain surrounding of P at any previous time, say t—τ.

If τ is large, that is, if that previous time lies far back, it may be necessary to know the previous situation for a wide domain around P. But the 'domain of influence' becomes smaller and smaller as τ becomes smaller, and becomes infinitesimal as τ becomes infinitesimal. Or, in plain, though less precise, words: what happens anywhere at a given moment depends only and unambiguously on what has been going on in the immediate neighbourhood 'just a moment earlier.' Classical physics rested entirely on this principle. The mathematical instrument to implement it was in all cases a system of partial differential equations—so-called field equations.

Obviously, if the ideal of continuous, 'gap-less,' description breaks down, this precise formulation of the principle of causality breaks down. And we must not be astonished to meet in this order of ideas with new, unprecedented difficulties as regards causation. We even meet (as you know) with the statement that there are gaps or flaws in strict causation. Whether this is the last word or not it is difficult to say. Some people believe that the question is by no manner of means settled (among them, by the way, is Albert Einstein). I shall tell you a little later about the 'emergency exit,' used at present to escape from the delicate situation. For the moment I wish to attach some further remarks to the classical ideal of continuous description.

THE INTRICACY OF THE CONTINUUM

However painful its loss may be, by losing it we probably lose something that is very well worth losing. It seems simple to us, because the idea of the continuum seems simple to us. We have somehow lost sight of the difficulties it implies. That is due to a suitable conditioning in early childhood. Such an idea as 'all the numbers between 0 and 1' or 'all the numbers between 1 and 2' has become quite familiar to us. We just think of them geometrically as the distance of any point like P or Q from 0 (see Figure 3).

Among the points like Q there is also the $\sqrt{2}$ ($= 1.414 \ldots$). We are told that such a number as $\sqrt{2}$ worried Pythagoras and his school almost to exhaustion. Being used to such queer numbers from early childhood,

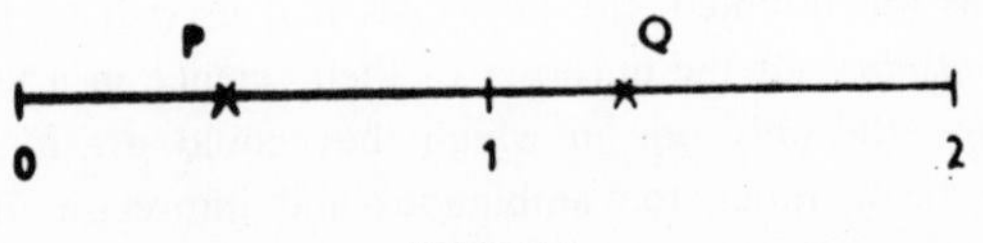

FIGURE 3

we must be careful not to form a low idea of the mathematical intuition of these ancient sages. Their worry was highly creditable. They were aware of the fact that no fraction can be indicated of which the square is exactly 2. You can indicate close approximations, as for instance $^{17}/_{12}$, whose square, $^{289}/_{144}$, is very near to $^{288}/_{144}$, which is 2. You can get closer by contemplating fractions with larger numbers than 17 and 12, but you will never get *exactly* 2.

The idea of a *continuous range*, so familiar to mathematicians in our days, is something quite exorbitant, an enormous extrapolation of what is really accessible to us. The idea that you should *really* indicate the exact values of any physical quantity—temperature, density, potential, field strength, or whatever it might be—for *all* the points of a continuous range, say between zero and 1, is a bold extrapolation. We *never* do anything

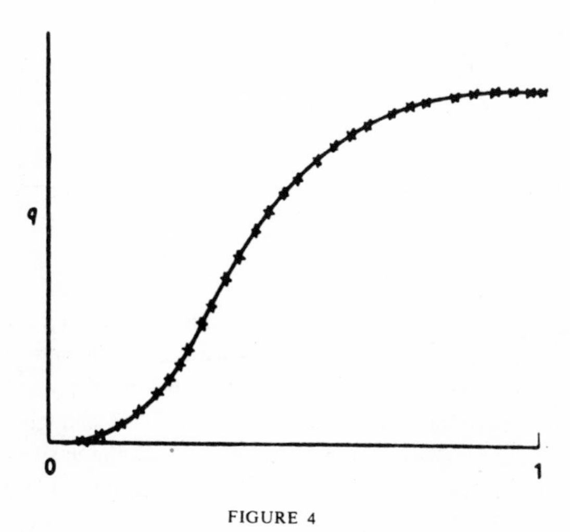

FIGURE 4

else than determine the quantity approximately for a very limited number of points and then 'draw a smooth curve through them.' This serves us well for many practical purposes, but from the epistemological point of view, from the point of view of the theory of knowledge, it is totally different from a supposed exact continual description. I might add that even in classical physics there were quantities—as, for instance, temperature or density—which avowedly did not admit of an exact continuous description. But this was due to the conception these terms represent—they have, even in classical physics, only a statistical meaning. However I shall not go into details about this at the moment, it would create confusion.

The demand for continuous description was encouraged by the fact that the mathematician claims to be able to indicate simple continuous descriptions of some of his simple mental constructions. For example, take again the range $0 \rightarrow 1$, call the variable in this range x, we claim to have an unambiguous idea of, say x^2 or $\sqrt{x}$.

The curves are pieces of parabolas (mirror images of each other). We

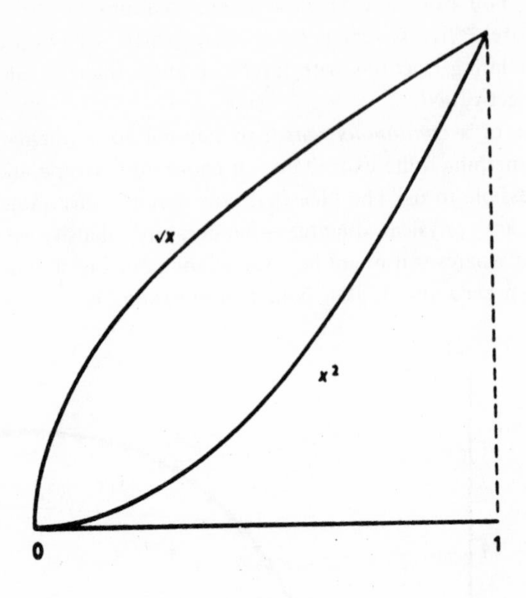

FIGURE 5

claim to have full knowledge of every point of such a curve, or rather, *given* the horizontal distance (abscissa) we are able to indicate the height (ordinate) *with any required precision*. But behold the words 'given' and 'with any required precision.' The first means 'we can give the answer, when it comes to it'—we cannot possibly have all the answers in store for you in advance. The second means 'even so, we cannot as a rule give you an absolutely precise answer.' You must tell us the precision you require, e.g., up to 1000 decimal places.

0 $\frac{1}{9}$ $\frac{2}{9}$ $\frac{1}{3}$ $\frac{2}{3}$ $\frac{7}{9}$ $\frac{8}{9}$ 1

FIGURE 6

Then we can give you the answer—if you leave us time.

Physical dependences can always be approximated by this simple kind of functions (the mathematician calls them 'analytical,' which means something like 'they can be analysed'). But to assume that physical dependence *is* of this simple type, is a bold epistemological step, and probably an inadmissible step.

However, the chief conceptual difficulty is the enormous number of 'answers' that are required, due to the enormous number of points contained in even the smallest continuous range. This quantity—the number of points between 0 and 1, for example—is so fabulously great that it is hardly diminished even if you take 'nearly all of them' away. Allow me to illustrate this by an impressive example.

Envisage again the line $0 \rightarrow 1$. I wish to describe a certain set of points that is *left over*, when you take some of them away, bar them, exclude them, make them inaccessible—or whatever you wish to call it. I shall use the word 'take away.'

First take away the whole middle third including its left border point, thus the points from ⅓ to ⅔ (but you *leave* ⅔). Of the remaining two thirds you again take away the 'middle thirds,' including *their* left border points, but leaving their right border points. With the remaining 'four ninths' you proceed in the same way. *And so on.*

If you actually try to continue for only a few steps you will soon get the impression that 'nothing is left over.' Indeed at every step we take away a third of the remaining length. Now supposing the Income Tax Inspector charged you first 6*s.* 8*d.* in the £, and of the remainder again 6*s.* 8*d.* in the £, and so on, *ad infinitum*, you agree you would not retain much.

We shall now analyse our case, and you will be astonished how many of our numbers or points are left. I regret that this needs a little preparation. A number between zero and one can be represented by a decadic fraction, as

$$0{\cdot}470802\ldots$$

and you know this means

$$\frac{4}{10}+\frac{7}{10^2}+\frac{0}{10^3}+\frac{8}{10^4}+\ldots.$$

That we habitually use here the number 10 is a pure accident, due to the fact that we have 10 fingers. We can use any other number, 8, 12, 3, 2. . . . We need, of course, different figure-symbols for all the numbers up to the chosen 'basis.' In our decadic system we need ten, 0, 1, 2, . . . 9. If we used 12 as our basis, we should have to invent single symbols for 10 and 11. If we used the basis 8, the symbols for 8 and 9 would become supernumerary.

Non-decadic fractions have not altogether been ousted by the decimal system. Dyadic fractions, that is those which use the basis 2, are quite popular, particularly with the British. When I asked my tailor the other day how much material I should get him for the flannel trousers I had just ordered, he answered—to my amazement—1⅜ yards. This is easily seen to be the *dyadic* fraction

$$1{\cdot}011,$$

meaning

$$1+\frac{0}{2}+\frac{1}{4}+\frac{1}{8}.$$

In the same way some stock exchanges quote shares not in shillings and pence but in dyadic fractions of a pound, for example £$^{13}/_{16}$, which in *dyadic* notation would read

$$0\cdot 1101,$$

meaning

$$\frac{1}{2}+\frac{1}{4}+\frac{0}{8}+\frac{1}{16}.$$

Notice that in a dyadic fraction only two symbols, viz. 0 and 1, occur.

For our present purpose we first need *triadic* fractions, which have the basis 3 and use only the symbols 0, 1, 2. Here, for instance, the notation

$$0\cdot 2012\ldots$$

means

$$\frac{2}{3}+\frac{0}{9}+\frac{1}{27}+\frac{2}{81}+\ldots.$$

(By adding dots we intentionally admit fractions that run to infinity, as for example the square root of 2). Now let us return to the problem of describing the 'almost vanishing' set of numbers that is left over in the construction illustrated by our figure. A little careful thinking will shew you that the points we have *taken away* are all those which in *triadic* representation contain a figure 1 *somewhere*. Indeed, by first cutting out the middle third we cut out all the numbers whose triadic fraction begins thus:

$$0\cdot 1\ldots.$$

At the second step we cut out all those whose triadic fraction begins

$$\text{either } 0\cdot 01\ldots \text{ or } 0\cdot 21\ldots.$$

And so on.—This consideration shews that there is something left, namely all those whose triadic fractions contain *no* 1, but only 0 and 2, as for instance

$$0\cdot 22000202\ldots$$

(where the dots stand for any sequence of 0s and 2s only). Among them are, of course, the *right* border points (as $0\cdot 2 = 2/3$ or $0\cdot 22 = 2/3 + 2/9 = 8/9$) of the excluded intervals; we had decided to let those border points stand. But there are a lot more, for instance the *periodic* dyadic fraction $0\cdot\dot{2}\dot{0}$, meaning $0\cdot 20202020\ldots$ *ad infinitum*. This is the infinite series

$$\frac{2}{3}+\frac{2}{3^3}+\frac{2}{3^5}+\frac{2}{3^7}+\ldots.$$

To find its value, think you multiply it by the square of 3, which is 9. Then the first term gives 18/3, that is, 6, while the remaining terms give the same series again. Hence *eight* times our series is 6, and our number is 6/8 or 3/4.

Still, recalling again that the intervals we have 'taken away' tend to cover the *whole* interval between 0 and 1, one is inclined to think that, compared with the original set (containing *all* numbers between 0 and 1), the remaining set must be 'exceedingly scarce.' But now comes the amazing turn: in a certain sense the remaining set is still just as vast as the original one. Indeed we can associate their respective members in pairs, by monogamously mating, as it were, each number of the original set with a definite number of the remaining set, without any number being left over on either side (the mathematician calls this a 'one-to-one correspondence'). This is so perplexing that, I am sure, many a reader will at first think he *must* have misunderstood the words, though I have taken pains to set them as unambiguously as possible.

How is this done? Well, the 'remaining set' is represented by *all* the *triadic* fractions containing only 0s and 2s; we gave the general example

$$0 \cdot 22000202 \ldots$$

(the dots standing for any sequence of 0s and 2s only). Associate with this *triadic* fraction the *dyadic* fraction

$$0 \cdot 11000101 \ldots$$

obtained from the former by replacing every figure 2 by the figure 1. Vice versa you can, from *any* dyadic fraction, by changing its 1s into 2s, obtain the *triadic* representation of a definite number in what we called 'the remaining set.' Since now any member of the original set, that is, any number between 0 and 1, is represented by one and only one [1] definite dyadic fraction, there is actually a perfect one-to-one mating between the members of the two sets.

[It may be useful to illustrate the 'mating' by examples. For instance the dyadic number that my tailor used

$$\frac{3}{8} = \frac{0}{2} + \frac{1}{4} + \frac{1}{8} = 0 \cdot 011$$

would lead to the triadic counterpart

$$0 \cdot 022 = \frac{0}{3} + \frac{2}{9} + \frac{2}{27} = \frac{8}{27};$$

that is to say, 3/8 of the original set corresponds to 8/27 in the remaining set.

[1] We have tacitly disregarded such trivial duplications as are instanced, in the decimal system, by $0.1 = 0.0\dot{9}$ or $0.8 = 0.7\dot{9}$.

Inversely, take out triadic 0.20, meaning, as we made out, ¾. The corresponding dyadic 0.10 means the infinite series

$$\frac{1}{2}+\frac{1}{2^3}+\frac{1}{2^5}+\frac{1}{2^7}+\frac{1}{2^9}+\ldots.$$

If you multiply this by the square of 2, which is 4, you get: 2 + *the same series*. In other words, *three* times our series equals 2, the series equals ⅔; that is to say, the number ¾ of the 'remaining set' corresponds (or 'is mated') to the number ⅔ in the original set.]

The remarkable fact about our 'remaining set' is that, though it covers no measurable interval, yet it has still the vast extension of any continuous range. This astonishing combination of properties is, in mathematical language, expressed by saying that our set has still the 'potency' of the continuum, although it is 'of measure zero.'

I have brought this case before you, in order to make you feel that there is something mysterious about the continuum and that we must not be too astonished at the apparent failure of our attempts to use it for a precise description of nature.

THE MAKESHIFT OF WAVE MECHANICS

Now I shall try to give you an idea of the way in which physicists at present endeavour to overcome this failure. One might term it an 'emergency exit,' though it was not intended as such, but as a new theory. I mean, of course, wave mechanics. (Eddington called it 'not a physical theory but a dodge—and a very good dodge too.')

The situation is about as follows. The observed facts (about particles and light and all sorts of radiation and their mutual interaction) appear to be *repugnant* to the classical ideal of a continuous description in space and time. (Let me explain myself to the physicist by hinting at one example: Bohr's famous theory of spectral lines in 1913 had to assume that the atom makes a *sudden* transition from one state into another state, and that in doing so it emits a train of light waves several feet long, containing hundreds of thousands of waves and requiring for its formation a considerable time. No information about the atom during this transition can be offered.)

So the facts of observation are irreconcilable with a continuous description in space and time; it just seems impossible, at least in many cases. On the other hand, from an incomplete description—from a picture with gaps in space and time—one cannot draw clear and unambiguous conclusions; it leads to hazy, arbitrary, unclear thinking—and that is the thing we must avoid at all costs! What is to be done? The method adopted at present may seem amazing to you. It amounts to this: we do give a complete description, continuous in space and time without leaving any gaps, conforming to the classical ideal—a description *of something*. But we do not claim that this 'something' is the observed or observable facts; and still less do we claim that we thus describe what nature (matter, radiation, etc.) really *is*. In fact we use this picture (the so-called wave picture) in full knowledge that it is *neither*.

There is no gap in this picture of wave mechanics, also no gap as regards *causation*. The wave picture conforms with the classical demand for complete determinism, the mathematical method used is that of field-equations, though sometimes they are a highly generalized type of field-equations.

But what is the use of such a description, which, as I said, is not believed to describe observable facts or what nature really is like? Well, it is believed to give us *information* about observed facts and their mutual dependence. There is an optimistic view, viz. that it gives us *all* the information obtainable about observable facts and their interdependence. But this view—which may or may not be correct—is *optimistic* only inasmuch as it may flatter our pride to possess in principle all obtainable information. It is pessimistic in another respect, we might say epistemologically pessimistic. *For the information we get as regards the causal dependence of observable facts is incomplete.* (The cloven hoof must show up *somewhere*!) The gaps, eliminated from the wave picture, have withdrawn to the connection between the wave picture and the observable facts. The latter are *not* in one-to-one correspondence with the former. Plenty of ambiguity remains, and, as I said, some optimistic pessimists or pessimistic optimists believe that this ambiguity is essential, it cannot be helped.

This is the logical situation at present. I believe I have depicted it correctly, though I am quite aware that without examples the whole discussion has remained a little bloodless—just purely logical. I am also afraid that I have given you too unfavourable an impression of the wave theory of matter. I ought to amend both points. The wave theory is not of yesterday and not of 25 years ago. It made its first appearance as the wave theory of light (Huygens 1690). For the better part of 100 years [2] light waves were regarded as an incontrovertible reality, as something of which the real existence had been proved beyond all doubt by experiments on the diffraction and interference of light. I do not think that even today many physicists—certainly not experimentalists—are ready to endorse the statement that 'light waves do not really exist, they are only waves of knowledge' (free quotation from Jeans).

If you observe a narrow luminous source L, a glowing Wollaston wire, a few thousandths of a millimetre thick, by a microscope whose objective

[2] Not the immediately following hundred years. Newton's authority eclipsed Huygens' theory for about a century.

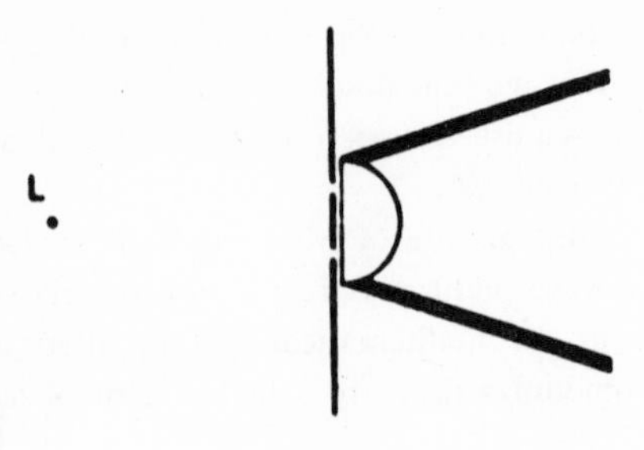

FIGURE 7

lens is covered by a screen with a couple of parallel slits, you find (in the image plane conjugate to L) a system of coloured fringes which conform exactly and quantitatively to the idea that light of a given colour is a wave motion of a certain small wave-length, shortest for violet, about twice as long for red light. This is one out of dozens of experiments that clinch the same view. Why, then, has this *reality* of the waves become doubtful? For two reasons:

(a) Similar experiments have been performed with beams of cathode rays (instead of light); and cathode rays—so it is said—*manifestly* consist of single electrons, which yield 'tracks' in the Wilson cloud chamber.

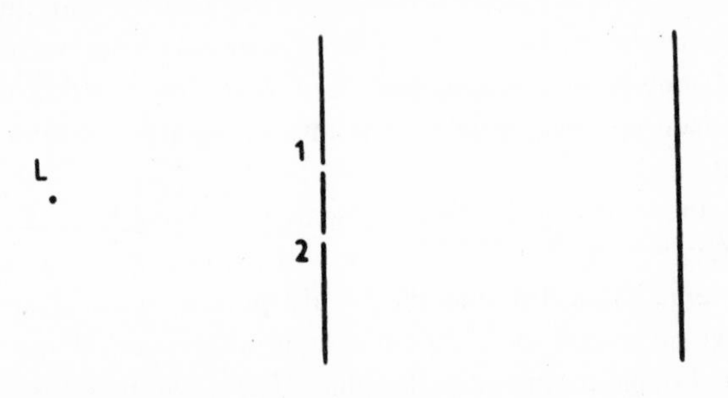

FIGURE 8

(b) There are reasons to assume that light itself also consists of single particles—called photons (from the Greek φῶς = light).

Against this one may argue that nevertheless in *both* cases the concept of waves is unavoidable, if you wish to account for the interference fringes. And one may also argue that the particles are not identifiable objects, they might be regarded as explosion like events within the wave-front—just the events by which the wave-front manifests itself to observation. These events—so one might say—are to a certain extent fortuitous, and that is why there is no strict causal connection between observations.

Let me explain in some detail why the phenomena, both in the case of light and in the case of cathode rays, cannot possibly be understood by the concept of single, individual, *permanently existing* corpuscles. This will also afford an example of what I call the 'gaps' in our description and of what I call the 'lack of individuality' of the particles. For the sake

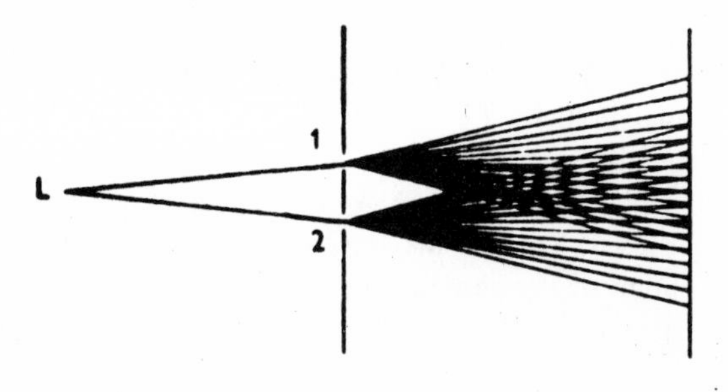

FIGURE 9

of argument we simplify the experimental arrangement to the utmost. We consider a small, almost point-like source which emits corpuscles in all directions, and a screen with two small holes, with shutters, so that we can open first only the one, then only the other, then both. Behind the screen we have a photographic plate which collects the corpuscles that emerge from the openings. After the plate has been developed, it shows, let me assume, the marks of the single corpuscles that have hit it, each rendering a grain of silver-bromide developable, so that it shows as a black speck after developing. (This is very near the truth.)

Now let us first open only one hole. You might expect that after exposing for some time we get a close cluster around one spot. This is not so. Apparently the particles are deflected from their straight path at the

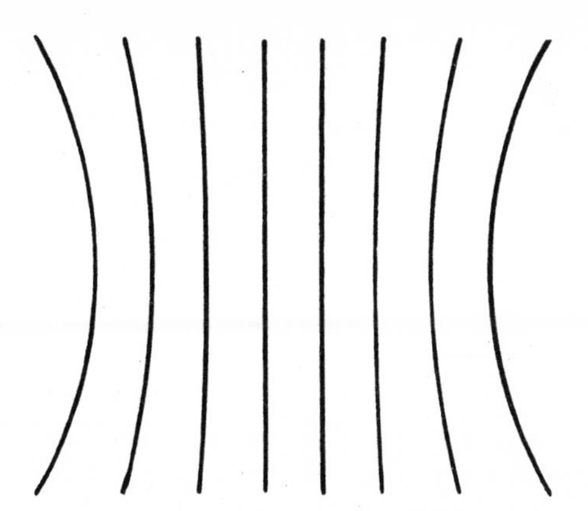

FIGURE 10—The lines indicate the places where there are few or no spots, while midway between any two lines the spots would be most frequent. The two straight lines in the middle are parallel to the slits.

opening. You get a fairly wide spreading of black specks, though they are densest in the middle, becoming rarer at greater angles. If you open the second hole alone, you clearly get a similar pattern, only around a different centre.

Now let us open both holes at the same time and expose the plate just as long as before. What would you expect—if the idea was correct, that single individual particles fly from the source to one of the holes, are deflected there, then continue along another straight line until they are caught by the plate? Clearly you would expect to get the two former patterns superposed. Thus in the region where the two fans overlap, if near a given point of the pattern you had, say, 25 spots per unit area in the first experiment and 16 more in the second, you would expect to find $25 + 16 = 41$ in the third experiment. This is not so. Keeping to these numbers (and *disregarding chance-fluctuations*, for the sake of argument), you may find anything between 81 and only 1 spot, this depending on the precise place on the plate. It is decided by the difference of its distances from the holes. The result is that in the overlapping part we get dark fringes separated by fringes of scarcity.
(N.B. The numbers 1 and 81 are obtained as

$$(\sqrt{25} \pm \sqrt{16})^2 = (5 \pm 4)^2 = 81, 1.)$$

If one wanted to keep up the idea of single individual particles flying continuously and independently either through one or through the other slit one would have to assume something quite ridiculous, namely that in some places on the plate the particles destroy each other to a large extent, while at other places they 'produce offspring.' This is not only ridiculous but can be refuted by experiment. (Making the source extremely weak and exposing for a very long time. This does not change the pattern!) The only other alternative is to assume that a particle flying through the opening No. 1 is influenced also by the opening No. 2, and that in an extremely mysterious fashion.

We must, so it seems, give up the idea of tracing back to the source the history of a particle that manifests itself on the plate by reducing a grain of silver-bromide. *We cannot tell where the particle was before it hit the plate*. We cannot tell through which opening it has come. This is one of the typical gaps in the description of observable events, and very characteristic of the lack of individuality in the particle. We must *think* in terms of spherical waves emitted by the source, parts of each wave-front passing through both openings, and producing our interference pattern on the plate—but this pattern manifests itself to *observation* in the form of single particles.

COMMENTARY ON

SIR ARTHUR STANLEY EDDINGTON

"I BELIEVE there are 15,747,724,136,275,002,577,605,653,961,181, 555,468,044,717,914,527,116,709,366,231,425,076,185,631,031,296 protons in the universe and the same number of electrons." Thus Eddington began one of the Tarner lectures in 1938.[1] It is improbable that even the drowsiest member of his audience failed to respond to this opening sentence. That it tells us much about the universe is doubtful, but it tells us a good deal about Eddington. He was the greatest astronomer of his day, commanding a superb mathematical technique and inclined strongly to philosophical speculation. The inventiveness of his imagery, the clarity and felicitousness of his prose in explaining science to the layman were unsurpassed. He was excruciatingly shy but had the self-confidence and boldness of the mystic in advancing his theories. The rather large number above represents the culmination of years of effort by Eddington to formulate a comprehensive theory of the universe. Multiply the number by 2 and you get the total number of particles: $2 \times 136 \times 2^{256}$. Eddington called this N, the *Cosmical Number*; he regarded it as his supreme achievement. Few scientists today subscribe to his theory, but none depreciates its intellectual grandeur.

Arthur Stanley Eddington was born in 1882 at Kendal, Westmorland, of Quaker parents. His father, headmaster of the local Friends school, died of typhoid fever when Arthur was two years old; his mother was left poor, but managed nevertheless to send him to private schools. It is reported that he was an exceptional child who could do the 24×24 multiplication table at the age of five and was experienced in the use of the telescope before he was ten. A variety of scholarships won in competitive examinations saw Eddington through Owens College, Manchester, and Trinity at Cambridge. In 1906 he was appointed Chief Assistant at the Royal Observatory at Greenwich and in 1907 was elected a Fellow of Trinity. The Plumian chair of astronomy was tendered him when he was only thirty. He held this post until his death, of cancer, in 1944.

Eddington's first major contribution to astronomy was an extension of Schwarzschild's theory of the radiative equilibrium of a star's atmosphere to the star's interior.[2] In a paper published on the subject in 1917 Ed-

[1] Sir Arthur Eddington, *The Philosophy of Physical Science*, Cambridge, 1939, p. 170.

[2] This synopsis of Eddington's scientific researches follows several accounts: J. G. Crowther, *British Scientists of the Twentieth Century*, London, 1952, pp. 140–196;

dington provided a beautiful explanation of the general features of stellar structure and stellar evolution. The heart of the problem to which he addressed himself was the relation between the mass and the brightness of a star. A giant star of low density could be supposed to be extremely bright because its "ether-waves," from X-rays to light-rays, are forced to flow out by the steep temperature gradient. Yet there was a huge discrepancy, which conventional theory could not account for, between the star's expected and observed brightness: it fell short of the expected luminosity by a factor of many millions. Eddington showed that the radiation waves are "hindered and turned back by their adventures with the atoms and electrons" and thus the leakage to the surface of the star is considerably reduced. He showed also that the radiation pressure balances the gravitational forces and supports the enormous weight of the star's upper layers. On the basis of this explanation of stellar structure, he demonstrated a surprising correlation, for all ordinary stars, between masses and luminosities. The more massive the star, the more energy it pours out.[3] The sun's material, he asserted, in spite of being denser than water, really is a perfect gas. Even more incredible is the figure of a mean density of 53,000 for a star like Sirius. This "foretold an Einstein shift of spectral lines," actually confirmed by observation shortly thereafter. In the *Internal Constitution of the Stars* (1926) were summarized the preceding fifteen years of brilliant astronomical investigations.[4]

From the first Eddington was interested in relativity. It appealed to his cosmological bent and he was one of the few quickly to master its mathematical difficulties and to apprehend the full significance of the theory. A copy of one of Einstein's papers on the General Theory was sent to him by the Dutch physicist De Sitter in 1915; for some time it was the only copy available in England. As a Quaker, Eddington was sitting out the war and thus found time to apply himself to the new ideas. His *Report on the Relativity Theory of Gravitation* made for the Physical Society of London in 1918, contains only ninety-one pages "and is a masterpiece of

Sir Harold Spencer Jones, E. A. Milne, E. N. daC. Andrade, Obituaries of Eddington in *Nature*, vol. 154, Dec. 16, 1944; H. C. Plummer in *Obituary Notices of Fellows of the Royal Society*, vol. 5, pp. 113 *et seq.*; Sir Edmund Whittaker, *Eddington's Principle in the Philosophy of Science*, Cambridge (England), 1951.

[3] The energy output is proportional to something between the cube and the fourth power of the mass. Thus, a star only twice the mass of the sun has twelve times its output.

[4] Interesting discussions of Eddington's researches on the radiation equilibrium of gaseous stars, of his scientific controversies with Jeans, and of related matters appear in E. A. Milne, *Sir James Jeans, A Biography*, Cambridge (England), 1952, *passim.* Milne says that Eddington, in calculating the state of equilibrium of a given mass of gas, and the rate of radiation from its surface, performed a first-class piece of scientific work. But he failed to realize exactly the nature of his achievements. What he found was the condition for a star to be gaseous throughout. What he claimed to have found—erroneously, says Milne—was the luminosity of the existing stars and their internal opacity. The dispute must be left to the judgment of experts. I abandon it gladly.

concise and elegant exposition." He "not only restated Einstein's work and De Sitter's expositions, he joyfully took wing in flights of physical and mathematical thought and fancy of his own." [5] In 1919 he was one of the leaders of an expedition to the Isle of Principe in the Gulf of Guiana which, during a solar eclipse, tested and confirmed Einstein's prediction of the bending of light rays by matter. A semipopular account of the theory was presented in his famous book *Space, Time and Gravitation* (1920)—from which an excerpt is given below. In *The Mathematical Theory of Relativity* (1923) he offers both a comprehensive technical analysis and, as his own contribution to the subject, a generalization of Herman Weyl's theory of the electromagnetic and gravitational fields "based on the notion of parallel displacement."

Eddington continued his researches on the structure and composition of the stars, studying such problems as ionization and capture phenomena attending energy interchanges in a diffuse gas, and the pulsations of the Cepheid variables. But in later years he focused his labors on developing the cosmological aspects of relativity theory and on unifying the quantum theory and relativity. What he tried to find was a grand embracing principle derived from a supposed relationship between certain important numbers such as "the radius curvature of the earth, the recession velocity constant of the external galaxies, the number of particles in the universe, and the physical constants such as the ratio of the mass of the proton to that of the electron, the ratio of the gravitational to the electrical force between a proton and an electron, the fine structure constant and the velocity of light." Sir Edmund Whittaker presents an admirably succinct description of the principle which Eddington wished to introduce into the philosophy of science and which is the central theme of his last book, *Fundamental Theory*.[6] One may distinguish between two kinds of assertions in physics: *quantitative* assertions, e.g., "The masses of the electron, the μ meson, the π meson, the τ meson, and the proton, are approximately in the ratios 1, 200, 300, 1000, 1836"; or "The ratio of the electron to the gravitational force between a proton and an electron is 2.2714×10^{39}"; and *qualitative* assertions, e.g., "The velocity of light is independent of the motion of its source," or, "It is impossible to derive mechanical effect from any portion of matter by cooling it below the temperature of the coldest of the surrounding objects." Eddington claimed a relationship between these quantitative and qualitative assertions as follows: "All the quantitative propositions of physics, that is, the exact values of the pure numbers that are constants of science, may be deduced by logical reasoning from qualitative assertions, without making any use of quantitative data derived from observation." [7] A more dramatic statement of the prin-

[5] Crowther, *op. cit.*, p. 161.
[6] Whittaker, *op. cit.*, pp. 1–3.
[7] Whittaker, *op. cit.*, p. 3.

ciple is given by Eddington: "An intelligence unacquainted with our universe but acquainted with the system of thought by which the human mind interprets to itself the content of its sensory experience, should be able to attain all the knowledge of physics that we have attained by experiment. He would not deduce the particular events and objects of our experience but he would deduce the generalizations we have based on them. For example, he would infer the existence and properties of radium but not the dimensions of the earth."

His conclusions, it is agreed, have not yet "carried general conviction." The theory is abstruse and complex; many students are repelled by its philosophical framework. Eddington allowed no compromise. He believed his constants to be absolutely precise, his theory as unexceptionable as the syllogism. In a letter he wrote shortly before his death to the astronomer Herbert Dingle, he rejected the criticism of his procedure as "obscure." Einstein, he said, "was once considered obscure . . . I cannot seriously believe that I ever attain the obscurity that Dirac does. But in the case of Einstein and Dirac people have thought it worth while to penetrate the obscurity. I believe they will understand me all right when they realize they have got to do so—and when it becomes the fashion 'to explain Eddington.' "

I am drawn to Milne's opinion. Some of the unified theory he pronounced "beautiful"; some, "palpable fudge." Altogether it was evidently "the production of a man of genius, not excluding tiresomeness; whether or not it will survive as a great scientific work, it is certainly a notable work of art." [8]

* * * * *

Eddington was a tall, handsome man, pale, with striking, deep-set eyes. He was inward but not unfriendly or unapproachable. He liked good books, detective stories, crossword puzzles, golf, and lonely bicycling jaunts, often conducted at breakneck speed. He was so shy, so slow of speech that it was torture for him to speak extemporaneously in public. A distinguished foreign astronomer, eager for years to meet him, finally got his wish. He said afterwards, "I was never more surprised. He can say 'yes' and he can say 'no' and that is all he can say." But his prepared addresses were unrivaled. Crowther describes him as "the master of the fable, the simile, the paradox, the epigram, the pun." Like Jeans, he knew how to make the listener or reader gape by presenting dramatic contrasts of magnitudes; but his repertory was much broader than Jeans', more poetic, and leavened with humor. The atom consists of a heavy nucleus with a "girdle or crinoline" of electrons. Stellar atoms, stripped of their crinolines by collisions are "nude savages innocent of the class distinctions of our fully arrayed terrestrial atoms." He devised the fable of the

[8] Quoted in Crowther, *op. cit.*, p. 194.

race of flat fish swimming in curved paths around a mound on the ocean floor (a mound they cannot see because they are two-dimensional) to explain space-time curvature; his story of the typing monkeys was an illustration of probability. He remarked that calcium is so prominent in the chromosphere because its atoms, though heavy, have mastered "the art of riding sunbeams"; he characterized reality as "a child which cannot survive without its nurse illusion." As an example of the process of abstraction used in the exact sciences, he cited the problem of an elephant sliding down a grassy hillside. This gay exploit, he complained, is transformed into an exercise involving a *mass* of two tons, an *angle* of 60°, and a coefficient of friction, so that "the poetry fades out of the problem . . . and we are left only with pointer readings." These are a few of the more famous Eddington images; they were an indispensable part of the technique by which he awakened a wide audience to the excellence, rare fancy and revolutionary implications of modern science.[9] Eddington's half-ironic appraisal of the new cosmology—of his own theories no less than the theories of others—is reflected in the lines from *Paradise Lost*, appearing as preface to one of his books:

Perhaps to move
His laughter at their quaint opinions wide
Hereafter, when they come to model heaven
And calculate the stars: how they will wield
The mighty frame: how build, unbuild, contrive
To save appearances.

[9] As I have indicated elsewhere in these pages (see article on Jeans, p. 2278), Eddington was strongly criticized by Susan Stebbing, and by others, for his metaphysics, his anthropomorphic imagery, his personification, and his mysticism. I think he was less of an offender in these matters than Jeans; I agree with Professor Stebbing that his fuzzy statements were not rare and that at their fuzziest they were pretty bad. I do not follow her all the way in her criticism, partly because not all his metaphysics strikes me as gibberish, and partly because even his gibberish can be enthralling.

Seek simplicity, and distrust it. —ALFRED NORTH WHITEHEAD

20 The Constants of Nature

By SIR ARTHUR STANLEY EDDINGTON

THE UNIVERSE AND THE ATOM

See Mystery to Mathematics fly!—Pope, *Dunciad.*

I

I HAVE explained in the previous chapters that theory led us to expect a systematic motion of recession of remote objects, and that by astronomical observation the most remote objects known have been found to be receding rapidly. The weak point in this triumph is that theory gave no indication how large a velocity of recession was to be expected. It is as though an explorer were given instructions to look out for a creature with a trunk; he has brought home an elephant—perhaps a *white elephant.* The conditions would equally well have been satisfied by a fly, with much less annoyance to his next-door neighbour the time-grabbing evolutionist. So there is great argument about it.

I think the only way to remove the cloud of doubt is to supplement the original prediction, and show that physical theory demands not merely a recession but a particular speed of recession. The theory of relativity alone will not give any more information; but we have other resources. I refer to the second great modern development of physics—the quantum theory, or (in its most recent form) wave-mechanics. By combining the two theories we can make the desired theoretical calculation of the speed of recession.

This is a new adventure, and I do not wish to insist on the accuracy or finality of the first attempt. I cannot see how there can be anything seriously wrong with it; but then one never does see these faults until some new circumstance arises or some ingenious person comes along to show us how blind we have been. But there are two kinds of scientific misadventure; we may start off on a false trail altogether, or we may make temporary blunders in following the true path. I am content if in this chapter I can justify my belief that at any rate we are not committing the first error.

According to the argument here developed we can calculate by pure theory what ought to be the speed of recession of the spiral nebulæ. (This is subject to the reservation that the restraining effect of their mutual gravitational attraction is relatively unimportant, a condition which appears to be satisfied in the present state of the universe.) Since certain small factors in the formulæ are at present left in suspense, there is a temporary indefiniteness; but we can say provisionally that the result is between 500 and 1,000 km. per sec. per megaparsec. No astronomical observations of any kind are used in this calculation, all the data being found in the laboratory. Therefore when we turn our telescopes and spectroscopes on the distant nebulæ and find them to be receding at a speed within these limits the confirmation is striking.

The original prediction of de Sitter and Lemaître gave no indication whether the phenomenon would first become perceptible at nebular distances or at distances 10^6 or 10^{60} times greater. We had not the faintest idea how large an effect would appear. By the new investigation, however, the amount is so closely defined that there can be little doubt as to the correspondence of the theoretical and observed effects. Our astronomical explorer cannot be accused of having brought home an elephant in mistake for a fly; and (if I may further complicate the zoological metaphor) even if it is a white elephant it is not a mare's nest.

Any theoretical step requires testing in as many directions as possible. If the theoretical ideas here employed had had only one application, viz. to calculate the recession of the nebulæ, there might be a certain amount of room for "fudging." As a matter of fact the danger of unconscious fudging is greatly exaggerated; there is an artistry in these fundamental equations of physics which one cannot trifle with. But it naturally strengthens our confidence if the same step also leads to the solution of another problem. This happens in the present case, the associated problem being the relation of the proton to the electron and in particular the ratio of their masses. Here a very delicate observational test of the theory is possible.

Thus we are not dealing with an isolated problem, but with a theory which determines at the same time two of the leading constants of physics, viz. the *cosmical constant* and through it the recession of the nebulæ, and the *mass-ratio* of the proton and electron.

I cannot give here the mathematical part of the argument. I want rather to show that all the necessary physical ideas present themselves naturally, and are waiting for the mathematician to express them in symbols and work out the answer. By a preliminary attempt at the latter task we gain fair assurance that no serious difficulty is likely to arise.

II

We have been contemplating the system of the galaxies—phenomena on the grandest scale yet imagined. I want now to turn to the other end of the scale and look into the interior of an atom.

The connecting link is the cosmical constant. Hitherto we have encountered it as the source of a scattering force, swelling the universe and driving the nebulæ far and wide. In the atom we shall find it in a different capacity, regulating the scale of construction of the system of satellite electrons. I believe that this wedding of great and small is the key to the understanding of the behaviour of electrons and protons.

The cosmical constant is equal in value to $1/R_e^2$ or to $3/R_s^2$, so that it is really a measure of world-curvature; and in place of it we can consider the initial radius of the universe R_e, or better the steady radius of curvature of empty space R_s. In the present chapter the unqualified phrase "radius of curvature" or the symbol R will be understood to refer to R_s. Being the radius *in vacuo* it has the same kind of pre-eminence in physical equations that the velociy of light *in vacuo* has. I will first explain why the radius of curvature is expected to play an essential part in the theory of the atom.

Length is relative. That is one of the principles of Einstein's theory that has now become a commonplace of physics. But it was a far from elementary kind of relativity that Einstein considered; according to him length is relative to a frame of reference moving with the observer, so that as reckoned by an observer moving with one star or planet it is not precisely equal to the length reckoned by an observer moving with another star. But besides this there is a much more obvious way in which length is relative. Reckoning of length always implies comparison with a standard of length, so that length is relative to a comparison standard. It is only the ratio of extensions that enters into experience. Suppose that every length in the universe were doubled; nothing in our experience would be altered. We cannot even attach a meaning to the supposed change. It is an empty form of words—as though an international conference should decree that the pound should henceforth be reckoned as two pounds, the dollar two dollars, the mark two marks, and so on.

In *Gulliver's Travels* the Lilliputians were about six inches high, their tallest trees about seven feet, their cattle, houses, cities in corresponding proportion. In Brobdingnag the folk appeared as tall as an ordinary spire-steeple; the cat seemed about three times larger than an ox; the corn grew forty feet high. Intrinsically Lilliput and Brobdingnag were just the same; that indeed was the principle on which Swift worked out his story. It needed an intruding Gulliver—an extraneous standard of length—to create a difference.

It is commonly stated in physics that all hydrogen atoms in their normal state have the same size, or the same spread of electric charge. But what do we mean by their having the same size? Or to put the question the other way round—What would it mean if we said that two normal hydrogen atoms were of different sizes, similarly constructed but on different scales? That would be Lilliput and Brobdingnag over again; to give meaning to the difference we need a Gulliver.

The Gulliver of physics is generally supposed to be a certain bar of metal called the International Metre. But he is not much of a traveller:

I do not think he has ever been away from Paris. We have, as it were, our Gulliver, but have left out his travels; and the travels are, as Prof. Weyl was the first to show, an essential part of the story.

It is evident that the metre bar in Paris is not the real Gulliver. It is one of those practical devices which serve a useful purpose, but dim the clear light of theoretical understanding. The real Gulliver must be ubiquitous. So I adopt the principle that when we come across the metre (or constants based on the metre) in the present fundamental equations of physics, our aim must be to eject it and to substitute the natural ubiquitous standard. The equations put into terms of the real standard will then reveal how they have arisen.

It is not difficult to find the ubiquitous standard. As a matter of fact Einstein told us what it was when he gave us the law of gravitation $G_{\mu\nu} = \lambda g_{\mu\nu}$. Some years ago I showed that this law could be stated in the form, "What we call a metre at any place and in any direction is a constant fraction ($\sqrt{\tfrac{1}{3}\lambda}$) of the radius curvature of space-time for that place and direction." In other words the metre is just a practically convenient sub-multiple of the radius of curvature at the place considered; so that measurement in terms of the metre is equivalent to measurement in terms of radius of curvature.

The radius of world-curvature is the real Gulliver. It is ubiquitous. Everywhere the radius of curvature exists as a comparison standard indicating, if they exist, such differences as Gulliver found between Lilliput and Brobdingnag. If we like we can use its sub-multiple the metre, remembering, however, that the metre is ubiquitous only in its capacity as a sub-multiple of the radius. We should, if possible, try to forget that in certain localities we have crystallised this metre into metallic bars for practical convenience.

We can now give a direct meaning to the statement that two normal hydrogen atoms in any part of the universe have the same size. We mean that the extent of each of them is the same fraction of the radius of curvature of space-time at the place where it lies. The atom here is a certain fraction of the radius here, and the atom on Sirius is the same fraction of the radius at Sirius. Whether the length of the radius here is absolutely the same as that of the radius at Sirius does not arise; and indeed I believe that such a comparison would be without meaning. We say that it is always the same number of metres; but we mean no more by that than when we say that the metre is always the same number of centimetres.

Thus it appears that in all our measures we are really comparing lengths and distances with the radius of world curvature at the spot. Provided that the law of gravitation is accepted, this is not a hypothesis; *it is the translation of the law from symbols into words*. It is not merely a suggestion for an ideal way of measuring lengths; it reveals the basis of the system which we have actually adopted, and to which the mechanical and optical laws assumed in practical measurements and triangulations are referred.

It is not difficult to see how it happens that our practical standard (the metre bar) is a crystallisation of the ideal standard (the radius of curvature, or a sub-unit thereof). Since the radius of curvature is the unit referred to in our fundamental physical equations, anything whose extension is determined by constant physical equations will have a constant length in terms of that unit. Thus the physical theory that provides that the normal hydrogen atom shall have the same size in terms of the radius of curvature wherever it may be, will also provide that a solid bar in a specified state shall have the same size in terms of the radius of curvature wherever it may be. The fact that the atom has a constant size in terms of the practical metre is a case of "things which are in a constant ratio to the same thing are in a constant ratio to one another." [1]

The simplification obtained by using the actual radius of curvature as unit of length (instead of using a sub-unit) is that all lengths will then become angles in our world-picture. The measure of any length will be the "tilt of space" in passing from one extremity to the other. It is true that these angles are not in actual space but in fictitious dimensions added for the purpose of obtaining a picture; but the justification of the picture is that it illustrates the analytical relations, and these angles will behave analogously to spatial angles in the mathematical equations.

To sum up this first stage of our inquiry: If in the most fundamental equations of physics we adopt the radius of curvature R_s as unit instead of the present arbitrary units, we shall have at least made the first step towards reducing them to a simpler form. We know that many equations are simplified when velocities are expressed in terms of the velocity of light *in vacuo*; we expect a corresponding simplification when lengths are expressed in terms of the radius of world-curvature *in vacuo*. When the equation is in this way freed from irrelevant complications it should be easier to detect its true significance. We cannot make this change of unit so long as the ratio of R_s to our ordinary unit is unknown; but observation of the spiral nebulæ has provided us with what we provisionally assume to be an approximate value of R_s, so that it is now possible to go ahead with our plan.

III[2]

In elementary geometry we generally think of space as consisting of infinitely many points. We approach nearer to the physical meaning of

[1] For a fuller explanation see *The Nature of the Physical World*, Chapter VII.

[2] This section is mainly an additional commentary on the principles explained under II. If found too difficult, it can be omitted. The main argument is resumed in IV.

space if we think of it as a network of distances. But this does not go far enough, for we have seen that it is only the ratios of distances which enter into physical experience. In order that a space may correspond exactly to physical actuality it must be capable of being built up out of ratios of distances.

The pure geometer is not bound by such considerations, and he freely invents spaces consisting only of points without distances, or spaces built up out of absolute distances. In adapting his work for application to space in the physical universe, we have to select that part of it which conforms to the above requirement. For that reason we must reject his first offer—flat space. Flat space cannot be constructed without absolute lengths, or at least without a conception of *a priori* comparability of lengths at a distance which can scarcely be distinguished from the conception of absolute length.[3]

Flat space, being featureless, does not contain within itself the requirement for reckoning length and size, viz. a ubiquitous comparison standard. But what is the use of a space which does not fulfil the functions of space, namely to constitute a scheme of reference for all those physical relations—length, distance, size—which are counted as spatial? Since it does not constitute a frame of reference for length, the name "space" is a misnomer. Whatever definition the pure geometer may adopt, the physicist must *define* space as something characterised at every point by an intrinsic magnitude which can be used as a standard for reckoning the size of objects placed there.

No question can arise as to whether the comparison unit for reckoning of lengths and distances is a magnitude intrinsic in space, or in some other physical quality of the universe, or is an absolute standard outside the universe. For whatever embodies this comparison unit is *ipso facto* the space of physics. Physical space therefore cannot be featureless. As a matter of geometrical terminology features of space are described as curvatures (including hypercurvatures); as already explained, no meta-

[3] In pre-relativity theory, and in the original form of Einstein's theory, "comparison of lengths at a distance" was assumed to be axiomatic; that is to say, there was a real difference of height between the Lilliputian and the Brobdingnagian irrespective of any physical connection between the islands. The fact that they were in the same universe—phenomena accessible to the same consciousness—had nothing to do with the comparison. Such a conception of unlimited comparability is scarcely distinguishable from the conception of absolute length. In a geometry based on this axiom, space only does half its proper work; the purpose of a field-representation of the relationships of objects is frustrated, if we admit that the most conspicuous spatial relationship, ratio of size, exists *a priori* and is not analysable by field-theory in the way that other relationships are. Weyl's theory rejected the axiom of comparability at a distance, and it was at first thought that such comparability could not exist in his scheme. But both in Weyl's theory and in the author's extension of it (affine field-theory) it is possible to compare lengths at a distance, not as an extra-geometrical *a priori* conception, but by the aid of the field which supplies the ubiquitous standard necessary.

physical implication of actual bending in new dimensions is intended. We have therefore no option but to look for the natural standard of length among the radii of curvature or hypercurvature of space-time.

To the pure geometer the radius of curvature is an incidental characteristic—like the grin of the Cheshire cat. To the physicist it is an indispensable characteristic. It would be going too far to say that to the physicist the cat is merely incidental to the grin. Physics is concerned with interrelatedness such as the interrelatedness of cats and grins. In this case the "cat without a grin" and the "grin without a cat" are equally set aside as purely mathematical phantasies.

When once it is admitted that there exists everywhere a radius of curvature ready to serve as comparison standard, and that spatial distances are directly or indirectly expressed in terms of this standard, the law of gravitation ($G_{\mu\nu} = \lambda g_{\mu\nu}$) follows without further assumption; and accordingly the existence of the cosmical constant λ with the corresponding force of cosmical repulsion is established. Being in this way based on a fundamental necessity of physical space,[4] the position of the cosmical constant seems to me impregnable; and if ever the theory of relativity falls into disrepute the cosmical constant will be the last stronghold to collapse. *To drop the cosmical constant would knock the bottom out of space.*

It would be a truism to say that space is not an ultimate conception; for in the relativity view of physics every conception is an intermediary between other conceptions. As in a closed universe where the galaxies form a system having no centre and no outside, so the conceptions of physics link into a system with no boundary; our goal is not to reach an ultimate conception but to complete the full circle of relationship. We have concluded that the ubiquitous comparison standard must be a characteristic of space, because it is the function of space to afford such a standard, but we can inquire further how space and the standard contained in it themselves originate.

The space in which the atom is pictured as having position and size is an intermediary conception used to relate the atom to the "rest of the universe." It is therefore no contradiction if we say sometimes that the extension of the atom is controlled by the curvature of space, and sometimes that it is controlled by forces of interaction proceeding from the rest of the universe. It must be remembered that we are only aware of an atom or any other object in so far as it interacts with the rest of the universe,

[4] The requirement is that the comparison standard shall be a magnitude intrinsic in the space—for whatever the standard is intrinsic in, that *ipso facto* is space. Space can have other characteristic magnitudes besides the radius of curvature—for example, magnitudes measuring various kinds of hypercurvature. Although the suggestion seems far-fetched, it is, I suppose, conceivable that one of these might be substituted. That would give a different law of gravitation; but there is still a cosmical constant, depending on the ratio of the metre to the natural comparison standard. In fact the cosmical term $\lambda g_{\mu\nu}$ remains unchanged; it is $G_{\mu\nu}$ which is modified.

and thereby gives rise to phenomena which ultimately reach our senses. The position and dimensions which we attribute to an atom are symbols associated with interaction effects; for there is no meaning in saying that an atom is at A rather than at B unless it makes some difference to something that it is at A not B. In considering this interaction it is not necessary to deal separately with every particle and every element of energy in the rest of the universe; if it were, progress in physics would be impracticable. For the most part it is sufficient to take averages. The multitudinous particles of the universe admit of an almost uncountable variety of change of configuration; in considering their interaction with the atom we need preserve only a few broad types of average change. The "rest of the universe" is thus idealised into something possessing only a few types of variation or degrees of freedom. This is illustrated in electrical theory where the interactions of myriads of electrical particles are replaced by the interaction of an *electric field* which is specified uniquely by six numbers. In the same way another part of the interaction of the rest of the universe on the atom is idealised into interaction of a *metrical field*, or—to give it its usual name—*space*. The few broad types of variation which are not smoothed out by averaging are retained in the curvatures of space.

We must distinguish in conception between space which for certain purposes replaces the rest of the universe and space which is occupied by the rest of the universe, although the two spaces ultimately become identical. The distinction is easier if we use the term "metrical field" instead of "space"; for (by analogy with electrical fields) we recognise that a field has a dual relation to matter, viz. it is produced by matter and it acts on matter.

The remainder of our task is to try to discover the details of this idealisation of the "rest of the universe" into a metrical field containing a radius of curvature.

IV

One of the most fundamental equations of physics is the wave-equation for a hydrogen atom, that is to say for a proton and electron. The equation determines the size of the atom or the spread of its electric charge. Clearly the ubiquitous standard of length R must come into this equation.

Now R does not appear in the equation as ordinarily written. That is because the equation has been reached through experiment, and is expressed in terms of quantities such as the charge of an electron, Planck's constant, the velocity of light, etc. The radius R though present is in disguise. We must try to penetrate the disguise.

At first sight a formidable obstacle appears. The radius of the hydrogen atom is of order 10^{-8} cm., and the natural unit R is of order 10^{27} cm.; thus the radius of the hydrogen atom in terms of the natural unit is of order 10^{-35}. Our idea was that by introducing the natural unit we should obtain a simplified equation; but can it be a very simple equation if its solution is 10^{-35}? Clearly it must contain an enormous numerical coefficient in one or more of its terms. If the equation is really in its most elementary form, every coefficient ought to have some simple meaning—some obviously appropriate reason for being what it is. We should not be surprised to see the 4π type of coefficient, which has a simple geometrical meaning; or a coefficient equal to the number of dimensions or degrees of freedom concerned in the problem, which arises from summing together a number of symmetrical terms. But what simple meaning can be attached to an enormously large number like 10^{35}?

I can think of only one large number which is in any way relevant to the problem, viz. the number of particles (electrons or protons) in the universe. Indeed there seems to be no other way of putting a large number into the structure of the physical world. I refer, of course, to pure numbers, not to the kind of number that we arbitrarily introduce by our centimetre-gram-second system of reckoning. We shall find presently that there are direct reasons for assuming that the number of particles in the universe N will occur in the coefficients of the wave-equation; but even without these reasons the enormous magnitude of the coefficients would be a sufficient indication of the occurrence of N.

Another aspect of the same large ratio appears when we compare the electric force between a proton and electron with the gravitational force between them. According to classical theory the ratio is $2 \cdot 3.10^{39}$. I have long thought that this must be related to the number of electrons and protons in the universe [5] and I expect that the same view has been entertained by others. Since N is about 10^{79} the above ratio is of the order $\sqrt{N}$.

The direct reason for the appearance of N is that N is actually an effective number of degrees of freedom of the universe. On classical theory the number would be greater than N, because each of the N particles would have several degrees of freedom; but there is a well-known exclusion principle which limits the freedom of a particle by forbidding it to go into an orbit already occupied by another particle. Wave-mechanics therefore approaches the problem from the other end and defines N as the number of independent wave-systems existing in the universe, and therefore equal to the number of separate constituents of the energy of the universe. It is quite possible that the number N approached in this way will be found, not to be arbitrary, but to have some definite theoretical foundation; but that is pure conjecture, and for the

[5] *Mathematical Theory of Relativity*, p. 167.

present we regard it as the one arbitrary element in the design of the actual universe.

Our atom is situated in and interacting with a universe containing N degrees of freedom. We idealise and simplify the problem by picturing it as situated in and interacting with a space (or metrical field) of radius of curvature R possessing a comparatively small number of degrees of freedom, say n. In this simplified form "the rest of the universe" comes into the equation of the hydrogen atom through the quantity R. I think we must expect that the numbers N and n will also occur in the equation, as a memorandum of the substitution of a space with n degrees of freedom for a universe with N degrees of freedom. For four-dimensional space-time the number n is found to be 10. We shall lose sight of it for the time being, but it will turn up later.

Having decided that N and R will enter into the coefficients of the equation of the hydrogen atom, we next inquire in what kind of association they will occur. The factor N, of course, arises from adding together equal contributions from each of the particles or wave-systems; the question is, What is the nature of the contributions to be added and how do they contain R? I do not profess to have achieved the necessary physical insight to settle the question. It will no doubt be much more satisfactory when we have a picture in which we can, as it were, see these entities adding themselves together, just as we can see a hundred centimetres adding themselves to form a metre. But when this kind of insight fails, we are not without a guide. As conduct may be guided by ethics or by "good form," so this kind of investigation can be guided by physical insight or by analytical form. Both wave-mechanics and relativity theory are *very strict on good form*. Only certain kinds of entities are allowed to be added together. To add anything else would be a solecism. "It isn't done."

In relativity theory the only things that are additive are action-invariants.[6] The action-invariant containing R is the Gaussian curvature, which is proportional to $1/R^2$. In quantum theory the entities which may be added are the squares of momenta, or as they are written symbolically $\partial^2/\partial x^2$. To construct a quantity of the same dimensions out of R we must take $1/R^2$.[7] I take it therefore that the entities to be added are, or are proportional to $1/R^2$; so that the required combination is N/R^2.

This gives us what may be called an "adjusted natural standard of length," viz. $R/\sqrt{N}$. By using $R/\sqrt{N}$ instead of R as our unit we absorb the factor N, so that it will not trouble us any more. The length of the ad-

[6] Other tensors may only be added if they are at the same point of space—a condition which is obviously not fulfilled here.

[7] The guidance of quantum theory is less obvious than that of relativity theory because the former commonly adopts a mixed system of units (dynamical and geometrical). Relativity theory being purely geometrical avoids the complication.

justed standard is about 3.10^{-13} cm., so that it is not unsuitable for dealing with phenomena of electrons.

Now we can go back to our problem, which was to discover how the natural standard of length is disguised in the familiar wave-equation. But this time we look for the adjusted standard $R/\sqrt{N}$ instead of the original standard R.

I think I have identified the adjusted standard in the wave-equation, disguised as the expression e^2/mc^2. Here e is the charge of an electron or proton, m the mass of an electron and c the velocity of light. This expression is well known to be of the dimensions of a length; in fact $\frac{2}{3}e^2/mc^2$ used to be called the "radius of an electron" in the days when the electron was conceived more substantially than it is now. The identification accordingly gives the equation

$$\frac{R}{\sqrt{N}} = \frac{e^2}{mc^2}.$$

I cannot enter here into the jurisdiction of this identification, which would lead deeply into the principles of quantum theory. But I may mention that the identification is a very simple one. The expression e^2/mc^2, or rather its reciprocal, stands rather disconsolately by itself in the wave-equation, forming a separate term. Investigators, who are busy transforming, explaining, theorising on the other terms, leave it alone; it has just been accepted as ballast. It calls out for identification.

It may be asked, Is not a straight identification too simple? Granting the identification in principle will there not be a numerical factor—say ½, or 2π, or perhaps something more complicated—the type of factor which usually appears when we reach the same entity by different routes? Perhaps there is; but at present the simple identification looks to me to be correct.[8] I should add, however, that I am uncertain whether in this formula N is to be taken as the number of electrons or the number of electrons *and* protons, so that a factor $\sqrt{2}$ is left in suspense. For definiteness I here take N to be the number of electrons only; the number of protons must be approximately, and probably exactly, the same.

By the relativity theory of the expanding universe, we have

$$\frac{N}{R} = \frac{\pi}{2\sqrt{3}}\frac{c^2}{Gm_p},$$

where m_p is the mass of a proton.*

Thus by relativity theory we find N/R, and by wave-mechanics we find

[8] Except that there are certain corrections amounting altogether to less than 1 per cent. which are explained under V. The mathematical arguments on which the identification is based are given in *Proceedings of the Royal Society*, vol. cxxxiii A, p. 605, and *Monthly Notices of the R.A.S.*, vol. xcii, p. 3.

* This follows from formulae given earlier by Eddington. ED.

$\sqrt{N}/R$. Combining the two results we find N and R separately. The resulting value of N is about 10^{79}. From R the limiting speed of recession of the galaxies c/R is found immediately. All the constants involved have been measured in the laboratory. The agreement of the result obtained in this way with the observed recession of the nebulæ has already been described (pp. 1074–1075).

The following summary of the theory is due to Prof. Dingle:

> He thought he saw electrons swift
> Their charge and mass combine.
> He looked again and saw it was
> The cosmic sounding line.
> The population then, said he,
> Must be 10^{79}.

V

Having, as we think, detected the adjusted natural standard in the terms of the wave-equation, we have next to inquire how our result affects the theory of protons and electrons. For in identifying the standard with e^2/mc^2 we have taken a step which links the universe to the atom; and we ought to verify the observational consequences not only in the astronomical universe but also within the atom.

In wave-mechanics the momentum of a particle is usually stated to be

$$\frac{ih}{2\pi}\frac{\partial}{\partial\chi}.$$

The factor $h/2\pi$ is an unnecessary complication due to our haphazard choice of units of length and mass. We shall adopt instead a natural unit of mass, which is related to the unit of length in such a way that the momentum is simply $i\partial/\partial\chi$. The meaning of i (literally the square root of -1) in an equation is that the two sides of the equation represent waves which, though equal in amplitude, are a quarter-period different in phase. When the mass of an electron is expressed in terms of this natural unit we shall denote by m_e. Making the change of unit, the identification on p. 1084 becomes

$$\frac{hc}{2\pi e^2}\,\mathrm{m}_e = \frac{\sqrt{N}}{R} \qquad \ldots\ldots(\mathrm{A}).$$

We have taken the opportunity to turn both sides upside down, since that is the way they actually appear in the wave-equation.

The coefficient $hc/2\pi e^2$, which is sometimes called the *fine-structure constant*, is a pure number; and it is well known that its value is close to 137. For my own part I think that its value is exactly 137, that being the number of degrees of freedom associated with the wave-function for a pair of charges. There has been much discussion whether the true value is 137·0 or 137·3; both values claim to be derived from observation. The latter, called the "spectroscopic value," is preferred by many physicists. It is, however, misleading to call these determinations *observational values*, for the observations are only a substratum; the spectroscopic value in particular is based on a rather complex theory and is certainly not to be treated as a "hard fact" of observation.

Although I believe $hc/2\pi e^2$ to be 137, I shall take the actual coefficient of m_e to be 136. This means that I slightly amend the original identification by inserting a factor 136/137. The reason for this change is that one of the 137 degrees of freedom, viz. that corresponding to radial displacement, occurs in some problems but not in others, and this appears to be a problem in which it will not occur. The fine-structure constant is introduced in the problem of interaction of two electric charges, and there the 137 degrees of freedom are all in play; a change of distance between the two charges is recognisable because a comparison standard for distances is furnished by the radius of curvature R. But now we are considering a formula for the mass of an electron, which arises from its interaction with the "rest of the universe." The N particles of the universe have been virtually simplified down to one particle by introducing the adjusting factor $\sqrt{N}$, so that the problem is not dissimilar to that of two interacting particles; but there is no longer an extraneous comparison standard of length. Tracing the analogy between the two problems, we find that the analogue of change of distance between the two electrons would be change of the radius of curvature of space. But by its very nature R cannot vary, since it is the standard unit of distance. There is therefore no analogue to the 137th degree of freedom; and we conclude that our first identification, which did not enter into such minutiæ, ought to be amended so as to show the correct number of degrees of freedom.

One might hesitate to introduce so odd-looking a factor as 136/137 were it not that we know of another case in which the radial degree of freedom is inhibited, and there the factor has been verified by observation. This occurs when a proton enters into the almost rigid helium nucleus. Its mass or energy is found to be reduced in a ratio which is very nearly 136/137; the reduction is called the *packing-fraction*. The disappearance of a degree of freedom is essentially the same in the helium nucleus and in the metrical field; the former cannot expand radially because it is rigid, the latter cannot expand because its radius is the standard of length.

Thus at present our result stands

$$136 m_e = \frac{\sqrt{N}}{R} \qquad \ldots\ldots(B).$$

But to this there is a serious objection. The result shows an unfair discrimination in favour of the electron, the proton not being mentioned. The proton is presumably as fundamental as the electron. But what can we put in place of $\sqrt{N}/R$ which would give an equally fundamental equation for the mass m_p of a proton?

With an electron and proton calling out for equal treatment the only way to satisfy their claims impartially is to make the fundamental equation a quadratic, so that there is one root for each. We do not want to alter the part we have already got, after taking so much trouble to justify it bit by bit; so we assume that

$$136m - \sqrt{N}/R = 0 \qquad \ldots\ldots(C).$$

gives correctly the last two terms of the equation, but there is a term in m^2 to come on at the beginning.

It is well known that we can learn something about the roots of a quadratic equation, even if only the last two terms are given. The ratio of the last two coefficients is the sum of the roots divided by the product of the roots. Since the equation is to have roots m_e and m_p, we must have

$$\frac{m_e + m_p}{m_e m_p} = \frac{136R}{\sqrt{N}}$$

or

$$\frac{136 m_e m_p}{m_p + m_e} = \frac{\sqrt{N}}{R} \qquad \ldots\ldots(D).$$

This is another change in the identification equation; but this time it is a very small change numerically. Comparing (D) with (B) we see that a factor $m_p \div (m_p + m_e)$ has been inserted. We know that m_p is about 1,847 times m_e, so that the factor is 1,847/1,848 or ·99946. Numerically the change is insignificant; but the proton no longer has any cause of complaint, for proton and electron receive perfectly impartial treatment in (D).

The next step is to complete the quadratic equation of which the last two terms are given in (C). Since we have finished with the problem of the identification of the adjusted standard (our final equation giving it in terms of known experimental quantities being (D)) we may as well now adopt it as our unit of length. As already explained this choice of unit ought to reduce the equations to their simplest possible form. This means that $R/\sqrt{N}$ can now be taken as unity. The two terms given in (C) are therefore $136m - 1 = 0$, and the completed quadratic is

$$?\, m^2 - 136m + 1 = 0.$$

What number must we put in place of the query? You may remember that there was a number $n = 10$, which we promised to bring into the wave-equation [9] sometime. Here is our chance. We take the equation to be

$$10\text{m}^2 - 136\text{m} + 1 = 0. \qquad \ldots\ldots(\text{E}).$$

For reference we write down the same equation, reintroducing the centimetre as the unit of length. It is

$$10\text{m}^2 - 136\text{m}\frac{\sqrt{N}}{R} + \frac{N}{R^2} = 0 \qquad \ldots\ldots(\text{F}).$$

You see that the number $n = 10$ occurs in the first term as a counterweight to N in the last term, which is evidently their proper relation.

Although we are not yet able to give a clear-cut theory of this equation, the argument is farther advanced than might appear from this superficial sketch. I think it will be agreed that, since the coefficient 136 represents number of degrees of freedom, it is extremely probable that the remaining coefficient in (E) will also represent number of degrees of freedom. Presumably each degree of freedom possesses a concealed energy or cyclic momentum similar to that provided by the ignoration of coordinates in ordinary dynamics. Approaching the problem from this dynamical point of view one arrives almost immediately at the term 10m^2, but there is more difficulty in seeing the reason for the term 136m.

Thus, broadly speaking, we have two lines of approach. They have not yet met and coalesced, as they ultimately must do. But since one line of approach gives the linear term and the other the squared term, apparently quite definitely, the progress seems to be already sufficient to give the correct equation.

We find the ratio of the two roots m_p, m_e by solving equation (E) or (F). The result is 1,847·60. The observed value [10] of the mass-ratio of the proton and electron is stated to be 1,847·0 with a probable error of half a unit. Thus the agreement is complete.

The solution of equation (F) gives

$$\left.\begin{aligned} 135\cdot 9264\text{m}_e &= \sqrt{N}/R \\ 0\cdot 073569\text{m}_p &= \sqrt{N}/R \end{aligned}\right\} \qquad \ldots\ldots(\text{G}).$$

This is another form of our final identification of the adjusted standard $R/\sqrt{N}$. The preliminary identification was $137\text{m}_e = \sqrt{N}/R$; so that there

[9] The wave-equation is formed by replacing the mass m by a differential operator.

[10] There is also a spectroscopic value about 10 units lower, obtained by adopting 137·3 instead of 137 for the fine-structure constant; but this is irrelevant. We are already committed to the value 137 at an earlier stage, so we *do not want* our computation to agree with a result which would only be true if the value were not 137. To put it another way—Naturally our theory cannot agree with both deflection values and spectroscopic values, since these differ; but it ought to agree consistently with one set, and not sometimes with one and sometimes with the other.

has not been any change worth considering in the numerical magnitude of $\sqrt{N}/R$, derived for the purpose of predicting the speed of recession of the nebulæ. The formulæ will work either way. Normally we apply them to calculate the astronomical results, using m_e or m_p determined by physicists; but we can also use them to give an astronomical method of measuring the mass of an electron or proton.

To measure the mass of an electron, a suitable procedure is to make astronomical observations of the distances and velocities of spiral nebulæ! The result, corrected if necessary for the mutual attraction of the galaxies is, let us say, 600 km. per sec. per megaparsec. This is c/R; and since the velocity of light c is 300,000 km. per sec., we have $R = 500$ megaparsecs, which is equal to $1 \cdot 54.10^{27}$ cm. The remaining steps, which involve a little algebraic handling of the equations, need not be described in detail; as soon as R is known, they become soluble and we can find N, and hence $\sqrt{N}/R$. The masses of the electron and proton are then given by equations (G). They are there expressed in terms of the natural unit of mass; we can convert them into grams if we know Planck's constant h. I am afraid that the accuracy attainable by this method would not satisfy the modern physicist, but we shall not be out by more than a factor 2 or thereabouts.

Perhaps you will object that this is not really *measuring* the mass of an electron; even supposing it to be right, it is a highly circuitous inference. But do you suppose that a physicist puts an electron in the scales and weighs it? If you will read an account of how the spectroscopic value of the mass has been determined, you will not think my method unduly complicated. But, of course, I do not seriously put it forward in rivalry; I only want to make vivid the wide interrelatedness of things.

We can show that the two roots of the quadratic represent electric charges of opposite sign. To test this an electric field must be introduced into the problem. Following Dirac's theory, this is done by adding to the equation a constant term (i.e. a term not involving the differential operator $m = id/ds$) depending on the electrical potential. Since this changes only the third term of the quadratic, the sum of the roots $m_e + m_p$ is unchanged. In other words, the mass or energy added by the field to m_e is equal and opposite to the mass or energy added by the field to m_p. But that is the definition of equal and opposite charges—in the same electric field they have equal and opposite potential energy.

Our conclusion that the fundamental wave-equation is really a quadratic has recently received unexpected support by the discovery of the neutron. It has been supposed that the wave-equation for two charges (a proton and electron [11]) was a linear equation first given by Dirac. The

[11] Note that our quadratic equation (E) is for a proton *or* electron—not for two charges.

complete set of solutions was found to represent (approximately, if not exactly) a hydrogen atom in its various possible states. But from recent experiments it has been discovered that there exists another state or group of states in which the proton and electron are much closer together and form a very minute kind of atom. This is called a neutron. Clearly the present wave-equation for a proton and electron cannot be correct, since its solutions do not give the neutron states. Just as for one charge we require a quadratic wave-equation whose two sets of solutions correspond to electrons and protons, so for two charges we require a quadratic wave-equation whose two sets of solutions correspond to hydrogen atoms and neutrons. I expect that this continues in more complicated systems, the two solutions then corresponding to extra-nuclear and nuclear binding of the charges.

The support to the present theory is twofold. Firstly it indicates that the theory of two charges will come into line with our theory of one charge as regards the general form of the equation. Secondly, since the "spectroscopic values" of the physical constants are based on an incomplete theory of two charges, we should not attach overmuch importance to a slight discrepancy between them and the values found in our theory.

It is perhaps still more important that we can see other problems ahead. Thus it will be necessary to investigate the theory of the additional term in the equation for the hydrogen atom and neutron, and to show that the extra solutions agree with the observed properties of the neutron. This will give further opportunities of testing, and if necessary revising, the present conclusions.

VI

To those whose interest in modern science is directed chiefly to the philosophical implications, the theory of the expanding universe does not, I think, bring any particularly new revelation. Except for one lapse, I have avoided questions savouring of philosophy, I have rather taken it for granted that the reader's attention, like my own, is fixed on the strictly scientific progress of the inquiry, and that he will suspend all questions as to how the physical scheme, which is here being developed, can be made to fit in with the general outlook of life and consciousness. It would be unfortunate to prejudice the inquiry by dragging in such questions prematurely.

We may perhaps emerge with the uneasy feeling that we no longer have vast domains of space and vast periods of time to dispose of. But the complaint has often been made that astronomical measures are too vast to conceive; and if we have never been able to conceive them, it scarcely affects our general outlook to have a few 0's lopped off here and there. In fact I consider the man who is dissatisfied with a universe containing ten thousand million million million stars rather grasping. That is, if he wants them merely for comfortable philosophic contemplation; in connection with scientific investigations the cuts may, of course, be serious, as we have seen in the case of the time-scale.

The new theory contains no obvious suggestion that the world will come to an end sooner than we had been expecting. The cosmical dispersal ignores the smaller scale aggregations like our galaxy. We anticipate that there will ultimately be a complete running-down of the universe by the slow degradation of energy into unavailable form; but that far distant day is not brought noticeably nearer by the existence of cosmical repulsion.

It would seem that the expansion of the universe is another one-day process parallel with the thermodynamical running-down. One cannot help thinking that the two processes are intimately connected; but, if so, the connection has not yet been found.

The position with regard to the thermodynamical running-down of the universe has not materially altered since I discussed it four years ago.[12] The impression has got abroad that the conclusions have been shaken by recent work on cosmic rays. That would be impossible, so far as I am concerned; for the theory of cosmic rays that is being urged in this connection happens to be the one that I was advocating at the time of writing, viz. that the cosmic rays give evidence of the building up of higher elements out of hydrogen in distant regions occupied by diffuse matter.[13] I am not at all sure that the more recent evidence should be interpreted as favourable to it; but if it is favourable, as Dr. Millikan maintains, I have the less reason to change my views. The coming together of electric particles to form a complex atom, and the consequent dispersal of some of the energy in cosmic rays, is clearly a step in the same direction as other energy-dissipating processes—for example, the coming together of nebulous matter to form a star, and the consequent dispersal of energy as radiant heat. It is one more contributor to the general running-down towards an ultimate state of thermodynamic equilibrium. Millikan has sometimes called the atom-building process a "winding-up" of the universe; but "up" and "down" are relative terms, and a transformation of axes may be needed in comparing his descriptions with mine.

It may be desirable to remind the philosophical reader of the reason why the scientist indulges in these extrapolations of our present imperfect knowledge to regions remote from our own experience—why he writes about the beginning and end of the world. It seems to be gratuitously courting disaster to expose our theories to conditions in which any slight weakness is likely to become magnified without limit. But that is just the

[12] *The Nature of the Physical World*, Chapter IV (1928).
[13] *Internal Constitution of the Stars*, p. 317 (1926).

principle of testing. "The real justification for making such forecasts is not that they are likely to be realised; but that they throw light upon the state of contemporary science, and may indicate where it requires supplementing." [14]

The test of extrapolation to the most distant future does not, I think, disclose any definite weakness in the present system of science—in particular, in the second law of thermodynamics on which physical science so largely relies. It is true that the extrapolation foretells that the material universe will some day arrive at a state of dead sameness and so virtually come to an end; to my mind that is a rather happy avoidance of a nightmare of eternal repetition. It is the opposite extrapolation towards the past which gives real cause to suspect a weakness in the present conceptions of science. The beginning seems to present insuperable difficulties unless we agree to look on it as frankly supernatural. We may have to let it go at that. But I have referred elsewhere to the danger of limiting scientific investigation to a bounded domain. Instead of honestly facing the intricacies of our problem, we may be led to think that its difficulties have been solved when they have only been swept over the boundary. Sweeping them back and back, the pile increases until it forms an unclimbable barrier. Perhaps it is this barrier that we call "the beginning."

VII

Now I have told you "everything right as it fell out."

How much of the story are we to believe?

Science has its showrooms and its workshops. The public to-day, I think rightly, is not content to wander round the showrooms where the tested products are exhibited; the demand is to see what is going on in the workshops. You are welcome to enter; but do not judge what you see by the standard of the showroom.

We have been going round a workshop in the basement of the building of science. The light is dim, and we stumble sometimes. About us is confusion and mess which there has not been time to sweep away. The workers and their machines are enveloped in murkiness. But I think that something is being shaped here—perhaps something rather big. I do not quite know what it will be when it is completed and polished for the showroom. But we can look at the present designs and the novel tools that are being used in its manufacture; we can contemplate too the little successes which make us hopeful.

A slight reddening of the light of distant galaxies, an adventure of the mathematical imagination in spherical space, reflections on the underlying principles implied in all measurement, nature's curious choice of certain numbers such as 137 in her scheme—these and many other scraps have

[14] Prof. H. T. H. Piaggio.

come together and formed a vision. As when the voyager sights a distant shore, we strain our eyes to catch the vision. Later we may more fully resolve its meaning. It changes in the mist; sometimes we seem to focus the substance of it, sometimes it is rather a vista leading on and on till we wonder whether aught can be final.

Once more I have recourse to Bottom the weaver—

> I have had a most rare vision. I have had a dream, past the wit of man to say what dream it was: man is but an ass, if he go about to expound this dream. . . . Methought I was,—and methought I had,—but man is but a patched fool, if he will offer to say what methought I had. . . .
>
> It shall be called Bottom's Dream, because it hath no bottom.

When Newton saw an apple fall, he found . . .
A mode of proving that the earth turn'd round
In a most natural whirl, called gravitation;
And thus is the sole mortal who could grapple
Since Adam, with a fall or with an apple. —BYRON

Nature and Nature's laws lay hid in night:
God said, "Let Newton be!" and all was light.
—ALEXANDER POPE (*Epitaph on Newton*)

It did not last: the Devil howling "Ho!
Let Einstein be!" restored the status quo. —SIR JOHN COLLINGS SQUIRE

21 The New Law of Gravitation and the Old Law

By SIR ARTHUR STANLEY EDDINGTON

I don't know what I may seem to the world, but, as to myself, I seem to have been only as a boy playing on the sea-shore, and diverting myself in now and then finding a smoother pebble or a prettier shell than ordinary, whilst the great ocean of truth lay all undiscovered before me.
SIR ISAAC NEWTON

WAS there any reason to feel dissatisfied with Newton's law of gravitation?

Observationally it had been subjected to the most stringent tests, and had come to be regarded as the perfect model of an exact law of nature. The cases, where a possible failure could be alleged, were almost insignificant. There are certain unexplained irregularities in the moon's motion; but astronomers generally looked—and must still look—in other directions for the cause of these discrepancies. One failure only had led to a serious questioning of the law; this was the discordance of motion of the perihelion of Mercury. How small was this discrepancy may be judged from the fact that, to meet it, it was proposed to amend *square* of the distance to the 2·00000016 power of the distance. Further it seemed possible, though unlikely, that the matter causing the zodiacal light might be of sufficient mass to be responsible for this effect.

The most serious objection against the Newtonian law as an exact law was that it had become ambiguous. The law refers to the product of the masses of the two bodies; but the mass depends on the velocity—a fact unknown in Newton's day. Are we to take the variable mass, or the mass reduced to rest? Perhaps a learned judge, interpreting Newton's statement like a last will and testament, could give a decision; but that is scarcely the way to settle an important point in scientific theory.

Further *distance*, also referred to in the law, is something relative to an observer. Are we to take the observer travelling with the sun or with the other body concerned, or at rest in the aether or in some gravitational medium? . . .

It is often urged that Newton's law of gravitation is much simpler than Einstein's new law. That depends on the point of view; and from the point of view of the four-dimensional world Newton's law is far more complicated. Moreover, it will be seen that if the ambiguities are to be cleared up, the statement of Newton's law must be greatly expanded.

Some attempts have been made to expand Newton's law on the basis of the restricted principle of relativity alone. This was insufficient to determine a definite amendment. Using the principle of equivalence, or relativity of force, we have arrived at a definite law proposed in the last chapter. Probably the question has arisen in the reader's mind, why should it be called the law of gravitation? It may be plausible as a law of nature; but what has the degree of curvature of space-time to do with attractive forces, whether real or apparent?

A race of flat-fish once lived in an ocean in which there were only two dimensions. It was noticed that in general fishes swam in straight lines, unless there was something obviously interfering with their free courses. This seemed a very natural behaviour. But there was a certain region where all the fish seemed to be bewitched; some passed through the region but changed the direction of their swim, others swam round and round indefinitely. One fish invented a theory of vortices, and said that there were whirlpools in that region which carried everything round in curves. By-and-by a far better theory was proposed; it was said that the fishes were all attracted towards a particular large fish—a sun-fish—which was lying asleep in the middle of the region; and that was what caused the deviation of their paths. The theory might not have sounded particularly plausible at first; but it was confirmed with marvellous exactitude by all kinds of experimental tests. All fish were found to possess this attractive power in proportion to their sizes; the law of attraction was extremely simple, and yet it was found to explain all the motions with an accuracy never approached before in any scientific investigations. Some fish grumbled that they did not see how there could be such an influence at a distance; but it was generally agreed that the influence was communicated through the ocean and might be better understood when more was known about the nature of water. Accordingly, nearly every fish who wanted to explain the attraction started by proposing some kind of mechanism for transmitting it through the water.

But there was one fish who thought of quite another plan. He was impressed by the fact that whether the fish were big or little they always took the same course, although it would naturally take a bigger force to deflect the bigger fish. He therefore concentrated attention on the courses rather than on the forces. And then he arrived at a striking explanation of the whole thing. There was a mound in the world round about where the sun-fish lay. Flat-fish could not appreciate it directly because they were two-dimensional; but whenever a fish went swimming over the slopes of the mound, although he did his best to swim straight on, he got turned round a bit. (If a traveller goes over the left slope of a mountain, he must consciously keep bearing away to the left if he wishes to keep to his original direction relative to the points of the compass.) This was the secret of the mysterious attraction, or bending of the paths, which was experienced in the region.

The parable is not perfect, because it refers to a hummock in space alone, whereas we have to deal with hummocks in space-time. But it illustrates how a curvature of the world we live in may give an illusion of attractive force, and indeed can only be discovered through some such effect. How this works out in detail must now be considered.

In the form $G_{\mu\nu} = 0$, Einstein's law expresses conditions to be satisfied in a gravitational field produced by any arbitrary distribution of attracting matter. An analogous form of Newton's law was given by Laplace in his celebrated expression $\nabla^2 V = 0$. A more illuminating form of the law is obtained if, instead of putting the question what kinds of space-time can exist under the most general conditions in an empty region, we ask what kind of space-time exists in the region round a single attracting particle? We separate out the effect of a single particle, just as Newton did. . . .

We need only consider space of two dimensions—sufficient for the so-called plane orbit of a planet—time being added as the third dimension. The remaining dimension of space can always be added, if desired, by conditions of symmetry. The result of long algebraic calculations is that, round a particle

$$ds^2 = -\frac{1}{\gamma}\,dr^2 - r^2 d\theta^2 + \gamma dt^2 \qquad \ldots\ldots\ldots(6)$$

where

$$\gamma = 1 - \frac{2m}{r}.$$

The quantity m is the gravitational mass of the particle—but we are not supposed to know that at present. r and θ are polar coordinates, or rather they are the nearest thing to polar coordinates that can be found in space which is not truly flat.

The fact is that this expression for ds^2 is found in the first place simply as a particular solution of Einstein's equations of the gravitational field; it is a variety of hummock (apparently the simplest variety) which is not curved beyond the first degree. There *could* be such a state of the world under suitable circumstances. To find out what those circumstances are, we have to trace some of the consequences, find out how any particle moves when ds^2 is of this form, and then examine whether we know of any case in which these consequences are found observationally. It is only after having ascertained that this form of ds^2 does correspond to the leading observed effects attributable to a particle of mass m at the origin that we have the right to identify this particular solution with the one we hoped to find.

It will be a sufficient illustration of this procedure, if we indicate how the position of the matter causing this particular solution is located. Wherever the formula (6) holds good there can be no matter, because the law which applies to empty space is satisfied. But if we try to approach the origin ($r = 0$), a curious thing happens. Suppose we take a measuring-rod, and, laying it radially, start marking off equal lengths with it along a radius, gradually approaching the origin. Keeping the time t constant, and $d\theta$ being zero for radial measurements, the formula (6) reduces to

$$ds^2 = -\frac{1}{\gamma}\, dr^2$$

or

$$dr^2 = -\gamma ds^2.$$

We start with r large. By-and-by we approach the point where $r = 2m$. But here, from its definition, γ is equal to 0. So that, however large the measured interval ds may be, $dr = 0$. We can go on shifting the measuring-rod through its own length time after time, but dr is zero; that is to say, we do not reduce r. There is a magic circle which no measurement can bring us inside. It is not unnatural that we should picture something obstructing our closer approach, and say that a particle of matter is filling up the interior.

The fact is that so long as we keep to space-time curved only in the first degree, we can never round off the summit of the hummock. It must end in an infinite chimney. In place of the chimney, however, we round it off with a small region of greater curvature. This region cannot be empty because the law applying to empty space does not hold. We describe it therefore as containing matter—a procedure which practically amounts to a definition of matter. Those familiar with hydrodynamics may be reminded of the problem of the irrotational rotation of a fluid; the conditions cannot be satisfied at the origin, and it is necessary to cut out a region which is filled by a vortex-filament.

A word must also be said as to the co-ordinates r and t used in (6). They correspond to our ordinary notion of radial distance and time—as well as any variables in a non-Euclidean world can correspond to words which, as ordinarily used, presuppose a Euclidean world. We shall thus call r and t, distance and time. But to give names to coordinates does not give more information—and in this case gives considerably less information—than is already contained in the formula for ds^2. If any question arises as to the exact significance of r and t it must always be settled by reference to equation (6).

The want of flatness in the gravitational field is indicated by the deviation of the coefficient γ from unity. If the mass $m = 0$, $\gamma = 1$, and space-time is perfectly flat. Even in the most intense gravitational fields known, the deviation is extremely small. For the sun, the quantity m, called the gravitational mass, is only $1 \cdot 47$ kilometres, for the earth it is 5 millimetres. In any practical problem the ratio $2m/r$ must be exceedingly small. Yet it is on the small corresponding difference in γ that the whole of the phenomena of gravitation depend. . . .

The mathematical reader should find no difficulty in proving that for a particle with small velocity the acceleration towards the sun is approximately m/r^2, agreeing with the Newtonian law. . . .

The result that the expression found for the geometry of the gravitational field of a particle leads to Newton's law of attraction is of great importance. It shows that the law $G_{\mu\nu} = 0$, proposed on theoretical grounds, agrees with observation at least approximately. It is no drawback that the Newtonian law applies only when the speed is small; all planetary speeds are small compared with the velocity of light, and the considerations mentioned at the beginning of this chapter suggest that some modification may be needed for speeds comparable with that of light.

Another important point to notice is that the attraction of gravitation is simply a geometrical deformation of the straight tracks. It makes no difference what body or influence is pursuing the track, the deformation is a general discrepancy between the "mental picture" and the "true map" of the portion of space-time considered. Hence light is subject to the same disturbance of path as matter. This is involved in the Principle of Equivalence; otherwise we could distinguish between the acceleration of a lift and a true increase of gravitation by optical experiments; in that case the observer for whom light-rays appear to take straight tracks might be described as absolutely unaccelerated and there could be no relativity theory. Physicists in general have been prepared to admit the likelihood of an influence of gravitation on light similar to that exerted on matter; and the problem whether or not light has "weight" has often been considered.

The appearance of γ as the coefficient of dt^2 is responsible for the main features of Newtonian gravitation; the appearance of $1/\gamma$ as the coefficient of dr^2 is responsible for the principal deviations of the new law from the old. This classification seems to be correct; but the Newtonian law is ambiguous and it is difficult to say exactly what are to be regarded as discrepancies from it. Leaving aside now the time-term as sufficiently discussed, we consider the space-terms alone [1]

$$ds^2 = \frac{1}{\gamma} dr^2 + r^2 d\theta^2.$$

The expression shows that space considered alone is non-Euclidean in the neighbourhood of an attracting particle. This is something entirely outside the scope of the old law of gravitation. Time can only be explored by something moving, whether a free particle or the parts of a clock, so that the non-Euclidean character of space-time can be covered up by introducing a field of force, suitably modifying the motion, as a convenient fiction. But space can be explored by static methods; and theoretically its non-Euclidean character could be ascertained by sufficient precise measures with rigid scales.

If we lay our measuring scale transversely and proceed to measure the circumference of a circle of nominal radius r, we see from the formula that the measured length ds is equal to $r\,d\theta$, so that, when we have gone right round the circle, θ has increased by 2π and the measured circumference is $2\pi r$. But when we lay the scale radially the measured length ds is equal to $dr/\sqrt{\gamma}$, which is always greater than dr. Thus, in measuring a diameter, we obtain a result greater than $2r$, each portion being greater than the corresponding change of r.

Thus if we draw a circle, placing a massive particle near the centre, so as to produce a gravitational field, and measure with a rigid scale the circumference and the diameter, the ratio of the measured circumference to the measured diameter will not be the famous number

$$\pi = 3 \cdot 14159265358979323846264338327 9\ldots$$

but a little smaller. Or if we inscribe a regular hexagon in this circle its sides will not be exactly equal to the radius of the circle. Placing the particle near, instead of at, the centre, avoids measuring the diameter *through* the particle, and so makes the experiment a practical one. But though practical, it is not practicable to determine the non-Euclidean character of space in this way. Sufficient refinement of measures is not attainable. If the mass of a ton were placed inside a circle of five yards radius, the defect in the value of π would only appear in the twenty-fourth or twenty-fifth place of decimals.

It is of value to put the result in this way, because it shows that the relativist is not talking metaphysics when he says that space in the gravitational field is non-Euclidean. His statement has a plain physical meaning,

[1] We change the sign of ds^2, so that ds, when real, means measured space instead of measured time.

which we may some day learn how to test experimentally. Meanwhile we can test it by indirect methods. . . .

[A body passing near a massive particle has its path bent owing to the non-Euclidean character of space.]

This bending of the path is additional to that due to the Newtonian force of gravitation which depends on the second appearance of γ in the formula. As already explained it is in general a far smaller effect and will appear only as a minute correction to Newton's law. The only case where the two rise to equal importance is when the track is that of a light wave, or of a particle moving with a speed approaching that of light; for then dr^2 rises to the same order of magnitude as dt^2.

To sum up, a ray of light passing near a heavy particle will be bent, firstly, owing to the non-Euclidean character of the combination of time with space. This bending is equivalent to that due to Newtonian gravitation, and may be calculated in the ordinary way on the assumption that light has weight like a material body. Secondly, it will be bent owing to the non-Euclidean character of space alone, and this curvature is additional to that predicted by Newton's law. If then we can observe the amount of curvature of a ray of light, we can make a crucial test of whether Einstein's or Newton's theory is obeyed. . . .

It is not difficult to show that the total deflection of a ray of light passing at a distance r from the centre of the sun is (in circular measure) $\frac{4m}{r}$, whereas the deflection of the same ray calculated on the Newtonian theory would be $\frac{2m}{r}$. For a ray grazing the surface of the sun the numerical value of this deflection is

$$1''\cdot 75 \text{ (Einstein's theory)},$$
$$0''\cdot 87 \text{ (Newtonian theory)}. \ . \ . \ .$$

The bending affects stars seen near the sun, and accordingly the only chance of making the observation is during a total eclipse when the moon cuts off the dazzling light. Even then there is a great deal of light from the sun's corona which stretches far above the disc. It is thus necessary to have rather bright stars near the sun, which will not be lost in the glare of the corona. Further the displacements of these stars can only be measured relatively to other stars, preferably more distant from the sun and less displaced; we need therefore a reasonable number of outer bright stars to serve as reference points.

In a superstitious age a natural philosopher wishing to perform an important experiment would consult an astrologer to ascertain an auspicious moment for the trial. With better reason, an astronomer to-day consulting the stars would announce that the most favourable day of the year for weighing light is May 29. The reason is that the sun in its annual journey round the ecliptic goes through fields of stars of varying richness, but on May 29 it is in the midst of a quite exceptional patch of bright stars—part of the Hyades—by far the best star-field encountered. Now if this problem had been put forward at some other period of history, it might have been necessary to wait some thousands of years for a total eclipse of the sun to happen on the lucky date. But by strange good fortune an eclipse did happen on May 29, 1919. Owing to the curious sequence of eclipses a similar opportunity will recur in 1938; we are in the midst of the most favourable cycle. It is not suggested that it is impossible to make the test at other eclipses; but the work will necessarily be more difficult.

Attention was called to this remarkable opportunity by the Astronomer Royal in March, 1917; and preparations were begun by a Committee of the Royal Society and Royal Astronomical Society for making the observations. Two expeditions were sent to different places on the line of totality to minimise the risk of failure by bad weather. Dr. A. C. D. Crommelin and Mr. C. Davidson went to Sobral in North Brazil; Mr. E. T. Cottingham and the writer went to the Isle of Principe in the Gulf of Guinea, West Africa. . . .

It will be remembered that Einstein's theory predicts a deflection of $1''\cdot 74$ at the edge of the sun,[2] the amount falling off inversely as the distance from the sun's centre. The simple Newtonian deflection is half this, $0''\cdot 87$. The final results (reduced to the edge of the sun) obtained at Sobral and Principe with their "probable accidental errors" were

$$\begin{array}{ll} \text{Sobral} & 1''\cdot 98 \pm 0''\cdot 12, \\ \text{Principe} & 1''\cdot 61 \pm 0''\cdot 30. \end{array}$$

It is usual to allow a margin of safety of about twice the probable error on either side of the mean. The evidence of the Principe plates is thus just about sufficient to rule out the possibility of the "half-deflection," and the Sobral plates exclude it with practical certainty. The value of the material found at Principe cannot be put higher than about one-sixth of that at Sobral; but it certainly makes it less easy to bring criticism against this confirmation of Einstein's theory seeing that it was obtained independently with two different instruments at different places and with different kinds of checks.

The best check on the results obtained with the 4-inch lens at Sobral is the striking internal accordance of the measures for different stars. The theoretical deflection should vary inversely as the distance from the sun's

[2] The predicted deflection of light from infinity to infinity is just over $1''\cdot 745$, from infinity to the earth it is just under.

centre; hence, if we plot the mean radial displacement found for each star separately against the inverse distance, the points should lie on a straight line. This is shown in Figure 1 where the broken line shows the theoretical prediction of Einstein, the deviations being within the accidental errors of the determinations. A line of half the slope representing the half-deflection would clearly be inadmissible. . . .

We have seen that the swift-moving light-waves possess great advantages as a means of exploring the non-Euclidean property of space. But there is an old fable about the hare and the tortoise. The slow-moving planets have qualities which must not be overlooked. The light-wave traverses the region in a few minutes and makes its report; the planet plods on and on for centuries, going over the same ground again and again. Each time it goes round it reveals a little about the space, and the knowledge slowly accumulates.

According to Newton's law a planet moves round the sun in an ellipse, and if there are no other planets disturbing it, the ellipse remains the same for ever. According to Einstein's law the path is very nearly an ellipse, but it does not quite close up; and in the next revolution the path has advanced slightly in the same direction as that in which the planet was moving. The orbit is thus an ellipse which very slowly revolves.

The exact prediction of Einstein's law is that in one revolution of the planet the orbit will advance through a fraction of a revolution equal to $3v^2/C^2$, where v is the speed of the planet and C the speed of light. The earth has 1/10,000 of the speed of light; thus in one revolution (one year) the point where the earth is at greatest distance from the sun will move on 3/100,000,000 of a revolution, or 0″·038. We could not detect this difference in a year, but we can let it add up for a century at least. It would then be observable but for one thing—the earth's orbit is very blunt, very nearly circular, and so we cannot tell accurately enough which way it is pointing and how its sharpest apses move. We can choose a planet with higher speed so that the effect is increased, not only because v^2 is increased, but because the revolutions take less time; but, what is perhaps more important, we need a planet with a sharp elliptical orbit, so that it is easy to observe how its apses move round. Both these conditions are fulfilled in the case of Mercury. It is the fastest of the planets, and the predicted advance of the orbit amounts to 43″ per century; further the eccentricity of its orbit is far greater than that of any of the other seven planets.

Now an unexplained advance of the orbit of Mercury had long been known. It had occupied the attention of Le Verrier, who, having successfully predicted the planet Neptune from the disturbances of Uranus, thought that the anomalous motion of Mercury might be due to an interior planet, which was called Vulcan in anticipation. But, though thor-

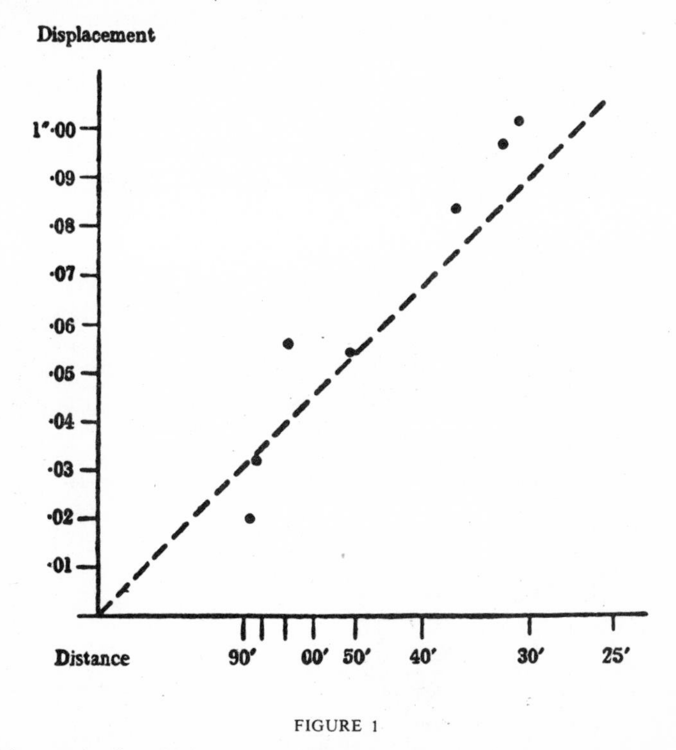

FIGURE 1

oughly sought for, Vulcan has never turned up. Shortly before Einstein arrived at his law of gravitation, the accepted figures were as follows. The actual observed advance of the orbit was 574″ per century; the calculated perturbations produced by all the known planets amounted to 532″ per century. The excess of 42″ per century remained to be explained. Although the amount could scarcely be relied on to a second of arc, it was at least thirty times as great as the probable accidental error.

The big discrepancy from the Newtonian gravitational theory is thus in agreement with Einstein's prediction of an advance of 43″ per century. . . .

The theory of relativity has passed in review the whole subject-matter of physics. It has unified the great laws, which by the precision of their formulation and the exactness of their application have won the proud place in human knowledge which physical science holds to-day. And yet, in regard to the nature of things, this knowledge is only an empty shell—a form of symbols. It is knowledge of structural form, and not knowledge of content. All through the physical world runs that unknown content, which must surely be the stuff of our consciousness. Here is a hint of aspects deep within the world of physics, and yet unattainable by the methods of physics. And, moreover, we have found that where science has progressed the farthest, the mind has but regained from nature that which the mind has put into nature.

We have found a strange foot-print on the shores of the unknown. We have devised profound theories, one after another, to account for its origin. At last, we have succeeded in reconstructing the creature that made the foot-print. And Lo! it is our own.

COMMENTARY ON

The Theory of Relativity

THE theory of relativity is fifty years old but we have not got used to it. In half a century it has not succeeded in changing the habits of our thought. For a long time many persons regarded the theory as a philosopher's fairy-tale; others looked upon it as the sort of hopeless abstraction on which mathematicians spend their lives. Lately we have come to realize that the ideas involved in Einstein's work have consequences. This has increased our respect for the ideas without, however, helping us to understand them. To be sure, we no longer stare at relativity, in Luther's classic phrase, like cows in front of a new gate, but this is more a sign of resignation than of comprehension.

The fact that the theory is still a stranger is not due to neglect. Many books about relativity have been written for ordinary readers. Leading scientists, among them the most gifted popularizers, have constructed ingenious analogies to make clear the physical and philosophical aspects of the theory. Einstein has tried his hand at this kind of treatment. Eddington and Jeans wrote brilliantly on the subject, but were sometimes carried away by their own metaphors. *The ABC of Relativity* by Bertrand Russell is a readable account; and I recommend highly two small, unpretentious volumes, now out of print and quite forgotten: *Relativity* by James Rice [1] and *The Idea of Einstein's Theory* by the noted Austrian physicist, J. H. Thirring.[2] Yet even the best of these expositions fails to satisfy the reader who is honest with himself that he has a firm grasp of the rudiments of relativity. The ideas and the paradoxes are carefully set forth; the paraphernalia of measuring rods, light signals and temperamental clocks are displayed: the effect is that of a conjurer's show. The tricks are made familiar to the onlooker, but he is not made familiar with them. He is entertained, perhaps impressed, but certainly not enlightened.

Is one then to conclude that the theory is too hard for simple language? I think not. It is revolutionary, but no more so than the theories of Copernicus and Galileo. It is abstract, but less so than the concepts of negative numbers and free enterprise. It is against common sense, but so at first were the ideas of vaccination and of men living upside down in the Antipodes.[3] The relativity primers have failed, I think, because of a dual misconception: the popularizer is convinced that he can make the subject plain without mathematics; and the reader is convinced that the subject

[1] Robert M. McBride Company, New York, no date.
[2] Robert M. McBride Company, New York, 1922.
[3] See J. E. Turner, "Relativity without Paradox," *The Monist*, January, 1930, for an interesting discussion of these points.

can never be made plain with mathematics. They are both mistaken, but the reader at least has Einstein on his side. When his teacher of mathematics, Hermann Minkowski, built up the special theory of relativity into a system of "world-geometry," Einstein remarked, "Since the mathematicians have invaded the theory of relativity, I do not understand it myself any more." [4]

There is a story about a student at Oxford who, when asked after an examination how he had proved the binomial theorem, replied cheerfully that, while he had not been able to prove the theorem, he had made it seem pretty plausible. The popular accounts of relativity do no more. They make the ideas seem plausible but avoid the precision essential to conviction and understanding. Relativity without mathematics has been compared to "painless dentistry," "skiing without falling," "reading without tears." I quote these comparisons from the preface to the volume from which the following selection has been taken. The author, Clement V. Durell, is a British mathematician who wrote textbooks, served as Senior Mathematical Master at Winchester College and had a high reputation as a teacher. His *Readable Relativity* is by far the best book on the subject for the nonspecialist. Durell does not suggest that Einstein's ideas can be traversed on a downhill glide. The theory of relativity imposes a discipline on those who seek to understand it. It demands concentration, venturesomeness and hospitality to concepts contradicting cherished beliefs. But understanding the theory does not presuppose extensive mathematical training. "This book," Durell says, "attempts to secure as high a degree of definition as is compatible with the standard of mathematical knowledge of the average person. The limitations this imposes are obvious, but inevitable if the subject is to lie within the sphere of general education. Given, however, this small amount of mathematical capacity and preferably also a willingness to work out a few numerical examples to test appreciation of the ideas peculiar to the subject, it should be possible to make Einstein's view of the universe as much a part of the intellectual equipment of ordinary people as is that of Newton."

Readable Relativity explains both the special and the general theory, using only elementary—truly elementary—algebra and geometry. I can promise even those readers who prefer to read about mathematics without having to practice it that, if they will suspend their prejudices, the next few pages will admit them to an exceptionally satisfying adventure in ideas. That they will then be able to say they understand the theory of relativity is of secondary importance.

[4] Arnold Sommerfeld, "To Albert Einstein's Seventieth Birthday" in *Albert Einstein, Philosopher-Scientist,* edited by Paul Arthur Schilpp, Evanston, 1949.

From henceforth, space by itself, and time by itself, have vanished into the merest shadows and only a kind of blend of the two exists in its own right.
—HERMAN MINKOWSKI

If my theory of relativity is proven successful, Germany will claim me as a German and France will declare that I am a citizen of the world. Should my theory prove untrue, France will say that I am a German and Germany will declare that I am a Jew.—ALBERT EINSTEIN (*Address at the Sorbonne, Paris*)

22 The Theory of Relativity

By CLEMENT V. DURELL

ALICE THROUGH THE LOOKING-GLASS

" 'I can't believe *that,*' said Alice.

" 'Can't you?' the Queen said in a pitying tone. 'Try again: draw a long breath and shut your eyes.'

"Alice laughed: 'There's no use trying,' she said; 'one can't believe impossible things.'

" 'I daresay you haven't had much practice,' said the Queen. 'When I was younger, I always did it for half an hour a day. Why, sometimes I've believed as many as six impossible things before breakfast.' "

—*Through the Looking-Glass.*

CAN NATURE DECEIVE?

THE scientists, in playing their game with Nature, are meeting an opponent on her own ground, who has not only made the rules of the game to suit herself, but may have even queered the pitch or cast a spell over the visiting team. If space possesses properties which distort our vision, deform our measuring-rods, and tamper with our clocks, is there any means of detecting the fact? Can we feel hopeful that eventually cross-examination will break through the disguise? *Professor Garnett,*[1] making use of *Lewis Carroll's* ideas, has given a most instructive illustration of a way in which Nature could mislead us, seemingly without any risk of exposure.

Ultimately, we can only rely on the evidence of our senses, checked and clarified of course by artificial apparatus, repeated experiment, and exhaustive inquiry. Observations can often be interpreted unwisely, as an anecdote told by Sir George Greenhill illustrates:

At the end of a session at the Engineering College, Coopers' Hill, a reception was held and the science departments were on view. A young lady, entering the physical laboratory and seeing an inverted image of herself in a large concave mirror, naïvely remarked to her companion: "They have hung that looking-glass upside down." Had the lady advanced past the focus of the mirror, she would have seen that the workmen were

[1] *Mathematical Gazette,* May 1918.

not to blame. If Nature deceived her, it was at least a deception which further experiment would have unmasked.

THE CONVEX LOOKING-GLASS

We shall now follow some of the adventures of Alice in a convex looking-glass world, as described by Professor Garnett. As a preliminary, it is necessary to enumerate some of the properties of reflection in a convex mirror. For the sake of any reader who wishes to see how they can be obtained, their proofs, which involve only the use of similar triangles and some elementary algebra, are indicated in footnote 2.

[2] *Exercise*

1. BAC is a convex mirror whose radius AO is large compared with the object PQ. The ray of light from P parallel to the axis meets the mirror at N, and is then

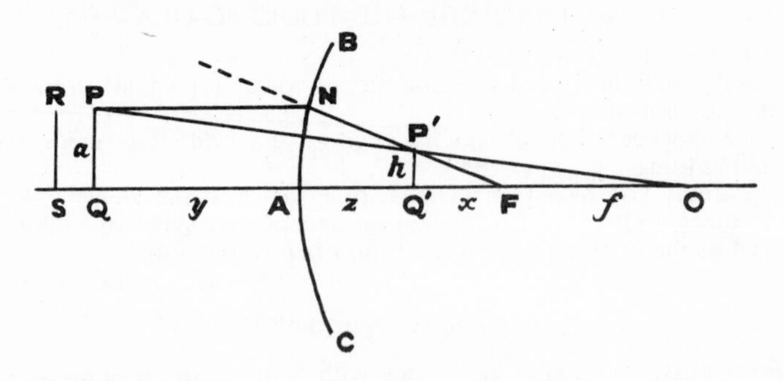

FIGURE 1

reflected along the line joining N to the focus F; the ray of light from P towards the centre O of the mirror is not changed in direction by the mirror; therefore the image of P is at the intersection P′ of NF and PO. Draw P′Q′ perpendicular to OA, then P′Q′ is the image of PQ. The focus F is at the mid-point of OA. Since the mirror has a large radius, its curvature is small, and NA can be treated as a straight line perpendicular to OA.

(i) Prove that $\dfrac{FO}{PN} = \dfrac{FP'}{P'N} = \dfrac{FQ'}{Q'A}$.

(ii) Hence, with the notation of p. 1109, prove that

$$\frac{1}{z} = \frac{1}{y} + \frac{1}{f}.$$

(iii) Prove that $\dfrac{P'Q'}{NA} = \dfrac{Q'F}{AF}$ and hence prove that $h = \dfrac{ax}{f}$.

(iv) Hence show that $x = f - z = \dfrac{fz}{y}$ and that $\dfrac{z}{y} = \dfrac{x}{f}$.

2. If in the figure PQ moves along the axis a *short* distance to RS, and if the image of RS is R′S′, and if AS $= y_1$, AS′ $= z_1$, use the formula $\dfrac{1}{z} = \dfrac{1}{y} + \dfrac{1}{f}$ to prove that

$$\frac{Q'S'}{QS} = \frac{z_1 - z}{y_1 - y} = \frac{zz_1}{yy_1} \simeq \frac{z^2}{y^2} = \frac{x^2}{f^2} = \left(\frac{h}{a}\right)^2.$$

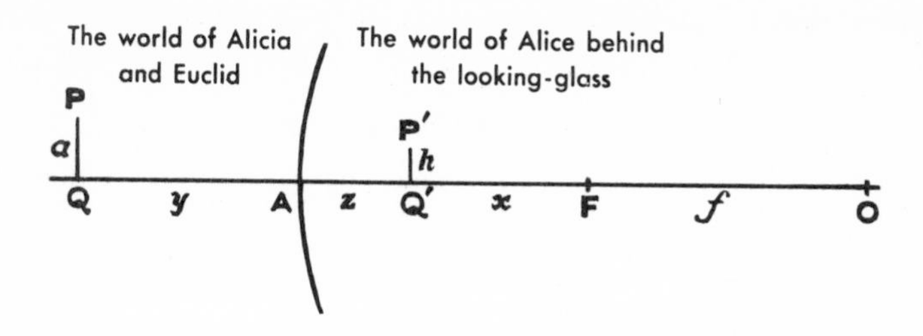

FIGURE 2

A is the apex of a convex mirror, of large radius; O is its centre, OA is the central radius or axis; the mid-point F of OA is the *focus*. PQ is an object outside the mirror, perpendicular to the axis and of height a feet, P′Q′ is the image of PQ in the mirror. Denote the various lengths as follows, in feet:

$$\mathrm{OF} = \mathrm{FA} = f;\ \mathrm{FQ'} = x;\ \mathrm{Q'A} = z;\ \mathrm{AQ} = y;\ \mathrm{P'Q'} = h.$$

We have the following formulæ:

$$\frac{1}{z} = \frac{1}{y} + \frac{1}{f};\ h = \frac{ax}{f};\ x = f - z.$$

The general consequences of these formulæ are easy to appreciate.

Since $\frac{1}{z} = \frac{1}{y} + \frac{1}{f}$, we see that $\frac{1}{z} > \frac{1}{y}$, so that $y > z$.

and that $\frac{1}{f} < \frac{1}{z}$, so that $z < f$ or $z < \mathrm{AF}$.

The image P′Q′ is therefore always nearer the mirror than the object PQ is, and is never as far from the mirror as F is.

Again, since $h = \frac{ax}{f}$, the height of the image is proportional to x, its

---

What do you deduce about the longitudinal contraction-ratio? Those acquainted with the calculus should show that $\frac{\delta z}{z^2} = \frac{\delta y}{y^2}$, and interpret the result.

3. What does Alicia think of the movement of the hands of Alice's watch (i) when held facing the mirror (ii) when laid flat on the ground?
4. Alice spins a top so that its axis is vertical; what is there unusual about it, according to Alicia?
5. The axis of the mirror A → O points due east. Alice, whose height has become only half that of Alicia, turns and walks north-east. What is her direction as measured by Alicia?
6. Alice believes she has proved two given triangles congruent by the method of superposition. Does Alicia agree with her?

distance from F, and therefore the nearer P′Q′ approaches F the smaller the length of P′Q′, the height of the image, becomes.

LIFE BEHIND THE LOOKING-GLASS

" 'He's dreaming now,' said Tweedledee: 'and what do you think he's dreaming about?'

" 'Nobody can guess that,' said Alice.

" 'Why, about you!' Tweedledee exclaimed. 'And if he left off dreaming about you, where do you suppose you'd be?'

" 'Where I am now, of course,' said Alice.

" 'Not you!' Tweedledee retorted contemptuously. 'You'd be nowhere. Why, you're only a sort of thing in his dream!'

" 'If that there King was to wake,' added Tweedledum, 'you'd go out—bang!—just like a candle!'

" 'I *am* real!' said Alice, and began to cry.

" 'You won't make yourself a bit realer by crying,' Tweedledee remarked."

We shall now treat Alice not as a thing in a dream, but as the image in a convex looking-glass of a pseudo-Alice who is moving about in our own world. Alice will insist as vehemently as she did to Tweedledee that she is a free agent with an independent existence, but we, looking from outside, will see that she conforms to the movements and amusements of this pseudo-Alice, whom we will call *Alicia*. We proceed to compare our (or Alicia's) observations with Alice's own ideas about her mode of life.

ALICE'S LIFE

Alicia is 4 feet tall and 1 foot broad. She starts at A with her back against the mirror, so that she and Alice are back to back, exactly the same size. Alicia is carrying a foot-rule which she holds against the mirror so that it touches and coincides with, and therefore equals, the corresponding foot-rule which Alice has.

Alicia now walks at a steady rate of 1 foot per second away from the mirror, along the axis. What happens to Alice?

Suppose the radius of the mirror is 40 feet, so that $AF = FO = f = 20$ feet. Then $a = PQ =$ Alicia's height $= 4$ ft.

After (say) 5 seconds, $AQ = y = 5$.

$$\text{Then } \frac{1}{z} = \frac{1}{y} + \frac{1}{f} = \frac{1}{5} + \frac{1}{20} = \frac{5}{20} = \frac{1}{4}.$$

$$\therefore AQ' = z = 4 \text{ and } x = Q'F = f - z = 20 - 4 = 16.$$

$$\therefore P'Q' = h = \frac{ax}{f} = \frac{4 \times 16}{20} = \frac{32}{10} = 3{\cdot}2 \text{ feet.}$$

If at this moment Alicia looks round, she will notice that Alice has only moved 4 feet compared with her own 5 feet, and that Alice's height has shrunk to 3·2 feet.

Alice's foot-rule, held vertically, has also shrunk: its length in fact is

now $\dfrac{1 \times x}{f} = \dfrac{1 \times 16}{20} = 0{\cdot}8$ foot.

Alice repudiates the idea that she has grown smaller, and to convince Alicia she takes her foot-rule and uses it to measure herself, and shows triumphantly that she is still exactly four foot-rules high ($3{\cdot}2 \div 0{\cdot}8 = 4$).

Alicia also notices that Alice is not so broad as she was; her breadth, in fact, has now dwindled to 0·8 foot: true, the breadth is still equal to Alice's foot-rule, but that foot-rule in any position *perpendicular* to the axis is now only 0·8 foot long.

CONTRACTION-RATIO PERPENDICULAR TO THE AXIS

We see that Alice continues to contract as she moves farther away from the mirror. The contraction-ratio in any direction perpendicular to the axis is $\dfrac{h}{a}$ which equals $\dfrac{x}{f}$ and is therefore proportional to x, Alice's distance from the focus F.

It is evident that Alice cannot detect this contraction, because Alice's ruler contracts in just the same proportion as Alice's body and Alice's clothes. In fact everything in Alice's world, regardless of what it is made, behaves in exactly the same way. We therefore call this contraction a property of space, not a property of matter. It is a form of influence which the space exercises on all things alike which enter it. And we say that *one of the laws of space in Alice's world is an automatic contraction-ratio which for any direction perpendicular to the axis, is proportional to* x, *the distance from the focus.*

CONTRACTION-RATIO ALONG THE AXIS

Alicia now lays her foot-rule down along the axis, and of course Alice imitates her.

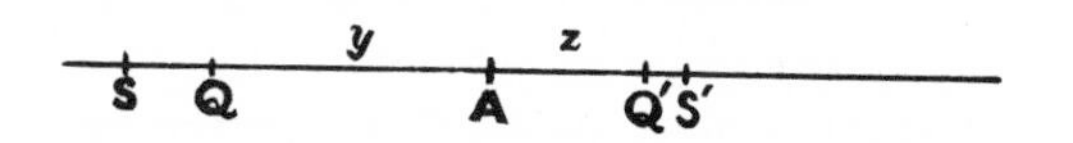

FIGURE 3

Q, S are two successive points of division on Alicia's foot-rule, such that AQ = 5 feet, QS = 0·1 foot, so that AS = 5·1 feet. The corresponding marks on Alice's foot-rule are Q′, S′. We have already proved that if

AQ $= y = 5$, then AQ$' = z = 4$. Further, if AS $= y = 5{\cdot}1$, then AS$' = z$ is given by

$$\frac{1}{z} = \frac{1}{y} + \frac{1}{f} = \frac{1}{5{\cdot}1} + \frac{1}{20} = \frac{20 + 5{\cdot}1}{5{\cdot}1 \times 20} = \frac{25{\cdot}1}{102}.$$

$$\therefore \text{AS}' = z = \frac{102}{25\cdot 1} = 4\cdot 064, \text{ approximately.}$$

$$\therefore \text{Q}'\text{S}' = \text{AS}' - \text{AQ}' = 0\cdot 064 \text{ foot.}$$

$\therefore$ the contraction-ratio along the axis at Q$'$ is

$$\frac{\text{Q}'\text{S}'}{\text{QS}} = \frac{0\cdot 064}{0\cdot 1} = 0\cdot 64.$$

But the contraction-ratio perpendicular to the axis at Q$'$ has been shown to be $0\cdot 8$.

Since $(0\cdot 8)^2 = 0\cdot 64$, this suggests that the contraction-ratio along the axis is the square of the contraction-ratio perpendicular to the axis, at the same place.

FIGURE 4

Alicia therefore notices that Alice becomes thinner from front to back as she moves away from the mirror, and the rate of getting thinner is more rapid than the rate of getting shorter.

For example, if Alice turns sideways and stretches out her left arm towards the mirror and her right arm away from it, the fingers of her left hand will be longer and fatter than those of her right hand, but the general effect will be to make the fingers of her right hand appear puffier—a chilblain effect—because these fingers have shortened more than they have thinned, compared with the other hand.

ALICE'S GEOMETRY

While Alicia is walking away from the mirror at a uniform speed, taking steps of equal length, Alice is also walking away in the opposite direction; but (according to Alicia) Alice's steps get shorter and shorter, and so she advances more and more slowly; and in fact, however far Alicia travels, Alice herself can never get as far as F. Alice of course imagines that there is no limit to the distance she can travel, and what Alicia calls the point F, Alice calls a point at infinity. If Alice walks along level ground, she imagines that the tip of her head and the soles of her feet are moving along parallel lines; indeed the lines along which they move express Alice's idea of parallelism. Alicia sees that such lines actually

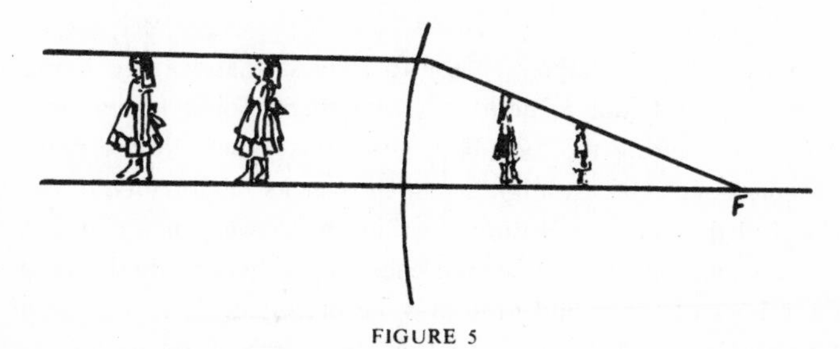

FIGURE 5

meet at F. If Alice lays a railway track along the axis, the railway lines will behave in the same way.

Suppose Alicia is riding on a bicycle down the axis away from the mirror, how do the wheels of Alice's bicycle behave?

The contraction is greater along the axis than at right angles to it. Consequently, not only are the wheels of Alice's machine smaller than those of Alicia's, but Alice's front wheel is smaller than her back wheel:

FIGURE 6

moreover, each wheel is approximately elliptical in shape, its vertical diameter being greater than its horizontal diameter, and although the wheels are turning round, the spokes which are vertical always appear to be longer than any of the others, and the spokes which are horizontal always appear to be shorter than any of the others; the consequence is that the spokes appear to expand as they revolve from the horizontal to the vertical, and then to contract as they revolve from the vertical to the horizontal.

Alice herself, after careful measurement, is satisfied that the machine is quite normal, but Alicia will think it most unsteady. Space does not permit of any inquiry into the mechanics of Alice's life: for this, reference should be made to Professor Garnett's article, mentioned above.

INNOCENTS AT PLAY

The object of this chapter is not to suggest that we are living in a looking-glass world, but to point out that there would appear to be no method of discovering the fact, if it were true. When Nature makes her laws of space, she can cast a binding spell over its inhabitants, if she cares to do so. But the fact that Nature is willing to answer some of the experimental questions which scientists put encourages them to think that gradually these laws of space and time may be disclosed. The purpose of this chapter will be served if it suggests that the search is not simple, and the results may be surprising.[3]

THE VELOCITY OF LIGHT

"The first thing to realise about the ether is its absolute continuity. A deep-sea fish has probably no means of apprehending the existence of water; it is too uniformly immersed in it: and that is our condition in regard to the ether." —SIR OLIVER LODGE, *Ether and Reality*

[3] [In this exercise, Alicia is supposed to be 4 feet high and her waist measurement is 1 foot broad, 6 inches thick; also $f = 20$ ft.]

1. When Alicia has moved 10 feet from the mirror, show that Alice has only moved 6 feet 8 inches, and is now 2 feet 8 inches tall and 8 inches broad. What happens when Alice measures her height with her own foot-rule?
2. When Alicia has moved 20 feet from the mirror, show that Alice has only moved 10 feet. What is Alice's height and breadth in this position?
3. Where is Alicia, when Alice's height is reduced to 1 foot? What is then Alice's breadth? What is the length of her foot-rule, held vertically? How many foot-rules high is Alice?
4. With the data of No. 1, find the thickness of Alice's waist. What is Alice's measurement of it?
5. With the data of No. 2, what is the contraction-ratio for Alice along the axis? What is the connection between the contraction-ratios along and perpendicular to the axis?
6. Alicia is 20 feet from the mirror and holds a 1-inch cube with its edges parallel or perpendicular to the axis. What is Alice holding?
7. When Alice's height shrinks to 1 foot, what are her waist measurements?
8. Does Alice's shape, as well as her actual size, alter as she moves away from the mirror?

THE ETHER

Those who have engaged in physical research during the last hundred years have been rewarded by discoveries of far-reaching importance and interest. Light, electricity, magnetism, and matter have been linked together so closely that it now appears that each must be interpreted in terms of a single medium, the ether.

What the ether is will, no doubt, remain a subject of acute controversy between physicists of rival schools for many years to come. Its existence was first postulated to serve as a vehicle for light. Experiment showed that light travels through space at approximately 300,000 km. per second, thus light takes about 8⅓ minutes to reach the Earth from the Sun.

According to the undulatory theory, light is propagated in the form of a wave motion through the ether. The work of Weber, Faraday, Clerk-Maxwell, and others established the remarkable fact that electro-magnetic radiation is a wave motion propagated with exactly the same velocity as light, and therefore presumably it uses the same medium as a vehicle. More recent research has shown that electric charges are discontinuous, and that the atoms of which matter is composed may themselves each be resolved into a group of electric particles; the group which constitutes each atom contains both negatively charged particles called electrons and positively charged particles called protons; atoms differ from each other according to the number and grouping of these particles. Each atom may be regarded as a miniature ultra-microscopic solar system, in which the electrons describe orbits round a central nucleus.

The ether is to be regarded as a *continuous* medium, filling the whole of space, and may indeed be identified with space: matter, electricity, etc., are *discontinuous*. But while identifying ether with space, we must also attribute to it some physical qualities in order that it may serve as a vehicle for physical phenomena. Some physicists credit it with weight and density. But if Einstein's view is accepted, it has no mechanical properties of this nature. Einstein holds that the idea of motion in connection with the ether is meaningless; the ether is everywhere and always. He does not say that the ether is at rest, but that the property of rest or motion can no more be applied to the ether than the property of mass can be applied to a man's reflection in a mirror, although the light-rays by which we perceive the reflection may and indeed do possess mass.

ABSOLUTE MOTION

If a passenger in a train observes another train moving past him, and if the motion is uniform and if there are no landmarks in view, it is impossible for him to determine whether his own or the other train, or both, are really in motion. This is a familiar experience. There is no difficulty in measuring the *relative* velocity of the two trains, but without a glimpse of the ground to act as a reference for measurement it is impossible to find what we tend to call the actual velocity of the train.

Again, suppose two balloons are drifting past each other above the clouds: an observer in one balloon tends to think of himself at rest and the other balloon as moving past him. Even when he obtains an accurate ground-observation, he can only calculate his velocity relative to the earth. An astronomer might continue the work and tell him the velocity of the observed point of the ground relative to the Sun and then the velocity of the Sun relative to one of the "fixed stars." But even all that will not enable him to find his actual or *absolute velocity*. What reason is there to consider any of the stars as fixed? we know that they also move relatively to each other; what indeed can the word "fixed" mean at all? Is there anything in the Universe we can mark down as really fixed? Scientists did not like the idea that all measurements of motion must be relative; it seemed like building a structure of the mechanics of the Universe on a shifting sand. When, therefore, physical research demanded the existence of a medium filling the whole of space, the ether was welcomed not only for what it could do for light and electricity but because it appeared to offer a standard of reference for the measurement of absolute velocity. Scientists, therefore, set to work to measure the velocity of the Earth through the ether. The fundamental experiment which had this object, and which may be taken as the basis for describing Einstein's (restricted) theory of Relativity, was performed in 1887 by *Michelson* and *Morley*. We shall in future refer to it as the M.-M. experiment. The idea of that experiment may be easily understood by taking a simple analogy.

ROWING ON RUNNING WATER

A stream is flowing at 4 feet per second between straight parallel banks 90 feet apart. Two men start from a point A on one bank; one of them T rows straight across the stream to the opposite bank at B and returns to A, the other L rows to a point C 90 feet downstream and then rows

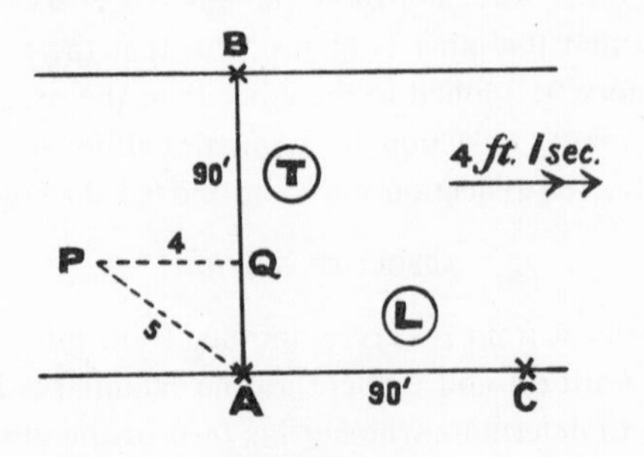

FIGURE 7

back to A. Each of them rows at 5 feet per second relatively to the water. Compare their times.

We have called the oarsmen T and L because T rows transversely to the stream and L rows longitudinally, in the line of the stream. Now T, in order to reach B, must point his boat upstream along a line AP such that if AP = 5 feet (*i.e.* the distance he moves relative to the water in 1 second), the water will carry him down 4 feet from P to Q (*i.e.* the distance the stream runs in 1 second), where Q is a point on AB and $\angle$AQP is a right angle.

$$\text{By Pythagoras, } AQ^2 + 4^2 = 5^2, \therefore AQ^2 = 25 - 16 = 9,$$
$$\therefore AQ = 3 \text{ feet.}$$

Therefore in each second the boat makes 3 feet headway along AB.

$$\therefore \text{the boat takes } \frac{90}{3} = 30 \text{ seconds to get from A to B.}$$

Similarly it takes 30 seconds to return from B to A.

$$\therefore \text{the total time across and back} = 60 \text{ seconds.}$$

Now L, on his journey to C, is moving relatively to the water at 5 feet per second, and the water carries him forwards 4 feet per second; therefore he advances at the rate of $5 + 4 = 9$ feet per second.

$$\therefore \text{the time from A to C} = \frac{90}{9} = 10 \text{ seconds.}$$

But when returning against the stream from C, his advance is only $5 - 4 = 1$ foot per second.

$$\therefore \text{the time from C to A} = \frac{90}{1} = 90 \text{ seconds.}$$

$$\therefore \text{the total time downstream and up} = 10 + 90 = 100 \text{ seconds.}$$

$$\therefore \frac{\text{time of L down and up}}{\text{time of T across and back}} = \frac{100}{60} = \frac{5}{3}.$$

It therefore takes longer to go down and up than an equal distance across and back. But the working of this example also shows that if we know the rate of rowing relative to the water, and if we know the ratio of the times taken for equal journeys in the two directions, we can calculate the velocity of the stream.

THE MICHELSON-MORLEY EXPERIMENT

It is known by experiment that light always travels through the ether at a constant rate of 300,000 km. per second. Let us suppose that at a

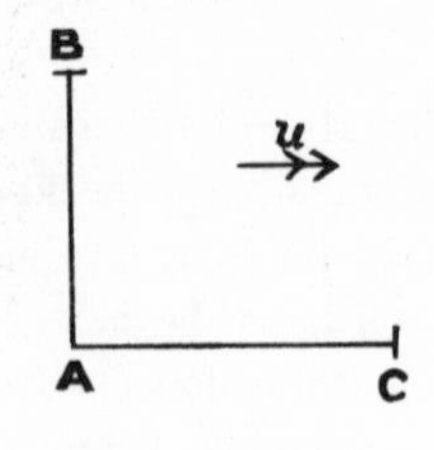

FIGURE 8

certain moment the Earth is moving through the ether at a speed of u km. per second in the direction C → A, then from the point of view of a man on the Earth the ether is streaming past A in the direction A → C at u km. per second. AC and AB are two rigid equal and perpendicular rods, with mirrors attached at C, B, so as to face A. At the same moment rays of light are dispatched from A, one along AC and the other along AB; these rays impinge on the mirrors and are reflected back to A. The motion of these rays corresponds to the motion of the boats in the example given above. Each ray travels at 300,000 km. per second relative to the ether-stream, since its mode of propagation is a wave in the ether, just as each boat moves at 5 feet per second relative to the water. Also the ether-stream is itself moving in the direction A → C at u km. per second, just as the water-stream is flowing at 4 feet per second.

Now it takes longer to go any given distance downstream and back than to go the same distance across the stream and back. Consequently the ray from C should arrive back at A later than the ray from B. If then we measure the ratio of the times taken by the two rays, we can calculate the speed u km. per second of the ether-stream, and this is equal and opposite to the velocity of the Earth through the ether.

The M.-M. experiment was designed to measure the ratio of these times: for a detailed account of the apparatus employed, reference may be made to any standard modern text-book on light. To the astonishment of the experimenters, the race proved to be a dead heat, the ray from C arriving back at A simultaneously with the ray from B.

Now the Earth is describing its orbit round the Sun at a speed approximately of 30 km. per second, consequently there is a difference of speed after a six months' interval of about 60 km. per second, so that even if

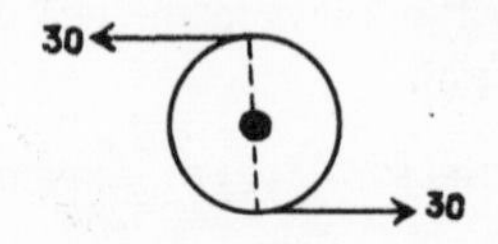

FIGURE 9

the Earth should happen to be at rest in the ether at one moment, it could not be still at rest six months later. But the repetition of the experiment after a six months' interval still gave a dead heat.

Also, in order to guard against any error arising from an inequality of the lengths of the arms AB and AC, the experiment was repeated after rotating the arms so that AB lay along the supposed stream and AC across it; but no difference of time was detected. Further, different directions were tried for AB, but without any result. The experiment has been carried out more recently with such added refinements that as small a speed as 1/5 km. per second would have been detected. Here, then, was an experimental result which contradicted a conclusion obtained by theory. Clearly there was something wrong with the theory. Scientists were compelled to look for some explanation or some modification of the theory which would reconcile calculation with observation.

WHAT IS THE ANSWER TO THE RIDDLE?

Let us return to the illustration of the boats which correspond to the light-rays in the M.-M. experiment. The two boats start together under the conditions stated on pp. 1116–1117, and every one is then amazed to see them arrive back simultaneously. How is this to be reconciled with the conclusions reached by calculation?

The first suggestion is that L rowed faster relatively to the water than T; this, however, must be rejected because the speed of rowing through the water corresponds in the M.-M. experiment to the velocity of light

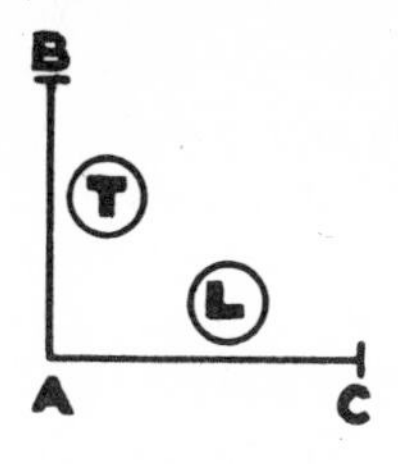

FIGURE 10

through the ether, which we know is a constant, 300,000 km. per second.

The next suggestion is that the courses are marked out incorrectly, and that the length of AC is less than that of AB, owing to careless measurement. But this view is untenable, because in the M.-M. experiment when the rigid arms AB and AC were interchanged, there was still no difference of time.

Fitzgerald then suggested that the arms AB and AC were unequal, not through faulty measurement, but because the shifting of a bar from a position across the stream to a position along the stream caused automatically a contraction in the length of the bar. The adventures of Alice have shown us that such a contraction could not be discovered by measurement, because the foot-rule with which AC is measured contracts in just the same proportion as the arm.

Suppose in the example of the boats the rule which measures 1 foot across the stream contracts to $\frac{3}{5}$ foot when in the line of the stream. When we measure out 90 foot-rule lengths to obtain AC, the outsider (Alicia) will say that AC is really $\frac{3}{5} \times 90 = 54$ feet, instead of 90 feet. L will then take $\frac{54}{9} = 6$ seconds to go downstream and $\frac{54}{1} = 54$ seconds to return upstream, so that his total time will be $6 + 54 = 60$ seconds, which is precisely the time taken by T.

This hypothetical phenomenon is called the "*Fitzgerald Contraction.*" Its value depends, of course, on the velocity of the stream; when the stream runs at 4 feet per second and the speed of the rowing is 5 feet per second, the contraction ratio has been shown to be $\sqrt{\left(1 - \frac{4^2}{5^2}\right)} = \sqrt{\left(\frac{9}{25}\right)} = \frac{3}{5}$.

The reader will see, from this way of writing it, what its value would be in other cases.

In 1905, an alternative explanation was offered by Einstein.

EINSTEIN'S HYPOTHESIS

Einstein lays down two general principles or axioms:

(i) It is impossible to detect uniform motion through the ether.

(ii) In all forms of wave motion, the velocity of propagation of the wave is independent of the velocity of the source.

Let us consider what these axioms mean.

(i) There is no difficulty in measuring the velocity of one body relative to another: all our ideas of velocity are essentially ideas of relative velocity, either velocities of other things relative to ourselves or our own velocity relative to something else—*e.g.* a man who looks at the road along which he is driving his car is probably estimating his own velocity relative to the road. But it is meaningless to inquire what our velocity is relative to the ether; no part of the ether can be distinguished from any other part: it may be possible to identify matter in the ether, but the ether itself defies identification. And if the ether cannot be (so to speak) labelled anywhere, the statement that a body is moving through it carries no information with it, or in other words has no meaning attaching to it.

(ii) The second axiom is perhaps more tangible. Imagine an engine moving at a uniform rate along a straight railway line on a perfectly calm day. If the engine-driver throws a stone forwards, a man on the line will observe that the velocity of the stone is equal to the velocity given it by the thrower + the velocity of the engine. The faster the engine is moving, the faster the stone will move, although the thrower exerts the same effort as before. The velocity of the stone relative to the air, therefore, depends on the velocity of the source, namely, the man on the engine.

Suppose now the engine whistles, and is heard by a man farther down the line. We know that the sound travels in the form of a wave through the air at approximately 1100 feet per second. But the motion of the sound-wave is a different type of motion from that of the stone: its velocity of propagation through the air does not depend on the velocity of the engine at the moment it whistled, *i.e.* it does not depend on the velocity of the source. The speed of the train will affect the pitch of the sound-wave, its musical note; but the time the wave takes to reach the man is not affected by the rate at which the engine is moving. If, then, a particle in motion sends out a beam of light, the rate of propagation of the light-wave through the ether has no connection with the velocity of the particle which emitted the beam of light.

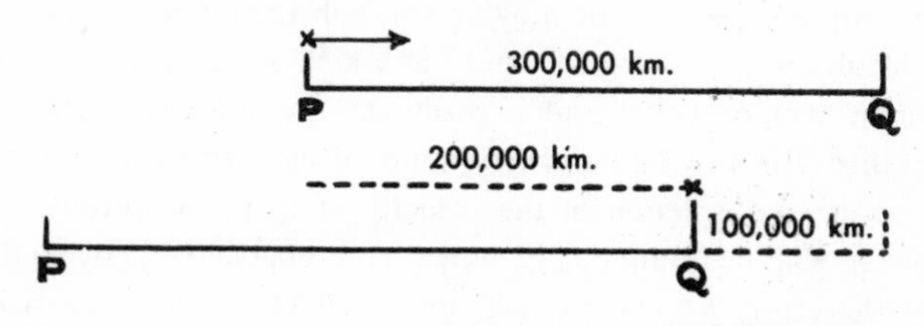

FIGURE 11

P and Q represent two places 300,000 km. apart and rigidly connected together. I take up my position at P and send a ray of light along PQ and measure the time it takes to reach Q. If PQ is fixed in the ether (assuming for the moment this phrase has a meaning), the time will be 1 second. If, however, my observations give the time as (say) only $\frac{2}{3}$ second, I can calculate that the ray itself only advanced $\frac{2}{3} \times 300{,}000 = 200{,}000$ km. through the ether, and that therefore Q must have advanced $300{,}000 - 200{,}000 = 100{,}000$ km. to meet it in the same time, $\frac{2}{3}$ second. Consequently the rigid bar PQ is moving at the rate of $100{,}000 \div \frac{2}{3} = 150{,}000$ km. per second. But as I remain at P, I deduce that my velocity through the ether is also 150,000 km. per second. This, however, contradicts Axiom (i), which lays down that a discovery of this nature is impossible. We are, therefore, forced to conclude that the measurement of the time of flight over this distance will always under all conditions be 1 second. Einstein's two axioms taken together, therefore, involve the following important result:

Any one who measures experimentally the velocity of light in a vacuum will always obtain the same result (within, of course, the limits of error imposed by the experiment). The velocity of light in a vacuum, as determined by every individual, is an absolute constant.

This conclusion may well cause a shock to any one who considers carefully what it implies; and the shock will not be diminished by examining its bearing on the problem of the boats.

It is, of course, important to notice the fundamental distinction between light-waves in the ether and sound-waves in the air. If an observer, when measuring the velocity of sound, obtains an answer which does not agree with the standard answer (about 1100 feet per second), he can at once calculate his velocity through the air. Nor is there any reason why he should not be able to do so; and he can compare his result with that obtained by other methods. But with the ether it is otherwise; the inability of an observer to measure his velocity through the ether involves the fact that his measure of the velocity of a light-wave must agree with the standard measurement.

THE APPLICATION OF EINSTEIN'S HYPOTHESIS

No one can be conscious of moving through the ether. An onlooker O has no difficulty in measuring the speed at which a man L is moving away from him; it will be equal and opposite to the velocity with which L calculates that O is moving away from him: if each expresses his measure of this velocity as a fraction of the velocity of light, the results obtained by O and L will be numerically equal and opposite in sign. Relative velocities, therefore, present no difficulty. But O and L alike will each consider himself at rest in the ether and will make his own measurements on that assumption. They, therefore, must be regarded as looking at the world from different points of view.

To explain the enigma of the boats, we must, therefore, consider separately the standpoint of each of the actors in the drama, the oarsmen T and L and an onlooker O, whom we will regard as poised just above T and L at the moment they start rowing. To make the analogy with the M.-M. experiment closer, imagine that the river-banks have disappeared, and that all we can see is an expanse of water devoid of all features or landmarks—that is what the ether-idea requires.

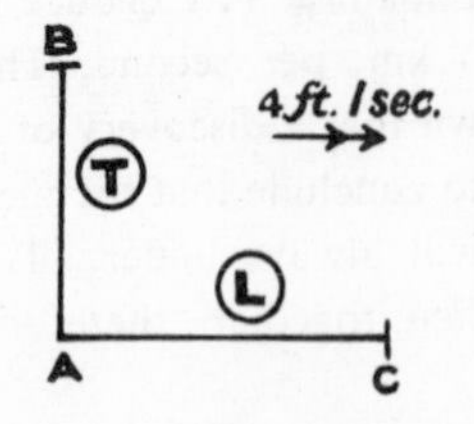

FIGURE 12

O says that this featureless ocean is moving in the direction A → C at 4 feet per second; in proof of this statement, he places a piece of cork on the water and observes that it at once moves away from him in the direction AC at 4 feet per second. T and L say that the water is motionless; each, sitting in his boat, places a piece of cork on the water and it remains where they have placed it; O, of course, says that these pieces of cork are drifting at the same rate as the boats. Further, T and L agree that O is moving away from them in the direction C → A at the rate of 4 feet per second; they say that the piece of cork which O has dropped remains stationary, and that it is O who is moving away from it.

The statement that the velocity of light is an absolute constant, or that each person who measures it obtains the same answer as any one else, when applied to the boats, means that T, L, and O will each obtain the same result when they measure the speed at which each boat is rowed through the water, because the boats are replacing the light-rays in the M.-M. experiment. We have taken this common measure of the speed as 5 feet per second.

Under these circumstances, our problem is to explain a definite experimental observation, namely, the fact that the boats (*i.e.* the light-rays) do return to A at the same moment.

Regard AB and AC as rigid planks of wood floating on the water. T and L believe that these planks are at rest, just as they believe the water is at rest; O believes that the planks are drifting with the water, just as T and L are doing.

T, L, and O each have a foot-rule; T holds his along AB, and L holds his along AC. O compares his foot-rule with T's by actual superposition, and they note that the two rules agree. As long as T's rule is kept perpendicular to the stream, it will remain identical with O's rule; but we shall see that when L, after comparing his rule with T's, places it along AC in the line of the stream, *O will consider* it to contract although *both L and T* are unconscious that it does so, and must remain unconscious of this fact, because they can have no knowledge of the existence of any stream carrying them along.

With the data of the problem, T and L satisfy themselves by direct measurement that AB and AC are each 90 foot-rules long. T and L, neither of whom recognise the existence of a stream, then calculate that their times to B, C respectively and back to A will be in each case $\frac{2 \times 90}{5} = 36$ seconds. And their clocks must bear this out when the trips have been made, for otherwise they could infer the influence of a stream and calculate its velocity.

O now times T's trip. By the argument on pp. 1116–1117, he sees that T makes a headway of 3 feet per second along AB and back, and therefore takes $\frac{2 \times 90}{3} = 60$ seconds for the whole journey. Consequently O says that T's clock only registers 36 seconds when it should register 60 seconds; therefore, according to O, T's clock loses.

Now L and T take precisely equal times for their trips. Therefore, by O's clock, L also takes 60 seconds. But by the argument on pp. 1116–1117, O sees that L advances at 9 feet per second from A to C, and returns from C to A at 1 foot per second. Therefore if AC = 90 feet, the total time = $\frac{90}{9} + \frac{90}{1} = 10 + 90 = 100$ seconds; but the total time according to O is only 60 seconds.

$$\therefore \text{ according to O, the length of AC is only } \frac{60}{100} \times 90 = 54 \text{ feet.}$$

$$\left[\text{As a check, note that } \frac{54}{9} + \frac{54}{1} = 6 + 54 = 60 \text{ seconds.}\right]$$

It is true that L marked out AC by taking 90 of his foot-rule lengths; therefore O is forced to conclude that L's foot-rule is only $\frac{54}{90} = \frac{3}{5}$ foot long, and so O says that the stream causes L's foot-rule, when placed along it, to contract to $\frac{3}{5}$ foot.

Further, as L also records the time of his trip as 36 seconds, O says that L's clock loses at just the same rate as T's clock.

We may summarise these results as follows:

O says that (i) clocks in the world of T and L lose time; they register an interval which is really 5 minutes long as only 3 minutes (60 : 36 = 5 : 3);

(ii) a foot-rule in the world of T and L measures 1 foot when placed along AB at right angles to the stream, but only measures $\frac{3}{5}$ foot when placed along AC in the line of the stream.

T and L say that (i) their clocks keep normal time;

(ii) their foot-rules remain 1 foot long, in whatever position they are placed.

WHO IS RIGHT?

It seems absurd to suggest that all of them are right. Let us, however, inquire what L thinks about O. Suppose that O and his brother O′ mark out two courses, AB and AD, each of length 90 feet, along CA produced and AB, in the air just above the ocean.

Then L says there is a current in the air of 4 feet per second which carries O and O′ in the direction C → A; O and O′, of course, say that the air is at rest and that L is drifting with the water in the direction A → C at 4 feet per second.

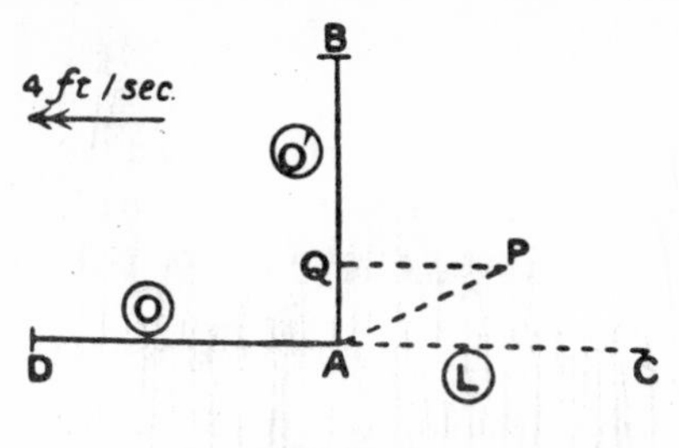

FIGURE 13

Suppose now that O and O′ fly through the air at 5 feet per second (*i.e.* relatively to the air). O′ flies to B and back again to A, O starts at the same time as O′ and flies to D and back again to A. They both arrive back at A at the same moment. This is the experimental fact established in the M.-M. experiment, which needs explanation.

It is clear that L's views (or T's views) about O and O′ are precisely the same as those which O formed about T and L. The arithmetical calculations are identical and need not be repeated. The results may be expressed as follows:

L and T say that (i) clocks in the world of O and O′ lose time; they register an interval which is really 5 minutes long as only 3 minutes;
(ii) a foot-rule in the world of O and O′ measures 1 foot when placed along AB at right angles to the current, but only measures ⅗ foot when placed along AD in the line of the current.

O and O′ say that (i) their clocks keep normal time;
(ii) their foot-rules remain 1 foot long, in whatever position they are placed.

It is clear, therefore, that any argument that can be used to support the views of O or O′ can be applied with equal force to support the views of T and L. We must, therefore, regard both views as equally true. We are therefore forced to conclude that each world, the world of O, O′, and the world of T, L, has its own standard of time-measurement and its own standard of length-measurement. If one world is moving relatively to another world, their standards of time and space automatically become different.

Suppose two people come together and compare their clocks to make sure they run at the same rate, and compare their foot-rules to make sure they agree, and suppose that afterwards they separate at a uniform rate, one from the other, along a line AC. Now imagine two explosions to take place at different times at different places somewhere on AC. Each observer, making proper allowance for the time sound takes to travel, can measure the time-interval between the two events and the distance-interval of the spots at which they occurred. But their measurements will not agree, either as regards time-interval or as regards distance-interval, for they have different standards of time and different standards of length.

There is indeed one measurement about which they will agree, namely, the velocity of a ray of light: each of them, using his own clock and his own rule, will find experimentally that a light-wave travels at 300,000 km. per second.[4]

NOTE.—The statement (see p. 1119) that further repetitions of the M.-M. experiment have confirmed the conclusion that no ether-stream can be detected requires some qualification. In 1925, Professor *Dayton Miller* announced that he had obtained results which showed the existence of a drift which varied from zero at sea-level up to 10 km. per second at the summit of Mount Wilson. This interpretation of Professor Miller's experiments has not, however, received support, and the conclusions inferred from the Michelson-Morley experiment remain generally accepted.

CLOCKS

"Alice looked round in great surprise. 'Why, I do believe we've been under this tree all the time! Everything's just as it was!'

" 'Of course it is,' said the Queen; 'what would you have it?'

---

[4] *Exercise*

1. A man's foot-rule is really only 10 inches long; what is the true length of a fence which the man measures as 12 yards? What will the man say is the length of a fence whose true length is 20 yards?
2. A foot-rule contracts to ¾ of its proper length. What is the true length of a line which according to this foot-rule is y feet? If the foot-rule is used to measure the length of a line whose true length is z feet, what result is obtained?
3. O says that two events occurred at an interval of 12 seconds at places 18 feet apart. What measurements are given by L, if his clock only registers 45 minutes for each hour of O's clock, and if his foot-rule only measures 8 inches according to O's rule?
4. A stream flows at 3 feet per second, and a man can row at 5 feet per second through the water. The width of the stream is 40 feet. Find the times taken to row (i) straight across the stream and back, (ii) 40 feet downstream and back.
5. With the data of No. 4, find how far the man can row downstream and back if he takes the same time as he would to go straight across and back.
6. It is found that a bullet from a rifle travels 1100 feet in the first second of motion. The bullet is fired along a railway line from a train at a moment when a man is 1100 feet away in the line of fire. There is no wind. Does the bullet or the noise of the explosion reach the man first if the train (i) is moving towards the man, (ii) is at rest, (iii) is moving away from the man?
7. A stream flows at u feet per second, and a man can row at c feet per second through the water. The width of the stream is x feet, and the man can row straight across and back in the same time that he can row x_1 feet downstream and back. Prove that—

$$\text{(i)}\quad \frac{2x}{\sqrt{c^2-u^2}}=\frac{x_1}{c+u}+\frac{x_1}{c-u}.$$

$$\text{(ii)}\quad x_1=x\sqrt{\left(1-\frac{u^2}{c^2}\right)}$$

" 'Well, in *our* country,' said Alice, 'you'd generally get to somewhere else if you ran very fast for a long time, as we've been doing.'

" 'A slow sort of country!' said the Queen. 'Now *here*, you see, it takes all the running you can do to keep in the same place. If you want to get somewhere else, you must run at least twice as fast as that!' "

—*Through the Looking-Glass.*

OBSERVATIONS AT DIFFERENT PLACES

If several observers, who are recording the times of occurrence of a series of events, wish to exchange their results, it is necessary for them to compare their clocks. Preferably the clocks should be synchronised, but it would be sufficient to note the differencce between each clock and some standard clock. The standard British clock registers what we call "Greenwich time."

Synchronising is a simple matter if the observers and their clocks are all at one place, but if the observation stations are far apart direct comparison is impossible, and we are forced to rely on indirect methods which may not be proof against criticism. To transport a clock from one station to another is not a reliable method, because the journey itself may set up an error in the running of the clock. The best method is to send signals from a standard station to all other stations, and use these signals to synchronise the various clocks or record their errors; this in fact is done each day by the wireless signals sent out at noon from Greenwich. Wireless signals travel with the velocity of light, and therefore, for such comparatively small distances as we are concerned with on the Earth, the time of transit of the signal is usually negligible. But for large distances, such as the distance of the Sun from the Earth, the time taken by the signal is material, and allowance must be made for it in setting the clock. We shall see, however, that the process involves another difficulty which we are powerless to remove. This is best illustrated by a numerical example. In order to avoid big numbers and to bring the arithmetic of this chapter into line with that of the last, we shall introduce (temporarily) a new unit of length:

$$60{,}000 \text{ km.} = 1 \textit{ leg.}$$

The velocity of light is therefore 5 legs per second.

SYNCHRONISING TWO CLOCKS

Suppose that two observers A and C, relatively at rest to each other, are at a distance of 75 legs apart, as measured by their own rules; this distance is about twelve times as much as the distance of the Moon from the Earth. We shall examine the process by which A and C attempt to synchronise their clocks.

Since light travels at 5 legs per second, A and C calculate that a ray of light sent by either to the other will take $75/5 = 15$ seconds to travel across the space separating them. It is agreed that, at the instant when A's clock

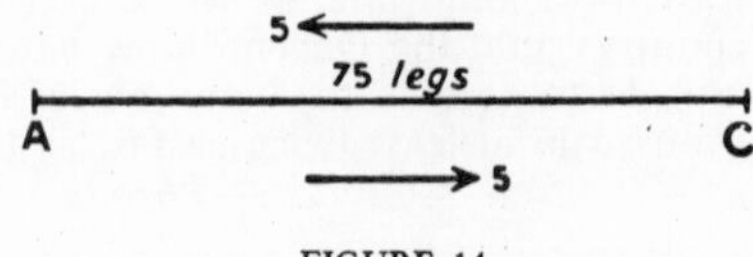

FIGURE 14

records zero hour, A shall send a light-signal to C, and that C, immediately he receives it, shall reflect it back to A.

C therefore *sets* his clock at 15 *seconds past zero*, but does *not start it* until the signal from A arrives. Immediately C receives the signal, he starts his clock and believes that it now agrees with A's clock. This opinion is shared by A, who, when he sees his clock indicating 15 seconds past zero, says to himself, "At this moment C is receiving my signal." From A's point of view, the fact appears to be established beyond doubt when the return signal from C reaches A at the instant his (A's) clock registers 30 seconds past zero. We know that A's receipt of the return reflected signal must occur at this instant, because otherwise A would be able to calculate his velocity through the ether (compare p. 1121), and this, as we have seen, is impossible. In the same way, C, when he sees his clock indicating 30 seconds past zero, says to himself, "At this moment A is receiving the return signal," and this opinion is confirmed by the fact that, if A then reflects the signal back to C, it will reach C when C's clock indicates 45 seconds past zero, for the same reason as before.

Now there can be no ambiguity as regards the time indicated by C's clock of an event happening to C, nor as regards the time indicated by A's clock of an event happening to A. But we shall see that there is unfortunately a great deal of uncertainty as to the time indicated by A's clock of an event happening to C, or *vice versa*. If the clocks of A and C are genuinely synchronised, this uncertainty would not exist. But if there are grounds for suspecting that A and C are mistaken in their belief that they have succeeded in synchronising their clocks, there is no direct method of either of them ascertaining the time by his own clock of an event which is happening to the other. Although, when A sees that his clock reads 15 seconds past zero, he says that at this moment his signal is arriving at C, yet he has no direct method of making sure that this statement is true. And, by enlisting the evidence of an eyewitness, we shall show that there are different, but equally trustworthy, opinions of the time recorded by A's clock of the arrival of the signal at C.

AN ONLOOKER'S OPINION

We now introduce an onlooker O, who considers that the world of A and C is moving away from him in the direction A $\rightarrow$ C at 4 legs per second.

Each individual acts on the supposition that he himself is at rest. In the following inquiry into O's opinions, we must therefore regard O as at rest and A, C as moving away from O. But if we had to inquire into the views of A or C, we should have to regard them as at rest and O as moving away from each of them in the opposite direction.

Suppose that A is passing O at the moment when A sends out his first light signal, and that O also sets his clock at zero at this instant. We can connect O with the world of A, C most easily by imagining that A combines his time-signal with a performance of the M.-M. experiment.

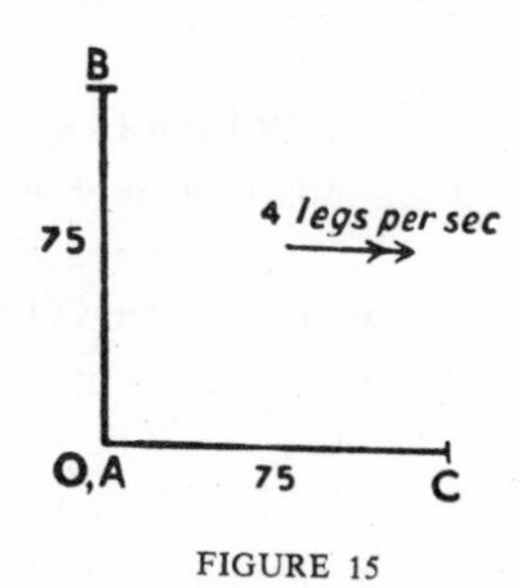

FIGURE 15

A marks out a track AB at right angles to AC and makes it 75 legs long by his rule, and places a mirror at B in the usual way. At the same time as he sends his light signal to C he sends another to B, and, as we know, both rays, reflected back, return to A at the same moment.

Now O and A agree that the length of AB is 75 legs, because for lengths across the stream their rules are identical. Also O, A, and C all agree that light travels at 5 legs per second through the ether.

Figure 16 represents *O's idea* of the path pursued by the light signal, which is directed to the mirror at B.

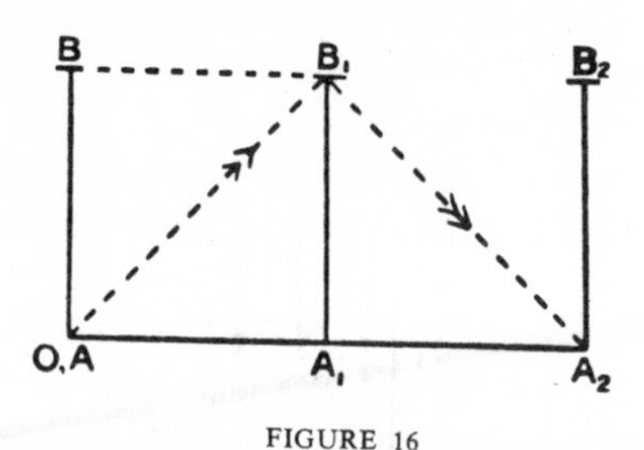

FIGURE 16

By the time the light signal impinges on the mirror at B, the arm AB has moved to the position A_1B_1, so that the signal starts from O, A and impinges on the mirror at B_1, and therefore travels along OB_1; by the time it returns to A the arm AB has moved to the position A_2B_2, so that the return path is B_1A_2.

The arm AB is advancing at 4 legs per second and the light signal travels along OB_1 at 5 legs per second. Suppose the outward journey takes t seconds. Then $OB_1 = 5t$ legs, $BB_1 = 4t$ legs, $OB = 75$ legs,

$$\therefore \text{ by Pythagoras, } (5t)^2 = (4t)^2 + 75^2.$$
$$\therefore 25t^2 - 16t^2 = 75^2 \text{ or } 9t^2 = 75^2 \text{ or } 3t = 75.$$
$$\therefore t = 75/3 = 25 \text{ seconds.}$$

$\therefore$ the total time out and back *according to O's clock* is $2 \times 25 = 50$ seconds.

But by the M.-M. experiment the ray returns to A from C at the same moment as the ray from B.

$\therefore$ *by O's clock*, the ray returns to A from C at 50 seconds past zero.

But *A's clock* registers 30 seconds past zero when the ray returns to A from C.

$\therefore$ the arrival of the ray back at A is said by O to occur at 50 seconds past zero by O's clock, and to occur at 30 seconds past zero by A's clock.

Therefore, although O's clock and A's clock agreed at zero hour, they do not agree afterwards: we may therefore say that the synchronisation between O and A has disappeared.

Let us next ascertain O's opinion as to the time when the first signal reached C.

O says that the ray from A to C is advancing at 5 legs per second towards a target C which is retreating at 4 legs per second: the ray therefore gains on the target C at $5 - 4 = 1$ leg per second. But on the return journey the ray advances at 5 legs per second towards a target A which advances to meet it at 4 legs per second: the ray therefore gains on the target A at the rate of $5 + 4 = 9$ legs per second. The distance which the ray has to gain on its target is the same on each journey (A and C believe this distance is 75 legs; O does not agree with them: but we need not stop to ascertain O's estimate of the distance), therefore the outward journey $A \to C$ takes 9 times as long as the journey back from C to A, so that $9/10$ths of the total time is spent on the outward journey, and $1/10$th of the total time on the return journey. Now the total time, out and back, by O's clock is 50 seconds.

$\therefore$ O says the outward journey, $A \to C$, takes $9/10$th of $50 = 45$ seconds, and the journey back from C to A takes $1/10$th of $50 = 5$ seconds.

$\therefore$ *O says* that the ray arrives at C at 45 seconds past zero *by O's clock.*

Also since A's clock records the total time, out and back, as 30 seconds, we see in the same way that *O says* that the ray arrives at C at $9/10$th of $30 = 27$ seconds past zero by *A's clock.*

Further, when the ray arrives at C, we know that C's clock registers 15 seconds past zero and is set going at this instant.

The occurrence of the event consisting of the arrival of the ray at C is therefore *registered by O* as follows:

O's Clock.	A's Clock.	C's Clock.
45 seconds past zero.	27 seconds past zero.	15 seconds past zero.

This is O's opinion of the operation. A of course does not agree with O; when A's clock registers 27 seconds past zero, A says that it is long past the time of C's receipt of the signal.

O, however, says that C's clock has been set $27 - 15 = 12$ second-spaces behind A's clock.

We can easily continue this process of calculating the times registered by O of further events. Consider the arrival of the ray back at A from C.

C dispatches the ray to A and receives it back again reflected from A after a total interval of $\frac{2 \times 75}{5} = 30$ seconds, by C's clock.

Now O says that the time from C to A is only $\frac{1}{10}$th of the total time C → A and A → C.

Therefore O says that the ray takes $\frac{1}{10}$th of $30 = 3$ seconds, by C's clock, to travel from C to A. But the time on C's clock when the ray left C was 15 seconds past zero; therefore the time on *C's clock* when the ray arrives at A is $15 + 3 = 18$ seconds past zero, *according to O.*

The occurrence of the event consisting of the arrival of the ray back at A is therefore *registered by O* as follows:

O's Clock.	A's Clock.	C's Clock.
50 seconds past zero.	30 seconds past zero.	18 seconds past zero.

It is worth comparing these two events as *recorded by O.*

	O's Clock.	A's Clock.	C's Clock.
Event I. (arrival at C)	45	27	15 seconds past zero.
Event II. (return to A)	50	30	18 seconds past zero.
Time-interval between the events	5	3	3 seconds.

O therefore says that A's clock and C's clock run at the same rate (each registers the *interval* between the two events as 3 seconds), but both their clocks lose time (each records an interval as 3 seconds which is really 5 seconds long) and C's clock has been set 12 seconds behind A's clock.

WHAT OTHER ONLOOKERS THINK

Now these calculations which O has made have depended on the fact that the world of A, C is moving away from O at 4 legs per second. Suppose that there is another onlooker P, who notes that the world of A, C is moving away from him in the direction A → C at (say) 3 legs per second. Then the same argument which has been used to obtain O's records may be used to obtain P's records of the various events, but the arithmetic will be different, and P's opinion about the behaviour of the clocks of A and C will not agree numerically with O's opinion. P will say that A and C have failed to synchronise, but will form a different estimate of the amount C's clock is behind A's clock, and will assess at a different figure the rate at which both A's clock and C's clock lose. It is left to the reader to make the necessary calculations, see Footnote 5. Each onlooker, therefore, has his own standard of time; and his judgment of the time-interval separating two events will differ from that formed by another observer moving relatively to him. This agrees with what has been said in the previous chapter. But the example we have just taken shows also that it is impossible to synchronise two clocks which are situated at different places. For, although the inhabitants of the world in which the clocks are at rest believe that they have secured synchronisation, the observers in other worlds not only deny that they have done so, but disagree amongst themselves as to the amount of the difference between the clocks. No setting of the clocks can therefore ever secure general approval, or indeed approval by the inhabitants of more than one world.

SIMULTANEOUS EVENTS

If after A and C believe they have synchronised their clocks, an event takes place at A and another event takes place at C, and if each records the time of the event which has happened to himself, and if these two records are the same, then A and C will say that the two events happened simultaneously. But with the data of our example, we see that O will say that the event at A took place before the event at C, for according to O when A's clock reads 27 seconds past zero C's clock reads 15 seconds past zero. Therefore if A and C both say that the times of the events are 27 seconds past zero, O says that, when the event occurs at A, C's clock has only got as far as 15 seconds past zero and therefore the event at C, timed as 27 seconds past zero at C, has not yet occurred. In fact the time-interval between these two events is $27 - 15 = 12$ seconds as measured by the clock-rate of A or C, which is equivalent to 20 seconds as measured by the clock-rate of O, for 5 of O's seconds are the same as 3 of A's seconds or C's seconds. O will therefore say that the event at A took place 20 seconds (by O's clock) before the event at C.

A and C therefore call two events simultaneous which O considers occur at a definite time-interval, and other onlookers will agree with O in saying that the events are not simultaneous, but will disagree with O as to the length of the time-interval between them. It is therefore impossible to attach any meaning to a general statement that two events at different places occurred at the same time. If the time-standard of one world makes them simultaneous, the time-standard of other worlds require a time-interval between the events. Since there is no reason to prefer the opinion of one onlooker to that of any other, we cannot say that any one opinion is more correct than any other. The bare statement that two events at different places were simultaneous is, therefore, devoid of meaning, unless we also specify the world in which this time-measure has been made.

UNION OF SPACE AND TIME

Time by itself ceases to be an absolute idea; it is a property of the world in which it is measured, and each world has its own standard.

Each individual has, of course, his own time-rule and his own distance-rule which he thinks of as absolute, because he thinks of his own world as at rest. But in a sense this is a delusion, because a transference to another world will modify each of them; a change in time-measure is bound up with a change in distance-measure. As we have already seen, the onlookers do not agree with A, C, or each other as to the distance between the two places where the events occurred, any more than they agree with the time-interval between the events. To quote the celebrated phrase of *Minkowski*: "From now onwards space and time sink to the position of mere shadows, and only a sort of union of both can claim an independent or absolute existence"—*i.e.* an existence to which all onlookers will give equal recognition and apply equal standards of measurement. We shall see later what form this union takes.[5]

[5] *Exercise*

1. A and C measure their distance apart as 50 legs; an onlooker P notes that the world of A, C is moving away from him in the direction A → C at 3 legs per second. A passes P at zero hour by A's clock and P's clock, at which moment A sends a light signal to C in order to synchronise with C; this signal is reflected back to A. What is P's estimate of the times recorded on the three clocks of (i) the arrival of the signal at C, and (ii) the return of the signal to A?
2. Repeat No. 1, if A and C are 75 legs apart.
3. With the data of No. 1, D is a person in the world of A and C at a distance of 100 legs from A and on the other side of A from C. If A and D synchronise, find the difference between their clocks according to P in terms of second-spaces (i) on A's clock, (ii) on P's clock.

ALGEBRAIC RELATIONS BETWEEN TWO WORLDS

"The progress of Science consists in observing interconnections and in showing with a patient ingenuity that the events of this ever-shifting world are but examples of a few general relations, called laws. To see what is general in what is particular, and what is permanent in what is transitory, is the aim of scientific thought."—A. N. WHITEHEAD, *An Introduction to Mathematics.*

GENERALISATIONS

We have shown in the previous discussion, by means of numerical examples, that any eye-witness will consider that standards of measurement of distance and time vary from one world to another. The real nature of these variations cannot be appreciated unless we pass on from numerical illustrations to general formulæ. We therefore shall now proceed to express in algebraic form the relations between two worlds which are moving with *uniform* velocity relatively one to another. These formulæ may then be utilised for solving special numerical cases.

It will simplify the work if we introduce a new unit of length:

300,000 km. (*i.e.* 5 legs) = 1 *lux.*

The velocity of light is therefore 1 lux per second.

STATEMENT OF THE PROBLEM

It may assist the reader if we state in great detail the problem proposed for solution in this section.

The world of A and C is moving away from O in the direction A → C at a uniform velocity of u luxes per second; at the instant when A passes O, both A and O set their clocks at zero hour. A and C are at rest relatively to each other, and they measure their distance apart as x_1 luxes. A and C believe they have synchronised their clocks.

An event (Event I.) occurs at A at zero hour by A's clock; another event (Event II.) occurs at C at t_1 seconds past zero by C's clock. Therefore in the world of A and C the *distance-interval* between the two events is x_1 luxes, and the *time-interval* between the two events is t_1 seconds. There is complete agreement between A and C as to both of these interval

4. Repeat No. 1, assuming that the world of A, C is moving away from P in the direction C → A at 3 legs per second, P and A being as before at the same place at zero hour.
5. With the data of No. 1, if an Event I. occurs at A and an Event II. occurs at C, and if A and C describe these events as simultaneous as recorded by their own clocks, which event will P consider to have occurred first? Repeat this problem with the data of No. 4.
6. Two events, I., II., occur simultaneously at different places in the world of A and C. An onlooker O says I. occurred before II. Would it ever be possible for some other onlooker to say that II. occurred before I.?
7. With the data of No. 1, find P's estimate of the distance of A from C.

measurements. Each regards both himself and the other as at rest in the ether. Their distance-measures agree because they can use the same rule to measure out AC; their time-measures agree, because otherwise they could deduce the velocity of their common world through the ether.

Next consider O's point of view. He says that Event I. occurs at O at zero hour, and that Event II. occurs at C at (say) t seconds past zero by his own (O's) clock. O regards himself as at rest and A, C as moving away from him. O therefore takes the distance-interval between the two events as the distance of C from him at the moment when Event II. takes place. Suppose that O's measure of this distance is x luxes. Then *O says* that the distance-interval between the two events is x luxes and the time-interval between the two events is t seconds.

In short, the interval between the two events is registered by A or C as x_1 luxes, t_1 seconds, and by O as x luxes, t seconds.

What are the formulæ which connect x, t with x_1, t_1?

Before tackling this general problem, we shall ascertain O's opinion about the measuring-rule used by A or C, the running of their clocks, and their attempts to synchronise.

MEASURING-RULES

A marks out a length AC of x_1 luxes along the line of motion of A's world relatively to O. What is the length of AC, according to O?

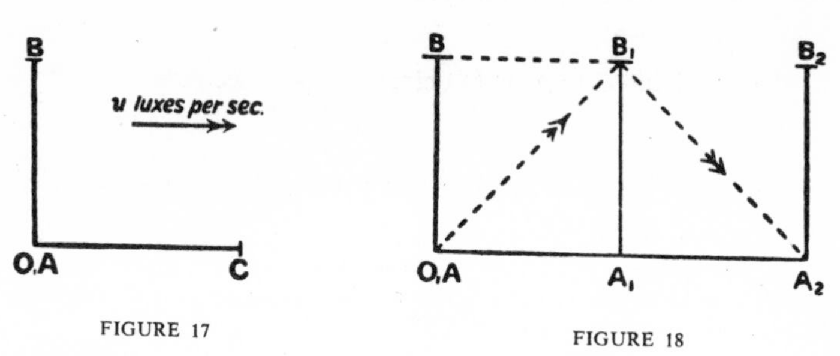

FIGURE 17 FIGURE 18

Suppose that O is watching A performing the M.-M. experiment. A and C agree that the lengths of AC and AB are each x_1 luxes, O agrees that the length of AB is x_1 luxes, but says that the length of AC is different, say z luxes.

O says that the arm AB moves away from him at u luxes per second, so that the ray sent towards the mirror at B impinges on it when AB has moved into the position A_1B_1; the path of the ray is therefore AB_1. Similarly, the ray returns to A when AB has moved into the position A_2B_2, so that the return path is B_1A_2.

Suppose the time from A to B_1 or from B_1 to A_2, is k seconds by O's clock. O makes the following calculations:

$AB_1 = k$ luxes (light travels along AB_1 at 1 lux per second).
$BB_1 = ku$ luxes (AB advances at u luxes per second).
$AB = x_1$ luxes (O agrees with A's measurements across the stream).

$$\therefore \text{ by Pythagoras, } k^2 = k^2u^2 + {x_1}^2$$

$$\therefore k^2 - k^2u^2 = {x_1}^2 \text{ or } k^2(1 - u^2) = {x_1}^2$$

$$\therefore k^2 = \frac{{x_1}^2}{1 - u^2}.$$

The total time by O's clock from A to B and back is $2k$ seconds.
$\therefore$ the total time by O's clock from A to C and back is $2k$ seconds.
But O can also reason as follows:
From A to C, the ray travels at 1 lux per second towards a target C, z luxes away, which is retreating at u luxes per second. Therefore the ray gains on the target at $(1 - u)$ luxes per second.

$$\therefore \text{ the time from A to C by O's clock is } \frac{z}{1 - u} \text{ seconds.}$$

Similarly from C to A the ray travels at 1 lux per second towards a target A, z luxes away, which is advancing at u luxes per second. Therefore the ray gains on the target at $(1 + u)$ luxes per second.
But the total time by O's clock is $2k$ seconds.

$$\therefore \text{ the time from C to A by O's clock is } \frac{z}{1 + u} \text{ seconds}$$

$$\therefore \text{ the total time from A to C and back is } \frac{z}{1 - u} + \frac{z}{1 + u} \text{ seconds}$$

$$= \frac{z(1 + u) + z(1 - u)}{(1 - u)(1 + u)} = \frac{z + zu + z - zu}{1 - u^2} = \frac{2z}{1 - u^2} \text{ seconds.}$$

But the total time by O's clock is $2k$ seconds.

$$\therefore \frac{2z}{1 - u^2} = 2k \text{ or } z = k(1 - u^2)$$

$$\therefore z^2 = k^2(1 - u^2)^2, \text{ but } k^2 = \frac{{x_1}^2}{1 - u^2}$$

$$\therefore z^2 = \frac{{x_1}^2}{1 - u^2}(1 - u^2)^2 = {x_1}^2(1 - u^2)$$

$$\therefore z = x_1\sqrt{(1 - u^2)}.$$

Therefore *O says* that a length in the direction of motion which A and C measure as x_1 luxes is really $x_1\sqrt{(1-u^2)}$ luxes:

Or, in proportion, what A and C measure as 1 lux is in O's opinion really $\sqrt{(1-u^2)}$ luxes.

Now $\sqrt{(1-u^2)}$ must be less than 1; consequently *O says* that the measuring-rule used by A and C, when placed along the line of motion, *contracts*; and the contraction-ratio is $\sqrt{(1-u^2)}$.

COMPARISON BY O OF A'S CLOCK AND C'S CLOCK WITH O'S CLOCK

Our numerical examples have shown that every one will agree that A's clock runs at the same rate as C's clock. The reason for this may be stated as follows:

An essential feature in every argument is that each individual regards himself as at rest in the ether and that all the observations he makes must bear this out. He cannot make any measurement which will reveal his velocity through the ether. A and C agree that their distance apart is x_1 luxes: they therefore argue that a time signal sent from either to the other and reflected back will return after $2x_1$ seconds, and their clocks must bear this out. But the experiment in which A sends a signal to C and receives it back again is identical with the experiment in which C sends a signal to A and receives it back again. Both A's clock and C's clock record the time of this experiment as performed by each of them as $2x_1$ seconds. Therefore A's clock and C's clock must run at the same rate. We have seen from numerical examples that O admits this, but says that both clocks lose and that they have not been synchronised. Let us now calculate O's estimate of the time-difference between A's clock and C's clock.

In order to synchronise the clocks, A proposes at zero hour by his clock to send a light-ray to C. As they agree that AC is x_1 luxes, they calculate that the signal will take x_1 seconds to reach C. Consequently C sets his clock at x_1 seconds past zero and starts it at the instant the light-ray arrives.

Now we have just seen that by O's clock the time from A to C is $\dfrac{z}{1-u}$ seconds, and the total time out and back is $\dfrac{2z}{1-u^2}$ seconds. O therefore says that the fraction of the total time out and back occupied by the journey out is

$$\frac{z}{1-u} \div \frac{2z}{1-u^2} = \frac{z}{1-u} \times \frac{(1+u)(1-u)}{2z} = \frac{1+u}{2}.$$

Now A's clock registers $2x_1$ seconds for the total time out and back. Therefore O says that, when the ray arrives at C, A's clock registers

$\dfrac{1+u}{2} \times 2x_1 = x_1(1+u)$ seconds past zero. But at this instant C's clock starts off at x_1 seconds past zero.

$$\therefore \text{A's clock is ahead of C's clock by } x_1(1+u) - x_1 \text{ seconds}$$
$$= x_1 + x_1u - x_1 = x_1u \text{ seconds.}$$

Therefore when A and C think they have synchronised their clocks, O says that A's clock is x_1u second-spaces ahead of C's clock.

The difference between the clocks depends on the value of x_1, the length of AC. Therefore the farther C is away from A in the direction of motion of the world of A, C from O, the more A's clock is ahead of C's clock, according to O. Suppose, for example, the world of A, C is moving due east away from O. Then A's clock is ahead of any clock east of A and is behind any clock west of A. Both these results are expressed in the statement given above, because, if C is west of A, x_1 is negative, and a clock which is a negative number of seconds ahead of another clock is, of course, a positive number of seconds behind it.

We must therefore regard each place on the line of motion of AC as having its own clock: the inhabitants of the world of A, C think all these clocks are synchronised, but O says each registers a local time whose difference from that of A is given by the formula above. We may express the facts by a diagram showing the local time, according to O, of the instant when A is passing O, which is taken as zero hour both by A and O.

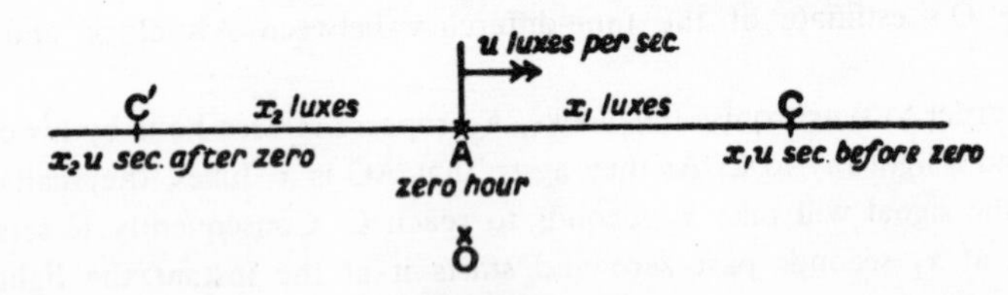

FIGURE 19

The distances indicated in the diagram represent A's or C's measurements.

CLOCK-RATES

A and C record the time-interval between two events as 1 second. What is O's estimate of this time-interval by his own clock?

With our previous notation, we know that O says that the time from A to C and back is $2k$ seconds by O's clock, where $k^2 = \dfrac{x_1^2}{1-u^2}$ or $k = \dfrac{x_1}{\sqrt{(1-u^2)}}$.

But A says that the time from A to C and back is $2x_1$ seconds by A's clock and O must agree with him.

$\therefore$ O says that $2x_1$ seconds on A's clock measures the same time-interval

$$\text{as } 2k \text{ seconds} = \frac{2x_1}{\sqrt{(1-u^2)}} \text{ seconds on O's clock.}$$

$\therefore$ O says that (in proportion) 1 second on A's clock measures the same

$$\text{time-interval as } \frac{1}{\sqrt{(1-u^2)}} \text{ seconds on O's clock.}$$

It is important to remember that this is a statement of *O's view* about the behaviour of A's clock.

Since $\sqrt{(1-u^2)}$ is less than 1, $\dfrac{1}{\sqrt{(1-u^2)}}$ is greater than 1, and therefore O says that A's clock loses. But, of course, A equally says that O's clock loses. Our results always depend on the point of view from which the progress of events is being observed.

TIME AND DISTANCE INTERVALS BETWEEN TWO EVENTS

The data which determine the two events have been stated in great detail on pp. 1134–1135. The diagram represents *O's view* of the events.

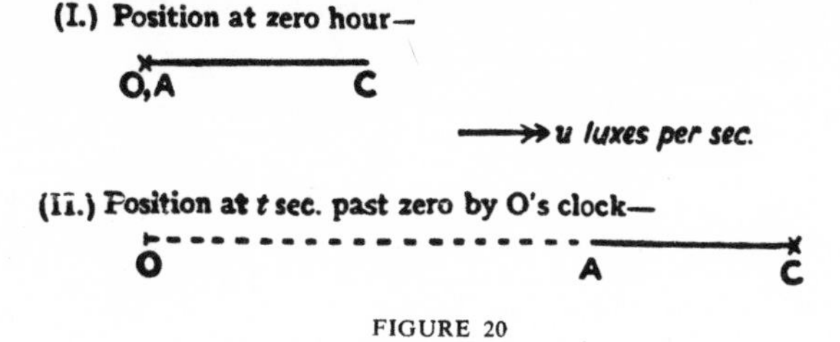

FIGURE 20

Event I. occurs at zero hour at A at the instant when A is passing O. Event II. occurs at C at t seconds past zero by O's clock. A and C say that the distance-interval between the two events is x_1 luxes, *i.e.* their measure of AC is x_1 luxes. O says that Event I. occurred at O, and that Event II. occurred at the position of C at t seconds past zero by O's clock. O therefore says that the distance-interval between the two events is x luxes, which is his measure of the length of OC in (II.). O also says that the measure of OA in (II.) is ut luxes, because A is moving away from him at u luxes per second.

$\therefore$ O says that by his rule $AC = x - ut$ luxes.

Now A measures AC as x_1 luxes and O says that a measurement of 1 lux by A is really $\sqrt{(1-u^2)}$ luxes (see p. 1137). Therefore O says that AC is really $x_1\sqrt{(1-u^2)}$ luxes.

$$\therefore x_1\sqrt{(1-u^2)} = x - ut$$

$$\therefore x_1 = \frac{x - ut}{\sqrt{(1-u^2)}}.$$

This relation is very important.

Again, suppose that the time of Event II. at C is recorded by C's clock as t_1 seconds past zero. Then A and C must agree that the time-interval between the events is t_1 seconds.

Now O says that A's clock is x_1u seconds ahead of C's clock (see p. 1138). Therefore O says that, when Event II. occurs, the time on A's clock is $t_1 + x_1u$ seconds past zero. But we know that 1 second on A's clock measures the same time-interval as $\dfrac{1}{\sqrt{(1-u^2)}}$ seconds on O's clock.

$\therefore$ when Event II. occurs, the time on O's clock is $\dfrac{t_1 + x_1u}{\sqrt{(1-u^2)}}$ seconds past zero; but the time on O's clock is t seconds past zero.

$$\therefore t = \frac{t_1 + x_1u}{\sqrt{(1-u^2)}}$$

$$\therefore t_1 + x_1u = t\sqrt{(1-u^2)}; \text{ now } x_1 = \frac{x - ut}{\sqrt{(1-u^2)}}$$

$$\therefore t_1 = t\sqrt{(1-u^2)} - \frac{u(x-ut)}{\sqrt{(1-u^2)}} = \frac{t(1-u^2) - u(x-ut)}{\sqrt{(1-u^2)}}$$

$$= \frac{t - u^2t - ux + u^2t}{\sqrt{(1-u^2)}}$$

$$\therefore t_1 = \frac{t - ux}{\sqrt{(1-u^2)}}.$$

This relation is also of great importance. It may be a help to the reader if we re-state what has been established.

Two events occur at distance-interval x_1 luxes and time-interval t_1 seconds according to A, C, and at distance-interval x luxes and time-interval t seconds according to O. The world of A, C is moving away from the world of O at u luxes per second, and distances are measured as positive in the direction of motion of A from O. Then A's records are connected with O's records by the formulæ:

$$x_1 = \frac{x - ut}{\sqrt{(1 - u^2)}}; \quad t_1 = \frac{t - ux}{\sqrt{(1 - u^2)}}.$$

If then we know the distance interval and the time-interval between two events as recorded in one world, we can calculate the distance and time-intervals between these two events as recorded in any other world moving with uniform velocity relative to the former, along the line joining the two events.

A'S OPINION OF O'S RECORDS

It has been pointed out frequently in previous chapters that there is no observer whose records are entitled to more respect than those of any other observer. It is therefore essential to show that the formulæ just obtained are consistent with this view. Using the same notation and axes as before, A says that O is moving away from him at $(-u)$ luxes per second. Now A says that the distance and time-intervals between the events are x_1 luxes and t_1 seconds.

∴ the formulæ just obtained show that

$$\text{O's distance-interval should} = \frac{x_1 - (-u)t_1}{\sqrt{(1 - u^2)}} = \frac{x_1 + ut_1}{\sqrt{(1 - u^2)}}$$

$$\text{and O's time-interval should} = \frac{t_1 - (-u)x_1}{\sqrt{(1 - u^2)}} = \frac{t_1 + ux_1}{\sqrt{(1 - u^2)}}$$

∴ the formulæ just obtained should be equivalent to

$$x = \frac{x_1 + ut_1}{\sqrt{(1 - u^2)}} \text{ and } t = \frac{t_1 + ux_1}{\sqrt{(1 - u^2)}}.$$

Unless they are, there is not that reciprocal relation between O and A which the Theory of Relativity requires. We may state the problem as follows:

$$\textit{Given that } x_1 = \frac{x - ut}{\sqrt{(1 - u^2)}} \text{ and } t_1 = \frac{t - ux}{\sqrt{(1 - u^2)}}$$

$$\textit{Prove that } x = \frac{x_1 + ut_1}{\sqrt{(1 - u^2)}} \text{ and } t = \frac{t_1 + ux_1}{\sqrt{(1 - u^2)}}$$

(i) We have $x_1 + ut_1 = \dfrac{x - ut}{\sqrt{(1 - u^2)}} + \dfrac{u(t - ux)}{\sqrt{(1 - u^2)}}$

$$= \frac{x - ut + ut - u^2x}{\sqrt{(1 - u^2)}} = \frac{x(1 - u^2)}{\sqrt{(1 - u^2)}}$$

$$= x\sqrt{(1 - u^2)}$$

$$\therefore x = \frac{x_1 + ut_1}{\sqrt{(1 - u^2)}}.$$

(ii) We have $t_1 + ux_1 = \dfrac{t - ux}{\sqrt{(1 - u^2)}} + \dfrac{u(x - ut)}{\sqrt{(1 - u^2)}}$

$$= \frac{t - ux + ux - u^2t}{\sqrt{(1 - u^2)}} = \frac{t(1 - u^2)}{\sqrt{(1 - u^2)}}$$

$$= t\sqrt{(1 - u^2)}$$

$$\therefore t = \frac{t_1 + ux_1}{\sqrt{(1 - u^2)}}.$$

We therefore see that the relations which express A's opinion of O's world are consistent with, and can be deduced from, the relations which express O's opinion of A's world.

THE VELOCITY OF LIGHT

The formulæ which connect the two worlds introduce the expression $\sqrt{(1 - u^2)}$, which is imaginary if $u > 1$, *i.e.* if the velocity of one world relatively to the other is greater than the velocity of light. We therefore say that we can have no experience of a body moving with a velocity greater than that of light. And in all our results u must stand for a fraction between $+1$ and -1. It is customary to represent $\sqrt{(1 - u^2)}$ by $\dfrac{1}{\beta}$ or to put $\beta = \dfrac{1}{\sqrt{(1 - u^2)}}$ so that $\beta > 1$. In this case, the standard formulæ may be written:

$$x_1 = \beta(x - ut);\ t_1 = \beta(t - ux)$$

or

$$x = \beta(x_1 + ut_1);\ t = \beta(t_1 + ux_1)$$

$$\text{where } \beta = \frac{1}{\sqrt{(1 - u^2)}} > 1.$$

And the results on pp. 1138–1139 may be stated as follows:

(i) O says that the length of a line in the direction of motion which A measures as 1 lux is $\frac{1}{\beta}$ luxes.

(ii) O says that a time-interval which A's clock records as 1 second is β seconds.[6]

[6] *Exercise*

1. The world of A is moving at 3/5 lux per second due east from O. What is O's opinion about (i) the length of A's foot-rule, (ii) the rate of running of A's clock? What is A's opinion about O's foot-rule and O's clock?
2. A and C, who are relatively at rest at a distance apart of 5 luxes, have synchronised their clocks; the world of A, C is moving away from O in the direction A → C at 7/25 lux per second. A passes O at zero hour by O's clock and A's clock. What does O say is the difference between A's clock and C's clock? D is a place in the world of A, C, such that DA = 10 luxes, DC = 15 luxes. What does O say is the difference between D's clock and A's clock? What does O say is the time recorded by the clocks of A, C, D when O's clock records 25 seconds past zero?
3. With the data of No. 1, A records two events as happening at an interval of 5 seconds and at a distance apart of 3 luxes, the second event being due east of the first event. What are the time and distance-intervals of the events as recorded by O?
4. With the data of No. 3, solve the question if the second event is due west of the first event.
5. Given that $x_1 = \dfrac{x - ut}{\sqrt{1 - u^2}}$ and $x = \dfrac{x_1 + ut_1}{\sqrt{1 - u^2}}$

   Prove that $t_1 = \dfrac{t - ux}{\sqrt{1 - u^2}}$ and $t = \dfrac{t_1 + ux_1}{\sqrt{1 - u^2}}$.
6. If Event I. is the dispatch of a light-signal by A and Event II. is the receipt of the light-signal by C, show that with the usual notation (i) $x_1 = t_1$, (ii) $x = t$. What does this mean in terms of O's opinion?
7. Using the equations on pp. 1141–1142, prove that $x^2 - t^2$ is always equal to $x_1^2 - t_1^2$.

PART VI

Mathematics and Social Science

1. Gustav Theodor Fechner *by* EDWIN G. BORING
2. Classification of Men According to Their Natural Gifts *by* SIR FRANCIS GALTON
3. Mathematics of Population and Food *by* THOMAS ROBERT MALTHUS
4. Mathematics of Value and Demand *by* AUGUSTIN COURNOT
5. Theory of Political Economy *by* WILLIAM STANLEY JEVONS
6. Mathematics of War and Foreign Politics *by* LEWIS FRY RICHARDSON
7. Statistics of Deadly Quarrels *by* LEWIS FRY RICHARDSON
8. The Theory of Economic Behavior *by* LEONID HURWICZ
9. Theory of Games *by* S. VAJDA
10. Sociology Learns the Language of Mathematics *by* ABRAHAM KAPLAN

COMMENTARY ON

The Founder of Psychophysics

GUSTAV FECHNER'S contribution to psychology was to introduce measurement as a tool. While there is some doubt as to what it was that he measured, the importance of his innovation of method is beyond question. For a long time it was believed that mental behavior, being linked to the soul, lay outside the crass reach of arithmetic, let alone of crude physical instruments such as rods and counters. (The prejudice is not entirely without merit; one might wish that the intrepid little band of surgeons known as lobotomists would succumb to it.) As long as psychology was classed as a branch of philosophy it was a subject for speculation, perhaps for description, but certainly not for experiment. In the nineteenth century this well-established position came under the attack of two advancing fronts of thought. One was led by physiologists—Charles Bell, Johannes Müller, E. H. Weber—who investigated the psychological aspects of physiological phenomena. They looked into such matters as the relation between stimuli and their corresponding sensations, the specific quality imposed on the mind by different kinds of nerves, the capacity to discriminate among stimuli. The other front was generated by the growing philosophical belief in a "scientific or physiological psychology." Lotze [1] and Bain,[2] each in his own way, supported this point of view; Herbart,[3] though opposed to the use of experiments, gave psychology status as a separate branch of learning. "He took it out of both philosophy and physiology and sent it forth with a mission of its own."

But it was Fechner who finally demolished the old edifice. He was a physicist, also a fervent and prolific philosopher; in these spheres, however, his accomplishments were minor. He is remembered for psychophysical experiments that were essentially a side line of his career. His results have not stood the test of research; his famous law is mainly of

[1] Rudolf Hermann Lotze (1817–1881) was a German philosopher who in his writings (e.g., *Medizinische Psychologie oder Physiologie der Seele*) stressed the view that the mind though an immaterial principal could only act on the body and be acted on by it through mechanical means. His opinions were said to have contributed to the destruction of the "phantom of Hegelian wisdom" and to have "vindicated the independent position of empirical philosophy." (*Encyclopaedia Britannica*, Eleventh Edition.)

[2] Alexander Bain (1818–1903) was a Scottish philosopher, educationist and psychologist. He was the first in Great Britain to emphasize the necessity of "clearing psychology of metaphysics, of applying the methods of the exact sciences to psychological phenomena and of referring these phenomena to their correlates in the nerves and brain."

[3] Johann Friedrich Herbart (1776–1841) was a famous German philosopher and educator who stressed the importance of framing educational methods on the basis of ethics and psychological knowledge. Psychology, he said, would help explain the mind to be educated; ethics would define the social goals of education, such as the cultivation of good will.

historical interest; but his thumbprint is on every page of modern psychological experiment. Fechner extended to psychology Kelvin's famous dictum that in science you cannot talk about a thing until you can measure it. In our century, to be sure, this dictum has come to sound a little glib. Mathematics has its paradoxes, astronomy its uncertainties (about what is being measured), physics having suffered certain metaphysical relapses can survive only by swallowing entire jugs of wholly contradictory measurements. As for psychology, its most brilliant and its most scandalous success has been in a realm of theory in which measurement is as welcome as Macduff at Dunsinane.

It appears that in psychology, as in other social studies, two distinct methods can be made to yield fruitful results. One is intuitive and non-quantitative—as in psychoanalysis; the other is analytic, formal, experimental, model-making—as in the physical sciences. In the last half century much has been learned about human behavior by developing and applying both methods, but no one has yet constructed a convincing unified field theory in psychology whereby the intuitive and qualitative approach can be reduced to the approach based on mechanism, microstructure and measurement. The divorcement of psychology from metaphysics undoubtedly marked an important step forward; yet it is doubtful that the decree was ever made absolute and final. At any rate one great branch of psychology has taken a new methodological partner much more closely related to philosophy than to physics. To put theories which can be tested by experiment, to discipline experiment by measurement, to evaluate measurement by mathematical apparatus, to advance new theories or models on the basis of what has been learned: these may not represent the only possible steps by which psychological knowledge can be broadened, psychological disorders diagnosed and treated, social behavior better understood. Nevertheless these steps constitute a creative and an indispensable sequence in the study of psychology and it is to the practice of this sequence that Fechner's ideas gave a major impetus.

The author of the following selection, Edwin Garrigues Boring, is a noted American scientist who since 1928 has been professor of psychology at Harvard University. His principal researches and writings have been in the fields of theoretical psychology, history of psychology, perception and sensation. Dr. Boring, born in Philadelphia in 1886, was educated at Cornell University from which he received his doctorate; he served for a time as professor of experimental psychology at Clark University, and for twenty-five years (1924–1949) was director of Harvard's Psychological Laboratory. He has been president of the American Psychological Society and his academic honors include membership in the National Academy of Sciences. The essay on Fechner is a chapter from Boring's standard work, *A History of Experimental Psychology.*

Fechner is a curiosity. His eyelids are strangely fringed and he has had a number of holes, square and round, cut, Heaven knows why, in the iris of each eye—and is altogether a bundle of oddities in person and manners. He has forgotten all the details of his "Psychophysik"; and is chiefly interested in theorizing how knots can be tied in endless strings, and how words can be written on the inner side of two slates sealed together.

—G. S. HALL *in a letter (1879) to William James*

1 Gustav Theodor Fechner

By EDWIN G. BORING

WE come at last to the formal beginning of experimental psychology, and we start with Fechner: not with Wundt, thirty-one years Fechner's junior, who published his first important but youthful psychological study two years after Fechner's epoch-making work; not with Helmholtz, twenty years younger, who was primarily a physiologist and a physicist but whose great genius extended to include psychology; but with Fechner, who was not a great philosopher nor at all a physiologist, but who performed with scientific rigor those first experiments which laid the foundation for the new psychology and still lie at the basis of its methodology. There had been, as we have seen, a psychological physiology: Johannes Müller, E. H. Weber. There had been, as we have also seen, the development of the philosophical belief in a scientific or a physiological psychology: Herbart, Lotze; Hartley, Bain. Nothing is new at its birth. The embryo had been maturing and had already assumed, in all great essentials, its later form. With Fechner it was born, quite as old, and also quite as young, as a baby.

THE DEVELOPMENT OF FECHNER'S IDEAS

Gustav Theodor Fechner (1801–1887) was a versatile man. He first acquired modest fame as professor of physics at Leipzig, but in later life he was a physicist only as the spirit of the *Naturforscher* penetrated all his work. In intention and ambition he was a philosopher, especially in his last forty years of life, but he was never famous, or even successful, in this fundamental effort that is, nevertheless, the key to his other activities. He was a humanist, a satirist, a poet in his incidental writings and an estheticist during one decade of activity. He is famous, however, for his psychophysics, and this fame was rather forced upon him. He did not wish his name to go down to posterity as a psychophysicist. He did not, like Wundt, seek to found experimental psychology. He might have been content to let experimental psychology as an independent science remain in the womb of time, could he but have established his spiritualistic *Tagesansicht* as a substitute for the current materialistic *Nachtansicht* of the universe. The world, however, chose for him; it seized upon the psychophysical experiments, which Fechner meant merely as contributory to his philosophy, and made them into an experimental psychology. A very interesting life to us, who are inquiring how psychologists are made!

Fechner was born in 1801 in the parsonage of a little village in southeastern Germany, near the border between Saxony and Silesia. His father had succeeded his grandfather as village pastor. His father was a man of independence of thought and of receptivity to new ideas. He shocked the villagers by having a lightning-rod placed upon the church tower, in the days when this precaution was regarded as a lack of faith in God's care of his own, and by preaching—as he urged that Jesus must also have done—without a wig. One can thus see in the father an anticipation of Fechner's own genius for bringing the brute facts of scientific materialism to the support of a higher spiritualism, but there can have been little, if any, direct influence of this sort, for the father died when Fechner was only five years old. Fechner, with his brother and mother, spent the next nine years with his uncle, also a preacher. Then he went for a short time to a *Gymnasium* and then for a half year to a medical and surgical academy. At the age of sixteen he was matriculated in medicine at the university in Leipzig, and at Leipzig he remained for the rest of his long life—for seventy years in all.

We are so accustomed to associating Fechner's name with the date 1860, the year of the publication of the *Elemente der Psychophysik*, and with the later years when he lived in Leipzig while Wundt's laboratory was being got under way, that we are apt to forget how old he was and how long ago he was beginning his academic life. In 1817, when Fechner went to Leipzig, Lotze was not even born. Herbart had just published his *Lehrbuch*, but his *Psychologie als Wissenschaft* was still seven years away in the future. In England, James Mill had barely completed the *History of India* and presumably had not even thought of writing a psychology. John Stuart Mill was eleven years old; Bain was not born. Phrenology had only just passed its first climax, and Gall was still writing on the functions of the brain. Flourens had not yet begun his researches on the brain. Bell, but not Magendie, had discovered the Bell-Magendie law. It was really, as the history of psychology goes, a very long time ago that Fechner went as a student to Leipzig.

It happened that E. H. Weber, the Weber after whom Fechner named "Weber's Law," went to Leipzig in the same year as *Dozent* in the faculty of medicine and was made in the following year *ausserordentlicher Professor* of comparative anatomy. After five years of study, Fechner took his degree in medicine, in 1822. Already, however, the humanistic side of the man was beginning to show itself. His first publication (1821), *Beweiss, dass der Mond aus Jodine bestehe*, was a satire on the current use of iodine as a panacea. The next year he wrote a satirical panegyric on modern medicine and natural history. Both these papers appeared under the *nom de plume* 'Dr. Mises,' and 'Dr. Mises' was reincarnated in ironical bursts altogether fourteen times from 1821 to 1876. Meanwhile Fechner's association with A. W. Volkmann had begun. Volkmann came to Leipzig as a student in medicine in 1821 and remained, later as *Dozent* and professor, for sixteen years.

After he had taken his degree, Fechner's interest shifted from biological science to physics and mathematics, and he settled down in Leipzig, at first without official appointment, for study in these fields. His means were slender, and he undertook to supplement them by the translation into German of certain French handbooks of physics and chemistry. This work must have been very laborious, for by 1830 he had translated more than a dozen volumes and nearly 9,000 pages; but it was work that brought him into prominence as a physicist. He was also appointed in 1824 to give lectures in physics at the university, and in addition he undertook physical research of his own. It was a very productive period. By 1830 he had published, including the translations, over forty articles in physical science. At this time the properties of electric currents were just beginning to become known. Ohm in 1826 had laid down the famous law that bears his name, the law that states the relation between current, resistance and electromotive force in a circuit. Fechner was drawn into the resulting problem, and in 1831 he published a paper of great importance on quantitative measurements of direct currents (*Massbestimmungen über die galvanische Kette*), a paper which made his reputation as a physicist.

The young Fechner in his thirties was a member of a delightful intellectual group in the university community at Leipzig. Volkmann, until he went to Dorpat in 1837, was also a member of this group, and it was Volkmann's sister whom Fechner married in 1833. The year after his marriage, the year in which, as we have already seen, Lotze came to Leipzig as a student, Fechner was appointed professor of physics. It must have seemed that his career was already determined. He was professor of physics at only thirty-three, with a program of work ahead of him and settled in a congenial social setting at one of the most important universities. We shall see presently how far wrong the obvious prediction would have been. Fechner for the time being kept on with his physical research, throughout the still very fertile decade of his thirties. 'Dr. Mises,' the humanistic Fechner, appeared as an author more than half a dozen times. Toward the end of this period there is, in Fechner's research, the first indication of a quasi-psychological interest: two papers on complementary colors and subjective colors in 1838, and the famous paper on subjective after-images in 1840. In general, however, Fechner was a promising younger physicist with the broad intellectual interests of the *deutscher Gelehrter.*

Fechner, however, had overworked. He had developed, as James diagnosed the disease, a 'habit-neurosis.' He had also injured his eyes in the research on after-images by gazing at the sun through colored glasses. He was prostrated, and resigned, in 1839, his chair of physics. He suffered great pain and for three years cut himself off from every one. This event seemed like a sudden and incomprehensible ending to a career so vividly begun. Then Fechner unexpectedly began to recover, and, since his malady was so little understood, his recovery appeared miraculous. This period is spoken of as the 'crisis' in Fechner's life, and it had a profound effect upon his thought and after-life.

The primary result was a deepening of Fechner's religious consciousness and his interest in the problem of the soul. Thus Fechner, quite naturally for a man with such an intense intellectual life, turned to philosophy, bringing with him a vivification of the humanistic coloring that always had been one of his attributes. His forties were, of course, a sterile decade as regards writing. 'Dr. Mises' published a book of poems in 1841 and several other papers later. The first book that showed Fechner's new tendency was *Nanna oder das Seelenleben der Pflanzen*, published in 1848. (Nanna was the Norse goddess of flowers.) For Fechner, in the materialistic age of science, to argue for the mental life of plants, even before Darwin had made the mental life of animals a crucial issue, was for him to court scientific unpopularity, but Fechner now felt himself possessed of a philosophic mission and he could not keep silence. He was troubled by materialism, as his *Büchlein vom Leben nach dem Tode* in 1836 had shown. His philosophical solution of the spiritual problem lay in his affirmation of the identity of mind and matter and in his assurance that the entire universe can be regarded as readily from the point of view of its consciousness, a view that he later called the *Tagesansicht*, as it can be viewed as inert matter, the *Nachtansicht*. Yet the demonstration of the consciousness of plants was but a step in a program.

Three years later (1851) a more important work of Fechner's appeared: *Zend-Avesta, oder über die Dinge des Himmels und des Jenseits.* Oddly enough this book contains Fechner's program of psychophysics and thus bears an ancestral relation to experimental psychology. We shall return to this matter in a moment. Fechner's general intent was that the book should be a new gospel. The title means practically "a revelation of the word." Consciousness, Fechner argued, is in all and through all. The earth, "our mother," is a being like ourselves but very much more perfect than ourselves. The soul does not die, nor can it be exorcised by the priests of materialism when all being is conscious. Fechner's argument was not rational; he was intensely persuasive and developed his theme by way of plausible analogies, which, but for their seriousness, resemble somewhat the method of Dr. Mises' satire, *Vergleichende Anatomie der Engel* (1825), where Fechner argued that the angels, as the most perfect beings, must be spherical, since the sphere is the most perfect form. Now, however, Fechner was in dead earnest. He said later in *Ueber die Seelenfrage* (1861) that he had then called four times to a sleeping public which had not yet been aroused from its bed. "I now," he went on, "say a fifth time, *'Steh' auf!'* and, if I live, I shall yet call a sixth and a seventh time, *'Steh' auf!'* and always it will be but the same *'Steh' auf!'* "

We need not go further into Fechner's philosophy. He did call, or at least so Titchener thought, a sixth and a seventh time, and these seven books with their dates show the persistence and the extent of Fechner's belief in his own gospel. They are: *Das Büchlein vom Leben nach dem Tode*, 1836; *Nanna*, 1848; *Zend-Avesta*, 1851; *Professor Schleiden und der Mond*, 1856; *Ueber die Seelenfrage*, 1861; *Die drei Motive und Gründe des Glaubens*, 1863; *Die Tagesansicht gegenüber der Nachtansicht*, 1879. As it happened, the public never "sprang out of bed," not even at the seventh call, as Fechner had predicted it would. His philosophy received some attention; many of these books of his have been reprinted in recent years; but Fechner's fame is as a psychophysicist and not as a philosopher with a mission.

His psychophysics, the sole reason for Fechner's inclusion in this book, was a by-product of his philosophy. We return to it.

It was one thing to philosophize about mind and matter as two alternative ways of regarding everything in the universe, and another thing to give the idea such concrete empirical form that it might carry weight with the materialistic intellectualism of the times or even be satisfactory to Fechner, the one-time physicist. This new philosophy, so Fechner thought, needed a solid scientific foundation. It was, as he tells us, on the morning of October 22, 1850, while he was lying in bed thinking about this problen., that the general outlines of the solution suggested themselves to him. He saw that the thing to be done was to make "the relative increase of bodily energy the measure of the increase of the corresponding mental intensity," and he had in mind just enough of the facts of this relationship to think that an arithmetic series of mental intensities might correspond to a geometric series of physical energies, that a given absolute increase of intensity might depend upon the ratio of the increase of bodily force to the total force. Fechner said that the idea was not suggested by a knowledge of Weber's results. This statement may seem strange, for Weber was in Leipzig and had published the *Tastsinn und Gemeingefühl* in 1846, and it was important enough to be separately reprinted in 1851. We must remember, however, that Weber himself had not pointed out the general significance of his law and may have seen its most general meaning only vaguely. He had hinted at generality in his manner of talking about ratios as if they were increments of stimulus, and in extending his finding for touch to visual extents and to tones. He had formulated no specific law. It was Fechner who realized later that his own principle was essentially what Weber's results showed, and it was Fechner who gave the empirical relationship mathematical form and called it "Weber's Law." In recent times there has been a tendency to correct Fechner's generosity, and to give the name *Fechner's Law* to what Fechner called "Weber's Law," reserving the latter term for Weber's simple statement that the just noticeable difference in a stimulus bears a constant ratio to the stimulus. (See formulas 1 and 6 *infra*, pp. 1159, 1160.)

The immediate result of Fechner's idea was the formulation of the program of what he later called psychophysics. This program, as we have already observed, was worked out in the *Zend-Avesta* of 1851. There was still, however, the program to carry out, and Fechner set about it. The methods of measurement were developed, the three psychophysical methods which are still fundamental to much psychological research. The mathematical form both of the methods and of the exposition of the general problem of measurement was established. The classical experiments on lifted weights, on visual brightnesses and on tactual and visual distances were performed. Fechner the philosopher proved to have lost none of the experimental care of Fechner the physicist. His friend and brother-in-law, A. W. Volkmann, then at Halle, helped with many of the experiments. Other data, notably the classification of the stars by magnitude, were brought forth to support the central thesis. For seven years Fechner published nothing of all this. Then in 1858 and 1859 two short anticipatory papers appeared, and then in 1860, full grown, the *Elemente der Psychophysik*, a text of the "exact science of the functional relations or relations of dependency between body and mind."

It would not be fair to say that the book burst upon a sleeping world. Fechner was not popular. *Nanna, Zend-Avesta* and similar writings had caused the scientists to look askance at him, and he was never accepted as a philosopher. No one suspected at the time what importance the book would come to have. There was no furor; nevertheless the work was scholarly and well grounded on both the experimental and mathematical sides, and, in spite of philosophical prejudice, it commanded attention in the most important quarter of all, namely, with the other scientists who were concerned with related problems. Even before the book itself appeared, the paper of 1858 had attracted the attention of Helmholtz and of Mach. Helmholtz proposed a modification of Fechner's fundamental formula in 1859. Mach began in 1860 tests of Weber's law in the time-sense and published in 1865. Wundt, in his first psychological publications in 1862 and again in 1863, called attention to the importance of Fechner's work. A. W. Volkmann published psychophysical papers in 1864. Aubert challenged Weber's law in 1865. Delbœuf, who later did so much for the development of psychophysics, began his experiments on brightness in 1865, inspired by Fechner. Vierordt similarly undertook in 1868 his study of the time-sense in the light of the *Elemente*. Bernstein, who had just divided with Volkmann the chair of anatomy and physiology at Halle, published in 1868 his irradiation theory, a theory that is based remotely on Herbart's law of the limen, but directly on Fechner's discussion. The *Elemente* did not take the world by the ears, but it got just the kind of attention that was necessary to give it a basic position in the new psychology.

Fechner, however, had now accomplished his purpose. He had laid the scientific foundation for his philosophy and was ready to turn to other matters, keeping always in mind the central philosophical theme. Moreover, he had reached his sixties, the age when men begin to be dominated more by their interests and less by their careers. The next topic, then, that caught the attention of this versatile man was esthetics, and, just as he had spent ten years on psychophysics, so now he spent a decade (1865–1876) on esthetics, a decade that was terminated when Fechner was seventy-five years old.

If Fechner 'founded' psychophysics, he also 'founded' experimental esthetics. His first paper in this new field was on the golden section and appeared in 1865. A dozen more papers came out from 1866 to 1872, and most of these had to do with the problem of the two Holbein Madonnas. Both Dresden and Darmstadt possessed Madonnas, very similar although different in detail, and both were reputed to have been painted by Holbein. There was much controversy about them, and Fechner plunged into it. There were several mooted points. The Darmstadt Madonna showed the Christ-child. The Dresden Madonna showed instead a sick child and might have been a votive picture, painted at the request of a family with the image of a child who had died. There was the general question of the significance of the pictures, and there was also the question of authenticity. Which was Holbein's and which was not? Experts disagreed. Fechner, maintaining the judicial attitude, was inclined to believe that they might both be authentic, that if Holbein had sought to portray two similar but different ideas he would have painted two similar but different pictures. And finally, of course, there was the question as to which was the more beautiful. These two latter questions were related in human judgment, for almost every one would be likely to believe that the authentic Madonna must be the more beautiful. Some of these questions Fechner sought to have answered 'experimentally' by a public opinion poll on the auspicious occasion when the two Madonnas were exhibited together. He placed an album by the pictures and asked visitors to record their judgments; but the experiment was a failure. Out of over 11,000 visitors, only 113 recorded their opinions, and most of these answers had to be rejected because they did not follow the instructions or were made by art critics or others who knew about the pictures and had formed judgments. Nevertheless the idea had merit and has been looked upon as the beginning of the use of the method of impression in the experimental study of feeling and esthetics.

In 1876 Fechner published the *Vorschule der Aesthetik*, a work that closed his active interest in that subject and laid the foundation for experimental esthetics. It goes into the various problems, methods and principles with a thoroughness that rivals the psychophysics, but is too far afield for detailed consideration in this book.

There is little doubt that Fechner would never have returned either to psychophysics or to esthetics, after the publication of his major book in each subject-matter, had the world let him be. The psychophysics, however, had immediately stimulated both research and criticism and, while Fechner was working on esthetics, was becoming important in the new psychology. In 1874, the year of the publication of Wundt's *Grundzüge der physiologischen Psychologie*, Fechner had been aroused to a brief criticism of Delbœuf's *Étude psychophysique* (1873). The next year Wundt came to Leipzig. The following year Fechner finished with esthetics and turned again to psychophysics, publishing in 1877 *In Sachen der Psychophysik*, a book which adds but little to the doctrine of the *Elemente*. Fechner was getting to be an old man, and his philosophical mission was still in his mind. In 1879, the year of Wundt's founding of the Leipzig psychological laboratory, Fechner issued *Die Tagesansicht gegenüber der Nachtansicht*, his seventh and last call to the somnolent world. He was then seventy-eight years old. Finally, in 1882, he published the *Revision der Hauptpunkte der Psychophysik*, a very important book, in which he took account of his critics and sought to meet the unexpected demand of experimental psychology upon him. In the following years there were half a dozen psychophysical articles by him, but actually this work was done. He died in 1887 at the age of eighty-six in Leipzig, where for seventy years he had lived the quiet life of the learned man, faring forth, while keeping his house, on these many and varied great adventures of the mind.

This then was Fechner. He was for seven years a physiologist (1817–1824); for fifteen a physicist (1824–1839); for a dozen years an invalid (1839 to about 1851); for fourteen years a psychophysicist (1851–1865); for eleven years an experimental estheticist (1865–1876); for at least two score years throughout this period, recurrently and persistently, a philosopher (1836–1879); and finally, during his last eleven years, an old man whose attention had been brought back by public acclaim and criticism to psychophysics (1876–1887)—all told three score years and ten of varied intellectual interest and endeavor. If he founded experimental psychology, he did it incidentally and involuntarily, and yet it is hard to see how the new psychology could have advanced as it did without an *Elemente der Psychophysik* in 1860. It is to this book, therefore, that we must now turn our attention.

PSYCHOPHYSICS

When Fechner began work on what was eventually to become the *Elemente der Psychophysik*, he had—beside his philosophical problem, his experience in physical research and his habits of careful experimentation—Herbart's psychology as a background. From Herbart he obtained the conception that psychology should be science, the general idea of mental measurement, the related notion of the application of mathematics to the study of the mind, the concept of the limen (which Herbart got from Leibnitz), the idea of mental analysis by way of the facts of the limen, and probably also a sensationistic cast to all of his work, a cast which resembles Herbart's intellectualism. When Fechner wrote the *Zend-Avesta*, Lotze had not published his psychology. There was really no psychology at all except the very influential psychology of Herbart and the psychological physiology of Johannes Müller and E. H. Weber. Fechner was, however, too much of an experimentalist to accept Herbart's metaphysical approach or to admit the validity of his denial of the psychological experiment. Instead he set himself to correct Herbart by an experimental measurement of mind. All this, we must not forget, was done in the interests of his philosophical attack upon materialism.

There is also to be mentioned Fechner's mathematical background. It will be recalled that Fechner had turned in part to the study of mathematics after he had obtained his doctor's degree. Fechner himself acknowledges debts to "Bernoulli (Laplace, Poisson), Euler (Herbart, Drobisch), Steinheil (Pogson)." He was thinking, however, more of the mathematical and experimental demonstration of Weber's Law. Steinheil had shown that stellar magnitudes follow this law; Euler, that tonal pitch follows it. It is plain, however, that Fechner placed the name of Daniel Bernoulli (1700–1782) first with reason. Bernoulli's interest in the theory of probabilities as applied to games of chance has led to the discussion of *fortune morale* and *fortune physique*, mental and physical values which he believed (1738) to be related to each other in such a way that a change in the amount of the 'mental fortune' varies with the ratio that the change in the physical fortune has to the total fortune of its possessor. (Thus in gambling with even stakes, one stands to lose more than one gains, for a given loss after the event bears a larger ratio to the reduced total fortune than would the same physical gain to an increased total fortune—a conclusion with a moral!) In this way *fortune morale* and *fortune physique* became mental and physical quantities, mathematically related, quantities that correspond exactly, both in kind and in relationship, to mind and body in general and to sensation and bodily energy in particular, the terms that Fechner sought to relate, in the interests of his philosophy, by way of Weber's law.

On the purely mathematical side, Fechner is less clear as to his background, but it is plain that Bernoulli, Laplace and Poisson were important. Nowadays we are apt to think especially of Fechner's use of the normal law of error as representing his mathematical interest. Fechner's method of constant stimuli makes use of this law, and the method has assumed importance because it is closely related to the biological and psychological statistical methods that also make use of normal distributions. The method of constant stimuli was, however, only one of Fechner's three fundamental methods.

Nevertheless, it is interesting to answer the question that arises about Fechner's use of the normal law. The principles were all contained in the earlier mathematicians' work on the theory of probabilities, work of which Bernoulli's is representative. Laplace, whom Fechner specially mentioned, developed the general law. Gauss gave it its more usual form, and the law ordinarily bears his name. Fechner refers to Gauss in his use of it, but Gauss seems to have been less important than Laplace. There is nothing new in making this practical application of the theory of probabilities. Since 1662 there had been attempts to apply it to the expectation of life, to the evaluation of human testimony and human innocence, to birth-rates and sex-ratios, to astronomical observations, to the facts of marriages, smallpox and inoculation, to weather forecasts, to annuities, to elections, and finally (Laplace and Gauss) to errors of scientific observation in general. It was in 1835 that Quetelet first thought of using the law of error to describe the distribution of human traits, as if nature, in aiming at an ideal average man, *l'homme moyen*, missed the mark and thus created deviations on either side of the average. It was Quetelet who gave Francis Galton the idea of the mathematical treatment of the inheritance of genius (1869), but Fechner had nothing of this sort in mind. The older tradition, however, he must have known, at least in part, and it is from it that he took for the method of constant stimuli the normal law of error, now so important to psychologists. It was easier to assume then than it is now that the normal law, as indeed its name implies, is a law of nature which applies whenever variability is uncontrolled.

Beside this general background and knowledge, Fechner brought to the problem of psychophysics several very definite things. First, there was the fact of the limen, made familiar by Herbart but also obvious enough in other ways, as, for example, in the invisibility of the stars in daylight. Second, there was Weber's law, a factual principle which, if not verified, could still be expected to persist in modified form. Third, there was the experimental method, which was equally fundamental and which derived from Fechner's own temperament in defiance of Herbart. Fourth, there was Fechner's clear conception of the nature of psychophysics as "an exact science of the functional relations or the relations of dependency between body and mind." This conception was the *raison d'être* for the entire undertaking. Finally, there was Fechner's very wise conclusion that he could not attempt the entire program of psychophysics and that he would therefore limit himself, not only to sensation, but further to the intensity of sensation, so that a final proof of his view in one field might, because of its finality, have the weight to lead later to extensions into other fields.

We must pause here to note that Fechner's view of the relation of *mind and body* was not that of psychophysical parallelism, but what has been called the *identity hypothesis* and also *panpsychism*. The writing of an equation between the mind and the body in terms of Weber's law seemed to him virtually a demonstration both of their identity and of their fundamental psychic character. Nevertheless, Fechner's psychophysics has played an important part in the history of psychophysical parallelism for the reason that mind and body, sensation and stimulus, have to be regarded as separate entities in order that each can be measured and the relation between the two determined. Fechner's psychology therefore, like so much of the psychology that came after him, seems at first to be dualistic. It is true that he began with a dualism, but we must remember that he thought he had shown that the dualism is not real and is made to disappear by the writing of the true equation between the two terms.

It is so easy nowadays to think that the Weber-Fechner law represents the functional relation between the measured magnitude of stimulus and the measured magnitude of sensation, that it is hard to realize what difficulty the problem presented to Fechner. It seemed plain to him, however, that sensation, a mental magnitude, could not be measured directly and that his problem was therefore to get at its measure indirectly. He began by turning to *sensitivity*.

Sensation, Fechner argued, we cannot measure; all we can observe is that a sensation is present or absent, or that one sensation is greater than, equal to, or less than another sensation. Of the absolute magnitude of a sensation we know nothing directly. Fortunately, however, we can measure stimuli, and thus we can measure the stimulus values necessary to give rise to a particular sensation or to a difference between two sensations; that is to say, we can measure threshold values of the stimulus. When we do this we are also measuring sensitivity, which is the inverse of the threshold value. Fechner distinguished between absolute and differential sensitivity, which correspond respectively to the absolute and differential limens. He recognized the importance of variability in this subject-matter and the necessity of dealing with averages, extreme values, the laws of averages and the laws of variability about the averages—in short, the necessity of using statistical methods.

Since Fechner believed that the stimulus, and hence sensitivity, can be measured directly but that sensation can not, he knew that he must measure sensation itself indirectly, and he hoped to do it by way of its differential increments. In determining the differential limen we have two sensations that are just noticeably different, and we may take the just noticeable difference (the jnd) as the unit of sensation, counting up jnd to determine the magnitude of a sensation. There was a long argument later as to whether every liminal increment of sensation (δS) equals every other one, but Fechner assumed that $\delta S =$ the jnd, and that the sensed differences, being all just noticeably different, are equal and therefore constitute a proper unit.

One does not in practice count up units for large magnitudes. One works mathematically on the general case for the general function which can, perhaps, later be applied in measurement. Fechner went to work in the following manner. In expounding him we shall use the familiar English abbreviations instead of Fechner's symbols: S for the magnitude of the sensation and R for the magnitude of the stimulus (*Reiz*).

Weber's experimental finding may be expressed:

$$\frac{\delta R}{R} = \text{constant, for the jnd.} \qquad \textit{Weber's Law} \ (1)$$

This fact ought to be called "Weber's law," since it is what Weber found. Fechner, however, used the phrase for his final result.

He assumed that, if (1) holds for the jnd, it must also hold for any small increment of S, δS, and that he could thus express the functional relation between S and R by writing:

$$\delta S = c \frac{\delta R}{R} \qquad \textit{Fundamental formula} \ (2)$$

where $c =$ a constant of proportionality. This was Fechner's *Fundamentalformel*, and we must note that the introduction of δS into the equation is the mathematical equivalent of Fechner's conclusion that all δS's are equal and can be treated as units. One has only to integrate to accomplish the mathematical counterpart of counting up units to perform a measurement.

If we can write the fundamental formula, we can certainly measure sensation. Fechner, therefore, integrated the equation, arriving at the result

$$S = c \log_e R + C \qquad (3)$$

where C = the constant of integration and e = the base of natural logarithms. In formula (3) we really have the desired result, since it gives the magnitudes of S for any magnitude of R, when the two constants are known. Fechner had thus demonstrated the fundamental point of his philosophy. Nevertheless this formula was unsatisfactory because of the unknown constants, and Fechner undertook to eliminate C by reference to other known facts. He let r = the threshold value of the stimulus, R, a value at which S, by definition, = 0. Thus:

When R = r, S = 0

Substituting these values of S and R in (3), we get:

$$0 = c \log_e r + C$$
$$C = -c \log_e r$$

Now we can substitute for C in (3):

$$\begin{aligned} S &= c \log_e R - c \log_e r \\ &= c(\log_e R - \log_e r) \\ &= c \log_e \frac{R}{r} \qquad (4) \end{aligned}$$

We can shift to common logarithms from natural logarithms by an appropriate change of the constant from c to, let us say, k:

$$S = k \log \frac{R}{r} \qquad \textit{Measurement formula} \ (5)$$

This is the formula for measurement, Fechner's *Massformel.* The scale of S is the number of jnd that the sensation is above zero, its value at the limen. Beyond this point Fechner went one more step. He suggested that we might measure R by its relation to its liminal value; that is to say, we might take r as the unit of R. If r be the unit of R, then:

$$S = k \log R. \qquad \textit{Fechner's Law} \ (6)$$

This last formula, (6), Fechner called "Weber's Law." It is only as we view the matter now that we see that formula (1) is really Weber's law and that formula (6) should be called Fechner's law. We must remember that S = k log R is true only when the unit of R is the liminal value of the stimulus and in so far as it is valid to integrate S and to assume that S = 0 at the limen. Furthermore, the entire conclusion depends on the validity of Weber's finding, formula (1), a generalization that further experimentation has verified only approximately in some cases, but not exactly nor for the entire range of stimuli.

About this claim of Fechner's that he had measured sensation vigorous controversy raged for forty years or more; and two of the fundamental objections are of sufficient interest to deserve brief mention here.

One argument was that Fechner had assumed the *equality of all jnd* without sufficient warrant and that he had thus in a sense begged the question, since there is no meaning to the statement that one δS equals another unless S is measurable. There is certainly some force to this criticism, but it can be met in two ways.

It was actually met in part by Delbœuf's notion of the sense-distance and the experiments on supraliminal sense-distances. Delbœuf pointed out that we can judge the size of the interval between two sensations immediately and directly. For example, we can say of three sensations, A, B and C, whether the distance AB is greater than, equal to or less than the distance BC. Thus we perform a mental measurement immediately, and the question is not begged. Now suppose AB = BC psychologically, and suppose that we find that the stimulus for B is the geometric mean of the stimuli for A and C. Then we have shown that the *Fundamentalformel* holds for a large S like AB, and, if the same law holds for large distances judged equal and for jnd, we may assume the jnd must also be equal. As a matter of fact, Weber's law has not been shown to hold generally nor exactly. It depends on what arbitrary scale of stimulus units is being used and it is apt to be wrong for the low values of any convenient measure of stimulus intensity. Modern findings show that the assumption of equality for intensitive jnd is often inconsistent with the direct judgmental comparison of supraliminal intensitive differences. There is evidence, for instance, that jnd for the pitch of tones are equal in this sense, but that jnd for the loudness of tones are not.

The other way to meet the objection that all jnd are not equal is to say frankly that the equality of units must be an assumption. Certainly one jnd is equivalent to another in that both are jnd. The issue can be met thus on purely logical grounds, though this solution leaves open still the question of the exact sense in which jnd as such are equal. So it must be with all units, and even Delbœuf's sense-distances are not more satisfactory in this regard. The obvious fact is, nevertheless, that the Fechner Law states a relationship between two entities that are not identical. S must be something, and it is not R. Something other than the stimulus has been measured.

The other important criticism of Fechner has been called the *quantity objection*. It was argued that it is patent to introspection that sensations do not have magnitude. "Our feeling of pink," said James, "is surely not

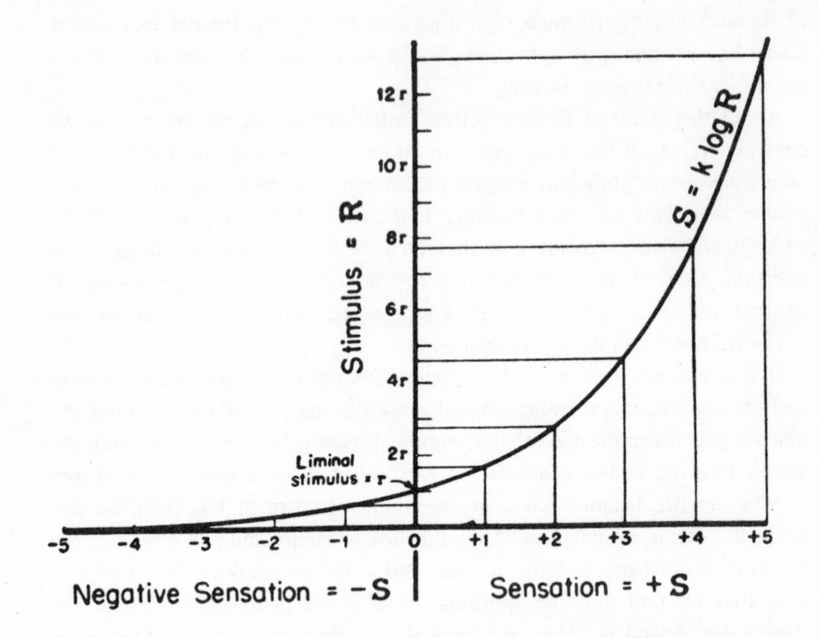

FIGURE 1—Fechner's Law: S = k log R. The positions of the equally spaced vertical ordinates represent an arithmetic series of S; their successive heights the corresponding geometric series of R. Thus the curve shows how a logarithmic function represents a correlation between an arithmetic and a geometric series. It also shows why the function requires the theoretical existence of negative sensations, for, when S = 0, R = a finite value, r, the limen; and S passes through an infinite number of negative values when R varies between r and 0. In this diagram R is plotted with r as the unit and k is arbitrarily chosen as 4·5 for common logarithms.

a portion of our feeling of scarlet; nor does the light of an electric arc seem to contain that of a tallow candle within itself." "This sensation of 'gray,'" Külpe remarked, "is not two or three of that other sensation of gray." Must not Fechner have tricked us when he proved by his figures something that we all can see is not true? The criticism is not valid, yet Fechner himself was to blame for this turn that criticism took. As we have seen, Fechner had said that stimuli can be measured directly and that sensations can not, that sensations must be measured indirectly by reference to the stimulus and by way of sensitivity. No wonder the critics accused Fechner of measuring the stimulus and calling it sensation. No wonder they argued that his own statement that sensation can not be measured directly is equivalent to saying that it can not be measured at all.

Actually the 'quantity objection' was met by being ignored. The experimentalists went on measuring sensation while the objectors complained, or at least they went on measuring whatever Fechner's S is. There are, however, two remarks that can be made about this matter. (1) Sensation can indeed be measured as directly as is the stimulus. You can compare directly in judgment two sensory differences. You can say that the difference AB is greater than, or less than or equal to the difference BC, when A, B and C are serial intensities or qualities or extents or durations. Such judgments boil down to the crucial judgment of *equal* or *not-different*. Such a comparison is quite as direct for the sensation as it is for the stimulus. Similarly, to compare weights you use a balance and form the judgment *equal* when the scalepans are *not-different* in height. Or for length you note on a tape the mark that is *not-different* in position from the end of the measured object. (2) Contrariwise, we may say that the stimulus is just as unitary and simple as the sensation. A meter is not made up of 100 parts which are called centimeters, or of a thousand parts which are called millimeters, or of 39·37 parts which are called inches. A meter in itself is just as unitary as a scarlet. The magnitude of neither implies complexity but simply a relationship to other objects that is got by the conventional methods of measurement.

We must now turn to certain matters that are connected with Fechner's name: Inner psychophysics, the limen of consciousness, negative sensations and the psychophysical methods.

Fechner distinguished *inner psychophysics* from outer psychophysics. Outer psychophysics, he said, deals with the relation between mind and stimulus, and it is in outer psychophysics that the actual experiments are to be placed. Inner psychophysics, however, is the relation between mind and the excitation most immediate to it and thus deals most immediately with the relationship in which Fechner was primarily interested. $S = k \log R$ is a relationship in outer psychophysics. Between R and S, excitation, E, is interposed. Just where is the locus of this logarithmic relationship, between R and E or between E and S? It is possible that S is simply proportional to E and that the true law is $E = k \log R$, a statement which means that Weber's law does not solve the problem of mind and body as Fechner hoped it would. Fechner, however, maintained that E is probably proportional to R and that Weber's law is the fundamental law of inner psychophysics, $S = k \log E$.

This view Fechner supported with five arguments. (1) In the first place, he said in the *Elemente*, it would be inconceivable that a logarithmic relation should exist between R and E. Such a statement is hardly an argument, and Fechner took it back in the *Revision*. (2) Then he observed that the magnitude of S does not change when sensitivity is reduced, whereas it should if $S = kE$ and E is involved in the change of sensitivity. (3) Further he noted that Weber's law holds for tonal pitch, and that it would be impossible for the vibrations of E to have other than a proportional relation to the vibrations of R. (Of course he was in error in supposing that nervous excitation is vibratory.) (4) Next he pointed out that a subliminal S probably has an E, that the invisible stars in daytime probably give rise to excitation which is below the limen of consciousness. Such a fact could be true only if $S = k \log E$. (5) Finally, he appealed to the distinction between sleep and waking, and between inattention and attention, as indicating the existence of a limen of consciousness rather than a limen of excitation. This last argument is the most cogent. Certainly the mere fact of the selectivity of attention seems to mean that there are many excitations, all prepotent for consciousness, of which only a few become conscious. However, Fechner's entire argument would not be taken very seriously at the present time. It is important for us merely to see why Fechner, working in outer psychophysics, thought he was solving the problem, all-important to him, of inner psychophysics.

From this discussion we see how important the fact of the *limen of consciousness* was to Fechner. What Fechner called Weber's Law is based upon the limen, for, if $S = k \log R$, then, when $S = 0$, R is some finite quantity, a liminal value. Herbart's limen of consciousness is thus simply a corollary of this law. In fact, Fechner was further consistent with Herbart in relating the limen to attention: when consciousness is already occupied with other sensations, a new sensation can not enter until it overcomes the "mixture limen."

The psychology that depends upon this law also requires the existence of *negative sensations.* Figure 3 shows graphically the logarithmic curve that gives the relationship of S to R for "Weber's Law." The function requires that $R = r$, the limen, when $S = 0$, and thus it gives negative sensations for subliminal values of R, for theoretically when $R = 0$, S is negative and infinite. Fechner believed that "the representation of unconscious psychical values by negative magnitude is a fundamental point for psychophysics," and by way of this mathematical logic he came to hold a doctrine of the unconscious not unlike that of his predecessors, Leibnitz and Herbart.

Fechner's claim to greatness within psychology does not, however, derive from these psychological conceptions of his, nor even from the formulation of his famous law. The great thing that he accomplished was a new kind of measurement. The critics may debate the question as to what it was that he measured; the fact stands that he conceived, developed and established new *methods of measurement,* and that, whatever interpretation may later be made of their products, these methods are essentially the first methods of mental measurement and thus the beginning of quantitative experimental psychology. Moreover, the methods have stood the test of time. They have proven applicable to all sorts of psychological problems and situations that Fechner never dreamed of, and they are all still used with only minor modifications in the greater part of quantitative work in the psychological laboratory today.

There were three fundamental methods: (1) the *method of just noticeable differences*, later called the *method of limits*; (2) the *method of right and wrong cases*, later called the *method of constant stimuli* or simply the *constant method*; and (3) the *method of average error*, later called the *method of adjustment* and the *method of reproduction*. Each of these methods is both an experimental procedure and a mathematical treatment. Each has special forms. The constant method has been much further developed by G. E. Müller and F. M. Urban. More recently the method of adjustment has shown certain advantages over the others. Changes and development, however, add to Fechner's distinction as the inventor. There are few other men who have done anything of equal importance for scientific psychology.

The storm of criticism that Fechner's work evoked was in general a compliment, but there were also those psychologists who were unable to see anything of value in psychophysics. But three years after Fechner's death, James wrote: "Fechner's book was the starting point of a new department of literature, which it would perhaps be impossible to match for the qualities of thoroughness and subtlety, but of which, in the humble opinion of the present writer, the proper psychological outcome is just *nothing*." Elsewhere he gave his picture of Fechner and his psychophysics:

> The Fechnerian *Massformel* and the conception of it as the ultimate 'psychophysic law' will remain an 'idol of the den,' if ever there was one. Fechner himself indeed was a German *Gelehrter* of the ideal type, at once simple and shrewd, a mystic and an experimentalist, homely and daring, and as loyal to facts as to his theories. But it would be terrible if even such a dear old man as this could saddle our Science forever with his patient whimsies, and, in a world so full of more nutritious objects of attention, compel all future students to plough through the difficulties, not only of his own works, but of the still drier ones written in his refutation. Those who desire this dreadful literature can find it; it has a 'disciplinary value'; but I will not even enumerate it in a foot-note. The only amusing part of it is that Fechner's critics should always feel bound, after smiting his theories hip and thigh and leaving not a stick of them standing, to wind up by saying that nevertheless to him belongs the *imperishable glory*, of first formulating them and thereby turning psychology into an *exact science*,
>
> " 'And everybody praised the duke
> Who this great fight did win.'
> 'But what good came of it at last?'
> Quoth little Peterkin.
> 'Why, that I cannot tell,' said he,
> 'But 'twas a famous victory!' "

It is plain to the reader that the present author does not agree with James. Of course, it is true that, without Fechner or a substitute which the times would almost inevitably have raised up, there might still have been an experimental psychology. There would still have been Wundt—and Helmholtz. There would, however, have been little of the breadth of science in the experimental body, for we hardly recognize a subject as scientific if measurement is not one of its tools. Fechner, because of what he did and the time at which he did it, set experimental quantitative psychology off upon the course which it has followed. One may call him the 'founder' of experimental psychology, or one may assign that title to Wundt. It does not matter. Fechner had a fertile idea which grew and brought forth fruit abundantly.

COMMENTARY ON

SIR FRANCIS GALTON

SIR FRANCIS GALTON was not merely an eminent Victorian but a thoroughly good and attractive man. It has been said of him that he was essentially a social reformer. The implied disparagement of his scientific abilities is unjust. He was imbued, it is true, with a strong sense of social responsibility and of mission; he was neither erudite nor philosophically profound. Yet, judged both by what he accomplished and by the impetus he gave to the researches of others, he was a great scientist. His kindly feelings and earnest social convictions are the more to be valued because they were carried over into his work and shaped its goals.

Karl Pearson in his monumental biography of Galton calls him the "master builder" of the modern theory of statistics.[1] Much of his work was handmade and primitive, many of his results were wrong, his theories of heredity and eugenics were grossly oversimplified; yet one is struck by the pioneer character of his labors, by their extraordinary suggestiveness even when, as Pearson says, his methods are "the crude extemporizations of the first settler." His outstanding merit as an investigator was that he blazed the trail.[2]

Galton was born near Birmingham on February 16, 1822. The same year saw the birth of Gregor Mendel, with whom Galton said he always felt himself "sentimentally connected."[3] On his father's side he came of a line of prosperous Quaker businessmen who combined a talent for practical affairs with scientific and statistical interests. His grandfather, Samuel John, was an intimate of Priestley, Watt and Boulton, and a Fellow of the Lunar Society, a provincially famous scientific-philosophical group. His father, Samuel Tertius, wrote a book on currency, filled his home with telescopes, barometers, solar microscopes and similar scientific instruments, was devoted to literature and still found time to be a successful banker. On his mother's side, Galton was related to the eminent Darwin family. Charles Darwin was his cousin and the two men were always on excellent terms. Galton credits his grandmother Darwin, who lived to be eighty-five, for the exceptional longevity of his family.[4]

[1] Karl Pearson, *The Life, Letters and Labours of Francis Galton*, Cambridge, 1914–1930, Vol. II, p. 424.

[2] Pearson, *op. cit.*, Vol. II, p. v.

[3] Francis Galton, *Memories of My Life*, London, 1908, p. 308.

[4] "My mother died just short of ninety, my eldest brother at eighty-nine, two sisters, as already mentioned, at ninety-three and ninety-seven respectively; my surviving brother is ninety-three and in good health. My own age is now only eighty-six, but may possibly be prolonged another year or more. I find old age thus far to be a very happy time, on the condition of submitting frankly to its many limitations." *Ibid.*, p. 7.

Galton's father was eager to give him an excellent education and also to make him self-reliant at the earliest possible age. The result was a higgledy-piggledy course of schooling. At the age of eight he was bundled off to a wretched boarding school at Boulogne, where he remained for two years. Galton remembered it as a fine place for birchings and bullying. He was then sent to small private schools at which, though he was a most gifted child,[5] he claims to have learned nothing. "I had craved for what was denied, namely an abundance of good English reading, well-taught mathematics, and solid science. Grammar and the dry rudiments of Latin and Greek were abhorrent to me." Both his parents wanted him to take up medicine and he was therefore apprenticed at sixteen to a leading physician. A short time later he became an "indoor pupil" at the Birmingham General Hospital. He completed his education at Kings College, London, and at Trinity College, Cambridge. His scholastic record was undistinguished.

The death of his father made Galton financially independent at the age of twenty-two. He thereupon abandoned medicine and set forth on a course of travel. In 1845 he visited Egypt, the Sudan and Syria. This and other journeys are vividly described in his *Memories*. For a time he cultivated, in moderation, the pleasures of sporting life; he summered in the Shetlands, went seal-shooting, collected sea-birds, essayed yachting and ballooning. He also published the first of an almost uncountable [illegible] scientific papers.[6] It describes an invention of his, the Telot[illegible], which was designed "to print telegraphic messages and to govern heavy machinery by an extremely feeble force." [7] In 1850 he left England for a two-year sojourn in South Africa. His explorations in South-West Africa, conducted under the usual dangerous and harassing conditions, brought out his best qualities as a man and as a scientist. He penetrated for more than 1,000 miles into unexplored territory and recorded a great deal of valuable information about the lands through which he passed and their people. For his geographical discoveries and astronomical observations he was awarded the gold medal of the Royal Geographical Society; and in 1856, at the age of thirty-four, was made a Fellow of the Royal Society. His scientific reputation was now established.

After his return from Africa Galton was, as he said, "rather used up in health." It happened quite a few times in his life that he suffered physical

[5] Galton learned to read at the age of two-and-a-half years, and wrote a letter before he was four. By the age of five he could read "almost any English book," could do multiplication, and could tell time. For an account of his precocity see Lewis M. Terman, "The Intelligence Quotient of Francis Galton in Childhood," *American Journal of Psychology* (1917), Vol. 28, p. 209; also by Terman, "The Psychological Approaches to the Biography of Genius," *Science* (1940), Vol. XCII, p. 264.

[6] Pearson refers to at least fifteen books by Galton, and lists more than 220 papers. He concedes that the register is probably incomplete.

[7] Galton, *op. cit.*, p. 119.

breakdowns and periods of severe depression and giddiness—what he called a "sprained brain"—during which he was incapable of working. Invariably, a change of habits, a tour abroad, "plenty of outdoor exercise," completely restored him.[8] Except for these "tours and cures" the story of the remaining fifty-five years of Galton's life "is to be told less in terms of his movements than of his thoughts." [9]

Galton was not a mathematician but he was mathematically minded. One of his maxims was: "Whenever you can, count"; he himself was almost obsessed by the need to count and to measure. In his laboratory he measured heads, noses, arms, legs, color of eyes and hair, breathing power, "strength of pull and of squeeze," keenness of sight and of hearing, reaction time, height, weight and so on. He compiled statistics of the weather, of the properties of identical twins, of the frequency of yawns, of the sterility of heiresses, of life span, of the inheritance of physical and mental characters. He counted the number of "fidgets" per minute among persons attending lectures; the purpose of this observation was apparently to derive a coefficient of boredom. Middle-aged persons, he found, are medium fidgets, "children are rarely still, while elderly philosophers will sometimes remain rigid for minutes together." He made a "Beauty Map" of the British Isles, classifying the girls he passed in the streets of various towns as "attractive, indifferent or repellent." The method he employed was to prick holes in a piece of paper, "torn rudely into a cross with a long leg," which he concealed within his pocket. London ranked highest; Aberdeen, lowest. When 800 visitors to a cattle exhibition at Plymouth tried to guess the weight of an ox, he tabulated the estimates and observed that the "*vox populi* was correct to within 1 per cent of the real value." This result might be construed, he felt, as evidence of the "trustworthiness of a democratic judgment" except that the proportion of voters capable of assessing the weight of an ox "undoubtedly surpassed that of the voters in ordinary elections who are versed in politics." [10]

Among his most incredible achievements was persuading Herbert Spencer to accompany him to the Derby. Spencer was mildly bored but Galton enjoyed every moment of the excursion. He took the opportunity of recording a scientific observation, viz., the change in "prevalent tint

[8] "I was blessed with an abundance of animal spirits and hopefulness, though they were dashed temporarily over and over again by the great readiness with which my brain became overtaxed; however I always recuperated quickly." Galton, *op. cit.*

[9] C. P. Blacker, *Eugenics, Galton and After*, Cambridge (Mass.), 1952, p. 37. This is a very agreeable and accurate account of Galton's life, of the work he did in eugenics and of later developments in the field.

[10] "The judgments were unbiassed by passion and uninfluenced by oratory and the like . . . The average competitor was probably as well fitted for making a just estimate of the dressed weight of the ox, as an average voter is to judge the merits of most political issues on which he votes." Galton, "Vox Populi," *Nature*, Vol. LXXV, pp. 450–451, March 7, 1907.

of the faces in the great stand," under the flush of excitement as the horses approached the finish line. To a benign reasonableness and disinterestedness Galton added an amazing innocence and suggestibility. His "passionate desire to subjugate the body to the spirit" led him to subject himself to some hair-raising experiments; the practice of slow self-suffocation had a strange fascination for him. Having invented an optical underwater device he amused himself by reading while submerged in his bath; on several occasions he almost drowned because he "forgot that [he] was nearly suffocating." He describes various adventures in auto-suggestion, such as pretending that a comic figure in *Punch* possessed divine attributes or that everyone he met while walking in Piccadilly was a spy. Invariably he managed in these experiments to frighten himself half to death. But they expressed truly his unquenchable curiosity; every experience was acceptable "for one wants to know the very worst of everything as well as the very best."

The range of Galton's scientific interests and pursuits is too broad to describe in this space. One of his major efforts was concentrated on the study of heredity; it was the subject of his most celebrated book, *Hereditary Genius* (1869), published when he was forty-seven. But this in a sense was the focus and synthesis of other studies. The earlier years were devoted to geography, ethnology and anthropology; later he turned to anthropometry and genetics.

Galton formulated the profoundly influential principle of correlation. Others in his century "hovered on the verge of the discovery," and the idea was greatly elaborated and refined by Pearson, Edgeworth and Weldon.[11] But Galton was the pioneer in evolving the conception of a correlated system of variates, the representation "by a single numerical quantity of the degree of relationship, or of partial causality, between the different variables of our ever-changing universe." [12] The connection between this concept and the problems of inheritance is self-evident; indeed it was his study of heredity that led Galton to correlation. How, for example, did a character possessed by the father influence the like character in the son? It was, of course, a contributing factor, but only one of many, derived from the son's mother, from his grandparents and other forebears. The method of expressing such relations of multiple causality in a single formula came to Galton one morning while waiting at a roadside station for a train, "poring over a small diagram in my notebook." [13]

[11] Helen M. Walker, *Studies in the History of Statistical Method*, Baltimore, 1931, p. 92. For a general discussion of Galton's work on correlation, see Pearson, *op. cit.*, Vol. III (a), Chapter XIV.

[12] Pearson, as in preceding note, p. 2.

[13] Galton's explanation of the correlation concept is a model of clarity and deserves reprinting: "It had appeared from observation, and it was fully confirmed by this theory, that such a thing existed as an 'Index of Correlation'; that is to say, a fraction, now commonly written *r*, that connects with closer approximation every value

It was an inspiration that changed the course of modern social studies.

An amateur in many fields, Galton was never a mere dabbler or dilettante. He deserves to be remembered for his researches in fingerprinting, his studies of imagery, of synesthesia and of color association, his discovery of number-forms (i.e., the images by which we represent numbers to ourselves), his contributions to meteorology (he invented the word "anticyclone"), his valuable experiments in blood transfusion, his origination of composite photography.

The selection below is a chapter from *Hereditary Genius.* The book exemplifies Galton's statistical approach and his primary interest in the problem of mental inheritance, and of improvement of the race by eugenic practices. The selection itself analyzes the distribution of mental ability and suggests that the pattern resembles that of the distribution of physical traits. Galton's interest in this subject culminated in *Inquiries into Human Faculty and Its Development* (1883), a book regarded "as the beginning of scientific individual psychology and of the mental tests." But as Boring points out in his excellent *History of Experimental Psychology,* Galton's intention regarding the book was different.[14] The effect on his mind of the publication of his cousin's *Origin of Species* was "to demolish a multitude of dogmatic barriers by a single stroke and to arouse a spirit of rebellion against all ancient authorities . . . contradicted by modern science." *Hereditary Genius, Human Faculty,* his writings on eugenics, even his passion for classification and measurement may be considered as Galton's attempts to further this rebellion. He believed in evolutionary progress and strove to prove that it was at once a more rational and a more promising faith than any offered by current religious dogmas. It was a faith for self-respecting, free men who dared to think they could do better for themselves and their ancestors than to rely on the efficacy of prayer. Galton held up "as the goal of human effort, not heaven, but the superman."[15] His studies of heredity and genius place excessive importance on biological factors as determinants of personality and achievement. Yet the nature-nurture controversy is far from settled, and no thoughtful person would deny the significance of Galton's work or the soundness of many of his principles of eugenics.

---

of deviation [from the median] on the part of the subject, with the *average* of all the associated deviations of the Relative as already described. Therefore the closeness of any specified kinship admits of being found and expressed by a single term. If a particular individual deviates so much, the *average* of the deviations of all his brothers will be a definite fraction of that amount; similarly as to sons, parents, first cousins, etc. Where there is no relationship at all, r becomes equal to 0; when it is so close that Subject and Relative are identical in value, then $r = 1$. Therefore the value of r lies in every case somewhere between the extreme limits of 0 and 1. Much more could be added, but not without using technical language, which would be inappropriate here." *Memories*, p. 303.

[14] Edwin G. Boring, *A History of Experimental Psychology*, New York, Second Edition, 1950, p. 483.

[15] Boring, *op. cit.*, p. 483.

Galton died on January 17, 1911, at his home at Haslemere. Until the last few days he was gay, socially active, busy with correspondence. He had a long and an uncommonly satisfying life. His marriage which lasted for forty-four years (his wife died in 1897) had been very happy, but there were no children. It is true of Galton, as of few men, that he did with his life exactly what he wanted to do and that if he had had it to live over he would not have varied its course. One envies and admires him. He had courage, honesty, imagination, tolerance, sympathy and a sense of humor. He labored and created, he gave friendship and he inspired it—both personally and through his work. This must be my excuse for this perhaps inexcusably long preface to his essay.

Many of the modern buildings in Italy are historically known to have been built out of the pillaged structures of older days. Here we may observe a column or a lintel serving the same purpose a second time. . . . I will pursue this rough simile just one step further, which is as much as it will bear. Suppose we were building a house with second-hand materials carted from a dealer's yard, we should often find considerable portions of the same old house to be still grouped together. . . . So in the process of transmission by inheritance, elements derived from the same ancestor are apt to appear in large groups. —SIR FRANCIS GALTON

2 Classification of Men According to Their Natural Gifts

By SIR FRANCIS GALTON

I HAVE no patience with the hypothesis occasionally expressed, and often implied, especially in tales written to teach children to be good, that babies are born pretty much alike, and that the sole agencies in creating differences between boy and boy, and man and man, are steady application and moral effort. It is in the most unqualified manner that I object to pretensions of natural equality. The experiences of the nursery, the school, the University, and of professional careers, are a chain of proofs to the contrary. I acknowledge freely the great power of education and social influences in developing the active powers of the mind, just as I acknowledge the effect of use in developing the muscles of a blacksmith's arm, and no further. Let the blacksmith labour as he will, he will find there are certain feats beyond his power that are well within the strength of a man of herculean make, even although the latter may have led a sedentary life. Some years ago, the Highlanders held a grand gathering in Holland Park, where they challenged all England to compete with them in their games of strength. The challenge was accepted, and the well-trained men of the hills were beaten in the foot-race by a youth who was stated to be a pure Cockney, the clerk of a London banker.

Everybody who has trained himself to physical exercises discovers the extent of his muscular powers to a nicety. When he begins to walk, to row, to use the dumb bells, or to run, he finds to his great delight that his thews strengthen, and his endurance of fatigue increases day after day. So long as he is a novice, he perhaps flatters himself there is hardly an assignable limit to the education of his muscles; but the daily gain is soon discovered to diminish, and at last it vanishes altogether. His maximum performance becomes a rigidly determinate quantity. He learns to an inch, how high or how far he can jump, when he has attained the highest state of training. He learns to half a pound, the force he can exert on the dynamometer, by compressing it. He can strike a blow against the machine used to measure impact, and drive its index to a certain graduation, but no further. So it is in running, in rowing, in walking, and in every other form of physical exertion. There is a definite limit to the muscular powers of every man, which he cannot by any education or exertion overpass.

This is precisely analogous to the experience that every student has had of the working of his mental powers. The eager boy, when he first goes to school and confronts intellectual difficulties, is astonished at his progress. He glories in his newly-developed mental grip and growing capacity for application, and, it may be, fondly believes it to be within his reach to become one of the heroes who have left their mark upon the history of the world. The years go by; he competes in the examinations of school and college, over and over again with his fellows, and soon finds his place among them. He knows he can beat such and such of his competitors; that there are some with whom he runs on equal terms, and others whose intellectual feats he cannot even approach. Probably his vanity still continues to tempt him, by whispering in a new strain. It tells him that classics, mathematics, and other subjects taught in universities, are mere scholastic specialties, and no test of the more valuable intellectual powers. It reminds him of numerous instances of persons who had been unsuccessful in the competitions of youth, but who had shown powers in after-life that made them the foremost men of their age. Accordingly, with newly furbished hopes, and with all the ambition of twenty-two years of age, he leaves his University and enters a larger field of competition. The same kind of experience awaits him here that he has already gone through. Opportunities occur—they occur to every man—and he finds himself incapable of grasping them. He tries, and is tried in many things. In a few years more, unless he is incurably blinded by self-conceit, he learns precisely of what performances he is capable, and what other enterprises lie beyond his compass. When he reaches mature life, he is confident only within certain limits, and knows, or ought to know, himself just as he is probably judged of by the world, with all his unmistakable weakness and all his undeniable strength. He is no longer tormented into hopeless efforts by the fallacious promptings of overweening vanity, but he limits his undertakings to matters below the level of his reach, and finds true moral repose in an honest conviction that he is engaged in as much good work as his nature has rendered him capable of performing.

There can hardly be a surer evidence of the enormous difference between the intellectual capacity of men, than the prodigious differences in the numbers of marks obtained by those who gain mathematical honours at Cambridge. I therefore crave permission to speak at some length upon this subject, although the details are dry and of little general interest.

There are between 400 and 450 students who take their degrees in each year, and of these, about 100 succeed in gaining honours in mathematics, and are ranged by the examiners in strict order of merit. About the first forty of those who take mathematical honours are distinguished by the title of wranglers, and it is a decidedly creditable thing to be even a low wrangler; it will secure a fellowship in a small college. It must be carefully borne in mind that the distinction of being the first in this list of honours, or what is called the senior wrangler of the year, means a vast deal more than being the foremost mathematician of 400 or 450 men taken at hap-hazard. No doubt the large bulk of Cambridge men are taken almost at hap-hazard. A boy is intended by his parents for some profession; if that profession be either the Church or the Bar, it used to be almost requisite, and it is still important, that he should be sent to Cambridge or Oxford. These youths may justly be considered as having been taken at hap-hazard. But there are many others who have fairly won their way to the Universities, and are therefore selected from an enormous area. Fully one-half of the wranglers have been boys of note at their respective schools, and, conversely, almost all boys of note at schools find their way to the Universities. Hence it is that among their comparatively small number of students, the Universities include the highest youthful scholastic ability of all England. The senior wrangler, in each successive year, is the chief of these as regards mathematics, and this, the highest distinction, is, or was, continually won by youths who had no mathematical training of importance before they went to Cambridge. All their instruction had been received during the three years of their residence at the University. Now, I do not say anything here about the merits or demerits of Cambridge mathematical studies having been directed along a too narrow groove, or about the presumed disadvantages of ranging candidates in strict order of merit, instead of grouping them, as at Oxford, in classes, where their names appear alphabetically arranged. All I am concerned with here are the results; and these are most appropriate to my argument. The youths start on their three years' race as fairly as possible. They are then stimulated to run by the most powerful inducements, namely, those of competition, of honour, and of future wealth (for a good fellowship is wealth); and at the end of the three years they are examined most rigorously according to a system that they all understand and are equally well prepared for. The examination lasts five and a half hours a day for eight days. All the answers are carefully marked by the examiners, who add up the marks at the end and range the candidates in strict order of merit. The fairness and thoroughness of Cambridge examinations have never had a breath of suspicion cast upon them.

Unfortunately for my purposes, the marks are not published. They are not even assigned on a uniform system, since each examiner is permitted to employ his own scale of marks, but whatever scale he uses, the results as to proportional merit are the same. I am indebted to a Cambridge examiner for a copy of his marks in respect to two examinations, in which the scales of marks were so alike as to make it easy, by a slight proportional adjustment, to compare the two together. This was, to a certain degree, a confidential communication, so that it would be improper for me to publish anything that would identify the years to which these marks refer. I simply give them as groups of figures, sufficient to show the enormous differences of merit. The lowest man in the list of honours gains less than 300 marks, the lowest wrangler gains about 1,500 marks; and the senior wrangler, in one of the lists now before me, gained more than 7,500 marks. Consequently, the lowest wrangler has more than five times the merit of the lowest junior optime, and less than one-fifth the merit of the senior wrangler.

SCALE OF MERIT AMONG THE MEN WHO OBTAIN MATHEMATICAL HONOURS AT CAMBRIDGE

The results of two years are thrown into a single table.
The total number of marks obtainable in each year was 17,000.

Number of marks obtained by candidates.	*Number of candidates in the two years, taken together, who obtained those marks.*
Under 500	24 [1]
500 to 1,000	74
1,000 to 1,500	38
1,500 to 2,000	21
2,000 to 2,500	11
2,500 to 3,000	8
3,000 to 3,500	11
3,500 to 4,000	5
4,000 to 4,500	2
4,500 to 5,000	1
5,000 to 5,500	3
5,500 to 6,000	1
6,000 to 6,500	0
6,500 to 7,000	0
7,000 to 7,500	0
7,500 to 8,000	1
	200

The precise number of marks obtained by the senior wrangler in the more remarkable of these two years was 7,634; by the second wrangler in the same year, 4,123; and by the lowest man in the list of honours, only 237. Consequently, the senior wrangler obtained nearly twice as

[1] I have included in this table only the first 100 men in each year. The omitted residue is too small to be important. I have omitted it lest, if the precise numbers of honour men were stated, those numbers would have served to identify the years. For reasons already given, I desire to afford no data to serve that purpose.

many marks as the second wrangler, and more than thirty-two times as many as the lowest man. I have received from another examiner the marks of a year in which the senior wrangler was conspicuously eminent. He obtained 9,422 marks, whilst the second in the same year—whose merits were by no means inferior to those of second wranglers in general—obtained only 5,642. The man at the bottom of the same honour list had only 309 marks, or one-thirtieth the number of the senior wrangler. I have some particulars of a fourth very remarkable year, in which the senior wrangler obtained no less than ten times as many marks as the second wrangler, in the "problem paper." Now, I have discussed with practised examiners the question of how far the numbers of marks may be considered as proportionate to the mathematical power of the candidate, and am assured they are strictly proportionate as regards the lower places, but do not afford full justice to the highest. In other words, the senior wranglers above mentioned had more than thirty, or thirty-two times the ability of the lowest men on the lists of honours. They would be able to grapple with problems more than thirty-two times as difficult; or when dealing with subjects of the same difficulty, but intelligible to all, would comprehend them more rapidly in perhaps the square root of that proportion. It is reasonable to expect that marks would do some injustice to the very best men, because a very large part of the time of the examination is taken up by the mechanical labour of writing. Whenever the thought of the candidate outruns his pen, he gains no advantage from his excess of promptitude in conception. I should, however, mention that some of the ablest men have shown their superiority by comparatively little writing. They find their way at once to the root of the difficulty in the problems that are set, and, with a few clean, apposite, powerful strokes, succeed in proving they can overthrow it, and then they go on to another question. Every word they write tells. Thus, the late Mr. H. Leslie Ellis, who was a brilliant senior wrangler in 1840, and whose name is familiar to many generations of Cambridge men as a prodigy of universal genius, did not even remain during the full period in the examination room: his health was weak, and he had to husband his strength.

The mathematical powers of the last man on the list of honours, which are so low when compared with those of a senior wrangler, are mediocre, or even above mediocrity, when compared with the gifts of Englishmen generally. Though the examination places 100 honour men above him, it puts no less than 300 "poll men" below him. Even if we go so far as to allow that 200 out of the 300 refuse to work hard enough to get honours, there will remain 100 who, even if they worked hard, could not get them. Every tutor knows how difficult it is to drive abstract conceptions, even of the simplest kind, into the brains of most people—how feeble and hesitating is their mental grasp—how easily their brains are mazed—how incapable they are of precision and soundness of knowledge. It often occurs to persons familiar with some scientific subject to hear men and women of mediocre gifts relate to one another what they have picked up about it from some lecture—say at the Royal Institution, where they have sat for an hour listening with delighted attention to an admirably lucid account, illustrated by experiments of the most perfect and beautiful character, in all of which they expressed themselves intensely gratified and highly instructed. It is positively painful to hear what they say. Their recollections seem to be a mere chaos of mist and misapprehension, to which some sort of shape and organization has been given by the action of their own pure fancy, altogether alien to what the lecturer intended to convey. The average mental grasp even of what is called a well-educated audience, will be found to be ludicrously small when rigorously tested.

In stating the differences between man and man, let it not be supposed for a moment that mathematicians are necessarily one-sided in their natural gifts. There are numerous instances of the reverse, of whom the following will be found, as instances of hereditary genius, in the appendix to my chapter on "Science." I would especially name Leibnitz, as being universally gifted, but Ampere, Arago, Condorcet, and D'Alembert, were all of them very far more than mere mathematicians. Nay, since the range of examination at Cambridge is so extended as to include other subjects besides mathematics, the differences of ability between the highest and lowest of the successful candidates, is yet more glaring than what I have already described. We still find, on the one hand, mediocre men, whose whole energies are absorbed in getting their 237 marks for mathematics; and, on the other hand, some few senior wranglers who are at the same time high classical scholars and much more besides. Cambridge has afforded such instances. Its lists of classical honours are comparatively of recent date, but other evidence is obtainable from earlier times of their occurrence. Thus, Dr. George Butler, the Head Master of Harrow for very many years, including the period when Byron was a schoolboy (father of the present Head Master, and of other sons, two of whom are also head masters of great public schools), must have obtained that classical office on account of his eminent classical ability; but Dr. Butler was also senior wrangler in 1794, the year when Lord Chancellor Lyndhurst was second. Both Dr. Kaye, the late Bishop of Lincoln, and Sir E. Alderson, the late judge, were the senior wranglers and the first classical prizemen of their respective years. Since 1824, when the classical tripos was first established, the late Mr. Goulburn (brother of Dr. Goulburn, Dean of Norwich, and son of the well-known Serjeant Goulburn) was second wrangler in 1835, and senior classic of the same year. But in more recent times, the necessary labour of preparation, in order to acquire the highest mathematical places, has become so enormous that there has been a wider differentiation of studies. There is no longer time for a man to acquire the necessary knowledge to succeed to the first place in more than one subject. There are, therefore, no instances of a man being absolutely first in both examinations, but a few can be found of high eminence in both classics and mathematics, as a reference to the lists published in the "Cambridge Calendar" will show. The best of these more recent degrees appears to be that of Dr. Barry, late Principal of Cheltenham, and now Principal of King's College, London (the son of the eminent architect, Sir Charles Barry, and brother of Mr. Edward Barry, who succeeded his father as architect). He was fourth wrangler and seventh classic of his year.

In whatever way we may test ability, we arrive at equally enormous intellectual differences. Lord Macaulay had one of the most tenacious of memories. He was able to recall many pages of hundreds of volumes by various authors, which he had acquired by simply reading them over. An average man could not certainly carry in his memory one thirty-second—ay, or one hundredth—part as much as Lord Macaulay. The father of Seneca had one of the greatest memories on record in ancient times. Porson, the Greek scholar, was remarkable for this gift, and, I may add, the "Porson memory" was hereditary in that family. In statesmanship, generalship, literature, science, poetry, art, just the same enormous differences are found between man and man, and numerous instances recorded in this book, will show in how small degree, eminence, either in these or any other class of intellectual powers, can be considered as due to purely special powers. They are rather to be considered in those instances as the result of concentrated efforts, made by men who are widely gifted. People lay too much stress on apparent specialities, thinking overrashly that, because a man is devoted to some particular pursuit, he could not possibly have succeeded in anything else. They might just as well say that, because a youth had fallen desperately in love with a brunette, he could not possibly have fallen in love with a blonde. He may or may not have more natural liking for the former type of beauty than the latter, but it is as probable as not that the affair was mainly or wholly due to a general amorousness of disposition. It is just the same with special pursuits. A gifted man is often capricious and fickle before he selects his occupation, but when it has been chosen, he devotes himself to it with a truly passionate ardour. After a man of genius has selected his hobby, and so adapted himself to it as to seem unfitted for any other occupation in life, and to be possessed of but one special aptitude, I often notice, with admiration, how well he bears himself when circumstances suddenly thrust him into a strange position. He will display an insight into new conditions, and a power of dealing with them, with which even his most intimate friends were unprepared to accredit him. Many a presumptuous fool has mistaken indifference and neglect for incapacity; and in trying to throw a man of genius on ground where he was unprepared for attack, has himself received a most severe and unexpected fall. I am sure that no one who has had the privilege of mixing in the society of the abler men of any great capital, or who is acquainted with the biographies of the heroes of history, can doubt the existence of grand human animals, of natures pre-eminently noble, of individuals born to be kings of men. I have been conscious of no slight misgiving that I was committing a kind of sacrilege whenever, in the preparation of materials for this book, I had occasion to take the measurement of modern intellects vastly superior to my own, or to criticize the genius of the most magnificent historical specimens of our race. It was a process that constantly recalled to me a once familiar sentiment in bygone days of African travel, when I used to take altitudes of the huge cliffs that domineered above me as I travelled along their bases, or to map the mountainous landmarks of unvisited tribes, that loomed in faint grandeur beyond my actual horizon.

I have not cared to occupy myself much with people whose gifts are below the average, but they would be an interesting study. The number of idiots and imbeciles among the twenty million inhabitants of England and Wales is approximately estimated at 50,000, or as 1 in 400. Dr. Segiun, a great French authority on these matters, states that more than thirty per cent of idiots and imbeciles, put under suitable instruction, have been taught to conform to social and moral law, and rendered capable of order, of good feeling, and of working like the third of an average man. He says that more than forty per cent have become capable of the ordinary transactions of life, under friendly control; of understanding moral and social abstractions, and of working like two-thirds of a man. And, lastly, that from twenty-five to thirty per cent come nearer and nearer to the standard of manhood, till some of them will defy the scrutiny of good judges, when compared with ordinary young men and women. In the order next above idiots and imbeciles are a large number of milder cases scattered among private families and kept out of sight, the existence of whom is, however, well known to relatives and friends; they are too silly to take a part in general society, but are easily amused with some trivial, harmless occupation. Then comes a class of whom the Lord Dundreary of the famous play may be considered a representative; and so, proceeding through successive grades, we gradually ascend to mediocrity. I know two good instances of hereditary silliness short of imbecility, and have reason to believe I could easily obtain a large number of similar facts.

To conclude, the range of mental power between—I will not say the highest Caucasian and the lowest savage—but between the greatest and least of English intellects, is enormous. There is a continuity of natural ability reaching from one knows not what height, and descending to one can hardly say what depth. I propose in this chapter to range men according to their natural abilities, putting them into classes separated by equal degrees of merit, and to show the relative number of individuals included in the several classes. Perhaps some persons might be inclined to make an offhand guess that the number of men included in the several classes would be pretty equal. If he thinks so, I can assure him he is most egregiously mistaken.

The method I shall employ for discovering all this, is an application of the very curious theoretical law of "deviation from an average." First, I will explain the law, and then I will show that the production of natural intellectual gifts comes justly within its scope.

The law is an exceedingly general one. M. Quetelet, the Astronomer-Royal of Belgium, and the greatest authority on vital and social statistics, has largely used it in his inquiries. He has also constructed numerical tables, by which the necessary calculations can be easily made, whenever it is desired to have recourse to the law. Those who wish to learn more than I have space to relate, should consult his work, which is a very readable octavo volume, and deserves to be far better known to statisticians than it appears to be. Its title is *Letters on Probabilities*, translated by Downes. Layton and Co. London: 1849.

So much has been published in recent years about statistical deductions, that I am sure the reader will be prepared to assent freely to the following hypothetical case:—Suppose a large island inhabited by a single race, who intermarried freely, and who had lived for many generations under constant conditions; then the average height of the male adults of that population would undoubtedly be the same year after year. Also—still arguing from the experience of modern statistics, which are found to give constant results in far less carefully-guarded examples—we should undoubtedly find, year after year, the same proportion maintained between the number of men of different heights. I mean, if the average stature was found to be sixty-six inches, and if it was also found in any one year that 100 per million exceeded seventy-eight inches, the same proportion of 100 per million would be closely maintained in all other years. An equal constancy of proportion would be maintained between any other limits of height we pleased to specify, as between seventy-one and seventy-two inches; between seventy-two and seventy-three inches; and so on. Statistical experiences are so invariably confirmatory of what I have stated would probably be the case, as to make it unnecessary to describe analogous instances. Now, at this point, the law of deviation from an average steps in. It shows that the number per million whose heights range between seventy-one and seventy-two inches (or between any other limits we please to name) can be predicted from the previous datum of the average, and of any one other fact, such as that of 100 per million exceeding seventy-eight inches.

The diagram on Figure 1 will make this more intelligible. Suppose a million of the men to stand in turns, with their backs against a vertical board of sufficient height, and their heights to be dotted off upon it. The

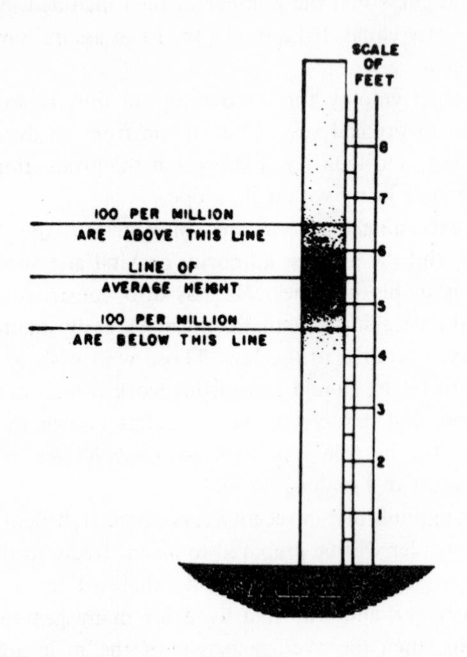

FIGURE 1

board would then present the appearance shown in the diagram. The line of average height is that which divides the dots into two equal parts, and stands, in the case we have assumed, at the height of sixty-six inches. The dots will be found to be ranged so symmetrically on either side of the line of average, that the lower half of the diagram will be almost a precise reflection of the upper. Next, let a hundred dots be counted from above downwards, and let a line be drawn below them. According to the conditions, this line will stand at the height of seventy-eight inches. Using the data afforded by these two lines, it is possible, by the help of the law of deviation from an average, to reproduce, with extraordinary closeness, the entire system of dots on the board.

M. Quetelet gives tables in which the uppermost line, instead of cutting off 100 in a million, cuts off only one in a million. He divides the intervals between that line and the line of average, into eighty equal divisions, and gives the number of dots that fall within each of those divisions. It is easy, by the help of his tables, to calculate what would occur under any other system of classification we pleased to adopt.

This law of deviation from an average is perfectly general in its application. Thus, if the marks had been made by bullets fired at a horizontal line stretched in front of the target, they would have been distributed according to the same law. Wherever there is a large number of similar events, each due to the resultant influences of the same variable conditions, two effects will follow. First, the average value of those events will be constant; and, secondly, the deviations of the several events from the average, will be governed by this law (which is, in principle, the same as that which governs runs of luck at a gaming-table).

The nature of the conditions affecting the several events must, I say, be the same. It clearly would not be proper to combine the heights of men belonging to two dissimilar races, in the expectation that the compound results would be governed by the same constants. A union of two dissimilar systems of dots would produce the same kind of confusion as if half the bullets fired at a target had been directed to one mark, and the other half to another mark. Nay, an examination of the dots would show to a person, ignorant of what had occurred, that such had been the case, and it would be possible, by aid of the law, to disentangle two or any moderate number of superimposed series of marks. The law may, therefore, be used as a most trustworthy criterion, whether or no the events of which an average has been taken are due to the same or to dissimilar classes of conditions.

I selected the hypothetical case of a race of men living on an island and freely intermarrying, to ensure the conditions under which they were all supposed to live, being uniform in character. It will now be my aim to show there is sufficient uniformity in the inhabitants of the British Isles to bring them fairly within the grasp of this law.

For this purpose, I first call attention to an example given in Quetelet's book. It is of the measurements of the circumferences of the chests of a large number of Scotch soldiers. The Scotch are by no means a strictly uniform race, nor are they exposed to identical conditions. They are a mixture of Celts, Danes, Anglo-Saxons, and others, in various proportions, the Highlanders being almost purely Celts. On the other hand, these races, though diverse in origin, are not very dissimilar in character. Consequently, it will be found that their deviations from the average, follow theoretical computations with remarkable accuracy. The instance is as follows. M. Quetelet obtained his facts from the thirteenth volume of the *Edinburgh Medical Journal*, where the measurements are given in respect to 5,738 soldiers, the results being grouped in order of magnitude, proceeding by differences of one inch. Professor Quetelet compares these results with those that his tables give, and here is the result. The marvellous accordance between fact and theory must strike the most unpractised eye. I should say that, for the sake of convenience, both the measurements and calculations have been reduced to per thousandths:—

Measures of the chest in inches	*Number of men per 1,000, by experience*	*Number of men per 1,000, by calculation*	*Measures of the chest in inches*	*Number of men per 1,000, by experience*	*Number of men per 1,000, by calculation*
33	5	7	41	1,628	1,675
34	31	29	42	1,148	1,096
35	141	110	43	645	560
36	322	323	44	160	221
37	732	732	45	87	69
38	1,305	1,333	46	38	16
39	1,867	1,838	47	7	3
40	1,882	1,987	48	2	1

I will now take a case where there is a greater dissimilarity in the elements of which the average has been taken. It is the height of 100,000 French conscripts. There is fully as much variety in the French as in the English, for it is not very many generations since France was divided into completely independent kingdoms. Among its peculiar races are those of Normandy, Brittany, Alsatia, Provence, Bearne, Auvergne—each with their special characteristics; yet the following table shows a most striking agreement between the results of experience compared with those derived by calculation, from a purely theoretical hypothesis.

The greatest differences are in the lowest ranks. They include the men who were rejected from being too short for the army. M. Quetelet boldly ascribes these differences to the effect of fraudulent returns. It certainly seems that men have been improperly taken out of the second rank and

Height of Men	*Number of Men*	
	Measured	*Calculated*
Under 61.8	28,620	26,345
61.8 to 62.9	11,580	13,182
62.9 to 63.9	13,990	14,502
63.9 to 65.0	14,410	13,982
65.0 to 66.1	11,410	11,803
66.1 to 67.1	8,780	8,725
67.1 to 68.2	5,530	5,527
68.2 to 69.3	3,190	3,187
Above 69.3	2,490	2,645

put into the first, in order to exempt them from service. Be this as it may, the coincidence of fact with theory is, in this instance also, quite close enough to serve my purpose.

I argue from the results obtained from Frenchmen and from Scotchmen, that, if we had measurements of the adult males in the British Isles, we should find those measurements to range in close accordance with the law of deviation from an average, although our population is as much mingled as I described that of Scotland to have been, and although Ireland is mainly peopled with Celts. Now, if this be the case with stature, then it will be true as regards every other physical feature—as circumference of head, size of brain, weight of grey matter, number of brain fibres, &c; and thence, by a step on which no physiologist will hesitate, as regards mental capacity.

This is what I am driving at—that analogy clearly shows there must be a fairly constant average mental capacity in the inhabitants of the British Isles, and that the deviations from that average—upwards towards genius, and downwards towards stupidity—must follow the law that governs deviations from all true averages.

I have, however, done somewhat more than rely on analogy. I have tried the results of those examinations in which the candidates had been derived from the same classes. Most persons have noticed the lists of successful competitors for various public appointments that are published from time to time in the newspapers, with the marks gained by each candidate attached to his name. These lists contain far too few names to fall into such beautiful accordance with theory, as was the case with Scotch soldiers. There are rarely more than 100 names in any one of these examinations, while the chests of no less than 5,700 Scotchmen were measured. I cannot justly combine the marks of several independent examinations into one fagot, for I understand that different examiners are apt to have different figures of merit; so I have analysed each examination separately. I give a calculation I made on the examination last before me; it will do as well as any other. It was for admission into the Royal Military College at Sandhurst, December 1868. The marks obtained were clustered most thickly about 3,000, so I take that number as representing the average ability of the candidates. From this datum, and from the fact that no candidate obtained more than 6,500 marks, I computed the column B in the following table [see p. 1186] by the help of Quetelet's numbers. It will be seen that column B accords with column A quite as closely as the small number of persons examined could have led us to expect.

The symmetry of the descending branch has been rudely spoilt by the conditions stated at the foot of column A. There is, therefore, little room for doubt, if everybody in England had to work up some subject and then to pass before examiners who employed similar figures of merit, that

Number of Marks obtained by the Candidates	*Number of Candidates who obtained those marks*			
	A *According to fact*		*B* *According to theory*	
6,500 and above	0		0	
5,800 to 6,500	1		1	
5,100 to 5,800	3		5	
4,400 to 5,100	6	73	8	72
3,700 to 4,400	11		13	
3,000 to 3,700	22		16	
2,300 to 3,000	22		16	
1,600 to 2,300	8		13	
1,100 to 1,600	Either did not ven-		8	
400 to 1,100	ture to complete, or		5	
below 400	were plucked.		1	

their marks would be found to range, according to the law of deviation from an average, just as rigorously as the heights of French conscripts, or the circumferences of the chests of Scotch soldiers.

The number of grades into which we may divide ability is purely a matter of option. We may consult our convenience by sorting Englishmen into a few large classes, or into many small ones. I will select a system of classification that shall be easily comparable with the numbers of eminent men, as determined in the previous chapter. We have seen that 250 men per million become eminent; accordingly, I have so contrived the classes in the following table [p. 1187] that the two highest, F and G, together with X (which includes all cases beyond G, and which are unclassed), shall amount to about that number—namely, to 248 per million:—

It will, I trust, be clearly understood that the numbers of men in the several classes in my table depend on no uncertain hypothesis. They are determined by the assured law of deviations from an average. It is an absolute fact that if we pick out of each million the one man who is naturally the ablest, and also the one man who is the most stupid, and divide the remaining 999,998 men into fourteen classes, the average ability in each being separated from that of its neighbours by equal grades, then the numbers in each of those classes will, on the average of many millions, be as is stated in the table. The table may be applied to special, just as truly as to general ability. It would be true for every examination that brought out natural gifts, whether held in painting, in music, or in statesmanship. The proportions between the different classes would be identical in all these cases, although the classes would be made up of different individuals, according as the examination differed in its purport.

It will be seen that more than half of each million is contained in the two mediocre classes a and A; the four mediocre classes a, b, A, B, con-

Classification of Men According to Their Natural Gifts

Grades of natural ability, separated by equal intervals		*Numbers of men comprised in the several grades of natural ability, whether in respect to their general powers, or to special aptitudes*							
				In total male population of the United Kingdom, viz. 15 millions, of the undermentioned ages:—					
Below average	*Above average*	*Propor-tionate, viz. one in*	*In each million of the same age*	*20–30*	*30–40*	*40–50*	*50–60*	*60–70*	*70–80*
a	A	4	256,791	651,000	495,000	391,000	268,000	171,000	77,000
b	B	6	162,279	409,000	312,000	246,000	168,000	107,000	48,000
c	C	16	63,563	161,000	123,000	97,000	66,000	42,000	19,000
d	D	64	15,696	39,800	30,300	23,900	16,400	10,400	4,700
e	E	413	2,423	6,100	4,700	3,700	2,520	1,600	729
f	F	4,300	233	590	450	355	243	155	70
g	G	79,000	14	35	27	21	15	9	4
x	X								
all grades below g	all grades above G	1,000,000	1	3	2	2	2	—	—
On either side of average......			500,000	1,268,000	964,000	761,000	521,000	332,000	149,000
Total, both sides.............			1,000,000	2,536,000	1,928,000	1,522,000	1,042,000	664,000	298,000

The proportions of men living at different ages are calculated from the proportions that are true for England and Wales. (Census 1861, Appendix, p. 107.)

Example.—The class F contains 1 in every 4,300 men. In other words, there are 233 of that class in each million of men. The same is true of class f. In the whole United Kingdom there are 590 men of class F (and the same number of f) between the ages of 20 and 30; 450 between the ages of 30 and 40; and so on.

tain more than four-fifths, and the six mediocre classes more than nineteen-twentieths of the entire population. Thus, the rarity of commanding ability, and the vast abundance of mediocrity, is no accident, but follows of necessity, from the very nature of these things.

The meaning of the word "mediocrity" admits of little doubt. It defines the standard of intellectual power found in most provincial gatherings, because the attractions of a more stirring life in the metropolis and elsewhere, are apt to draw away the abler classes of men, and the silly and the imbecile do not take a part in the gatherings. Hence, the residuum that forms the bulk of the general society of small provincial places, is commonly very pure in its mediocrity.

The class C possesses abilities a trifle higher than those commonly possessed by the foreman of an ordinary jury. D includes the mass of men who obtain the ordinary prizes of life. E is a stage higher. Then we reach F, the lowest of those yet superior classes of intellect, with which this volume is chiefly concerned.

On descending the scale, we find by the time we have reached f, that we are already among the idiots and imbeciles. We have seen that there are 400 idiots and imbeciles, to every million of persons living in this country; but that 30 per cent. of their number, appear to be light cases, to whom the name of idiot is inappropriate. There will remain 280 true idiots and imbeciles, to every million of our population. This ratio coincides very closely with the requirements of class f. No doubt a certain proportion of them are idiotic owing to some fortuitous cause, which may interfere with the working of a naturally good brain, much as a bit of dirt may cause a first-rate chronometer to keep worse time than an ordinary watch. But I presume, from the usual smallness of head and absence of disease among these persons, that the proportion of accidental idiots cannot be very large.

Hence we arrive at the undeniable, but unexpected conclusion, that eminently gifted men are raised as much above mediocrity as idiots are depressed below it; a fact that is calculated to considerably enlarge our ideas of the enormous differences of intellectual gifts between man and man.

I presume the class F of dogs, and others of the more intelligent sort of animals, is nearly commensurate with the f of the human race, in respect to memory and powers of reason. Certainly the class G of such animals is far superior to the g of humankind.

COMMENTARY ON

THOMAS ROBERT MALTHUS

IT WAS in 1798 that Thomas Robert Malthus (1766–1834) published anonymously *An Essay on the Principle of Population As It Affects the Future Improvement of Society*. The main argument of this famous pamphlet was that the world's population increases in a geometrical, and food only in an arithmetical, ratio. Since an arithmetic progression is no match for a geometric progression, man must forever be outstripping his food supplies. Thus, said Malthus, population is necessarily limited by the "checks" of vice and misery.

The pamphlet aroused bitter controversy. It offended those who believed in the perfectability of man and the gradual advent of the "happy society," and it annoyed conservatives who regarded Malthus as a mischief maker, a remembrancer of unpleasant facts. On the other hand, his views were received with favor in certain conservative circles because of an impression "very welcome to the higher ranks of society, that they tended to relieve the rich and powerful of responsibility for the condition of the working classes, by showing that the latter had chiefly themselves to blame, and not either the negligence of their superiors or the institutions of the country." [1]

In 1803, having traveled in Germany, Sweden, Norway, Finland and Russia to collect further information, Malthus issued a second edition of his essay, which was "substantially a new book." Without receding from his central principle, based on the unmistakably persuasive "postulata" that "food is necessary to the existence of man" and that "the passion between the sexes is necessary, and will remain nearly in its present state," he took the position that the checks to population were not to be regarded as "insuperable obstacles" to social advance but "as defining the dangers which must be avoided if improvement is to be achieved."

Malthus' doctrine, despite the furor it created, was not new. Condorcet, the eminent French philosopher, mathematician and revolutionist, had anticipated his ideas as to population, but took the cheerful view that as mankind improved it would avert misery by the practice of birth control. Malthus also acknowledged his indebtedness to Robert Wallace (*Various Prospects of Mankind, Nature and Providence*, 1761) and to J. P. Süssmilch (*Göttliche Ordnung*) from whom he got many of his statistical facts.[2] Nonetheless, the importance and influence of the *Essay* must not be underestimated. It gave the first full, systematic exposition to a power-

[1] *Encyclopaedia Britannica*, Eleventh Edition, article on Malthus, Vol. 17, p. 516.
[2] *Dictionary of National Biography*, article on Malthus.

ful doctrine, and not the least of its accomplishments was in stimulating the thoughts of Darwin. For as he himself took pains to point out, the phrase "struggle for existence," used by Malthus in relation to social competition, suggested the operation of this principle, with its corollary of survival of the fittest, as the determining factor of evolution in all forms of organic life.

For a time in the nineteenth century, as new lands and resources were developed, the Malthusian doctrine fell into disfavor. Today, with the world's population sharply rising and with a scarcity of new frontiers, the truth of his basic assertion is again very much in vogue. It is known that food production can be greatly increased by improvements in agricultural methods, by food synthesis, by ingenuities beyond present conception. It is known also that an unchecked population must soon outrun food supplies however much they are augmented. The effect of scientific progress is in some respects to aggravate the difficulty. Epidemic diseases have been largely brought under control; the practice of "vice" is no longer so widespread or so lethal as to assure the virtuous of more to eat; even wars, though much more efficient than in Malthus' day, have not yet succeeded in substantially reducing the number of the hungry. The Malthusian governors are a grim company and should obviously be eliminated. But this does not solve the problem of food shortages. The only permanent solution is voluntary birth control, coupled with a rational, worldwide eugenics policy, but many persons despair of its ever being adopted.[3] In this, as in other aspects of human affairs, man is his own principal adversary.

I add a brief biographical note. Malthus was born in Surrey in 1766. His father was well off, a cultured and thoughtful man who "lived quietly among his books." He was a friend and disciple of Rousseau and was said to have been his executor. Malthus was educated at Cambridge, where he studied history, poetry and the classics, and distinguished himself in mathematics. In 1798 he took "holy orders," and the same year published the *Essay*, which was to some extent a challenge to the social views held by his father, who shared the optimistic theories of Condorcet and Godwin. After spending five years on the revision of the *Essay*, Malthus became professor of history and political economy at the newly founded college of Haileybury. In this post he spent the rest of his life, publishing papers on the corn laws, rent and various current economic topics, and several larger treatises on political economy. He was elected a Fellow of the Royal Society in 1819, and an associate of the Royal Society of Literature in 1824.

[3] For an unmixedly gloomy but extraordinarily readable discussion of man's future prospects, in light of Malthusian principles, see Sir Charles Galton Darwin, *The Next Million Years*, New York, 1953. Another valuable study is L. Dudley Stamp, *Land for Tomorrow*, Indiana University Press, 1952.

Malthus carried on an extensive correspondence with Ricardo, and was a friend of James Mill, George Grote and other influential thinkers of his period. Politically he was a Whig, moderate in his opinions, favoring most of the contemporary reform legislation yet not altogether convinced of its value. Like other men who have advanced controversial theories, Malthus has been belabored, at times, as if he had personally created the distressing predicament he described. Fortunately he was a "serene and cheerful man," "singularly amiable"—as Harriet Martineau described him—without malice and with a keen sense of humor. Miss Martineau also tells us that although he had "a defect in the palate" which made his speech "hopelessly imperfect," he was the only friend whom she could hear without her trumpet. The world has been hearing him clearly for more than a century. His predictions have not been vindicated by population trends but the central point of his analysis has yet to be disproved.

The following selection is from the sixth edition of *An Essay on the Principle of Population.*

1, 2, 3, 4, 5, 6, 7, 8 . . . —ARITHMETIC PROGRESSION

1, 2, 4, 8, 16, 32, 64, 128 . . . —GEOMETRIC PROGRESSION

3 Mathematics of Population and Food

By THOMAS ROBERT MALTHUS

IN an inquiry concerning the improvement of society the mode of conducting the subject which naturally presents itself is,

1. To investigate the causes that have hitherto impeded the progress of mankind towards happiness; and

2. To examine the probability of the total or partial removal of these causes in the future.

To enter fully into this question and to enumerate all the causes that have hitherto influenced human improvement would be much beyond the power of an individual. The principal object of the present essay is to examine the effects of one great cause intimately united with the very nature of man; which, though it has been constantly and powerfully operating since the commencement of society, has been little noticed by the writers who have treated this subject. The facts which establish the existence of this cause have, indeed, been repeatedly stated and acknowledged; but its natural and necessary effects have been almost totally overlooked; though probably among these effects may be reckoned a very considerable portion of that vice and misery, and of that unequal distribution of the bounties of nature, which it has been the unceasing object of the enlightened philanthropist in all ages to correct.

The cause to which I allude is the constant tendency in all animated life to increase beyond the nourishment prepared for it.

It is observed by Dr. Franklin that there is no bound to the prolific nature of plants or animals but what is made by their crowding and interfering with each other's means of subsistence. Were the face of the earth, he says, vacant of other plants, it might be gradually sowed and overspread with one kind only, as for instance with fennel; and were it empty of other inhabitants, it might in a few ages be replenished from one nation only, as for instance with Englishmen.

This is incontrovertibly true. Through the animal and vegetable kingdoms nature has scattered the seeds of life abroad with the most profuse and liberal hand, but has been comparatively sparing in the room and the nourishment necessary to rear them. The germs of existence contained in this earth, if they could freely develop themselves, would fill millions of worlds in the course of a few thousand years. Necessity, that imperious, all-pervading law of nature, restrains them within the prescribed bounds. The race of plants and the race of animals shrink under this great restrictive law; and man cannot by any efforts of reason escape from it.

In plants and irrational animals the view of the subject is simple. They are all impelled by a powerful instinct to the increase of their species; and this instinct is interrupted by no doubts about providing for their offspring. Wherever therefore there is liberty, the power of increase is exerted; and the superabundant effects are repressed afterwards by want of room and nourishment.

The effects of this check on man are more complicated. Impelled to the increase of his species by an equally powerful instinct, reason interrupts his career, and asks him whether he may not bring beings into the world for whom he cannot provide the means of support. If he attend to this natural suggestion, the restriction too frequently produces vice. If he hear it not, the human race will be constantly endeavoring to increase beyond the means of subsistence. But as, by that law of our nature which makes food necessary to the life of man, population can never actually increase beyond the lowest nourishment capable of supporting it, a strong check on population, from the difficulty of acquiring food, must be constantly in operation. This difficulty must fall somewhere, and must necessarily be severely felt in some or other of the various forms of misery, or the fear of misery, by a large portion of mankind.

That population has this constant tendency to increase beyond the means of subsistence, and that it is kept to its necessary level by these causes will sufficiently appear from a review of the different states of society in which man has existed. But before we proceed to this review the subject will, perhaps, be seen in a clearer light, if we endeavor to ascertain what would be the natural increase of population if left to exert itself with perfect freedom, and what might be expected to be the rate of increase in the productions of the earth under the most favorable circumstances of human industry.

It will be allowed that no country has hitherto been known where the manners were so pure and simple, and the means of subsistence so abundant, that no check whatever has existed to early marriages from the difficulty of providing for a family, and that no waste of the human species has been occasioned by vicious customs, by towns, by unhealthy occupations, or too severe labor. Consequently, in no state that we have yet known has the power of population been left to exert itself with perfect freedom.

Whether the law of marriage be instituted or not, the dictates of nature and virtue seem to be an early attachment to one woman; and where there were no impediments of any kind in the way of an union to which such an attachment would lead, and no causes of depopulation afterwards, the increase of the human species would be evidently much greater than any increase which has hitherto been known.

In the Northern States of America, where the means of subsistence have been more ample, the manners of the people more pure, and the checks to early marriages fewer than in any of the modern states of Europe, the population has been found to double itself, for above a century and a half successively, in less than twenty-five years. Yet, even during these periods, in some of the towns the deaths exceeded the births, a circumstance which clearly proves that, in those parts of the country which supplied this deficiency, the increase must have been much more rapid than the general average.

In the back settlements, where the sole employment is agriculture, and vicious customs and unwholesome occupations are little known, the population has been found to double itself in fifteen years. Even this extraordinary rate of increase is probably short of the utmost power of population. Very severe labor is requisite to clear a fresh country; such situations are not in general considered as particularly healthy; and the inhabitants, probably, are occasionally subject to the incursions of the Indians, which may destroy some lives, or at any rate diminish the fruits of industry.

According to a table of Euler, calculated on a mortality of one to thirty-six, if the births be to the deaths in the proportion of three to one, the period of doubling will be only twelve years and four fifths. And this proportion is not only a possible supposition, but has actually occurred for short periods in more countries than one.

Sir William Petty supposes a doubling possible in so short a time as ten years.

But, to be perfectly sure that we are far within the truth, we will take the slowest of these rates of increase, a rate in which all concurring testimonies agree, and which has been repeatedly ascertained to be from procreation only.

It may safely be pronounced, therefore, that population, when unchecked, goes on doubling itself every twenty-five years, or increases in a geometrical ratio.

The rate according to which the productions of the earth may be supposed to increase it will not be so easy to determine. Of this, however, we may be perfectly certain—that the ratio of their increase in a limited territory must be of a totally different nature from the ratio of the increase of population. A thousand millions are just as easily doubled every twenty-five years by the power of population as a thousand. But the food to support the increase from the greater number will by no means be obtained with the same facility. Man is necessarily confined in room. When acre has been added to acre till all the fertile land is occupied, the yearly increase of food must depend upon the melioration of the land already in possession. This is a fund which, from the nature of all soils instead of increasing, must be gradually diminishing. But population, could it be supplied with food, would go on with unexhausted vigor; and the increase of one period would furnish the power of a greater increase the next, and this without any limit.

From the accounts we have of China and Japan, it may be fairly doubted whether the best-directed efforts of human industry could double the produce of these countries even once in any number of years. There are many parts of the globe, indeed, hitherto uncultivated and almost unoccupied, but the right of exterminating, or driving into a corner where they must starve, even the inhabitants of these thinly-peopled regions, will be questioned in a moral view. The process of improving their minds and directing their industry would necessarily be slow; and during this time, as population would regularly keep pace with the increasing produce, it would rarely happen that a great degree of knowledge and industry would have to operate at once upon rich unappropriated soil. Even where this might take place, as it does sometimes in new colonies, a geometrical ratio increases with such extraordinary rapidity that the advantage could not last long. If the United States of America continue increasing, which they certainly will do, though not with the same rapidity as formerly, the Indians will be driven further and further back into the country, till the whole race is ultimately exterminated and the territory is incapable of further extension.

These observations are, in a degree, applicable to all the parts of the earth where the soil is imperfectly cultivated. To exterminate the inhabitants of the greatest part of Asia and Africa is a thought that could not be admitted for a moment. To civilize and direct the industry of the various tribes of Tartars and Negroes would certainly be a work of considerable time, and of variable and uncertain success.

Europe is by no means so fully peopled as it might be. In Europe there is the fairest chance that human industry may receive its best direction. The science of agriculture has been much studied in England and Scotland, and there is still a great portion of uncultivated land in these countries. Let us consider at what rate the produce of this island might be supposed to increase under circumstances the most favorable to improvement.

If it be allowed that by the best possible policy, and great encouragement to agriculture, the average produce of the island could be doubled in the first twenty-five years, it will be allowing, probably, a greater increase than could with reason be expected.

In the next twenty-five years it is impossible to suppose that the produce could be quadrupled. It would be contrary to all our knowledge of the properties of land. The improvement of the barren parts would be a work of time and labor; and it must be evident, to those who have the slightest acquantance with agricultural subjects, that in proportion as cultivation is extended, the additions that could yearly be made to the former average produce must be gradually and regularly diminishing. That we may be the better able to compare the increase of population and food, let us make a supposition which, without pretending to accuracy, is clearly more favorable to the power of production in the earth than any experience we have had of its qualities will warrant.

Let us suppose that the yearly additions which might be made to the former average produce, instead of decreasing, which they certainly would do, were to remain the same; and that the produce of this island might be increased every twenty-five years by a quantity equal to what it at present produces. The most enthusiastic speculator cannot suppose a greater increase than this. In a few centuries it would make every acre of land in the island like a garden.

If this supposition be applied to the whole earth, and if it be allowed that the subsistence for man which the earth affords might be increased every twenty-five years by a quantity equal to what it at present produces, this will be supposing a rate of increase much greater than we can imagine that any possible exertions of mankind could make it.

It may be fairly pronounced, therefore, that considering the present average state of the earth, the means of subsistence, under circumstances the most favorable to human industry, could not possibly be made to increase faster than in an arithmetical ratio.

The necessary effects of these two different rates of increase, when brought together, will be very striking. Let us call the population of this island eleven millions; and suppose the present produce equal to the easy support of such a number. In the first twenty-five years the population would be twenty-two millions, and the food being also doubled, the means of subsistence would be equal to this increase. In the next twenty-five years the population would be forty-four millions, and the means of subsistence only equal to the support of thirty-three millions. In the next period the population would be eighty-eight millions and the means of subsistence just equal to the support of half that number. And at the conclusion of the first century the population would be a hundred and seventy-six millions, and the means of subsistence only equal to the support of fifty-five millions, leaving a population of a hundred and twenty-one millions totally unprovided for.

Taking the whole earth instead of this island, emigration would of course be excluded; and, supposing the present population equal to a thousand millions, the human species would increase as the numbers, 1, 2, 4, 8, 16, 32, 64, 128, 256, and subsistence as 1, 2, 3, 4, 5, 6, 7, 8, 9. In two centuries the population would be to the means of subsistence as 256 to 9; in three centuries, as 4096 to 13; and in two thousand years the difference would be almost incalculable.

In this supposition no limits whatever are placed to the produce of the earth. It may increase forever, and be greater than any assignable quantity; yet still, the power of population being in every period so much superior, the increase of the human species can only be kept down to the level of the means of subsistence by the constant operation of the strong law of necessity, acting as a check upon the greater power.

OF THE GENERAL CHECKS TO POPULATION, AND THE MODE OF THEIR OPERATION

The ultimate check to population appears then to be a want of food arising necessarily from the different ratios according to which population and food increase. But this ultimate check is never the immediate check, except in cases of actual famine.

The immediate check may be stated to consist in all those customs, and all those diseases, which seem to be generated by a scarcity of the means of subsistence; and all those causes, independent of this scarcity, whether of a moral or physical nature, which tend prematurely to weaken and destroy the human frame.

These checks to population, which are constantly operating with more or less force in every society, and keep down the number to the level of the means of subsistence, may be classed under two general heads—the preventive, and the positive checks.

The preventive check, as far as it is voluntary, is peculiar to man, and arises from that distinctive superiority in his reasoning faculties which enables him to calculate distant consequences. The checks to the indefinite increase of plants and irrational animals are all either positive, or, if preventive, involuntary. But man cannot look around him and see the distress which frequently presses upon those who have large families; he cannot contemplate his present possessions or earnings, which he now nearly consumes himself, and calculate the amount of each share, when with very little addition they must be divided, perhaps, among seven or eight, without feeling a doubt whether, if he follow the bent of his inclinations, he may be able to support the offspring which he will probably bring into the world. In a state of equality, if such can exist, this would be the simple question. In the present state of society other considerations occur. Will he not lower his rank in life, and be obliged to give up in great measure his former habits? Does any mode of employment present itself by which he may reasonably hope to maintain a family? Will he not at any rate subject himself to greater difficulties and more severe labor than in his single state? Will he not be unable to transmit to his children the same advantages of education and improvement that he had himself possessed? Does he even feel secure that, should he have a large family, his utmost exertions can save them from rags and squalid poverty, and their consequent degradation in the community? And may he not be reduced to the grating necessity of forfeiting his independence, and of being obliged to the sparing hand of charity for support?

These considerations are calculated to prevent, and certainly do prevent, a great number of persons in all civilized nations from pursuing the dictate of nature in an early attachment to one woman.

If this restraint does not produce vice, it is undoubtedly the least evil that can arise from the principle of population. Considered as a restraint on a strong natural inclination, it must be allowed to produce a certain degree of temporary unhappiness, but evidently slight compared with the evils which result from any of the other checks to population, and merely of the same nature as many other sacrifices of temporary to permanent gratification, which it is the business of a moral agent continually to make.

When this restraint produces vice, the evils which follow are but too conspicuous. A promiscuous intercourse to such a degree as to prevent the birth of children seems to lower, in the most marked manner, the dignity of human nature. It cannot be without its effect on men, and nothing can be more obvious than its tendency to degrade the female character and to destroy all its most amiable and distinguishing characteristics. Add to which, that among those unfortunate females with which all great towns abound more real distress and aggravated misery are, perhaps, to be found, than in any other department of human life.

When a general corruption of morals with regard to the sex pervades all the classes of society, its effects must necessarily be to poison the springs of domestic happiness, to weaken conjugal and parental affection, and to lessen the united exertions and ardor of parents in the care and education of their children—effects which cannot take place without a decided diminution of the general happiness and virtue of the society; particularly as the necessity of art in the accomplishment and conduct of intrigues and in the concealment of their consequences necessarily leads to many other vices.

The positive checks to population are extremely various, and include every cause, whether arising from vice or misery, which in any degree contributes to shorten the natural duration of human life. Under this head, therefore, may be enumerated all unwholesome occupations, severe labor and exposure to the seasons, extreme poverty, bad nursing of children, great towns, excesses of all kinds, the whole train of common diseases and epidemics, wars, plague, and famine.

On examining these obstacles to the increase of population which I have classed under the heads of preventive and positive checks, it will appear that they are all resolvable into moral restraint, vice and misery.

Of the preventive checks, the restraint from marriage which is not followed by irregular gratifications may properly be termed moral restraint.[1]

Promiscuous intercourse, unnatural passions, violations of the marriage bed, and improper arts to conceal the consequences of irregular connections are preventive checks that clearly come under the head of vice.

Of the positive checks, those which appear to arise unavoidably from the laws of nature may be called exclusively misery, and those which we obviously bring upon ourselves, such as wars, excesses, and many others which it would be in our power to avoid, are of a mixed nature. They are brought upon us by vice, and their consequences are misery.

The sum of all these preventive and positive checks taken together forms the immediate check to population; and it is evident that in every country where the whole of the procreative power cannot be called into action, the preventive and the positive checks must vary inversely as each other; that is, in countries either naturally unhealthy or subject to a great morality, from whatever cause it may arise, the preventive check will prevail very little. In those countries, on the contrary, which are naturally healthy, and where the preventive check is found to prevail with considerable force, the positive check will prevail very little, or the mortality be very small.

In every country some of these checks are with more or less force in constant operation; yet notwithstanding their general prevalence, there are few states in which there is not a constant effort in the population to increase beyond the means of subsistence. This constant effort as constantly tends to subject the lower classes of society to distress, and to prevent any great permanent melioration of their condition.

[1] It will be observed that I here use the term *moral* in its most confined sense. By moral restraint I would be understood to mean a restraint from marriage from prudential motives, with a conduct strictly moral during the period of this restraint; and I have never intentionally deviated from this sense. When I have wished to consider the restraint from marriage unconnected with its consequences, I have either called it prudential restraint, or a part of the preventive check, of which indeed it forms the principal branch. In my review of the different stages of society I have been accused of not allowing sufficient weight in the prevention of population to moral restraint; but when the confined sense of the term, which I have here explained, is adverted to, I am fearful that I shall not be found to have erred much in this respect. I should be very glad to believe myself mistaken.

COMMENTARY ON

COURNOT, JEVONS, and the Mathematics of Money

MATHEMATICAL economics is old enough to be respectable, but not all economists respect it. It has powerful supporters and impressive testimonials, yet many capable economists deny that mathematics, except as a shorthand or expository device, can be applied to economic reasoning. There have even been rumors that mathematics is used in economics (and in other social sciences) either for the deliberate purpose of mystification or to confer dignity upon commonplaces—as French was once used in diplomatic communications.

The value of graphs and symbols for expressing simple facts of economics is widely acknowledged and needs no discussion. A more important use of mathematics is the application of its methods to restricted economic problems (for example, the theory of partial equilibria) selected because of their "particular susceptibility to mathematical treatment." [1] An exponent of this approach was the famous economist Alfred Marshall.[2] The most sweeping and most controversial development in mathematical economics has been the organization of the "whole body of economic theory into an interdependent set of propositions stated in mathematical terms." Alfredo Pareto was among the leaders of this ambitious school of thought which sought to formulate a theory of general equilibrium in which all the economic unknowns—such as prices of consumption goods and of productive services, costs, market quantities of outputs and inputs—could be represented in simultaneous equations, analogous to the procedures of classical physics. "The problem is determined when there can be obtained as many equations as there are unknowns."

To be sure, mathematics can be extended to any branch of knowledge, including economics, provided the concepts are so clearly defined as to permit accurate symbolic representation. That is only another way of saying that in some branches of discourse it is desirable to know what you are talking about. But this extension does not in itself guarantee fruit-

[1] See Oskar Morgenstern, "Article on Mathematical Economics," in *Encyclopaedia of the Social Sciences*, 1931, Vol. 5, pp. 364–368, an admirable summary from which the quoted material and other points in this introduction have been taken.

[2] ". . . the most helpful applications of mathematics to economics are those which are short and simple, which employ few symbols, and which aim at throwing a bright light on some small part of the great economic movement rather than at representing its endless complexities." *Memorials of Alfred Marshall*, ed. by A. C. Pigou, London, 1925, p. 313. Quoted by Morgenstern.

ful results. Wittgenstein has described the service mathematics performs for science in a brilliant passage: "Mathematics is a logical method . . . Mathematical propositions express no thoughts. In life it is never a mathematical proposition which we need, but we use mathematical propositions *only* in order to infer from propositions which do not belong to mathematics to others which equally do not belong to mathematics." [3] Mathematical methods facilitate inference in some branches of economic thought; their usefulness is therefore beyond dispute. On the other hand, it has not been established—at least for the techniques customarily employed—that mathematics is helpful in handling every type of economic problem, much less that a fruitful mathematization of the entire body of economic theory is possible.[4] Recently, however, Von Neumann and Morgenstern have introduced into economics a novel analytical apparatus, employing the tools of modern logic, that may profoundly alter the perspective and give new impetus to the process of mathematical-economic generalization.[5]

The first book dealing with the application of mathematics to economics was written by an Italian engineer, Giovanni Ceva, in 1711. Half a century later, one Cesare Beccaria "employed algebra effectively in an essay on the hazards and profits of smuggling": *Tentativo analytico sui contrabbandi.* Other minor writings followed, but the first treatise to explore the subject systematically was Augustin Cournot's masterpiece, *Recherches sur les principes mathématiques de la théorie des richesses* (*Researches into the Mathematical Principles of the Theory of Wealth*), an immensely influential work published in 1838.[6] Cournot (1801–1877) was a mathematician, a philosopher and a student of probability theory. He served as a university official at Lyons and Grenoble, held high positions in the French government and wrote and translated a large number of books on philosophy, mathematics, statistics and economics. His treatise, from which two selections appear below, enunciates principles—for example, as to the law of diminishing returns—which have become classic. It presents the original interpretation of supply and demand as "functions or schedules." Cournot's book "seemed a failure when first published. It was too far in advance of the times. Its methods were too

[3] Ludwig Wittgenstein, *Tractatus Logico Philosophicus*, New York, 1922, p. 169.

[4] "The problem that presents itself in connection with the mathematical method, so long as the question of its applicability is settled, is one of preference: is it of greater utility than other methods, and are there cases in which it is the only method possible. So far there have been found very few instances in which mathematics is absolutely necessary, but there are more examples in which the application of mathematics facilitates the prosecution of the argument." Morgenstern, *op. cit.*, p. 364.

[5] *See* selection by Leonid Hurwicz, "The Theory of Economic Behavior," pp. 1267–1284.

[6] In 1897 it was translated into English by Irving Fisher, a teacher at Yale for forty-five years and one of the first exponents of mathematical economics in the United States. The selection used here is from his translation.

strange, its reasoning too intricate for the crude and confident notions of political economy then current." [7]

Among the first to recognize the virtues of Cournot's work was William Stanley Jevons (1835–1882), an English economist and logician noted for developing the theory of utility and for his chief writings, *The Theory of Political Economy* and *The Principles of Science.* As a student at University College in London, Jevons specialized in mathematics, biology, chemistry and metallurgy—an unusual but apparently stimulating training for an economist. He worked for five years in Australia as assayer to the mint, and then returned to England to devote himself to investigations in economics and logic. The scientific character of his studies in economics, statistics and logic was recognized by his election in 1872 as a Fellow of the Royal Society; he was the first economist so elected, as John Maynard Keynes points out, since Sir William Petty. In introducing the material I have taken from Jevons,[8] I cannot do better than quote a passage from a lecture given by Keynes on the centenary of Jevons' birth:

> Jevons' *Theory of Political Economy* and the place it occupies in the history of the subject are so well known that I need not spend time in describing its content. It was not as uniquely original in 1871 as it would have been in 1862. For, leaving on one side the precursors Cournot, Gossen Dupuit, Von Thünen and the rest, there were several economists, notably Walras and Marshall, who by 1871, were scribbling equations with X's and Y's, big Deltas and little d's. Nevertheless, Jevons' *Theory* is the first treatise to present in a finished form the theory of value based on subjective valuations, the marginal principle and the now familiar technique of the algebra and the diagrams of the subject. The first modern book on the subject, it has proved singularly attractive to all bright minds newly attacking the subject;—simple, lucid, unfaltering, chiselled in stone where Marshall knits in wool.[9]

[7] Irving Fisher, in a review of Cournot's work published in *Quarterly Journal of Economics*, January 1898, reprinted in Henry William Spiegel, *The Development of Economic Thought*, New York, 1952, pp. 459–469. Cournot himself, it may be noted, took a modest view of his essay. ". . . I believe [he says in his preface], if this essay is of any practical value, it will be chiefly in making clear how far we are from being able to solve, with full knowledge of the case, a multitude of questions which are boldly decided every day."

[8] From *The Theory of Political Economy*, Chaps. I, III, IV.

[9] John Maynard Keynes, *Essays in Biography*, New York, 1951, p. 284.

What we might call, by way of eminence, the dismal science.

—THOMAS CARLYLE

Professor Planck of Berlin, the famous originator of the quantum theory, once remarked to me that in early life he had thought of studying economics, but had found it too difficult! Professor Planck could easily master the whole corpus of mathematical economics in a few days. He did not mean that! But the amalgam of logic and intuition and the wide knowledge of facts, most of which are not precise, which is required for economic interpretation in its highest form, is, quite truly, overwhelmingly difficult for those whose gift consists mainly in the power to imagine and pursue to their furthermost points, the implications and prior conditions of comparatively simple facts, which are known with a high degree of precision.

—JOHN MAYNARD KEYNES (*Biography of Alfred Marshall*)

I happened to sit next to Keynes at the High Table of King's College a day or two after Planck had made this observation, and Keynes told me of it. Lowes Dickinson was sitting opposite. "That's funny," he said, "because Bertrand Russell once told me that in early life he had thought of studying economics, but had found it too easy"!

—R. F. HARROD (*Life of John Maynard Keynes*)

4 Mathematics of Value and Demand

By AUGUSTIN COURNOT

OF CHANGES IN VALUE, ABSOLUTE AND RELATIVE

WHENEVER there is occasion to go back to the fundamental conceptions on which any science rests, and to formulate them with accuracy, we almost always encounter difficulties, which come, sometimes from the very nature of these conceptions, but more often from the imperfections of language. For instance, in the writings of economists, the definition of *value*, and the distinction between absolute and relative value, are rather obscure: a very simple and strikingly exact comparison will serve to throw light on this.

We conceive that a body moves when its situation changes with reference to other bodies which we look upon as fixed. If we observe a system of material points at two different times, and find that the respective situations of these points are not the same at both times, we necessarily conclude that some, if not all, of these points have moved; but if besides this we are unable to refer them to points of the fixity of which we can be sure, it is, in the first instance, impossible to draw any conclusions as to the motion or rest of each of the points in the system.

However, if all of the points in the system, except one, had preserved their relative situation, we should consider it very probable that this single point was the only one which had moved, unless, indeed, all the other points were so connected that the movement of one would involve the movement of all.

We have just pointed out an extreme case, viz., that in which all except one had kept their relative positions; but, without entering into details, it is easy to see that among all the possible ways of explaining the change in the state of the system there may be some much simpler than others, and which without hesitation we regard as much more probable.

If, without being limited to two distinct times, observation should follow the system through its successive states, there would be hypotheses as to the absolute movements of the different points of the system, which would be considered preferable for the explanation of their relative movements. Thus, without reference to the relative size of the heavenly bodies and to knowledge of the laws of gravitation, the hypothesis of Copernicus would explain the apparent motions of the planetary system more simply and plausibly than those of Ptolemy or Tycho.

In the preceding paragraph we have only looked on motion as a geometric relation, a change of position, without reference to any idea of cause or motive power or any knowledge of the laws which govern the movements of matter. From this new point of view other considerations of probability will arise. If, for instance, the mass of the body A is considerably greater than that of the body B, we judge that the change in the relative situation of the bodies A and B is more probably due to the displacement of B than of A.

Finally, there are some circumstances which may make it certain that relative or apparent movements come from the displacement of one body and not of another.[1] Thus the appearance of an animal will show by unmistakable signs whether it is stopping or starting. Thus, to return to the preceding example, experiments with the pendulum, taken in connection with the known laws of mechanics, will prove the diurnal motion of the earth; the phenomenon of the aberration of light will prove its annual motion; and the hypothesis of Copernicus will take its place among established truths.

Let us now examine how some considerations perfectly analogous to those which we have just considered, spring from the idea of exchangeable values.

Just as we can only assign situation to a point by reference to other points, so we can only assign value to a commodity[2] by reference to

[1] See Newton, *Principia*, Book I, at the end of the preliminary definitions.

[2] It is almost needless to observe that for conciseness the word *commodity* is used in its most general sense, and that it includes the rendering of valuable services, which can be exchanged either for other services or for commodities proper, and which, like such commodities, have a definite price or a value in exchange. We shall not repeat this remark in the future, as it can easily be supplied from the context.

other commodities. In this sense there are only relative values. But when these relative values change, we perceive plainly that the reason of the variation may lie in the change of one term of the relation or of the other or of both at once; just as when the distance varies between two points, the reason for the change may lie in the displacement of one or the other or both of the two points. Thus again when two violin strings have had between them a definite musical interval, and when after a certain time they cease to give this interval, the question is whether one has gone up or the other gone down, or whether both of these effects have joined to cause the variation of the interval.

We can therefore readily distinguish the relative changes of value manifested by the changes of relative values from the absolute changes of value of one or another of the commodities between which commerce has established relations.

Just as it is possible to make an indefinite number of hypotheses as to the absolute motion which causes the observed relative motion in a system of points, so it is also possible to multiply indefinitely hypotheses as to the absolute variations which cause the relative variations observed in the values of a system of commodities.

However, if all but one of the commodities preserved the same relative values, we should consider by far the most probable hypothesis, the one which would assign the absolute change to this single article; unless there should be manifest such a connection between all the others, that one cannot vary without involving proportional variations in the values of those which depend on it.

For instance, an observer who should see by inspection of a table of statistics of values from century to century, that the value of money fell about four-fifths towards the end of the sixteenth century, while other commodities preserved practically the same relative values, would consider it very probable that an absolute change had taken place in the value of money, even if he were ignorant of the discovery of mines in America. On the other hand, if he should see the price of wheat double from one year to the next without any remarkable variation in the price of most other articles or in their relative values, he would attribute it to an absolute change in the value of wheat, even if he did not know that a bad grain harvest had preceded the high price.

Without reference to this extreme case, where the disturbance of the system of relative values is explained by the movement of a single article, it is evident that among all the possible hypotheses on absolute variations some explain the relative variations more simply and more probably than others.

If, without being limited to consideration of the system of relative values at two distinct periods, observation follows it through its intermediate states, a new set of data will be provided to determine the most probable law of absolute variations, from all possibilities for satisfying the observed law of relative variations.

Let

$$p_1, p_2, p_3, \text{etc.},$$

be the values of certain articles, with reference to a gram of silver; if the standard of value is changed and a myriagram of wheat is substituted for the gram of silver, the values of the same articles will be given by the expressions

$$\frac{1}{a}p_1, \frac{1}{a}p_2, \frac{1}{a}p_3, \text{etc.},$$

a being the price of the myriagram of wheat, or its value with reference to a gram of silver. In general, whenever it is desired to change the standard of value, it will suffice to multiply the numerical expressions of individual values by a constant factor, greater or less than unity; just as with a system of points conditioned to remain in a straight line, it would suffice to know the distances from these points to any one of their number, to determine by the addition of a constant number, positive or negative, their distances referred to another point of the system, taken as the new origin.

From this there results a very simple method of expressing by a mathematical illustration the variations which occur in the relative values of a system of articles. It is sufficient to conceive of a system composed of as many points arranged in a straight line as there are articles to be compared, so that the distances from one of these points to all the others constantly remain proportional to the logarithms of the numbers which measure the values of all these articles with reference to one of their number. All the changes of distance which occur by means of addition and subtraction, from the relative and absolute motions of such a system of movable points, will correspond perfectly to the changes by means of multiplication and division in the system of values which is being compared: from which it follows that the calculations for determining the most probable hypothesis as to the absolute movements of a system of points, can be applied, by going from logarithms back to numbers, to the determination of the most probable hypothesis for the absolute variations of a system of values.

But, in general, such calculations of probability, in view of the absolute ignorance in which we would be of the causes of variation of values, would be of very slight interest. What is really important is to know the laws which govern the variation of values, or, in other words, the theory of wealth. This theory alone can make it possible to prove to what absolute variations are due the relative variations which come into the field of observation; in the same manner (if it is permissible to compare the most exact of sciences with the one nearest its cradle) as the theory of the laws of motion, begun by Galileo and completed by Newton, alone makes it possible to prove to what real and absolute motions are due the relative and apparent motions of the solar system.

To sum up, there are only relative values; to seek for others is to fall into a contradiction with the very idea of *value in exchange*, which necessarily implies the idea of a ratio between two terms. But also an accomplished change in this ratio is a relative effect, which can and should be explained by absolute changes in the terms of the ratio. There are no absolute values, but there are movements of absolute rise and fall in values.

Among the possible hypotheses on the absolute changes which produce the observed relative changes, there are some which the general laws of probability indicate as most probable. Only knowledge of the special laws of the matter in question can lead to the substitution of an assured decision for an opinion as to probability.

If theory should indicate one article incapable of absolute variation in its value, and should refer to it all others, it would be possible to immediately deduce their absolute variations from their relative variations; but very slight attention is sufficient to prove that such a fixed term does not exist, although certain articles approach much more nearly than others to the necessary conditions for the existence of such a term.

The monetary metals are among the things which, under ordinary circumstances and provided that too long a period is not considered, only experience slight absolute variations in their value. If it were not so, all transactions would be disturbed, as they are by paper money subject to sudden depreciation.[3]

On the other hand, articles such as wheat, which form the basis of the food supply, are subject to violent disturbances; but, if a sufficient period is considered, these disturbances balance each other, and the average value approaches fixed conditions, perhaps even more closely than the monetary metals. This will not make it impossible for the value so determined to vary, nor prevent it from actually experiencing absolute variations on a still greater scale of time. Here, as in astronomy, it is necessary to recognize *secular* variations, which are independent of *periodic* variations.

Even the wages of that lowest grade of labour, which is only considered as a kind of mechanical agent, the element often proposed as the

[3] What characterizes a contract of sale, and distinguishes it essentially from a contract of exchange, is the invariability of the absolute value of the monetary metals, at least for the lapse of time covered by an ordinary business transaction. In a country where the absolute value of the monetary tokens is perceptibly variable, there are, properly speaking, no contracts of sale. This distinction should affect some legal questions.

standard of value, is subject like wheat to periodic as well as secular variations; and, if the periodic oscillations of this element have generally been less wide than those of wheat, on the other hand we may suspect that in future the progressive changes in the social status will cause it to suffer much more rapid secular variations.

But if no article exists having the necessary conditions for perfect fixity, we can and ought to imagine one, which, to be sure, will only have an abstract existence.[4] It will only appear as an auxiliary term of comparison to facilitate conception of the theory, and will disappear in the final applications.

In like manner, astronomers imagine a mean sun endowed with a uniform motion, and to this imaginary star they refer, as well the real sun as the other heavenly bodies, and so finally determine the actual situation of these stars with reference to the real sun.

It would perhaps seem proper to first investigate the causes which produce absolute variations in the value of the monetary metals, and, when these are accounted for, to reduce to the corrected value of money the variations which occur in the value of other articles. *This corrected money* would be the equivalent of the mean sun of astronomers.

But, on one hand, one of the most delicate points in the theory of wealth is just this analysis of the causes of variation of the value of the monetary metals used as means of circulation, and on the other hand it is legitimate to admit, as has been already said, that the monetary metals do not suffer notable variations in their values except as we compare very distant periods, or else in case of sudden revolutions, now very improbable, which would be caused by the discovery of new metallurgical processes, or of new mineral deposits. It is, to be sure, a common saying, that the price of money is steadily diminishing, and fast enough for the depreciation of value of coin to be very perceptible in the course of a generation; but by going back to the cause of this phenomenon, as we have shown how to do in this chapter, it is plain that the relative change is chiefly due to an absolute upward movement of the prices of most of the articles which go directly for the needs or pleasures of mankind, an ascending movement produced by the increase in population and by the progressive developments of industry and labour. Sufficient explanations on this doctrinal point can be found in the writings of most modern economists.

Finally, in what follows, it will be the more legitimate to neglect the absolute variations which affect the value of the monetary metals, as we do not have numerical applications directly in view. If the theory were sufficiently developed, and the data sufficiently accurate, it would be easy to go from the value of an article in terms of a fictitious and invariable

[4] Montesquieu, *Esprit des Lois*, Book XXII, Chap. 8.

modulus, to its monetary value. If the value of an article, in terms of this fictitious modulus, was p, at a time when that of the monetary metal was π, and if at another time these quantities had taken other values, p' and π', it is evident that the monetary value of the article would have varied in the ratio of

$$\frac{p}{\pi} \text{ to } \frac{p'}{\pi'}.$$

If the absolute value of the monetary metals during long periods only suffers slow variations, which are hardly perceptible throughout the commercial world, the relative values of these very metals suffer slight variations from one commercial centre to another, which constitute what is known as the *rate of exchange*, and of which the mathematical formula is very simple.

OF THE LAW OF DEMAND

To lay the foundations of the theory of exchangeable values, we shall not accompany most speculative writers back to the cradle of the human race; we shall undertake to explain neither the origin of property nor that of exchange or division of labour. All this doubtless belongs to the history of mankind, but it has no influence on a theory which could only become applicable at a very advanced state of civilization, at a period when (to use the language of mathematicians) the influence of the *initial* conditions is entirely gone.

We shall invoke but a single axiom, or, if you prefer, make but a single hypothesis, *i.e.* that each one seeks to derive the greatest possible value from his goods or his labour. But to deduce the rational consequences of this principle, we shall endeavour to establish better than has been the case the elements of the data which observation alone can furnish. Unfortunately, this fundamental point is one which theorists, almost with one accord, have presented to us, we will not say falsely, but in a manner which is really meaningless.

It has been said almost unanimously that "the price of goods is in the inverse ratio of the quantity offered, and in the direct ratio of the quantity demanded." It has never been considered that the statistics necessary for accurate numerical estimation might be lacking, whether of the quantity offered or of the quantity demanded, and that this might prevent deducing from this principle general consequences capable of useful application. But wherein does the principle itself consist? Does it mean that in case a double quantity of any article is offered for sale, the price will fall one-half? Then it should be more simply expressed, and it should only be said that the price is in the inverse ratio of the quantity offered. But the principle thus made intelligible would be false; for, in general, that 100 units of an article have been sold at 20 francs is no reason that 200 units would sell at 10 francs in the same lapse of time and under the same circumstances. Sometimes less would be marketed; often much more.

Furthermore, what is meant by the quantity demanded? Undoubtedly it is not that which is actually marketed at the demand of buyers, for then the generally absurd consequence would result from the pretended principle, that the more of an article is marketed the dearer it is. If by demand only a vague desire of possession of the article is understood, without reference to the *limited price* which every buyer supposes in his demand, there is scarcely an article for which the demand cannot be considered indefinite; but if the price is to be considered at which each buyer is willing to buy, and the price at which each seller is willing to sell, what becomes of the pretended principle? It is not, we repeat, an erroneous proposition—it is a proposition devoid of meaning. Consequently all those who have united to proclaim it have likewise united to make no use of it. Let us try to adhere to less sterile principles.

The cheaper an article is, the greater ordinarily is the demand for it. The sales or the demand (for to us these two words are synonymous, and we do not see for what reason theory need take account of any demand which does not result in a sale)—the sales or the demand generally, we say, increases when the price decreases.

We add the word *generally* as a corrective; there are, in fact, some objects of whim and luxury which are only desirable on account of their rarity and of the high price which is the consequence thereof. If any one should succeed in carrying out cheaply the crystallization of carbon, and in producing for one franc the diamond which to-day is worth a thousand, it would not be astonishing if diamonds should cease to be used in sets of jewellery, and should disappear as articles of commerce. In this case a great fall in price would almost annihilate the demand. But objects of this nature play so unimportant a part in social economy that it is not necessary to bear in mind the restriction of which we speak.

The demand might be in the inverse ratio of the price; ordinarily it increases or decreases in much more rapid proportion—an observation especially applicable to most manufactured products. On the contrary, at other times the variation of the demand is less rapid; which appears (a very singular thing) to be equally applicable both to the most necessary things and to the most superfluous. The price of violins or of astronomical telescopes might fall one-half and yet probably the demand would not double; for this demand is fixed by the number of those who cultivate the art or science to which these instruments belong; who have the disposition requisite and the leisure to cultivate them and the means to pay teachers and to meet the other necessary expenses, in consequences of which the price of the instruments is only a secondary question. On the contrary, firewood, which is one of the most useful articles, could probably double in price, from the progress of clearing land or increase in population, long before the annual consumption of fuel would be halved; as a large number of consumers are disposed to cut down other expenses rather than get along without firewood.

Let us admit therefore that the sales or the annual demand D is, for each article, a particular function $F(p)$ of the price p of such article. To know the form of this function would be to know what we call *the law of demand* or *of sales*. It depends evidently on the kind of utility of the article, on the nature of the services it can render or the enjoyments it can procure, on the habits and customs of the people, on the average wealth, and on the scale on which wealth is distributed.

Since so many moral causes capable of neither enumeration nor measurement affect the law of demand, it is plain that we should no more expect this law to be expressible by an algebraic formula than the law of mortality, and all the laws whose determination enters into the field of statistics, of what is called social arithmetic. Observation must therefore be depended on for furnishing the means of drawing up between proper limits a table of the corresponding values of D and p; after which, by the well-known methods of interpolation or by graphic processes, an empiric formula or a curve can be made to represent the function in question; and the solution of problems can be pushed as far as numerical applications.

But even if this object were unattainable (on account of the difficulty of obtaining observations of sufficient number and accuracy, and also on account of the progressive variations which the law of demand must undergo in a country which has not yet reached a practically stationary condition), it would be nevertheless not improper to introduce the unknown law of demand into analytical combinations, by means of an indeterminate symbol; for it is well known that one of the most important functions of analysis consists precisely in assigning determinate relations between quantities to which numerical values and even algebraic forms are absolutely unassignable.

Unknown functions may none the less possess properties or general characteristics which are known; as, for instance, to be indefinitely increasing or decreasing, or periodical, or only real between certain limits. Nevertheless such data, however imperfect they may seem, by reason of their very generality and by means of analytical symbols, may lead up to relations equally general which would have been difficult to discover without this help. Thus without knowing the law of decrease of the capillary forces, and starting solely from the principle that these forces are inappreciable at appreciable distances, mathematicians have demonstrated the general laws of the phenomena of capillarity, and these laws have been confirmed by observation.

On the other hand, by showing what determinate relations exist between unknown quantities, analysis reduces these unknown quantities to the smallest possible number, and guides the observer to the best observations for discovering their values. It reduces and coördinates statistical documents; and it diminishes the labour of statisticians at the same time that it throws light on them.

For instance, it is impossible *a priori* to assign an algebraic form to the law of mortality; it is equally impossible to formulate the function expressing the subdivision of population by ages in a stationary population; but these two functions are connected by so simple a relation, that, as soon as statistics have permitted the construction of a table of mortality, it will be possible, without recourse to new observations, to deduce from this table one expressing the proportion of the various ages in the midst of a stationary population, or even of a population for which the annual excess of deaths over births is known.[5]

Who doubts that in the field of social economy there is a mass of figures thus mutually connected by assignable relations, by means of which the easiest to determine empirically might be chosen, so as to deduce all the others from it by means of theory?

We will assume that the function $F(p)$, which expresses the law of demand or of the market, is a *continuous* function, *i.e.* a function which does not pass suddenly from one value to another, but which takes in passing all intermediate values. It might be otherwise if the number of consumers were very limited: thus in a certain household the same quantity of firewood will possibly be used whether wood costs 10 francs or 15 francs the stere,[6] and the consumption may suddenly be diminished if the price of the stere rises above the latter figure. But the wider the market extends, and the more the combinations of needs, of fortunes, or even of caprices, are varied among consumers, the closer the function $F(p)$ will come to varying with p in a continuous manner. However little may be the variation of p, there will be some consumers so placed that the slight rise or fall of the article will affect their consumptions, and will lead them to deprive themselves in some way or to reduce their manufacturing output, or to substitute something else for the article that has grown dearer, as, for instance, coal for wood or anthracite for soft coal. Thus the "exchange" is a thermometer which shows by very slight variations of rates

[5] The *Annuaire du Bureau des Longitudes* contains these two tables, the second deduced from the first, as above, and calculated on the hypothesis of a stationary population.

The work by Duvillard, entitled *De l'influence de la petite vérole sur la mortalité*, contains many good examples of mathematical connections between essentially em- [illegible]al functions.

1 stere = 1 M^3 = 35.3 cu. ft. = 5⁄18 cord.—TRANSLATOR.]

the fleeting variations in the estimate of the chances which affect government bonds, variations which are not a sufficient motive for buying or selling to most of those who have their fortunes invested in such bonds.

If the function $F(p)$ is continuous, it will have the property common to all functions of this nature, and on which so many important applications of mathematical analysis are based: *the variations of the demand will be sensibly proportional to the variations in price so long as these last are small fractions of the original price.* Moreover, these variations will be of opposite signs, *i.e.* an increase in price will correspond with a diminution of the demand.

Suppose that in a country like France the consumption of sugar is 100 million kilograms when the price is 2 francs a kilogram, and that it has been observed to drop to 99 millions when the price reached 2 francs 10 centimes. Without considerable error, the consumption which would correspond to a price of 2 francs 20 centimes can be valued at 98 millions, and the consumption corresponding to a price of 1 franc 90 centimes at 101 millions. It is plain how much this principle, which is only the mathematical consequence of the continuity of functions, can facilitate applications of theory, either by simplying analytical expressions of the laws which govern the movement of values, or in reducing the number of data to be borrowed from experience, if the theory becomes sufficiently developed to lend itself to numerical determinations.

Let us not forget that, strictly speaking, the principle just enunciated admits of exceptions, because a continuous function may have interruptions of continuity in some points of its course; but just as friction wears down roughnesses and softens outlines, so the wear of commerce tends to suppress these exceptional cases, at the same time that commercial machinery moderates variations in prices and tends to maintain them between limits which facilitate the application of theory.

To define with accuracy the quantity D, or the function $F(p)$ which is the expression of it, we have supposed that D represented the quantity sold *annually* throughout the extent of the country or of the market [7] under consideration. In fact, the year is the natural unit of time, especially for researches having any connection with social economy. All the wants of mankind are reproduced during this term, and all the resources which mankind obtains from nature and by labour. Nevertheless, the price of an article may vary notably in the course of a year, and, strictly speaking, the law of demand may also vary in the same interval, if the country experiences a movement of progress or decadence. For greater accuracy, therefore, in the expression $F(p)$, p must be held to denote the annual

[7] It is well known that by *market* economists mean, not a certain place where purchases and sales are carried on, but the entire territory of which the parts are so united by the relations of unrestricted commerce that prices there take the same level throughout, with ease and rapidity.

average price, and the curve which represents function F to be in itself an average of all the curves which would represent this function at different times of the year. But this extreme accuracy is only necessary in case it is proposed to go on to numerical applications, and it is superfluous for researches which only seek to obtain a general expression of average results, independent of periodical oscillations.

Since the function $F(p)$ is continuous, the function $pF(p)$, which expresses the total value of the quantity annually sold, must be continuous also. This function would equal zero if p equals zero, since the consumption of any article remains finite even on the hypothesis that it is absolutely free; or, in other words, it is theoretically always possible to assign to the symbol p a value so small that the product $pF(p)$ will vary imperceptibly from zero. The function $pF(p)$ disappears also when p becomes infinite, or, in other words, theoretically a value can always be assigned to p so great that the demand for the article and the production of it would cease. Since the function $pF(p)$ at first increases, and then decreases as p increases, there is therefore a value of p which makes this function a maximum, and which is given by the equation,

$$F(p) + pF'(p) = 0, \tag{1}$$

in which F', according to Lagrange's notation, denotes the differential coefficient of function F.

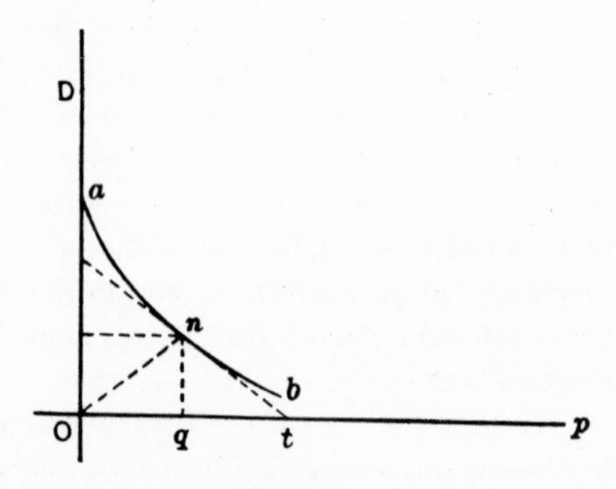

FIGURE 1

If we lay out the curve *anb* (Figure 1) of which the abscissas *oq* and the ordinates *qn* represent the variables p and D, the root of equation (1) will be the abscissa of the point *n* from which the triangle *ont*, formed by the tangent *nt* and the radius vector *on*, is isosceles, so that we have $oq = qt$.

We may admit that it is impossible to determine the function $F(p)$ empirically for each article, but it is by no means the case that the same obstacles prevent the approximate determination of the value of p which satisfies equation (1) or which renders the product $pF(p)$ a maximum.

The construction of a table, where these values could be found, would be the work best calculated for preparing for the practical and rigorous solution of questions relating to the theory of wealth.

But even if it were impossible to obtain from statistics the value of p which should render the product $pF(p)$ a maximum, it would be easy to learn, at least for all articles to which the attempt has been made to extend commercial statistics, whether current prices are above or below this value. Suppose that when the price becomes $p + \Delta p$, the annual consumption as shown by statistics, such as customhouse records, becomes $D - \Delta D$. According as

$$\frac{\Delta D}{\Delta p} < \text{or} > \frac{D}{p},$$

the increase in price, Δp, will increase or diminish the product $pF(p)$; and, consequently, it will be known whether the two values p and $p + \Delta p$ (assuming Δp to be a small fraction of p) fall above or below the value which makes the product under consideration a maximum.

Commercial statistics should therefore be required to separate articles of high economic importance into two categories, according as their current prices are above or below the value which makes a maximum of $pF(p)$. We shall see that many economic problems have different solutions, according as the article in question belongs to one or the other of these two categories.

We know by the theory of maxima and minima that equation (1) is satisfied as well by the values of p which render $pF(p)$ a minimum as by those which render this product a maximum. The argument used at the beginning of the preceding article shows, indeed, that the function $pF(p)$ necessarily has a maximum, but it might have several and pass through minimum values between. A root of equation (1) corresponds to a maximum or a minimum according as

$$2\,F'(p) + pF''(p) < \text{or} > 0,$$

or, substituting for p its value and considering the essentially negative sign of $F'(p)$,

$$2\,[F'(p)]^2 - F(p) \times F''(p) > \text{or} < 0.$$

In consequence, whenever $F''(p)$ is negative, or when the curve $D = F(p)$ turns its concave side to the axis of the abscissas, it is impossible that there should be a minimum, nor more than one maximum. In the contrary case, the existence of several maxima or minima is not proved to be impossible.

But if we cease considering the question from an exclusively abstract standpoint, it will be instantly recognized how improbable it is that the function $pF(p)$ should pass through several intermediate maxima and minima inside of the limits between which the value of p can vary; and as it is unnecessary to consider maxima which fall beyond these limits, if any such exist, all problems are the same as if the function $pF(p)$ only admitted a single maximum. The essential question is always whether, for the extent of the limits of oscillation of p, the function $pF(p)$ is increasing or decreasing for increasing values of p.

Any demonstration ought to proceed from the simple to the complex: the simplest hypothesis for the purpose of investigating by what laws prices are fixed, is that of monopoly, taking this word in its most absolute meaning, which supposes that the production of an article is in one man's hands. This hypothesis is not purely fictitious: it is realized in certain cases; and, moreover, when we have studied it, we can analyze more accurately the effects of competition of producers.

The age of chivalry is gone. That of sophisters, economists and calculators has succeeded.

—EDMUND BURKE (*Reflections on the Revolution in France*)

The effort of the economist is to "see," to picture the interplay of economic elements. The more clearly cut these elements appear in his vision, the better; the more elements he can grasp and hold in his mind at once, the better. The economic world is a misty region. The first explorers used unaided vision. Mathematics is the lantern by which what before was dimly visible now looms up in firm, bold outlines. The old phantasmagoria disappear. We see better. We also see further.

—IRVING FISHER (*Transactions of Conn. Academy, 1892*)

5 Theory of Political Economy

By WILLIAM STANLEY JEVONS

MATHEMATICAL CHARACTER OF THE SCIENCE

IT IS clear that Economics, if it is to be a science at all, must be a mathematical science. There exists much prejudice against attempts to introduce the methods and language of mathematics into any branch of the moral sciences. Many persons seem to think that the physical sciences form the proper sphere of mathematical method, and that the moral sciences demand some other method—I know not what. My theory of Economics, however, is purely mathematical in character. Nay, believing that the quantities with which we deal must be subject to continuous variation, I do not hesitate to use the appropriate branch of mathematical science, involving though it does the fearless consideration of infinitely small quantities. The theory consists in applying the differential calculus to the familiar notions of wealth, utility, value, demand, supply, capital, interest, labour, and all the other quantitative notions belonging to the daily operations of industry. As the complete theory of almost every other science involves the use of that calculus, so we cannot have a true theory of Economics without its aid.

To me it seems that *our science must be mathematical, simply because it deals with quantities*. Wherever the things treated are capable of being *greater or less*, there the laws and relations must be mathematical in nature. The ordinary laws of supply and demand treat entirely of quantities of commodity demanded or supplied, and express the manner in which the quantities vary in connection with the price. In consequence of this fact the laws *are* mathematical. Economists cannot alter their nature by denying them the name; they might as well try to alter red light by calling it blue. Whether the mathematical laws of Economics are stated in words, or in the usual symbols, x, y, z, p, q, etc., is an accident, or a matter of mere convenience. If we had no regard to trouble and prolixity, the most complicated mathematical problems might be stated in ordinary language, and their solution might be traced out by words. In fact, some distinguished mathematicians have shown a liking for getting rid of their symbols, and expressing their arguments and results in language as nearly as possible approximating to that in common use. In his *Système du Monde*, Laplace attempted to describe the truths of physical astronomy in common language; and Thomson and Tait interweave their great *Treatise on Natural Philosophy* with an interpretation in ordinary words, supposed to be within the comprehension of general readers.

These attempts, however distinguished and ingenious their authors, soon disclose the inherent defects of the grammar and dictionary for expressing complicated relations. The symbols of mathematical books are not different in nature from language; they form a perfected system of language, adapted to the notions and relations which we need to express. They do not constitute the mode of reasoning they embody; they merely facilitate its exhibition and comprehension. If, then, in Economics, we have to deal with quantities and complicated relations of quantities, we must reason mathematically; we do not render the science less mathematical by avoiding the symbols of algebra—we merely refuse to employ, in a very imperfect science, much needing every kind of assistance, that apparatus of appropriate signs which is found indispensable in other sciences.

CONFUSION BETWEEN MATHEMATICAL AND EXACT SCIENCES

Many persons entertain a prejudice against mathematical language, arising out of a confusion between the ideas of a mathematical science and an exact science. They think that we must not pretend to calculate unless we have the precise data which will enable us to obtain a precise answer to our calculations; but, in reality, there is no such thing as an exact science, except in a comparative sense. Astronomy is more exact than other sciences, because the position of a planet or star admits of close measurement; but, if we examine the methods of physical astronomy, we find that they are all approximate. Every solution involves hypotheses which are not really true: as, for instance, that the earth is a smooth, homogeneous spheroid. Even the apparently simpler problems in statics or dynamics are only hypothetical approximations to the truth.

We can calculate the effect of a crowbar, provided it be perfectly inflexible and have a perfectly hard fulcrum—which is never the case. The data are almost wholly deficient for the complete solution of any one problem in natural science. Had physicists waited until their data were perfectly precise before they brought in the aid of mathematics, we should have still been in the age of science which terminated at the time of Galileo.

When we examine the less precise physical sciences, we find that physicists are, of all men, most bold in developing their mathematical theories in advance of their data. Let any one who doubts this examine Airy's "Theory of the Tides," as given in the *Encyclopaedia Metropolitana*; he will there find a wonderfully complex mathematical theory which is confessed by its author to be incapable of exact or even approximate application, because the results of the various and often unknown contours of the seas do not admit of numerical verification. In this and many other cases we have mathematical theory without the data requisite for precise calculation.

The greater or less accuracy attainable in a mathematical science is a matter of accident, and does not affect the fundamental character of the science. There can be but two classes of sciences—those which are *simply logical*, and *those which, besides being logical, are also mathematical.* If there be any science which determines merely whether a thing be or be not—whether an event will happen, or will not happen—it must be a purely logical science; but if the thing may be greater or less, or the event may happen sooner or later, nearer or farther, then quantitative notions enter, and the science must be mathematical in nature, by whatever name we call it.

CAPABILITY OF EXACT MEASUREMENT

Many will object, no doubt, that the notions which we treat in this science are incapable of any measurement. We cannot weigh, nor gauge, nor test the feelings of the mind; there is no unit of labour, or suffering, or enjoyment. It might thus seem as if a mathematical theory of Economics would be necessarily deprived for ever of numerical data.

I answer, in the first place, that nothing is less warranted in science than an uninquiring and unhoping spirit. In matters of this kind, those who despair are almost invariably those who have never tried to succeed. A man might be despondent had he spent a lifetime on a difficult task without a gleam of encouragement; but the popular opinions on the extension of mathematical theory tend to deter any man from attempting tasks which, however difficult, ought, some day, to be achieved.

* * * * *

MEASUREMENT OF FEELING AND MOTIVES

Many readers may, even after reading the preceding remarks, consider it quite impossible to create such a calculus as is here contemplated, because we have no means of defining and measuring quantities of feeling, like we can measure a mile, or a right angle; or any other physical quantity. I have granted that we can hardly form the conception of a unit of pleasure or pain, so that the numerical expression of quantities of feeling seems to be out of the question. But we only employ units of measurement in other things to facilitate the comparison of quantities; and if we can compare the quantities directly, we do not need the units.

* * * * *

LOGICAL METHOD OF ECONOMICS

* * * * *

To return, however, to the topic of the present work, the theory here given may be described as *the mechanics of utility and self-interest.* Oversights may have been committed in tracing out its details, but in its main features this theory must be the true one. Its method is as sure and demonstrative as that of kinematics or statics, nay, almost as self-evident as are the elements of Euclid, when the real meaning of the formulae is fully seized.

I do not hesitate to say, too, that Economics might be gradually erected into an exact science, if only commercial statistics were far more complete and accurate than they are at present, so that the formulae could be endowed with exact meaning by the aid of numerical data. These data would consist chiefly in accurate accounts of the quantities of goods possessed and consumed by the community, and the prices at which they are exchanged. There is no reason whatever why we should not have those statistics, except the cost and trouble of collecting them, and the unwillingness of persons to afford information. The quantities themselves to be measured and registered are most concrete and precise. In a few cases we already have information approximating to completeness, as when a commodity like tea, sugar, coffee, or tobacco is wholly imported. But when articles are untaxed, and partly produced within the country, we have yet the vaguest notions of the quantities consumed. Some slight success is now, at last, attending the efforts to gather agricultural statistics; and the great need felt by men engaged in the cotton and other trades to obtain accurate accounts of stocks, imports, and consumption, will probably lead to the publication of far more complete information than we have hitherto enjoyed.

The deductive science of Economics must be verified and rendered useful by the purely empirical science of Statistics. Theory must be invested with the reality and life of fact. But the difficulties of this union are immensely great, and I appreciate them quite as much as does Cairnes in his admirable lectures "On the Character and Logical Method of Political Economy." I make hardly any attempt to employ statistics in this work, and thus I do not pretend to any numerical precision. But, before we attempt any investigation of facts, we must have correct theoretical notions; and of what are here presented, I would say, in the words of Hume, in his *Essay on Commerce*, "If false, let them be rejected: but no one has a right to entertain a prejudice against them merely because they are out of the common road."

RELATION OF ECONOMICS TO ETHICS

I wish to say a few words, in this place, upon the relation of Economics to Moral Science. The theory which follows is entirely based on a calculus of pleasure and pain; and the object of Economics is to maximise happiness by purchasing pleasure, as it were, at the lowest cost of pain. The language employed may be open to misapprehension, and it may seem as if pleasures and pains of a gross kind were treated as the all-sufficient motives to guide the mind of man. I have no hesitation in accepting the Utilitarian theory of morals which does uphold the effect upon the happiness of mankind as the criterion of what is right and wrong. But I have never felt that there is anything in that theory to prevent our putting the widest and highest interpretation upon the terms used.

Jeremy Bentham put forward the Utilitarian theory in the most uncompromising manner. According to him, whatever is of interest or importance to us must be the cause of pleasure or of pain; and when the terms are used with a sufficiently wide meaning, pleasure and pain include all the forces which drive us to action. They are explicitly or implicitly the matter of all our calculations, and form the ultimate quantities to be treated in all the moral sciences. The words of Bentham on this subject may require some explanation and qualification, but they are too grand and too full of truth to be omitted. "Nature," he says, "has placed mankind under the governance of two sovereign masters—*pain* and *pleasure*. It is for them alone to point out what we ought to do, as well as to determine what we shall do. On the one hand the standard of right and wrong, on the other the chain of causes and effects, are fastened to their throne. They govern us in all we do, in all we say, in all we think: every effort we can make to throw off our subjection will serve but to demonstrate and confirm it. In words a man may pretend to abjure their empire; but, in reality, he will remain subject to it all the while. The *principle of utility* recognises this subjection, and assumes it for the foundation of that system, the object of which is to rear the fabric of felicity by the hands of reason and of law. Systems which attempt to question it deal in sounds instead of sense, in caprice instead of reason, in darkness instead of light."

* * * * *

THE THEORY OF UTILITY

UTILITY IS NOT AN INTRINSIC QUALITY

My principal work now lies in tracing out the exact nature and conditions of utility. It seems strange indeed that economists have not bestowed more minute attention on a subject which doubtless furnishes the true key to the problem of economics.

In the first place, utility, though a quality of things, is *no inherent quality*. It is better described as *a circumstance of things* arising out of their relation to man's requirements. As Senior most accurately says, "Utility denotes no intrinsic quality in the things which we call useful; it merely expresses their relations to the pains and pleasures of mankind." We can never, therefore, say absolutely that some objects have utility and others have not. The ore lying in the mine, the diamond escaping the eye of the searcher, the wheat lying unreaped, the fruit ungathered for want of consumers, have no utility at all. The most wholesome and necessary kinds of food are useless unless there are hands to collect and mouths to eat them sooner or later. Nor, when we consider the matter closely, can we say that all portions of the same commodity possess equal utility. Water, for instance, may be roughly described as the most useful of all substances. A quart of water per day has the high utility of saving a person from dying in a most distressing manner. Several gallons a day may possess much utility for such purposes as cooking and washing; but after an adequate supply is secured for these uses, any additional quantity is a matter of comparative indifference. All that we can say, then, is that water, up to a certain quantity, is indispensable; that further quantities will have various degrees of utility; but that beyond a certain quantity the utility sinks gradually to zero; it may even become negative, that is to say, further supplies of the same substance may become inconvenient and hurtful.

Exactly the same considerations apply more or less clearly to every other article. A pound of bread per day supplied to a person saves him from starvation, and has the highest conceivable utility. A second pound per day has also no slight utility; it keeps him in a state of comparative plenty, though it be not altogether indispensable. A third pound would begin to be superfluous. It is clear, then, that *utility is not proportional to commodity*: the very same articles vary in utility according as we already possess more or less of the same article. The like may be said of other things. One suit of clothes per annum is necessary, a second convenient, a third desirable, a fourth not unacceptable, but we sooner or later reach a point at which further supplies are not desired with any perceptible force unless it be for subsequent use.

LAW OF THE VARIATION OF UTILITY

Let us now investigate this subject a little more closely. Utility must be considered as measured by, or even as actually identical with, the addition made to a person's happiness. It is a convenient name for the

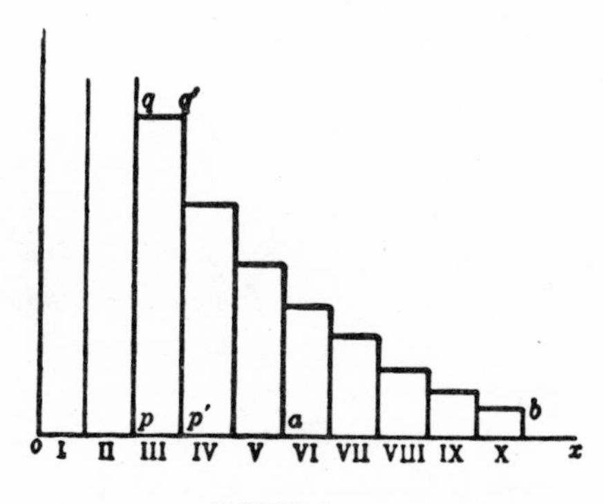

FIGURE 1

aggregate of the favorable balance of feeling produced,—the sum of the pleasure created and the pain prevented. We must now carefully discriminate between the *total utility* arising from any commodity and the utility attaching to any particular portion of it. Thus the total utility of the food we eat consists in maintaining life, and may be considered as infinitely great; but if we were to subtract a tenth part from what we eat daily, our loss would be but slight. We should certainly not lose a tenth part of the whole utility of food to us. It might be doubtful whether we should suffer any harm at all.

Let us imagine the whole quantity of food which a person consumes on an average during twenty-four hours to be divided into ten equal parts. If his food be reduced by the last part, he will suffer but little; if a second tenth part be deficient, he will feel the want distinctly; the subtraction of the third tenth part will be decidedly injurious; with every subsequent subtraction of a tenth part his sufferings will be more and more serious, until at length he will be upon the verge of starvation. Now, if we call each of the tenth parts *an increment*, the meaning of these facts is, that each increment of food is less necessary, or possesses less utility, than the previous one. To explain this variation of utility we may make use of space representations, which I have found convenient in illustrating the laws of economics in my college lectures during fifteen years past (Figure 1).

Let the line *ox* be used as a measure of the quantity of food, and let it be divided into ten equal parts to correspond to the ten portions of food mentioned above. Upon these equal lines are constructed rectangles and the area of each rectangle may be assumed to represent the utility of the increment of food corresponding to its base. Thus the utility of the last increment is small, being proportional to the small rectangle on *x*. As we approach towards *o*, each increment bears a larger rectangle, that standing upon III being the largest complete rectangle. The utility of the next increment, II, is undefined, as also that of I, since these portions of food would be indispensable to life, and their utility, therefore, infinitely great.

We can now form a clear notion of the utility of the whole food, or of any part of it, for we have only to add together the proper rectangles. The utility of the first half of the food will be the sum of the rectangles standing on the line *oa*; that of the second half will be represented by the sum of the smaller rectangles between *a* and *b*. The total utility of the food will be the whole sum of the rectangles, and will be infinitely great.

The comparative utility of the several portions is, however, the most important. Utility may be treated as *a quantity of two dimensions*, one dimension consisting in the quantity of the commodity, and another in the intensity of the effect produced upon the consumer. Now the quantity of the commodity is measured on the horizontal line *ox*, and the intensity of utility will be measured by the length of the upright lines, or *ordinates*. The intensity of utility of the third increment is measured either by *pq*, or *p′q′*, and its utility is the product of the units in *pp′* multiplied by those in *pq*.

But the division of the food into ten equal parts is an arbitrary supposition. If we had taken twenty or a hundred or more equal parts, the same general principle would hold true, namely, that each small portion would be less useful and necessary than the last. The law may be considered to hold true theoretically, however small the increments are made; and in this way we shall at last reach a figure which is undistinguishable from a continuous curve. The notion of infinitely small quantities of food may seem absurd as regards the consumption of one individual; but when we consider the consumption of a nation as a whole, the consumption may well be conceived to increase or diminish by quantities which are, practically speaking, infinitely small compared with the whole consumption. The laws which we are about to trace out are to be conceived as theoretically true of the individual; they can only be practically verified as regards the aggregate transactions, productions, and consumptions of a large body of people. But the laws of the aggregate depend of course upon the laws applying to individual cases.

The law of the variation of the degree of utility of food may thus be represented by a continuous curve *pbq*, and the perpendicular height of each point at the curve above the line *ox* represents the degree of utility of the commodity when a certain amount has been consumed. (See Figure 2.)

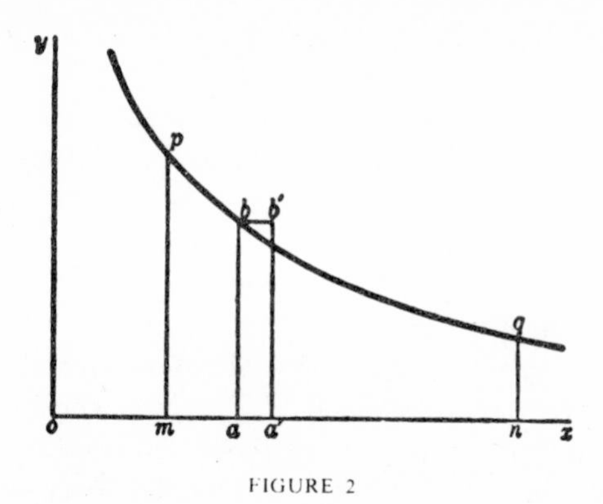

FIGURE 2

Thus, when the quantity *oa* has been consumed, the degree of utility corresponds to the length of the line *ab*; for if we take a very little more food, *aa′*, its utility will be the product of *aa′* and *ab* very nearly, and more nearly the less is the magnitude of *aa′*. The degree of utility is thus properly measured by the height of a very narrow rectangle corresponding to a very small quantity of food, which theoretically ought to be infinitely small.

TOTAL UTILITY AND DEGREE OF UTILITY

We are now in a position to appreciate perfectly the difference between the *total utility* of any commodity and the *degree of utility* of the commodity at any point. These are, in fact, quantities of altogether different kinds, the first being represented by an area, and the second by a line. We must consider how we may express these notions in appropriate mathematical language.

Let x signify, as is usual in mathematical books, the quantity which varies independently—in this case the quantity of commodity. Let u denote the *whole utility* proceeding from the consumption of x. Then u will be, as mathematicians say, *a function of* x; that is, it will vary in some continuous and regular, but probably unknown, manner, when x is made to vary. Our great object at present, however, is to express the *degree of utility*.

Mathematicians employ the sign Δ prefixed to a sign of quantity, such as x, to signify that a quantity of the same nature as x, but small in proportion to x, is taken into consideration. Thus Δx means a small portion of x, and $x + \Delta x$ is therefore a quantity a little greater than x. Now when x is a quantity of commodity, the utility of $x + \Delta x$ will be more than that of x as a general rule. Let the whole utility of $x + \Delta x$ be denoted by $u + \Delta u$; then it is obvious that the increment of utility Δu belongs to the increment of commodity Δx; and if, for the sake of argument, we suppose the degree of utility uniform over the whole of Δx, which is nearly true, owing to its smallness, we shall find the corresponding degree of utility by dividing Δu by Δx.

We find these considerations fully illustrated by the last figure, in which *oa* represents *x*, and *ab* is the degree of utility at the point *a*. Now, if we increase *x* by the small quantity *aa′*, or Δx, the utility is increased by the small rectangle *abb′a′*, or Δu; and since a rectangle is the product of its sides, we find that the length of the line *ab*, the degree of utility, is represented by the fraction $\Delta u/\Delta x$.

As already explained, however, the utility of a commodity may be considered to vary with perfect continuity, so that we commit a small error in assuming it to be uniform over the whole increment Δx. To avoid this, we must imagine Δx to be reduced to an infinitely small size, Δu decreasing with it. The smaller the quantities are the more nearly we shall have a correct expression for *ab*, the degree of utility at the point *a*. Thus the *limit* of this fraction $\Delta u/\Delta x$, or, as it is commonly expressed, du/dx, is the degree of utility corresponding to the quantity of commodity *x*. *The degree of utility is*, in mathematical language, *the differential coefficient of u considered as a function of x*, and will itself be another function of *x*.

We shall seldom need to consider the degree of utility except as regards the last increment which has been consumed, or, which comes to the same thing, the next increment which is about to be consumed. I shall therefore commonly use the expression *final degree of utility*, as meaning the degree of utility of the last addition, or the next possible addition of a very small, or infinitely small, quantity to the existing stock. In ordinary circumstances, too, the final degree of utility will not be great compared with what it might be. Only in famine or other extreme circumstances do we approach the higher degrees of utility. Accordingly we can often treat the lower portions of the curves of variation (*pbq*) which concern ordinary commercial transactions, while we leave out of sight the portions beyond *p* or *q*. It is also evident that we may know the degree of utility at any point while ignorant of the total utility, that is, the area of the whole curve. To be able to estimate the total enjoyment of a person would be an interesting thing, but it would not be really so important as to be able to estimate the additions and subtractions to his enjoyment which circumstances occasion. In the same way a very wealthy person may be quite unable to form any accurate statement of his aggregate wealth, but he may nevertheless have exact accounts of income and expenditure, that is, of additions and subtractions.

VARIATION OF THE FINAL DEGREE OF UTILITY

The final degree of utility is that function upon which the theory of economics will be found to turn. Economists, generally speaking, have failed to discriminate between this function and the total utility, and from this confusion has arisen much perplexity. Many commodities which are most useful to us are esteemed and desired but little. We cannot live without water, and yet in ordinary circumstances we set no value on it. Why is this? Simply because we usually have so much of it that its final degree of utility is reduced nearly to zero. We enjoy every day the almost infinite utility of water, but then we do not need to consume more than we have. Let the supply run short by drought, and we begin to feel the higher degrees of utility, of which we think but little at other times.

The variation of the function expressing the final degree of utility is the all-important point in economic problems. We may state, as a general law, that *the degree of utility varies with the quantity of commodity, and ultimately decreases as that quantity increases.* No commodity can be named which we continue to desire with the same force, whatever be the quantity already in use or possession. All our appetites are capable of *satisfaction* or *satiety* sooner or later, in fact, both these words mean, etymologically, that we have had *enough*, so that more is of no use to us. It does not follow, indeed, that the degree of utility will always sink to zero. This may be the case with some things, especially the simple animal requirements, such as food, water, air, etc. But the more refined and intellectual our needs become, the less are they capable of satiety. To the desire for articles of taste, science, or curiosity, when once excited, there is hardly a limit.

* * * * *

DISUTILITY AND DISCOMMODITY

A few words will suffice to suggest that as utility corresponds to the production of pleasure, or, at least, a favorable alteration in the balance of pleasure and pain, so negative utility will consist in the production of pain, or the unfavorable alteration of the balance. In reality we must be almost as often concerned with the one as with the other; nevertheless, economists have not employed any distinct technical terms to express that production of pain which accompanies so many actions of life. They have fixed their attention on the more agreeable aspect of the matter. It will be allowable, however, to appropriate the good English word *discommodity*, to signify any substance or action which is the opposite of *commodity*, that is to say, *anything which we desire to get rid of*, like ashes or sewage. Discommodity is, indeed, properly an abstract form signifying inconvenience, or disadvantage; but as the noun *commodities* has been used in the English language for four hundred years at least as a concrete term, so we may now convert discommodity into a concrete term, and speak of *discommodities* as substances or things which possess the quality of causing inconvenience or harm. For the abstract notion, the opposite or negative of utility, we may invent the term *disutility*, which will mean something different from inutility, or the absence of utility. It is obvious that utility passes through inutility before changing into disutility, these notions being related as +, 0, and −.

DISTRIBUTION OF COMMODITY IN DIFFERENT USES

The principles of utility may be illustrated by considering the mode in which we distribute a commodity when it is capable of several uses. There are articles which may be employed for many distinct purposes: thus, barley may be used either to make beer, spirits, bread, or to feed cattle; sugar may be used to eat, or for producing alcohol; timber may be used in construction, or as fuel; iron and other metals may be applied to many different purposes. Imagine, then, a community in the possession of a certain stock of barley; what principles will regulate their mode of consuming it? Or, as we have not yet reached the subject of exchange, imagine an isolated family, or even an individual, possessing an adequate stock, and using some in one way and some in another. The theory of utility gives, theoretically speaking, a complete solution of the question.

Let s be the whole stock of some commodity, and let it be capable of two distinct uses. Then we may represent the two quantities appropriated to these uses by x_1 and y_1, it being a condition that $x_1 + y_1 = s$. The person may be conceived as successively expending small quantities of the commodity; now it is the inevitable tendency of human nature to choose that course which appears to offer the greatest advantage at the moment. Hence, when the person remains satisfied with the distribution he has made, it follows that no alteration would yield him more pleasure, which amounts to saying that an increment of commodity would yield exactly as much utility in one use as in another. Let Δu_1, Δu_2 be the increments of utility which might arise respectively from consuming an increment of commodity in the two different ways. When the distribution is completed, we ought to have $\Delta u_1 = \Delta u_2$; or at the limit we have the equation

$$\frac{du_1}{dx} = \frac{du_2}{dy},$$

which is true when x, y are respectively equal to x_1, y_1. We must, in other words, have the *final degrees of utility* in the two uses equal.

The same reasoning which applies to uses of the same commodity will evidently apply to any two uses, and hence to all uses simultaneously, so that we obtain a series of equations less numerous by a unit than the number of ways of using the commodity. The general result is that commodity, if consumed by a perfectly wise being, must be consumed with a maximum production of utility.

We should often find these equations to fail. Even when x is equal to $^{99}/_{100}$ of the stock, its degree of utility might still exceed the utility attaching to the remaining $^{1}/_{100}$ part in either of the other uses. This would mean that it was preferable to give the whole commodity to the first use. Such a case might perhaps be said to be not the exception but the rule; for whenever a commodity is capable of only one use, the circumstance is theoretically represented by saying that the final degree of utility in this employment always exceeds that in any other employment.

Under peculiar circumstances great changes may take place in the consumption of a commodity. In a time of scarcity the utility of barley as food might rise so high as to exceed altogether its utility, even as regards the smallest quantity, in producing alcoholic liquors; its consumption in the latter way would then cease. In a besieged town the employment of articles becomes revolutionized. Things of great utility in other respects are ruthlessly applied to strange purposes. In Paris a vast stock of horses was eaten, not so much because they were useless in other ways, as because they were needed more strongly as food. A certain stock of horses had, indeed, to be retained as a necessary aid to locomotion, so that the equation of the degrees of utility never wholly failed.

THE LAW OF INDIFFERENCE

When a commodity is perfectly uniform or homogeneous in quality, any portion may be indifferently used in place of an equal portion: hence, in the same market, and at the same moment, all portions must be exchanged at the same ratio. There can be no reason why a person should treat exactly similar things differently, and the slightest excess in what is demanded for one over the other will cause him to take the latter instead of the former. In nicely balanced exchanges it is a very minute scruple which turns the scale and governs the choice. A minute difference of quality in a commodity may thus give rise to preference, and cause the ratio of exchange to differ. But where no difference exists at all, or where no difference is known to exist, there can be no ground for preference whatever. If, in selling a quantity of perfectly equal and uniform barrels of flour, a merchant arbitrarily fixed different prices on them, a purchaser would of course select the cheaper ones; and where there was absolutely no difference in the thing purchased, even an excess of a penny in the price of a thing worth a thousand pounds would be a valid ground of choice. Hence follows what is undoubtedly true, with proper explanations, that *in the same open market, at any one moment, there cannot be two prices for the same kind of article.* Such differences as may practically occur arise from extraneous circumstances, such as the defective credit of the purchasers, their imperfect knowledge of the market, and so on.

The principle above expressed is a general law of the utmost importance in Economics, and I propose to call it *The Law of Indifference*, meaning that, when two objects or commodities are subject to no important difference as regards the purpose in view, they will either of them be taken instead of the other with perfect indifference by a purchaser. Every such act of indifferent choice gives rise to an equation of degrees of utility, so that in this principle of indifference we have one of the central pivots of the theory.

Though the price of the same commodity must be uniform at any one moment, it may vary from moment to moment, and must be conceived as in a state of continual change. Theoretically speaking, it would not usually be possible to buy two portions of the same commodity *successively* at the same ratio of exchange, because, no sooner would the first portion have been bought than the conditions of utility would be altered. When exchanges are made on a large scale, this result will be verified in practice. If a wealthy person invested £100,000 in the funds in the morning, it is hardly likely that the operation could be repeated in the afternoon at the same price. In any market, if a person goes on buying largely, he will ultimately raise the price against himself. Thus it is apparent that extensive purchases would best be made gradually, so as to secure the advantage of a lower price upon the earlier portions. In theory this effect of exchange upon the ratio of exchange must be conceived to exist in some degree, however small may be the purchases made. Strictly speaking, the ratio of exchange at any moment is that of dy to dx, of an infinitely small quantity of one commodity to the infinitely small quantity of another which is given for it. The ratio of exchange is really a differential coefficient. The quantity of any article purchased is a function of the price at which it is purchased, and the ratio of exchange expresses the rate at which the quantity of the article increases compared with what is given for it.

We must carefully distinguish, at the same time, between the Statics and Dynamics of this subject. The real condition of industry is one of perpetual motion and change. Commodities are being continually manufactured and exchanged and consumed. If we wished to have a complete solution of the problem in all its natural complexity, we should have to treat it as a problem of motion—a problem of dynamics. But it would surely be absurd to attempt the more difficult question when the more easy one is yet so imperfectly within our power. It is only as a purely statical problem that I can venture to treat the action of exchange. Holders of commodities will be regarded not as continuously passing on these commodities in streams of trade, but as possessing certain fixed amounts which they exchange until they come to equilibrium.

It is much more easy to determine the point at which a pendulum will come to rest than to calculate the velocity at which it will move when displaced from that point of rest. Just so, it is a far more easy task to lay down the conditions under which trade is completed and interchange ceases, than to attempt to ascertain at what rate trade will go on when equilibrium is not attained.

The difference will present itself in this form: dynamically we could not treat the ratio of exchange otherwise than as the ratio of dy and dx, infinitesimal quantities of commodity. Our equations would then be regarded as differential equations, which would have to be integrated. But in the statical view of the question we can substitute the ratio of the finite quantities y and x. Thus, from the self-evident principle, stated earlier, that there cannot, in the same market, at the same moment, be two different prices for the same uniform commodity, it follows that *the last increments in an act of exchange must be exchanged in the same ratio as the whole quantities exchanged.* Suppose that two commodities are bartered in the ratio of x for y; then every m^{th} part of x is given for the mth part of y, and it does not matter for which of the mth parts. No part of the commodity can be treated differently from any other part. We may carry this division to an indefinite extent by imagining m to be constantly increased, so that, at the limit, even an infinitely small part of x must be exchanged for an infinitely small part of y, in the same ratio as the whole quantities. This result we may express by stating that the increments concerned in the process of exchange must obey the equation.

$$\frac{dy}{dx} = \frac{y}{x}.$$

The use which we shall make of this equation will be seen in the next section.

THE THEORY OF EXCHANGE

The keystone of the whole Theory of Exchange, and of the principal problems of Economics, lies in this proposition—*The ratio of exchange of any two commodities will be the reciprocal of the ratio of the final degrees of utility of the quantities of commodity available for consumption after the exchange is completed.* When the reader has reflected a little upon the meaning of this proposition, he will see, I think, that it is necessarily true, if the principles of human nature have been correctly represented in previous pages.

Imagine that there is one trading body possessing only corn, and another possessing only beef. It is certain that, under these circumstances, a portion of the corn may be given in exchange for a portion of the beef with a considerable increase of utility. How are we to determine at what point the exchange will cease to be beneficial? This question must involve both the ratio of exchange and the degrees of utility. Suppose, for a moment, that the ratio of exchange is approximately that of ten pounds of corn for one pound of beef: then if, to the trading body which possesses corn, ten pounds of corn are less useful than one of beef, that body will desire to carry the exchange further. Should the other body possessing beef find one pound less useful than ten pounds of corn, this body will also be desirous to continue the exchange. Exchange will thus go on until each party has obtained all the benefit that is possible, and loss of utility would result if more were exchanged. Both parties, then, rest in satisfaction and equilibrium, and the degrees of utility have come to their level, as it were.

This point of equilibrium will be known by the criterion, that an infinitely small amount of commodity exchanged in addition, at the same rate, will bring neither gain nor loss of utility. In other words, if increments of commodities be exchanged at the established ratio, their utilities will be equal for both parties. Thus, if ten pounds of corn were of exactly the same utility as one pound of beef, there would be neither harm nor good in further exchange of this ratio.

It is hardly possible to represent this theory completely by means of a diagram, but the accompanying figure may, perhaps, render it clearer. Suppose the line *pqr* to be a small portion of the curve of utility of one commodity, while the broken line *p′qr′* is the like curve of another com-

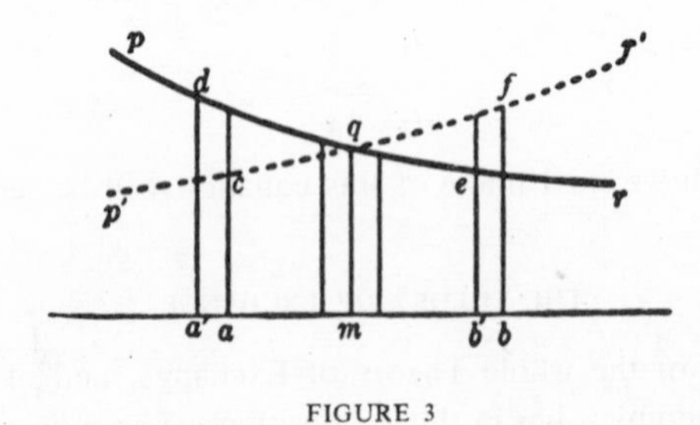

FIGURE 3

modity which has been reversed and superposed on the other. Owing to this reversal, the quantities of the first commodity are measured along the base line from *a* towards *b*, whereas those of the second must be measured in the opposite direction. Let units of both commodities be represented by equal lengths: then the little line of *a′a* indicates an increase of the first commodity, and a decrease of the second. Assume the ratio of exchange to be that of unit for unit, or 1 to 1: then, by receiving the commodity *a'a* the person will gain the utility *ad*, and lose the utility *a'c*; or he will make a net gain of the utility corresponding to the mixtilinear figure *cd*. He will, therefore, wish to extend the exchange. If he were to go up to the point *b'*, and were still proceeding, he would, by the next small exchange, receive the utility *be*, and part with *b'f*; or he would have a net loss of *ef*. He would, therefore, have gone too far; and it is pretty obvious that the point of intersection, *q*, defines the place where he would stop with the greatest advantage. It is there that a net gain is converted into a net loss, or rather where, for an infinitely small quantity, there is neither gain nor loss. To respresent an infinitely small quantity, or even an exceedingly small quantity, on a diagram is, of course, impossible; but on either side of the line *mq* I have represented the utilities of a small quantity of commodity more or less, and it is apparent that the net gain or loss upon the exchange of these quantities would be trifling.

SYMBOLIC STATEMENT OF THE THEORY

To represent this process of reasoning in symbols, let Δx denote a small increment of corn, and Δy a small increment of beef exchanged for it. Now our Law of Indifference comes into play. As both the corn and the beef are homogeneous commodities, no parts can be exchanged at a different ratio from other parts in the same market: hence, if x be the whole quantity of corn given for y the whole quantity of beef received, Δy must have the same ratio to Δx as y to x; we have then,

$$\frac{\Delta y}{\Delta x} = \frac{y}{x}, \text{ or } \Delta y = \frac{y}{x}\Delta x.$$

In a state of equilibrium, the utilities of these increments must be equal in the case of each party, in order that neither more nor less exchange would be desirable. Now the increment of beef, Δy, is $\frac{y}{x}$ times as great as the increment of corn, Δx, so that, in order that their utilities shall be equal, the degree of utility of beef must be $\frac{x}{y}$ times as great as the degree of utility of corn. Thus we arrive at the principle that *the degrees of utility of commodities exchanged will be in the inverse proportion of the magnitudes of the increments exchanged.*

Let us now suppose that the first body, A, originally possessed the quantity a of corn, and that the second body, B, possessed the quantity b of beef. As the exchange consists in giving x of corn for y of beef, the state of things after exchange will be as follows:—

A holds $a - x$ of corn, and y of beef,
B holds x of corn, and $b - y$ of beef.

Let $\phi_1(a - x)$ denote the final degree of utility of corn to A, and $\phi_2 x$ the corresponding function for B. Also let $\psi_1 y$ denote A's final degree of utility for beef, and $\psi_2(b - y)$ B's similar function. Then, as explained previously A will not be satisfied unless the following equation holds true:—

$$\phi_1(a - x) \cdot dx = \psi_1 y \cdot dy;$$

$$\text{or } \frac{\phi_1(a - x)}{\psi_1 y} = \frac{dy}{dx}.$$

Hence, substituting for the second member by the equation given previously we have

$$\frac{\phi_1(a - x)}{\psi_1 y} = \frac{y}{x}.$$

What holds true of A will also hold true of B, *mutatis mutandis.* He must also derive exactly equal utility from the final increments, otherwise it will be for his interest to exchange either more or less, and he will disturb the conditions of exchange. Accordingly the following equation must hold true:

$$\psi_2(b - y) \cdot dy = \phi_2 x \cdot dx;$$

or, substituting as before,

$$\frac{\phi_2 x}{\psi_2(b - y)} = \frac{y}{x}.$$

We arrive, then, at the conclusion, that whenever two commodities are exchanged for each other, and *more or less can be given or received in infinitely small quantities*, the quantities exchanged satisfy two equations, which may be thus stated in a concise form—

$$\frac{\phi_1(a - x)}{\psi_1 y} = \frac{y}{x} = \frac{\phi_2 x}{\psi_2(b - y)}.$$

The two equations are sufficient to determine the results of exchange; for there are only two unknown quantities concerned, namely, x and y, the quantities given and received.

A vague notion has existed in the minds of economical writers, that the conditions of exchange may be expressed in the form of an equation. Thus, J. S. Mill has said:[1] "The idea of a *ratio*, as between demand and

[1] *Principles of Political Economy*, book iii, chap. ii, sec. 4.

supply, is out of place, and has no concern in the matter: the proper mathematical analogy is that of an *equation*. Demand and supply, the quantity demanded and the quantity supplied, will be made equal." Mill here speaks of an equation as only a proper mathematical *analogy*. But if Economics is to be a real science at all, it must not deal merely with analogies; it must reason by real equations, like all the other sciences which have reached at all a systematic character. Mill's equation, indeed, is not explicitly the same as any at which we have arrived above. His equation states that the quantity of a commodity given by A is equal to the quantity received by B. This seems at first sight to be a mere truism, for this equality must necessarily exist if any exchange takes place at all. The theory of value, as expounded by Mill, fails to reach the root of the matter, and show how the amount of demand or supply is caused to vary. And Mill does not perceive that, as there must be two parties and two quantities to every exchange, there must be two equations.

Nevertheless, our theory is perfectly consistent with the laws of supply and demand; and if we had the functions of utility determined, it would be possible to throw them into a form clearly expressing the equivalence of supply and demand. We may regard x as the quantity demanded on one side and supplied on the other; similarly, y is the quantity supplied on the one side and demanded on the other. Now, when we hold the two equations to be simultaneously true, we assume that the x and y of one equation equal those of the other. The laws of supply and demand are thus a result of what seems to me the true theory of value or exchange.

* * * * *

THEORY OF LABOUR

QUANTITATIVE NOTIONS OF LABOUR

Let us endeavour to form a clear notion of what we mean by amount of labour. It is plain that duration will be one element of it; for a person labouring *uniformly* during two months must be allowed to labour twice as much as during one month. But labour may vary also in intensity. In the same time a man may walk a greater or less distance; may saw a greater or less amount of timber; may pump a greater or less quantity of water; in short, may exert more or less muscular and nervous force. Hence amount of labour will be a quantity of two dimensions, the product of intensity and time when the intensity is uniform, or the sum represented by the area of a curve when the intensity is variable.

But intensity of labour may have more than one meaning; it may mean the quantity of work done, or the painfulness of the effort of doing it. These two things must be carefully distinguished, and both are of great importance for the theory. The one is the reward, the other the penalty, of labour. Or rather, as the produce is only of interest to us so far as it possesses utility, we may say that there are three quantities involved in the theory of labour—the amount of painful exertion, the amount of produce, and the amount of utility gained. The variation of utility, as depending on the quantity of commodity possessed, has already been considered; the variation of the amount of produce will be treated in the next chapter; we will here give attention to the variation of the painfulness of labour.

Experience shows that as labour is prolonged the effort becomes as a general rule more and more painful. A few hours' work per day may be considered agreeable rather than otherwise; but so soon as the overflowing energy of the body is drained off, it becomes irksome to remain at work. As exhaustion approaches, continued effort becomes more and more intolerable. Jennings has so clearly stated this law of the variation of labour, that I must quote his words. "Between these two points, the point of incipient effort and the point of painful suffering, it is quite evident that the degree of toilsome sensations endured does not vary directly as the quantity of work performed, but increases much more rapidly, like the resistance offered by an opposing medium to the velocity of a moving body." [2]

* * * * *

There can be no question of the general truth of the above statement, although we may not have the data for assigning the exact law of the

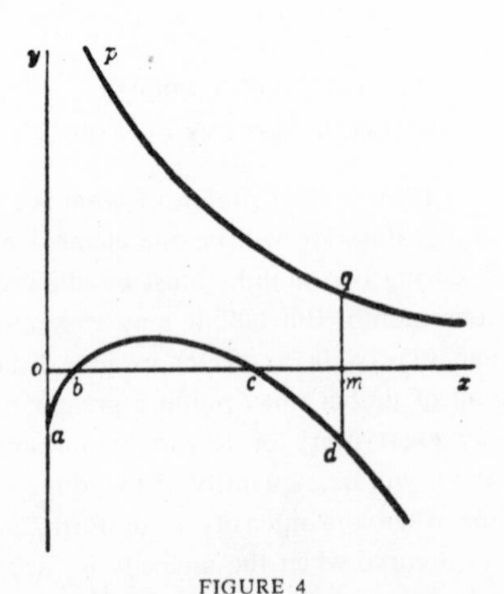

FIGURE 4

variation. We may imagine the painfulness of labour in proportion to produce to be represented by some such curve as *abcd* (see figure above).

[2] "*Natural Elements of Political Economy,*" p. 119.

In this diagram the height of points above the line *ox* denotes pleasure, and depths below it pain. At the moment of commencing labour it is usually more irksome than when the mind and body are well bent to the work. Thus, at first, the pain is measured by *oa*. At *b* there is neither pain nor pleasure. Between *b* and *c* an excess of pleasure is represented as due to the exertion itself. But after *c* the energy begins to be rapidly exhausted, and the resulting pain is shown by the downward tendency of the line *cd*.

We may at the same time represent the degree of utility of the produce by some such curve as *pq*, the amount of produce being measured along the line *ox*. Agreeably to the theory of utility, already given, the curve shows that, the larger the wages earned, the less is the pleasure derived from a further increment. There will, of necessity, be some point *m* such that $qm = dm$, that is to say, such that the pleasure gained is exactly equal to the labour endured. Now, if we pass the least beyond this point, a balance of pain will result: there will be an ever-decreasing motive in favour of labour, and an ever-increasing motive against it. The labourer will evidently cease, then, at the point *m*. It would be inconsistent with human nature for a man to work when the pain of work exceeds the desire of possession, including all the motives for exertion.

We must consider the duration of labour as measured by the number of hours' work per day. The alternation of day and night on the earth has rendered man essentially periodic in his habits and actions. In a natural and wholesome condition a man should return each twenty-four hours to exactly the same state; at any rate, the cycle should be closed within the seven days of the week. Thus the labourer must not be supposed to be either increasing or diminishing his normal strength. But the theory might also be made to apply to cases where special exertion is undergone for many days or weeks in succession, in order to complete work, as in collecting the harvest. Adequate motives may lead to and warrant overwork, but, if long continued, excessive labour reduces the strength and becomes insupportable; and the longer it continues the worse it is, the law being somewhat similar to that of periodic labour.

COMMENTARY ON

A Distinguished Quaker and War

LEWIS FRY RICHARDSON (1881–1953) was a British physicist who for twenty years devoted himself to research in the psychology of peace and war. His earlier work lay in physics, mathematics and meteorology. He wrote a famous paper "The Supply of Energy from and to Atmospheric Eddies" (1920) dealing with the transformations of radiation in the atmosphere and atmospheric turbulence; he investigated the formidably difficult problems of weather prediction, and though the equations he invented were not altogether successful, "a mighty vision had been displayed"; he was the author of a notable volume, *Weather Prediction by Numerical Process* (1922).[1] These scientific labors gained wide recognition, including a fellowship in the Royal Society and other academic honors. The shift in interest to social studies, evidenced among others by his acquiring a degree in psychology in 1929, at the age of fifty, had its origins during the First World War. Dr. Richardson, a Quaker, served in the Friends' Ambulance Unit attached to the French Army. Thus for two years while being "not paid to think," as he says, he had abundant opportunities for meditation.[2] The publication of Bertrand Russell's *War, The Offspring of Fear*, further stimulated his mind. The result was an essay, *The Mathematical Psychology of War*, which appeared in 1919.[3] A subsequent period of "comparative tranquillity" in international affairs turned his mind from the problem, but the failure of the Disarmament Conference in 1935 led him to reconsider his theory. His conclusions were published first as communications to *Nature* [4] and were later reprinted in a monograph which drew considerable attention.[5]

Richardson was one of no more than a handful of serious scholars who have in recent years attempted a quantitative treatment of the causes of war, the mechanics of foreign politics, the effects of armament races and kindred matters. His undertaking was bold and unorthodox; its political implications were often unwelcome. Accordingly, it has not been warmly received by the average specialist in social studies. I suspect that, while Richardson is respected for his "more solid" scientific achievements, his work in the field of international brawling is regarded much as was the

[1] A most interesting sketch of Richardson's meteorological ideas is given by O. G. Sutton in "Methods of Forecasting the Weather," *The Listener*, March 25, 1954, pp. 522–24.

[2] Preface to *Generalized Foreign Politics: A Study in Group Psychology*, by Lewis F. Richardson, The *British Journal of Psychology*, Monograph Supplement XXII, Cambridge, 1939.

[3] Oxford, 1919 (William Hunt).

[4] May 18 and December 25, 1935.

[5] Lewis F. Richardson, *op. cit.*

psychic research of Sir Oliver Lodge. In his monumental treatise, *A Study of War*, Professor Quincy Wright—an uncommonly able student of the problem—acknowledges the importance of Richardson's contribution. He analyzes the theory and calls it "suggestive, though not in all respects convincing." [6] That is a fair appraisal.

The selections below furnish a clear summary of Richardson's results. The presentation is for a general audience and the mathematics is not only held to a minimum but explained simply, step by step. In the first essay, "Threats and Security," Richardson expresses the main thesis that the more a nation arms—even granting that those in power sincerely desire to preserve peace—the better are its chances for war. If the best way to preserve peace is to prepare for war, it is not clear, as C. E. M. Joad has pointed out, "why all nations should regard the armaments of other nations as a menace to peace. However, they do so regard them and are accordingly stimulated to increase their armaments to overtop the armaments by which they conceive themselves to be threatened." [7] This ruinous, deadly competition is what Richardson examines. Among the Kwakiutl Indians, competition between tribes took the form of each attempting to outdo the other in destroying its own property. But this curious practice, roughly analogous to that of more civilized peoples, was held within prudent limits. A chief might destroy a copper pot, bought for "four thousand blankets," in order to vanquish his rival by wasting more than the other could afford to waste. But he "was not free to destroy property to the utter impoverishment of his people or to engage in contests ruinous to them. Overdoing was always dangerous. . . ." [8]

The second essay, "Statistics of Deadly Quarrels," deals with such topics as the distribution of wars in time, the number of big and little wars since the sixteenth century, the nations most involved, the possibility of forecasting future wars on the basis of past records and so on. It is full of fresh insights. Richardson wrote with humor and becoming diffidence; he advanced no sweeping claims for his theories but neither did he undervalue them in false modesty. He made no effort to conceal his hatred of war. I suppose the time has come when such behavior must be counted as remarkable.

[6] Quincy Wright, *A Study of War*, Chicago, 1942, pp. 1482 *et seq.* See also Wright's interesting exposition, some of it mathematical, of his own theories of the probability, causes and prevention of war. (Chapters 36, 37, 39, 40, *op. cit.*)

[7] *Why War*, Penguin Special, Harmondsworth, 1939, p. 69; quoted by Richardson.

[8] Ruth Benedict, *Patterns of Culture*, New York, 1934.

Though this may be play to you, 'tis death to us.—SIR ROGER L'ESTRANGE

Man, a child in understanding of himself, has placed in his hands physical tools of incalculable power. He plays with them like a child. . . . The instrumentality becomes a master and works fatally . . . not because it has a will but because man has not. —JOHN DEWEY

6 Mathematics of War and Foreign Politics

By LEWIS FRY RICHARDSON

THE DIVERSE EFFECTS OF THREATS

THE reader has probably heard a mother say to her child: "Stop that noise, or I'll smack you." Did the child in fact become quiet?

A threat from one person, or group of people, to another person or group has occasionally produced very little immediate effect, being received with contempt. Effects, when conspicuous, may be classified as submission at one extreme, negotiation or avoidance in the middle, and retaliation at the other extreme. The following incidents are classified in that manner; otherwise they are purposely miscellaneous.

Contempt

EXAMPLE 1. About fifty states, organized as the League of Nations, tried in 1935 and 1936, by appeals and by cutting off supplies, to restrain Mussolini's Italy from making war on Abyssinia. At the time Mussolini disregarded the League, and went on with the conquest of Abyssinia. He did not however forget. Four years later in his speech to the Italian people on the occasion of Italy's declaration of war against Britain and France, Mussolini said, "The events of quite recent history can be summarized in these words—half promises, constant threats, blackmail and finally, as the crown of this ignoble edifice, the League siege of the fifty-two States." [1]

Submission

EXAMPLE 2.[2] In 1906 the British Government had a disagreement with the Sultan of Turkey about the exact location of the frontier between Egypt and Turkish Palestine. A British battleship was sent on 3rd May with an ultimatum, and thereupon the Sultan accepted the British view.

[1] *Glasgow Herald*, 11th June, 1940.

[2] *Ency. Brit.* XIV, ed. 14, **2**, 156. Grey, Viscount, 1925, *Twenty-Five Years*. Hodder and Stoughton, London.

Some resentment may perhaps have lingered, for after the First World War had been going on for three months Turkey joined the side opposite to Britain.

EXAMPLE 3. In August 1945 the Japanese, having suffered several years of war, having lost by defeat their Italian and German allies, being newly attacked by Russia, having had two atomic bombs dropped on them, and being threatened with more of the same, surrendered unconditionally.

Negotiation Followed by Submission

EXAMPLE 4.[3] After the Germans annexed Austria on 13th March, 1938, the German minority in Czechoslovakia began to agitate for self-government, and both the German and Czech Governments moved troops towards their common frontier. On 13th August there began German Army manœuvres on an unprecedented scale. On 6th September France called up reservists. In September the cession of the Sudetenland by Czechoslovakia to Germany was discussed, to the accompaniment of threats by Hitler on 12th, the partial mobilization of France on 24th, and the mobilization of Czechoslovakia and of the British Navy on 27th. Finally, at Munich on the 29th the French and British agreed to advise the Czechs to submit to the German demand for the Sudetenland, partly because its population spoke German, and partly because of the German threat to take it by armed force. Intense resentment at this humiliation lingered.

Negotiation Followed by a Bargain

EXAMPLE 5.[4] In the spring of 1911 French troops entered Fez, in Morocco, the German Government protesting. On 1st July, 1911, the German Government notified those of France and Britain that a German gunboat was being dispatched to Agadir on the southern coast of Morocco in order there to protect some German firms from the local tribesmen. The French and British interpreted the movement of the gunboat as a threat against themselves, like a 'thumping of the diplomatic table.' On 21st July Mr. Lloyd George made a speech containing the sentence "I say emphatically that peace at that price would be humiliation intolerable for a great country like ours to endure." Much indignation was expressed in the newspapers of France, Britain and Germany. Negotiations ensued. By 4th November France and Germany had agreed on a rearrangement of their West African territories and rights. The arms race continued.

[3] *Keesing's Contemporary Archives.* Bristol.

[4] *Ency. Brit.* XIV, ed. 14, **23,** 349, 352. Morel, E. D., 1915, *Ten Years of Secret Diplomacy*, National Labour Press, London. Grey, Viscount, 1925, *Twenty-Five Years*, Hodder and Stoughton, London.

Avoidance

EXAMPLE 6. The normal behaviour of the armed personnel guarding any frontier in time of peace is to avoid crossing the frontier lest they should be attacked.

EXAMPLE 7. During 1941 British shipping mostly went eastward via the Cape of Good Hope avoiding the Mediterranean where the enemy threat was too strong.

EXAMPLE 8. Criminals are usually said to avoid the police: that is, when the criminals are decidedly outnumbered by the police.

Retaliation

EXAMPLE 9. After the Agadir incident had been settled by the Franco-German Agreement of 4th November, 1911, the 'defence' expenditures of both France and Germany nevertheless continued to increase. See the tabular statement on page 1246.

EXAMPLE 10. Within six weeks after the Munich Agreement of 29th September, 1938, rearmament was proceeding more rapidly in France,[5] Britain,[6] Germany[7] and U.S.A.[8]

EXAMPLE 11. On 5th November, 1940, the British armed merchantship *Jervis Bay*, being charged with the defence of a convoy, saw a German pocket-battleship threatening it. The *Jervis Bay*, although of obviously lesser power, attacked the battleship, and continued to fight until sunk, thus distracting the battleship's attention from the convoy, and giving the latter a chance to escape.

EXAMPLE 12. On 10th November, 1941, Mr. Winston Churchill warned the Japanese that "should the United States become involved in war with Japan the British declaration will follow within the hour." This formidable threat did not deter the Japanese from attacking Pearl Harbour within a month.

These miscellaneous illustrations may serve to remind the reader of many others. Is there any understandable regularity about the wider phenomena which they represent? The present question is about what happens in fact; the other very important question as to whether the fact is ethically good or bad, is here left aside. Some conclusions emerge:

(*a*) People, when threatened, do not always behave with coldly calculated self-interest. They sometimes fight back, taking extreme risks. (Examples 11 and 12.)

(*b*) There is a notable distinction between fresh and tired nations, in the sense that a formidable threat to a fresh nation was followed by

[5] Athlone broadcast, 5th October and 12th October.
[6] *Parliamentary Debates*, 3rd to 6th October.
[7] *Glasgow Herald*, 10th October.
[8] *Glasgow Herald*, 7th November.

retaliation (Example 12) whereas an even more severe threat to the same nation, when tired, produced submission. (Example 3.)

(*c*) A group of people, having a more or less reasonable claim, has sometimes quickly obtained by a threat of violence, more than it otherwise would. But that may not have been the end of the matter. (Examples 2 and 4.)

(*d*) There have often been two contrasted effects, one immediate, the other delayed. An immediate effect of contempt or submission or negotiation or avoidance has been followed by resentful plans for retaliation at some later opportunity. (Example 1, Example 2, Examples 4 and 10, Examples 5 and 9.)

(*e*) What nowadays is euphemistically called national 'defence,' in fact always includes preparations for attack, and thus constitutes a threat to some other group of people. This type of 'defence' is based on the assumption that threats directed towards other people will produce in them either submission, or negotiation, or avoidance; and it neglects the possibility that contempt or retaliation may be produced instead. Yet in fact the usual effect between comparable nations is retaliation by counter-preparations, thus leading on by way of an arms race towards another war.

SCHISMOGENESIS

In his study of the Iatmul tribe in New Guinea, Gregory Bateson[9] noticed a custom whereby, at a meeting in the ceremonial hall two men would boast alternately, each provoking the other to make bolder claims, until they reached extravagant extremes.

He also noticed a process whereby a man would have some control over a woman. Then her acceptance of his leadership would encourage him to become domineering. This in turn made her submissive. Then he became more domineering and she became more abject, the process running to abnormal extremes.

Bateson called both these processes 'schismogenesis,' which may be translated as 'the manner of formation of cleavages.' When both parties developed the same behaviour, for example, both boasting, Bateson called the schismogenesis 'symmetrical.' When the parties developed contrasted behaviour, say, one domineering and the other submissive, he called the schismogenesis 'complementary.' In this terminology an arms race between two nations is properly described as a case of symmetrical schismogenesis.

In the year 1912 Germany was allied with Austria-Hungary, while France was allied with Czarist Russia. Britain was loosely attached to the latter group, thus forming the Triple Entente, while Italy was nominally

[9] Bateson, G., 1935, in the periodical *Man*, p. 199. Bateson, G., 1936, *Naven*, University Press, Cambridge.

attached to the former group, thus making the Triple Alliance. The warlike preparations of the Alliance and of the Entente were both increasing. The usual explanation was then, and perhaps still is, that the motives of the two sides were quite different, for we were only doing what was right, proper and necessary for our own defence, whilst they were disturbing the peace by indulging in wild schemes and extravagant ambitions. There are several distinct contrasts in that omnibus statement. Firstly that their conduct was morally bad, ours morally good. About so national a dispute it would be difficult to say anything that the world as a whole would accept. But there is another alleged contrast as to which there is some hope of general agreement. It was asserted in the years 1912–14 that their motives were fixed and independent of our behaviour, whereas our motives were a response to their behaviour and were varied accordingly. In 1914 Bertrand Russell[10] (now Earl Russell) put forward the contrary view that the motives of the two sides were essentially the same, for each was afraid of the other; and it was this fear which caused each side to increase its armaments as a defence against the other. Russell's pamphlet came at a time when a common boast in the British newspapers was that the British people 'knew no fear.' Several conspicuous heroes have since explained that they achieved their aims in spite of fear. When we analyse arms races it is, however, unnecessary to mention fear, or any other emotion; for an arms race can be recognized by the characteristic outward behaviour, which is shown in the diagram on page 1246. The valuable part of Russell's doctrine was not his emphasis on fear, but his emphasis on mutual stimulation.

This view has been restated by another philosopher, C. E. M. Joad:[11]

> . . . if, as they maintain, the best way to preserve peace is to prepare war, it is not altogether clear why all nations should regard the armaments of other nations as a menace to peace. However, they do so regard them and are accordingly stimulated to increase their armaments to overtop the armaments by which they conceive themselves to be threatened. . . . These increased arms being in their turn regarded as a menace by nation A whose allegedly defensive armaments have provoked them, are used by nation A as a pretext for accumulating yet greater armaments wherewith to defend itself against the menace. These yet greater armaments are in their turn interpreted by neighbouring nations as constituting a menace to themselves and so on. . . .

This statement is, I think, a true and very clear description but needs two amendments. The competition is not usually between every nation and every other nation, but rather between two sides; so that a nation looks with moderate favour on the armaments of other nations on its own side, and with strong dislike on those of the opposite side. Joad's descrip-

[10] Russell, B. A. W., 1914, "War the Offspring of Fear," Union of Democratic Control, London.

[11] Joad, C. E. M., 1939, *Why War?* Penguin Special, p. 69.

tion applies to an arms race which has become noticeable. Motives other than defence may have been important in starting the arms race.

It may be well to translate these ideas into the phraseology of 'operational research' which began to be used during the Second World War. Professor C. H. Waddington [12] explains that "The special characteristic which differentiates operational research from other branches of applied science is that it takes as the phenomenon to be studied the whole executive problem and not the individual technical parts. . . ." Surely the maintenance of world peace is an executive problem large enough to be called an operation and to require an appropriate background of operational research. Sir Charles Goodeve,[13] in a survey of operational research, distinguishes between 'self-compensating and self-aggravating systems,' and he mentions, as an example of the latter, the system composed of the public and of the store-keepers; a system such that a rumour of scarcity can make a real scarcity. In this phraseology it can be said that a system of two great powers, not in the presence of any common enemy, is a 'self-aggravating system' such that a rumour of war can make a real war.

It will be shown in the next section that arms races are best described in quantitative terms; but, for those who do not like mathematics, Bateson's word 'schismogenesis' may serve as an acceptable summary of a process which otherwise requires a long verbal description such as those given by Russell, Bateson, or Joad.

THE QUANTITATIVE THEORY OF ARMS RACES

The facts for the years 1909 to 1914 are interesting. The 'defence' budgets of France, Germany and Russia were taken from a digest by Per Jacobsson; [14] those for Austria-Hungary from the *Statesman's Year Books*. To make them comparable, they were all reduced to sterling. In those years the exchange rates between national currencies were held steady by the shipment of gold, so that the conversion to sterling is easy and definite.

Because France was allied to Russia it is reasonable to consider the total of their 'defence' expenditures. Let it be U. For a similar reason let V be the total for Germany and Austria-Hungary. In the accompanying diagram the rate of increase $(U + V)$ is plotted against $(U + V)$. See Figure 1.

The accuracy with which the four observed points are fitted by a straight line is remarkable, especially as one of the co-ordinates is a difference. Similar diagrams drawn for other years, for other countries, and from other sources of information, are not so straight; but still they are straight enough to suggest that the explanation of the phenomenon is

[12] Waddington, C. H., in *Nature*, Vol. CLXI, p. 404.

[13] Goodeve, Sir Charles in *Nature*, Vol. CLXI, p. 384.

[14] Jacobsson, Per (1929?). *Armaments Expenditures of the World*, published by the *Economist*, London.

hardly likely to be found in the caprice of a few national leaders; the financial facts suggest either regular planning, or the regularity which results from the average of many opinions.

The main feature shown by the diagram is that the more these nations spent, the more rapidly did they increase their expenditure. Athletic races are not like that, for in them the speed of the contestants does not increase so markedly with the distance that they have run.

TABLE. THE ARMS RACE OF 1909–14

Defence budgets expressed in millions of £ sterling

	1909		*1910*		*1911*		*1912*		*1913*
France	48·6		50·9		57·1		63·2		74·7
Russia	66·7		68·5		70·7		81·8		92·0
Germany	63·1		62·0		62·5		68·2		95·4
Austria-Hungary	20·8		23·4		24·6		25·5		26·9
Total $= U + V$	199·2		204·8		214·9		238·7		289·0
Time rate $= \Delta(U + V)/\Delta t$		5·6		10·1		23·8		50·3	
$(U + V)$ at same date		202·0		209·8		226·8		263·8	

Here Δ signifies 'take the annual increase of' whatever symbol follows next.
From Monog. Supplt. No. 23 of *Brit. Journ. Psychol.*, by permission of the British Psychological Society.

The sloping line when produced backwards cuts the horizontal, where $\Delta(U + V)/\Delta t$ vanishes, at the point where $U + V = 194$ million £. This point may suitably be called a point of equilibrium. To explain how it could be a point of equilibrium we can suppose that the total expenditure of 194 million was regarded as that which would have been so ordinary as not to constitute any special threat.

It was a theory [15] which led L. F. Richardson to make a diagram

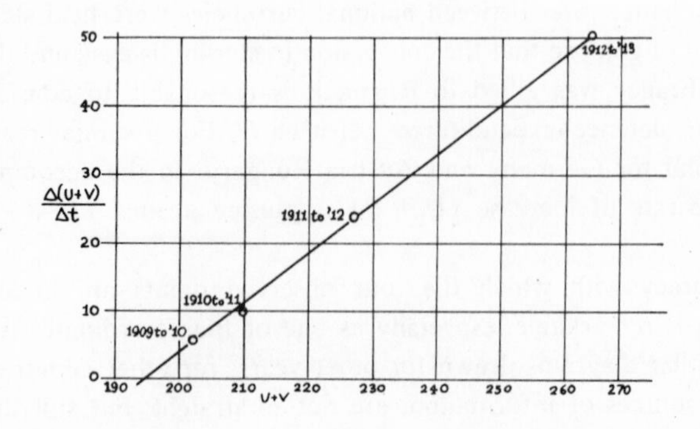

FIGURE 1—The financial facts in the table above are here plotted.

[15] Richardson, L. F., 1919, *Mathematical Psychology of War*. In British copyright libraries. The diagram first appeared in *Nature* of 1938, Vol. CXLII, p. 792.

having those co-ordinates $U + V$ and $\Delta(U + V)/\Delta t$. This theory will now be explained. The opening phase of the First World War afforded a violent illustration of Russell's doctrine of mutuality, for it was evident that warlike activity, and the accompanying hatred, were both growing by tit for tat, alias mutual reprisals. Tit for tat is a jerky alternation; but apart from details, the general drift of mutual reprisals was given a smoothed quantitative expression in the statement that the rate of increase of the warlike activity of each side was proportional to the warlike activity of the other side. This statement is equivalent to the following pair of simultaneous differential equations

$$\frac{dx}{dt} = ky, \qquad \frac{dy}{dt} = lx \qquad (1), (2)$$

where t is the time, x is the warlike activity of one side, y that of the other side, k and l are positive constants, dx/dt is the excellent notation of Leibniz for the time rate of increase of x, and dy/dt is the time rate of increase of y. In accordance with modern custom, the fraction-line is set sloping when it occurs in a line of words. If k were equal to l, then the relation of x to y would be the same as the relation of y to x, so that the system of x and y would be strictly mutual. Strict mutuality is, however, not specially interesting. The essential idea is that k and l, whether equal or not, are both positive. They are called '*defence coefficients*' because they represent a pugnacious response to threats.

The reader may here object that anything so simple as the pair of equations (1) and (2) is hardly likely to be true description of anything so complicated as the politics of an arms race. In reply appeal must be made to a working rule known as Occam's Razor whereby the simplest possible descriptions are to be used until they are proved to be inadequate.

The meaning of (1) and (2) will be further illustrated by deducing from them some simple consequences. If at any time both x and y were zero, it follows according to (1) and (2) that x and y would always remain zero. This is a mathematical expression of the idea of permanent peace by all-round total disarmament. Criticism of that idea will follow, but for the present let us continue to study the meaning of equations (1) and (2). Suppose that x and y being zero, the tranquillity were disturbed by one of the nations making some very slightly threatening gesture, so that y became slightly positive. According to (1) x would then begin to grow. According to (2) as soon as x had become positive, y would begin to grow further. The larger x and y had become the faster would they increase. Thus the system defined by (1) and (2) represents a possible equilibrium at the point where x and y are both zero, but this equilibrium is unstable, because any slight deviation from it tends to increase. If any historian or politician reads these words, I beg him or her to notice that in the mechanical sense, which is used here, stability is not the same as equilibrium; for on the contrary stable and unstable are adjectives qualifying equilibrium. Thus an equilibrium is said to be stable, or to have stability, if a small disturbance tends to die away; whereas an equilibrium is said to be unstable, or to have instability, if a small disturbance tends to increase. In this mechanical sense the system defined by (1) and (2) has instability. It describes a schismogenesis. "It is an old proverb," wrote William Penn in 1693: "*Maxima bella ex levissimis causis:* The greatest Feuds have had the smallest Beginnings."

One advantage of expressing a concept in mathematics is that deductions can then be made by reliable techniques. Thus in (1) and (2) the nations appear as entangled with one another, for each equation involves both x and y. These variables can, however, be separated by repeating the operation d/dt which signifies 'take the time rate of.'

Thus from (1) it follows that $\frac{d}{dt}\left(\frac{dx}{dt}\right) = k\frac{dy}{dt}$

Simultaneously from (2) $k\frac{dy}{dt} = klx$

On elimination of dy/dt between these two equations there remains an equation which does not involve y, namely $\frac{d}{dt}\left(\frac{dx}{dt}\right) = klx.$ (3)

Similarly $\frac{d}{dt}\left(\frac{dy}{dt}\right) = kly$ (4)

In (3) and (4) each nation appears as if sovereign and independent, managing its own affairs, until we notice that the constant kl is a property of the two nations jointly.

Another advantage of a mathematical statement is that it is so definite that it might be definitely wrong; and if it is found to be wrong, there is a plenteous choice of amendments ready in the mathematicians' stock of formulæ. Some verbal statements have not this merit; they are so vague that they could hardly be wrong, and are correspondingly useless.

The formulæ (1) and (2) do indeed require amendment, for they contain no representation of any restraining influences; whereas it is well known that, after a war, the victorious side, no longer feeling threatened by its defeated enemy, proceeds to reduce its armed forces in order to save expenditure, and because the young men are desired at home. The simplest mathematical representation of disarmament by a victor is

$$\frac{dx}{dt} = -\alpha x \tag{5}$$

where α is a positive constant. For, so long as x is positive, equation (5) asserts that dx/dt is negative, so that x is decreasing. Equation (5) is commonly used in physics to describe fading away. In accountancy, depreciation at a fixed annual percentage is a rule closely similar to (5). As a matter of fact [16] (5) is a good description of the disarmaments of Britain, France, U.S.A., or Italy during the years just after the First World War. In equation (5), which represents disarmament of the victor, there is no mention of y, because the defeated nation no longer threatens. It seems reasonable to suppose that restraining influences of the type represented by (5) are also felt by both of the nations during an arms race, so that equations (1) and (2) should be amended so as to become

$$\frac{dx}{dt} = ky - \alpha x, \qquad \frac{dy}{dt} = lx - \beta y \qquad (6), (7)$$

in which β is another positive constant. At first [17] α and β were called 'fatigue and expense coefficients,' but a shorter and equally suitable name is *restraint coefficients.* These restraining influences may, or may not, be sufficient to render the equilibrium stable. The interaction is easily seen in the special, but important, case of similar nations, such that $\alpha = \beta$, and $k = l$. For then the subtraction of (7) from (6) gives

$$\frac{d(x-y)}{dt} = -(k+\alpha)(x-y) \qquad (8)$$

In this $(k + \alpha)$ is always positive. If at any time $(x - y)$ is positive, equation (8) shows that then $(x - y)$ is decreasing; and that moreover $(x - y)$ will continue to decrease until it vanishes, leaving $x = y$. If on the contrary $(x - y)$ is initially negative, (8) shows that $(x - y)$ will increase towards zero. Thus there is a stable drift from either side towards equality of x with y. That is more or less in accord with the historical facts about arms races between nations which can be regarded as similar.

To see the other aspect, let (7) be added to (6) giving

$$\frac{d(x+y)}{dt} = (k-\alpha)(x+y) \qquad (9)$$

The meaning of (9) can be discussed in the same manner as that of (8). The result is that $(x + y)$ will drift towards zero if $(k - \alpha)$ is negative, that is if $\alpha > k$. We may then say that restraint overpowers 'defence,' and that the system is thoroughly stable. Unfortunately that is not what has happened in Europe in the present century. The other case is that in which $k > \alpha$ so that 'defence' overpowers restraint, and $(x + y)$ drifts away from zero. That is like an arms race. When $k > \alpha$, the system is

[16] (See footnote 21.)
[17] Richardson, L. F., in *Nature* of 18th May, 1935, p. 830.

stable as to $(x - y)$, but unstable as to $(x + y)$. It is the instability which has the disastrous consequences. The owner of a ship which has capsized by rolling over sideways can derive little comfort from the knowledge that it was perfectly stable for pitching fore and aft. People who trust in the balance of power should note this combination of stability with instability.

If at any time x and y were both zero, it would follow from (6) and (7) that x and y would always remain zero. So that the introduction of the restraining terms still leaves the theoretical possibility of permanent peace by universal total disarmament. Small-scale experiments on absence of armament have been tried with success between Norway and Sweden, between Canada and U.S.A., between the early settlers in Pennsylvania and the Red Indians. The experiment of a general world-wide absence of arms has never been tried. Many people doubt if it would result in permanent peace; for, they say, grievances and ambitions would cause various groups to acquire arms in order to assert their rights, or to domineer over their unarmed neighbours. The theory is easily amended to meet this objection. Let two constants g and h be inserted respectively into (6) and (7) thus

$$\frac{dx}{dt} = ky - \alpha x + g; \qquad \frac{dy}{dt} = lx - \beta y + h \qquad (10), (11)$$

If x and y were at any time both zero, then would $dx/dt = g$, and $dy/dt = h$, which do not indicate a permanent condition.

There may be still an equilibrium, but it is not at the point $x = 0$, $y = 0$. To find the new point of equilibrium, let $dx/dt = 0$, and $dy/dt = 0$. Then by (10) and (11)

$$0 = ky - \alpha x + g; \qquad 0 = lx - \beta y + h \qquad (12), (13)$$

These equations represent two straight lines in the plane of x and y. If these lines are not parallel, their intersection is the point of equilibrium. It may be stable or unstable.

The assertion that the defence-coefficients k and l are positive is equivalent to supposing that the effect of threats is always retaliation. The reader may object that in the opening section of this chapter other effects were also mentioned, namely, contempt, submission, negotiation, or avoidance. The most important of these objections relates to submission, because it is the direct opposite of retaliation. The answer is that the scope of the present theory is restricted to the interaction of groups which style themselves powers, which are proud of their so-called sovereignty and independence, are proud of their armed might, and are not exhausted by combat. This theory is not about victory and defeat. In different circumstances k or l might be negative. A theory of submissiveness showing this has been published.[18] As to contempt, negotiation, or avoidance, they have sometimes gone on concurrently with an arms race, as in Examples (1), (5) and (9), (4) and (10) of the diverse effects of threats.

Let us now return to the 'defence' budgets of France, Russia, Germany and Austria-Hungary. The diagram on page 1246 relates to the total $U + V$ of the warlike expenditures of the two opposing sides. The equilibrium point, at $U + V = 194$ millions sterling, presumably represents the expenditure which was excused as being customary for the maintenance of internal order, and harmless in view of the treaty situation. In the theory the treaty situation is represented by the constants g and h. Their effects can be regarded as included in the 194, together with the general good-will between the nations. It appears suitable therefore to compare fact with theory by setting simultaneously $x + y = U + V - 194$, together with $g = 0$ and $h = 0$. (14), (15), (16)

From (14) one derives

$$\frac{d(x + y)}{dt} = \frac{d(U + V)}{dt} \tag{17}$$

The two opposing alliances were about of equal size and civilization, so that it seems permissible to simplify the formulæ by setting $\alpha = \beta$ and $k = l$. Addition of the formulæ (10) and (11) then gives, as before in (9),

$$\frac{d(x + y)}{dt} = (k - \alpha)(x + y) \tag{18}$$

Now $x + y$ can be thoroughly eliminated from (18); from its first member by (17), and from its second member by (14), with the result that

$$\frac{d(U + V)}{dt} = (k - \alpha)(U + V - 194) \tag{19}$$

This is a statement about the expenditures of the nations in a form comparable with Figure 1 on page 1246. For $d(U + V)/dt$ is a close approximation to $\Delta(U + V)/\Delta t$. The assumed constancy of $k - \alpha$ agrees with the fact that the sloping line on the diagram is straight. Moreover the absence from (6) and (7) of squares, reciprocals, or other more complicated functions, which was at first excused on the plea of simplicity, is now seen to be so far justified by comparison with historical fact. The slope of the line on the diagram, when compared with (19) gives

$$k - \alpha = 0 \cdot 7 \text{ per year} \tag{20}$$

[18] See footnotes 19, 21.

Further investigations of this sort have dealt with demobilization,[19, 21] with the arms race of 1929–39 between nine nations,[19, 21] with war-weariness and its fading,[21] with submissiveness in general,[19, 21] and with the submission of the defeated in particular.[20, 21]

All those investigations had to do with warlike preparations, an outward or behaviouristic manifestation. The best measure of it was found to be a nation's expenditure on defence divided, in the same currency, by the annual pay of a semi-skilled engineer. This conception may be called 'war-finance per salary,' a phrase which can be packed into the new word 'warfinpersal.'

MOODS, FRIENDLY OR HOSTILE, PRIOR TO A WAR

What of the inner thoughts, emotions, and intentions which accompany the growth of warfinpersal? Lloyd George, who was Chancellor of the Exchequer in 1914, describes some of them in his *War Memoirs*, revised in 1938. He relates that although there had been naval rivalry between Britain and Germany during the previous six years, yet as late as 24th July, 1914, only a very small minority of Britons wished for war with Germany. Eleven days later the British nation had changed its mind. Another brilliant description of the moods occurs in H. G. Wells's novel *Mr. Britling Sees It Through*. The contrast between the comparatively slow growth of irritation over years, and the sudden outbreak of war, can be explained by the well-established concept of the subconscious. Suppose, for simplicity, that in a person there are only two mental levels, the overt and the subconscious, and that the moods in these levels are not necessarily the same. In Britain in the year 1906 the prevailing mood towards Germany was friendly openly, and friendly also in the subconscious. The arms race during 1908 to 1913 did not prevent the King from announcing annually to Parliament that "My relations with foreign powers continue to be friendly"; and the majority of British citizens continued to speak in friendly terms of their German acquaintances. It is reasonable to suppose, however, that during those same years there was a growing hostility to Germany in the British subconscious mind, caused by the arms race and by diplomatic crises. The hostile mood, having been thus slowly prepared in the subconscious, was ready suddenly to take open control at the beginning of August 1914. A quantitative theory of such changes of mood is offered by L. F. Richardson.[22] Here is a simplified specimen of that theory:

$$d\eta_1/dt = C_{12}\xi_1\eta_2$$

[19] Richardson, L. F., 1939, *Generalized Foreign Politics.* Monog. Supplt. No. 23 of the *British Journal of Psychology*.

[20] Richardson, L. F., 1944, letter in *Nature* of 19th August, p. 240.

[21] Richardson, L. F., 1947, *Arms and Insecurity* on 35 mm. punched safety microfilm, sold by the author.

[22] Richardson, L. F., 1948, *Psychometrika*, Vol. XIII, pp. 147–74 and 197–232.

where η_1 is the fraction of the British population that was eager for war at time t, while η_2 is the corresponding fraction for Germany, C_{12} is a constant, and ξ_1 is the fraction of the British population that was in the susceptible mood: overtly friendly, but subconsciously hostile. An equation of this type is used in the theory of epidemics of disease.[23] Eagerness for war can be regarded analogously as a mental disease infected into those in a susceptible mood by those who already have the disease in the opposing country. In this theory, as in Russell's *War the Offspring of Fear*, the relations between the two nations are regarded as mutual. Accordingly the same letter, ξ say, is used in relation to either, but is distinguished by suffix 1 for the British, suffix 2 for the German quantity. Also there is another equation obtainable from that above by interchanging the suffixes.

CONCLUSION

This chapter is not about wars and how to win them, but is about attempts to maintain peace by a show of armed strength. Is there any escape from the disastrous mutual stimulation by threat and counter-threat? Jonathan Griffin [24] argued that each nation should confine itself to pure defence which did not include any preparation for attack, while aggressive weapons should be controlled by a supranational authority. The difficulty is that, when once a war has started, attack is more effective than defence. Gandhi's remarkable discipline and strategy of non-violent resistance is explained and discussed by Gregg.[25] The pacifying influence of intermarriage has been considered by Richardson.[26]

[23] Kermack, W. O., and McKendrick, A. G., 1927, *Proc. Roy. Soc. Lond. A* Vol. CXV, pp. 700-22.

[24] Griffin, J. 1936, *Alternative to Rearmament*, Macmillan, London.

[25] Gregg, R. B., 1936, *The Power of Non-Violence*, Routledge, London.

[26] Richardson, L. F., 1950, *The Eugenics Review*, Vol. XLII, pp. 25-36.

They be farre more in number, that love to read of great Armies, bloudy Battels, and many thousands slaine at once, then that minde the Art by which the Affairs, both of Armies, and Cities, be conducted to their ends.
—THOMAS HOBBES (*Preface to Thucydides*)

Men grow tired of sleep, love, singing and dancing sooner than of war.
—HOMER (*Iliad*)

The success of a war is gauged by the amount of damage it does.
—VICTOR HUGO

There mustn't be any more war. It disturbs too many people.
—AN OLD FRENCH PEASANT WOMAN (*To Aristide Briand, 1917*)

7 Statistics of Deadly Quarrels

By LEWIS FRY RICHARDSON

THERE are many books by military historians dealing in one way or another with the general theme 'wars, and how to win them.' The theme of the present chapter is different, namely, 'wars, and how to take away the occasions for them,' as far as this can be done by inquiring into general causes. But is there any scope for such an inquiry? Can there be any general causes that are not well known, and yet of any importance? Almost every individual in a belligerent nation explains the current war quite simply by giving particulars of the abominable wickedness of his enemies. Any further inquiry into general causes appears to a belligerent to be futile, comic, or disloyal. Of course an utterly contradictory explanation is accepted as obviously true by the people on the other side of the war; while the neutrals may express chilly cynicism. This contradiction and variety of explanation does provide a prima-facie case for further investigation. Any such inquiry should be so conducted as to afford a hope that critical individuals belonging to all nations will ultimately come to approve of it. National alliances and enmities vary from generation to generation. One obvious method of beginning a search for general causes is therefore to collect the facts from the whole world over a century or more. Thereby national prejudices are partly eliminated.

COLLECTIONS OF FACTS FROM THE WHOLE WORLD

Professor Quincy Wright [1] has published a collection of *Wars of Modern Civilization* extending from A.D. 1482 to A.D. 1940, and including 278 wars, together with their dates of beginning and ending, the name of any treaty of peace, the names of the participating states, the number of battles, and a classification into four types of war. This extensive summary

[1] Wright, Q., 1942, *A Study of War*, Chicago University Press, Chicago.

of fact is very valuable, for it provides a corrective to those frequent arguments which are based on the few wars which the debater happens to remember, or which happen to support his theory. Wright explains his selection by the statement that his list

"is intended to include all hostilities involving members of the family of nations, whether international, civil, colonial, or imperial, which were recognized as states of war in the legal sense or which involved over 50,000 troops. Some other incidents are included in which hostilities of considerable but lesser magnitude, not recognized at the time as legal states of war, led to important legal results such as the creation or extinction of states, territorial transfers, or changes of government."

Another world-wide collection has been made by L. F. Richardson for a shorter time interval, only A.D. 1820 onwards, but differently selected and classified. No attention was paid to legality or to important legal results, such concepts being regarded as varying too much with opinion. Instead attention was directed to deaths caused by quarrelling, with the idea that these are more objective than the rights and wrongs of the quarrel. The wide class of 'deadly quarrels' includes any quarrel that caused death to humans. This class was subdivided according to the number of deaths. For simplicity the deaths on the opposing sides were added together. The size of the subdivisions had to be suited to the uncertainty of the data. The casualties in some fightings are uncertain by a factor of three. It was found in practice that a scale which proceeded by factors of ten was suitable, in the sense that it was like a sieve which retained the reliable part of the data, but let the uncertainties pass through and away. Accordingly the first notion was to divide deadly quarrels into those which caused about 10,000,000 or 1,000,000 or 100,000 or 10,000 or 1,000 or 100 or 10 or 1 deaths. These numbers are more neatly written respectively as 10^7, 10^6, 10^5, 10^4, 10^3, 10^2, 10^1, 10^0 in which the index is the logarithm of the number of deaths. The subsequent discussion is abbreviated by the introduction of a technical term. *Let the 'magnitude' of any deadly quarrel be defined to be the logarithm, to the base ten, of the number of persons who died because of that quarrel.* The middles of the successive classes are then at magnitudes 7, 6, 5, 4, 3, 2, 1, 0. To make a clean cut between adjacent classes it is necessary to specify not the middles of the classes, but their edges. Let these edges be at 7·5, 6·5, 5·5, 4·5, 3·5, 2·5 . . . on the scale of magnitude. For example magnitude 3·5 lies between 3,162 and 3,163 deaths, magnitude 4·5 lies between 31,622 and 31,623 deaths, magnitude 5·5 lies between 316,227 and 316,228 deaths, and so on.

THE DISTRIBUTION OF WARS IN TIME

This aspect of the collections is taken first, not because it is of the most immediate political interest, but almost for the opposite reason, namely, that it is restfully detached from current controversies.

Before beginning to build, I wish to clear three sorts of rubbish away from the site.

1. There is a saying that "If you take the date of the end of the Boer War and add to it the sum of the digits in the date, you obtain the date of the beginning of the next war, thus $1902 + 1 + 9 + 0 + 2 = 1914$." Also $1919 + 1 + 9 + 1 + 9 = 1939$. These are merely accidental coincidences. If the Christian calendar were reckoned from the birth of Christ in 4 B.C. then the first sum would be $1906 + 1 + 9 + 0 + 6 = 1922$, not $1914 + 4$.

2. There is a saying that "Every generation must have its war." This is an expression of a belief, perhaps well founded, in latent pugnacity. As a statistical idea, however, the duration of a generation is too vague to be serviceable.

3. There is an assertion of a fifty-year period in wars which is attributed by Wright (1942, p. 230) to Mewes in 1896. Wright mentions an explanation by Spengler of this supposed period, thus: "The warrior does not wish to fight again himself and prejudices his son against war, but the grandsons are taught to think of war as romantic." This is certainly an interesting suggestion, but it contradicts the other suggestion that "Every generation must have its war." Moreover the genuineness of the fifty-year period is challenged. Since 1896, when Mewes published, the statisticians have developed strict tests for periodicity (*see* for example Kendall's *Advanced Theory of Statistics*, Part II, 1946). These tests have discredited various periods that were formerly believed. In particular the alleged fifty-year period in wars is mentioned by Kendall [2] as an example of a lack of caution.

Having thus cleared the site, let us return to Wright's collection as to a quarry of building material.

The Distribution of Years in Their Relation to War and Peace

A list was made of the calendar years. Against each year was set a mark for every war that began in that year. Thus any year was characterized by the number, x, of wars that began in it. The number, y, of years having the character x was then counted. The results were as follows.[3]

YEARS FROM A.D. 1500, TO A.D. 1931 INCLUSIVE. WRIGHT'S COLLECTION.

Number, x, of outbreaks in a year .	0	1	2	3	4	>4	Totals
Number, y, of such years	223	142	48	15	4	0	432
Y, as defined below .	216·2	149·7	51·8	12·0	2·1	0·3	432·1

It is seen that there is some regularity about the progression of the numbers y. Moreover they agree roughly with the numbers Y. These are

[2] Kendall, M. G., 1945, *J. Roy. Statistical Soc.*, **108,** 122.
[3] Richardson, L. F., 1945, *J. Roy. Statistical Soc.*, **107,** 242.

of interest because they are calculated from a well-known formula, called by the name of its discoverer the 'Poisson Distribution' and specified thus

$$Y = \frac{N\lambda^x}{(2\cdot7183)^\lambda\, x!}$$

in which N is the whole number of years, λ is the mean number of outbreaks per year, and $x!$ is called 'factorial x' and is equal respectively to 1, 1, 2, 6, 24, when x equals 0, 1, 2, 3, 4. Similar results were obtained from Richardson's collection both for the beginnings and for the ends of fatal quarrels in the range of magnitude extending from 3·5 to 4·5, thus:

YEARS A.D. 1820 TO 1929 INCLUSIVE

x outbreaks in a year	0	1	2	3	4	>4	Total
y for war	65	35	6	4	0	0	110
Poisson	64·3	34·5	9·3	1·7	0·2	0·0	110·0
y for peace	63	35	11	1	0	0	110
Poisson	63·8	34·8	9·5	1·7	0·2	0·0	110·0

The numbers in the rows beginning with the word 'Poisson' were calculated from the formula already given, in which N and λ have the same *verbal* definitions as before, and therefore have appropriately altered *numerical* values. Such adjustable constants are called parameters.

If every fatal quarrel had the same duration, then the Poisson distribution for their beginnings would entail a Poisson distribution for their ends; but in fact there is no such rigid connection. The durations are scattered: Spanish America took fourteen years to break free from Spain, but the siege of Bharatpur was over in two months. Therefore the Poisson distributions for war and for peace may reasonably be regarded as separate facts.

Observed numbers hardly ever agree perfectly with the formulae that are accepted as representing them. In the paper cited [4] the disagreement with Wright's collection is examined by the χ^2 test and is shown to be unimportant. It should be noted, however, that the application of this standard χ^2 test involves the tacit assumption that there is such a thing as chance in history.

There is much available information about the Poisson distribution; about the theories from which it can be derived; and about the phenomena which are approximately described by it.[5] The latter include the distribution of equal time intervals classified according to the number of alpha particles emitted during each by a film of radioactive substance.

[4] *J. Roy. Statistical Soc.*, **107,** 242.

[5] Jeffreys, H., 1939, *Theory of Probability*, Oxford University Press. Kendall, M. G., 1943, *The Advanced Theory of Statistics*, Griffin, London. Shilling, W., 1947, *J. Amer. Statistical Assn.*, **42,** 407–24. Cramér, H., 1946, *Mathematical Methods of Statistics*, Princeton University Press.

In order to bring the idea home, an experiment in cookery may be suggested. Take enough flour to make N buns. Add λN currants, where λ is a small number such as 3. Add also the other usual ingredients, and mix all thoroughly. Divide the mass into N equal portions, and bake them. When each bun is eaten, count carefully and record the number of currants which it contains. When the record is complete, count the number y of buns, each of which contains exactly x currants. Theory would suggest that y will be found to be nearly equal to Y, as given by the Poisson formula. I do not know whether the experiment has been tried.

A more abstract, but much more useful, summary of the relations, is to say that the Poisson distribution of years, follows logically from the hypothesis that there is the same very small probability of an outbreak of war, or of peace, somewhere on the globe on every day. In fact there is a seasonal variation, outbreaks of war having been commoner in summer than in winter, as Q. Wright shows. But when years are counted as wholes, this seasonal effect is averaged out; and then λ is such that the probability of a war beginning, or ending, during any short time dt years is λdt.

This explanation of the occurrence of wars is certainly far removed from such explanations as ordinarily appear in newspapers, including the protracted and critical negotiations, the inordinate ambition and the hideous perfidy of the opposing statesmen, and the suspect movements of their armed personnel. The two types of explanation are, however, not necessarily contradictory; they can be reconciled by saying that each can separately be true as far as it goes, but cannot be the whole truth. A similar diversity of explanation occurs in regard to marriage: on the one hand we have the impersonal and moderately constant marriage rate; on the other hand we have the intense and fluctuating personal emotions of a love-story; yet both types of description can be true.

Those who wish to abolish war need not be discouraged by the persistent recurrence which is described by the Poisson formula. The regularities observed in social phenomena are seldom like the unalterable laws of physical science. The statistics, if we had them, of the sale of snuff or of slaves, would presumably show a persistence during the eighteenth century; yet both habits have now ceased. The existence of a descriptive formula does not necessarily indicate an absence of human control, especially not when the agreement between formula and fact is imperfect. Nevertheless, the Poisson distribution does suggest that the abolition of war is not likely to be easy, and that the League of Nations and its successor the United Nations have taken on a difficult task. In some other fields of human endeavour there have been long lags between aspiration and achievement. For example Leonardo da Vinci drew in detail a flying machine of graceful appearance. But four centuries of mechanical research intervened before flight was achieved. Much of the research that afterwards was applied to aeroplanes was not at first made specifically for that object. So it may be with social science and the abolition of war.

The Poisson distribution is not predictive; it does not answer such questions as 'when will the present war end?' or 'when will the next war begin?' On the contrary the Poisson distribution draws attention to a persistent probability of change from peace to war, or from war to peace. Discontent with present weather has been cynically exaggerated in a comic rhyme:

> As a rule a man's a fool:
> When it's hot he wants it cool,
> When it's cool he wants it hot,
> Always wanting what is not.

A suggestion made by the Poisson law is that discontent with present circumstances underlies even the high purposes of peace and war. There is plenty of psychological evidence in support. This is not the place to attempt a general review of it; but two illustrations may serve as pointers. In 1877 Britain had not been engaged in any considerable war since the end of the conflict with China in 1860. During the weeks of national excitement in 1877 preluding the dispatch of the British Mediterranean squadron to Gallipoli, in order to frustrate Russian designs on Constantinople, a bellicose music-hall song with the refrain:

'We don't want to fight, but, by Jingo, if we do:

We've got the men, we've got the ships, we've got the money too.'

was produced in London and instantly became very popular.[6]

Contrast this with the behaviour of the governments of Britain, China, USA, and USSR in 1944, after years of severe war, but with victory in sight, who then at Dumbarton Oaks officially described themselves as 'peace-loving.'[7]

Chance in history. The existence of a more or less constant λ, a probability per time of change, plainly directs our attention to chance in history. Thus the question which statisticians are accustomed to ask about any sample of people or things, namely "whether the sample is large enough to justify the conclusions which have been drawn from it" must also be asked about any set of wars.

Have wars become more frequent? In particular the discussion of any alleged trend towards more or fewer wars is a problem in sampling. No definite conclusion about trend can be drawn from the occurrence of two world wars in the present century, because the sample is too small. When,

[6] *Ency. Brit.*, XIV, ed. **13,** 69.
[7] H.M. Stationery Office, London, Cmd. 6666.

however, the sample was enlarged by the inclusion of all the wars in Wright's collection, and the time was divided into two equal intervals, the following result was obtained.

Dates of beginning	A.D. 1500 to 1715	A.D. 1716 to 1931
Numbers of wars	143	156

The increase from 143 to 156 can be explained away as a chance effect.

This was not so for all subdivisions of the time. When the interval from A.D. 1500 to A.D. 1931 was divided into eight consecutive parts of fifty-four years each, it was found that the fluctuation, from part to part, of the number of outbreaks in Wright's collection was too large to be explained away as chance. The extremes were fifty-four outbreaks from A.D. 1824 to 1877, and sixteen outbreaks from A.D. 1716 to 1769. Other irregular fluctuations of λ were found, although less definitely, for parts of twenty-seven and nine years.[8] All these results may, of course, depend on Wright's selection rules. The problem has been further studied by Moyal.[9]

THE LARGER, THE FEWER

When the deadly quarrels in Richardson's collection were counted in units ranges of magnitude, the following distribution was found.[10] The numbers are those of deadly quarrels which ended from A.D. 1820 to 1929 inclusive.

Ends of range of magnitude	$7 \pm \frac{1}{2}$	$6 \pm \frac{1}{2}$	$5 \pm \frac{1}{2}$	$4 \pm \frac{1}{2}$
Quarrel-dead at centre of range	10,000,000	1,000,000	100,000	10,000
Number of deadly quarrels	1	3	16	62

Although Wright's list is not classified by magnitudes, yet some support for the observation that the smaller incidents were the more numerous is provided by his remark (p. 636) that "A list of all revolutions, insurrections, interventions, punitive expeditions, pacifications, and explorations involving the use of armed force would probably be more than ten times as long as the present list." Deadly quarrels that cause few deaths are not in popular language called wars. The usage of the word 'war' is variable and indefinite; but perhaps on the average the customary boundary may be at about 3,000 deaths. From the scientific point of view it would be desirable to extend the above tabular statement to the ranges of magnitude ending at $3 \pm \frac{1}{2}$, $2 \pm \frac{1}{2}$, $1 \pm \frac{1}{2}$, by collecting the corresponding

[8] *J. Roy. Statistical Soc.*, **107**, 246–7.
[9] Moyal, J. E., 1950. *J. Roy. Statistical Soc.*, **112**, 446–9.
[10] Letter in *Nature* of 15th November, 1941.

numbers of deadly quarrels from the whole world. There is plenty of evidence that such quarrels, involving about 1,000 or 100 or 10, deaths, have existed in large numbers. They are frequently reported in the radio news. Wright alludes to them in the quotation above. Many are briefly mentioned in history books. But it seems not to have been anyone's professional duty to record them systematically. For the range of magnitude between 3·5 and 2·5 I have made a card index for the years A.D. 1820 to 1929 which recently contained 174 incidents, but was still growing. This number 174, though an underestimate, notably exceeds 62 fatal quarrels in the next unit range of larger magnitude, and is thus in accordance with 'the larger the fewer.'

Between magnitudes 2·5 and 0·5 the world totals are unknown. Beyond this gap in the data are those fatal quarrels which caused 3, 2, or 1 deaths, which are mostly called murders, and which are recorded in criminal statistics. For the murders it is possible to make a rough estimate of the world total in the following manner. Different countries are first compared by expressing the murders per million of population during a year. This 'murder rate' has varied from 610 for Chile [11] in A.D. 1932, to 0·3 for Denmark [12] A.D. 1911–20. The larger countries had middling rates. From various sources, including a governmental report [13] it was estimated that the murder rate for the whole world was of the order of 32 in the interval A.D. 1820 to 1929. As the world population [14] averaged about 1,358 million for the same interval, it follows that the whole number of murders in the world was about

$$110 \times 32 \times 1358 = 5 \text{ million}$$

This far exceeds the number of small wars in the whole world during the same 110 years. Thus 'the larger, the fewer' is a true description of all the known facts about world totals of fatal quarrels.

In the gap where world totals are lacking there are local samples: one of banditry in Manchukuo,[15] and one of ganging in Chicago.[16] Before these can be compared with the world totals it is essential that they should be regrouped according to equal ranges of quarrel-dead or of magnitude; for the maxim 'the larger the fewer' relates to statistics arranged in either of those manners. When thus transformed the statistics of banditry and of ganging fit quite well with the gradation of the world totals, on certain assumptions. A thorough statistical discussion will be found elsewhere.[17]

[11] *Keesing's Contemporary Archives*, p. 1052, Bristol. Corrected by a factor of ten.

[12] Calvert, E. R., 1930, *Capital Punishment in the Twentieth Century*, Putnam's, London.

[13] *Select Committee on Capital Punishment*, 1931, H.M.S.O., London, for reference to which I am indebted to Mr. John Paton.

[14] Carr-Saunders, A. M., 1936, *World Population*, Clarendon Press, Oxford.

[15] *Japan and Manchukuo Year Book*, 1938, Tokio.

[16] Thrasher, F. M., 1927, *The Gang*, Chicago University Press.

[17] Richardson, L. F., 1948, *Journ. Amer. Statistical Assn.*, Vol. XLIII, pp. 523–46.

The suggestion is that deadly quarrels of all magnitudes, from the world wars to the murders, are suitably considered together as forming one wide class, gradated as to magnitude and as to frequency of occurrence. This is a statistical chapter; and for that reason the other very important gradations, legal, social, and ethical, between a world war and a murder are not discussed here.

WHICH NATIONS WERE MOST INVOLVED?

This section resembles quinine: it has a bitter taste, but medicinal virtues. The participation of some well-known states in the 278 'wars of modern civilization' as listed by Wright is summarized and discussed by him.[18]

Over the whole time interval from A.D. 1480 to 1941 the numbers of wars in which the several nations participated were as follows: England (Great Britain) 78, France 71, Spain 64, Russia (USSR) 61, Empire (Austria) 52, Turkey 43, Poland 30, Sweden 26, Savoy (Italy) 25, Prussia (Germany) 23, Netherlands 23, Denmark 20, United States 13, China 11, Japan 9.

It may be felt that the year 1480 has not much relevance to present-day affairs. So here are the corresponding numbers for the interval A.D. 1850 to 1941, almost within living memory: Great Britain 20, France 18, Savoy (Italy) 12, Russia (USSR) 11, China 10, Spain 10, Turkey 10, Japan 9, Prussia (Germany) 8, USA 7, Austria 6, Poland 5, Netherlands 2, Denmark 2, Sweden 0.

It would be difficult to reconcile these numbers of wars in which the various nations have participated, with the claim made in 1945 by the Charter of the United Nations [19] to the effect that Britain, France, Russia, China, Turkey, and USA, were 'peace-loving' in contrast with Italy, Japan, and Germany. Some special interpretation of peace-lovingness would be necessary: such as either 'peace-lovingness' at a particular date; or else that 'peace-loving' states participated in many wars in order to preserve world peace.

It would be yet more difficult to reconcile the participations found by Wright with the concentration of Lord Vansittart's invective against Germans, as though he thought that Germans were the chief, and the most persistent, cause of war.[20]

In fact no one nation participated in a majority of the wars in Wright's list. For the greatest participation was that of England (Great Britain) namely in seventy-eight wars; leaving 200 wars in which England did not

[18] Wright, Q., 1942, *A Study of War*, Chicago University Press, pp. 220 3 and 650.
[19] H.M. Stationery Office, London, Cmd. 6666, Articles 3 and 4 together with the list of states represented at the San Francisco Conference.
[20] Vansittart, Sir Robert (now Lord), 1941, *Black Record*, Hamish Hamilton, London.

participate. The distinction between aggression and defence is usually controversial. Nevertheless, it is plain that a nation cannot have been an aggressor in a war in which it did not participate. The conclusion is, therefore, that no one nation was the aggressor in more than 28 per cent of the wars in Wright's list. Aggression was widespread. This result for wars both civil and external agrees broadly with Sorokin's findings after his wide investigation of internal disturbance. He attended to Ancient Greece, Ancient Rome, and to the long interval A.D. 525 to 1925 in Europe. Having compared different nations in regard to internal violence, Sorokin concluded that 'these results are enough to dissipate the legend of "orderly" and "disorderly" peoples.' . . . 'All nations are orderly and disorderly according to the times.' [21]

There does not appear to be much hope of forming a group of permanently peace-loving nations to keep the permanently aggressive nations in subjection; for the reason that peace-lovingness and aggressiveness are not permanent qualities of nations. Instead the facts support Ranyard West's [22] conception of an international order in which a majority of momentarily peace-loving nations, changing kaleidoscopically in its membership, may hope to restrain a changing minority of momentarily aggressive nations.

[21] Sorokin, Pitirim A., 1937, *Social and Cultural Dynamics*, American Book Co.
[22] West, R., 1942, *Conscience and Society*, Methuen, London.

COMMENTARY ON

The Social Application of Mathematics

IN 1928 John von Neumann reported a curious discovery to the Mathematical Society of Göttingen: he had worked out a rational strategy for matching pennies. This may not strike you as a momentous achievement, but it was the beginning of a new branch of science. The Theory of Games is today regarded as the most promising mathematical tool yet devised for the analysis of man's social relations.

Von Neumann's proof, which extended to other and more polite amusements such as chess, cards, backgammon, showed that there is in each case a mathematically determinable "best-possible" method of play. The "best-possible" or "rational" strategy is that which assures a player the maximum advantage, regardless of what his opponents may do. It does not promise him good fortune; it does not insure him against ruin. Its best office is to minimize the maximum loss he can sustain "not in every play of the game, but in the long run." To be sure, the rational strategy indicated for a given game may not always be practicable. Von Neumann's version of chess, for example, is a no-move contest in which the opponents work out their secret strategies in advance, each player specifying the moves he will make under all possible circumstances, and then leaving it to an umpire, on the basis of these schedules, to declare the winner. The calculations required for this dismal affair might take centuries and wear out batteries of electronic computers. Even a simplified version of poker, involving a three-card deck, a one-card, no-draw hand, and two participants, would require for its strategic determination the performance of at least two billion multiplications and additions.[1] But these limitations are of secondary concern. What is important is that in each case an optimum strategy exists, that the game has what might be called a "solution." By demonstrating this fact Von Neumann introduced several novel and far-reaching concepts to the field of mathematics.

The name of this branch of science is perhaps too restricted. The Theory of Games has as much to do with players as with games. Von Neumann was not interested in helping people to play winning bridge nor in proving that only croupiers can safely depend on roulette for their living. He was led to his researches by the conjecture that an analysis of the general structure of games would in itself be of mathematical value;

[1] Oskar Morgenstern, "The Theory of Games," *Scientific American*, May 1949, p. 23.

but beyond that, he hoped that the solution of certain problems of games might throw light on problems of economic theory. It is evident that games of strategy have many elements in common with "real-life" situations. Decisions must be faced and choices must be made; some issues can be solved in advance by pure reasoning (e.g. in chess), others involve chance elements (e.g. in poker) and require a different treatment; rarely does a player have sole control of the variables determining the final outcome. Games vary as to the information available to each participant regarding the past actions and resources of his opponents. It may be essential for a player to conceal his strategy; in that case, he must be prepared for the contingency that his plans will become known. Above all, games are pervaded by conflicts of interest: one player cannot win unless another loses and it is plausible to assume that most players want to win. These and other resemblances justify the belief that the study of "rational behavior" in games is a fruitful approach to an understanding of rational behavior in social and economic processes.

The classic work on the subject is Von Neumann and Morgenstern's great treatise *Theory of Games and Economic Behavior.* Its aim is not merely to demonstrate an analogy between the competitive dynamics of games and economics; but to prove "that the typical problems of economic behavior [are] strictly identical with the mathematical notions of suitable games of strategy." [2] The authors do not claim to have formulated a full-fledged mathematical theory of society; their book treats only of economic problems, and even in this area both theory and applications are in infancy. But it is a vigorous and promising infancy; the theory of games can fairly be said to have laid the foundation for a systematic and penetrating mathematical treatment of a vast range of problems in social science.

The three papers which follow illustrate different aspects of this exciting subject. The first is a review of the Von Neumann book by Leonid Hurwicz, Research Professor of Economics and Mathematical Statistics at the University of Illinois. Hurwicz's essay, which appeared in the *American Economic Review*, is an admirable simplification of some of the main ideas presented by Von Neumann and Morgenstern. The second selection discusses a few elementary applications of the theory to a few simple games—matching pennies, the three-boxes game and so on. The author, Dr. S. Vajda, is a mathematician and physicist, now on the staff of the British Admiralty as a member of the Royal Naval Scientific Service. Dr. Abraham Kaplan, head of the department of philosophy of the University of California at Los Angeles, presents a survey of recent attempts to extend the uses of mathematics to social phenomena. It is a

[2] John von Neumann and Oskar Morgenstern, *Theory of Games and Economic Behavior, Princeton*, 1947, p. 2.

readable and succinct essay, affording the reader an opportunity to compare several different approaches, of which the theory of games is obviously the most fruitful. The theory of probability and statistics, which made earlier important contributions in this sphere, also originates in the study of games. Pastimes are apparently an inexhaustible source of knowledge about the outside world and society.

I warne yow wel, it is no childes pley. —CHAUCER

8 The Theory of Economic Behavior[1]

By LEONID HURWICZ

HAD it merely called to our attention the existence and exact nature of certain fundamental gaps in economic theory, the *Theory of Games and Economic Behavior* by von Neumann and Morgenstern would have been a book of outstanding importance. But it does more than that. It is essentially constructive: where existing theory is considered to be inadequate, the authors put in its place a highly novel analytical apparatus designed to cope with the problem.

It would be doing the authors an injustice to say that theirs is a contribution to economics only. The scope of the book is much broader. The techniques applied by the authors in tackling economic problems are of sufficient generality to be valid in political science, sociology, or even military strategy. The applicability to games proper (chess and poker) is obvious from the title. Moreover, the book is of considerable interest from a purely mathematical point of view. This review, however, is in the main confined to the purely economic aspects of the *Theory of Games and Economic Behavior.*

To a considerable extent this review is of an expository nature. This seems justified by the importance of the book, its use of new and unfamiliar concepts and its very length which some may find a serious obstacle.

The existence of the gap which the book attempts to fill has been known to the economic theorists at least since Cournot's work on duopoly, although even now many do not seem to realize its seriousness. There is no adequate solution of the problem of defining "rational economic behavior" on the part of an individual when the very rationality of his actions depends on the probable behavior of other individuals: in the case of oligopoly, other sellers. Cournot and many after him have attempted to sidetrack the difficulty by assuming that every individual has a definite idea as to what others will do under given conditions. Depending on the nature of this expected behavior of other individuals, we have the special, well-known solutions of Bertrand and Cournot, as well as the more general Bowley concept of the "conjectural variation."[2] Thus, the individual's

[1] The tables and figures used in this article were drawn by Mrs. D. Friedlander of the University of Chicago.

[2] More recent investigations have led to the idea of a kinked demand curve. This, however, is a special—though very interesting—case of the conjectural variation.

"rational behavior" is determinate *if* the pattern of behavior of "others" can be assumed *a priori* known. But the behavior of "others" cannot be known *a priori* if the "others," too, are to behave rationally! Thus a logical *impasse* is reached.

The way, or at least *a* way,[3] out of this difficulty had been pointed out by one of the authors[4] over a decade ago. It lies in the rejection of a narrowly interpreted maximization principle as synonymous with rational behavior. Not that maximization (of utility[5] or profits) would not be desirable if it were feasible, but there can be no true maximization when only one of the several factors which decide the outcome (of, say, oligopolistic competition) is controlled by the given individual.

Consider, for instance, a duopolistic situation[6] where each one of the duopolists A and B is *trying* to maximize his profits. A's profits will depend not only on his behavior ("strategy") but on B's strategy as well. Thus, *if* A could control (directly or indirectly) the strategy to be adopted by B, he would select a strategy for himself and one for B so as to maximize his own profits. But he cannot select B's strategy. Therefore, he can in no way make sure that by a proper choice of his own strategy his profits will actually be unconditionally maximized.

It might seem that in such a situation there is no possibility of defining rational behavior on the part of the two duopolists. But it is here that the novel solution proposed by the authors comes in. An example will illustrate this.

Suppose each of the duopolists has three possible strategies at his disposal.[7] Denote the strategies open to duopolist A by A_1, A_2, and A_3, and those open to duopolist B by B_1, B_2, and B_3. The profit made by A, to be denoted by a, obviously is determined by the choices of strategy made by the two duopolists. This dependence will be indicated by subscripts attached to a, with the first subscript referring to A's strategy and the second subscript to that of B; thus, *e.g.*, a_{13} is the profit which will be made by A if he chooses strategy A_1 while B chooses the strategy B_3. Similarly, b_{13} would denote the profits by B under the same circumstances.

[3] *Cf.* reference to von Stackelberg in footnote 16 and some of the work quoted by von Stackelberg, *op. cit.*

[4] J. von Neumann, "Zur Theorie der Gesellschaftsspiele," *Math. Annalen* (1928).

[5] A side-issue of considerable interest discussed in the *Theory of Games* is that of measurability of the utility function. The authors need measurability in order to be able to set up tables of the type to be presented later in the case where utility rather than profit is being maximized. The proof of measurability is not given; however, an article giving the proof is promised for the near future and it seems advisable to postpone comment until the proof appears. But it should be emphasized that the validity of the core of the *Theory of Games* is by no means dependent on measurability or transferability of the utilities and those who feel strongly on the subject would perhaps do best to substitute "profits" for "utility" in most of the book in order to avoid judging the achievements of the *Theory of Games* from the point of view of an unessential assumption.

[6] It is assumed that the buyer's behavior may be regarded as known.

[7] Actually the number of strategies could be very high, perhaps infinite.

TABLE 1A

A's Profits

B's choice of strategies / A's choice of strategies	B_1	B_2	B_3
A_1	a_{11}	a_{12}	a_{13}
A_2	a_{21}	a_{22}	a_{23}
A_3	a_{31}	a_{32}	a_{33}

TABLE 1B

B's Profits

B's choice of strategies / A's choice of strategies	B_1	B_2	B_3
A_1	b_{11}	b_{12}	b_{13}
A_2	b_{21}	b_{22}	b_{23}
A_3	b_{31}	b_{32}	b_{33}

The possible outcomes of the "duopolistic competition" may be represented in the following two tables:

Table 1A shows the profits A will make depending on his own and B's choice of strategies. The first row corresponds to the choice of A_1, etc.; columns correspond to B's strategies. Table 1B gives analogous information regarding B's profits.

In order to show how A and B will make decisions concerning strategies we shall avail ourselves of a numerical example given in Tables 2A and 2B.

TABLE 2A

A's Profits

B's choice of strategies / A's choice of strategies	B_1	B_2	B_3
A_1	2	8	1
A_2	4	3	9
A_3	5	6	7

TABLE 2B

B's Profits

B's choice of strategies / A's choice of strategies	B_1	B_2	B_3
A_1	11	2	20
A_2	9	15	3
A_3	8	7	6

Now let us watch A's thinking processes as he considers his choice of strategy. First of all, he will notice that by choosing strategy A_3 he will be sure that his profits cannot go down below 5, while either of the remaining alternatives would expose him to the danger of going down to 3 or even to 1. But there is another reason for his choosing A_3. Suppose there is a danger of a "leak": B might learn what A's decision is before he makes his own. Had A chosen, say, A_1, B—if he knew about this—would obviously choose B_3 so as to maximize his own profits; this would leave

A with a profit of only 1. Had A chosen A_2, B would respond by selecting B_2, which again would leave A with a profit below 5 which he could be sure of getting if he chose A_3.

One might perhaps argue whether A's choice of A_3 under such circumstances is the only way of defining rational behavior, but it certainly is *a* way of accomplishing this and, as will be seen later, a very fruitful one. The reader will verify without difficulty that similar reasoning on B's part will make him choose B_1 as the optimal strategy. Thus, the outcome of

TABLE 3A

A's Profits

B's choice of strategies / A's choice of strategies	B_1	B_2	B_3
A_1	2	8	1
A_2	4	3	9
A_3	5	6	7

TABLE 3B

B's Profits

B's choice of strategies / A's choice of strategies	B_1	B_2	B_3
A_1	8	2	9
A_2	6	7	1
A_3	5	4	3

the duopolistic competition is determinate and can be described as follows: A will choose A_3, B will choose B_1, A's profit will be 5, B's 8.

An interesting property of this solution is that neither duopolist would be inclined to alter his decision, even if he were able to do so, after he found out what the other man's strategy was.

To see this, suppose B has found out that A's decision was in favor of strategy A_3. Looking at the third row of Table 2B, he will immediately see that in no case could he do better than by choosing B_1, which gives him the highest profit consistent with A's choice of A_3. The solution arrived at is of a very stable nature, independent of finding out the other man's strategy.

But the above example is artificial in several important respects. For one thing, it ignores the possibility of a "collusion" or, to use a more neutral term, coalition between A and B. In our solution, yielding the strategy combination (A_3, B_1), the joint profits of the two duopolists amount to 13; they could do better than that by acting together. By agreeing to choose the strategies A_1 and B_3 respectively, they would bring their joint profits up to 21; this sum could then be so divided that both would be better off than under the previous solution.

A major achievement of the *Theory of Games* is the analysis of the conditions and nature of coalition formation. How that is done will be shown below. But, for the moment, let us eliminate the problem of coalitions by considering a case which is somewhat special but nevertheless of great theoretical interest: the case of *constant sum* profits. An example of such a case is given in Tables 3A and 3B.

Table 3A is identical with Table 2A. But figures in Table 3B have been selected in such a manner that the joint profits of the two duopolists always amount to the same (10), no matter what strategies have been chosen. In such a case, A's gain is B's loss and *vice versa.* Hence, it is intuitively obvious (although the authors take great pains to show it rigorously) that no coalition will be formed.

The solution can again be obtained by reasoning used in the previous case and it will again turn out to be (A_3, B_1) with the respective profits 5 and 5 adding up to 10. What was said above about stability of solution and absence of advantage in finding the opponent [8] out still applies.

There is, however, an element of artificiality in the example chosen that is responsible for the determinateness of the solution. To see this it

TABLE 4

A's Profits

B's choice of strategies / A's choice of strategies	B_1	B_2	B_3
A_1	2	8	1
A_2	4	3	9
A_3	6	5	7

will suffice to interchange 5 and 6 in Table 3A. The changed situation is portrayed in Table 4 which gives A's profits for different choices of strategies.[9]

There is no solution now which would possess the kind of stability found in the earlier example. For suppose A again chooses A_3; then if B should find that out, he would obviously "play" B_2 which gives him the highest possible profit consistent with A_3. But then A_3 would no longer be A's optimum strategy: he could do much better by choosing A_1; but if he does so, B's optimum strategy is B_3, not B_2, etc. There is no solution

[8] In this case the interests of the two duopolists are diametrically opposed and the term "opponents" is fully justified; in the previous example it would not have been.

[9] The table for B's profits is omitted because of the constant sum assumption. Clearly, in the constant sum case, B may be regarded as minimizing A's profits since this implies maximization of his own.

which would not give at least one of the opponents an incentive to change his decision if he found the other man out! There is no stability.[10]

What is it in the construction of the table that insured determinateness in the case of Table 3 and made it impossible in Table 4? The answer is that Table 3 has a *saddle point* ("minimax") while Table 4 does not.

The saddle point has the following two properties: it is the highest of all the row minima and at the same time it is lowest of the column maxima. Thus, in Table 3a the row minima are respectively 1, 3, and 5, the last one being highest among them (*Maximum Minimorum*); on the other hand, the column maxima are respectively 5, 8, and 9 with 5 as the lowest (*Minimum Maximorum*). Hence the combination (A_3, B_1) yields both the highest row minimum and the lowest column maximum, and, therefore, constitutes a saddle point. It is easy to see that Table 4 does *not* possess a saddle point. Here 5 is still the *Maximum Minimorum*, but the *Minimum Maximorum* is given by 6; the two do not coincide, and it is the absence of the saddle point that makes for indeterminateness in Table 4.

Why is the existence of a unique saddle point necessary (as well as sufficient) to insure the determinateness of the solution? The answer is inherent in the reasoning used in connection with the earlier examples: if A chooses his strategy so as to be protected in case of any leakage of information concerning his decision, he will choose the strategy whose row in the table has the highest minimum value, *i.e.*, the row corresponding to the *Maximum Minimorum*—A_3 in case of Table 4—for then he is sure he will not get less than 5, even if B should learn of this decision. B, following the same principle, will choose the column (*i.e.*, strategy) corresponding to the *Minimum Maximorum*—B_1 in Table 4—thus making sure he will get at least 4, even if the information does leak out.

In this fashion both duopolists are sure of a certain minimum of profit —5 and 4, respectively. But this adds up to only 9. The residual—1— is still to be allocated and this allocation depends on outguessing the opponent. It is this residual that provides an explanation, as well as a measure, of the extent of indeterminacy. Its presence will not surprise economists familiar with this type of phenomenon from the theory of bilateral monopoly. But there are cases when this residual does equal zero, that is, when the *Minimum Maximorum* equals the *Maximum Minimorum*, which (by definition) implies the existence of the saddle point and complete determinacy.

At this stage the authors of the *Theory of Games* had to make a choice.

[10] There is, however, a certain amount of determinateness, at least in the negative sense, since certain strategy combinations are excluded: *e.g.* (A_2, B_1); A would never choose A_2 if he knew B had chosen B_1, and *vice versa*.

They could have accepted the fact that saddle points do not always exist so that a certain amount of indeterminacy would, in general, be present. They preferred, however, to get rid of the indeterminacy by a highly ingenious modification of the process which leads to the choice of appropriate strategy.

So far our picture of the duopolist making a decision on strategy was that of a man reasoning out which of the several possible courses of action is most favorable ("*pure strategy*"). We now change this picture and put in his hands a set of dice which he will throw to determine the strategy to be chosen. Thus, an element of chance is introduced into decision making ("mixed strategy").[11] But not everything is left to chance. The duopolist A must in advance formulate a rule as to what results of the

TABLE 5

A's Profits

B's choice of strategies / A's choice of strategies	B_1	B_2	ROW MINIMA
A_1	5	3	3 (MAXIMUM MINIMORUM)
A_2	1	5	1
COLUMN MAXIMA	5	5	

MINIMUM MAXIMORUM

throw—assume that just one die is thrown—would make him choose a given strategy. In order to illustrate this we shall use a table that is somewhat simpler, even if less interesting than those used previously. In this new table (Table 5)[12] each duopolist has only two strategies at his disposal.

[11] The authors' justification for introducing "mixed strategies" is that leaving one's decision to chance is an effective way of preventing "leakage" of information since the individual making the decision does not himself know which strategy he will choose.

[12] In Table 5 there is no saddle point.

An example of a rule A might adopt would be:

> If the result of the throw is 1 or 2, choose A_1;
> if the result of the throw is 3, 4, 5, or 6, choose A_2.

If this rule were followed, the probability that A will chose A_1 is ⅓, that of his choosing A_2 is ⅔. If a different rule had been decided upon (say, one of choosing A_1 whenever the result of the throw is 1, 2, or 3), the probability of choosing A_1 would have been ½. Let us call the fraction giving the probability of choosing A_1 A's *chance coefficient*; in the two examples, A's chance coefficients were ⅓ and ½ respectively.[13]

As a special case the value of the chance coefficient might be zero (meaning, that is, definitely choosing strategy A_2) or one (meaning that A is definitely choosing strategy A_1); thus in a sense "pure strategies" may be regarded as a special case of mixed strategies. However, this last statement is subject to rather important qualifications which are of a complex nature and will not be given here.

TABLE 6

Mathematical Expectations of A's Profits

B's chance coefficients / A's chance coefficients	0	$\frac{1}{3}$	$\frac{2}{3}$	1	ROW MINIMA
0	5	$3\frac{2}{3}$	$2\frac{1}{3}$	1	1
$\frac{1}{3}$	$4\frac{1}{3}$	$3\frac{2}{3}$	3	$2\frac{1}{3}$	$2\frac{1}{3}$
$\frac{2}{3}$	$3\frac{2}{3}$	$3\frac{2}{3}$	$3\frac{2}{3}$	$3\frac{2}{3}$	$3\frac{2}{3}$ } MAXIMUM MINIMORUM
1	3	$3\frac{2}{3}$	$4\frac{1}{3}$	5	3
COLUMN MAXIMA	5	$3\frac{2}{3}$ (MINIMUM MAXIMORUM)	$4\frac{1}{3}$	5	

Now instead of choosing one of the available strategies the duopolist A must choose the optimal (in a sense not yet defined) chance coefficient.

[13] Since the probability of choosing A_2 is always equal to one minus that of choosing A_1, specification of the probability of choosing A_1 is sufficient to describe a given rule. However, when the number of available strategies exceeds two, there are several such chance coefficients to be specified.

How is the choice of the chance coefficient made? The answer lies in constructing a table which differs in two important respects from those used earlier. Table 6 provides an example. Each row in the table now corresponds to a possible value of A's chance coefficient; similarly, columns correspond to possible values of B's chance coefficient. Since the chance coefficient may assume any value between zero and one (including the latter two values), the table is to be regarded merely as a "sample." This is indicated by spaces between rows and between columns.

The numbers entered in the table are the average values (mathematical expectations) corresponding to the choice of chance coefficients indicated by the row and column.[14] (One should mention that Table 6 is only an expository device: the actual procedures used in the book are algebraic and much simpler computationally.)

If we now assume with the authors that each duopolist is trying to maximize the mathematical expectation of his profits (Table 6) rather than the profits themselves (Table 5), it might seem that the original source of difficulty remains if a saddle point does not happen to exist. But the mixed strategies were not introduced in vain! It is shown (the theorem was originally proved by von Neumann in 1928) that in the table of mathematical expectations (like Table 6) a saddle point *must* exist; the problem is always determinate.[15]

The reader who may have viewed the introduction of dice into the decision-making process with a certain amount of suspicion will probably agree that this is a rather spectacular result. Contrary to the initial impression, it *is* possible to render the problem determinate. But there is a

TABLE 7

COMPUTATION OF THE MATHEMATICAL EXPECTATION FOR THE 2ND ROW, 3RD COLUMN IN TABLE 6

A's choice of strategies \ B's choice of strategies	A's chance coefficients \ B's chance coefficients	B_1	B_2
		$\frac{2}{3}$	$\frac{1}{3}$
A_1	$\frac{1}{3}$	5	3
A_2	$\frac{2}{3}$	1	5

$$\tfrac{1}{3} \times \tfrac{2}{3} \times 5 + \tfrac{1}{3} \times \tfrac{1}{3} \times 3 + \tfrac{2}{3} \times \tfrac{2}{3} \times 1 + \tfrac{2}{3} \times \tfrac{1}{3} \times 5 = \tfrac{27}{9} = 3$$

[14] To see this we shall show how, *e.g.*, we have obtained the value in the second row and third column of Table 5 (*viz.*, 3).

We construct an auxiliary table valid only for this particular combination of chance coefficients (A's ⅓, B's ⅔).

This table differs from Table 5 only by the omission of row maxima and column minima and by the insertion of the probabilities of choosing the available strategies corresponding to the second row third column of Table 6. The computation of the mathematical expectation is indicated in Table 6.

[15] In Table 6 the saddle point is in the third row second column; it is to be stressed that Table 5 has no saddle point.

price to be paid: acceptance of mixed strategies, assumption that only the mathematical expectation of profit (not its variance, for instance) matters, seem to be necessary. Many an economist will consider the price too high. Moreover, one might question the need for introducing determinateness into a problem of this nature. Perhaps we should consider as the "solution" the interval of indeterminacy given by the two critical points: the *Minimum Maximorum* and *Maximum Minimorum*.

As indicated earlier in this review, one should not ignore, in general, the possibility of a collusion. This is especially evident when more complex economic situations are considered.

We might, for instance, have a situation where there are two sellers facing two buyers. Here a "coalition" of buyers, as well as one of sellers, may be formed. But it is also conceivable that a buyer would bribe a seller into some sort of coöperation against the other two participants. Several other combinations of this type can easily be found.

When only *two* persons enter the picture, as in the case of duopoly (where the rôle of buyers was ignored), it was seen that a coalition would not be formed if the sum of the two persons' profits remained constant. But when the number of participants is *three* or more, subcoalitions can profitably be formed even if the sum of all participants' profits is constant; in the above four-person example it might pay the sellers to combine against the buyers even if (or, perhaps, especially if) the profits of all four always add to the same amount.

Hence, the formation of coalitions may be adequately treated without abandoning the highly convenient constant-sum assumption. In fact, when the sum is known to be non-constant, it is possible to introduce (conceptually) an additional fictitious participant who, by definition, loses what all the real participants gain and *vice versa*. In this fashion a non-constant sum situation involving, say, three persons may be considered as a special case of a constant-sum four-person situation. This is an additional justification for confining most of the discussion (both in the book and in the review) to the constant-sum case despite the fact that economic problems are as a rule of the non-constant sum variety.

We shall now proceed to study the simplest constant-sum case which admits coalition formation, that involving three participants. The technique of analysis presented earlier in the two-person case is no longer adequate. The number of possibilities increases rapidly. Each of the participants may be acting independently; or else, one of the three possible two-person coalitions (A and B *vs.* C, A and C *vs.* B, B and C *vs.* A) may be formed. Were it not for the constant-sum restriction, there would be the additional possibility of the coalition comprising all three participants.

Here again we realize the novel character of the authors' approach to the problem. In most[16] of traditional economic theory the formation—or absence—of specific coalitions is *postulated*. Thus, for instance, we discuss the economics of a cartel without rigorously investigating the necessary and sufficient conditions for its formation. Moreover, we tend to exclude *a priori* such phenomena as collusion between buyers and sellers even if these phenomena are known to occur in practice. The *Theory of Games*, though seemingly more abstract than economic theory known to us, approaches reality much more closely on points of this nature. A complete solution to the problems of economic theory requires an answer to the question of coalition formation, bribery, collusion, etc. This answer is now provided, even though it is of a somewhat formal nature in the more complex cases; and even though it does not always give sufficient insight into the actual workings of the market.

Let us now return to the case of three participants. Suppose two of them are sellers, one a buyer. Traditional theory would tell us the quantity sold by each seller and the price. But we know that in the process of bargaining one of the sellers might bribe the other one into staying out of the competition. Hence the seller who refrained from market operations would make a profit; on the other hand, the nominal profit made by the man who did make the sale would exceed (by the amount of bribe) the actual gain made.

It is convenient, therefore, to introduce the concept of *gain*: the bribed man's gain is the amount of the bribe, the seller's gain is the profit made on a sale minus the bribe, etc. A given distribution of gains among the participants is called an *imputation*. The imputation is not a number: it is a set of numbers. For instance, if the gains of the participants in a given situation were g_A, g_B, g_C, it is the set of these three g's that is called the imputation. The imputation summarizes the outcome of the economic process. In any given situation there are a great many possible imputations. Therefore, one of the chief objectives of economic theory is that of finding those among all the possible imputations which will actually be observed under rational behavior.

In a situation such as that described (three participants, constant-sum) each man will start by asking himself how much he could get acting independently, even if the worst should happen and the other two formed a coalition against him. He can determine this by treating the situation as a two-person case (the opposing coalition regarded as one person) and finding the relevant *Maximum Minimorum*, or the saddle point, if that

[16] In his *Grundlagen einer reinen Kostentheorie* (Vienna, 1932) H. von Stackelberg does point out (p. 89) that "the competitors [duopolists] must somehow unite; they must supplement the economic mechanics, which in this case is inadequate, by economic politics." But no rigorous theory is developed for such situations (although an outline of possible developments is given). This is where the *Theory of Games* has made real progress.

point does exist; the saddle point would, of course, exist if "mixed strategies" are used. Next, the participant will consider the possibility of forming a coalition with one of the other two men. Now comes the crucial question: under what conditions might such a coalition be formed?

Before discussing this in detail, let us summarize, in Table 8, all the relevant information.

It will be noted that under imputation #1, B and C are each better off than if they had been acting individually: they get respectively 8.3 and 10.2 instead of 7 and 10. Hence, there is an incentive for B and C to form

TABLE 8

I.	If A acts alone, he can get	5
	If B acts alone, he can get	7
	If C acts alone, he can get	10.
II.	If A and B form a coalition, they can get	15
	If A and C form a coalition, they can get	18
	If B and C form a coalition, they can get	20.
III.	If A, B, and C act together, they can get	25.

Among the many possible imputations, let us now consider the three given in Table 9.

TABLE 9

	A	B	C
#1	6.5	8.3	10.2
#2	5.0	9.5	10.5
#3	4.0	10.0	11.0

a coalition since without such a coalition imputation #1 would not be possible. But once the coalition is formed, they can do better than under #1; *viz.*, under #2, where each gets more (9.5 and 10.5 instead of 8.3 and 10.2, respectively). In such a case we say that imputation #2 *dominates* imputation #1. It might seem that #3, in turn, dominates #2 since it promises still more to both B and C. But it promises too much: the sum of B's and C's gains under #3 is 21, which is more than their coalition could get (*cf.* Table 8)! Thus #3 is ruled out as unrealistic and cannot be said to dominate any other imputation.

Domination is an exceptionally interesting type of relation. For one thing, it is not transitive: we may have an imputation i_1 dominating the imputation i_2 and i_2 dominating i_3, without thereby implying that i_1 dominates i_3; in fact, i_1 might be dominated by i_3.[17] Moreover, it is easy to construct examples of, say, two imputations, neither of which dominates the other one.[18]

To get a geometric picture of this somewhat unusual situation one may turn to Figure 1, where points on the circle represent different possible imputations. (The reader must be cautioned that this is merely a geometrical analogy, though a helpful one.) Let us now say that point #1 dominates point #2 if #2 is less than 90° (clockwise) from #1. It is easy to see in Figure 1 that #1 dominates #2 and #2 dominates #3, but in spite of that, #1 does not dominate #3.

This geometrical picture will help define the very fundamental concept of a *solution.*

Consider the points (imputations) #1, 3, 5 and 7 in Figure 1. None of them dominates any other since any two are either *exactly* or more than 90° apart. But any other point on the circle is dominated by at least (in this case: exactly) one of them: all points between #1 and #3 are dominated by #1, etc. There is no point on the circle which is not dominated

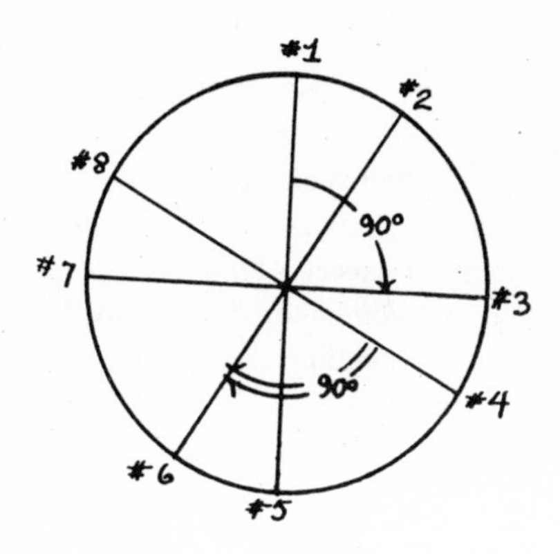

FIGURE 1

by one of the above four points. Now we *define* a solution as a set of points (imputations) with two properties: (1) no element of the set dominates any other element of the set, and (2) any point outside the set must be dominated by at least one element within the set.

We have seen that the points #1, 3, 5, 7 do have both of these proper-

[17] *I.e.*, domination may be a *cyclic* relation. For instance, consider the following three imputations in the above problem: #1 and #2 as in Table 9, and #4, where

	A	B	C
#4	6.0	7.0	12.0.

Here #2 (as shown before) dominates #1 (for coalition B, C), #4 dominates #2 (for coalition A, C), but at the same time #1 dominates #4 (for the coalition A, B): the cycle is completed.

[18] For instance, #2 and #3 in Table 9.

ties; hence, the four points together form a solution. It is important to see that none of the individual points by itself can be regarded as a solution. In fact, if we tried to leave out any one of the four points of the set, the remaining three would no longer form a solution; for instance, if #1 were left out, the points between #1 and #3 are not dominated by any of the points #3, 5, 7. This violates the second property required of a solution and the three points by themselves are not a solution. On the other hand, if a fifth point were added to #1, 3, 5, 7, the resulting five element set would not form a solution either; suppose #2 is the fifth point chosen; we note that #2 is dominated by #1 and it also dominates #3. Thus, the first property of a solution is absent.

Contrary to what would be one's intuitive guess, an element of the solution may be dominated by points outside the solution: #1 is dominated by #8, etc.

There can easily be more than one solution. The reader should have no trouble verifying the fact that #2, 4, 6, 8 also form a solution, and it is clear that infinitely many other solutions exist.

Does there always exist at least one solution? So far this question remains unanswered. Among the cases examined by the authors none has been found without at least one solution. But it has not yet been proved that there must always be a solution. To see the theoretical possibility of a case without a solution we shall redefine slightly our concept of domination (*cf.* Figure 2): #1 dominates #2 if the angle between them (measured clockwise) does not exceed 180°.

Hence, in Figure 2 point #1 dominates #3, but not #4, etc. It can now be shown that in this case *no* solution exists. For suppose there is one; then we may, without loss of generality, choose #1 as one of its points. Clearly, #1 by itself does not constitute a solution, for there are points

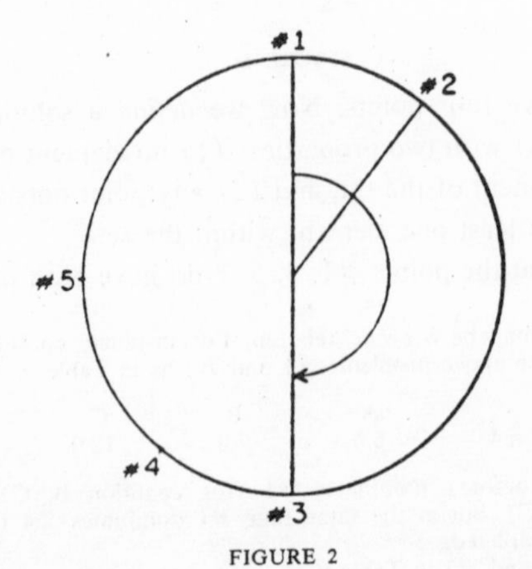

FIGURE 2

on the circle (*e.g.*, #4) not dominated by #1; thus the solution must have at least two points. But any other point on the circle either is dominated by #1 (*e.g.*, #2), or it dominates #1 (*e.g.*, #4), or both (#3), which contradicts the first requirement for the elements of a solution. Hence there is no solution consisting of two points either. *A fortiori*, there are no solutions containing more than two points. Hence we have been able to construct an example without a solution. But whether this type of situation could arise in economics (or in games, for that matter) is still an open question.

Now for the economic interpretation of the concept of solution. Within the solution there is no reason for switching from one imputation to another since they do not dominate each other. Moreover, there is never a good reason for going outside a given solution: any imputation outside the solution can be "discredited" by an imputation within the solution which dominates the one outside. But, as we have seen, the reverse is also usually true: imputations within the solution may be dominated by those outside. If we are to assume that the latter consideration is ignored, the given solution acquires an institutional, if not accidental, character. According to the authors, a solution may be equivalent to what one would call the "standards of behavior" which are accepted by a given community.

The multiplicity of solutions can then be considered as corresponding to alternative institutional setups; for a given institutional framework only one solution would be relevant. But even then a large number of possibilities remains since, in general, a solution contains more than one imputation. More indeterminacy yet would be present if we had refrained from introducing mixed strategies.

It would be surprising, therefore, if in their applications von Neumann and Morgenstern should get no more than the classical results without discovering imputations hitherto neglected or ignored. And there are some rather interesting "unorthodox" results pointed out, especially in the last chapter of the book.

In one case, at least, the authors' claim to generality exceeding that of economic theory is not altogether justified in view of the more recent literature. That is the case of what essentially corresponds to bilateral monopoly (p. 564, proposition 61:C). The authors obtain (by using their newly developed methods) a certain interval of indeterminacy for the price; this interval is wider than that indicated by Böhm-Bawerk, because (as the authors themselves point out) of the dropping of Böhm-Bawerk's assumption of a unique price. But this assumption has been abandoned, to give only one example, in the theories of consumer's surplus, with analogous extension of the price interval.

It will stand repeating, however, that the *Theory of Games* does offer a greater generality of approach than could be attained otherwise. The existence of "discriminatory" solutions, discovered by purely analytical methods, is an instance of this. Also, the possibility of accounting for various types of deals and collusions mentioned earlier in connection with the three-person and four-person cases go far beyond results usually obtained by customarily used methods and techniques of economic theory.

The potentialities of von Neumann's and Morgenstern's new approach seem tremendous and may, one hopes, lead to revamping, and enriching in realism, a good deal of economic theory. But to a large extent they are only potentialities: results are still largely a matter of future developments.

The difficulties encountered in handling, even by the more powerful mathematical methods, the situations involving more than three persons are quite formidable. Even the problems of monopoly and monopsony are beyond reach at the present stage of investigation. The same is true of perfect competition, though it may turn out that the latter is not a "legitimate" solution since it excludes the formation of coalitions which may dominate the competitive imputations. A good deal of light has been thrown on the problem of oligopoly, but there again the results are far from the degree of concreteness desired by the economic theorist.

The reviewer therefore regards as somewhat regrettable some of the statements made in the initial chapter of the book attacking (rather indiscriminately) the analytical techniques at present used by the economic theorists. True enough, the deficiencies of economic theory pointed out in the *Theory of Games* are very real; nothing would be more welcome than a model giving the general properties of a system with, say, m sellers and n buyers, so that monopoly, duopoly, or perfect competition could simply be treated as special cases of the general analysis. Unfortunately, however, such a model is not yet in sight. In its absence less satisfactory, but still highly useful, models have been and no doubt will continue to be used by economic theorists. One can hardly afford to ignore the social need for the results of economic theory even if the best is rather crude. The fact that the theory of economic fluctuations has been studied as much as it has is not a proof of "how much the attendant difficulties have been underestimated" (p. 5). Rather it shows that economics cannot afford the luxury of developing in the theoretically most "logical" manner when the need for the results is as strong as it happens to be in the case of the ups and downs of the employment level!

Nor is it quite certain, though of course conceivable, that, when a rigorous theory developed along the lines suggested by von Neumann and Morgenstern is available, the results obtained in the important problems will be sufficiently remote from those obtained with the help of the current (admittedly imperfect) tools to justify some of the harsher accusations to be found in the opening chapter of the book. It must not be forgotten, for instance, that, while theoretical derivation of coalitions to be formed is of great value, we do have empirical knowledge which can be used as a substitute (again imperfect) for theory. For example, cartel formation may be so clearly "in the cards" in a given situation that the economic theorist will simply include it as one of his assumptions while von Neumann and Morgenstern would (at least in principle) be able to *prove* the formation of the cartel without making it an additional (and logically unnecessary) assumption.

The authors criticize applications of the mathematical methods to economics in a way which might almost, in spite of protests to the contrary, mislead some readers into thinking that von Neumann and Morgenstern are not aware of the amount of recent progress in many fields of economic theory due largely to the use of mathematical tools. They also seem to ignore the fact that economics developed in literary form is, implicitly, based on the mathematical techniques which the authors criticize. (Thus it is not the methods of mathematical economics they are really questioning, but rather those elements of economic theory which literary and mathematical economics have in common.) While it is true that even mathematical treatment is not always sufficiently rigorous, it is as a rule more so than the corresponding literary form, even though the latter is not infrequently more realistic in important respects.

There is little doubt in the reviewer's mind that nothing could have been further from the authors' intentions than to give aid and comfort to the opponents of rigorous thinking in economics or to increase their complacency. Yet such may be the effect of some of the vague criticisms contained in the first chapter; they hardly seem worthy of the constructive achievements of the rest of the book.

Economists will probably be surprised to find so few references to more recent economic writings. One might almost form the impression that economics is synonymous with Böhm-Bawerk plus Pareto. Neither the nineteenth century pioneers (such as Cournot) nor the writers of the last few decades (Chamberlin, Joan Robinson, Frisch, Stackelberg) are even alluded to. But, perhaps, the authors are entitled to claim exemption from the task of relating their work to that of their predecessors by virtue of the tremendous amount of constructive effort they put into their opus. One cannot but admire the audacity of vision, the perseverance in details, and the depth of thought displayed on almost every page of the book.

The exposition is remarkably lucid and fascinating, no matter how involved the argument happens to be. The authors made an effort to avoid the assumption that the reader is familiar with any but the more elementary parts of mathematics; more refined tools are forged "on the spot" whenever needed.

One should also mention, though this transcends the scope of the review, that in the realm of strategic games proper (chess, poker) the results obtained are more specific than some of the economic applications. Those interested in the nature of determinacy of chess, in the theory of "bluffing" in poker, or in the proper strategy for Sherlock Holmes in his famous encounter with Professor Moriarty, will enjoy reading the sections of the book which have no direct bearing on economics. The reader's views on optimum military or diplomatic strategies are also likely to be affected.

Thus, the reading of the book is a treat as well as a stage in one's intellectual development. The great majority of economists should be able to go through the book even if the going is slow at times; it is well worth the effort. The appearance of a book of the caliber of the *Theory of Games* is indeed a rare event.

Heads I win; tails you lose. —ENGLISH SAYING, *17th century.*

It is a silly game where nobody wins. —THOMAS FULLER (*Gnomologia*)

It has long been an axiom of mine that the little things are infinitely the most important.—SIR ARTHUR CONAN DOYLE (*The Adventures of Sherlock Holmes. A Case of Identity*)

9 Theory of Games

By S. VAJDA

FROM a variety of considerations which, not very long ago, would have seemed in no way to be connected, a new branch of science has emerged: the Theory of Games. The origins of this theory go back to 1928, when John von Neumann read a paper to the Mathematical Society of Göttingen, introducing new and unorthodox concepts. During the last war Professors von Neumann and Morgenstern published their monumental and painstaking treatise *Theory of Games and Economic Behavior*, and scientists concerned with operational research on both sides of the Atlantic investigated and developed various theoretical aspects of tactics and strategy. The theory does not help one to become proficient in any specific games. It does not tabulate chess openings or advise on poker biddings. It is concerned, not with any particular game, but rather with general aspects applicable to all games, and with processes which obtain a special significance when a long succession of plays is being considered. Such conclusions as that you cannot win at Monte Carlo in the long run may emerge as a trivial consequence of the theory, but statistics is not of its essence. Being a mathematical theory, it draws on the results of several branches of mathematics, such as algebra and measure theory, but entirely new concepts had to be created as well, and it is in these that the particular attraction of the subject lies.

The essential feature of any game is the fact that one has to do with one or more opponents and that, therefore, only some of the relevant variables are under the control of any single player. It is clear that a theory which takes account of this peculiarity can be applied to the analysis of warfare, to economic problems, and even to decisions which must be taken where no specific opponent appears to exist, but where all variables outside one's one control are dependent on 'Chance' or on 'Laws of Nature.' It is the aim of this article to introduce the reader to some of the new concepts and to show how they emerge naturally from a consideration of typical features of games.

MATCHING PENNIES

To begin with, let us consider one of the simplest games imaginable, that of 'Matching Pennies.' Two players put down a penny each, either head or tail up, unknown to the opponent. They then uncover their coins and A receives his own and also B's penny, if both coins show the same side. Otherwise B collects both pennies. Clearly it is A's aim to show the same side of the penny as B does, and all his decisions connected with the game depend on this aim of his and on his ignorance of B's procedure.

It is convenient to introduce here a few definitions. There are two persons to play Matching Pennies and one person wins what the other loses. In other words, the gains and losses of the two players are balanced. Such a game is called a zero-sum two-person game. By a 'game' we understand in this context the aggregate of rules which set out the possible behaviour of the players and their gains and losses. Thus 'game' is an abstract concept. One particular instance of it, as it is actually performed, is called a 'play,' consisting of a set of 'moves.' Matching Pennies exhibits an extreme simplicity because it finishes after only one move of each player.

In more complicated games one imagines, as a rule, that the players decide on their own moves as the play develops. This is, however, not essential. One could equally well imagine that each player decides before the play begins what 'strategy' he wants to apply. Such a strategy determines in advance what moves the player would make under all conceivable circumstances. One could imagine that each player chooses his strategy unknown to the other player and informs an umpire of his choice. The latter would then consult a list containing all possible pairs of strategies of the two players and would read out the result. It would not be necessary to play any more. Fortunately the number of possible strategies in such sophisticated games as, for instance, chess is so enormously large that there is no danger that their complete enumeration and evaluation will ever kill the interest in the game.

Reverting to Matching Pennies, we can construct a table showing the payments which A receives and how they depend on the possible outcomes of a play. Such a table is referred to as the 'pay-off table' (for A) and looks as follows:

TABLE I: PAY-OFF TABLE FOR MATCHING PENNIES

		B's coin	
		Head	Tail
A's coin	Head	1	−1
	Tail	−1	1

It is clearly impossible for either player to choose his move or his strategy (the two concepts are equivalent in this particular game) in such a way that he can be sure of winning. On the other hand, any player who knows the opponent's move can win. It would be very bad policy to stick to any particular move in a succession of games, since this would soon be noticed. It is therefore natural to suggest that a player should change his strategies at random, i.e., in a manner that does not give any indication of the strategy to be used in the next play. He can still decide on the over-all proportions in which his strategies should be used in the long run. It is, of course, also possible that the opponent finds out what these proportions are. Each player should therefore choose them in such a way that his position does not become worse when he is found out. It is clear that if *A* chooses his originally given 'pure strategies' in the proportion ½:½, then, whatever *B* does, *A*'s gains will equal his losses in the long run. But this is, in fact, the best *A* can hope to achieve. If he chooses his strategies in any other proportion, say once head to three times tail, then *B*, provided he knew it, would be shrewd enough to choose always head, because then *A* would lose more often in the long run than he would win. Reversing the argument, it is seen that *B* should, in ignorance of *A*'s strategy, once again choose head and tail with equal frequencies, in order to view without concern the danger of being found out. (This principle has been described as expecting the best in the worst possible world. We leave it to the reader's temperament to decide whether this description, if justified, should induce the player to gamble excessively rather than to play safe.) If, then, both players use their pure strategies in the proportion ½:½, a certain stability will be reached in the sense that no player will change his behaviour, even if he finds out what his adversary does, since by doing so he would not improve his prospects. The pair of frequencies, which leads to such a stability, is called a 'solution,' and the decision of a player to use his (pure) strategies in given proportions is itself a strategy and is called 'mixed.' We have just seen that in Matching Pennies no pure, but only a mixed, strategy leads to a solution. However, this is by no means always so, as can be shown by considering a simple modification of the game.

MODIFIED MATCHING PENNIES

We assume now that any player is allowed to 'call off' the play and that he receives, if he is the only one to do this, a halfpenny from his opponent, whereas no payment is made if both players call off. The corresponding pay-off table (again for *A*) is shown as Table II.

An inspection of the table shows that now both players can choose pure strategies which give a solution. To achieve this, they must both call off.

TABLE II: PAY-OFF TABLE FOR MODIFIED MATCHING PENNIES

		B's coin			
		Head	Tail	Call off	Row minima
A's coin	Head	1	−1	−½	−1
	Tail	−1	1	−½	−1
	Call off	½	½	0	0
Column maxima		1	1	0	

In any other case the player who finds out his opponent's intention can win a penny.

The entry 0, corresponding to the pair of strategies which form the solution, is called the value of the game. It is the smallest number in its row and the highest in its column. Therefore its position is called a 'saddle point,' by an obvious analogy. It can easily be proved that if there are more saddle points in a pay-off table, then the numbers in all of them must be equal. A simple way of finding a saddle point, if one exists, is to write down the column maxima and the row minima, as we have done in the margins of the table above. In this example the maximum of the row minima equals the minimum of the column maxima, viz. 0 (the value of the game). In more abbreviated terms, the 'maximin' equals the 'minimax.' The common value indicates a saddle point, and the corresponding solution is called a 'Minimax Solution.'

Now one can easily prove that the maximin can never exceed the minimax, but as we have already seen (see Matching Pennies), the two need not be equal. However, it has been proved that the maximin of all possible strategies, including mixed ones, is always equal to the minimax of all possible strategies of the opponent, again including mixed ones. Once more, the resulting pay-off is called the value of the game. The equality of the maximin and the minimax, extended to refer to mixed strategies, was proved first by John von Neumann in 1928. He called it the Main Theorem. His proof has since been simplified in various ways, but it is still too involved to be included in this merely expository article.

THE THREE BOXES GAME

The game of Matching Pennies led to an obvious best mixed strategy and its modification had a Minimax Solution of pure strategies. We shall now deal with a game whose solution is again given by mixed strategies, but not by such obvious ones. Let three boxes marked 1, 2, and 3 be given, containing these numbers of shillings respectively. The 'banker' removes the bottom of one of the boxes, but this is not discernible from the outside. A 'player' puts into two of the boxes the amounts of shillings marked on them. He then receives all the money in those two boxes. He will, of course, lose the money that he happened to put into a box without bottom. The pay-off table for the player is easily constructed (Table III). For instance, if the bottom of box 3 is missing and if the player puts 1s and 2s into the appropriate boxes, then he wins 3s altogether. However, if he puts his money into boxes 1 and 3, then he wins 1s and loses 3s, so that, on balance, he suffers a loss of 2s. In this way the following table has been constructed:

TABLE III: PAY-OFF TABLE FOR THE THREE BOXES GAME

		Bottom removed from box			Row minima
		1	2	3	
Money put into boxes	1, 2	1	−1	3	−1
	1, 3	2	4	−2	−2
	2, 3	5	1	−1	−1
Column maxima		5	4	3	

This table has no saddle point, since the maximum of the row minima is -1, whereas the minimum of the column maxima is 3. Therefore the solution (which is known to exist by virtue of the Main Theorem) must contain mixed strategies.

In order to find the solution, one could first try to see whether three positive numbers a, b, and c, adding up to unity, can be found such that every row gives the same value if the banker follows a mixed strategy in which the bottoms of the three boxes are removed in those proportions. This is the condition that the player will gain no benefit from discovering the strategy used by the banker. We could then repeat the procedure for every column, and hope in that way to obtain the player's mixed strategy such that the banker gains nothing from its discovery.

In the present case, the conditions for the banker's mixed strategy are (for the rows) that $a + b + c = 1$, $a - b + 3c = 2a + 4b - 2c = 5a + b - c$. If the banker chooses the proportions $a = 5/22$, $b = 8/22$, and $c = 9/22$, which are consistent with these conditions, then the player wins 12/11, whatever he does, whereas if the banker chose any other proportions, the player could win more than this amount. By the same argument, we must try to find three numbers x, y, and z, say, so that $x + y + z = 1$ and (applying these proportions to the columns) $x + 2y + 5z = -x + 4y + z = 3x - 2y - z$. However, this results in $x = 7/11$, $y = 5/11$, and $z = -1/11$. Since z is negative, this is not a possible frequency, and we must find some other method for finding a solution. It would be too long to explain here how this is done systematically. The solution turns out to be (3/5, 2/5, 0) for the player and (0, 1/2, 1/2) for the banker.

When the banker uses the proportions (0, 1/2, 1/2), the gains of the player for his pure strategies are $-1/2 + 3/2 = 1$, $4/2 - 2/2 = 1$, and $1/2 - 1/2 = 0$ respectively. The latter will, therefore, be wise to use mixed strategies, in which only two pure strategies are combined—those in which he puts money into the 1s and 2s, and the 1s and 3s boxes respectively—since by using the 2s, 3s strategy as well he would invariably reduce his expected average gain. The proportions in which he should use those strategies are 3 : 2, as already stated. Doing this, he gains 1 and the banker loses 1. (We may assume that the latter will demand a fee amounting to 1, the value of the game, for participation.) The player cannot improve on his gain of 1, as long as the banker adheres to the proportions (0, 1/2, 1/2) on his part. For any other choice of the banker's the player could gain more. Thus, for instance, if the banker used (0, 1/3, 2/3), the player could obtain a gain of 5/3, by choosing the pure strategy (1s, 2s). Conversely, if the player chose proportions different from those given for him, he could be made to gain less than 1, if he were found out. Thus, for instance, if he chose (4/5, 1/5, 0), then the banker might choose pure strategy 2, and the gain of the player would thereby be reduced, in the long run, to zero.

So far we have always assumed that all proceedings depend only on the decisions of the players. However, this would not cover all possible games. We imagine now a slightly modified Three Boxes Game, where the banker does not decide on his own moves, but where they are chosen for him by a chance mechanism, which makes every pure strategy equally likely. In other words, his mixed strategy is (1/3, 1/3, 1/3), and this is known to the player, who must choose the pure strategy of always putting his money in the 2s and 3s boxes to gain in the proportion 5 : 3 over a series of plays; every other choice would give him less in the long run. This is an example of a game with so-called 'chance moves.' Obviously, these can also occur in combination with deliberate moves, and they may be available to both players. Card games, for instance, contain always both types of moves, because at least some of the cards are dealt at random (or should be).

In the Three Boxes Game both participants make their choice without knowledge of the other's moves, as was also the case in Matching Pennies, where indeed otherwise the game would have been senseless. However, games exist where the moves occur one at a time alternately and where, whenever a move is about to be made, all players know all previous moves, whether their own or those of the opponent (though not, of course, the opponent's strategy, which includes also all his future intentions for any possible contingency). Such games are said to have 'perfect information,' whereas Matching Pennies, for instance, is a game with 'imperfect information.' It has been proved by von Neumann that a game with perfect information has always a saddle point and that, in a probability sense, this is also true of games with chance moves (and perfect information). With regard to games without chance moves, the theorem has, in effect, also been proved.[1] We shall now introduce a 'Modified Three Boxes Game' with perfect information which will, therefore, have a solution consisting of pure strategies for both players.

MODIFIED THREE BOXES GAME

The rules of the Modified Three Boxes Game are as follows: (1) the player puts his money into one of the boxes, (2) the banker, who has watched him, removes the bottom of one of the boxes, (3) the player puts his money into one of the two boxes not previously chosen.

The player has thus two moves. As to the banker, who has only one move, we can take it for granted that his best strategy is to remove the bottom of that box into which the player has put the money on his first move. Thus the move of the banker is known to the player in advance. The latter, however, has a choice of the following six strategies, with the gains as given (provided the banker uses his best strategy, as he will):

First move of the player	1	1	2	2	3	3
Second move of the player	2	3	1	3	1	2
Gain	1	2	−1	1	−2	−1

It emerges, then, that the player should first put 1s into box 1 (knowing that he will lose his money) and then 3s into box 3, obtaining an overall gain of 2s. This is, of course, also clear from common-sense considerations.

The Modified Three Boxes Game, besides being somewhat artificial, is also of rather restricted generality in that the banker has only one move and that, if this is known, his whole strategy is thereby known as well. But it will be clear by now that the same methods are applicable also to more complicated games. It is perhaps worth mentioning that there is no connexion between the division of games according to information and according to chance moves. This can be seen from Table IV:

[1] [By D. W. Davies in an article on "A Theory of Chess and Noughts and Crosses" in *Science News 16* (Penguin Books, Harmondsworth). ED.]

TABLE IV: THE DIVISION OF GAMES ON THE BASIS OF INFORMATION AND OF CHANCE MOVES

	Games with perfect information	Games with imperfect information
Games with chance moves	Backgammon	Poker
Games without chance moves	Chess [2]	Matching Pennies

THE TWO GENERALS

So far, we have considered games with saddle points, or games where the solution consists of a single mixed strategy for both players. We shall now introduce a game which has more than one single solution. The example will also illustrate how the theory of games may be applied to military considerations. We imagine the following simple strategic situation.

Two generals, *A* and *B*, face one another. *A* has three companies at his disposal and *B* has four. It is the aim of *A* to reach a town on one of two possible roads. *B* tries to prevent him from doing it. General *A* can send all three companies on the same road, or he may split up their number, but he may not divide up any single company. General *B* again, may send all his companies on the same road, or on different roads, but he must keep any single company undivided. We assume that *A* has achieved his object if he has on any road a number of companies exceeding that of his adversary. If he reaches the town, he wins (indicated by 1), if he does not, the issue remains undecided (indicated by 0). The pay-off table (for *A*) is shown here as Table V:

TABLE V: PAY-OFF TABLE FOR THE 'TWO GENERALS' GAME

		Number of companies of *B* on the two roads, respectively				
		0 and 4	1 and 3	2 and 2	3 and 1	4 and 0
Number of companies of *A* on the two roads, respectively	0 and 3	0	0	1	1	1
	1 and 2	1	0	0	1	1
	2 and 1	1	1	0	0	1
	3 and 0	1	1	1	0	0

[2] It is assumed that the strategies of chess-players are chosen when it is known who plays White.

The row minima are all 0, the column maxima are all 1. Hence this table has no saddle point.

It is now possible to see, by inspection, that if A uses strategies mixed in any of the following proportions: $S_1 = (\frac{1}{2}, 0, \frac{1}{2}, 0)$, or $S_2 = (\frac{1}{2}, 0, 0, \frac{1}{2})$, or $S_3 = (0, \frac{1}{2}, 0, \frac{1}{2})$, he will gain on the average at least ½ in the long run, and that B can keep him down to this if he uses strategy $(0, \frac{1}{2}, 0, \frac{1}{2}, 0) = T$, say, but cannot reduce A's gain below ½. On the other hand, if B uses strategy T, then A cannot obtain more than ½, whatever he does. This means that any of the following pairs of mixed strategies, (S_1, T), (S_2, T), and (S_3, T), is a solution. The value of the game is the same for all solutions (namely ½), and it can be proved generally that whenever a game has more than one solution, the value is the same for all of them.

General A has the three 'best' strategies S_1, S_2, and S_3 at his disposal, and it is obvious that, if he uses any combination of these mixed strategies, the resulting mixed strategy will also form together with T for General B, a solution in the sense (as before) that neither can lose from the discovery by the other of their respective strategies. Thus if r, s, and t are three positive numbers which add up to 1, but are otherwise quite arbitrary, then A may use the following mixed strategy: $(r + s)/2$, $t/2$, $r/2$, $(s + t)/2$. For example, taking $r = s = t = 1/3$, the following strategy emerges: $(1/3, 1/6, 1/6, 1/3)$. Together with $(0, \frac{1}{2}, 0, \frac{1}{2}, 0)$ for B, we obtain again the value of the game, namely one-half.

MORE PERSON GAMES

Games in which more than two players take part have not been so intensively studied as two-person games, though the great book by J. von Neumann and O. Morgenstern contains chapters on this subject as well. Pay-off tables can again be constructed, but no theorem similar to the Main Theorem exists. An essentially new situation arises from the fact that two players can now combine in 'coalitions' and play against the remaining player or players. This is quite a common situation in real-life problems, and sometimes, as at an auction, the coalition may be agreed upon in advance between some of the players. Only if this condition is satisfied can the effects of coalitions be treated by the theory in its present form.

Coalitions can also arise spontaneously in many-person games, which ostensibly are all-against-all, at a stage when some one individual is too evidently winning. Coalitions of the latter kind are unpredictable, and may be changed and re-formed, according to the mood of the players, as the play proceeds. They cannot be treated by the theory, as at present developed, and illustrate one type of limitation in its application to real-life problems.

Never too late to learn. —SCOTTISH PROVERB

With just enough of learning to misquote. —BYRON

10 Sociology Learns the Language of Mathematics

By ABRAHAM KAPLAN

A TROUBLING question for those of us committed to the widest application of intelligence in the study and solution of the problems of men is whether a general understanding of the social sciences will be possible much longer. Many significant areas of these disciplines have already been removed by the advances of the past two decades beyond the reach of anyone who does not know mathematics; and the man of letters is increasingly finding, to his dismay, that the study of mankind proper is passing from his hands to those of technicians and specialists. The aesthetic effect is admittedly bad: we have given up the belletristic "essay on man" for the barbarisms of a technical vocabulary, or at best the forbidding elegance of mathematical syntax. What have we gained in exchange?

To answer this question we must be able to get at the content of the new science: But when it is conveyed in mathematical formulas, most of us are in the position of the medieval layman confronted by clerical Latin —with this difference: mathematical language cannot be forced to give way to a vernacular. Mathematics, if it has a function at all in the sciences, has an indispensable one; and now that so much of man's relation to man has come to be treated in mathematical terms, it is impossible to ignore or escape the new language any longer. There is no completely satisfactory way out of this dilemma. All this article can do is to grasp either horn, sometimes oversimplifying, sometimes taking the reader out of his depth: but hoping in the end to suggest to him the significance of the growing use of mathematical language in social science.

To complicate matters even further, the language has several dialects. "Mathematics" is a plural noun in substance as well as form. Geometry, algebra, statistics, and topology use distinct concepts and methods, and are applicable to characteristically different sorts of problems. The role of mathematics in the new social science cannot be discussed in a general way: as we shall see, everything depends on the kind of mathematics being used.

I

The earliest and historically most influential of the mathematical sciences is geometry. Euclid's systematization of the results of Babylonian astronomy and Egyptian surveying set up a model for scientific theory that remained effective for two thousand years. Plato found in Euclid's geometry the guide to the logical analysis of all knowledge, and it was the Renaissance's "rediscovery" of Plato's insistence on the fundamentally geometric structure of reality that insured the triumph of the modern world view inaugurated by Copernicus. Scientists like Kepler accepted the Copernican hypothesis because it made the cosmos more mathematically elegant than Ptolemy's cumbersome epicycles had been able to do.

The study of man—to say nothing of God!—enthusiastically availed itself of mathematical method: witness Spinoza's *Ethics*, which claimed that it "demonstrated according to the geometrical manner." But Spinoza's *Ethics* failed, as demonstrative science, because the 17th century did not clearly understand the geometry it was applying with such enthusiasm. The discovery of non-Euclidean geometries two hundred years later revealed that the so-called axioms of geometry are not *necessary* truths, as Spinoza and his fellow rationalists had always supposed, but merely postulates: propositions put forward for the sake of subsequent inquiry. It is only by deducing their consequences and comparing these with the perceived facts that we can decide whether or not the postulates are true. Geometry is a fully developed example of a set of undefined terms, prescribed operations, and the resulting postulates and theorems which make up a *postulational system*; it is in this form that it is influential in some of the recent developments in the social sciences.

Perhaps the most elaborate postulational system for dealing with the data of social and psychological science was constructed in 1940 by Clark Hull and associates of the Yale Institute of Human Relations (C. L. Hull et al., *Mathematico-Deductive Theory of Rote Learning*, Yale University Press, 1940). "Rote learning" is a very specialized branch of psychology that studies the learning of series of nonsense syllables: presumably, this tells us something about the act of learning in its "purest" form, with no admixture of influence from the thing learned.

The problems of the field revolve around ways of explaining the patterns of learning that are in fact discovered; why the first syllables in any series are learned first, the last soon after, and why the syllables a little past the middle of the given series take longest to memorize, and so on. There is a vast number of observations of this sort to be made, and Hull's ideal was to set up a postulational system which would allow observed patterns of learning to be logically deduced from relatively few postulates.

The system consists of 16 undefined terms, 86 definitions, and 18 postulates. From these, 54 theorems are deduced. The deductions can, in principle, be checked against the results of direct observation, in experimental situations or elsewhere. In many cases, as the book points out, existing evidence is as yet inadequate to determine whether the theorems hold true; in the great majority of cases, experimental evidence is in agreement with logically deduced theorems; in others, there is undoubted disagreement between deduced expectation and observed fact. Such disagreements point to basic defects in the postulate system, which, however, can be progressively improved.

The authors consider their book to be principally important as an example of the proper scientific method to be used in the study of behavior. And certainly, as a formal demonstration of the handling of definitions, postulates, and theorems, the book is unexceptionable. However, science prides itself on its proper method because of the fruitfulness of its results; and it is to the fruitfulness of this effort that we must address ourselves.

One example of the method may suggest better than general criticism the problem raised. Hull proves in one of his theorems that the greater the "inhibitory potential" at a given point in the learning of a rote series. the greater will be the time elapsing between the stimulus (the presentation of one nonsense syllable in the list) and the reaction (the pronouncing of the next memorized nonsense syllable). "Inhibitory potential" is one of the undefined terms of the system; it denotes the inability of the subject, before his involvement in the learning process, to pronounce the appropriate syllable on the presentation of the appropriate stimulus. (It may be pictured as a force that wanes as the stimuli are repeated and the syllable to be uttered is learned.)

Now this theorem certainly follows logically from three postulates of the system (they involve too many special terms to be enlightening if quoted). However, on examining these postulates, the theorem is seen to be so directly implied by them that one wonders what additional knowledge has been added by formally deducing it. A certain amount must have been known about rote learning to justify the selection of those postulates in the first place. To deduce this theorem from them has added very little—if anything—to what was already known. In short: the geometric method used by Hull, correct as it is formally, does not, for this reader, extend significantly what we already knew about rote learning from his and others' work.

In the course of Hull's book "qualitative postulates," by which is meant the "unquantified" ideas of thinkers like Freud and Darwin, are condemned because they have "so little deductive fertility"—because so few theorems may be deduced from them. In the narrowest logical sense of the phrase, this may be true. But fertility in the sense of yielding precisely determinable logical consequences is one thing; in the sense of yielding further insights into the subject matter—whether or not these can be presented as strict consequences of a system of postulates—it is quite another. The ideas of Darwin and Freud can hardly be condemned as lacking in fertility, even though they leave much to be desired from the standpoint of logical systematization.

This is not to deny that the postulational method can play a valuable role in science. But it is a question of the scientific context in which the method is applied. Newton and Euclid both had available to them a considerable body of fairly well-established knowledge which they could systematize, and in the process derive other results not apparent in the disconnected materials. Hull recognizes this condition, but supposes it to be already fulfilled in the area of learning theory. The results of the particular system he has constructed raise serious doubts that his supposition is true.

Science, basically, does not proceed by the trial-and-error method to which Hull, as a student of animal learning, is so much attached. It employs insight, itself subject to logical analysis, but too subtle to be caught in the coarse net of any present-day system of postulates. The geometric method in the new social science can be expected to increase in value as knowledge derived from other methods grows. But for the present, it is an elegantly written check drawn against insufficient funds.

II

If the 17th century was the age of geometry, the 18th was that of algebra. The essential idea of algebra is to provide a symbolism in which relations between quantities can be expressed *without a specification of the magnitudes of the quantities.* An equation simply formulates the equality between two quantities in a way that shows how the magnitude of one depends on the magnitude of certain constituents of the other.

The characterization of mathematics as a language is nowhere more appropriate than in algebra; the notation is everything. The power of algebra consists in that it allows the symbolism to think for us. Thought enters into the formulation of the equations and the establishing of the rules of manipulation by which they are to be solved, but the rest is mechanical—so much so that more and more the manipulation is being done, literally, by machines. The postulational method characteristic of classical geometry no longer plays as important a part here. Derivation is replaced by calculation; we proceed, not from postulates of uncertain truth, but from arithmetical propositions whose truth is a matter of logic following necessarily from the definitions of the symbols occurring in them. The equations that express relations between real quantities are, of course, empirical hypotheses; but *if* the equations correctly formulate the function relating two quantities, then certain numerical values for one necessarily imply certain numerical values for the other. Again, as in geometry, the facts are the final test.

In this spirit, the mathematicians of the 18th century replaced Newton's geometrizing by the methods of algebra, and the culmination was Laplace's system of celestial mechanics. Laplace's famous superman, given the position and momentum of every particle in the universe, could compute, it was thought, the entire course of world history, past and future. The development of quantum mechanics made this program unrealizable even in principle, just as the non-Euclidean geometries were fatal to the aspirations of the 17th-century rationalists. Nevertheless, this scientific ideal, so nearly realized in physics, has exerted enormous influence on the study of man, and much of the new social science is motivated by the attempt to take advantage of the powerful resources of algebra.

Among the most ambitious, but also most misguided, of such attempts is a 900-page treatise by the sociologist Stuart C. Dodd, *Dimensions of Society* (Macmillan, 1942). The author's ambition, declared in the subtitle, is to provide "a quantitative systematics for the social sciences." What is misguided is the failure to realize that a system is not provided by a symbolism alone: it is necessary also to have something to say in the symbolism that is rich enough to give point to the symbolic manipulation. Dodd presents a dozen symbols of what he regards as basic sociological concepts, together with four others for arithmetical operations. They stand for such ideas as space, time, population, and characteristics (the abstract idea "characteristics," and not the specific characteristics to be employed in sociological theory). In addition to these sixteen basic symbols, there are sixteen auxiliary symbols, compounds or special cases of the basic symbols, or supplements to them—for instance, the question mark, described as one of four "new operators," used to denote a hypothesis, or a questioned assertion. With this notation, every situation studied by sociologists is to be defined by an expression of the form:

$$S = {}^{s}_{s}(T\,;I\,;L\,;P)^{s}_{s}$$

The capital "S" stands for the *situation*, and the letters within the parentheses for *time*, indicators of the *characteristics* specified, *length* or *spatial regions*, and *populations*, *persons*, or *groups*. The semicolon stands for an unstated form of mathematical combination, and the small "s" for various ways of particularizing the four major kinds of characterizations. Thus "T^0" stands for a *date*, "T^1" for a *rate of change*, both of these being different sorts of time specifications. Instructions for the use of the notation require one hundred distinct rules for their formulation.

But this whole notational apparatus is, as Dodd recognizes, "a systematic way of expressing societal data, and not, directly, a system of the functionings of societal phenomena." "Facts," however, are data only *for* hypotheses; without the latter, they are of no scientific significance. Certainly a notational system can hardly be called a "theory," as Dodd constantly designates it, unless it contains some statements *about* the facts. But *Dimensions of Society* contains only one social generalization: "This theory . . . generalizes societal phenomena in the statement: 'People, Environments, and Their Characteristics May Change.' This obvious generalization becomes even more obvious if the time period in which the change is observed is prolonged." The last sentence may save Dodd, but not his "theory," from hopeless naivety.

Dodd's hope that his system of "quantic classification" will "come to play a role for the social sciences comparable to the classification of the chemical atoms in Mendelyeev's periodic table" is groundless. The periodic table, after all, told us something about the elements; more, it suggested that new elements that we knew nothing of existed, and told us what their characteristics would be when discovered. The fundamental point is that we have, in the case of Dodd, only a *notation*; when he speaks of the "verification" of his "theory," he means only that it is possible to formulate societal data with his notation.

Dodd's basic error is his failure to realize that after "Let x equal such-and-such," the important statement is still to come. The question is how to put *that* statement into mathematical form.

An answer to this question is provided in two books of a very different sort from Dodd's, both by the biophysicist N. Rashevsky: *Mathematical Theory of Human Relations* (The Principia Press, 1947) and *Mathematical Biology of Social Behavior* (University of Chicago Press, 1951). In these two books the author does not merely talk *about* mathematics; he actually *uses* it. In the earlier one, the methods of mathematical biology are applied to certain social phenomena on the basis of formal postulates—more simply, assumptions—about these phenomena. In the later book, these assumptions are interpreted in terms of neurobiophysical theory, and are derived as first approximations from that theory. The results, according to Rashevsky, are "numerous suggestions as to how biological measurements, made on large groups of individuals, may lead to the prediction of some social phenomena."

As a natural scientist, Rashevsky is not seduced, like so many aspirants to a social science, by the blandishments of physics. Scientific method is the same everywhere: it is the method of logical inference from data provided and tested by experience. But the specific techniques of science are different everywhere, not only as between science and science but even as between problem and problem in the same field. The confusion of method and technique, and the resultant identification of scientific method with the techniques of physics (and primarily 19th-century physics at that) has hindered the advance of the social sciences not a little. For the problems of sociology *are* different from those of physics. There are no concepts in social phenomena comparable in simplicity and fruitfulness to the space, time, and mass of classical mechanics; experiments are difficult to perform, and even harder to control; measurement in sociological situations presents special problems from which physics is relatively free. Yet none of these differences, as Rashevsky points out, prevents a scientific study of man. That social phenomena are complex means only that techniques must be developed of corresponding complexity: today's schoolboy can solve mathematical problems beyond the reach of Euclid or Archimedes. Difficulties in the way of experimentation have not prevented astronomy from attaining full maturity as a science. And the allegedly "qualitative" character of social facts is, after all, only an allegation; such facts have their quantitative aspects too. And what is perhaps more to the point, mathematics can also deal with qualitative characteristics, as we shall see.

Rashevsky addresses himself to specific social subject matters: the formation of closed social classes, the interaction of military and economic factors in international relations, "individualistic" and "collectivistic" tendencies, patterns of social influence, and many others. But though the problems are concrete and complex, he deals with them in a deliberately abstract and oversimplified way. The problems are real enough, but their formulation is idealized, and the equations solved on the basis of quite imaginary cases. Both books constantly repeat that the treatment is intended to "illustrate" how mathematics is applicable "in principle" to social science: for actual solutions, the theory is admitted to be for the most part too crude and the data too scarce.

What this means is that Rashevsky's results cannot be interpreted as actual accounts of social phenomena. They are, rather, ingenious elaborations of what *would* be the case if certain unrealistic assumptions were granted. Yet this is not in itself objectionable. As he points out in his own defense, physics made great strides by considering molecules as rigid spheres, in complete neglect of the complexity of their electronic and nuclear internal structures. But the critical question is whether Rashevsky's simplifications are, as he claims, "a temporary expedient." An idealization is useful only if an assumption that is approximately true yields a solution that is approximately correct; or at any rate, if we can point out the ways in which the solution must be modified to compensate for the errors in the assumptions.

It is in this respect that Rashevsky's work is most questionable. Whatever the merits of his idealizations from the standpoint of "illustrating," as he says, the potentialities of mathematics, from the standpoint of the study of man they are so idealized as almost to lack all purchase on reality.

Rashevsky's treatment of individual freedom, for example, considers it in two aspects: economic freedom and freedom of choice. The former is defined mathematically as the fraction obtained when the amount of work a man must actually do is subtracted from the maximum amount of work of which he is capable, and this remainder is divided by the original maximum. A person's economic freedom is *0* when he is engaged in hard labor to the point of daily exhaustion; it is *1* when he does not need to work at all. This definition equates increase in economic freedom with shortening of the hours of work; and an unemployed worker, provided he is kept alive on a dole, enjoys complete economic freedom. Such critical elements of economic freedom as real wages, choice of job, and differences in level of aspirations are all ignored.

Freedom of choice, the other aspect of individual freedom, is analyzed as the proportion borne by the amount of time an individual is not in contact with others who might interfere with his choices, to the time he spends alone plus time that is spent in the company of others with the same preferences. This makes freedom of choice decrease with increasing population, so that by this definition one is freer in a small village than a large city. Nothing is said about prying neighbors, or the presence or absence of a secret police, a most important determinant of "freedom of choice." The whole matter of the "introjection" of other persons' standards, as discussed for instance in Erich Fromm's *Escape from Freedom*, is ignored, as are such fundamental considerations as knowledge of the choices available, or the opportunity to cultivate skills and tastes.

On current social issues Rashevsky betrays that he suffers from the same confusions and rationalizations as afflict students without the advantages of a mathematical training. To explain the Lysenko case, for example, he suggests that it is possible that the facts of genetics "may be *interpreted* from two different points of view," thus naively substituting a scientific question (if there be one) for the real issue, which is the political control of science. His assumptions encourage him to attempt the conclusion, "even at the present state of our knowledge," that after World War II peace will be most strongly desired by the Soviet Union, least by the United States, with England and Germany in between. And he confesses that he finds it "difficult to understand why the Soviet Union insists on repatriating individuals who left the Soviet Union during World War II and do not desire to return." Mathematics is not yet capable of coping with the naivety of the mathematician himself.

III

The 19th century saw the rise of mathematical statistics. From its origins in the treatment of games of chance, it was expanded to cope with the new problems posed by the development of insurance, and finally came to be recognized as fundamental to every science dealing with repetitive phenomena. This means the whole of science, for knowledge everywhere rests on the cumulation of data drawn from sequences of situations on whose basis predictions can be made as to the recurrence of similar situations. Mathematical statistics is the theory of the treatment of repeated—or multiple—observations in order to obtain all and only those conclusions for which such observations are evidence. This, and not merely the handling of facts stated in numerical terms, is what distinguishes statistics from other branches of quantitative mathematics.

The application of statistics to social phenomena far exceeds, in fruitfulness as well as extent, the use of mathematics of a fundamentally geometrical—i.e., postulational—or algebraic character in the social sciences. Social scientists themselves have made important contributions to mathematical statistics—which is a good indication of its usefulness to them in their work. Only two of the most recent contributions in the application of statistics to social phenomena can be dealt with here.

The first is a rather remarkable book by a Harvard linguist, G. K. Zipf, *Human Behavior and the Principle of Least Effort* (Addison-Wesley Press, 1949). Its basic conception is that man is fundamentally a user of tools confronted with a variety of jobs to do. Culture can be regarded as constituting a set of tools, and human behavior can be analyzed as the use of such tools in accord with the principle of minimizing the probable rate of work which must be done to perform the jobs that arise. It is this principle that Zipf calls "the law of least effort." As a consequence of it, he claims, an enormous variety of social phenomena exhibit certain regularities of distribution, in accordance with the principle that the tools nearest to hand, easiest to manipulate, and adapted to the widest variety of purposes are those which tend to be used most frequently. These regularities often take the form, according to Zipf, of a constant rank-frequency relationship, according to which the tenth most frequently used word, for instance, is used one-tenth as often as the most frequently used one of all. This is the case, for example, with James Joyce's *Ulysses* as well as with clippings from American newspapers.

A large part of Zipf's book deals with linguistic phenomena, since he is most at home in this field, and it is there that his results seem most fully established. But an enormous range of other subjects is also treated: evolution, sex, schizophrenia, dreams, art, population, war, income, fads, and many others. Many of these topics are dealt with statistically, as likewise conforming to the law of least effort; and all are discussed with originality and insight. For example, the cities in any country, according to Zipf, tend to show the same regularity—the tenth most populous city will have one-tenth as many people as the most populous, the one-hundredth most populous city will have one-hundredth as many. Where this pattern does not prevail, we have an indication of serious potential conflict. It seems that starting about 1820 the growing divisions between Northern and Southern economies in the United States could be seen by a break in this pattern, which reached a peak of severity around 1840, and disappeared after the Civil War!

But while this breadth of topic endows the book with a distinctive fascination, it also makes the reader skeptical of the validity of the theory it puts forward. In the human sciences, the scope of a generalization is usually inversely proportional to its precision and usefulness—at any rate, in the present state of our knowledge. Zipf's law, or something like it, is well known in physics as Maupertuis' principle of least action. But Zipf applies it to a much wider field of phenomena than physical flows of energy. It is understandable that action will follow some sort of least-action pattern *if* economy enters into its motivation and *if* the action is sufficiently rational. But that the law of least effort should be manifested everywhere in human conduct, as Zipf holds—indeed, "in all living process"—is difficult to believe.

That a theory is incredible is, of course, no logical objection whatever. And Zipf does not merely speculate; he presents an enormous mass of empirical evidence. The question is what it proves. It does not show, as he claims, the existence of "natural social laws," but, at best, only certain regularities. Brute empiricism is not yet science. Unless observed regularities can be brought into logical relation with other regularities previously observed, we remain at the level of description rather than explanation; and without explanation, we cannot attach much predictive weight to description. As a collection of data that deserve the careful attention of the social scientist, Zipf's work will have interest for some time to come. But something more precise and less universal than his principle of least effort will be required to transform that data into a body of scientific knowledge.

The importance of clear conceptualization in giving scientific significance to observed fact is admirably expounded in the recently published fourth volume of the monumental *American Soldier* series (S. A. Stouffer, L. Guttman, et al., *Measurement and Prediction: Studies in Social Psychology in World War II*, Vol. IV, Princeton University Press, 1950). *Measurement and Prediction* deals with the study of attitudes. It is concerned with the development of methodological rather than substantive theory. It deals with the way in which attitudes—any attitudes—should be studied and understood, but says little about attitudes themselves. However, methodology here is not an excuse for irresponsibility in substantive assumptions, or for confusion as to the ultimate importance of a substantive theory of attitudes.

The major problem taken up is this: how can we tell whether a given set of characteristics is to be regarded as variations of some single underlying quality? Concretely, when do the responses to a questionaire constitute expressions of a single attitude, and when do they express a number of attitudes? If there is such a single quality, how can we measure how much of it is embodied in each of the characteristics? If all the items on a questionaire *do* deal with only one attitude, can we measure, in any meaningful sense, how favorable or unfavorable that attitude is, and how intensely it is held by the particular respondent?

This problem arises directly out of the practical—as well as the theoretical—side of opinion study. Consider the case of a poll designed to test the extent and intensity of anti-Semitism. Various questions are included: "Do you think the Jews have too much power?" "Do you think we should allow Jews to come into the country?" "Do you approve of what Hitler did to the Jews?" Some people give anti-Semitic answers to all the questions, some to a few, some to none. Is this because they possess varying amounts of a single quality that we may call "anti-Semitism"? Or is there really a mixture of two or more very different attitudes in the individual, present in varying proportions? If a person is against Jewish immigration, is this because he is against immigration or against Jews, and to what extent? And if there is a single quality such as anti-Semitism, what questions will best bring it out for study? It is problems such as these that the research reported on in *Measurement and Prediction* permits us to solve.

The approach taken stems from the work done in the past few decades by L. L. Thurstone and C. Spearman. Their problems were similar, but arose in a different field, the study of intelligence and other psychological characteristics. Their question was: do our intelligence tests determine a single quality called intelligence? Or do they actually tap a variety of factors, which though combining to produce the total intelligence score, are really quite different from each other? Thurstone and Spearman developed mathematical methods that in effect determined which items in a questionnaire were interdependent—that is, if a person answered *a*, he would tend to answer *b*, but not *c*. On the basis of such patterns, various factors of intelligence were discovered.

In opinion study, one inquires whether items of a questionnaire "hang together"—or, to use the technical term, whether they *scale*. A complex of attitudes is said to be *scalable* if it is possible to arrange the items in it in such a way that every person who answers "yes" to any item also answers "yes" to every item coming after it in the arrangement. In the case of anti-Semitism, we would consider the complex of anti-Semitism scalable—and therefore referring to a single factor in a person's attitudes, rather than including a few distinct attitudes—if we could order the questions in such a way that if someone answered any question in the list "anti-Semitically," he would answer all those following it "anti-Semitically." Attitudes on anti-Semitism would then have the same cumulative character as a series of questions on height—if a person answers "yes" to "Are you more than six feet tall?" we know he will answer "yes" to the question "Are you more than five and a half feet tall?" and all others down the list. This type of reduction of an apparently complex field of attitudes to the simple scheme of a series of cumulative questions is of great value. In *Measurement and Prediction* Louis Guttman describes how to determine whether a group of attitudes does "scale"—that is, does measure a single underlying reality.

Guttman developed, for example, such a scale for manifestations of fear in battle—vomiting, trembling, etc.: soldiers who vomit when faced with combat also report trembling, and so on down the scale; while those who report trembling do not necessarily report vomiting too. On the other hand, it turned out, when he studied paratroopers, various *kinds* of fear of very different types had to be distinguished, for the paratroopers' symptoms were not scalable in terms of a single variable.

One of the most direct applications of scaling methods is in the detection of spurious "causal" connections. We may find, for instance, that the attitude to the continuation of OPS by the government correlates closely with the attitude to farm subsidies. Scale analysis now makes it possible for us to provide an explanation for this fact by testing whether these two items do not in fact express a single attitude—say, the attitude to governmental controls.

Scale analysis permits us to handle another important problem. Suppose we find that 80 per cent of a group of soldiers tested agreed that "the British are doing as good a job as possible of fighting the war, everything considered," while only 48 per cent disagree with the statement that "the British always try to get other countries to do their fighting for them." How many soldiers are "favorable" toward the British? It is clear that we can get different percentages in answer to this question, depending on how we word our question. Scale analysis provides a method which yields what is called an "objective zero point": a division between the numbers of those "favorable" and those "unfavorable" that remains constant no matter what questions we ask about the British. The method demands that, besides asking a few questions testing the attitude, we also get a measure of the "intensity" with which the respondent holds his opinion—we ask for example, whether the respondent feels strongly, not so strongly, or is relatively indifferent about the matter. With this method, it turns out that if we asked a group of entirely different questions about the British, the application of the procedure for measuring the "objective zero point" would show the same result. This limited area of attitude comes to have the same objectivity as the temperature scale, which shows the same result whether we use an alcohol or a mercury thermometer.

Measurement and Prediction also presents, for the first time, a full description of Lazarsfeld's "latent structure" analysis. This is in effect a mathematical generalization of the scaling method of Guttman, which permits us to extend the type of inquiry that scale analysis makes possible into other areas. Scale analysis and latent structure analysis together form an important contribution to the development of more reliable methods of subjecting attitudes—and similar "qualities"—to precise and meaningful measurement.

The prediction part of *Measurement and Prediction* does not contain any comparable theoretical developments. For the most part, prediction in the social sciences today is not above the level of "enlightened common sense," as the authors recognize.

IV

The distinctive development in mathematics in the last one hundred years is the rise of a discipline whose central concept is neither number nor quantity, but *structure*. The mathematics of structure might be said to focus on qualitative differences. *Topology*, as the new discipline is called, is occupied with those properties of figures—conceived as sets of points—that are independent of mere differences of quantity. Squares, circles, and rectangles, of whatever size, are topologically indistinguishable from one another, or from any simple closed figure, however irregular. On the other hand, if a "hole" is inscribed in any of these figures, it thereby becomes qualitatively different from the rest.

Topology, more than most sectors of higher mathematics, deals with questions that have direct intuitive meaning. But intuition often misleads us as to the answers, when it does not fail us altogether. For instance, how many different colors are required to color *any* map so that no two adjoining countries have the same color? It is known, from topological analyses, that five colors are quite enough; but no one has been able to produce a map that needs more than four, or even to prove whether there could or could not be such a map.

It is paradoxical that the field of mathematics which deals with the most familiar subject matter is the least familiar to the non-mathematician. A smattering of geometry, algebra, and statistics is very widespread; topology is virtually unknown, even among social scientists sympathetic to mathematics. To be sure, the late Kurt Lewin's "topological psychology" has given the name much currency. But topology, for Lewin, provided only a notation. The rich content of his work bears no *logical* relation to the topological framework in which it is presented, as is clear from the posthumously published collection of his more important papers, *Field Theory in Social Science* (Harper, 1951). In these papers, talk about the "life space," and "paths," "barriers," and "regions" in it, are elaborately sustained metaphors. Such figures of speech are extraordinarily suggestive, but do not in themselves constitute a strict mathematical treatment of the "life space" in a topological geometry. The actual application of topology to psychology remained for Lewin a program and a hope.

One further development must be mentioned as playing an important role in the new approaches: the rise of symbolic logic. As in the case of topology, this discipline can be traced back to the 17th century, to Leibniz's ideas for a universal language; but not till the late 19th century did it undergo any extensive and precise development. Boolean algebra provided a mechanical method for the determination of the consequences of a given set of propositions (in another form, this is called the "calculus of propositions"). A few years later, De Morgan, Schroeder, and the founder of Pragmatism, Charles Peirce, investigated the formal properties of relations, leading to an elaborately abstract theory of relations. These results, together with work on the foundations of mathematics (like Peano's formulation of five postulates sufficing for the whole of arithmetic), were extended and systematized shortly after the turn of the century in Russell and Whitehead's monumental *Principia Mathematica.*

The word *cybernetics* (N. Wiener, *Cybernetics*, Wiley, 1948) is from the Greek for "steersman," from the same root as the word "governor." It is Wiener's coinage for the new science of "control and communication in the animal and the machine." The element of control is that exemplified in the mechanical device known as a governor, operating on the fundamental principle of the *feed-back:* the working of the machine at a given time is itself a stimulus for the modification of the future course of its working. Communication enters into cybernetics by way of an unorthodox and ingeniously precise concept of "information" as the degree to which the impulses entering a machine reduce the uncertainty among the set of alternatives available to it. Thus if a machine contains a relay which has two possible positions, an impulse which puts the relay into one of these two has conveyed to the machine exactly one "bit" of information. All communication can be regarded as made up of such bits, as is suggested by the dots and dashes of the Morse code, for example. Modern machines actually use a binary arithmetic, i.e., a notation in which all numbers are expressed in terms of powers of 2 (rather than 10), and are therefore formulable as strings of the cyphers for "zero" and "one." One bit of information is conveyed by each choice of either a zero or a one, the word "bit" being an abbreviation for "binary digit."

Working with these concepts of communication and control, cybernetics becomes relevant to the study of man because human behavior is paralleled in many respects by the communication machines. This parallel is no mere metaphor, but consists in a similarity of structure between the machine processes and those of human behavior. The Darwinian continuity between man and the rest of nature has now been carried to completion: man's rationality marks only a difference of degree from other animals, and fundamentally, no difference at all from the machine. For modern computers are essentially logical machines: they are designed to confront propositions and to draw from them their logical conclusions. With communication and control as the key, a similarity of structure can also be traced between an individual (whether human or mechanical) and a society (again, whether human or machines in a well-designed factory). The metaphors of Plato and Hobbes can now be given a literalist interpretation.

Thus cybernetics bears on the study of human behavior in a variety of ways: most directly, by way of neurology and physiological psychology; and by simple extension, to an improved understanding of functional mental disorders, which Wiener finds to be primarily diseases of memory, thus arriving at Freudian conclusions by a totally different route. Of particular interest are the implications, definitely present though not explicitly drawn, for such classical philosophical puzzles as those concerning free will and the mind-body relation. The analysis of mind and individual personality as a structure of certain information processes renders obsolete not only the "mind substance" of the idealist, but mechanistic materialism as well. Mind is a patterning of information and not spirit, matter, or energy.

In a more recent book (*The Human Use of Human Beings*, Houghton Mifflin, 1950) Wiener considers the relationship between cybernetics and the study of man at quite a different level. The sorts of control mechanisms with which cybernetics is concerned are creating, Wiener argues, a fundamental social transformation which he calls the "cybernetic revolution." This is the age of such mechanisms, he says, as the 18th century was of the clock and the 19th of the steam engine. It is now possible to construct machines for almost any degree of elaborateness of performance —chess-playing machines, for example, are no longer the hoaxes of Edgar Allan Poe's day. It is even possible to arrange machines so that they can communicate with one another, and in no merely figurative sense. And in addition to electronic brains, machines can be equipped with sensory receptors and efferent channels.

The social problems posed by the "cybernetic revolution" are basically those of the industrial one, but on an enlarged scale: whether the new technology is to be organized so as to produce leisure or unemployment, the ennoblement or degradation of man. *The Human Use of Human Beings* consists mainly in a forceful statement of this problem. The Industrial Revolution at least allowed for the localization of human dignity in man's reason, which still played an indispensable rôle in the operation of the technology, even though his muscle could be increasingly dispensed with. This last line of defense is in process of being undermined. Wiener accordingly devotes his book to a vigorous protest against "any use of a human being in which less is demanded of him and less is attributed to him than his full status." In the concrete, this inhuman use of man he finds in the increasing control over the individual in social organization, particularly in the control over information, as in scientific research under military sponsorship. But cybernetics itself only allows us to understand something of the technological developments that have posed this social problem; it does not contain, nor does it pretend to contain, any scientific theory—mathematical or otherwise—of how the problem can be resolved.

Among the recent applications to the study of man of this whole general body of ideas, one is especially celebrated; the theory of games presented by J. von Neumann and O. Morgenstern in *Theory of Games and Economic Behavior* (Princeton University Press, 1947). Here the focus is confined to problems of economics, but it is hoped that it will be extended to the whole range of man's social relations.

It may seem, superficially, that von Neumann and Morgenstern, in selecting games as a way of approaching the study of social organization, fall into the trap of oversimplification. But unlike Rashevsky, von Neumann and Morgenstern do not so much introduce simplifying assumptions in order to deal artificially with the whole complex social order as *select* relatively simple aspects of that order for analysis. Only after a theory adequate to the simple problems has been developed can the more complicated problems be attacked fruitfully. To be sure, the decision-maker cannot suspend action to satisfy the scruples of a scientific conscience; but neither can the scientist pretend to an advisory competence that he does not have, in order to satisfy practical demands.

While the theory of games does not deal with the social process in its full complexity, it is not merely a peripheral aspect of that process which it studies, but its very core. The aim is "to find the mathematically complete principles which define 'rational behavior' for the participants in a social economy, and to derive from them the general characteristics of that behavior." Games are analyzed because the pattern of rational behavior that they exhibit is the same as that manifested in social action, insofar as the latter does in fact involve rationality.

The theory is concerned with the question of what choice of strategy is rational when all the relevant probabilities are known and the outcome is not determined by one's choice alone. It is in the answer to this question that the new mathematics enters. And with this kind of mathematics, the social sciences finally abandon the imitation of natural science that has dogged so much of their history.

The authors first present a method of applying a numerical scale other than the price system to a set of human preferences. A man might prefer a concert to the theater, and either to staying at home. If we assign a utility of "1" to the last alternative, we know that going to the theater must be assigned a higher number, and the concert a higher one still. But how much higher? Suppose the decision were to be made by tossing two coins, the first to settle whether to go to the theater or not, the second (if necessary) to decide between the remaining alternatives. If the utility of the theater were very little different from that of staying at home, most of the time (three-fourths, to be exact) the outcome would be an unhappy one; similarly, if the theater and concert had comparable utilities, the outcome would be usually favorable. Just how the utilities compare, therefore, could be measured by allowing the second coin to be a loaded one. When it is a matter of indifference to the individual whether he goes to the theater or else tosses the loaded coin to decide whether he hears the concert or stays home, the loading of the coin provides a numerical measure of the utilities involved.

Once utility can be measured in a way that does not necessarily correspond to monetary units, a theory of rational behavior can be developed which takes account of other values than monetary ones. A game can be regarded as being played, not move by move, but on the basis of an overall *strategy* that specifies beforehand what move is to be made in every situation that could possibly arise in the game. Then, for every pair of strategies selected—one by each player in the game—the rules of the game determine a *value* for the game: namely, what utility it would then have for each player. An optimal strategy is one guaranteeing a certain value for the game even with best possible play on the part of the opponent. Rational behavior, that is to say, is characterized as the selection of the strategy which minimizes the maximum loss each player can sustain.

If a game has only two players and is "zero-sum"—whatever is won by one player being lost by the other, and vice versa—then, if each player has "perfect information" about all the previous moves in the game, there always exists such an optimal strategy for each player. The outcome of a rationally played game can therefore be computed in advance; in principle, chess is as predictable as ticktacktoe.

Not every game, however, is as completely open as chess. In bridge and poker, for example, we do not know what cards are in the other players' hands; and this is ordinarily the situation in the strategic interplay of management and labor, or in the relations among sovereign states. In such cases rationality consists in playing several strategies, each with a mathematically determined probability. Consider a type of matching pennies in which each player is permitted to place his coin as he chooses. If we were to select heads always, we should be quickly found out; even if we favored heads somewhat, our opponent (assuming it is he who wins when the coins match) could always select heads and thereby win more than half the time. But if we select heads half the time—not in strict alternation, to be sure, but at random—then, no matter what our opponent does, we cannot lose more than half the time. Rational play consists in what is actually done in the game: we toss the coin each time to determine, as it were, whether we select heads or tails. Of course, in more complex games our strategies are not always to be "mixed" in equal proportions. The fundamental theorem of the theory of games is that for every two-person zero-sum game, no matter how complex the rules or how great the range of possible strategies open to each player, there always exists some specific pattern of probabilities of strategies which constitutes rational play. It minimizes the maximum loss that each player can sustain, not in every play of the game, but in the long run. And there is a mathematical solution that tells us what this strategy is.

Unfortunately, many games are not "zero-sum": in the game between management and labor, utilities are created in the process of play; in war they are destroyed. It is simply not true in such cases that what one side loses the other gains, or vice versa. In such cases, the mathematics of the "zero-sum" game will not apply. Moreover, many games have more than two-players: a number of manufacturers may be competing for a single market. Here the mathematics of a two-person game will not hold. The first difficulty, however, can be absorbed into the second. A non-"zero-sum" two-person game can be regarded as a three-person game, where the third person wins or loses whatever is lost or won by the other two together.

But how are we to solve games of more than two persons? Only if coalitions are formed, in effect making the game a two-person one. This is not an unrealistic assumption, since, obviously, if two players can coordinate their strategies against a third, they will enjoy a special advantage: odd-man-out is the simple but fundamental principle in such situations. For such games, however, the theory does not provide a detailed solution, for it cannot determine what is a rational division of the spoils between the members of a coalition in the way it can determine what is a rational strategy for the coalition as a whole. And here, of course, is the great difficulty in politics. The United States and Russia may be conceived of, in this theory, as playing a two-person non-"zero-sum" game with nature as the third player: only nature wins from atomic destruction, only nature loses if resources need not be diverted to military purposes. But the coalition of men against nature still leaves open how the utilities acquired in the game are to be divided between the participants. And here conflicting interests stand in the way of the joint interests that would make rational a coalition strategy.

From our present standpoint, the important outcome of the theory is to show that there exists a rigorously mathematical approach to precisely those aspects of the study of man that have seemed in the past to be least amenable to mathematical treatment—questions of conflicting or parallel interest, perfect or imperfect information, rational decision or chance effect. Mathematics is of importance for the social sciences not merely in the study of those aspects in which man is assimilable to inanimate nature, but precisely in his most human ones. On this question the theory of games leaves no room for doubt.

But a mathematical theory of games is one thing, and a mathematical theory of society another. Von Neumann and Morgenstern, it must be said, never confuse the two. Many fundamental difficulties remain, even within the limitations of the games framework. The theory of games involving many players is in a very unsatisfactory state; there is no way at present of comparing the utilities of different persons; and the whole theory is so far a static one, unable to take account of the essential learning process which (it may be hoped) takes place in the course of the selection of real-life strategies. Yet the theory has already provided enormous stimulation to mathematical and substantive research, and much more can be expected from it. Above all, it has shown that the resources of the human mind for the exact understanding of man himself are by no means already catalogued in the existing techniques of the natural sciences.

Thus the application of mathematics to the study of man is no longer a programmatic hope, but an accomplished fact. The books we have surveyed are by no means the only mathematical approaches to problems of social science published even within the past few years. For instance, K. Arrow's *Social Choice and Individual Values* (Wiley, 1951) is a penetrating application of postulational method and the logical theory of relations to the problem of combining, in accord with democratic principles, individual preferences into a representative set of preferences for the group as a whole. Harold Lasswell and his associates report in *The Language of Politics* (G. W. Stewart, 1949) on the procedures and results of the application of the now widely familiar content-analysis techniques to political discourse, in order to objectify and quantify the role of ideologies and utopias in politics. Shannon and Weaver's *Mathematical Theory of Communication* (University of Illinois Press, 1950) reprints the classic papers in which Claude Shannon first developed the precise theory of information now being applied in cybernetics, linguistics, and elsewhere. There are a number of other books; and dozens of papers have appeared in the professional journals of a wide variety of disciplines, including such pre-eminently "qualitative" fields as social organizations, psychiatry, and even literary history.

But if the new language is widely spoken, there are also wide local differences in dialects, and many individual peculiarities of usage. Yet on the scientific scene, this is in itself scarcely objectionable. New problems call for new methods, and if for a time these remain *ad hoc*, it is only from such a rich variety of approaches that systematic and general theories can emerge. No such theories of society or even of part of the social process have yet been developed, though, as we have seen, there have been some premature attempts. But the situation is promising. If the mathematics employed is not merely notational, if it is not merely an "illustration" of an abstract methodology, if it does not outstrip the availability of data and especially of ideas, there is reason to hope that it will eventually contribute as much to the study of man as it already has to the understanding of the rest of nature.

PART VII

The Laws of Chance

1. Concerning Probability *by* PIERRE SIMON DE LAPLACE
2. The Red and the Black *by* CHARLES SANDERS PEIRCE
3. The Probability of Induction *by* CHARLES SANDERS PEIRCE
4. The Application of Probability to Human Conduct *by* JOHN MAYNARD KEYNES
5. Chance *by* HENRI POINCARÉ
6. The Meaning of Probability *by* ERNEST NAGEL

COMMENTARY ON
PIERRE SIMON DE LAPLACE

HISTORIANS of science and biographers have no difficulty pronouncing a verdict on the Marquis de Laplace. As a scientist—mathematical astronomer and mathematician—he was second only to Newton; as a person, his qualities were mixed. He was ambitious, but not unamiable; disposed to borrow without acknowledgment from the works of others, but not ungenerous. Above all he was a virtuoso in the art of rapid adaptation to a changing social and political environment. He is condemned for this—probably because it is such a common frailty that mere mention of it makes everyone uncomfortable.

Pierre Simon de Laplace was born at Beaumont-en-Auge, a Normandy village in sight of the English Channel, on March 23, 1749. The facts of his life, of the earlier years especially, are both sparse and in dispute.[1] Most of the original documents essential to an accurate account were burned in a fire which in 1925 destroyed the château of his great-great-grandson the Comte de Colbert-Laplace; others were lost during World War II in the bombardment of Caen. Many errors about Laplace's life have gained currency: that his father was a poor peasant, that he owed his education to the generosity of prosperous neighbors, that after he became famous he sought to conceal his "humble origins." Recent researches by the mathematician Sir Edmund Whittaker seem to show that whatever Laplace's reasons for reticence about his childhood, poverty of his parents was not among them. His father owned a small estate and was a syndic of the parish; his family belonged to the "good bourgeoisie of the land." One of Laplace's uncles was a surgeon, another a priest. The latter, a member of the teaching staff of the Benedictine Priory at Beaumont, where Laplace had his first schooling, is said to have awakened the boy's interest in mathematics. For a time it was thought that Laplace would follow his uncle's profession as a priest, but at the University of Caen, which he entered at the age of sixteen, he soon demonstrated his mathematical inclinations. He wrote a paper on the calculus of finite differences which was published in a journal edited by Joseph Louis

[1] The leading older biographical sources are Baron Fourier's *Éloge, Mémoires de l'institut*, X. LXXXI (1831); *Revue Encyclopédique*, XLIII (1829); S. D. Poisson's Funeral Oration (*Conn. des Temps*, 1830, p. 19); Francois Arago, *Biographies of Distinguished Scientific Men*, London, 1857; Agnes Mary Clerke, article on Laplace, Encyclopaedia Britannica, Eleventh Edition. Also brief readable accounts in E. T. Bell's *Men of Mathematics*, New York, 1937; Sir Oliver Lodge, *Pioneers of Science*, London, 1928; E. J. C. Morton, *Heroes of Science*, London (n.d.); Sir Robert S. Ball, *Great Astronomers*, London, 1901. The recent researches of the late Sir Edmund Whittaker are reported in an article by him in *The Mathematical Gazette*, XXXIII, February 1949. See also H. Andoyer, *L'Oeuvre scientifique de Laplace*, Paris, 1922.

Lagrange, the great mathematician, thirteen years Laplace's senior, with whom he was later to collaborate.

When Laplace was eighteen, he set out for Paris. He carried enthusiastic letters of recommendation to d'Alembert, the most prominent mathematician of France. D'Alembert ignored them; Laplace, not an easy fellow to put off, thereupon wrote him a letter on the general principles of mechanics, which made so strong an impression that d'Alembert at once sent for the precocious young man and said: "Monsieur, as you see, I pay little enough attention to recommendations; you had no need of them. You made your worth known; that is enough for me; my support is your due." A short while later d'Alembert procured for him an appointment as professor of mathematics in the *École Militaire* of Paris.

Laplace's rise was rapid and brilliant. His very first paper demonstrated a mastery of the calculus. Then, in one memoir after another submitted to the Academy of Sciences, he applied his formidable mathematical capabilities to the outstanding questions of planetary theory. "We have never seen," said a spokesman for the usually imperturbable savants of the Academy, "a man so young present in so short a time so many important memoirs on such diverse and difficult problems."

One of the main problems Laplace ventured to attack was the perturbations of the planets. The anomalies of their motion had long been known; the English astronomer Edmund Halley had noted, for instance, that Jupiter and Saturn over the centuries alternately lagged behind and were accelerated ahead of their expected places in a peculiar kind of orbital horse race. The application of Newton's theory of gravitation to the behavior of the planets and their satellites entailed fearful difficulties. The famous three-body problem (how three bodies behave when attracting one another under the inverse square law) is not completely solved today; Laplace tackled the much more complex problem of all the planets cross-pulling on one another and on the sun.

Newton had feared that the planetary melee would in time derange the solar system and that God's help would be needed to restore order. Laplace decided to look elsewhere for reassurance. In a memoir described as "the most remarkable ever presented to a scientific society," he demonstrated that the perturbations of the planets were not cumulative but periodic. He then set out to establish a comprehensive rule concerning these oscillations and the inclination of the planetary orbits.[2] This

[2] "Represent the masses of the several planets by m_1, m_2, etc.; their mean distances from the sun (or radii vectores) by r_1, r_2, etc.; the eccentricities of their orbits by e_1, e_2, etc.; and the obliquity of the planes of these orbits, reckoned from a single plane of reference or 'invariable plane,' by θ_1, θ_2, etc.; then all these quantities (except m) are liable to fluctuate; but however much they change, an increase for one planet will be accompanied by a decrease for some others; so that, taking all the planets into account, the sum of a set of terms like these, $m_1e_1{}^2\sqrt{r_1} + m_2e_2{}^2\sqrt{r_2}$ + etc., will remain always the same. This is summed up briefly in the following

work bore on the fate of the entire solar system. If it could be shown that disturbances in the machinery were gradually overcome and the status quo restored—a kind of self-healing and self-preserving process analogous

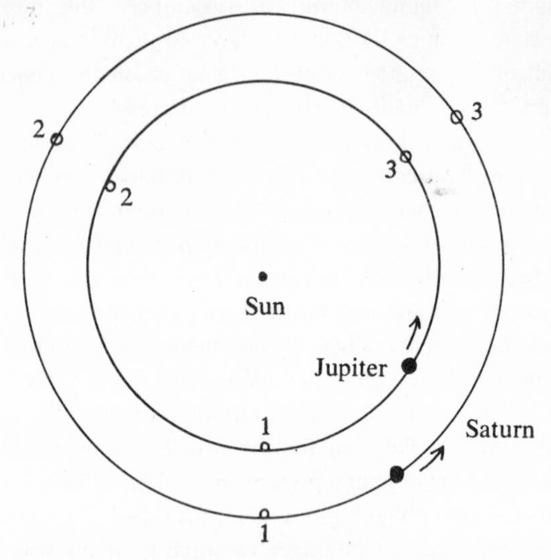

Perturbations of the planets were mathematically described by Laplace and Lagrange. When Jupiter and Saturn are in conjunction (on a line with the sun), one planet speeds up and the other slows down. Numbers indicate they are in conjunction three times in every five circuits of Jupiter.

to the physiological principle which Walter Cannon has called homeostasis—the future of the cosmic machine, and of its accidental passenger, man, was reasonably secure. If, however, the disturbances tended to accumulate, and each oscillation simply paved the way for a wilder successor, catastrophe was the inevitable end. Laplace worked out a theoretical solution which seemed to fit observation, showing that the outcome would be happy, that the changes of the solar system merely "repeat themselves at regular intervals, and never exceed a certain moderate amount." The period itself is of course tremendously long; the oscillations are those of "a great pendulum of eternity which beats ages as our pendulums beat seconds."

Thus Laplace's theorems gave assurance of the reliability of the stellar clockwork of the universe; its peculiar wobbles and other irregularities were seen to be minor, self-correcting blemishes which in no sense threatened the revolutions of the engine as a whole. Indeed, Laplace regarded

---

statement: $\Sigma(me^2\sqrt{r}) = $ constant. That is one law, and the other is like it, but with inclination of orbit instead of eccentricity, viz: $\Sigma(m\theta^2\sqrt{r}) = $ constant. The value of each of these two constants can at any time be calculated. At present their values are small. Hence they always were and always will be small; being in fact invariable. Hence neither e nor r nor θ can ever become infinite, nor can their average value for the system ever become zero." Lodge, *op. cit.*, p. 266.

the anomalies as a boon to astronomers. He wrote in the *Mécanique céleste:* "The irregularities of the two planets appeared formerly to be inexplicable by the law of universal gravitation; they now form one of its most striking proofs. Such has been the fate of this brilliant discovery, that each difficulty which has arisen has become for it a new subject of triumph—a circumstance which is the surest characteristic of the true system of nature."

Two reservations about this work have to be noted. Laplace's solution did not completely prove the stability of the solar system. His solution would be valid for an "idealized solar system undisturbed by tidal friction or other forces"; but the earth is now known, as it was not in Laplace's day, to be a nonrigid body subject to deformation by tidal friction, which thus acts as a brake on its motion. The effect is very small but acts always in one direction. There is no reason to anticipate an early dissolution of the system by collision or other gross accident, but there is also no basis for concluding, as Laplace did, that nature arranged the operations of the celestial machine "for an eternal duration, upon the same principles as those which prevail so admirably upon the earth, for the preservation of individuals and for the perpetuity of the species."

The second point concerns Laplace's failure to mention his indebtedness to Lagrange. Almost everything that Laplace accomplished in physical astronomy owes a debt to Lagrange's profound mathematical discoveries. It is impossible in many instances to separate their contributions; they "alternately surpassed and supplemented each other." [3] Lagrange was the greater mathematician; Laplace, for whom mathematics was only a means to an objective, was primarily a mathematical physicist and astronomer. And the reason for stressing Lagrange's share in these developments is that Laplace himself was not inclined to do so. It was his practice, says Bell, to "steal outrageously, right and left, wherever he could lay his hands on anything of his contemporaries and predecessors which he could use." Others have severely censured Laplace for his failure to acknowledge his collaborator's contributions, but Lagrange, obviously a saintly soul, did not; the two always remained on the best of terms.

Laplace's *Mécanique céleste* appeared in five immense volumes between 1799 and 1825. The distinguished historian Agnes Mary Clerke characterized it as a "record of unmixed triumph," a work which so completely fulfills the author's aims that it "hints at no unsatisfied ambitions." It sums up the labors of three generations of mathematicians in applying the law of gravitation to astronomical phenomena. It rightly earns for Laplace the title of the Newton of France. Laplace described its scope as follows:

"We have given, in the first part of this work, the general principles

[3] Clerke, *loc. cit.*

of the equilibrium and motion of bodies. The application of these principles to the motions of the heavenly bodies has conducted us, by geometrical reasoning, without any hypothesis, to the law of universal attraction; the action of gravity, and the motion of projectiles, being particular cases of this law. We have then taken into consideration a system of bodies subjected to this great law of nature; and have obtained, by a singular analysis, the general expressions of their motions, of their figures, and of the oscillations of the fluids which cover them. From these expressions we have deduced all the known phenomena of the flow and ebb of the tide; the variations of the degrees, and of the force of gravity at the surface of the earth; the precession of the equinoxes; the libration of the moon; and the figure and rotation of Saturn's rings. We have also pointed out the cause why these rings remain permanently in the plane of the equator of Saturn. Moreover, we have deduced, from the same theory of gravity, the principal equations of the motions of the planets; particularly those of Jupiter and Saturn, whose great inequalities have a period of above 900 years."

To mathematicians the work is especially memorable. The Irish mathematician William Rowan Hamilton is said to have begun his mathematical career by discovering a mistake in the *Mécanique céleste*. George Green, the English mathematician, derived from it a mathematical theory of electricity. Perhaps the greatest single contribution of the work was the famous Laplace equation:

$$\frac{\partial^2 u}{\partial x^2} + \frac{\partial^2 u}{\partial y^2} + \frac{\partial^2 u}{\partial z^2} = 0$$

Laplace's expression is a field equation, which is to say it can be used to describe what is happening at every instant of time at every point in a field produced by a gravitational mass, an electric charge, fluid flow and so on. Another way of saying this is that the equation deals with the value of a physical quantity, the potential, throughout a continuum. The potential function u, introduced in the first instance as a purely mathematical quantity, later acquired a physical meaning. The difference between the values of the potential function at two different points of a field measures the amount of work required to move a unit of matter from one of these points to the other; the rate of change of potential in any direction measures the force in that direction.

By giving u different meanings (e.g., temperature, velocity potential and so on) the equation is found to have an enormous range of applications in the theories of electrostatics, gravitation, hydrodynamics, magnetism, sound, light, conduction of heat. In hydrodynamics, where u is the velocity potential (distance squared divided by time), the rate of change of potential is the measure of the velocity of the fluid. The equation applies to a fluid which is incompressible and indestructible; if as much fluid flows out of any tiny element of volume as flows in, the potential function satisfies Laplace's equation. A rough explanation of why this equation serves as an almost universal solvent of physical problems is that it describes a characteristic economy of natural behavior—"a general tendency toward uniformity so that local inequalities tend to be smoothed out." Thus a metal rod heated at one end tends to become of uniform temperature throughout; a solute in a liquid tends to distribute itself evenly.

The *Mécanique céleste* is a book of formidable difficulties. Laplace made no concession to the reader. The style is extremely obscure and great gaps in the argument are bridged only by the infuriating phrase "it is easy to see." The great man himself, when required to reconstruct some of his reasoning, found it not at all *"aisé à voir,"* and often had to spend hours figuring out how his conclusions had been reached. The American mathematician Nathaniel Bowditch, who translated four of the volumes into English, said he never came across this expression "without feeling sure that I have hours of hard work before me to fill up the chasm." Nor is it a modest or entirely honorable writing. "Theorems and formulae are appropriated wholesale without acknowledgment, and a production which may be described as the organized result of a century of patient toil presents itself to the world as the offspring of a single brain." [4] When Napoleon protested to Laplace, on receiving a copy of the work, that in all its vast expanse God is not mentioned, the author replied that he had no need of that hypothesis. Napoleon, much amused, repeated the reply to Lagrange, who is said to have exclaimed: "Ah, but it is a beautiful hypothesis; it explains many things." (See selection by De Morgan, p. 2369.)

For those unable to follow the formidable abstractions of the *Mécanique* Laplace wrote in 1786 the *Exposition du système du monde*, one of the most charming popular treatises on astronomy ever published. In this masterpiece Laplace put forward his famous nebular hypothesis (which had been anticipated by Immanuel Kant in 1755). Its gist is that the solar system evolved from a rotating mass of gas, which condensed to form the sun and later threw off a series of gaseous rings that became the planets. While still in the gaseous state the planets threw off rings which became satellites. The hypothesis has had its ups and downs since Kant and Laplace advanced it. In Laplace's theory revolution in a retro-

[4] Agnes Mary Clerke, *loc. cit.* E. T. Bell's comment is also of interest: "From Lagrange, for example, he lifted the fundamental concept of the potential . . . from Legendre he took whatever he needed in the way of analysis; and finally, in his masterpiece, the *Mécanique céleste*, he deliberately omits references to the work of others incorporated in his own, with the intention of leaving posterity to infer that he alone created the mathematical theory of the heavens." *Op. cit.*, p. 174.

grade direction by a member of the solar system was impossible; yet before Laplace died Sir William Herschel found that the satellites of Uranus misbehaved in this way, and others have since been discovered. Yet the theory was an intellectual landmark, and much of its basic reasoning is still accepted by some cosmologists as valid for astronomical aggregates larger than the solar system.

Another subject upon which Laplace bestowed his attention, both as a mathematician and as a popularizer, is the theory of probability. His comprehensive treatise *Théorie analytique des probabilités* described a useful calculus for assigning a "degree of rational" belief to propositions about chance events. Its framework was the science of permutations and combinations, which might be called the mathematics of possibility. The theory of probability, said Laplace, is at bottom nothing more than common sense reduced to calculation. But his treatise seemed to indicate that the arithmetic of common sense is even more intricate than that of the planets. No less a mathematician than Augustus De Morgan described it as "by very much the most difficult mathematical work we have ever met with," exceeding in complexity the *Mécanique céleste.*

Laplace's contributions to probability are perhaps unequaled by any other single investigator; nevertheless the *Théorie analytique*, like the *Mécanique*, failed to acknowledge the labors of other mathematicians, on which many of its conclusions depended. De Morgan said of Laplace: "There is enough originating from himself to make any reader wonder that one who could so well afford to state what he had taken from others, should have set an example so dangerous to his own claims."

The companion to the main treatise, the *Essai philosophique sur les probabilités* (1814), from which the next selection has been taken, is a superbly lucid nontechnical introduction to the laws of chance. It contains, as you will see, what is regarded as the most perfect statement of the deterministic interpretation of the universe, a symbol of that happy and confident age which supposed that the past could be described and the future predicted from a single snapshot of the present.

Together with the great chemist Antoine Lavoisier, Laplace engaged in experiments to determine the specific heats of a number of substances. They designed the instrument known as Laplace's ice calorimeter, which measures heat by the amount of ice melted, a method employed earlier by the Scottish chemist Joseph Black and the German Johann Karl Wilke.

Laplace prospered financially and politically; Lavoisier died on the guillotine. In 1784 Laplace was appointed "examiner to the royal artillery," a lucrative post and one in which he had the good fortune to examine a promising sixteen-year-old candidate named Napoleon Bonaparte. The relationship was to blossom forth twenty years later, much to

Laplace's advantage. With Lagrange, Laplace taught mathematics at the École Normale, became a member and then president of the Bureau of Longitudes, aided in the introduction of the decimal system and suggested, in keeping with the reform spirit of the Revolution, the adoption of a new calendar based on certain astronomical calculations.

There is some reason to believe that for a brief period during the Revolution Laplace fell under suspicion; he was removed from the commission of weights and measures. But he managed not only to hold on to his head but to win new honors. He had a knack for riding the waves of his turbulent era. Under the Republic he was an ardent Republican and declared his "inextinguishable hatred to royalty." The day following the 18th Brumaire (November 9, 1799), when Napoleon seized power, he shed his Republicanism and formed an ardent attachment for the first consul, whom he had helped earlier to form a Commission for Egypt. Almost immediately Napoleon rewarded Laplace with the portfolio of the Interior. The evening of his appointment the new minister demanded a pension of 2,000 francs for the widow of the noted scholar Jean Bailly, executed during the Terror, and early the next morning Madame Laplace herself brought the first half-year's income to "this victim of the passions of the epoch." It was a "noble beginning," as Laplace's protégé François Arago wrote, but it is hard to discover any other noble accomplishment gracing Laplace's ministerial career. His tenure of office was brief—six weeks. Napoleon wrote tartly of Laplace's shortcomings in his St. Helena memoirs: "He was a worse than mediocre administrator who searched everywhere for subtleties, and brought into the affairs of government the spirit of the infinitely small." But to soothe the hurt of his dismissal the deposed minister was given a seat in the Senate and in 1803 became its Chancellor.

Historians have amused themselves describing Laplace's skill in running with the hare and hunting with the hounds. The neatest evidence appears in his introductions to successive editions of his books. He inscribed the first edition of the *Système du monde* in 1796 to the Council of Five Hundred, and in 1802 prefixed the third volume of the *Mécanique céleste* with a worshipful paean to Napoleon, who had dispersed the Council. Laplace dedicated the 1812 edition of the *Théorie analytique des probabilités* to "Napoleon the Great"; in the 1814 edition he suppressed this dedication and wrote "that the fall of empires which aspired to universal dominion could be predicted with very high probability by one versed in the calculus of chances." Napoleon had made Laplace a count; this gave him the opportunity to join in the 1814 decree of forfeiture banishing the man who had made him a count. When the Bourbons returned, Laplace was one of the first to fall at their feet; for this genuflection he received a marquisate.

Laplace was not really a bad man. He gave a hand up to many younger scientists. At his country home in Arcueil he surrounded himself with "adopted children of his thought": Arago, an astronomer and physicist; the physicist Jean Biot, noted for his investigations of the polarization of light; Baron Alexander von Humboldt, the celebrated German naturalist and traveler; Joseph Gay-Lussac, the great chemist and physicist; Siméon Poisson, the brilliant mathematician. Biot related that after he had read a paper on the theory of equations, Laplace took him aside and showed him "under a strict pledge of secrecy papers yellow with age in which he had long before obtained the same results." Having soothed his ego, Laplace told the young man to say nothing about the earlier work and to publish his own.

The almost universal admiration for Laplace's scientific genius did not mitigate the widespread distrust inspired by his political adaptability. The more tolerantly cynical of his contemporaries referred to his "suppleness." The stock appraisal is to compare him to the Vicar of Bray. The Vicar, an accommodating man who was twice a Papist and twice a Protestant, is said to have defended the charge of being a time-server by replying: "Not so, neither, for if I changed my religions, I am sure I kept true to my principle, which is to live and die the Vicar of Bray." Laplace could have made similar answer.[5]

About his family life and personal habits there is a strange lack of information. Laplace's marriage with Charlotte de Courty de Romanges, contracted in 1788, was apparently a happy one. They had a daughter and a son, Emile, who rose to the rank of general in the artillery. In later years Laplace passed much of his time at Arcueil, where he had a house next to the chemist Count de Berthollet. There in his study, where the portrait of Racine, his favorite author, hung opposite that of Newton, he pursued his studies with "unabated ardor" and received "distinguished visitors from all parts of the world." He died on March 5, 1827, a few days before his seventy-eighth birthday. Illustrious men are required to say deathless things on their deathbeds. Laplace is said to have departed after expressing the reasonable opinion, "What we know is very slight; what we don't know is immense." De Morgan, observing that "this looks like a parody on Newton's pebbles," claims on close authority to have learned Laplace's very last words: "Man follows only phantoms."

[5] *The Oxford Companion to English Literature*, Oxford, 1946, p. 822.

Fate, Time, Occasion, Chance and Change?—To these
All things are subject . . . —PERCY BYSSHE SHELLEY

1 Concerning Probability

By PIERRE SIMON DE LAPLACE

ALL events, even those which on account of their insignificance do not seem to follow the great laws of nature, are a result of it just as necessarily as the revolutions of the sun. In ignorance of the ties which unite such events to the entire system of the universe, they have been made to depend upon final causes or upon hazard, according as they occur and are repeated with regularity, or appear without regard to order; but these imaginary causes have gradually receded with the widening bounds of knowledge and disappear entirely before sound philosophy, which sees in them only the expression of our ignorance of the true causes.

Present events are connected with preceding ones by a tie based upon the evident principle that a thing cannot occur without a cause which produces it. This axiom, known by the name of *the principle of sufficient reason*, extends even to actions which are considered indifferent; the freest will is unable without a determinative motive to give them birth; if we assume two positions with exactly similar circumstances and find that the will is active in the one and inactive in the other, we say that its choice is an effect without a cause. It is then, says Leibnitz, the blind chance of the Epicureans. The contrary opinion is an illusion of the mind, which, losing sight of the evasive reasons of the choice of the will in indifferent things, believes that choice is determined of itself and without motives.

We ought then to regard the present state of the universe as the effect of its anterior state and as the cause of the one which is to follow. Given for one instant an intelligence which could comprehend all the forces by which nature is animated and the respective situation of the beings who compose it—an intelligence sufficiently vast to submit these data to analysis—it would embrace in the same formula the movements of the greatest bodies of the universe and those of the lightest atom; for it, nothing would be uncertain and the future, as the past, would be present to its eyes. The human mind offers, in the perfection which it has been able to give to astronomy, a feeble idea of this intelligence. Its discoveries in mechanics and geometry, added to that of universal gravity, have enabled it to comprehend in the same analytical expressions the past and future states of the system of the world. Applying the same method to some other objects of its knowledge, it has succeeded in referring to general laws observed phenomena and in foreseeing those which given circumstances ought to produce. All these efforts in the search for truth tend to lead it back continually to the vast intelligence which we have just mentioned, but from which it will always remain infinitely removed. This tendency, peculiar to the human race, is that which renders it superior to animals; and their progress in this respect distinguishes nations and ages and constitutes their true glory.

Let us recall that formerly, and at no remote epoch, an unusual rain or an extreme drought, a comet having in train a very long tail, the eclipses, the aurora borealis, and in general all the unusual phenomena were regarded as so many signs of celestial wrath. Heaven was invoked in order to avert their baneful influence. No one prayed to have the planets and the sun arrested in their courses: observation had soon made apparent the futility of such prayers. But as these phenomena, occurring and disappearing at long intervals, seemed to oppose the order of nature, it was supposed that Heaven, irritated by the crimes of the earth, had created them to announce its vengeance. Thus the long tail of the comet of 1456 spread terror through Europe, already thrown into consternation by the rapid successes of the Turks, who had just overthrown the Lower Empire. This star after four revolutions has excited among us a very different interest. The knowledge of the laws of the system of the world acquired in the interval had dissipated the fears begotten by the ignorance of the true relationship of man to the universe; and Halley, having recognized the identity of this comet with those of the years 1531, 1607, and 1682, announced its next return for the end of the year 1758 or the beginning of the year 1759. The learned world awaited with impatience this return which was to confirm one of the greatest discoveries that have been made in the sciences, and fulfill the prediction of Seneca when he said, in speaking of the revolutions of those stars which fall from an enormous height: "The day will come when, by study pursued through several ages, the things now concealed will appear with evidence; and posterity will be astonished that truths so clear had escaped us." Clairaut then undertook to submit to analysis the perturbations which the comet had experienced by the action of the two great planets, Jupiter and Saturn; after immense calculations he fixed its next passage at the perihelion toward the beginning of April, 1759, which was actually verified by observation. The regularity which astronomy shows us in the movements of the comets doubtless exists also in all phenomena.

The curve described by a simple molecule of air or vapor is regulated in a manner just as certain as the planetary orbits; the only difference between them is that due to our ignorance.

Probability is relative, in part to this ignorance, in part to our knowledge. We know that of three or a greater number of events a single one ought to occur; but nothing induces us to believe that one of them will occur rather than the others. In this state of indecision it is impossible for us to announce their occurrence with certainty. It is, however, probable that one of these events, chosen at will, will not occur because we see several cases equally possible which exclude its occurrence, while only a single one favors it.

The theory of chance consists in reducing all the events of the same kind to a certain number of cases equally possible, that is to say, to such as we may be equally undecided about in regard to their existence, and in determining the number of cases favorable to the event whose probability is sought. The ratio of this number to that of all the cases possible is the measure of this probability, which is thus simply a fraction whose numerator is the number of favorable cases and whose denominator is the number of all the cases possible.

The preceding notion of probability supposes that, in increasing in the same ratio the number of favorable cases and that of all the cases possible, the probability remains the same. In order to convince ourselves let us take two urns, A and B, the first containing four white and two black balls, and the second containing only two white balls and one black one. We may imagine the two black balls of the first urn attached by a thread which breaks at the moment when one of them is seized in order to be drawn out, and the four white balls thus forming two similar systems. All the chances which will favor the seizure of one of the balls of the black system will lead to a black ball. If we conceive now that the threads which unite the balls do not break at all, it is clear that the number of possible chances will not change any more than that of the chances favorable to the extraction of the black balls; but two balls will be drawn from the urn at the same time; the probability of drawing a black ball from the urn A will then be the same as at first. But then we have obviously the case of urn B with the single difference that the three balls of this last urn would be replaced by three systems of two balls invariably connected.

When all the cases are favorable to an event the probability changes to certainty and its expression becomes equal to unity. Upon this condition, certainty and probability are comparable, although there may be an essential difference between the two states of the mind when a truth is rigorously demonstrated to it, or when it still perceives a small source of error.

In things which are only probable the difference of the data, which each man has in regard to them, is one of the principal causes of the diversity of opinions which prevail in regard to the same objects. Let us suppose, for example, that we have three urns, A, B, C, one of which contains only black balls while the two others contain only white balls; a ball is to be drawn from the urn C and the probability is demanded that this ball will be black. If we do not know which of the three urns contains black balls only, so that there is no reason to believe that it is C rather than B or A, these three hypotheses will appear equally possible, and since a black ball can be drawn only in the first hypothesis, the probability of drawing it is equal to one third. If it is known that the urn A contains white balls only, the indecision then extends only to the urns B and C, and the probability that the ball drawn from the urn C will be black is one half. Finally this probability changes to certainty if we are assured that the urns A and B contain white balls only.

It is thus that an incident related to a numerous assembly finds various degrees of credence, according to the extent of knowledge of the auditors. If the man who reports it is fully convinced of it and if, by his position and character, he inspires great confidence, his statement, however extraordinary it may be, will have for the auditors who lack information the same degree of probability as an ordinary statement made by the same man, and they will have entire faith in it. But if some one of them knows that the same incident is rejected by other equally trustworthy men, he will be in doubt and the incident will be discredited by the enlightened auditors, who will reject it whether it be in regard to facts well averred or the immutable laws of nature.

It is to the influence of the opinion of those whom the multitude judges best informed and to whom it has been accustomed to give its confidence in regard to the most important matters of life that the propagation of those errors is due which in times of ignorance have covered the face of the earth. Magic and astrology offer us two great examples. These errors inculcated in infancy, adopted without examination, and having for a basis only universal credence, have maintained themselves during a very long time; but at last the progress of science has destroyed them in the minds of enlightened men, whose opinion consequently has caused them to disappear even among the common people, through the power of imitation and habit which had so generally spread them abroad. This power, the richest resource of the moral world, establishes and conserves in a whole nation ideas entirely contrary to those which it upholds elsewhere with the same authority. What indulgence should we not have then for opinions different from ours, when this difference often depends only upon the various points of view where circumstances have placed us! Let us enlighten those whom we judge insufficiently instructed; but first let us examine critically our own opinions and weigh with impartiality their respective probabilities.

The difference of opinions depends, however, upon the manner in which the influence of known data is determined. The theory of probabilities holds to considerations so delicate that it is not surprising that with the same data two persons arrive at different results, especially in very complicated questions. Let us examine now the general principles of this theory.

THE GENERAL PRINCIPLES OF THE CALCULUS OF PROBABILITIES

First Principle.—The first of these principles is the definition itself of probability, which, as has been seen, is the ratio of the number of favorable cases to that of all the cases possible.

Second Principle.—But that supposes the various cases equally possible. If they are not so, we will determine first their respective possibilities, whose exact appreciation is one of the most delicate points of the theory of chance. Then the probability will be the sum of the possibilities of each favorable case. Let us illustrate this principle by an example.

Let us suppose that we throw into the air a large and very thin coin whose two large opposite faces, which we will call heads and tails, are perfectly similar. Let us find the probability of throwing heads at least one time in two throws. It is clear that four equally possible cases may arise, namely, heads at the first and at the second throw; heads at the first throw and tails at the second; tails at the first throw and heads at the second; finally, tails at both throws. The first three cases are favorable to the event whose probability is sought; consequently this probability is equal to ¾; so that it is a bet of three to one that heads will be thrown at least once in two throws.

We can count at this game only three different cases, namely, heads at the first throw, which dispenses with throwing a second time; tails at the first throw and heads at the second; finally, tails at the first and at the second throw. This would reduce the probability to ⅔ if we should consider with d'Alembert these three cases as equally possible. But it is apparent that the probability of throwing heads at the first throw is ½, while that of the two other cases is ¼, the first case being a simple event which corresponds to two events combined: heads at the first and at the second throw, and heads at the first throw, tails at the second. If we then, conforming to the second principle, add the possibility ½ of heads at the first throw to the possibility ¼ of tails at the first throw and heads at the second, we shall have ¾ for the probability sought, which agrees with what is found in the supposition when we play the two throws. This supposition does not change at all the chance of him who bets on this event; it simply serves to reduce the various cases to the cases equally possible.

Third Principle.—One of the most important points of the theory of probabilities and that which tends the most to illusions is the manner in which these probabilities increase or diminish by their mutual combination. If the events are independent of one another, the probability of their combined existence is the product of their respective probabilities. Thus the probability of throwing one ace with a single die is ⅙; that of throwing two aces in throwing two dice at the same time is 1/36. Each face of the one being able to combine with the six faces of the other, there are in fact thirty-six equally possible cases, among which one single case gives two aces. Generally the probability that a simple event in the same circumstances will occur consecutively a given number of times is equal to the probability of this simple event raised to the power indicated by this number. Having thus the succesive powers of a fraction less than unity diminishing without ceasing, an event which depends upon a series of very great probabilities may become extremely improbable. Suppose then an incident be transmitted to us by twenty witnesses in such manner that the first has transmitted it to the second, the second to the third, and so on. Suppose again that the probability of each testimony be equal to the fraction 9/10; that of the incident resulting from the testimonies will be less than 1/8. We cannot better compare this diminution of the probability than with the extinction of the light of objects by the interposition of several pieces of glass. A relatively small number of pieces suffices to take away the view of an object that a single piece allows us to perceive in a distinct manner. The historians do not appear to have paid sufficient attention to this degradation of the probability of events when seen across a great number of successive generations; many historical events reputed as certain would be at least doubtful if they were submitted to this test.

In the purely mathematical sciences the most distant consequences participate in the certainty of the principle from which they are derived. In the applications of analysis to physics the results have all the certainty of facts or experiences. But in the moral sciences, where each inference is deduced from that which precedes it only in a probable manner, however probable these deductions may be, the chance of error increases with their number and ultimately surpasses the chance of truth in the consequences very remote from the principle.

Fourth Principle.—When two events depend upon each other, the probability of the compound event is the product of the probability of the first event and the probability that, this event having occurred, the second will occur. Thus in the preceding case of the three urns A, B, C, of which two contain only white balls and one contains only black balls, the probability of drawing a white ball from the urn C is 2/3, since of the three urns only two contain balls of that color. But when a white ball has been drawn from the urn C, the indecision relative to that one of the urns which contain only black balls extends only to the urns A and B; the probability of drawing a white ball from the urn B is 1/2; the product of 2/3 by 1/2, or 1/3, is then the probability of drawing two white balls at one time from the urns B and C.

We see by this example the influence of past events upon the probability of future events. For the probability of drawing a white ball from the urn B, which primarily is ⅔, becomes ½ when a white ball has been drawn from the urn C; it would change to certainty if a black ball had been drawn from the same urn. We will determine this influence by means of the following principle, which is a corollary of the preceding one.

Fifth Principle.—If we calculate *a priori* the probability of the occurred event and the probability of an event composed of that one and a second one which is expected, the second probability divided by the first will be the probability of the event expected, drawn from the observed event.

Here is presented the question raised by some philosophers touching the influence of the past upon the probability of the future. Let us suppose at the play of heads and tails that heads has occurred oftener than tails. By this alone we shall be led to believe than in the constitution of the coin there is a secret cause which favors it. Thus in the conduct of life constant happiness is a proof of competency which should induce us to employ preferably happy persons. But if by the unreliability of circumstances we are constantly brought back to a state of absolute indecision, if, for example, we change the coin at each throw at the play of heads and tails, the past can shed no light upon the future and it would be absurd to take account of it.

Sixth Principle.—Each of the causes to which an observed event may be attributed is indicated with just as much likelihood as there is probability that the event will take place, supposing the event to be constant. The probability of the existence of any one of these causes is then a fraction whose numerator is the probability of the event resulting from this cause and whose denominator is the sum of the similar probabilities relative to all the causes; if these various causes, considered *a priori*, are unequally probable, it is necessary, in place of the probability of the event resulting from each cause, to employ the product of this probability by the possibility of the cause itself. This is the fundamental principle of this branch of the analysis of chances which consists in passing from events to causes.

This principle gives the reason why we attribute regular events to a particular cause. Some philosophers have thought that these events are less possible than others and that at the play of heads and tails, for example, the combination in which heads occurs twenty successive times is less easy in its nature than those where heads and tails are mixed in an irregular manner. But this opinion supposes that past events have an influence on the possibility of future events, which is not at all admissible. The regular combinations occur more rarely only because they are less numerous. If we seek a cause wherever we perceive symmetry, it is not that we regard a symmetrical event as less possible than the others, but, since this event ought to be the effect of a regular cause or that of chance, the first of these suppositions is more probable than the second. On a table we see letters arranged in this order, *Constantinople*, and we judge that this arrangement is not the result of chance, not because it is less possible than the others, for if this word were not employed in any language we should not suspect it came from any particular cause, but this word being in use among us, it is incomparably more probable that some person has thus arranged the aforesaid letters than that this arrangement is due to chance.

This is the place to define the word *extraordinary*. We arrange in our thought all possible events in various classes; and we regard as *extraordinary* those classes which include a very small number. Thus at the play of heads and tails the occurrence of heads a hundred successive times appears to us extraordinary because of the almost infinite number of combinations which may occur in a hundred throws; and if we divide the combinations into regular series containing an order easy to comprehend, and into irregular series, the latter are incomparably more numerous. The drawing of a white ball from an urn which among a million balls contains only one of this color, the others being black, would appear to us likewise extraordinary, because we form only two classes of events relative to the two colors. But the drawing of the number 475813, for example, from an urn that contains a million numbers seems to us an ordinary event; because, comparing individually the numbers with one another without dividing them into classes, we have no reason to believe that one of them will appear sooner than the others.

From what precedes, we ought generally to conclude that the more extraordinary the event, the greater the need of its being supported by strong proofs. For those who attest it, being able to deceive or to have been deceived, these two causes are as much more probable as the reality of the event is less. We shall see this particularly when we come to speak of the probability of testimony.

Seventh Principle.—The probability of a future event is the sum of the products of the probability of each cause, drawn from the event observed, by the probability that, this cause existing, the future event will occur. The following example will illustrate this principle.

Let us imagine an urn which contains only two balls, each of which may be either white or black. One of these balls is drawn and is put back into the urn before proceeding to a new draw. Suppose that in the first two draws white balls have been drawn; the probability of again drawing a white ball at the third draw is required.

Only two hypotheses can be made here: either one of the balls is white and the other black, or both are white. In the first hypothesis the probability of the event observed is $\frac{1}{4}$; it is unity or certainty in the second. Thus in regarding these hypotheses as so many causes, we shall have from the sixth principle $\frac{1}{5}$ and $\frac{4}{5}$ for their respective probabilities. But if the first hypothesis occurs, the probability of drawing a white ball at the third draw is $\frac{1}{2}$; it is equal to certainty in the second hypothesis; multiplying then the last probabilities by those of the corresponding hypotheses, the sum of the products, or $\frac{9}{10}$, will be the probability of drawing a white ball at the third draw.

When the probability of a single event is unknown we may suppose it equal to any value from zero to unity. The probability of each of these hypotheses, drawn from the event observed, is, by the sixth principle, a fraction whose numerator is the probability of the event in this hypothesis and whose denominator is the sum of the similar probabilities relative to all the hypotheses. Thus the probability that the possibility of the event is comprised within given limits is the sum of the fractions comprised within these limits. Now if we multiply each fraction by the probability of the future event, determined in the corresponding hypothesis, the sum of the products relative to all the hypotheses will be, by the seventh principle, the probability of the future event drawn from the event observed. Thus we find that an event having occurred successively any number of times, the probability that it will happen again the next time is equal to this number increased by unity divided by the same number, increased by two units. Placing the most ancient epoch of history at five thousand years ago, or at 1826213 days, and the sun having risen constantly in the interval at each revolution of twenty-four hours, it is a bet of 1826214 to one that it will rise again to-morrow. But this number is incomparably greater for him who, recognizing in the totality of phenomena the principal regulator of days and seasons, sees that nothing at the present moment can arrest the course of it.

Buffon in his *Political Arithmetic* calculates differently the preceding probability. He supposes that it differs from unity only by a fraction whose numerator is unity and whose denominator is the number 2 raised to a power equal to the number of days which have elapsed since the epoch. But the true manner of relating past events with the probability of causes and of future events was unknown to this illustrious writer.

Lest men suspect your tale untrue,
Keep probability in view.

—JOHN GAY

2 The Red and the Black

By CHARLES SANDERS PEIRCE [1]

THE theory of probabilities is simply the science of logic quantitatively treated. There are two conceivable certainties with reference to any hypothesis, the certainty of its truth and the certainty of its falsity. The numbers *one* and *zero* are appropriated, in this calculus, to marking these extremes of knowledge; while fractions having values intermediate between them indicate, as we may vaguely say, the degrees in which the evidence leans toward one or the other. The general problem of probabilities is, from a given state of facts, to determine the numerical probability of a possible fact. This is the same as to inquire how much the given facts are worth, considered as evidence to prove the possible fact. Thus the problem of probabilities is simply the general problem of logic.

Probability is a continuous quantity, so that great advantages may be expected from this mode of studying logic. Some writers have gone so far as to maintain that, by means of the calculus of chances, every solid inference may be represented by legitimate arithmetical operations upon the numbers given in the premises. If this be, indeed, true, the great problem of logic, how it is that the observation of one fact can give us knowledge of another independent fact, is reduced to a mere question of arithmetic. It seems proper to examine this pretension before undertaking any more recondite solution of the paradox.

But, unfortunately, writers on probabilities are not agreed in regard to this result. This branch of mathematics is the only one, I believe, in which good writers frequently get results entirely erroneous. In elementary geometry the reasoning is frequently fallacious, but erroneous conclusions are avoided; but it may be doubted if there is a single extensive treatise on probabilities in existence which does not contain solutions absolutely indefensible. This is partly owing to the want of any regular method of procedure; for the subject involves too many subtilities to make it easy to put its problems into equations without such an aid. But, beyond this, the fundamental principles of its calculus are more or less in dispute. In regard to that class of questions to which it is chiefly applied for practical purposes, there is comparatively little doubt; but in regard to others to which it has been sought to extend it, opinion is somewhat unsettled.

[1] For a biographical essay on Peirce, see p. 1767.

This last class of difficulties can only be entirely overcome by making the idea of probability perfectly clear in our minds in the way set forth in our last paper.[2]

To get a clear idea of what we mean by probability, we have to consider what real and sensible difference there is between one degree of probability and another.

The character of probability belongs primarily, without doubt, to certain inferences. Locke explains it as follows: After remarking that the mathematician positively knows that the sum of the three angles of a triangle is equal to two right angles because he apprehends the geometrical proof, he thus continues: "But another man who never took the pains to observe the demonstration, hearing a mathematician, a man of credit, affirm the three angles of a triangle to be equal to two right ones, *assents* to it; i.e., receives it for true. In which case the foundation of his assent is the probability of the thing, the proof being such as, for the most part, carries truth with it; the man on whose testimony he receives it not being wont to affirm anything contrary to, or besides his knowledge, especially in matters of this kind." The celebrated *Essay concerning Human Understanding* contains many passages which, like this one, make the first steps in profound analyses which are not further developed. It was shown in the first of these papers that the validity of an inference does not depend on any tendency of the mind to accept it, however strong such tendency may be; but consists in the real fact that, when premises like those of the argument in question are true, conclusions related to them like that of this argument are also true. It was remarked that in a logical mind an argument is always conceived as a member of a *genus* of arguments all constructed in the same way, and such that, when their premises are real facts, their conclusions are so also. If the argument is demonstrative, then this is always so; if it is only probable, then it is for the most part so. As Locke says, the probable argument is "*such as* for the most part carries truth with it."

According to this, that real and sensible difference between one degree of probability and another, in which the meaning of the distinction lies, is that in the frequent employment of two different modes of inference, one will carry truth with it oftener than the other. It is evident that this is the only difference there is in the existing fact. Having certain premises, a man draws a certain conclusion, and as far as this inference alone is concerned the only possible practical question is whether that conclusion is true or not, and between existence and non-existence there is no middle term. "Being only is and nothing is altogether not," said Parmenides; and this is in strict accordance with the analysis of the conception of reality

[2] [The reference is to the essay, "How to Make Our Ideas Clear." ED.]

given in the last paper. For we found that the distinction of reality and fiction depends on the supposition that sufficient investigation would cause one opinion to be universally received and all others to be rejected. That presupposition, involved in the very conceptions of reality and figment, involves a complete sundering of the two. It is the heaven-and-hell idea in the domain of thought. But, in the long run, there is a real fact which corresponds to the idea of probability, and it is that a given mode of inference sometimes proves successful and sometimes not, and that in a ratio ultimately fixed. As we go on drawing inference after inference of the given kind, during the first ten or hundred cases the ratio of successes may be expected to show considerable fluctuations; but when we come into the thousands and millions, these fluctuations become less and less; and if we continue long enough, the ratio will approximate toward a fixed limit. We may, therefore, define the probability of a mode of argument as the proportion of cases in which it carries truth with it.

The inference from the premise, A, to the conclusion, B, depends, as we have seen, on the guiding principle, that if a fact of the class A is true, a fact of the class B is true. The probability consists of the fraction whose numerator is the number of times in which both A and B are true, and whose denominator is the total number of times in which A is true, whether B is so or not. Instead of speaking of this as the probability of the inference, there is not the slightest objection to calling it the probability that, if A happens, B happens. But to speak of the probability of the event B, without naming the condition, really has no meaning at all. It is true that when it is perfectly obvious what condition is meant, the ellipsis may be permitted. But we should avoid contracting the habit of using language in this way (universal as the habit is), because it gives rise to a vague way of thinking, as if the action of causation might either determine an event to happen or determine it not to happen, or leave it more or less free to happen or not, so as to give rise to an *inherent* chance in regard to its occurrence. It is quite clear to me that some of the worst and most persistent errors in the use of the doctrine of chances have arisen from this vicious mode of expression.[3]

But there remains an important point to be cleared up. According to what has been said, the idea of probability essentially belongs to a kind of inference which is repeated indefinitely. An individual inference must be either true or false, and can show no effect of probability; and, therefore, in reference to a single case considered in itself, probability can have no meaning. Yet if a man had to choose between drawing a card from a

[3] The conception of probability here set forth is substantially that first developed by Mr. Venn, in his *Logic of Chance*. Of course, a vague apprehension of the idea had always existed, but the problem was to make it perfectly clear, and to him belongs the credit of first doing this.

pack containing twenty-five red cards and a black one, or from a pack containing twenty-five black cards and a red one, and if the drawing of a red card were destined to transport him to eternal felicity, and that of a black one to consign him to everlasting woe, it would be folly to deny that he ought to prefer the pack containing the larger portion of red cards, although, from the nature of the risk, it could not be repeated. It is not easy to reconcile this with our analysis of the conception of chance. But suppose he should choose the red pack, and should draw the wrong card, what consolation would he have? He might say that he had acted in accordance with reason, but that would only show that his reason was absolutely worthless. And if he should choose the right card, how could he regard it as anything but a happy accident? He could not say that if he had drawn from the other pack, he might have drawn the wrong one, because an hypothetical proposition such as, "if A, then B," means nothing with reference to a single case. Truth consists in the existence of a real fact corresponding to the true proposition. Corresponding to the proposion, "if A, then B," there may be the fact that *whenever* such an event as A happens such an event as B happens. But in the case supposed, which has no parallel as far as this man is concerned, there would be no real fact whose existence could give any truth to the statement that, if he had drawn from the other pack, he might have drawn a black card. Indeed, since the validity of an inference consists in the truth of the hypothetical proposition that *if* the premises be true the conclusion will also be true, and since the only real fact which can correspond to such a proposition is that whenever the antecedent is true the consequent is so also, it follows that there can be no sense in reasoning in an isolated case, at all.

These considerations appear, at first sight, to dispose of the difficulty mentioned. Yet the case of the other side is not yet exhausted. Although probability will probably manifest its effect in, say, a thousand risks, by a certain proportion between the numbers of successes and failures, yet this, as we have seen, is only to say that it certainly will, at length, do so. Now the number of risks, the number of probable inferences, which a man draws in his whole life, is a finite one, and he cannot be absolutely *certain* that the mean result will accord with the probabilities at all. Taking all his risks collectively, then, it cannot be certain that they will not fail, and his case does not differ, except in degree, from the one last supposed. It is an indubitable result of the theory of probabilities that every gambler, if he continues long enough, must ultimately be ruined. Suppose he tries the martingale, which some believe infallible, and which is, as I am informed, disallowed in the gambling-houses. In this method of playing, he first bets say \$1; if he loses it he bets \$2; if he loses that he bets \$4; if he loses that he bets \$8; if he then gains he has lost $1 + 2 + 4 = 7$, and he has gained \$1 more; and no matter how many bets he loses, the first one he gains will make him $1 richer than he was in the beginning. In that way, he will probably gain at first; but, at last, the time will come when the run of luck is so against him that he will not have money enough to double, and must, therefore, let his bet go. This will *probably* happen before he has won as much as he had in the first place, so that this run against him will leave him poorer than he began; some time or other it will be sure to happen. It is true that there is always a possibility of his winning any sum the bank can pay, and we thus come upon a celebrated paradox that, though he is certain to be ruined, the value of his expectation calculated according to the usual rules (which omit this consideration) is large. But, whether a gambler plays in this way or any other, the same thing is true, namely, that if he plays long enough he will be sure some time to have such a run against him as to exhaust his entire fortune. The same thing is true of an insurance company. Let the directors take the utmost pains to be independent of great conflagrations and pestilences, their actuaries can tell them that, according to the doctrine of chances, the time must come, at last, when their losses will bring them to a stop. They may tide over such a crisis by extraordinary means, but then they will start again in a weakened state, and the same thing will happen again all the sooner. An actuary might be inclined to deny this, because he knows that the expectation of his company is large, or perhaps (neglecting the interest upon money) is infinite. But calculations of expectations leave out of account the circumstance now under consideration, which reverses the whole thing. However, I must not be understood as saying that insurance is on this account unsound, more than other kinds of business. All human affairs rest upon probabilities, and the same thing is true everywhere. If man were immortal he could be perfectly sure of seeing the day when everything in which he had trusted should betray his trust, and, in short, of coming eventually to hopeless misery. He would break down, at last, as every good fortune, as every dynasty, as every civilization does. In place of this we have death.

But what, without death, would happen to every man, with death must happen to some man. At the same time, death makes the number of our risks, of our inferences, finite, and so makes their mean result uncertain. The very idea of probability and of reasoning rests on the assumption that this number is indefinitely great. We are thus landed in the same difficulty as before, and I can see but one solution of it. It seems to me that we are driven to this, that logicality inexorably requires that our interests shall *not* be limited. They must not stop at our own fate, but must embrace the whole community. This community, again, must not be limited, but must extend to all races of beings with whom we can come into immediate or mediate intellectual relation. It must reach, however, vaguely, beyond this geological epoch, beyond all bounds. He who would not sacrifice his own soul to save the whole world, is, as it seems to me, illogical in all his inferences, collectively. Logic is rooted in the social principle.

To be logical men should not be selfish; and, in point of fact, they are not so selfish as they are thought. The willful prosecution of one's desires is a different thing from selfishness. The miser is not selfish; his money does him no good, and he cares for what shall become of it after his death. We are constantly speaking of *our* possessions on the Pacific, and of *our* destiny as a republic, where no personal interests are involved, in a way which shows that we have wider ones. We discuss with anxiety the possible exhaustion of coal in some hundreds of years, or the cooling-off of the sun in some millions, and show in the most popular of all religious tenets that we can conceive the possibility of a man's descending into hell for the salvation of his fellows.

Now, it is not necessary for logicality that a man should himself be capable of the heroism of self-sacrifice. It is sufficient that he should recognize the possibility of it, should perceive that only that man's inferences who has it are really logical, and should consequently regard his own as being only so far valid as they would be accepted by the hero. So far as he thus refers his inferences to that standard, he becomes identified with such a mind.

This makes logicality attainable enough. Sometimes we can personally attain to heroism. The soldier who runs to scale a wall knows that he will probably be shot, but that is not all he cares for. He also knows that if all the regiment, with whom in feeling he identifies himself, rush forward at once, the fort will be taken. In other cases we can only imitate the virtue. The man whom we have supposed as having to draw from the two packs, who if he is not a logician will draw from the red pack from mere habit, will see, if he is logician enough, that he cannot be logical so long as he is concerned only with his own fate, but that that man who should care equally for what was to happen in all possible cases of the sort could act logically, and would draw from the pack with the most red cards, and thus, though incapable himself of such sublimity, our logician would imitate the effect of that man's courage in order to share his logicality.

But all this requires a conceived identification of one's interests with those of an unlimited community. Now, there exist no reasons, and a later discussion will show that there can be no reasons, for thinking that the human race, or any intellectual race, will exist forever. On the other hand, there can be no reason against it; [4] and, fortunately, as the whole requirement is that we should have certain sentiments, there is nothing in

[4] I do not here admit an absolutely unknowable. Evidence could show us what would probably be the case after any given lapse of time; and though a subsequent time might be assigned which that evidence might not cover, yet further evidence would cover it.

the facts to forbid our having a *hope*, or calm and cheerful wish, that the community may last beyond any assignable date.

It may seem strange that I should put forward three sentiments, namely, interest in an indefinite community, recognition of the possibility of this interest being made supreme, and hope in the unlimited continuance of intellectual activity, as indispensable requirements of logic. Yet, when we consider that logic depends on a mere struggle to escape doubt, which, as it terminates in action, must begin in emotion, and that, furthermore, the only cause of our planting ourselves on reason is that other methods of escaping doubt fail on account of the social impulse, why should we wonder to find social sentiment presupposed in reasoning? As for the other two sentiments which I find necessary, they are so only as supports and accessories of that. It interests me to notice that these three sentiments seem to be pretty much the same as that famous trio of Charity, Faith, and Hope, which, in the estimation of St. Paul, are the finest and greatest of spiritual gifts. Neither Old nor New Testament is a textbook of the logic of science, but the latter is certainly the highest existing authority in regard to the dispositions of heart which a man ought to have.

He who has heard the same thing told by 12,000 eye-witnesses has only 12,000 probabilities, which are equal to one strong probability, which is far from certainty.
—VOLTAIRE

A reasonable probability is the only certainty.
—E. W. HOWE (*Sinner Sermons*)

There is a tradition of opposition between adherents of induction and of deduction. In my view it would be just as sensible for the two ends of a worm to quarrel.
—ALFRED NORTH WHITEHEAD

3 The Probability of Induction[1]

By CHARLES SANDERS PEIRCE[2]

WE have found[3] that every argument derives its force from the general truth of the class of inferences to which it belongs; and that probability is the proportion of arguments carrying truth with them among those of any *genus*. This is most conveniently expressed in the nomenclature of the medieval logicians. They called the fact expressed by a premiss an *antecedent*, and that which follows from it its *consequent*; while the leading principle, that every (or almost every) such antecedent is followed by such a consequent, they termed the *consequence*. Using this language, we may say that probability belongs exclusively to *consequences*, and the probability of any consequence is the number of times in which antecedent and consequent both occur divided by the number of all the times in which the antecedent occurs. From this definition are deduced the following rules for the addition and multiplication of probabilities:

Rule for the Addition of Probabilities.—Given the separate probabilities of two consequences having the same antecedent and incompatible consequents. Then the sum of these two numbers is the probability of the consequence, that from the same antecedent one or other of those consequents follows.

Rule for the Multiplication of Probabilities.—Given the separate probabilities of the two consequences, "If A, then B," and "If both A and B, then C." Then the product of these two numbers is the probability of the consequence, "If A, then both B and C."

Special Rule for the Multiplication of Independent Probabilities.—Given the separate probabilities of two consequences having the same antecedents, "If A, then B," and "If A, then C." Suppose that these con-

[1] This chapter, with Peirce's title, is the entire fourth paper of a series, *Popular Science Monthly* 1878; C. Hartshorne and P. Weiss, *Collected Papers of C.S.P.*, Cambridge, 1931–35.

[2] For a biographical essay on Peirce, see p. 1767.

[3] The reference is to Peirce's essay, "The Doctrine of Chances," part of which is given on page 1334.

sequences are such that the probability of the second is equal to the probability of the consequence, "If both A and B, then C." Then the product of the two given numbers is equal to the probability of the consequence, "If A, then both B and C."

To show the working of these rules we may examine the probabilities in regard to throwing dice. What is the probability of throwing a six with one die? The antecedent here is the event of throwing a die; the consequent, its turning up a six. As the die has six sides, all of which are turned up with equal frequency, the probability of turning up any one is $1/6$. Suppose two dice are thrown, what is the probability of throwing sixes? The probability of either coming up six is obviously the same when both are thrown as when one is thrown—namely, $1/6$. The probability that either will come up six when the other does is also the same as that of its coming up six whether the other does or not. The probabilities are, therefore, independent; and, by our rule, the probability that both events will happen together is the product of their several probabilities, or $1/6 \times 1/6$. What is the probability of throwing deuce-ace? The probability that the first die will turn up ace and the second deuce is the same as the probability that both will turn up sixes—namely, $1/36$; the probability that the *second* will turn up ace and the *first* deuce is likewise $1/36$; these two events—first, ace; second, deuce; and, second, ace; first, deuce—are incompatible. Hence the rule for addition holds, and the probability that either will come up ace and the other deuce is $1/36 + 1/36$, or $1/18$.

In this way all problems about dice, etc., may be solved. When the number of dice thrown is supposed very large, mathematics (which may be defined as the art of making groups to facilitate numeration) comes to our aid with certain devices to reduce the difficulties.

The conception of probability as a matter of *fact, i.e.*, as the proportion of times in which an occurrence of one kind is accompanied by an occurrence of another kind, is termed by Mr. Venn the materialistic view of the subject. But probability has often been regarded as being simply the degree of belief which ought to attach to a proposition; and this mode of explaining the idea is termed by Venn the conceptualistic view. Most writers have mixed the two conceptions together. They, first, define the probability of an event as the reason we have to believe that it has taken place, which is conceptualistic; but shortly after they state that it is the ratio of the number of cases favourable to the event to the total number of cases favourable or contrary, and all equally possible. Except that this introduces the thoroughly unclear idea of cases equally possible in place of cases equally frequent, this is a tolerable statement of the materialistic view. The pure conceptualistic theory has been best expounded by Mr. De Morgan in his *Formal Logic: or, the Calculus of Inference, Necessary and Probable.*

The great difference between the two analyses is, that the conceptualists refer probability to an event, while the materialists make it the ratio of frequency of events of a *species* to those of a *genus* over that *species*, thus *giving it two terms instead of one*. The opposition may be made to appear as follows:

Suppose that we have two rules of inference, such that, of all the questions to the solution of which both can be applied, the first yields correct answers to $^{81}/_{100}$, and incorrect answers to the remaining $^{19}/_{100}$; while the second yields correct answers to $^{93}/_{100}$, and incorrect answers to the remaining $^{7}/_{100}$. Suppose, further, that the two rules are entirely independent as to their truth, so that the second answers correctly $^{93}/_{100}$ of the questions which the first answers correctly, and also $^{93}/_{100}$ of the questions which the first answers incorrectly, and answers incorrectly the remaining $^{7}/_{100}$ of the questions which the first answers correctly, and also the remaining $^{7}/_{100}$ of the questions which the first answers incorrectly. Then, of all the questions to the solution of which both rules can be applied—

both answer correctly . . . . $\frac{93}{100}$ of $\frac{81}{100}$, or $\frac{93 \times 81}{100 \times 100}$;

the second answers correctly and the first incorrectly $\frac{93}{100}$ of $\frac{19}{100}$, or $\frac{93 \times 19}{100 \times 100}$;

the second answers incorrectly and the first correctly $\frac{7}{100}$ of $\frac{81}{100}$, or $\frac{7 \times 81}{100 \times 100}$;

and both answer incorrectly . . . . . . $\frac{7}{100}$ of $\frac{19}{100}$, or $\frac{7 \times 19}{100 \times 100}$;

Suppose, now, that, in reference to any question, both give the same answer. Then (the questions being always such as are to be answered by *yes* or *no*), those in reference to which their answers agree are the same as those which both answer correctly together with those which both answer falsely, or

$$\frac{93 \times 81}{100 \times 100} + \frac{7 \times 19}{100 \times 100}$$

of all. The proportion of those which both answer correctly out of those their answers to which agree is, therefore—

$$\frac{\dfrac{93 \times 81}{100 \times 100}}{\dfrac{93 \times 81}{100 \times 100} + \dfrac{7 \times 19}{100 \times 100}} \text{ or } \frac{93 \times 81}{(93 \times 81) + (7 \times 19)}.$$

This is, therefore, the probability that, if both modes of inference yield the same result, that result is correct. We may here conveniently make use of another mode of expression. *Probability* is the ratio of the favourable cases to all the cases. Instead of expressing our result in terms of this ratio, we may make use of another—the ratio of favourable to unfavourable cases. This last ratio may be called the *chance* of an event. Then the chance of a true answer by the first mode of inference is $^{81}/_{19}$ and by the second is $^{93}/_{7}$; and the chance of a correct answer from both, when they agree, is—

$$\frac{81 \times 93}{19 \times 7} \text{ or } \frac{81}{19} \times \frac{93}{7},$$

or the product of the chances of each singly yielding a true answer.

It will be seen that a chance is a quantity which may have any magnitude, however great. An event in whose favour there is an even chance, or $^1/_1$, has a probability of $^1/_2$. An argument having an even chance can do nothing toward reënforcing others, since according to the rule its combination with another would only multiply the chance of the latter by 1.

Probability and chance undoubtedly belong primarily to consequences, and are relative to premisses; but we may, nevertheless, speak of the chance of an event absolutely, meaning by that the chance of the combination of all arguments in reference to it which exist for us in the given state of our knowledge. Taken in this sense it is incontestable that the chance of an event has an intimate connection with the degree of our belief in it. Belief is certainly something more than a mere feeling; yet there is a feeling of believing, and this feeling does and ought to vary with the chance of the thing believed, as deduced from all the arguments. Any quantity which varies with the chance might, therefore, it would seem, serve as a thermometer for the proper intensity of belief. Among all such quantities there is one which is peculiarly appropriate. When there is a very great chance, the feeling of belief ought to be very intense. Absolute certainty, or an infinite chance, can never be attained by mortals, and this may be represented appropriately by an infinite belief. As the chance diminishes the feeling of believing should diminish, until an even chance is reached, where it should completely vanish and not incline either toward or away from the proposition. When the chance becomes less, then a contrary belief should spring up and should increase in intensity as the chance diminishes, and as the chance almost vanishes (which it can never quite do) the contrary belief should tend toward an infinite intensity. Now, there is one quantity which, more simply than any other, fulfills these conditions; it is the *logarithm* of the chance. But there is another consideration which must, if admitted, fix us to this choice for our thermometer. It is that our belief ought to be proportional to the weight of evidence, in this sense, that two arguments which are entirely independent, neither weakening nor strengthening each other, ought, when they concur, to produce a belief equal to the sum of the intensities of belief which either would produce separately. Now, we have seen that the chances of independent concurrent arguments are to be multiplied together to get the chance of their combination, and therefore the quantities which best express the intensities of belief should be such that they are to be *added* when the *chances* are multiplied in order to produce the quantity which corresponds to the combined chance. Now, the logarithm is the only quantity which fulfills this condition. There is a general law of sensibility, called Fechner's psycho-physical law. It is that the intensity of any sensation is proportional to the logarithm of the external force which produces it. It is entirely in harmony with this law that the feeling of belief should be as the logarithm of the chance, this latter being the expression of the state of facts which produces the belief.

The rule for the combination of independent concurrent arguments takes a very simple form when expressed in terms of the intensity of belief, measured in the proposed way. It is this: Take the sum of all the feelings of belief which would be produced separately by all the arguments *pro*, subtract from that the similar sum for arguments *con*, and the remainder is the feeling of belief which we ought to have on the whole. This is a proceeding which men often resort to, under the name of *balancing reasons*.

These considerations constitute an argument in favour of the conceptualistic view. The kernel of it is that the conjoint probability of all the arguments in our possession, with reference to any fact, must be intimately connected with the just degree of our belief in that fact; and this point is supplemented by various others showing the consistency of the theory with itself and with the rest of our knowledge.

But probability, to have any value at all, must express a fact. It is, therefore, a thing to be inferred upon evidence. Let us, then, consider for a moment the formation of a belief of probability. Suppose we have a large bag of beans from which one has been secretly taken at random and hidden under a thimble. We are now to form a probable judgment of the colour of that bean, by drawing others singly from the bag and looking at them, each one to be thrown back, and the whole well mixed up after each drawing. Suppose the first drawing is white and the next black. We conclude that there is not an immense preponderance of either colour, and that there is something like an even chance that the bean under the thimble is black. But this judgment may be altered by the next few drawings. When we have drawn ten times, if 4, 5, or 6, are white, we have more confidence that the chance is even. When we have drawn a thousand times, if about half have been white, we have great confidence in this result. We now feel pretty sure that, if we were to make a large number of bets upon the colour of single beans drawn from the bag, we could approximately insure ourselves in the long run by betting each time upon the white, a confidence which would be entirely wanting if, instead of sampling the bag by 1,000 drawings, we had done so by only two. Now, as the whole utility of probability is to insure us in the long run, and as that assurance depends, not merely on the value of the chance, but also on the accuracy of the evaluation, it follows that we ought not to have the same feeling of belief in reference to all events of which the chance is even. In short, to express the proper state of our belief, not *one* number but *two* are requisite, the first depending on the inferred probability, the second on the amount of knowledge on which that probability is based. It is true that when our knowledge is very precise, when we have made many drawings from the bag, or, as in most of the examples in the books, when the total contents of the bag are absolutely known, the number which expresses the uncertainty of the assumed probability and its liability to be changed by further experience may become insignificant, or utterly vanish. But, when our knowledge is very slight, this number may be even more important than the probability itself; and when we have no knowledge at all this completely overwhelms the other, so that there is no sense in saying that the chance of the totally unknown event is even (for what expresses absolutely no fact has absolutely no meaning), and what ought to be said is that the chance is entirely indefinite. We thus perceive that the conceptualistic view, though answering well enough in some cases, is quite inadequate.

Suppose that the first bean which we drew from our bag were black. That would constitute an argument, no matter how slender, that the bean under the thimble was also black. If the second bean were also to turn out black, that would be a second independent argument reënforcing the first. If the whole of the first twenty beans drawn should prove black, our confidence that the hidden bean was black would justly attain considerable strength. But suppose the twenty-first bean were to be white and that we were to go on drawing until we found that we had drawn 1,010 black beans and 990 white ones. We should conclude that our first twenty beans being black was simply an extraordinary accident, and that in fact the proportion of white beans to black was sensibly equal, and that it was an even chance that the hidden bean was black. Yet according to the rule of *balancing reasons*, since all the drawings of black beans are so many independent arguments in favour of the one under the thimble being black, and all the white drawings so many against it, an excess of twenty black beans ought to produce the same degree of belief that the hidden bean was black, whatever the total number drawn.

In the conceptualistic view of probability, complete ignorance, where the judgment ought not to swerve either toward or away from the hypothesis, is represented by the probability ½.

But let us suppose that we are totally ignorant what coloured hair the inhabitants of Saturn have. Let us, then, take a colour-chart in which all possible colours are shown shading into one another by imperceptible degrees. In such a chart the relative areas occupied by different classes of colours are perfectly arbitrary. Let us inclose such an area with a closed line, and ask what is the chance on conceptualistic principles that the colour of the hair of the inhabitants of Saturn falls within that area? The answer cannot be indeterminate because we must be in some state of belief; and, indeed, conceptualistic writers do not admit indeterminate probabilities. As there is no certainty in the matter, the answer lies between *zero* and *unity*. As no numerical value is afforded by the data, the number must be determined by the nature of the scale of probability itself, and not by calculation from the data. The answer can, therefore, only be one-half, since the judgment should neither favour nor oppose the hypothesis. What is true of this area is true of any other one; and it will equally be true of a third area which embraces the other two. But the probability for each of the smaller areas being one-half, that for the larger should be at least unity, which is absurd.

All our reasonings are of two kinds: 1. *Explicative, analytic,* or *deductive*; 2. *Amplificative, synthetic,* or (loosely speaking) *inductive.* In explicative reasoning, certain facts are first laid down in the premisses. These facts are, in every case, an inexhaustible multitude, but they may often be summed up in one simple proposition by means of some regularity which runs through them all. Thus, take the proposition that Socrates was a man; this implies (to go no further) that during every fraction of a second of his whole life (or, if you please, during the greater part of them) he was a man. He did not at one instant appear as a tree and at another as a dog; he did not flow into water, or appear in two places at once; you could not put your finger through him as if he were an optical image, etc. Now, the facts being thus laid down, some order among some of them, not particularly made use of for the purpose of stating them, may perhaps be discovered; and this will enable us to throw part or all of them into a new statement, the possibility of which might have escaped attention. Such a statement will be the conclusion of an analytic inference. Of this sort are all mathematical demonstrations. But synthetic reasoning is of another kind. In this case the facts summed up in the conclusion are not among those stated in the premisses. They are different facts, as when one sees that the tide rises m times and concludes that it will rise the next time. These are the only inferences which increase our real knowledge, however useful the others may be.

In any problem in probabilities, we have given the relative frequency of certain events, and we perceive that in these facts the relative frequency of another event is given in a hidden way. This being stated makes the solution. This is therefore mere explicative reasoning, and is evidently entirely inadequate to the representation of synthetic reasoning, which goes out beyond the facts given in the premisses. There is, therefore, a manifest impossibility in so tracing out any probability for a synthetic conclusion.

Most treatises on probability contain a very different doctrine. They state, for example, that if one of the ancient denizens of the shores of the Mediterranean, who had never heard of tides, had gone to the bay of Biscay, and had there seen the tide rise, say m times, he could know that there was a probability equal to

$$\frac{m+1}{m+2}$$

that it would rise the next time. In a well-known work by Quetelet, much stress is laid on this, and it is made the foundation of a theory of inductive reasoning.

But this solution betrays its origin if we apply it to the case in which the man has never seen the tide rise at all; that is, if we put $m = 0$. In this case, the probability that it will rise the next time comes out ½, or, in other words, the solution involves the conceptualistic principle that there is an even chance of a totally unknown event. The manner in which it has been reached has been by considering a number of urns all containing the same number of balls, part white and part black. One urn contains all white balls, another one black and the rest white, a third two black and the rest white, and so on, one urn for each proportion, until an urn is reached containing only black balls. But the only possible reason for drawing any analogy between such an arrangement and that of Nature is the principle that alternatives of which we know nothing must be considered as equally probable. But this principle is absurd. There is an indefinite variety of ways of enumerating the different possibilities, which, on the application of this principle, would give different results. If there be any way of enumerating the possibilities so as to make them all equal, it is not that from which this solution is derived, but is the following: Suppose we had an immense granary filled with black and white balls well mixed up; and suppose each urn were filled by taking a fixed number of balls from this granary quite at random. The relative number of white balls in the granary might be anything, say one in three. Then in one-third of the urns the first ball would be white, and in two-thirds black. In one-third of those urns of which the first ball was white, and also in one-third of those in which the first ball was black, the second ball would be white. In this way, we should have a distribution like that shown in the following table, where w stands for a white ball and b for a black one. The reader can, if he chooses, verify the table for himself.

wwww.					
wwwb.	wwbw.	wbww.	bwww.		
wwwb.	wwbw.	wbww.	bwww.		
wwbb.	wbwb.	bwwb.	wbbw.	bwbw.	bbww.
wwbb.	wbwb.	bwwb.	wbbw.	bwbw.	bbww.
wwbb.	wbwb.	bwwb.	wbbw.	bwbw.	bbww.
wwbb.	wbwb.	bwwb.	wbbw.	bwbw.	bbww.
wbbb.	bwbb.	bbwb.	bbbw.		
wbbb.	bwbb.	bbwb.	bbbw.		
wbbb.	bwbb.	bbwb.	bbbw.		
wbbb.	bwbb.	bbwb.	bbbw.		
wbbb.	bwbb.	bbwb.	bbbw.		
wbbb.	bwbb.	bbwb.	bbbw.		
wbbb.	bwbb.	bbwb.	bbbw.		
wbbb.	bwbb.	bbwb.	bbbw.		
bbbb.					
bbbb.					
bbbb.					
bbbb.					
bbbb.					
bbbb.					
bbbb.					
bbbb.					
bbbb.					
bbbb.					
bbbb.					
bbbb.					
bbbb.					
bbbb.					
bbbb.					
bbbb.					

In the second group, where there is one b, there are two sets just alike; in the third there are 4, in the fourth 8, and in the fifth 16, doubling every time. This is because we have supposed twice as many black balls in the granary as white ones; had we supposed 10 times as many, instead of

1, 2, 4, 8, 16

sets we should have had

1, 10, 100, 1000, 10000

sets; on the other hand, had the numbers of black and white balls in the granary been even, there would have been but one set in each group. Now suppose two balls were drawn from one of these urns and were found to be both white, what would be the probability of the next one being white? If the two drawn out were the first two put into the urns, and the next to be drawn out were the third put in, then the probability of this third being white would be the same whatever the colours of the first two, for it has been supposed that just the same proportion of urns has the third ball white among those which have the first two *white-white*, *white-black*, *black-white*, and *black-black*. Thus, in this case, the chance of the third ball being white would be the same whatever the first two were. But, by inspecting the table, the reader can see that in each group all orders of the balls occur with equal frequency, so that it makes no difference whether they are drawn out in the order they were put in or not. Hence the colours of the balls already drawn have no influence on the probability of any other being white or black.

Now, if there be any way of enumerating the possibilities of Nature so as to make them equally probable, it is clearly one which should make one arrangement or combination of the elements of Nature as probable as another, that is, a distribution like that we have supposed, and it, therefore, appears that the assumption that any such thing can be done, leads simply to the conclusion that reasoning from past to future experience is absolutely worthless. In fact, the moment that you assume that the chances in favour of that of which we are totally ignorant are even, the problem about the tides does not differ, in any arithmetical particular, from the case in which a penny (known to be equally likely to come up heads and tails) should turn up heads m times successively. In short, it would be to assume that Nature is a pure chaos, or chance combination of independent elements, in which reasoning from one fact to another would be impossible; and since, as we shall hereafter see, there is no judgment of pure observation without reasoning, it would be to suppose all human cognition illusory and no real knowledge possible. It would be to suppose that if we have found the order of Nature more or less regular in the past, this has been by a pure run of luck which we may expect is now at an end. Now, it may be we have no scintilla of proof to the contrary, but reason is unnecessary in reference to that belief which is of all the most settled, which nobody doubts or can doubt, and which he who should deny would stultify himself in so doing.

The relative probability of this or that arrangement of Nature is something which we should have a right to talk about if universes were as plenty as blackberries, if we could put a quantity of them in a bag, shake them well up, draw out a sample, and examine them to see what proportion of them had one arrangement and what proportion another. But, even in that case, a higher universe would contain us, in regard to whose arrangements the conception of probability could have no applicability.

We have examined the problem proposed by the conceptualists, which, translated into clear language, is this: Given a synthetic conclusion; required to know out of all possible states of things how many will accord, to any assigned extent, with this conclusion; and we have found that it is only an absurd attempt to reduce synthetic to analytic reason, and that no definite solution is possible.

But there is another problem in connection with this subject. It is this: Given a certain state of things, required to know what proportion of all synthetic inferences relating to it will be true within a given degree of approximation. Now, there is no difficulty about this problem (except for its mathematical complication); it has been much studied, and the answer is perfectly well known. And is not this, after all, what we want to know much rather than the other? Why should we want to know the probability that the fact will accord with our conclusion? That implies that we are interested in all possible worlds, and not merely the one in which we find ourselves placed. Why is it not much more to the purpose to know the probability that our conclusion will accord with the fact? One of these questions is the first above stated and the other the second, and I ask the reader whether, if people, instead of using the word probability without any clear apprehension of their own meaning, had always spoken of relative frequency, they could have failed to see that what they wanted was not to follow along the synthetic procedure with an analytic one, in order to find the probability of the conclusion; but on the contrary, to begin with the fact at which the synthetic inference aims, and follow back to the facts it uses for premisses in order to see the probability of their being such as will yield the truth.

As we cannot have an urn with an infinite number of balls to represent the inexhaustibleness of Nature, let us suppose one with a finite number, each ball being thrown back into the urn after being drawn out, so that there is no exhaustion of them. Suppose one ball out of three is white and the rest black, and that four balls are drawn. Then the table on p. 1349 represents the relative frequency of the different ways in which these balls might be drawn. It will be seen that if we should judge by these four balls of the proportion in the urn, 32 times out of 81 we should find it ¼, and 24 times out of 81 we should find it ½, the truth being ⅓. To extend this table to high numbers would be great labour, but the mathematicians have found some ingenious ways of reckoning what the numbers would be. It is found that, if the true proportion of white balls is p, and s balls are drawn, then the error of the proportion obtained by the induction will be—

half the time within	$0{\cdot}477\sqrt{\frac{2p(1-p)}{s}}$
9 times out of 10 within	$1{\cdot}163\sqrt{\frac{2p(1-p)}{s}}$
99 times out of 100 within	$1{\cdot}821\sqrt{\frac{2p(1-p)}{s}}$
999 times out of 1,000 within	$2{\cdot}328\sqrt{\frac{2p(1-p)}{s}}$
9,999 times out of 10,000 within	$2{\cdot}751\sqrt{\frac{2p(1-p)}{s}}$
9,999,999,999 times out of 10,000,000,000 within	$4{\cdot}77\sqrt{\frac{2p(1-p)}{s}}$

The use of this may be illustrated by an example. By the census of 1870, it appears that the proportion of males among native white children under one year old was 0·5082, while among coloured children of the same age the proportion was only 0·4977. The difference between these is 0·0105, or about one in 100. Can this be attributed to chance, or would the difference always exist among a great number of white and coloured children under like circumstances? Here p may be taken at ½; hence $2p(1-p)$ is also ½. The number of white children counted was near 1,000,000; hence the fraction whose square-root is to be taken is about $1/2000000$. The root is about $1/1400$, and this multiplied by 0·477 gives about 0·0003 as the probable error in the ratio of males among the whites as obtained from the induction. The number of black children was about 150,000 which gives 0·0008 for the probable error. We see that the actual discrepancy is ten times the sum of these, and such a result would happen, according to our table, only once out of 10,000,000,000 censuses, in the long run.

It may be remarked that when the real value of the probability sought inductively is either very large or very small, the reasoning is more secure. Thus, suppose there were in reality one white ball in 100 in a certain urn, and we were to judge of the number by 100 drawings. The probability of drawing no white ball would be $366/1000$; that of drawing one white ball would be $370/1000$; that of drawing two would be $185/1000$; that of drawing three would be $61/1000$; that of drawing four would be $15/1000$; that of drawing five would be only $3/1000$, etc. Thus we should be tolerably certain of not being in error by more than one ball in 100.

It appears, then, that in one sense we can, and in another we cannot, determine the probability of synthetic inference. When I reason in this way:

> Ninety-nine Cretans in a hundred are liars;
> But Epimenides is a Cretan;
> Therefore, Epimenides is a liar;

I know that reasoning similar to that would carry truth 99 times in 100. But when I reason in the opposite direction:

> Minos, Sarpedon, Rhadamanthus, Deucalion, and Epimenides, are all the Cretans I can think of;
> But these were all atrocious liars;
> Therefore, pretty much all Cretans must have been liars;

I do not in the least know how often such reasoning would carry me right. On the other hand, what I do know is that some definite proportion of Cretans must have been liars, and that this proportion can be probably approximated to by an induction from five or six instances. Even in the worst case for the probability of such an inference, that in which about half the Cretans are liars, the ratio so obtained would probably not be in error by more than 1/6. So much I know; but, then, in the present case the inference is that pretty much all Cretans are liars, and whether there may not be a special improbability in that I do not know.

Late in the last century, Immanuel Kant asked the question, "How are synthetical judgments *a priori* possible?" By synthetical judgments he meant such as assert positive fact and are not mere affairs of arrangement; in short, judgments of the kind which synthetical reasoning produces, and which analytic reasoning cannot yield. By *a priori* judgments he meant such as that all outward objects are in space, every event has a cause, etc., propositions which according to him can never be inferred from experience. Not so much by his answer to this question as by the mere asking of it, the current philosophy of that time was shattered and destroyed, and a new epoch in its history was begun. But before asking *that* question he ought to have asked the more general one, "How are any synthetical judgments at all possible?" How is it that a man can observe one fact and straightway pronounce judgment concerning another different fact not involved in the first? Such reasoning, as we have seen, has, at least in the usual sense of the phrase, no definite probability; how, then, can it add to our knowledge? This is a strange paradox; the Abbé Gratry says it is a miracle, and that every true induction is an immediate inspiration from on high. I respect this explanation far more than many a pedantic attempt to solve the question by some juggle with probabilities, with the forms of syllogism, or what not. I respect it because it shows an appreciation of the depth of the problem, because it assigns an adequate cause, and because it is intimately connected—as the true account should be—with a general philosophy of the universe. At the same time, I do not accept this explanation, because an explanation should tell *how* a thing is done, and to assert a perpetual miracle seems to be an abandonment of all hope of doing that, without sufficient justification.

It will be interesting to see how the answer which Kant gave to his question about synthetical judgments *a priori* will appear if extended to the question of synthetical judgments in general. That answer is, that synthetical judgments *a priori* are possible because whatever is universally true is involved in the conditions of experience. Let us apply this to a general synthetical reasoning. I take from a bag a handful of beans; they are all purple, and I infer that all the beans in the bag are purple. How can I do that? Why, upon the principle that whatever is universally true of my experience (which is here the appearance of these different beans) is involved in the condition of experience. The condition of this special experience is that all these beans were taken from that bag. According to Kant's principle, then, whatever is found true of all the beans drawn from the bag must find its explanation in some peculiarity of the contents of the bag. This is a satisfactory statement of the principle of induction.

When we draw a deductive or analytic conclusion, our rule of inference is that facts of a certain general character are either invariably or in a certain proportion of cases accompanied by facts of another general character. Then our premiss being a fact of the former class, we infer with certainty or with the appropriate degree of probability the existence of a fact of the second class. But the rule for synthetic inference is of a different kind. When we sample a bag of beans we do not in the least assume that the fact of some beans being purple involves the necessity or even the probability of other beans being so. On the contrary, the conceptualist method of treating probabilities, which really amounts simply to the deductive treatment of them, when rightly carried out leads to the result that a synthetic inference has just an even chance in its favour, or in other words is absolutely worthless. The colour of one bean is entirely independent of that of another. But synthetic inference is founded upon a classification of facts, not according to their characters, but according to the manner of obtaining them. Its rule is, that a number of facts obtained in a given way will in general more or less resemble other facts obtained in the same way; or, *experiences whose conditions are the same will have the same general characters.*

In the former case, we know that premisses precisely similar in form to those of the given ones will yield true conclusions, just once in a calculable number of times. In the latter case, we only know that premisses obtained under circumstances similar to the given ones (though perhaps themselves very different) will yield true conclusions, at least once in a calculable number of times. We may express this by saying that in the case of analytic inference we know the probability of our conclusion (if the premisses are true), but in the case of synthetic inferences we only know the degree of trustworthiness of our proceeding. As all knowledge comes from synthetic inference, we must equally infer that all human certainty consists merely in our knowing that the processes by which our knowledge has been derived are such as must generally have led to true conclusions.

Though a synthetic inference cannot by any means be reduced to deduction, yet that the rule of induction will hold good in the long run may be deduced from the principle that reality is only the object of the final opinion to which sufficient investigation would lead. That belief gradually tends to fix itself under the influence of inquiry is, indeed, one of the facts with which logic sets out.

COMMENTARY ON

LORD KEYNES

JOHN MAYNARD KEYNES has been called, I think rightly, the outstanding intellectual of the age. He was dazzlingly versatile and creative, enlightened and bold in outlook. He had a high capacity for abstract thinking coupled with a rare gift of translating ideas into action. Men of widely differing views fell under his spell and thus he became immensely influential. Without doubt he was the most important economist of his time, both for getting his theories adopted and for provoking policies designed to counteract them.

Keynes was born at Cambridge on June 5, 1883. He was a happy child in a happy and peaceful home, elevated in its intellectual and social climate. His father, a Fellow of Pembroke College and for many years registrar of the university, was the noted logician and political economist John Neville Keynes; his mother was a woman of literary talent; only recently, at ninety-two, she published a charming volume of family reminiscences, *Gathering Up the Threads*.[1] Keynes himself was educated at Eton and at King's College, Cambridge, and had a brilliant undergraduate career. At Eton he played fiercely at Rugby, rowed, frequented the rare-book stalls, wrote Latin hymns, was a pre-eminent mathematical scholar, played Malvolio in *Twelfth Night*, made a speech on the thesis that "women are more fitted to rule than men," won almost every calfbound prize volume that was offered and generally overawed his tutors as well as his fellows. Nevertheless, he made friends. At Cambridge the competition was stiffer. It soon became apparent that although Keynes was a fine mathematician he would never be a great one. His métier lay elsewhere, and while he searched for it he took full part in the rich life of the university: the seances of undergraduate societies, the debates at the Union, the sharp controversy with men of mettle on any and every subject, preferably deep. He was admitted to the select group known as the "Apostles." This secret society, founded in the 1820s, had had Alfred Tennyson, William Kingdon Clifford, William Harcourt, James Clerk Maxwell and Henry Sidgwick as members; shortly before Keynes was invited to join, Frederick Maitland, Walter Raleigh, J. M. McTaggart, Alfred North Whitehead and Lowes Dickinson had belonged. It was gov-

[1] Cambridge (England), 1950. The principal sources of this sketch are R. F. Harrod, *The Life of John Maynard Keynes*, New York, 1951 (unkeyed quotations are from Harrod); *John Maynard Keynes, 1883–1946, A Memoir prepared by the direction of the Council of King's College, Cambridge*, Cambridge, 1949; Obituary of Keynes in *The Economist*, 1946; articles in the *Times Literary Supplement*, May 20, 1949 (pp. 321–322), and February 23, 1951 (pp. 109–111). I have drawn heavily on one of my own pieces appearing in *Scientific American*, April 1951.

erned by common intellectual tastes, common studies, common literary aspirations. Clifford described its agenda as "solving the universe with delight." Keynes formed lasting friendships with Lytton Strachey, Duncan Grant, Leonard Woolf, Thoby Stephens and Clive Bell. G. E. Moore influenced his ethical thought.[2] He attended the lectures of Alfred Marshall, who decisively persuaded him "to give up everything for economics."

At Cambridge and forever after Keynes was hospitable to sweeping, revolutionary ideas. The timid, makeshift or ambiguous proposal irritated and disgusted him. " 'What *exactly* do you mean?' was the phrase most frequently on our lips. If it appeared under cross-examination that you did not mean *exactly* anything, you lay under a strong suspicion of meaning nothing." [3] There was a streak of iconoclasm in him, and an impish spirit. "To tease, to flout" (says Harrod), "finally perhaps to overthrow venerable authorities—that was a sport which had great appeal for him." His intellectual predisposition vied at times with his practical traits, but on the whole they harmonized and reinforced one another. He was once dubbed Pozzo, for the Corsican diplomat Pozzo di Borgo, "not a diplomat of evil motive or base conduct, but certainly a schemer and man of many facets."

For two years after leaving Cambridge Keynes served as a civil servant in the India Office. His first book, *Indian Currency and Finance*, is a work of prime quality in its practical application of Marshall's principles to the complex problems of Indian exchange. Even those who have no use for the later Keynesian theories acknowledge it as a classic. In 1909 he was elected to a fellowship in King's College, a post which, despite many other activities, he held until his death. From 1911 to 1937 he was Lecturer in Economics.

While still at the India Office, Keynes began the *Treatise on Probability*, a work which occupied all his "spare time" for five or six years (1906–1912). The subject intrigued him, no less in its applications than in the theory. Keynes was a gambler. Throughout his life he successfully played the market and also more amiable games. "One day a friend called on Keynes, Duncan Grant and Adrian Stephen. 'I have just returned from Ostend,' he said. 'They are playing roulette there without a zero.' Without a second's hesitation the future author of the *Treatise on Probability* rose, urged them all to pack, and the four left on the night boat to return with all expenses paid and well in pocket." [4] The *Treatise* (from which the next selection has been taken) is a fascinating book. It is learned, clear, provocative in its philosophical speculations. "The book as a whole," wrote Bertrand Russell, "is one which it is impossible to praise too highly . . ."

[2] Moore's influence on Keynes, and his early philosophical beliefs are described in a delightful small volume, *Two Memoirs* (by J. M. Keynes, London, 1949).

[3] "My Early Beliefs," in *Two Memoirs*, p. 88.

[4] *Times Literary Supplement*, February 23, 1951, p. 111.

It was greatly approved by Whitehead, though he had criticized it severely (and constructively) in manuscript. Writing under the joint influence of Moore's *Principia Ethica* and Whitehead and Russell's *Principia Mathematica*, Keynes made a sharp attack on the frequency theory of probability. He himself favored the old and almost forgotten inductive concept, proposed by Bernoulli and Laplace, according to which the purpose of theory was to judge hypotheses and to guide judgments on the basis of evidence. Following Russell's view Keynes adopted a proposition rather than an event "as that which carries the attribute of probability." Moore's basic doctrine that the "good" is an indefinable, intuitive concept finds an echo in Keynes' similar judgment regarding the concept of probability. Keynes was attracted to the skepticism of Hume, according to which it is impossible to prove that because two events, X and Y, occur in conjunction there is a causal connection between them; he found it necessary to invent a rescuing hypothesis, that of Limited Independent Variety. With this it is assumed "that the experienced properties of things arise out of a finite number of generator properties," whence there is a finite probability, however small, in favor of a connection between any two properties. Evidence for the validity of the hypothesis is to be found, Keynes argued, in experience. For all its merits the *Treatise* is not a profoundly original work, but it amply exhibits Keynes' erudition, his powers as a logician, his method of assault upon a problem by concerting theoretical and practical weapons. It is notable also for an elegance of style and wit not to be found in other treatises on this formidable subject—except for the writings of Laplace and Poincaré.

From 1915 to 1919 Keynes held a post at the Treasury. The book that first brought him fame, *The Economic Consequences of the Peace*, was based on his experiences as a British official at Versailles. It was an impeccably reasoned study but its main impact was polemic. The set pieces on Clemenceau, Woodrow Wilson and Lloyd George are unforgettable; the castigation of the Allies for their hypocrisy and shortsightedness made a worldwide impression. *The Economic Consequences of the Peace* was a work of somber prophecy, of bitterness, magnanimity and courage; it incurred "great odium in official circles" and cast Keynes "for many years in the wilderness." But the very fact that the official world turned its back on him was both emancipation and challenge. His greatest achievements, which lay ahead, were his response.

It is outside the scope of this introduction to say more than a few words about Keynes' economic doctrines. His major contributions to this sphere of thought are most fully and systematically expressed in *Treatise on Money* and the later *General Theory of Employment, Interest and Money*. The *Treatise* advanced as its central doctrine the theory that there is no necessary link between savings and investment. This challenged the sacred penny-saved, penny-earned precept. "Hitherto the economist, as such, had tended to encourage economy and thrift in all circumstances. If Keynes' doctrine was correct, it was most desirable to do so in times of incipient inflation—but not at all times. On the contrary, in times of depression and unemployment it was desirable to encourage spending and lavishness." Thrift, in other words, is not an absolute virtue for either individuals or governments; it may do good or it may do harm. Usually, Keynes said, it does harm. In the *General Theory* Keynes was primarily concerned with analyzing the causes of unemployment. Here again he departed from traditional economic theory in denying, subject to certain qualifications, that high wages caused unemployment or that lowering them would raise the level of employment. In these difficult theoretical works and in some of his more popular writings (e.g., *How to Pay for the War*) are to be found the ideas which have so vastly influenced students, practical men, theoreticians and governments; which have shaped financial policies and altered the circumstances of millions of men and women to whom Keynes' thought is incomprehensible and his very name unknown.

A dozen different themes ran through Keynes' life. He was an intimate of the "Bloomsbury Circle," a shrewd and conservative banker, a successful speculator, a mathematician, a civil servant, a pamphleteer, a Cambridge don, a college bursar, a government spokesman and adviser, a book collector, a political scientist, a biographer, essayist and editor. In each of these activities he gained distinction, in some, pre-eminence. He worked terribly hard but understood how to take his ease. "He labored early and late, changing from problem to problem, from advising a friend to deciding on affairs of state, with an agility which enabled him at any moment to appear, and indeed to become, a man of leisure: the very opposite of Chaucer's Sergeant of the Law who 'seemed busier than he was.' "[5] Keynes was devoted to the theater, painting, scenic design and the ballet; he inspired and supported various movements and institutions in these fields. But he was neither a dilettante nor merely a rich man basking in the role of Maecenas. He "believed in art as education because he was a Victorian, and believed in it as pure entertainment because he was ahead of his age."[6] He was himself a creative writer of first rank.

[5] *John Maynard Keynes, 1883–1946* (King's College Memoir), "The Arts," by G. H. W. Rylands, pp. 36–37.

[6] Even in his devotion to art Keynes' shrewd business sense peeped through. "Toward the end of the war of 1914–18, Duncan Grant showed Keynes the illustrated catalogue of the sale of the Degas collection and suggested (without hope) that the British government should take the opportunity to remedy the meager representation of French nineteenth-century painting in the National Gallery. To everybody's astonishment the Treasury allowed Keynes a limit of £20,000, and he at once set off for Paris in high excitement with Sir Charles Holmes to attend the sale, which was punctuated, to the advantage of buyers, by shell-bursts from Big Bertha. It was then that Keynes laid the foundation of his own collection with the celebrated 'Apples'

"In one art, certainly [T. S. Eliot wrote after his death] he had no reason to defer to any opinion: in expository prose he had the essential style of the clear mind which thinks structurally and respects the meaning of words. He had been both a classical and mathematical scholar: he had excelled under those two best disciplines, which, when imposed upon an uncommon mind capable of profiting by both, should co-operate to produce lucid thinking and correct expression. And, unlike some other brilliant scholars, he had continued throughout his life to feed and exercise his mind by wide reading in English and other literature." [7]

Savage in writing and debate, Keynes could also be magnanimous and delicate of understanding with "a kind of critical intuition only to be paralleled by that of some of our greatest historians and scientists." He was both rude and kind, a fierce controversialist and, as Henry Morgenthau said, "a gentle soul." He delighted in paradox for its own sake, in pretending to be omniscient, in playing the prophet. "No one in our age," Harrod says somewhat regretfully, "was cleverer than Keynes nor made less attempt to conceal it." When the necessary statistics were available, he had them at his fingertips; when they were not, he would guess at them, admitting that he did so with disarming candor. He respected statistics, but they also bored him; above all he despised "the mean statistical view of life." [8] To say of Keynes that he had the defects of his qualities is not merely to stress the obvious; it is, in a sense, the definition both of his uniqueness and his genius.

Keynes believed in the planning and regulation of economic forces to improve the condition of men everywhere, but he abhorred regimentation or any threat to personal freedom. He condemned the predominance of the money motive and made a fortune as a speculator. He was a passionate reformer but had no sympathy for socialism. He thought all nations "dishonorable, cruel and designing" but burned himself out in his country's service.

He was a full man and he had a full life. His marriage to the lovely ballerina, Lydia Lopokova, brought him much happiness and enlarged his sympathies and interests. "I like him," said Margot Asquith, "he is such a good man." In 1946 his heart gave out. He had not spared himself though he knew he was dangerously ill. He died too soon. "Ministers of good things," said Hooker, "are like torches, a light to others, waste and destruction to themselves."

---

of Cezanne and a magnificent drawing by Ingres." *John Maynard Keynes, 1883–1946* (King's College Memoir), "The Arts," by G. H. W. Rylands, p. 39.

[7] *John Maynard Keynes, 1883–1946* (King's College Memoir), *ibid.*, p. 38.

[8] *Times Literary Supplement*, February 23, 1951, p. 111.

Probability is the very guide of life. —BISHOP BUTLER

It is truth very certain that, when it is not in our power to determine what is true, we ought to follow what is most probable. —RENÉ DESCARTES

Life is a school of probability. —WALTER BAGEHOT

4 The Application of Probability to Conduct

By JOHN MAYNARD KEYNES

1. GIVEN as our basis what knowledge we actually have, the probable, I have said, is that which it is rational for us to believe. This is not a definition. For it is not rational for us to believe that the probable is true; it is only rational to have a probable belief in it or to believe it in preference to alternative beliefs. To believe one thing *in preference* to another, as distinct from believing the first true or more probable and the second false or less probable, must have reference to action and must be a loose way of expressing the propriety of *acting* on one hypothesis rather than on another. We might put it, therefore, that the probable is the hypothesis on which it is rational for us to act. It is, however, not so simple as this, for the obvious reason that of two hypotheses it may be rational to act on the less probable if it leads to the greater good. We cannot say more at present than that the probability of a hypothesis is one of the things to be determined and taken account of before acting on it.

2. I do not know of passages in the ancient philosophers which explicitly point out the dependence of the duty of pursuing goods on the reasonable or probable expectation of attaining them relative to the agent's knowledge. This means only that analysis had not disentangled the various elements in rational action, not that common sense neglected them. Herodotus puts the point quite plainly. "There is nothing more profitable for a man," he says, "than to take good counsel with himself; for even if the event turns out contrary to one's hope, still one's decision was right, even though fortune has made it of no effect: whereas if a man acts contrary to good counsel, although by luck he gets what he had no right to expect, his decision was not any the less foolish." [1]

3. The first contact of theories of probability with modern ethics appears in the Jesuit doctrine of probabilism. According to this doctrine one is justified in doing an action for which there is *any* probability, however small, of its results being the best possible. Thus, if any priest is willing

[1] Herod. vii. 10.

to permit an action, that fact affords some probability in its favour, and one will not be damned for performing it, however many other priests denounce it.[2] It may be suspected, however, that the object of this doctrine was not so much duty as safety. The priest who permitted you so to act assumed thereby the responsibility. The correct application of probability to conduct naturally escaped the authors of a juridical ethics, which was more interested in the fixing of responsibility for definite acts, and in the various specified means by which responsibility might be disposed of, than in the greatest possible sum-total of resultant good.

A more correct doctrine was brought to light by the efforts of the philosophers of the Port Royal to expose the fallacies of probabilism. "In order to judge," they say, "of what we ought to do in order to obtain a good and to avoid an evil, it is necessary to consider not only the good and evil in themselves, but also the probability of their happening and not happening, and to regard geometrically the proportion which all these things have, taken together." [3] Locke perceived the same point, although not so clearly.[4] By Leibniz this theory is advanced more explicitly; in such judgments, he says, "as in other estimates disparate and heterogeneous and, so to speak, of more than one dimension, the greatest of that which is discussed is in reason composed of both estimates (*i.e.* of goodness and of probability), and is like a rectangle, in which there are two considerations, viz. that of length and that of breadth. . . . Thus we should still need the art of thinking and that of estimating probabilities, besides the knowledge of the value of goods and evils, in order properly to employ the art of consequences." [5]

In his preface to the *Analogy* Butler insists on "the absolute and formal obligation" under which even a low probability, if it is the greatest, may lay us: "To us probability is the very guide of life."

4. With the development of a utilitarian ethics largely concerned with the summing up of consequences, the place of probability in ethical theory has become much more explicit. But although the general outlines of the problem are now clear, there are some elements of confusion not yet dispersed. I will deal with some of them.

In his *Principia Ethica* (p. 152) Dr. Moore argues that "the first

[2] Compare with this doctrine the following curious passage from Jeremy Taylor:— "We being the persons that are to be persuaded, we must see that we be persuaded reasonably. And it is unreasonable to assent to a lesser evidence when a greater and clearer is propounded: but of that every man for himself is to take cognisance, if he be able to judge; if he be not, he is not bound under the tie of necessity to know anything of it. That that is necessary shall be certainly conveyed to him: God, that best can, will certainly take care for that; for if he does not, it becomes to be not necessary; or if it should still remain necessary, and he be damned for not knowing it, and yet to know it be not in his power, then who can help it! There can be no further care in this business."

[3] *The Port Royal Logic* (1662), Eng. Trans., p. 367.

[4] *Essay concerning Human Understanding*, book ii. chap. xxi. § 66.

[5] *Nouveaux Essais*, book ii. chap. xxi.

difficulty in the way of establishing a probability that one course of action will give a better total result than another, lies in the fact that we have to take account of the effects of both throughout an infinite future. . . . We can certainly only pretend to calculate the effects of actions within what may be called an 'immediate future.' . . . We must, therefore, certainly have some reason to believe that no consequences of our action in a further future will generally be such as to reverse the balance of good that is probable in the future which we can foresee. This large postulate must be made, if we are ever to assert that the results of one action will be even probably better than those of another. Our utter ignorance of the far future gives us no justification for saying that it is even probably right to choose the greater good within the region over which a probable forecast may extend."

This argument seems to me to be invalid and to depend on a wrong philosophical interpretation of probability. Mr. Moore's reasoning endeavours to show that there is not even a *probability* by showing that there is not a *certainty*. We must not, of course, have reason to believe that remote consequences will *generally* be such as to reverse the balance of immediate good. But we need not be certain that the opposite is the case. If good is additive, if we have reason to think that of two actions one produces more good than the other in the near future, and if we have no means of discriminating between their results in the distant future, then by what seems a legitimate application of the Principle of Indifference we may suppose that there is a probability in favour of the former action. Mr. Moore's argument must be derived from the empirical or frequency theory of probability, according to which we must know for certain what will happen *generally* (whatever that may mean) before we can assert a probability.

The results of our endeavours are very uncertain, but we have a genuine probability, even when the evidence upon which it is founded is slight. The matter is truly stated by Bishop Butler: "From our short views it is greatly uncertain whether this endeavour will, in particular instances, produce an overbalance of happiness upon the whole; since so many and distant things must come into the account. And that which makes it our duty is that there is some appearance that it will, and no positive appearance to balance this, on the contrary side. . . ." [6]

The difficulties which exist are not chiefly due, I think, to our ignorance of the remote future. The possibility of our knowing that one thing rather than another is our duty depends upon the assumption that a greater goodness in any part makes, in the absence of evidence to the contrary,

[6] This passage is from the *Analogy*. The Bishop adds: ". . . and also that such benevolent endeavour is a cultivation of that most excellent of all virtuous principles, the active principle of benevolence."

a greater goodness in the whole more probable than would the lesser goodness of the part. We assume that the goodness of a part is *favourably* relevant to the goodness of the whole. Without this assumption we have no reason, not even a probable one, for preferring one action to any other on the whole. If we suppose that goodness is always *organic*, whether the whole is composed of simultaneous or successive parts, such an assumption is not easily justified. The case is parallel to the question whether physical law is organic or atomic.

Nevertheless we can admit that goodness is partly organic and still allow ourselves to draw probable conclusions. For the alternatives, that *either* the goodness of the whole universe throughout time is organic *or* the goodness of the universe is the arithmetic sum of the goodnesses of infinitely numerous and infinitely divided parts, are not exhaustive. We may suppose that the goodness of conscious persons is organic for each distinct and individual personality. Or we may suppose that, when conscious units are in conscious relationship, then the whole which we must treat as organic includes both units. These are only examples. We must suppose, in general, that the units whose goodness we must regard as organic and indivisible are not always larger than those the goodness of which we can perceive and judge directly.

5. The difficulties, however, which are most fundamental from the standpoint of the student of probability, are of a different kind. Normal ethical theory at the present day, if there can be said to be any such, makes two assumptions: first, that degrees of goodness are numerically measurable and arithmetically additive, and second, that degrees of probability also are numerically measurable. This theory goes on to maintain that what we ought to add together, when, in order to decide between two courses of action, we sum up the results of each, are the 'mathematical expectations' of the several results. 'Mathematical expectation' is a technical expression originally derived from the scientific study of gambling and games of chance, and stands for the product of the possible gain with the probability of attaining it.[7] In order to obtain, therefore, a measure of what ought to be our preference in regard to various alternative courses of action, we must sum for each course of action a series of terms made up of the amounts of good which may attach to each of its possible consequences, each multiplied by its appropriate probability.

[7] Priority in the conception of mathematical expectation can, I think, be claimed by Leibniz, *De incerti aestimatione*, 1678 (Couturat, *Logique de Leibniz*, p. 248). In a letter to Placcius, 1687 (Dutens, vi. i. 36 and Couturat, *op. cit.* p. 246) Leibniz proposed an application of the same principle to jurisprudence, by virtue of which, if two litigants lay claim to a sum of money, and if the claim of the one is twice as probable as that of the other, the sum should be divided between them in that proportion. The doctrine, seems sensible, but I am not aware that it has ever been acted on.

The first assumption, that quantities of goodness are duly subject to the laws of arithmetic, appears to me to be open to a certain amount of doubt. But it would take me too far from my proper subject to discuss it here, and I shall allow, for the purposes of further argument, that in some sense and to some extent this assumption can be justified. The second assumption, however, that degrees of probability are wholly subject to the laws of arithmetic, runs directly counter to the view which has been advocated in Part I. of this treatise. Lastly, if both these points be waived, the doctrine that the 'mathematical expectations' of alternative courses of action are the proper measures of our degrees of preference is open to doubt on two grounds—first, because it ignores what I have termed in Part I. the 'weights' of the arguments, namely, the amount of evidence upon which each probability is founded; and second, because it ignores the element of 'risk' and assumes that an even chance of heaven or hell is precisely as much to be desired as the certain attainment of a state of mediocrity. Putting on one side the first of these grounds of doubt, I will treat each of the others in turn.

6. I have argued that only in a strictly limited class of cases are degrees of probability numerically measurable.[8] It follows from this that the 'mathematical expectations' of goods or advantages are not always numerically measurable; and hence, that even if a meaning can be given to the sum of a series of non-numerical 'mathematical expectations,' not every pair of such sums are numerically comparable in respect of more and less. Thus even if we know the degree of advantage which might be obtained from each of a series of alternative courses of actions and know also the probability in each case of obtaining the advantage in question, it is not always possible by a mere process of arithmetic to determine which of the alternatives ought to be chosen. If, therefore, the question of right action is under all circumstances a determinate problem, it must be in virtue of an intuitive judgment directed to the situation as a whole, and not in virtue of an arithmetical deduction derived from a series of separate judgments directed to the individual alternatives each treated in isolation.

We must accept the conclusion that, if one good is greater than another, but the probability of attaining the first less than that of attaining the second, the question of which it is our duty to pursue may be indeterminate, unless we suppose it to be within our power to make direct quantitative judgments of probability and goodness jointly. It may be remarked, further, that the difficulty exists, whether the numerical indeterminateness of the probability is intrinsic or whether its numerical value is, as it is according to the Frequency Theory and most other theories, simply unknown.

[8] [In the earlier portion of the book. ED.]

7. The second difficulty, to which attention is called above, is the neglect of the 'weights' of arguments in the conception of 'mathematical expectation.' The significance of 'weight' has already been discussed.[9] In the present connection the question comes to this—if two probabilities are equal in degree, ought we, in choosing our course of action, to prefer that one which is based on a greater body of knowledge?

The question appears to me to be highly perplexing, and it is difficult to say much that is useful about it. But the degree of completeness of the information upon which a probability is based does seem to be relevant, as well as the actual magnitude of the probability, in making practical decisions. Bernoulli's maxim,[10] that in reckoning a probability we must take into account all the information which we have, even when reinforced by Locke's maxim that we must get all the information we can,[11] does not seem completely to meet the case. If, for one alternative, the available information is necessarily small, that does not seem to be a consideration which ought to be left out of account altogether.

8. The last difficulty concerns the question whether, the former difficulties being waived, the 'mathematical expectation' of different courses of action accurately measures what our preferences ought to be—whether, that is to say, the undesirability of a given course of action increases in direct proportion to any increase in the uncertainty of its attaining its object, or whether some allowance ought to be made for 'risk,' its undesirability increasing more than in proportion to its uncertainty.

In fact the meaning of the judgment, that we ought to act in such a way as to produce most probably the greatest sum of goodness, is not perfectly plain. Does this mean that we ought so to act as to make the sum of the goodnesses of each of the possible consequences of our action multiplied by its probability a maximum? Those who rely on the conception of 'mathematical expectation' must hold that this is an indisputable proposition. The justifications for this view most commonly advanced resemble that given by Condorcet in his "Réflexions sur la règle générale, qui prescrit de prendre pour valeur d'un événement incertain, la probabilité de cet événement multipliée par la valeur de l'événement en luimême," [12] where he argues from Bernoulli's theorem that such a rule will lead to satisfactory results if a very large number of trials be made. As, however, it will be shown that Bernoulli's theorem is not applicable in by

[9] [In an earlier chapter. ED.]

[10] *Ars Conjectandi*, p. 215: "Non sufficit expendere unum alterumve argumentum, sed conquirenda sunt omnia, quae in cognitionem nostram venire possunt, atque ullo modo ad probationem rei facere videntur."

[11] *Essay concerning Human Understanding*, book ii. chap. xxi. § 67: "He that judges without informing himself to the utmost that he is capable, cannot acquit himself of *judging amiss*."

[12] *Hist. de l'Acad.*, Paris, 1781.

any means every case, this argument is inadequate as a general justification.[13]

In the history of the subject, nevertheless, the theory of 'mathematical expectation' has been very seldom disputed. As D'Alembert has been almost alone in casting serious doubts upon it (though he only brought himself into disrepute by doing so), it will be worth while to quote the main passage in which he declares his scepticism: "Il me sembloit" (in reading Bernoulli's *Ars Conjectandi*) "que cette matière avoit besoin d'être traitée d'une manière plus claire; je voyois bien que l'espérance étoit plus grande, 1° que la somme espérée étoit plus grande, 2° que la probabilité de gagner l'étoit aussi. Mais je ne voyois pas avec la même évidence, et je ne le vois pas encore, 1° que la probabilité soit estimée exactement par les méthodes usitées; 2° que quand elle le seroit, l'espérance doive être proportionnelle à cette probabilité simple, plutôt qu'à une puissance ou même à une fonction de cette probabilité; 3° que quand il y a plusieurs combinaisons qui donnent différens avantages ou différens risques (qu'on regarde comme des avantages négatifs) il faille se contenter d'*ajouter* simplement ensemble toutes les espérances pour avoir l'espérance totale." [14]

In extreme cases it seems difficult to deny some force to D'Alembert's objection; and it was with reference to extreme cases that he himself raised it. Is it certain that a larger good, which is extremely improbable, is precisely equivalent ethically to a smaller good which is proportionately more probable? We may doubt whether the moral value of speculative and cautious action respectively can be weighed against one another in a simple arithmetical way, just as we have already doubted whether a good whose probability can only be determined on a slight basis of evidence can be compared by means merely of the magnitude of this probability with another good whose likelihood is based on completer knowledge.

There seems, at any rate, a good deal to be said for the conclusion that, other things being equal, that course of action is preferable which involves least risk, and about the results of which we have the most complete knowledge. In marginal cases, therefore, the coefficients of weight and risk as well as that of probability are relevant to our conclusion. It seems natural to suppose that they should exert some influence in other cases also, the only difficulty in this being the lack of any principle for the calculation of the degree of their influence. A high weight and the absence of risk increase *pro tanto* the desirability of the

[13] [The argument is made later by Keynes. For a further discussion of Bernoulli's theorem, see p. 1448 of the present book. ED.]

[14] *Opuscules mathématiques*, vol. iv., 1768 (extraits de lettres), pp. 284, 285. See also p. 88 of the same volume.

action to which they refer, but we cannot measure the amount of the increase.

The 'risk' may be defined in some such way as follows. If A is the amount of good which may result, p its probability $(p + q = 1)$, and E the value of the 'mathematical expectation,' so that $E = pA$, then the 'risk' is R, where $R = p(A - E) = p(1 - p)A = pqA = qE$. This may be put in another way: E measures the net immediate sacrifice which should be made in the hope of obtaining A; q is the probability that this sacrifice will be made in vain; so that qE is the 'risk.'[15] The ordinary theory supposes that the ethical value of an expectation is a function of E only and is entirely independent of R.

We could, if we liked, define a conventional coefficient c of weight and risk, such as $c = \dfrac{2pw}{(1 + q)(1 + w)}$, where w measures the 'weight,' which is equal to unity when $p = 1$ and $w = 1$, and to zero when $p = 0$ or $w = 0$, and has an intermediate value in other cases.[16] But if doubts as to the sufficiency of the conception of 'mathematical expectation' be sustained, it is not likely that the solution will lie, as D'Alembert suggests, and as has been exemplified above, in the discovery of some more complicated function of the probability wherewith to compound the proposed good. The judgment of goodness and the judgment of probability both involve somewhere an element of direct apprehension, and both are quantitative. We have raised a doubt as to whether the magnitude of the 'oughtness' of an action can be in all cases directly determined by simply multiplying together the magnitudes obtained in the two direct judgments; and a new direct judgment may be required, respecting the magnitude of the 'oughtness' of an action under given circumstances, which need not bear any simple and necessary relation to the two former.

The hope, which sustained many investigators in the course of the nineteenth century, of gradually bringing the moral sciences under the sway of mathematical reasoning, steadily recedes—if we mean, as they meant, by mathematics the introduction of precise numerical methods. The old assumptions, that all quantity is numerical and that all quantita-

[15] The theory of *Risiko* is briefly dealt with by Czuber, *Wahrscheinlichkeitsrechnung*, vol. i. pp. 219 *et seq.* If R measures the first insurance, this leads to a *Risiko* of the second order, $R_1 = qR = q^2E$. This again may be insured against, and by a sufficient number of such reinsurances the risk can be completely shifted:

$$E + R_1 + R_2 + \ldots = E(1 + q + q^2 + \ldots) = \frac{E}{1 - q} = \frac{E}{p} = A.$$

[16] If $pA = p'A'$, $w > w'$, and $q = q'$, then $cA > c'A'$; if $pA = p'A'$, $w = w'$, and $q < q'$, then $cA > c'A'$; if $pA = p'A'$, $w > w'$, and $q < q'$, then $cA > c'A'$; but if $pA = p'A'$, $w = w'$, and $q > q'$, we cannot in general compare cA and $c'A'$.

tive characteristics are additive, can be no longer sustained. Mathematical reasoning now appears as an aid in its symbolic rather than in its numerical character. I, at any rate, have not the same lively hope as Condorcet, or even as Edgeworth, "éclairer les Sciences morales et politiques par le flambeau de l'Algèbre." In the present case, even if we are able to range goods in order of magnitude, and also their probabilities in order of magnitude, yet it does not follow that we can range the products composed of each good and its corresponding probability in this order.

9. Discussions of the doctrine of Mathematical Expectation, apart from its directly ethical bearing, have chiefly centred round the classic Petersburg Paradox,[17] which has been treated by almost all the more notable writers, and has been explained by them in a great variety of ways. The Petersburg Paradox arises out of a game in which Peter engages to pay Paul one shilling if a head appears at the first toss of a coin, two shillings if it does not appear until the second, and, in general, 2^{r-1} shillings if no head appears until the rth toss. What is the value of Paul's expectation, and what sum must he hand over to Peter before the game commences, if the conditions are to be fair?

The mathematical answer is $\sum_1^n (\tfrac{1}{2})^r 2^{r-1}$, if the number of tosses is not in any case to exceed n in all, and $\sum_1^\infty (\tfrac{1}{2})^r 2^{r-1}$ if this restriction is removed. That is to say, Paul should pay $\frac{n}{2}$ shillings in the first case, and an infinite sum in the second. Nothing, it is said, could be more paradoxical, and no sane Paul would engage on these terms even with an honest Peter.

Many of the solutions which have been offered will occur at once to the reader. The conditions of the game *imply* contradiction, say Poisson and Condorcet; Peter has undertaken engagements which he cannot fulfil; if the appearance of heads is deferred even to the 100th toss, he will owe a mass of silver greater in bulk than the sun. But this is no answer. Peter has promised much and a belief in his solvency will strain our imagination; but it is imaginable. And in any case, as Bertrand points out, we may suppose the stakes to be, not shillings, but grains of sand or molecules of hydrogen.

D'Alembert's principal explanations are, first, that true expectation is not necessarily the product of probability and profit (a view which has

[17] For the history of this paradox see Todhunter. The name is due, he says, to its having first appeared in a memoir by Daniel Bernoulli in the *Commentarii* of the Petersburg Academy.

been discussed above), and second, that very long runs are not only very improbable, but do not occur at all.

The next type of solution is due, in the first instance, to Daniel Bernoulli, and turns on the fact that no one but a miser regards the desirability of different sums of money as directly proportional to their amount; as Buffon says, "L'avare est comme le mathématicien: tous deux estiment l'argent par sa quantité numérique." Daniel Bernoulli deduced a formula from the assumption that the importance of an increment is inversely proportional to the size of the fortune to which it is added. Thus, if x is the 'physical' fortune and y the 'moral' fortune,

$$dy = k\frac{dx}{x},$$

or $y = k \log \frac{x}{a}$, where k and a are constants.

On the basis of this formula of Bernoulli's a considerable theory has been built up both by Bernoulli [18] himself and by Laplace.[19] It leads easily to the further formula—

$$x = (a + x_1)p_1(a + x_2)p_2 \ . \ . \ . ,$$

where a is the initial 'physical' fortune, p_1, etc., the probabilities of obtaining increments x_1, etc., to a, and x the 'physical' fortune whose present possession would yield the same 'moral' fortune as does the expectation of the various increments, x_1, etc. By means of this formula Bernoulli shows that a man whose fortune is £1000 may reasonably pay a £6 stake in order to play the Petersburg game with £1 units. Bernoulli also mentions two solutions proposed by Cramer. In the first all sums greater than 2^{24} (16,777,116) are regarded as 'morally' equal; this leads to £13 as the fair stake. According to the other formula the pleasure derivable from a sum of money varies as the square root of the sum; this leads to £2 : 9s. as the fair stake. But little object is served by following out these arbitrary hypotheses.

As a solution of the Petersburg problem this line of thought is only partially successful: if increases of 'physical' fortune beyond a certain finite limit can be regarded as 'morally' negligible, Peter's claim for an infinite initial stake from Paul is, it is true, no longer equitable, but with any reasonable law of diminution for successive increments Paul's stake will still remain paradoxically large. Daniel Bernoulli's suggestion is, however, of considerable historical interest as being the first explicit attempt to take account of the important conception known to modern economists

[18] "Specimen Theoriae Novae de Mensura Sortis," *Comm. Acad. Petrop.* vol. v. for 1730 and 1731, pp. 175–192 (published 1738). See Todhunter, pp. 213 *et seq.*

[19] *Théorie analytique*, chap. x. "De l'espérance morale," pp. 432–445.

as the diminishing marginal utility of money,—a conception on which many important arguments are founded relating to taxation and the ideal distribution of wealth.

Each of the above solutions probably contains a part of the psychological explanation. We are unwilling to be Paul, partly because we do not believe Peter will pay us if we have good fortune in the tossing, partly because we do not know what we should do with so much money or sand or hydrogen if we won it, partly because we do not believe we ever should win it, and partly because we do not think it would be a rational act to risk an infinite sum or even a very large finite sum for an infinitely larger one, whose attainment is infinitely unlikely.

When we have made the proper hypotheses and have eliminated these elements of psychological doubt, the theoretic dispersal of what element of paradox remains must be brought about, I think, by a development of the theory of risk. It is primarily the great *risk* of the wager which deters us. Even in the case where the number of tosses is in no case to exceed a finite number, the risk R, as already defined, may be very great, and the relative risk $\frac{R}{E}$ will be almost unity. Where there is no limit to the number of tosses, the risk is infinite. A relative risk, which approaches unity, may, it has been already suggested, be a factor which must be taken into account in ethical calculation.

10. In establishing the doctrine, that all private gambling must be with certainty a losing game, precisely contrary arguments are employed to those which do service in the Petersburg problem. The argument that "you must lose if only you go on long enough" is well known. It is succinctly put by Laurent:[20] Two players A and B have a and b francs respectively. $f(a)$ is the chance that A will be ruined. Thus $f(a) = \frac{b}{a+b}$,[21] so that the poorer a gambler is, relatively to his opponent, the more likely he is to be ruined. But further, if $b = \infty$, $f(a) = 1$, *i.e.*, ruin is certain. The infinitely rich gambler is the public. It is against the public that the professional gambler plays, and his ruin is therefore certain.

Might not Poisson and Condorcet reply, The conditions of the game *imply* contradiction, for no gambler plays, as this argument supposes, forever?[22] At the end of any *finite* quantity of play, the player, even if he is not the public, *may* finish with winnings of any finite size. The gambler is in a worse position if his capital is smaller than his opponents'—at poker, for instance, or on the Stock Exchange. This is clear. But our

[20] *Calcul des probabilités*, p. 129.

[21] This would possibly follow from the theorem of Daniel Bernoulli. The reasoning by which Laurent obtains it seems to be the result of a mistake.

[22] Cf. also Mr. Bradley, *Logic*, p. 217.

desire for moral improvement outstrips our logic if we tell him that he *must* lose. Besides it is paradoxical to say that everybody individually must lose and that everybody collectively must win. For every individual gambler who loses there is an individual gambler or syndicate of gamblers who win. The true moral is this, that poor men should not gamble and that millionaires should do nothing else. But millionaires gain nothing by gambling with one another, and until the poor man departs from the path of prudence the millionaire does not find his opportunity. If it be replied that in fact most millionaires are men originally poor who departed from the path of prudence, it must be admitted that the poor man is not doomed with certainty. Thus the philosopher must draw what comfort he can from the conclusion with which his theory furnishes him, that millionaires are often fortunate fools who have thriven on unfortunate ones.[23]

11. In conclusion we may discuss a little further the conception of 'moral' risk, raised in § 8 and at the end of § 9. Bernoulli's formula crystallises the undoubted truth that the value of a sum of money to a man varies according to the amount he already possesses. But does the value of an amount of goodness also vary in this way? May it not be true that the addition of a given good to a man who already enjoys much good is less good than its bestowal on a man who has little? If this is the case, it follows that a smaller but relatively certain good is better than a greater but proportionately more uncertain good.

In order to assert this, we have only to accept a particular theory of organic goodness, applications of which are common enough in the mouths of political philosophers. It is at the root of all principles of equality, which do not arise out of an assumed diminishing marginal utility of money. It is behind the numerous arguments that an equal distribution of benefits is better than a very unequal distribution. If this is the case, it follows that, the sum of the goods of all parts of a community taken together being fixed, the organic good of the whole is greater the more equally the benefits are divided amongst the individuals. If the doctrine is to be accepted, more risks, like financial risks, must not be undertaken unless they promise a profit actuarially.

There is a great deal which could be said concerning such a doctrine, but it would lead too far from what is relevant to the study of Probability. One or two instances of its use, however, may be taken from the literature of Probability. In his essay, "Sur l'application du calcul des probabilités à l'inoculation de la petite vérole," [24] D'Alembert points out that the com-

[23] From the social point of view, however, this moral against gambling may be drawn—that those who start with the largest initial fortunes are most likely to win, and that a given increment to the wealth of these benefits them, on the assumption of a diminishing marginal utility of money, less than it injures those from whom it is taken.

[24] *Opuscules mathématiques*, vol. ii.

munity would gain on the average if, by sacrificing the lives of one in five of its citizens, it could ensure the health of the rest, but he argues that no legislator could have the right to order such a sacrifice. Galton, in his *Probability, the Foundation of Eugenics*, employed an argument which depends essentially on the same point. Suppose that the members of a certain class cause an average detriment M to society, and that the mischiefs done by the several individuals differ more or less from M by amounts whose average is D, so that D is the average amount of the individual deviations, all regarded as positive, from M; then, Galton argued, the smaller D is, the stronger is the justification for taking such drastic measures against the propagation of the class as would be consonant to the feelings, if it were known that each individual member caused a detriment M. The use of such arguments seems to involve a qualification of the simple ethical doctrine that right action should make the sum of the benefits of the several individual consequences, each multiplied by its probability, a maximum.

On the other hand, the opposite view is taken in the *Port Royal Logic* and by Butler, when they argue that everything ought to be sacrificed for the hope of heaven, even if its attainment be thought infinitely improbable, since "the smallest degree of facility for the attainment of salvation is of higher value than all the blessings of the world put together." [25] The argument is, that we ought to follow a course of conduct which may with the slightest probability lead to an infinite good, until it is logically disproved that such a result of our action is impossible. The Emperor who embraced the Roman Catholic religion, not because he believed it, but because it offered insurance against a disaster whose future occurrence, however improbable, he could not certainly disprove, may not have considered, however, whether the product of an infinitesimal probability and an infinite good might not lead to a finite or infinitesimal result. In any case the argument does not enable us to choose between different courses of conduct, unless we have reason to suppose that one path is *more* likely than another to lead to infinite good.

12. In estimating the risk, 'moral' or 'physical,' it must be remembered that we cannot necessarily apply to individual cases results drawn from the observation of a long series resembling them in some particular. I am thinking of such arguments as Buffon's when he names $1/_{10,000}$ as the limit, beyond which probability is negligible, on the ground that, being the chance that a man of fifty-six *taken at random* will die within a day, it is practically disregarded by a man of fifty-six *who knows his health to be*

[25] *Port Royal Logic* (Eng. trans.), p. 369: "It belongs to infinite things alone, as eternity and salvation, that they cannot be equalled by any temporal advantage; and thus we ought never to place them in the balance with any of the things of the world. This is why the smallest degree of facility for the attainment of salvation is of higher value than all the blessings of the world put together. . . ."

good. "If a public lottery," Gibbon truly pointed out, "were drawn for the choice of an immediate victim, and if our name were inscribed on one of the ten thousand tickets, should we be perfectly easy?"

Bernoulli's second axiom, that in reckoning a probability we must take everything into account, is easily forgotten in these cases of statistical probabilities. The statistical result is so attractive in its definiteness that it leads us to forget the more vague though more important considerations which may be, in a given particular case, within our knowledge. To a stranger the probability that I shall send a letter to the post unstamped may be derived from the statistics of the Post Office; for me those figures would have but the slightest bearing upon the question.

13. It has been pointed out already that no knowledge of probabilities, less in degree than certainty, helps us to know what conclusions are true, and that there is no direct relation between the truth of a proposition and its probability. Probability begins and ends with probability. That a scientific investigation pursued on account of its probability will generally lead to truth, rather than falsehood, is at the best only probable. The proposition that a course of action guided by the most probable considerations will generally lead to success, is not certainly true and has nothing to recommend it but its probability.

The importance of probability can only be derived from the judgment that it is *rational* to be guided by it in action; and a practical dependence on it can only be justified by a judgment that in action we *ought* to act to take some account of it. It is for this reason that probability is to us the "guide of life," since to us, as Locke says, "in the greatest part of our concernment, God has afforded only the Twilight, as I may so say, of Probability, suitable, I presume, to that state of Mediocrity and Probationership He has been pleased to place us in here."

COMMENTARY ON

An Absent-minded Genius and the Laws of Chance

HENRI POINCARE was a French savant who looked alarmingly like the popular image of a French savant. He was short and plump, carried an enormous head set off by a thick spade beard and splendid mustache, was myopic, stooped, distraught in speech, absent-minded and wore pince-nez glasses attached to a black silk ribbon. He resembled his cousin Raymond who was less absent-minded and became President of the French Republic. Poincaré was a man of high philosophic capacity, and the greatest mathematician of his day. E. T. Bell calls him the "last universalist" because of the breadth of his mathematical knowledge, the amazing variety of his researches in the pure and applied field. It is improbable that a future mathematician will attain the spacious grasp of a Poincaré or a Gauss. The subject is difficult and has grown very big.

Poincaré was born in 1854 at Nancy, where his ancestors had long been established. The name, Henri said, had been Pontcaré, which at least made better sense than Pointcaré "for one can imagine a square bridge but not a square point." [1] It turns out, however, that the original form in the fifteenth century was Poingquarré, which means "clenched fist." Poincaré's father was a physician and professor in the faculty of medicine at the university. He was of scholarly inclination and Darboux says that Henri passed his childhood in a circle of "savants, university men and polytechnicians." To this severe handicap to normal development were added the child's disturbing precocity, physical frailty and poor co-ordination. An attack of diphtheria when he was five left him for nine months with a paralyzed larynx; "feeble and timid," he was a suitable victim of the brutalities of children his own age.[2] After his early education at home under the supervision of his mother, a kindly woman of superior intelligence, Poincaré attended the Nancy lycée, the Ecole Polytechnique and the School of Mines, and received his doctorate of mathematical sciences in 1879 from the University of Paris. It is no surprise to learn that he made a remarkable record at school despite the fact that he drew so badly that even his geometric diagrams were undecipherable. He got a zero in

[1] Edwin E. Slosson, *Major Prophets of Today*, Boston, 1914, p. 131. One of the best biographical sources is the *Eloge Historique d'Henri Poincaré* (Paris, 1913) by the distinguished geometer Gaston Darboux, and an article reprinted in *Le Temps*, December 15, 1913. The chapter on Poincaré in E. T. Bell's *Men of Mathematics* is also very readable and accurate.

[2] Ganesh Prasad, *Some Great Mathematicians of the Nineteenth Century*, Benares City, 1934, Vol. II, p. 279.

drawing in the entrance examination for the Ecole Polytechnique and the examiners had to make an exception to admit him despite this deficiency. He was widely read, had a taste for languages and history and displayed exceptional mathematical aptitude when he was still in high school. His memory was fabulous and he never needed to read a book more than once to retain every detail. Like the geometer John Bolyai, who performed a similar feat at his university in Hungary, Poincaré followed the courses in mathematics at the Ecole Polytechnique "without taking a note and without the syllabus." [3] Poincaré's intuitive powers and fluency had their drawbacks. At school, as well as later, he had to guard against a tendency to carelessness because problems too hard for others were too easy for him. Throughout his life (said Darboux) "when one asked him to solve a difficulty the reply came like an arrow."

In 1879 Poincaré was appointed to the faculty at Caen; two years later he was made *Maître de Conférences* in mathematical analysis at Paris, and in 1886 was promoted to professor of mathematical physics and the calculus of probabilities. On the strength of researches described as "above ordinary praise," he was elected to the Academy of Sciences at the early age of thirty-two. Poincaré had settled, said the academician who nominated him, "questions which, before him, were unimagined." [4]

His scientific output, which began about 1879, was tremendous. In thirty-four years he published more than thirty books on mathematical physics and celestial mechanics and nearly 500 memoirs on mathematics, some of them of the very first order. In addition, he wrote two books of popular essays, and three volumes on the philosophy of science which are not only acknowledged as classics but provide the most delectable reading to be found in this branch of literature.[5]

One of Poincaré's principal discoveries was in the field of elliptic functions. Perhaps one or two paragraphs on this subject will give the reader with a liking for symbols an entry into Poincaré's thought.[6] Certain functions have the characteristic of being periodic, which is to say that when the value of the independent variable is increased by a regular quantity, the function recurs to an initial value. Thus the function Sin Z has the period 2π, e.g., $\sin(Z + 2\pi) = \sin(Z + 4\pi) = \sin(Z + 6\pi) \ldots = \sin Z$. An elliptic function, $E(Z)$ has *two* distinct periods, say p_1 and p_2, such that $E(z + p_1) = E(z)$, $E(z + p_2) = E(z)$. "Poincaré found that *periodicity* is merely a special case of a more general property: the value of certain functions is restored when the variable is replaced by any one

[3] George Bruce Halsted in the preface to his translation of Poincaré's *The Foundations of Science*, Lancaster, 1946.

[4] Bell, *op. cit.*, p. 541.

[5] If the *Dernières Pensées*, never translated, is counted, Poincaré wrote four volumes on the philosophy of science.

[6] Here I use E. T. Bell's lucid simplification of the work in his *Men of Mathematics.*

of a *denumerable* infinity [7] of linear fractional transformations of itself, and all these transformations form a group." The same statement in symbols reads as follows:

Let Z be replaced by $\dfrac{az+b}{cz+d}$. Then for a *denumerable infinity* of sets of values of a, b, c, d, there are uniform functions of z, say $F(z)$ is one of them, such that

$$F\left(\frac{az+b}{cz+d}\right) = F(Z).$$

Further, if a_1, b_1, c_1, d_1 and a_2, b_2, c_2, d_2 are any two of the sets of values of a, b, c, d, and if Z be replaced first by

$$\frac{a_1z+b_1}{c_1z+d_1},$$

and then, in this, z be replaced by

$$\frac{a_2z+b_2}{c_2z+d_2},$$

giving, say,

$$\frac{AZ+B}{CZ+D},$$

then not only do we have

$$F\left(\frac{a_1z+b_1}{c_1z+d_1}\right) = F(Z), F\left(\frac{a_2z+b_2}{c_2z+d_2}\right) = F(Z)$$

but also

$$F\left(\frac{Az+B}{Cz+D}\right) = F(z).$$

Further, the set of all the substitutions

$$Z \to \frac{az+b}{cz+d}$$

which leave the value of $F(Z)$ unchanged *form a group* [8] one of whose properties is that the result of two successive substitutions in the set

$$Z \to \frac{a_1z+b_1}{c_1z+d_1},\ Z \to \frac{a_2z+b_2}{c_2z+d_2}$$

is itself in the set.

[7] For the meaning of this term see the selection by Hans Hahn, p. 1593.

[8] For the group concept see selections by Keyser and Eddington.

Functions of this kind, which are said to be invariant under a denumerably infinite group of linear fractional transformations, were constructed by Poincaré in the 1880s and their properties described in a series of memorable papers. He called some of them "Fuchsian" functions (after the German mathematician Lazarus Fuchs), though they are today called *automorphic*. The Fuchsian functions are alluded to in Poincaré's famous essay on "Mathematical Creation," which is also included in this book. After weeks of intense concentration the concept came to him during a sleepless night. A little later, another inspiration which profoundly elaborated the original concept struck him as he was setting forth on a journey and placed his foot on the step of an omnibus.

In theoretical astronomy, Poincaré undertook a celebrated investigation to determine the form taken by a gravitating mass of fluid in rotation. He showed that a liquid sphere which rotates with increasing speed will first be transformed into a flattened spheroidal shape, then into an ellipsoid of revolution, then will develop a "lop-sided bulge" and finally "will become pear-shaped until at last the mass, hollowing out more and more at its 'waist,' will separate into two distinct and unequal bodies." Other researches dealt with the famous problem of three bodies, tides, light, electricity, capillarity, potential, thermodynamics, conduction of heat, elasticity, wireless telegraphy. Sir George Darwin, in presenting the gold medal of the Royal Astronomical Society to Poincaré in 1909, commented on the richness and number of his ideas, and on the "immense amplitude of generalizations" which was the dominant character of his work. Most scientists, said Darwin, raise themselves to the general aspects of a problem after examining simple and concrete cases. "I realize that M. Poincaré is destined to follow in his work another route than that, and that he finds it easier to consider at the outset the larger issues for descending towards the more special cases."

The selection which follows, an essay on chance, is an excellent example of Poincaré's profundity, of his wit and his supreme gift of clarity. "One of the many reasons for which he [Poincaré] will live," Jourdain wrote, "is because he made it possible for us to understand him as well as to admire him." [9] He made it possible for us to understand him by admitting us, as few men have been able to do, into his mind; by taking us along, step by step, on the side journeys as well as the main route of his thought. He had a wonderfully curt, flat way of disposing of nonsense. Except for the Gallic flavor of his sentences, his liquid, subtle style resembles Bertrand Russell's.

In his theory of knowledge and philosophy of science Poincaré might be called a pragmatist. The test of a scientific theory was its simplicity and

[9] Philip E. B. Jourdain, obituary of Poincaré in *The Monist*, Vol. 22, 1912, pp. 611 *et seq.*

convenience—his favorite word. Certain hypotheses are valuable because by experiment they can be verified or refuted. But there is another important class of hypotheses which, though they can neither be confirmed nor disproved, are yet indispensable to science because (in the words of Josiah Royce) they are "devices of the understanding whereby we give conceptual unity and an invisible connectedness to certain types of phenomenal facts which come to us in a discrete form and in a confused variety." I shall quote a few examples. "Masses," writes Poincaré, "are coefficients it is convenient to introduce into calculations. We could reconstruct all mechanics by attributing different values to all the masses. This new mechanics would not be in contradiction either with experience or with the general principles of dynamics. Only the equations of this new mechanics would be *less simple*." [10] We choose rules for ordering our experience, "not because they are true, but because they are most convenient." [11] "Time should be so defined that the equations of mechanics may be as simple as possible. . . . Of two watches, we have no right to say that one goes true, the other wrong: we can only say that it is advantageous to conform to the indications of the first." [12] "Behold then the rule we follow and the only one we can follow: when a phenomenon appears to us as the cause of another, we regard it as anterior. It is therefore by cause we define time." [13] ". . . by natural selection our mind has adapted itself to the conditions of the external world. It has adopted the geometry *most advantageous* to the species or, in other words, the *most convenient*. Geometry is not true, it is advantageous." [14] "Absolute space, that is to say, the mark to which it would be necessary to refer the earth to know whether it really moves, has no objective existence. . . . The two propositions: 'The earth turns round' and 'it is more convenient to suppose the earth turns round' have the same meaning; there is nothing more in the one than in the other." [15]

One or two further passages are irresistible. What is the business of the scientist? He must observe, of course, but merely to observe is not enough. "The scientist must set in order. Science is built up with facts, as a house is with stones. But a collection of facts is no more a science than a heap of stones is a house. And above all the scientist must foresee. Carlyle has somewhere said something like this: Nothing but facts are of importance. John Lackland passed by here. Here is something that is admirable. Here is a reality for which I would give all the theories in the world.' Carlyle was a fellow countryman of Bacon; but Bacon would not have said that.

[10] *Science and Hypothesis.*
[11] *The Value of Science.*
[12] *Ibid.*
[13] *Ibid.*
[14] *Science and Method.*
[15] *Science and Hypothesis.*

That is the language of the historian. The physicist would say rather: 'John Lackland passed by here; that makes no difference to me, for he will never pass this way again.' " [16]

Even before 1882 when the squaring of the circle was proved impossible, the Academy of Sciences "rejected without examination the alas! too numerous memoirs on the subject, that some unhappy madmen sent in every year. Was the Academy wrong? Evidently not, and it knew well that in acting thus it did not run the least risk of stifling a discovery of the moment. . . . If you had asked the Academicians, they would have answered: 'We have compared the probability that an unknown savant should have found out what has been vainly sought for so long, with the probability that there is one madman the more on the earth; the second appears to us the greater.' "

Poincaré was not neglected in his day. He won almost every medal and prize for which a scientist was eligible and gained membership in the most distinguished scientific bodies. The immortal J. J. Sylvester in his old age made a pilgrimage to Paris to see the new star of mathematics. He puffed up three flights of steep, narrow stairs on a hot summer day to Poincaré's "airy perch in the Rue Gay-Lussac." At the open door he paused in doubt and astonishment "at beholding a mere boy, 'so blond, so young,' as the author of the deluge of papers which had heralded the advent of a successor to Cauchy." [17] "In the presence," said Sylvester, "of that mighty pent-up intellectual force my tongue at first refused its office, and it was not until I had taken some time (it may be two or three minutes) to peruse and absorb as it were the idea of his external youthful lineaments that I found myself in a condition to speak." In 1904 Poincaré attended the International Congress of Arts and Science at St. Louis; his appearance is still vividly remembered by the survivors of that remarkable gathering. In 1908 he was anointed as one of the forty celestial objects of the French Academy. This accolade is the more remarkable because it was awarded him for the literary quality of his popular essays. He was taken ill while attending the international congress of mathematicians at Rome in 1908; the cause, enlargement of the prostate, was temporarily relieved by surgery. In 1912 a second operation was necessary. From this he seemed to recover, but on July 17 he died suddenly from an embolism.

[16] *Ibid.*

[17] E. T. Bell, *op. cit.*, p. 527; also Slosson, *op. cit.*, pp. 136–137.

I returned and saw under the sun that the race is not to the swift, nor the battle to the strong, neither yet bread to the wise, nor yet riches to men of understanding, nor yet favour to men of skill; but time and chance happeneth to them all.
—ECCLESIASTES

5 Chance

By HENRI POINCARÉ

I

"HOW dare we speak of the laws of chance? Is not chance the antithesis of all law?" So says Bertrand at the beginning of his *Calcul des probabilités*. Probability is opposed to certitude; so it is what we do not know and consequently it seems what we could not calculate. Here is at least apparently a contradiction, and about it much has already been written.

And first, what is chance? The ancients distinguished between phenomena seemingly obeying harmonious laws, established once for all, and those which they attributed to chance; these were the ones unpredictable because rebellious to all law. In each domain the precise laws did not decide everything, they only drew limits between which chance might act. In this conception the word chance had a precise and objective meaning; what was chance for one was also chance for another and even for the gods.

But this conception is not ours to-day. We have become absolute determinists, and even those who want to reserve the rights of human free will let determinism reign undividely in the inorganic world at least. Every phenomenon, however minute, has a cause; and a mind infinitely powerful, infinitely well-informed about the laws of nature, could have foreseen it from the beginning of the centuries. If such a mind existed, we could not play with it at any game of chance; we should always lose.

In fact for it the word chance would not have any meaning, or rather there would be no chance. It is because of our weakness and our ignorance that the word has a meaning for us. And, even without going beyond our feeble humanity, what is chance for the ignorant is not chance for the scientist. Chance is only the measure of our ignorance. Fortuitous phenomena are, by definition, those whose laws we do not know.

But is this definition altogether satisfactory? When the first Chaldean shepherds followed with their eyes the movements of the stars, they knew not as yet the laws of astronomy; would they have dreamed of saying that the stars move at random? If a modern physicist studies a new phenomenon, and if he discovers its law Tuesday, would he have said Monday that this phenomenon was fortuitous? Moreover, do we not often invoke what Bertrand calls the laws of chance, to predict a phenomenon? For example, in the kinetic theory of gases we obtain the known laws of Mariotte and of Gay-Lussac by means of the hypothesis that the velocities of the molecules of gas vary irregularly, that is to say at random. All physicists will agree that the observable laws would be much less simple if the velocities were ruled by any simple elementary law whatsoever, if the molecules were, as we say, *organized*, if they were subject to some discipline. It is due to chance, that is to say, to our ignorance, that we can draw our conclusions; and then if the word chance is simply synonymous with ignorance what does that mean? Must we therefore translate as follows?

"You ask me to predict for you the phenomena about to happen. If, unluckily, I knew the laws of these phenomena I could make the prediction only by inextricable calculations and would have to renounce attempting to answer you; but as I have the good fortune not to know them, I will answer you at once. And what is most surprising, my answer will be right."

So it must well be that chance is something other than the name we give our ignorance, that among phenomena whose causes are unknown to us we must distinguish fortuitous phenomena about which the calculus of probabilities will provisionally give information, from those which are not fortuitous and of which we can say nothing so long as we shall not have determined the laws governing them. For the fortuitous phenomena themselves, it is clear that the information given us by the calculus of probabilities will not cease to be true upon the day when these phenomena shall be better known.

The director of a life insurance company does not know when each of the insured will die, but he relies upon the calculus of probabilities and on the law of great numbers, and he is not deceived, since he distributes dividends to his stockholders. These dividends would not vanish if a very penetrating and very indiscreet physician should, after the policies were signed, reveal to the director the life chances of the insured. This doctor would dissipate the ignorance of the director, but he would have no influence on the dividends, which evidently are not an outcome of this ignorance.

II

To find a better definition of chance we must examine some of the facts which we agree to regard as fortuitous, and to which the calculus of probabilities seems to apply; we then shall investigate what are their common characteristics.

The first example we select is that of unstable equilibrium; if a cone rests upon its apex, we know well that it will fall, but we do not know toward what side; it seems to us chance alone will decide. If the cone were perfectly symmetric, if its axis were perfectly vertical, if it were acted upon by no force other than gravity, it would not fall at all. But the least defect in symmetry will make it lean slightly toward one side or the other, and if it leans, however little, it will fall altogether toward that side. Even if the symmetry were perfect, a very slight tremor, a breath of air could make it incline some seconds of arc; this will be enough to determine its fall and even the sense of its fall which will be that of the initial inclination.

A very slight cause, which escapes us, determines a considerable effect which we can not help seeing, and then we say this effect is due to chance. If we could know exactly the laws of nature and the situation of the universe at the initial instant, we should be able to predict exactly the situation of this same universe at a subsequent instant. But even when the natural laws should have no further secret for us, we could know the initial situation only *approximately*. If that permits us to foresee the subsequent situation *with the same degree of approximation*, this is all we require, we say the phenomenon has been predicted, that it is ruled by laws. But this is not always the case; it may happen that slight differences in the initial conditions produce very great differences in the final phenomena; a slight error in the former would make an enormous error in the latter. Prediction becomes impossible and we have the fortuitous phenomenon.

Our second example will be very analogous to the first and we shall take it from meteorology. Why have the meteorologists such difficulty in predicting the weather with any certainty? Why do the rains, the tempests themselves seem to us to come by chance, so that many persons find it quite natural to pray for rain or shine, when they would think it ridiculous to pray for an eclipse? We see that great perturbations generally happen in regions where the atmosphere is in unstable equilibrium. The meteorologists are aware that this equilibrium is unstable, that a cyclone is arising somewhere; but where they can not tell; one-tenth of a degree more or less at any point, and the cyclone bursts here and not there, and spreads its ravages over countries it would have spared. This we could have foreseen if we had known that tenth of a degree, but the observations were neither sufficiently close nor sufficiently precise, and for this reason all seems due to the agency of chance. Here again we find the same contrast between a very slight cause, unappreciable to the observer, and important effects, which are sometimes tremendous disasters.

Let us pass to another example, the distribution of the minor planets on the zodiac. Their initial longitudes may have been any longitudes whatever; but their mean motions were different and they have revolved for so long a time that we may say they are now distributed *at random* along the zodiac. Very slight initial differences between their distances from the sun, or, what comes to the same thing, between their mean motions, have ended by giving enormous differences between their present longitudes. An excess of the thousandth of a second in the daily mean motion will give in fact a second in three years, a degree in ten thousand years, an entire circumference in three or four million years, and what is that to the time which has passed since the minor planets detached themselves from the nebula of Laplace? Again therefore we see a slight cause and a great effect; or better, slight differences in the cause and great differences in the effect.

The game of roulette does not take us as far as might seem from the preceding example. Assume a needle to be turned on a pivot over a dial divided into a hundred sectors alternately red and black. If it stops on a red sector I win; if not, I lose. Evidently all depends upon the initial impulse I give the needle. The needle will make, suppose, ten or twenty turns, but it will stop sooner or not so soon, according as I shall have pushed it more or less strongly. It suffices that the impulse vary only by a thousandth or a two thousandth to make the needle stop over a black sector or over the following red one. These are differences the muscular sense can not distinguish and which elude even the most delicate instruments. So it is impossible for me to foresee what the needle I have started will do, and this is why my heart throbs and I hope everything from luck. The difference in the cause is imperceptible, and the difference in the effect is for me of the highest importance, since it means my whole stake.

III

Permit me, in this connection, a thought somewhat foreign to my subject. Some years ago a philosopher said that the future is determined by the past, but not the past by the future; or, in other words, from knowledge of the present we could deduce the future, but not the past; because, said he, a cause can have only one effect, while the same effect might be produced by several different causes. It is clear no scientist can subscribe to this conclusion. The laws of nature bind the antecedent to the consequent in such a way that the antecedent is as well determined by the consequent as the consequent by the antecedent. But whence came the error of this philosopher? We know that in virtue of Carnot's principle physical phenomena are irreversible and the world tends toward uniformity. When two bodies of different temperature come in contact, the warmer gives up heat to the colder; so we may foresee that the temperature will equalize. But once equal, if asked about the anterior state, what can we answer? We might say that one was warm and the other cold, but not be able to divine which formerly was the warmer.

And yet in reality the temperatures will never reach perfect equality. The difference of the temperatures only tends asymptotically toward zero. There comes a moment when our thermometers are powerless to make it known. But if we had thermometers a thousand times, a hundred thousand times as sensitive, we should recognize that there still is a slight difference, and that one of the bodies remains a little warmer than the other, and so we could say this it is which formerly was much the warmer.

So then there are, contrary to what we found in the former examples, great differences in cause and slight differences in effect. Flammarion once imagined an observer going away from the earth with a velocity greater than that of light; for him time would have changed sign. History would be turned about, and Waterloo would precede Austerlitz. Well, for this observer, effects and causes would be inverted; unstable equilibrium would no longer be the exception. Because of the universal irreversibility, all would seem to him to come out of a sort of chaos in unstable equilibrium. All nature would appear to him delivered over to chance.

IV

Now for other examples where we shall see somewhat different characteristics. Take first the kinetic theory of gases. How should we picture a receptacle filled with gas? Innumerable molecules, moving at high speeds, flash through this receptacle in every direction. At every instant they strike against its walls or each other, and these collisions happen under the most diverse conditions. What above all impresses us here is not the littleness of the causes, but their complexity, and yet the former element is still found here and plays an important rôle. If a molecule deviated right or left from its trajectory, by a very small quantity, comparable to the radius of action of the gaseous molecules, it would avoid a collision or sustain it under different conditions, and that would vary the direction of its velocity after the impact, perhaps by ninety degrees or by a hundred and eighty degrees.

And this is not all; we have just seen that it is necessary to deflect the molecule before the clash by only an infinitesimal, to produce its deviation after the collision by a finite quantity. If then the molecule undergoes two successive shocks, it will suffice to deflect it before the first by an infinitesimal of the second order, for it to deviate after the first encounter by an infinitesimal of the first order, and after the second hit, by a finite quantity. And the molecule will not undergo merely two shocks; it will undergo a very great number per second. So that if the first shock has multiplied the deviation by a very large number A, after n shocks it will be multiplied by A^n. It will therefore become very great not merely because A is large, that is to say because little causes produce big effects, but because the exponent n is large, that is to say because the shocks are very numerous and the causes very complex.

Take a second example. Why do the drops of rain in a shower seem to be distributed at random? This is again because of the complexity of the causes which determine their formation. Ions are distributed in the atmosphere. For a long while they have been subjected to air-currents constantly changing, they have been caught in very small whirlwinds, so that their final distribution has no longer any relation to their initial distribution. Suddenly the temperature falls, vapor condenses, and each of these ions becomes the center of a drop of rain. To know what will be the distribution of these drops and how many will fall on each paving-stone, it would not be sufficient to know the initial situation of the ions, it would be necessary to compute the effect of a thousand little capricious air-currents.

And again it is the same if we put grains of powder in suspension in water. The vase is ploughed by currents whose law we know not, we only know it is very complicated. At the end of a certain time the grains will be distributed at random, that is to say uniformly, in the vase; and this is due precisely to the complexity of these currents. If they obeyed some simple law, if for example the vase revolved and the currents circulated around the axis of the vase, describing circles, it would no longer be the same, since each grain would retain its initial altitude and its initial distance from the axis.

We should reach the same result in considering the mixing of two liquids or of two fine-grained powders. And to take a grosser example, this is also what happens when we shuffle playing cards. At each stroke the cards undergo a permutation (analogous to that studied in the theory of substitutions). What will happen? The probability of a particular permutation (for example, that bringing to the nth place the card occupying the $\phi(n)$th place before the permutation) depends upon the player's habits. But if this player shuffles the cards long enough, there will be a great number of successive permutations, and the resulting final order will no longer be governed by aught but chance; I mean to say that all possible orders will be equally probable. It is to the great number of successive permutations, that is to say to the complexity of the phenomenon, that this result is due.

A final word about the theory of errors. Here it is that the causes are complex and multiple. To how many snares is not the observer exposed, even with the best instrument! He should apply himself to finding out the largest and avoiding them. These are the ones giving birth to systematic errors. But when he has eliminated those, admitting that he succeeds, there remain many small ones which, their effects accumulating, may become dangerous. Thence come the accidental errors; and we attribute them to chance because their causes are too complicated and too numerous. Here again we have only little causes, but each of them would produce only a slight effect; it is by their union and their number that their effects become formidable.

V

We may take still a third point of view, less important than the first two and upon which I shall lay less stress. When we seek to foresee an event and examine its antecedents, we strive to search into the anterior situation. This could not be done for all parts of the universe and we are content to know what is passing in the neighborhood of the point where the event should occur, or what would appear to have some relation to it. An examination can not be complete and we must know how to choose. But it may happen that we have passed by circumstances which at first sight seemed completely foreign to the foreseen happening, to which one would never have dreamed of attributing any influence and which nevertheless, contrary to all anticipation, come to play an important rôle.

A man passes in the street going to his business; some one knowing the business could have told why he started at such a time and went by such a street. On the roof works a tiler. The contractor employing him could in a certain measure foresee what he would do. But the passer-by scarcely thinks of the tiler, nor the tiler of him; they seem to belong to two worlds completely foreign to one another. And yet the tiler drops a tile which kills the man, and we do not hesitate to say this is chance.

Our weakness forbids our considering the entire universe and makes us cut it up into slices. We try to do this as little artificially as possible. And yet it happens from time to time that two of these slices react upon each other. The effects of this mutual action then seem to us to be due to chance.

Is this a third way of conceiving chance? Not always; in fact most often we are carried back to the first or the second. Whenever two worlds usually foreign to one another come thus to react upon each other, the laws of this reaction must be very complex. On the other hand, a very slight change in the initial conditions of these two worlds would have been sufficient for the reaction not to have happened. How little was needed for the man to pass a second later or the tiler to drop his tile a second sooner.

VI

All we have said still does not explain why chance obeys laws. Does the fact that the causes are slight or complex suffice for our foreseeing, if not their effects *in each case*, at least what their effects will be, *on the average*? To answer this question we had better take up again some of the examples already cited.

I shall begin with that of the roulette. I have said that the point where the needle will stop depends upon the initial push given it. What is the probability of this push having this or that value? I know nothing about it, but it is difficult for me not to suppose that this probability is represented by a continuous analytic function. The probability that the push is comprised between α and $\alpha + \epsilon$ will then be sensibly equal to the probability of its being comprised between $\alpha + \epsilon$ and $\alpha + 2\epsilon$, *provided* ϵ *be very small.* This is a property common to all analytic functions. Minute variations of the function are proportional to minute variations of the variable.

But we have assumed that an exceedingly slight variation of the push suffices to change the color of the sector over which the needle finally stops. From α to $\alpha + \epsilon$ it is red, from $\alpha + \epsilon$ to $\alpha + 2\epsilon$ it is black; the probability of each red sector is therefore the same as of the following black, and consequently the total probability of red equals the total probability of black.

The datum of the question is the analytic function representing the probability of a particular initial push. But the theorem remains true whatever be this datum, since it depends upon a property common to all analytic functions. From this it follows finally that we no longer need the datum.

What we have just said for the case of the roulette applies also to the example of the minor planets. The zodiac may be regarded as an immense roulette on which have been tossed many little balls with different initial impulses varying according to some law. Their present distribution is uniform and independent of this law, for the same reason as in the preceding case. Thus we see why phenomena obey the laws of chance when slight differences in the causes suffice to bring on great differences in the effects. The probabilities of these slight differences may then be regarded as proportional to these differences themselves, just because these differences are minute, and the infinitesimal increments of a continuous function are proportional to those of the variable.

Take an entirely different example, where intervenes especially the complexity of the causes. Suppose a player shuffles a pack of cards. At each shuffle he changes the order of the cards, and he may change them in many ways. To simplify the exposition, consider only three cards. The cards which before the shuffle occupied respectively the places 123, may after the shuffle occupy the places

$$123, 231, 312, 321, 132, 213.$$

Each of these six hypotheses is possible and they have respectively for probabilities:

$$p_1, p_2, p_3, p_4, p_5, p_6.$$

The sum of these six numbers equals 1; but this is all we know of them; these six probabilities depend naturally upon the habits of the player which we do not know.

At the second shuffle and the following, this will recommence, and under the same conditions; I mean that p_4 for example represents always the probability that the three cards which occupied after the nth shuffle and before $n+1$th the places 123, occupy the places 321 after the $n+1$th shuffle. And this remains true whatever be the number n, since the habits of the player and his way of shuffling remain the same.

But if the number of shuffles is very great, the cards which before the first shuffle occupied the places 123 may, after the last shuffle, occupy the places

$$123, 231, 312, 321, 132, 213$$

and the probability of these six hypotheses will be sensibly the same and equal to $1/6$; and this will be true whatever be the numbers $p_1 \ldots p_6$ which we do not know. The great number of shuffles, that is to say the complexity of the causes, has produced uniformity.

This would apply without change if there were more than three cards, but even with three cards the demonstration would be complicated; let it suffice to give it for only two cards. Then we have only two possibilities 12, 21 with the probabilities p_1 and $p_2 = 1 - p_1$.

Suppose n shuffles and suppose I win one franc if the cards are finally in the initial order and lose one if they are finally inverted. Then, my mathematical expectation will be $(p_1 - p_2)^n$.

The difference $p_1 - p_2$ is certainly less than 1; so that if n is very great my expectation will be zero; we need not learn p_1 and p_2 to be aware that the game is equitable.

There would always be an exception if one of the numbers p_1 and p_2 was equal to 1 and the other naught. *Then it would not apply because our initial hypotheses would be too simple.*

What we have just seen applies not only to the mixing of cards, but to all mixings, to those of powders and of liquids; and even to those of the molecules of gases in the kinetic theory of gases.

To return to this theory, suppose for a moment a gas whose molecules can not mutually clash, but may be deviated by hitting the insides of the vase wherein the gas is confined. If the form of the vase is sufficiently complex the distribution of the molecules and that of the velocities will not be long in becoming uniform. But this will not be so if the vase is spherical or if it has the shape of a cuboid. Why? Because in the first case the distance from the center to any trajectory will remain constant; in the second case this will be the absolute value of the angle of each trajectory with the faces of the cuboid.

So we see what should be understood by conditions *too simple*; they are those which conserve something, which leave an invariant remaining. Are the differential equations of the problem too simple for us to apply the laws of chance? This question would seem at first view to lack precise meaning; now we know what it means. They are too simple if they conserve something, if they admit a uniform integral. If something in the initial conditions remains unchanged, it is clear the final situation can no longer be independent of the initial situation.

We come finally to the theory of errors. We know not to what are due the accidental errors, and precisely because we do not know, we are aware they obey the law of Gauss. Such is the paradox. The explanation is nearly the same as in the preceding cases. We need know only one thing: that the errors are very numerous, that they are very slight, that each may be as well negative as positive. What is the curve of probability of each of them? We do not know; we only suppose it is symmetric. We prove then that the resultant error will follow Gauss's law, and this resulting law is independent of the particular laws which we do not know. Here again the simplicity of the result is born of the very complexity of the data.

VII

But we are not through with paradoxes. I have just recalled the figment of Flammarion, that of the man going quicker than light, for whom time changes sign. I said that for him all phenomena would seem due to chance. That is true from a certain point of view, and yet all these phenomena at a given moment would not be distributed in conformity with the laws of chance, since the distribution would be the same as for us, who, seeing them unfold harmoniously and without coming out of a primal chaos, do not regard them as ruled by chance.

What does that mean? For Lumen, Flammarion's man, slight causes seem to produce great effects; why do not things go on as for us when we think we see grand effects due to little causes? Would not the same reasoning be applicable in his case?

Let us return to the argument. When slight differences in the causes produce vast differences in the effects, why are these effects distributed according to the laws of chance? Suppose a difference of a millimeter in the cause produces a difference of a kilometer in the effect. If I win in case the effect corresponds to a kilometer bearing an even number, my probability of winning will be ½. Why? Because to make that, the cause must correspond to a millimeter with an even number. Now, according to all appearance, the probability of the cause varying between certain limits will be proportional to the distance apart of these limits, provided this distance be very small. If this hypothesis were not admitted there would no longer be any way of representing the probability by a continuous function.

What now will happen when great causes produce small effects? This is the case where we should not attribute the phenomenon to chance and where on the contrary Lumen would attribute it to chance. To a difference of a kilometer in the cause would correspond a difference of a millimeter in the effect. Would the probability of the cause being comprised between two limits n kilometers apart still be proportional to n? We have no reason to suppose so, since this distance, n kilometers, is great. But the probability that the effect lies between two limits n millimeters apart will be precisely the same, so it will not be proportional to n, even though this distance, n millimeters, be small. There is no way therefore of representing the law of probability of effects by a continuous curve. This curve, understand, may remain continuous in the *analytic* sense of the word; to *infinitesimal* variations of the abscissa will correspond infinitesimal variations of the ordinate. But *practically* it will not be continuous, since *very small* variations of the ordinate would not correspond to very small variations of the abscissa. It would become impossible to trace the curve with an ordinary pencil; that is what I mean.

So what must we conclude? Lumen has no right to say that the probability of the cause (*his* cause, our effect) should be represented necessarily by a continuous function. But then why have we this right? It is because this state of unstable equilibrium which we have been calling initial is itself only the final outcome of a long previous history. In the course of this history complex causes have worked a great while: they have contributed to produce the mixture of elements and they have tended to make everything uniform at least within a small region; they have rounded off the corners, smoothed down the hills and filled up the valleys. However capricious and irregular may have been the primitive curve given over to them, they have worked so much toward making it regular that finally they deliver over to us a continuous curve. And this is why we may in all confidence assume its continuity.

Lumen would not have the same reasons for such a conclusion. For him complex causes would not seem agents of equalization and regularity, but on the contrary would create only inequality and differentiation. He would see a world more and more varied come forth from a sort of primitive chaos. The changes he could observe would be for him unforeseen and impossible to foresee. They would seem to him due to some caprice or another; but this caprice would be quite different from our chance, since it would be opposed to all law, while our chance still has its laws. All these points call for lengthy explications, which perhaps would aid in the better comprehension of the irreversibility of the universe.

VIII

We have sought to define chance, and now it is proper to put a question. Has chance thus defined, in so far as this is possible, objectivity?

It may be questioned. I have spoken of very slight or very complex causes. But what is very little for one may be very big for another, and what seems very complex to one may seem simple to another. In part I have already answered by saying precisely in what cases differential equations become too simple for the laws of chance to remain applicable. But it is fitting to examine the matter a little more closely, because we may take still other points of view.

What means the phrase 'very slight'? To understand it we need only go back to what has already been said. A difference is very slight, an interval is very small, when within the limits of this interval the probability remains sensibly constant. And why may this probability be regarded as constant within a small interval? It is because we assume that the law of probability is represented by a continuous curve, continuous not only in the analytic sense, but *practically* continuous, as already explained. This means that it not only presents no absolute hiatus, but that it has neither salients nor reentrants too acute or too accentuated.

And what gives us the right to make this hypothesis? We have already said it is because, since the beginning of the ages, there have always been complex causes ceaselessly acting in the same way and making the world tend toward uniformity without ever being able to turn back. These are the causes which little by little have flattened the salients and filled up the reentrants, and this is why our probability curves now show only gentle undulations. In milliards of milliards of ages another step will have been made toward uniformity, and these undulations will be ten times as gentle; the radius of mean curvature of our curve will have become ten times as great. And then such a length as seems to us to-day not very small, since on our curve an arc of this length can not be regarded as rectilineal, should on the contrary at that epoch be called very little, since the curvature will have become ten times less and an arc of this length may be sensibly identified with a sect.

Thus the phrase 'very slight' remains relative; but it is not relative to such or such a man, it is relative to the actual state of the world. It will change its meaning when the world shall have become more uniform, when all things shall have blended still more. But then doubtless men can no longer live and must give place to other beings—should I say far smaller or far larger? So that our criterion, remaining true for all men, retains an objective sense.

And on the other hand what means the phrase 'very complex'? I have already given one solution, but there are others. Complex causes we have said produce a blend more and more intimate, but after how long a time will this blend satisfy us? When will it have accumulated sufficient complexity? When shall we have sufficiently shuffled the cards? If we mix two powders, one blue, the other white, there comes a moment when the tint of the mixture seems to us uniform because of the feebleness of our senses; it will be uniform for the presbyte, forced to gaze from afar, before it will be so for the myope. And when it has become uniform for all eyes, we still could push back the limit by the use of instruments. There is no chance for any man ever to discern the infinite variety which, if the kinetic theory is true, hides under the uniform appearance of a gas. And yet if we accept Gouy's ideas on the Brownian movement, does not the microscope seem on the point of showing us something analogous?

This new criterion is therefore relative like the first; and if it retains an objective character, it is because all men have approximately the same senses, the power of their instruments is limited, and besides they use them only exceptionally.

IX

It is just the same in the moral sciences and particularly in history. The historian is obliged to make a choice among the events of the epoch he studies; he recounts only those which seem to him the most important. He therefore contents himself with relating the most momentous events of the sixteenth century, for example, as likewise the most remarkable facts of the seventeenth century. If the first suffice to explain the second, we say these conform to the laws of history. But if a great event of the seventeenth century should have for cause a small fact of the sixteenth century which no history reports, which all the world has neglected, then we say this event is due to chance. This word has therefore the same sense as in the physical sciences; it means that slight causes have produced great effects.

The greatest bit of chance is the birth of a great man. It is only by chance that meeting of two germinal cells, of different sex, containing precisely, each on its side, the mysterious elements whose mutual reaction must produce the genius. One will agree that these elements must be rare and that their meeting is still more rare. How slight a thing it would have required to deflect from its route the carrying spermatozoon. It would have sufficed to deflect it a tenth of a millimeter and Napoleon would not have been born and the destinies of a continent would have been changed. No example can better make us understand the veritable characteristics of chance.

One more word about the paradoxes brought out by the application of the calculus of probabilities to the moral sciences. It has been proven that no Chamber of Deputies will ever fail to contain a member of the opposition, or at least such an event would be so improbable that we might without fear wager the contrary, and bet a million against a sou.

Condorcet has striven to calculate how many jurors it would require to make a judicial error practically impossible. If we had used the results of this calculation, we should certainly have been exposed to the same disappointments as in betting, on the faith of the calculus, that the opposition would never be without a representative.

The laws of chance do not apply to these questions. If justice be not always meted out to accord with the best reasons, it uses less than we think the method of Bridoye. This is perhaps to be regretted, for then the system of Condorcet would shield us from judicial errors.

What is the meaning of this? We are tempted to attribute facts of this nature to chance because their causes are obscure; but this is not true chance. The causes are unknown to us, it is true, and they are even complex; but they are not sufficiently so, since they conserve something. We have seen that this it is which distinguishes causes 'too simple.' When men are brought together they no longer decide at random and independently one of another; they influence one another. Multiplex causes come into action. They worry men, dragging them to right or left, but one thing there is they can not destroy, this is their Panurge flock-of-sheep habits. And this is an invariant.

X

Difficulties are indeed involved in the application of the calculus of probabilities to the exact sciences. Why are the decimals of a table of logarithms, why are those of the number π distributed in accordance with the laws of chance? Elsewhere I have already studied the question in so far as it concerns logarithms, and there it is easy. It is clear that a slight difference of argument will give a slight difference of logarithm, but a great difference in the sixth decimal of the logarithm. Always we find again the same criterion.

But as for the number π, that presents more difficulties, and I have at the moment nothing worth while to say.

There would be many other questions to resolve, had I wished to attack them before solving that which I more specially set myself. When we reach a simple result, when we find for example a round number, we say that such a result can not be due to chance, and we seek, for its explanation, a non-fortuitous cause. And in fact there is only a very slight probability that among 10,000 numbers chance will give a round number; for example, the number 10,000. This has only one chance in 10,000. But there is only one chance in 10,000 for the occurrence of any other one number; and yet this result will not astonish us, nor will it be hard for us to attribute it to chance; and that simply because it will be less striking.

Is this a simple illusion of ours, or are there cases where this way of thinking is legitimate? We must hope so, else were all science impossible. When we wish to check a hypothesis, what do we do? We can not verify all its consequences, since they would be infinite in number; we content ourselves with verifying certain ones and if we succeed we declare the hypothesis confirmed, because so much success could not be due to chance. And this is always at bottom the same reasoning.

I can not completely justify it here, since it would take too much time; but I may at least say that we find ourselves confronted by two hypotheses, either a simple cause or that aggregate of complex causes we call chance. We find it natural to suppose that the first should produce a simple result, and then, if we find that simple result, the round number for example, it seems more likely to us to be attributable to the simple cause which must give it almost certainly, than to chance which could only give it once in 10,000 times. It will not be the same if we find a result which is not simple; chance, it is true, will not give this more than once in 10,000 times; but neither has the simple cause any more chance of producing it.

COMMENTARY ON

ERNEST NAGEL and the Laws of Probability

IT is three hundred years exactly (I am writing in 1954) since Pascal and Fermat conducted their famous exchange of letters dealing with a question proposed by the gambler Chevalier de Meré. The question concerned the division of stakes at games of dice, and the answer to it, contained in the correspondence of the two great French mathematicians, is generally regarded as the foundation of the theory of probability.[1] In the centuries that have elapsed since the theory was created no branch of mathematics has been more assiduously cultivated, none has found a wider range of application. It is remarkable, as Laplace wrote, that "a science which began with the considerations of play has risen to the most important objects of human knowledge." But it is in a sense even more remarkable that despite the attention bestowed on this science, and its enormous influence, mathematicians and philosophers are quite unable to agree on the meaning of probability. Their disagreement is less easily explained than that of the three men describing the elephant. For in this case the observers are not blind and the creature is of their own design.

There are three main interpretations of probability. The classic view, formulated by Laplace and De Morgan, holds that the notion refers to a state of mind.[2] None of our knowledge is certain; the degree or strength of our belief as to any given proposition is its probability. The mathematical theory of probability tells us how a measure can be assigned to each proposition, and how such measures can be combined in a calculus. Another view defines probability as an essentially unanalyzable, but intuitively understandable, logical relation between propositions. Accord-

[1] Recent historical researches have uncovered the fact that the Italian mathematician Jerome Cardan (Gerolamo Cardano), of Milan (1501–1576) was the real pioneer in this field. In his little gamblers' manual, the *Liber de Ludo Aleae*, Cardan exhibited remarkable insight about fundamental problems of probability. He discusses equiprobability, mathematical expectation, reasoning on the mean, frequency tables for dice probabilities, additive properties of probabilities, the power formula $p_n = p^n$ for obtaining n successes in n independent trials; there is even an adumbration of the so-called law of large numbers. The 15 folio-page *Liber de Ludo* did not appear in print until the publication of a ten-volume edition of Cardan's complete works in 1663 and even then it attracted scant attention. It cannot be said, therefore, that his work influenced the development of probability. Indeed it was not until 1953 that a proper evaluation of Cardan's achievements in mathematical probability was published. See Oystein Ore, *Cardano, The Gambling Scholar*, Princeton, 1953; also my review of this book in *Scientific American*, June, 1953.

[2] "I consider the word *probability*," De Morgan explained (*An Essay on Probability*, London, 1838) "as meaning the state of mind with respect to an assertion, a coming event, or any other matter on which absolute knowledge does not exist."

ing to John Maynard Keynes, a principal exponent of this interpretation, "we must have a 'logical intuition' of the probable relations between propositions." [3] Once we apprehend the existence of this relation between evidence and conclusion, the latter becomes a subject of "rational belief." The third view of probability rests on the statistical concept of relative frequency. This interpretation was developed during the last century by the Austrian philosopher and mathematician Bernhard Bolzano (1781–1841), the English logician John Venn (1834–1923), the French economist Cournot (see p. 1203), and by Charles Sanders Peirce; and in our time by R. A. Fisher (see p. 1512) and Richard von Mises (see p. 1723), among others. Statistical probability stems from the idea of the relative frequency of an event in a class of events. Thus for example when it is said the probability of surviving an attack of pneumonia if sulfa drugs are promptly administered is $^{11}/_{12}$, what is meant is that records show that 11 out of 12 persons who have had this disease and received this treatment have recovered. Most scientists today think of probability in this sense.

Probability exerts a peculiar fascination even over persons who care nothing for mathematics. It is rich in philosophical interest and of the highest scientific importance. But it is also baffling. Each of the three interpretations is represented by at least one selection in this book, and the reader who has followed carefully the various persuasive arguments supporting the different views may by now be thoroughly confused. I cannot claim that the next selection resolves the question but its great merit is that it explains clearly the main issues in dispute. It is a sharp, lucid, disinterested analysis and comparison of the conflicting theories by a keenly perceptive thinker. Ernest Nagel is a leader among contemporary American philosophers. He was born in Czechoslovakia in 1901 and came to the U.S. at the age of nine. He attended the College of the City of New York, taught in the New York City public schools, and later took his doctorate in philosophy at Columbia University. In 1931 he joined the staff of that institution and since 1946 has been professor of philosophy. Nagel's specialties are symbolic logic and the philosophy of science; his writings have gained him an international reputation. But this is an inadequate index of his capabilities. He masters the equipment for handling hard technical problems of logic and mathematics but his vision extends beyond them. He is a rounded philosopher, sensitive to the role of historical and cultural circumstance, extraordinarily free from every trace of humbug and pretension. That explains why he writes so well, both in his professional papers and in his reviews and more popular essays. With the late Morris Cohen he wrote *Introduction to Logic and*

[3] Ernest Nagel, *Principles of the Theory of Probability*, International Encyclopedia of Unified Science, Chicago, 1939, Vol. 1, No. 6, p. 49.

Scientific Method, generally regarded as a classic text. It is hard to imagine a more readable and instructive book. Nagel is a rarely gifted teacher, as I know from having studied with him. Apart from his command of the subject and fluency as a lecturer, he cares about his students, is unfailingly kind and forthcoming. It is a special pleasure to include three of his essays (see also pp. 1668 and 1878) in this anthology.[1]

[1] The present paper was presented at the 97th annual meeting of the American Statistical Association, New York, December 28, 1935, and was reprinted in the *Journal of the American Statistical Association*, March 1936, Vol. 31, pp. 10–30.

That which chiefly constitutes Probability is expressed in the Word Likely, i.e. like some Truth, or true Event; like it, in itself, in its Evidence, in some more or fewer of its Circumstances. For when we determine a thing to be probably true . . . 'tis from the Mind's remarking in it a Likeness to some other Event, which we have observed has come to pass.

—BISHOP BUTLER

6 The Meaning of Probability

By ERNEST NAGEL

THERE is at least the appearance of presumption in the title of this paper. The title could be taken to imply that my object was either to legislate what meaning the term "probability" *must* have, or to settle summarily the many difficult problems which an analysis of the meaning of the term faces. Reflection on the follies of other philosophers who have ruled out as "meaningless" the meanings which scientists have assigned to certain terms, prevents me from undertaking the former task; and limitations of space, to say nothing of my own sense of unsolved difficulties, makes it impossible to attempt the latter. All that I wish to do is to make explicit certain generally recognized methodological principles to which, I believe, all rational inquiry appeals, and to test by their means three analyses of the meaning of "probability" which have been offered. One of Dickens' characters composed an essay on Chinese metaphysics by combining the contents of the encyclopaedia article on metaphysics with the information obtained from the article on China. However, even if, like the essay in Pickwick Papers, the present paper is simply a mechanical juxtaposition of some reflections on methodology and others on probability, I hope it succeeds in showing the relevance of the former to the discussion of the meaning of the latter.

I

I begin with the methodological principles.

1. If the term "probability" were being introduced for the first time, or in some special technical sense, into every-day or scientific language, its meaning would be given by specifying the occasion when it is to be employed, and by stating the rules which would govern its occurrence and relations to other terms in the language. Such a definition would be nominal and arbitrary, because the meaning of the term would be established by a resolution or convention, so that questions involving truth or falsity would not be relevant concerning the definition. Thus, the expres-

sion "work done by a body" was introduced as a technical term into physics by Coriolis, who used it to mean distance multiplied by the component of force. There are often good practical reasons for introducing new terms into a language in this arbitrary way; and having once fixed the meaning of a term, it is of course no longer an arbitrary matter whether, for example, a given body is performing work or not. Nevertheless, the initial specification of meaning is arbitrary, because every dispute about the meaning would be decided by an appeal to a resolution made, and not by an appeal to matters of objective fact.

With respect to the term "probability" such is not the situation in which we find ourselves, except in those cases where a writer explicitly stipulates that he will employ it in some unique sense. The term is of great antiquity and is used in more or less determinate ways in various contexts. What it *does* mean in any given context cannot therefore be decided by an arbitrary resolution as to what it *shall* mean; and we cannot rule out some usages of the term as "meaningless" merely on the ground that such a usage does not conform to the usage we have decided to give it. Hence, whether the term "probability" has one meaning or many must be determined by a study of the different contexts in which it is employed. The univocality or equivocality of the term cannot be determined *prior* to such a study.

There seems to be but one reliable way of discovering what meaning or meanings the term "probability" does in fact have. For people assert propositions like "It is probable that it will snow to-night," "The probability that he will survive the operation is 3/5," "The photon has a probability of 3/4 of being reflected," "The helio-centric theory is more probable than the geo-centric one," and so on. Now we would discover what is meant by these propositions and consequently by "probability," if we could ascertain rules controlling the assertion of each, and therefore of the usage of the term. To put the matter in different language, if we can obtain an answer to the question "What *sort of evidence* would be regarded as relevant for propositions asserting something about probabilities?" we would obtain at the same time an answer to the question "What is meant by probability?"

This point may be made clearer by an illustration. Suppose a person *X* were to assert the proposition "Events *A* and *B* are simultaneous," and were to give as evidence for it the proposition that he saw them occurring within one specious moment. We would now know what *X* means by "simultaneity"; and we could legitimately conclude that he would judge a sudden increase in the brightness of each of two stars lying in an approximately straight line with his eyes as simultaneous events, if he were to observe the change in brightness within the same specious moment. On the other hand, if a person *Y* were to give as evidence for the proposition "Events *A* and *B* are simultaneous," the proposition that an instrument placed half-way between them registers light rays having *A* and *B* as their source within a small interval of time as measured by a clock on the instrument, we would also know what *Y* means by "simultaneity"; and we could conclude that what *X* means by the term is not in general identical with what *Y* means by it.

It is an inquiry of this sort which seems to me necessary for determining what the term "probability" means. Such an inquiry is more than a project in anthropology or linguistics. For consider the following imagined situation: In virtue of what the term "work" means in physics, and by using other physical principles and data, we may validly infer from the proposition that a given body has performed a certain quantity of work another proposition, for example, that some other body suffers specifiable deformations. We are thus able to predict empirical states of affairs not yet examined and to regulate our behavior accordingly. Now suppose a person *X*, unfamiliar with the specific meaning and function of the term "work" as employed in physics, nevertheless believes that when a doorman of a hotel is said to be "working," just the same sort of consequences can be validly inferred from this as in the preceding illustration. It would still be the case that the term "work" would have a meaning for *X*, though a confused one. Nevertheless, it would be proper to criticise his use of the term "work" on the ground that the consequences he draws are not valid ones.

Now I think the way "probability" is frequently employed illustrates confusions similar to this one; propositions are sometimes believed to be validly inferable from propositions about probabilities when in fact such inferences are not justifiable in the light of the meaning, whether confused or not, ascribed to the term "probability." Hence, while I do not believe it is the philosopher's task to legislate away any of the meanings which "probability" may have, I think it is his task to *distinguish* between different meanings. It is also his task to evaluate the cogency of arguments which claim that propositions about probabilities are adequate premises for certain kinds of conclusions, and to suggest what the term *ought* to mean, and what it *could not* mean if the conclusions are to be valid consequences.

2. The second point I wish to make concerns the conditions for an empirically significant theory. Every theory is required to be formulated in such a manner that determinate propositions may be inferred from it by logical means alone. And furthermore, it is requisite that among the logical consequences of a theory there must be propositions which are capable of empirical corroboration or refutation. In brief, theories must be verifiable.

It will be obvious that no theory can be established beyond every possibility of doubt by any finite number of observations. This is so because theories are universal propositions which are intended to express the constant relations between an indeterminate number of specific events, most of which lie in the future. But while the empirical evidence for a theory can never be complete, *some* empirical evidence there must be. This statement implies the proposition that not every state of affairs can be confirmatory evidence for a given theory, so that propositions about possibly observable states of affairs must be specifiable which would contradict the theory. Any theory for which this statement is not true is without empirical content. Laplace did not require God in his *Celestial Mechanics* to explain the motion of heavenly bodies, simply because the hypothesis of God leads to no consequences which are empirically refutable.

But a word of caution must be added. Not all theories are rejected simply because an observed fact seems to contradict some logical consequence of a theory. For any empirically testable consequence which is said to be drawn from a theory does not follow *simply* from the given theory. It follows from the theory *conjoined with* other theories and observational data. It is thus the whole system of our knowledge which is put to an empirical test. Hence an alleged contradiction between theory and observation can be eliminated by making suitable changes in other parts of the body of our knowledge, so that what was initially regarded as a logical consequence from the theory no longer is capable of being deduced from it. Nevertheless, while so-called crucial experiments for a theory are not finally decisive and are only relatively crucial, *in the context of a given set of assumptions* the propositions about possible observations which would contradict the theory must be determinate; in that context we must in principle be capable of deciding whether such propositions corroborate the theory or not. This methodological principle will be shown to play an important role in the discussion of the meaning of probability.

3. There remains one further point in methodology which I wish to make explicit. Every theory about a subject matter involves a selection of phases of behavior within that subject matter, and does not consider the interrelations of all its phases. Moreover, for various reasons, a theory will state only an idealized schema of the relations between the selected phases. These observations lead to the consequence that the confirmation of a theory by experiment is only approximate. Now the degree of approximation which must hold between the consequences of an acceptable theory and the observation propositions is not determined by the theory. The degree may vary for the different situations to which a theory is applied and may even be left unexpressed as a tacit rule of inquiry. It is important to note, however, that the degree of allowable approximation

is determined by various material considerations, such as the purpose for which the inquiry is undertaken, the kind of activity which the theory is intended to coördinate and foretell, or the character of the instruments by means of which the testing is carried on. In some domains of research, the ideal pursued may be the development of theories for which the degree of allowable approximation progressively diminishes; in other fields the pursuit of such an ideal may be a fatal obstacle to the achievement of the goal of an inquiry. From the point of view of an outsider, the limit of allowable approximation is arbitrary, conventional and "subjective." If such a person were to refuse to accept certain allegedly confirmatory evidence for a theory on the ground that the approximation was not close enough, neither logic nor matters of fact could force him to do so. Such a person would simply refuse to abide by the rules of the game which that special science or special inquiry agrees to follow.

These remarks will be shown to bear upon the interpretation of the meaning of probability, because they are relevant for the interpretation of any symbolic operation which is performed within a theory. Mathematical physics, for example, differentiates and integrates certain functions representing the distributions of material particles; nevertheless, it is safe to assert that the mathematical conditions required for these operations do not obtain for the actual distribution of particles. Similarly, infinite series are employed to calculate magnitudes of various kinds, even though the objects which have those magnitudes would not be regarded as capable of physical decomposition into an infinity of parts. Such symbolic operations upon infinite and continuous manifolds must clearly be supplemented by stipulations about the degree of approximation within which experimental findings will be regarded as corroborating the theory. These operations are effective tools for dealing intellectually and in a generalized way with a set of otherwise unrelated problems; they are so many different *façons de parler* for formulating and translating the invariant relations between what is directly observable. They are not in general "literal" statements about a subject matter, so that some indication must be supplied concerning the extent to which empirical traits may deviate from the idealized schema so that the latter will still serve the specific objectives of an inquiry.

II

With these methodological principles in mind, I turn now to an analysis of the meaning of probability. Lack of time, to say nothing of lack of competence, does not permit an adequate survey of the innumerable contexts in which propositions about probabilities occur. I wish, however, to distinguish five broad types of contexts in which they do occur, consider three analyses of the meaning of the term "probability," and indicate the bearings of the above methodological considerations upon each.

Statements involving probabilities are to be found in (1) every-day discourse, (2) in the field of applied statistics and measurements, (3) within the context of physical and biological theories, (4) in the comparison of theories with each other for their respective degrees of probability, and (5) in the branch of mathematics known as the calculus of probability. Examples of each type of statement will appear in due course, in the discussion of the three major interpretations of probability with which I wish to concern myself. To these I now turn.

The first interpretation is the classic one, associated with the historical development of the mathematical theory of probability. It has been expounded with vigor by the English logician and mathematician De Morgan. According to him, the word "probable" refers to the state of mind with respect to an assertion for which complete certainty or knowledge does not exist. Hence the degree of certainty of a proposition, its "probability," is the degree of belief with which it is held. For certainty has degrees, and all grades of "knowledge," it is claimed, are capable of being quantitatively conceived. It is possible, therefore, to apply the calculus of probability to the degrees or strength of belief, if probability is defined algebraically as the ratio of the number of alternatives "favorable" to an "event" to the total number of equiprobable alternatives. The transition from this definition to the previous one is mediated by the principle of sufficient reason or indifference, according to which two propositions are equally probable if the strength of our belief is equally divided between them.

The second interpretation of probability is professed by certain English logicians like the economist Keynes. According to it, any two propositions are related not only by the relations usually studied in traditional logic, such as implication, but also by a directly intuitable relation called probability. This relation is not analyzable, although it is capable of having degrees. However, while any two propositions will each have some degree of probability with respect to a third, the degrees of probability are not in general comparable or measurable. Hence it is not always possible to apply the calculus of probability in order to explore the implications of compound assertions of probability relations. But whenever the calculus is applicable, the application is carried on in terms of a modified principle of indifference.

The third interpretation of probability is already implicit in Aristotle, but has become prominent only within the last century as a consequence of applying the probability calculus to statistics and physics. Its central idea is that by the probability of a proposition or an "event" is meant the relative frequency of the "event" in an indefinite class of events. A more precise statement of this view will be given presently. It is sufficient at this point to emphasize the fact that on this interpretation every statement involving probabilities is a *material proposition* whose truth or falsity is to be discovered by examining objective relative frequencies.

We must now decide (1) whether the meaning of propositions about probabilities in any of the five contexts I have enumerated is adequately stated by any one of these interpretations; and (2) whether any of these interpretations correctly analyzes the meaning of the propositions in *all* the five contexts.

Now it is demonstrable that statements involving the term "probability" which occur in the mathematical calculus of probability, in no way depend upon any of these interpretations. For the calculus of probability is a branch of what is called "pure" mathematics, a discipline whose sole object is to discover whether something follows logically from something else. The premises or axioms of the calculus, as a branch of pure mathematics, are not propositions but *propositional functions*, i.e. expressions containing *free* variables, as in the statement "If p, q, are the probabilities of two exclusive alternatives x, y, respectively, then the probability of x or y is p plus q." If we examine this and analogous statements and the operations of the calculus carefully, we discover that the term "probability" is a *free* variable, which is defined only *implicitly* by the axioms of the calculus. As far as the calculus itself is concerned, the free variable may be interpreted in *any* manner whatsoever consistent with the axioms of the system. It may even be possible to interpret the free variable "probability" in each of the three ways suggested above, just as it is possible to interpret the letters a, b, in the formula $a + b = b + a$, as integers, fractions, or complex numbers although these are three different sorts of numbers.

The calculus of probability does not therefore determine the specific empirical content of the term "probability" which occurs in its statements. Its axioms may of course be used to define implicitly what probability is, and thus limit the range of possible interpretations of the term. But its primary function is to enable us to discover, given certain initial probabilities, what other probabilities are implied by them. It therefore follows that the use of a so-called principle of indifference *within* the calculus to *define*, for purposes of supplying hypotheses for deductions, the alternatives which are to be treated symmetrically, does *not* justify the use of that principle as a criterion for *deciding* which material propositions are "equiprobable" in any nonformal sense of the word.

There thus remain for consideration four types of propositions involving the term "probability." Now I think that when some people assert propositions in every-day discourse, such as "It is very probable he read it in some book," "It is not probable that he could have forgotten me," or "The real existence of Christ is most probable," they mean by "probability" just what De Morgan says they mean. For when we inquire what sort of evidence is believed by these people as relevant for these propositions they can do nothing but reassert their beliefs; while the rapid variation in the degrees of probability which they attribute to propositions makes it plausible to assume that a degree of probability is in some way an index of the "intensity" of their convictions. Moreover, I also think that when some people evaluate theories with respect to their truth, for example when they declare "The theory of evolution is just as probable as the theory of special creation," it is a subjective interpretation which seems to be the correct one. However, those who consistently use the term in this sense assert propositions about probability which have no verifiable consequences, since the implicit predictions which they think are involved in their statements can in no way be justified as logical consequences from them. In spite of the efforts of De Morgan, Stumpf, and others to assign an interpretation to the numerical value of a "probability" when the latter is understood as strength of belief, no unambiguous criteria have been specified for the equality, addition, or multiplication of probabilities. They are therefore not entitled to employ the calculus of probability upon their interpretation of what probability is, because the fundamental operations of the calculus are without any specified content. And finally, this interpretation of probability is completely irrelevant for propositions in applied statistics or physics like "The probability of a male birth in the U. S. is .52," or "The probability of a 10° deflection of an α-ray passing through a film is ¼." As Norman Campbell remarked apropos of the last statement, if anyone proposed to attribute to that probability any value other than that determined by frequency, he would convince us of nothing but his ignorance of physics.

It has been claimed for the Keynes interpretation of probability that it is the only adequate one for most uses of the term. Thus, it is claimed that for propositions such as the following: "On the evidence as to their moral character, it is more probable that witness *A* speaks the truth than witness *B*," "It is probable that Miss *C* would have won the beauty prize if she had entered the contest," or "Relative to experiment *D* the Einstein theory is more probable than relative to *E*," we must have a "rational insight" into the connections between the evidence and the conclusion said to be probable. For, it is alleged, in these cases the probability relation cannot be interpreted in terms of relative frequencies; for no numerical evaluation of the degrees of probability is possible since the relevant statistical information is completely lacking. Hence "probability" refers to a unique, directly intuitable relation between propositions.

I shall examine presently what a frequentist might say to this. I will admit, however, that probability relations are often asserted to hold on the basis of no specific statistical evidence such as the frequency interpretation seems to require. But I do not think this *proves* the existence of such a unique relation of probability which can be "rationally" intuited. In the first place, there are few people who claim for themselves the requisite intellectual faculty for intuiting these unique relations. In the second place, the alleged intuition is not controllable, and there seems no way of corroborating its findings. For the alleged unique probability relation is not definitely localizable, so that agreement both about its presence and degree cannot be obtained among competent students. And in the third place, on Keynes' view a degree of probability is assignable to a *single* proposition with respect to given evidence. But what *verifiable* consequences can be drawn from the statement that with respect to the evidence the proposition that on the *next* throw with a given pair of dice 7 will appear, has a probability of 1/6? For on the view that it is significant to predicate a probability to the single instance there is nothing to be verified or to be refuted. Hence Keynes' view, like that of De Morgan, violates the principle of verifiability required for all statements, and cannot be regarded as a satisfactory analysis of probability propositions in the sciences which claim to abide by this canon.

III

We must therefore examine the frequency interpretation of probability, to discover whether it offers an adequate account of the types of probability propositions I am considering and whether it is free from the objections which are fatal to the alternative interpretations. It seems to me, however, that the views of De Morgan and Keynes are correct in insisting upon two points, so that a frequency interpretation must be so formulated as to include them. The first is, that probability is not a property of facts or events, but of propositions. It follows that the analysis of the meaning of probability and the consideration of the evidence for probability propositions belong to logic, not to physics or metaphysics; and indeed I think that at no point in that analysis or consideration is a prior decision required about the universality of causal determination or the presence of absolute contingency. Thus, when we seem to talk about the probability of an event we are simply talking inaccurately although conveniently about the probability of a proposition stating the occurrence of that event. The second point is that probability is not a property of a *single* proposition, but is a relation between propositions. A proposition is either true or false, and is not, as such, probable; it is probable with respect to other propositions. It seems to me, therefore, that a frequency view will not quarrel with Keynes on the ground that "probability" refers to a relation between propositions, but only on the ground of what *sort* of relation it is.

The chief difficulty for the frequentist is to find a formulation for his view, or at least an interpretation of a formulation, which will make it adequate for the sort of situations where it is fairly clear that frequency considerations are relevant. I wish to consider the following one.

Let C_1 be a characteristic, e.g. being a coin with usual shape and construction tossed in the usual way, and C_2 be another characteristic, e.g. falling heads uppermost. Then "x_i is C_1" "x_i is C_2" and "x_i is C_1 and C_2" are propositional functions whose values are propositions, true or false, when constants are substituted for the variable x_i. Let T_i^n (x_i is C_1 and C_2) be the number of true propositions obtained when the constants replace the variables in the respective propositional functions, where n is the total number of propositions obtained in this way; and let $T_i^n(x_i \text{ is } C_1)$ be interpreted in a similar manner. Finally, form the fraction $F_n(C_1\ C_2) = T_i^n(x_i \text{ is } C_1 \text{ and } C_2)/T_i^n(x_i \text{ is } C_1)$, which will be the numerical value, in n cases, of the relative frequency with which propositions of the form "x_i is C_2" are true. If, now, the fraction F_n approaches p as a limit as n increases without limit, p will be called the probability with which, to speak loosely, a coin falls heads on the evidence that it is "fairly" constructed and "fairly" tossed.

This definition of "probability" incorporates the two important insights of the alternative interpretations already discussed, and adds a feature which they neglect—namely, relative frequency. Consequently, on this "truth-frequency" view, probability is a relation between propositions, but an *analyzable* relation; secondly every statement predicating probabilities is a *material* proposition for which empirical evidence is required; and thirdly, probability denotes a relation between *classes* of propositions, so that a statement about the probability of a single proposition is an elliptic way of asserting a relation between classes of propositions to one of which the given proposition belongs. Since probability is a fraction, addition and multiplication of probabilities are given intelligible meanings. A calculus of probability may therefore be employed to calculate from certain initial probabilities of propositions the probability values of other propositions related in definite ways to the given data.

I shall take for granted, what is generally admitted, that in such contexts as vital statistics and physics a frequency interpretation for propositions about probabilities is the relevant one. But this discussion of the meaning of probability would be sadly incomplete if I did not touch upon the question whether the formulation of a frequency interpretation I have reported is capable of meeting the serious objections which have been leveled against it. To these objections I now turn.

1. It has been argued that we cannot define probability as a limit of a

ratio in the mathematical sense. For if the probability of a proposition *were* such a limit, there would have to be a term in the series studied, *after which* the difference between the limit and the relative frequency empirically found would be less than any preassigned magnitude. But if there were such a term for a given preassigned magnitude, it would follow necessarily that the empirical frequencies could not deviate from the limit by more than this magnitude. This, however, contradicts well established theorems in the calculus of probability.

This objection seems to me to have a point only when we are dealing with what by some writers are called "normal" series, to which alone Bernoulli's theorem is applicable. It loses all force for a series which exhibits a certain kind of internal regularity, e.g. the series of numbers 1, 0, 1, 0, 1, 0, etc. In the second place, as Von Mises has shown, for "normal" series the objection fails to distinguish between the probability that a certain sequence of events will *occur* in a given series and the probability that such a sequence of events will occur at a *definite place* in a series. The latter probability involves not one series but a *whole class* of them, so that a whole set of preassigned magnitudes is required and not only one such magnitude. But in the third place, whatever force the objection has is directed against the dialectical elaboration of probability in terms of the calculus of probability. Now the limit definition of probability is required to develop such a calculus consistently and at the same time to find a way of formulating empirical frequencies so as to be amenable to convenient calculation. Nevertheless, as has already been indicated, the definition has to be supplemented by indicating the degree of approximation which would be allowed in practice between the empirical frequencies discovered and the theoretical limit stipulated. As long as the empirical relative frequencies remain within approximately assigned intervals, it is convenient to select *one* of the values which the empirical ratios take on within this interval, and to treat it, for the purposes of calculation, as the mathematical limit of the ratios. This assumption is permissible because it leads to no contradictions, and is advisable because of the dispatch it introduces into the mathematical operations. Since the range of allowable oscillation of the ratios is agreed upon on extra-logical grounds, the range as well as the limit may both be altered with the progress of inquiry. Hence even apart from the acute observations of Von Mises the alleged contradiction could always be obviated by conveniently altering the value of the limit. I think therefore that one virtue in this objection consists in calling our attention to the conventional moment in the value assigned to a probability.

2. A second objection is directed not only against a limit definition of probability, but against *any* frequency interpretation of the term. I shall state it, however for the limit definition. How can we ever discover the

probability value of a proposition on given evidence, it is objected, if this value is defined as the limit of an infinite sequence of ratios? For we do not, in general, know the law of the sequence, and we can never examine more than a finite number of its terms. Hence the value of a probability is a sort of "unknowable," so that this definition of probability cannot be a correct analysis of what we mean in making judgments of probability.

A two-fold answer may be offered: In the first place, while it is true that we can examine only a finite number of terms in an empirical series, we can employ the empirical ratios as hypotheses for the "true" probability value of the infinite series. These hypotheses are to be tested in the usual way, by comparing their logical consequences with the ratios subsequently observed empirically. And in the second place, a probability value may be deduced from some general theory already established, instead of being obtained from observation of a statistical series. For example, the probability of a "fair" coin falling heads uppermost may be computed by the aid of theoretical mechanics, although in this case the calculated probability is once more a hypothesis to be tested by empirical frequencies.

Before elaborating these answers I think it is necessary to note an important objection that is made to both of them. Suppose we entertain the hypothesis that the probability of getting a head with a coin thrown in the usual way is ½. This hypothesis must then have determinate consequences, in accordance with the principle of verifiability discussed above. But suppose that in a thousand throws heads turn up 900 times. Has the hypothesis been refuted? It is well known, however, that this empirical result is not incompatible with the hypothesis of ½ as the probability, since, indeed, this hypothesis requires us to expect it; for that hypothesis means simply that in an *indefinitely* long series the proportion of heads will be approximately ½, and does not therefore exclude a preponderance of heads within a finite segment of the series. It seems, therefore, that the hypothesis of ½ as the probability, like the hypothesis of a Providence, has no *refutable* consequence, and is without empirical content. It is no answer to this observation that no hypothesis can be *completely* verified. For the criticism doesn't challenge the hypothesis of ½ as the probability on the ground that it is not capable of *complete* confirmation, but on the ground that, in terms of the hypothesis itself, it does not seem capable of refutation no matter *what* the empirical findings are.

It seems to me, however, that this objection is fatal only if the frequentist fails to introduce the degree of allowable approximation into his predictions. For it is obvious that in practice we do decide, on the evidence of empirical frequencies in final series, whether the assumption we have made as to the value of a probability is corroborated or not. Thus, if a coin were thrown a thousand times and the head came up 490 times, we would regard this as supporting the hypothesis that the probability of its coming up is ½, where the hypothesis has been suggested by observing other finite series of throws; if it came up 520 times we would usually still regard this as confirmatory evidence; but if it came up only 400 times we would normally reject the hypothesis, suspect the coin was loaded, and propose a different value for the probability. We proceed in this manner because we have in mind a degree of allowable approximation, in this case lying somewhere between a deviation of twenty and one of a hundred. That we do not always explicitly define this degree is not important. We cannot specify this degree once for all, since it is a function of the specific context in which the coin is thrown. What is important, to my mind, is that there is a tacit acceptance of *some* degree of allowable deviation, however vaguely we may formulate it to ourselves.

But I think something further can be said about the way the specific context of inquiry controls this degree of allowable deviation, and about the way in which the testing of hypotheses about probabilities is conducted. Propositions about probabilities are not isolated propositions in the body of our knowledge. The evidence for them comes not merely from an examination of statistical frequencies, but also, and sometimes exclusively, from various theories and other data. Consequently, in the above case, if we have reason to believe that the coin is not loaded and is "fairly thrown," in any finite segment of an endless series a larger degree of deviation from the hypothetical frequencies will be allowable than if we did not possess this information. On the other hand, if a man is gambling with a limited amount of money, he will generally impose a smaller degree of permissible deviation from the hypothesis he entertains for the relative frequencies. Thus, although we cannot fix once for all the number of throws within which the hypothesis about the probability would be taken as sufficiently corroborated, in practice such a number is roughly indicated.

It must also be stressed that the method of rational inquiry is a self-corrective one, and in general we place greater reliance upon our rules of procedure and "long-run" results than upon particular conclusions obtained. Thus, suppose that in the case where we obtain 400 heads out of a thousand throws with a coin we are led to entertain the hypothesis of ⅖ for the probability of getting a head. It may be the case that for this particular coin the "true" probability is in fact ½, so that the hypothesis of ⅖ is false. Nevertheless, it would not be irrational to adopt the latter hypothesis in this instance, if it were the case that when we continually formulated hypotheses on evidence in the manner indicated, we did hit upon a good approximation to the relative frequencies more often than not. Indeed, while we are not in the position to assert that the empirical relative frequencies do approximate to a limit, nevertheless *if* they do we are bound to discover this approximate value by a repeated and systematic correction of the hypotheses suggested by the empirical samples we continue to examine.

IV

According to the frequency interpretation of probability it seems as if we ought to be able to associate a definite numerical value with every statement about probability. But it must be pointed out that the evidence for statements of probability may often be inadequate, because both statistical and theoretical information may be lacking. In such while it still may be true that what we *mean* by probability is stated by the frequency view, for lack of relevant evidence we rely on general impressions and "hunches." Thus, for the proposition "It is probable on the evidence that witness *A* speaks the truth," we do not possess any means for assigning a definite numerical value to the probability. It does not follow, however, that a frequency interpretation is illegitimate. For what may be meant by this statement is that the relative frequency with which a regular church-goer tells lies on important occasions is a small number considerably less than ½. For many purposes we do not require an actual numerical value for the probability, since in some contexts an indeterminate value which is greater (or less) than ½ is sufficient; and although we may lack adequate statistical material to support such statements, we may nevertheless possess considerable general experience to corroborate them. In these contexts I would admit that the more precise formulation given above is not a good picture of the meaning of probability statements; and that a frequency interpretation is satisfactory only if we take care not to define probability too rigorously.

Nevertheless, in the light of the preceding I think it is at least plausible that in the first three contexts enumerated earlier, a frequency interpretation of probability is a workable one, although whether it is satisfactory for any given instance of a probability proposition can be decided only by examining the instance itself. The remaining question I wish to touch upon is whether statements about the probability of theories can also be interpreted in the frequency sense.

Since statements about the probability of a proposition are material statements for which, in the light of the preceding, the evidence must be judged as incomplete, such statements are hypotheses about the probability in question. It is therefore not surprising to find statements like the following: "It is probable that the probability of getting heads with a coin is ½," and it is natural to ask whether in the first occurrence of the term "probable" means the same as in its second occurrence.

There is a way, I believe, in which both occurrences of the term "probability" in this statement can receive a frequency interpretation.

Suppose there are an indefinite number of coins *a*, *b*, *c*, etc., for each of which we seek the probability of getting heads by repeated tosses; we then would obtain a theoretically endless "square array" of propositions. Suppose further that after say a thousand trials with each coin, we formulate the series of propositions: "The probability that coin *a* falls heads is ½," "The probability that coin *b* falls heads is ½," etc. Each proposition is a hypothesis about the respective coin, and requires to be tested in the usual ways. Now suppose some of these hypotheses were confirmed by continuing the series of throws with the corresponding coins and some were not, but that nevertheless the ratio of verified hypotheses to the total number exceeds ½ say. We may express this by saying that on the evidence that these hypotheses have been tested to the extent indicated, the statement: "The probability that coin *x* falls heads is ½" where *x* is any coin, has a probability greater than ½. It seems to me, therefore, that in *some* cases where hypotheses are said to be probable, it is possible to interpret the probability of hypotheses in a frequency way.

But is it *always* correct to interpret such statements in this way? I think there are several cogent arguments for denying this.

1. It is possible to interpret the probability of *singular* propositions, like "This coin falls head uppermost" in a frequency manner, because singular propositions are *values* of propositional *functions* and because we can define the expression "the truth-frequency of a propositional function." But, it is urged, we cannot in general regard a theory as a *value* of some propositional function, and in any case we cannot attach a clear sense to the expression "the truth-frequency of a propositional function for which a theory is a determinate value." For consider such a complicated theory as Newtonian mechanics. Is it possible and is it practicable to regard this theory as some value of a propositional function? And is it really helpful to talk about the truth-value of this theory in any sense analogous to the way in which we discuss the truth value of the proposition "This coin falls heads uppermost?" It will follow that if the answer to these questions is in the negative we cannot assign a frequency meaning to a statement that the Newtonian theory is probable on given evidence.

2. It is sometimes thought that we can assign a numerical and therefor a frequency probability to a theory under the following circumstances. On the hypothesis that theory *A* is true the probability of an empirically confirmed proposition *p* is 1; on the hypothesis that theory *B* is true, the probability of *p* is ⅓. Hence, it is claimed, *A* is three times as probable as *B* on the evidence that *p* is true. But this argument must be rejected by a frequentist on the ground that it uses Bayes' theorem in the form in which it tacitly employs the assumption that the *antecedent* probabilities of *A* and *B* are equal. Such an assumption is incompatible with

the central thesis of the frequency view. The most that a frequentist can claim is that by adopting theory *A* he is employing a system of inferences which would lead him, *via* the theory, to empirically false conclusions not oftener than approximately once out of three times.

3. This last remark may be thought to provide a clue for interpreting in a frequency way the probability of theories. It has been suggested that such statements mean that the logical consequences of a theory which may be approximately confirmed by experiment are true with a certain relative frequency for the series of *all* the logical consequences of the theory capable of experimental testing. This suggestion, however, does not seem to be a way out, for according to it we would say that the probability of the Newtonian theory is $\frac{4}{5}$ if every fifth consequence of the theory were materially false. But in such a case the theory would ordinarily be rejected as false, and not accepted as highly probable.

4. And finally, it may be urged that what we mean by the probability of a theory is the ratio of its *actually* verified consequences to the total number of its *possible* consequences. In general, however, since this ratio would have the value zero, this cannot be taken as the measure of the probability of a theory.

It seems to me therefore that in general statements about the probability of theories do not refer to relative frequencies in any obvious way, although for some theories which are not too complex it is possible to provide a frequency interpretation for propositions about their probabilities. I do not see in what way it is possible, or for that matter in what way it would be useful, to compare, say, the "probability" of the gene theory in genetics on the evidence for it, with the "probability" of the relativity theory on the evidence for the latter, although on a frequency interpretation such comparison should be significant. I do not reject the frequency interpretation for statements about the probability of theories, as is sometimes done, on the ground that there is an essential difference between universal and singular propositions because of an alleged difference in the logic of verification required for them. Nor do I believe that the acceptance or rejection of a theory is a completely arbitrary matter, or that the evidence for one of two alternative theories may not be more complete or satisfactory than for the other. This is a matter which requires an independent study, and would lead, I think, to another meaning of the term "probability" into which frequency considerations enter in ways different from the way they enter the frequency view here discussed. But I do think that it is not very illuminating to place the interpretation of all statements about the probability of theories upon the Proscrustean bed of the frequency view I have considered. For while there are certain rough analogies between the way the term "probability" is used in the context of statements about theories and its use in other contexts, the differences in this case seem to me to outweigh the similarities.

I conclude, therefore, that the term "probability" is not a univocal term, for it has different meanings in different contexts. However, the frequency interpretation, when properly qualified, seems to be the most satisfactory one for analyzing the meaning of the term in the contexts of every-day discourse, applied statistics and measurement, and within many branches of the theoretical sciences. Furthermore, some statements about the probability of hypotheses can be interpreted in a frequency manner; but I do not find sufficient grounds for maintaining that such is the meaning of probability in statements about the probability of complicated theories like the corpuscular or wave theories of light.

GEORGE ALLEN & UNWIN LTD
London: 40 Museum Street, W.C.1

Auckland: 24 Wyndham Street
Sydney, N.S.W.: Bradbury House, 55 York Street
Bombay: 15 Graham Road, Ballard Estate, Bombay 1
Calcutta: 17 Chittaranjan Avenue, Calcutta 13
Cape Town: 109 Long Street
Karachi: Metherson's Estate, Wood Street, Karachi, 2
New Delhi: 13-14 Ajmeri Gate Extension, New Delhi 1
São Paulo: Avenida 9 de Julho 1138-Ap. 51
Singapore (South East Asia and Far East): 36c Prinsep Street
Toronto: 91 Wellington Street West